T0181706

Lecture Notes in Computer Science 13801

More information about this series at https://link.springer.com/bookseries/558

Leonid Karlinsky · Tomer Michaeli ·
Ko Nishino (Eds.)

Computer Vision – ECCV 2022 Workshops

Tel Aviv, Israel, October 23–27, 2022
Proceedings, Part I

 Springer

Editors
Leonid Karlinsky
IBM Research - MIT-IBM Watson AI Lab
Massachusetts, USA

Tomer Michaeli 🆔
Technion – Israel Institute of Technology
Haifa, Israel

Ko Nishino 🆔
Kyoto University
Kyoto, Japan

ISSN 0302-9743   ISSN 1611-3349 (electronic)
Lecture Notes in Computer Science
ISBN 978-3-031-25055-2   ISBN 978-3-031-25056-9 (eBook)
https://doi.org/10.1007/978-3-031-25056-9

This Springer imprint is published by the registered company Springer Nature Switzerland AG
The registered company address is: Gewerbestrasse 11, 6330 Cham, Switzerland

Foreword

Organizing the European Conference on Computer Vision (ECCV 2022) in Tel-Aviv during a global pandemic was no easy feat. The uncertainty level was extremely high, and decisions had to be postponed to the last minute. Still, we managed to plan things just in time for ECCV 2022 to be held in person. Participation in physical events is crucial to stimulating collaborations and nurturing the culture of the Computer Vision community.

There were many people who worked hard to ensure attendees enjoyed the best science at the 17th edition of ECCV. We are grateful to the Program Chairs Gabriel Brostow and Tal Hassner, who went above and beyond to ensure the ECCV reviewing process ran smoothly. The scientific program included dozens of workshops and tutorials in addition to the main conference and we would like to thank Leonid Karlinsky and Tomer Michaeli for their hard work. Finally, special thanks to the web chairs Lorenzo Baraldi and Kosta Derpanis, who put in extra hours to transfer information fast and efficiently to the ECCV community.

We would like to express gratitude to our generous sponsors and the Industry Chairs Dimosthenis Karatzas and Chen Sagiv, who oversaw industry relations and proposed new ways for academia-industry collaboration and technology transfer. It's great to see so much industrial interest in what we're doing!

Authors' draft versions of the papers appeared online with open access on both the Computer Vision Foundation (CVF) and the European Computer Vision Association (ECVA) websites as with previous ECCVs. Springer, the publisher of the proceedings, has arranged for archival publication. The final version of the papers is hosted by SpringerLink, with active references and supplementary materials. It benefits all potential readers that we offer both a free and citable version for all researchers, as well as an authoritative, citeable version for SpringerLink readers. Our thanks go to Ronan Nugent from Springer, who helped us negotiate this agreement. Last but not least, we wish to thank Eric Mortensen, our publication chair, whose expertise made the process smooth.

October 2022

Rita Cucchiara
Jiří Matas
Amnon Shashua
Lihi Zelnik-Manor

Preface

Welcome to the workshop proceedings of the 17th European Conference on Computer Vision (ECCV 2022). This year, the main ECCV event was accompanied by 60 workshops, scheduled between October 23–24, 2022. We received 103 workshop proposals on diverse computer vision topics and unfortunately had to decline many valuable proposals because of space limitations. We strove to achieve a balance between topics, as well as between established and new series. Due to the uncertainty associated with the COVID-19 pandemic around the proposal submission deadline, we allowed two workshop formats: hybrid and purely online. Some proposers switched their preferred format as we drew near the conference dates. The final program included 30 hybrid workshops and 30 purely online workshops. Not all workshops published their papers in the ECCV workshop proceedings, or had papers at all. These volumes collect the edited papers from 38 out of the 60 workshops. We sincerely thank the ECCV general chairs for trusting us with the responsibility for the workshops, the workshop organizers for their hard work in putting together exciting programs, and the workshop presenters and authors for contributing to ECCV.

October 2022

Tomer Michaeli
Leonid Karlinsky
Ko Nishino

Organization

General Chairs

Rita Cucchiara   University of Modena and Reggio Emilia, Italy
Jiří Matas   Czech Technical University in Prague,
     Czech Republic
Amnon Shashua   Hebrew University of Jerusalem, Israel
Lihi Zelnik-Manor   Technion – Israel Institute of Technology, Israel

Program Chairs

Shai Avidan   Tel-Aviv University, Israel
Gabriel Brostow   University College London, UK
Giovanni Maria Farinella   University of Catania, Italy
Tal Hassner   Facebook AI, USA

Program Technical Chair

Pavel Lifshits   Technion – Israel Institute of Technology, Israel

Workshops Chairs

Leonid Karlinsky   IBM Research - MIT-IBM Watson AI Lab, USA
Tomer Michaeli   Technion – Israel Institute of Technology, Israel
Ko Nishino   Kyoto University, Japan

Tutorial Chairs

Thomas Pock   Graz University of Technology, Austria
Natalia Neverova   Facebook AI Research, UK

Demo Chair

Bohyung Han   Seoul National University, South Korea

Social and Student Activities Chairs

Tatiana Tommasi   Italian Institute of Technology, Italy
Sagie Benaim   University of Copenhagen, Denmark

Diversity and Inclusion Chairs

Xi Yin   Facebook AI Research, USA
Bryan Russell   Adobe, USA

Communications Chairs

Lorenzo Baraldi   University of Modena and Reggio Emilia, Italy
Kosta Derpanis   York University and Samsung AI Centre Toronto,
                                Canada

Industrial Liaison Chairs

Dimosthenis Karatzas   Universitat Autònoma de Barcelona, Spain
Chen Sagiv   SagivTech, Israel

Finance Chair

Gerard Medioni   University of Southern California and Amazon,
                                USA

Publication Chair

Eric Mortensen   MiCROTEC, USA

Workshops Organizers

W01 - AI for Space

Tat-Jun Chin   The University of Adelaide, Australia
Luca Carlone   Massachusetts Institute of Technology, USA
Djamila Aouada   University of Luxembourg, Luxembourg
Binfeng Pan   Northwestern Polytechnical University, China
Viorela Ila   The University of Sydney, Australia
Benjamin Morrell   NASA Jet Propulsion Lab, USA
Grzegorz Kakareko   Spire Global, USA

W02 - Vision for Art

Alessio Del Bue   Istituto Italiano di Tecnologia, Italy
Peter Bell   Philipps-Universität Marburg, Germany
Leonardo L. Impett   École Polytechnique Fédérale de Lausanne
                                (EPFL), Switzerland
Noa Garcia   Osaka University, Japan
Stuart James   Istituto Italiano di Tecnologia, Italy

W03 - Adversarial Robustness in the Real World

Angtian Wang	Johns Hopkins University, USA
Yutong Bai	Johns Hopkins University, USA
Adam Kortylewski	Max Planck Institute for Informatics, Germany
Cihang Xie	University of California, Santa Cruz, USA
Alan Yuille	Johns Hopkins University, USA
Xinyun Chen	University of California, Berkeley, USA
Judy Hoffman	Georgia Institute of Technology, USA
Wieland Brendel	University of Tübingen, Germany
Matthias Hein	University of Tübingen, Germany
Hang Su	Tsinghua University, China
Dawn Song	University of California, Berkeley, USA
Jun Zhu	Tsinghua University, China
Philippe Burlina	Johns Hopkins University, USA
Rama Chellappa	Johns Hopkins University, USA
Yinpeng Dong	Tsinghua University, China
Yingwei Li	Johns Hopkins University, USA
Ju He	Johns Hopkins University, USA
Alexander Robey	University of Pennsylvania, USA

W04 - Autonomous Vehicle Vision

Rui Fan	Tongji University, China
Nemanja Djuric	Aurora Innovation, USA
Wenshuo Wang	McGill University, Canada
Peter Ondruska	Toyota Woven Planet, UK
Jie Li	Toyota Research Institute, USA

W05 - Learning With Limited and Imperfect Data

Noel C. F. Codella	Microsoft, USA
Zsolt Kira	Georgia Institute of Technology, USA
Shuai Zheng	Cruise LLC, USA
Judy Hoffman	Georgia Institute of Technology, USA
Tatiana Tommasi	Politecnico di Torino, Italy
Xiaojuan Qi	The University of Hong Kong, China
Sadeep Jayasumana	University of Oxford, UK
Viraj Prabhu	Georgia Institute of Technology, USA
Yunhui Guo	University of Texas at Dallas, USA
Ming-Ming Cheng	Nankai University, China

W06 - Advances in Image Manipulation

Radu Timofte	University of Würzburg, Germany, and ETH Zurich, Switzerland
Andrey Ignatov	AI Benchmark and ETH Zurich, Switzerland
Ren Yang	ETH Zurich, Switzerland
Marcos V. Conde	University of Würzburg, Germany
Furkan Kınlı	Özyeğin University, Turkey

W07 - Medical Computer Vision

Tal Arbel	McGill University, Canada
Ayelet Akselrod-Ballin	Reichman University, Israel
Vasileios Belagiannis	Otto von Guericke University, Germany
Qi Dou	The Chinese University of Hong Kong, China
Moti Freiman	Technion, Israel
Nicolas Padoy	University of Strasbourg, France
Tammy Riklin Raviv	Ben Gurion University, Israel
Mathias Unberath	Johns Hopkins University, USA
Yuyin Zhou	University of California, Santa Cruz, USA

W08 - Computer Vision for Metaverse

Bichen Wu	Meta Reality Labs, USA
Peizhao Zhang	Facebook, USA
Xiaoliang Dai	Facebook, USA
Tao Xu	Facebook, USA
Hang Zhang	Meta, USA
Péter Vajda	Facebook, USA
Fernando de la Torre	Carnegie Mellon University, USA
Angela Dai	Technical University of Munich, Germany
Bryan Catanzaro	NVIDIA, USA

W09 - Self-Supervised Learning: What Is Next?

Yuki M. Asano	University of Amsterdam, The Netherlands
Christian Rupprecht	University of Oxford, UK
Diane Larlus	Naver Labs Europe, France
Andrew Zisserman	University of Oxford, UK

W10 - Self-Supervised Learning for Next-Generation Industry-Level Autonomous Driving

Xiaodan Liang	Sun Yat-sen University, China
Hang Xu	Huawei Noah's Ark Lab, China

Fisher Yu	ETH Zürich, Switzerland
Wei Zhang	Huawei Noah's Ark Lab, China
Michael C. Kampffmeyer	UiT The Arctic University of Norway, Norway
Ping Luo	The University of Hong Kong, China

W11 - ISIC Skin Image Analysis

M. Emre Celebi	University of Central Arkansas, USA
Catarina Barata	Instituto Superior Técnico, Portugal
Allan Halpern	Memorial Sloan Kettering Cancer Center, USA
Philipp Tschandl	Medical University of Vienna, Austria
Marc Combalia	Hospital Clínic of Barcelona, Spain
Yuan Liu	Google Health, USA

W12 - Cross-Modal Human-Robot Interaction

Fengda Zhu	Monash University, Australia
Yi Zhu	Huawei Noah's Ark Lab, China
Xiaodan Liang	Sun Yat-sen University, China
Liwei Wang	The Chinese University of Hong Kong, China
Xiaojun Chang	University of Technology Sydney, Australia
Nicu Sebe	University of Trento, Italy

W13 - Text in Everything

Ron Litman	Amazon AI Labs, Israel
Aviad Aberdam	Amazon AI Labs, Israel
Shai Mazor	Amazon AI Labs, Israel
Hadar Averbuch-Elor	Cornell University, USA
Dimosthenis Karatzas	Universitat Autònoma de Barcelona, Spain
R. Manmatha	Amazon AI Labs, USA

W14 - BioImage Computing

Jan Funke	HHMI Janelia Research Campus, USA
Alexander Krull	University of Birmingham, UK
Dagmar Kainmueller	Max Delbrück Center, Germany
Florian Jug	Human Technopole, Italy
Anna Kreshuk	EMBL-European Bioinformatics Institute, Germany
Martin Weigert	École Polytechnique Fédérale de Lausanne (EPFL), Switzerland
Virginie Uhlmann	EMBL-European Bioinformatics Institute, UK

Peter Bajcsy   National Institute of Standards and Technology,
                                         USA
Erik Meijering   University of New South Wales, Australia

W15 - Visual Object-Oriented Learning Meets Interaction: Discovery, Representations, and Applications

Kaichun Mo   Stanford University, USA
Yanchao Yang   Stanford University, USA
Jiayuan Gu   University of California, San Diego, USA
Shubham Tulsiani   Carnegie Mellon University, USA
Hongjing Lu   University of California, Los Angeles, USA
Leonidas Guibas   Stanford University, USA

W16 - AI for Creative Video Editing and Understanding

Fabian Caba   Adobe Research, USA
Anyi Rao   The Chinese University of Hong Kong, China
Alejandro Pardo   King Abdullah University of Science and
                                         Technology, Saudi Arabia
Linning Xu   The Chinese University of Hong Kong, China
Yu Xiong   The Chinese University of Hong Kong, China
Victor A. Escorcia   Samsung AI Center, UK
Ali Thabet   Reality Labs at Meta, USA
Dong Liu   Netflix Research, USA
Dahua Lin   The Chinese University of Hong Kong, China
Bernard Ghanem   King Abdullah University of Science and
                                         Technology, Saudi Arabia

W17 - Visual Inductive Priors for Data-Efficient Deep Learning

Jan C. van Gemert   Delft University of Technology, The Netherlands
Nergis Tömen   Delft University of Technology, The Netherlands
Ekin Dogus Cubuk   Google Brain, USA
Robert-Jan Bruintjes   Delft University of Technology, The Netherlands
Attila Lengyel   Delft University of Technology, The Netherlands
Osman Semih Kayhan   Bosch Security Systems, The Netherlands
Marcos Baptista Ríos   Alice Biometrics, Spain
Lorenzo Brigato   Sapienza University of Rome, Italy

W18 - Mobile Intelligent Photography and Imaging

Chongyi Li   Nanyang Technological University, Singapore
Shangchen Zhou   Nanyang Technological University, Singapore

Ruicheng Feng   Nanyang Technological University, Singapore
Jun Jiang   SenseBrain Research, USA
Wenxiu Sun   SenseTime Group Limited, China
Chen Change Loy   Nanyang Technological University, Singapore
Jinwei Gu   SenseBrain Research, USA

W19 - People Analysis: From Face, Body and Fashion to 3D Virtual Avatars

Alberto Del Bimbo   University of Florence, Italy
Mohamed Daoudi   IMT Nord Europe, France
Roberto Vezzani   University of Modena and Reggio Emilia, Italy
Xavier Alameda-Pineda   Inria Grenoble, France
Marcella Cornia   University of Modena and Reggio Emilia, Italy
Guido Borghi   University of Bologna, Italy
Claudio Ferrari   University of Parma, Italy
Federico Becattini   University of Florence, Italy
Andrea Pilzer   NVIDIA AI Technology Center, Italy
Zhiwen Chen   Alibaba Group, China
Xiangyu Zhu   Chinese Academy of Sciences, China
Ye Pan   Shanghai Jiao Tong University, China
Xiaoming Liu   Michigan State University, USA

W20 - Safe Artificial Intelligence for Automated Driving

Timo Saemann   Valeo, Germany
Oliver Wasenmuller   Hochschule Mannheim, Germany
Markus Enzweiler   Esslingen University of Applied Sciences,
                               Germany
Peter Schlicht   CARIAD, Germany
Joachim Sicking   Fraunhofer IAIS, Germany
Stefan Milz   Spleenlab.ai and Technische Universität Ilmenau,
                               Germany
Fabian Hüger   Volkswagen Group Research, Germany
Seyed Ghobadi   University of Applied Sciences Mittelhessen,
                               Germany
Ruby Moritz   Volkswagen Group Research, Germany
Oliver Grau   Intel Labs, Germany
Frédérik Blank   Bosch, Germany
Thomas Stauner   BMW Group, Germany

W21 - Real-World Surveillance: Applications and Challenges

Kamal Nasrollahi   Aalborg University, Denmark
Sergio Escalera   Universitat Autònoma de Barcelona, Spain

Radu Tudor Ionescu   University of Bucharest, Romania
Fahad Shahbaz Khan   Mohamed bin Zayed University of Artificial
                                      Intelligence, United Arab Emirates
Thomas B. Moeslund   Aalborg University, Denmark
Anthony Hoogs   Kitware, USA
Shmuel Peleg   The Hebrew University, Israel
Mubarak Shah   University of Central Florida, USA

W22 - Affective Behavior Analysis In-the-Wild

Dimitrios Kollias   Queen Mary University of London, UK
Stefanos Zafeiriou   Imperial College London, UK
Elnar Hajiyev   Realeyes, UK
Viktoriia Sharmanska   University of Sussex, UK

W23 - Visual Perception for Navigation in Human Environments: The JackRabbot Human Body Pose Dataset and Benchmark

Hamid Rezatofighi   Monash University, Australia
Edward Vendrow   Stanford University, USA
Ian Reid   University of Adelaide, Australia
Silvio Savarese   Stanford University, USA

W24 - Distributed Smart Cameras

Niki Martinel   University of Udine, Italy
Ehsan Adeli   Stanford University, USA
Rita Pucci   University of Udine, Italy
Animashree Anandkumar   Caltech and NVIDIA, USA
Caifeng Shan   Shandong University of Science and Technology,
                                      China
Yue Gao   Tsinghua University, China
Christian Micheloni   University of Udine, Italy
Hamid Aghajan   Ghent University, Belgium
Li Fei-Fei   Stanford University, USA

W25 - Causality in Vision

Yulei Niu   Columbia University, USA
Hanwang Zhang   Nanyang Technological University, Singapore
Peng Cui   Tsinghua University, China
Song-Chun Zhu   University of California, Los Angeles, USA
Qianru Sun   Singapore Management University, Singapore
Mike Zheng Shou   National University of Singapore, Singapore
Kaihua Tang   Nanyang Technological University, Singapore

W26 - In-Vehicle Sensing and Monitorization

Jaime S. Cardoso	INESC TEC and Universidade do Porto, Portugal
Pedro M. Carvalho	INESC TEC and Polytechnic of Porto, Portugal
João Ribeiro Pinto	Bosch Car Multimedia and Universidade do Porto, Portugal
Paula Viana	INESC TEC and Polytechnic of Porto, Portugal
Christer Ahlström	Swedish National Road and Transport Research Institute, Sweden
Carolina Pinto	Bosch Car Multimedia, Portugal

W27 - Assistive Computer Vision and Robotics

Marco Leo	National Research Council of Italy, Italy
Giovanni Maria Farinella	University of Catania, Italy
Antonino Furnari	University of Catania, Italy
Mohan Trivedi	University of California, San Diego, USA
Gérard Medioni	Amazon, USA

W28 - Computational Aspects of Deep Learning

Iuri Frosio	NVIDIA, Italy
Sophia Shao	University of California, Berkeley, USA
Lorenzo Baraldi	University of Modena and Reggio Emilia, Italy
Claudio Baecchi	University of Florence, Italy
Frederic Pariente	NVIDIA, France
Giuseppe Fiameni	NVIDIA, Italy

W29 - Computer Vision for Civil and Infrastructure Engineering

Joakim Bruslund Haurum	Aalborg University, Denmark
Mingzhu Wang	Loughborough University, UK
Ajmal Mian	University of Western Australia, Australia
Thomas B. Moeslund	Aalborg University, Denmark

W30 - AI-Enabled Medical Image Analysis: Digital Pathology and Radiology/COVID-19

Jaime S. Cardoso	INESC TEC and Universidade do Porto, Portugal
Stefanos Kollias	National Technical University of Athens, Greece
Sara P. Oliveira	INESC TEC, Portugal
Mattias Rantalainen	Karolinska Institutet, Sweden
Jeroen van der Laak	Radboud University Medical Center, The Netherlands
Cameron Po-Hsuan Chen	Google Health, USA

Diana Felizardo	IMP Diagnostics, Portugal
Ana Monteiro	IMP Diagnostics, Portugal
Isabel M. Pinto	IMP Diagnostics, Portugal
Pedro C. Neto	INESC TEC, Portugal
Xujiong Ye	University of Lincoln, UK
Luc Bidaut	University of Lincoln, UK
Francesco Rundo	STMicroelectronics, Italy
Dimitrios Kollias	Queen Mary University of London, UK
Giuseppe Banna	Portsmouth Hospitals University, UK

W31 - Compositional and Multimodal Perception

Kazuki Kozuka	Panasonic Corporation, Japan
Zelun Luo	Stanford University, USA
Ehsan Adeli	Stanford University, USA
Ranjay Krishna	University of Washington, USA
Juan Carlos Niebles	Salesforce and Stanford University, USA
Li Fei-Fei	Stanford University, USA

W32 - Uncertainty Quantification for Computer Vision

Andrea Pilzer	NVIDIA, Italy
Martin Trapp	Aalto University, Finland
Arno Solin	Aalto University, Finland
Yingzhen Li	Imperial College London, UK
Neill D. F. Campbell	University of Bath, UK

W33 - Recovering 6D Object Pose

Martin Sundermeyer	DLR German Aerospace Center, Germany
Tomáš Hodaň	Reality Labs at Meta, USA
Yann Labbé	Inria Paris, France
Gu Wang	Tsinghua University, China
Lingni Ma	Reality Labs at Meta, USA
Eric Brachmann	Niantic, Germany
Bertram Drost	MVTec, Germany
Sindi Shkodrani	Reality Labs at Meta, USA
Rigas Kouskouridas	Scape Technologies, UK
Ales Leonardis	University of Birmingham, UK
Carsten Steger	Technical University of Munich and MVTec, Germany
Vincent Lepetit	École des Ponts ParisTech, France, and TU Graz, Austria
Jiří Matas	Czech Technical University in Prague, Czech Republic

W34 - Drawings and Abstract Imagery: Representation and Analysis

Diane Oyen	Los Alamos National Laboratory, USA
Kushal Kafle	Adobe Research, USA
Michal Kucer	Los Alamos National Laboratory, USA
Pradyumna Reddy	University College London, UK
Cory Scott	University of California, Irvine, USA

W35 - Sign Language Understanding

Liliane Momeni	University of Oxford, UK
Gül Varol	École des Ponts ParisTech, France
Hannah Bull	University of Paris-Saclay, France
Prajwal K. R.	University of Oxford, UK
Neil Fox	University College London, UK
Ben Saunders	University of Surrey, UK
Necati Cihan Camgöz	Meta Reality Labs, Switzerland
Richard Bowden	University of Surrey, UK
Andrew Zisserman	University of Oxford, UK
Bencie Woll	University College London, UK
Sergio Escalera	Universitat Autònoma de Barcelona, Spain
Jose L. Alba-Castro	Universidade de Vigo, Spain
Thomas B. Moeslund	Aalborg University, Denmark
Julio C. S. Jacques Junior	Universitat Autònoma de Barcelona, Spain
Manuel Vázquez Enríquez	Universidade de Vigo, Spain

W36 - A Challenge for Out-of-Distribution Generalization in Computer Vision

Adam Kortylewski	Max Planck Institute for Informatics, Germany
Bingchen Zhao	University of Edinburgh, UK
Jiahao Wang	Max Planck Institute for Informatics, Germany
Shaozuo Yu	The Chinese University of Hong Kong, China
Siwei Yang	Hong Kong University of Science and Technology, China
Dan Hendrycks	University of California, Berkeley, USA
Oliver Zendel	Austrian Institute of Technology, Austria
Dawn Song	University of California, Berkeley, USA
Alan Yuille	Johns Hopkins University, USA

W37 - Vision With Biased or Scarce Data

Kuan-Chuan Peng	Mitsubishi Electric Research Labs, USA
Ziyan Wu	United Imaging Intelligence, USA

W38 - Visual Object Tracking Challenge

Matej Kristan   University of Ljubljana, Slovenia
Aleš Leonardis   University of Birmingham, UK
Jiří Matas   Czech Technical University in Prague,
                                  Czech Republic
Hyung Jin Chang   University of Birmingham, UK
Joni-Kristian Kämäräinen   Tampere University, Finland
Roman Pflugfelder   Technical University of Munich, Germany,
                                  Technion, Israel, and Austrian Institute of
                                  Technology, Austria
Luka Čehovin Zajc   University of Ljubljana, Slovenia
Alan Lukežič   University of Ljubljana, Slovenia
Gustavo Fernández   Austrian Institute of Technology, Austria
Michael Felsberg   Linköping University, Sweden
Martin Danelljan   ETH Zurich, Switzerland

Contents – Part I

W03 - Adversarial Robustness in the Real World

W04 - Autonomous Vehicle Vision

W01 - AI for Space

W01 - AI for Space

The space sector is experiencing significant growth. Currently planned activities and utilization models also greatly exceed the scope, ambition, and/or commercial value of space missions in the previous century, e.g., autonomous spacecraft, space mining, and understanding the universe. Achieving these ambitious goals requires surmounting non-trivial technical obstacles. This workshop focuses on the role of AI, particularly computer vision and machine learning, in helping to solve those technical hurdles. The workshop aims to highlight the space capabilities that draw from and/or overlap significantly with vision and learning research, outline the unique difficulties presented by space applications to vision and learning, and discuss recent advances towards overcoming those obstacles.

October 2022

Tat-Jun Chin
Luca Carlone
Djamila Aouada
Binfeng Pan
Viorela Ila
Benjamin Morrell
Grzegorz Kakareko

Globally Optimal Event-Based Divergence Estimation for Ventral Landing

Sofia McLeod[1] ⓘ, Gabriele Meoni[2] ⓘ, Dario Izzo[2] ⓘ, Anne Mergy[2] ⓘ,
Daqi Liu[1] ⓘ, Yasir Latif[1(✉)] ⓘ, Ian Reid[1] ⓘ, and Tat-Jun Chin[1] ⓘ

[1] School of Computer Science, The University of Adelaide, Adelaide, Australia
{Sofia.McLeod,Daqi.Liu,Yasir.Latif,Ian.Reid,Tat-Jun.Chin}@adelaide.edu.au
[2] Advanced Concepts Team, European Space Research and Technology Centre,
Keplerlaan 1, 2201 AZ Noordwijk, The Netherlands
{Gabriele.Meoni,Dario.Izzo,Anne.Mergy}@esa.int

Abstract. Event sensing is a major component in bio-inspired flight guidance and control systems. We explore the usage of event cameras for predicting time-to-contact (TTC) with the surface during ventral landing. This is achieved by estimating divergence (inverse TTC), which is the rate of radial optic flow, from the event stream generated during landing. Our core contributions are a novel contrast maximisation formulation for event-based divergence estimation, and a branch-and-bound algorithm to exactly maximise contrast and find the optimal divergence value. GPU acceleration is conducted to speed up the global algorithm. Another contribution is a new dataset containing real event streams from ventral landing that was employed to test and benchmark our method. Owing to global optimisation, our algorithm is much more capable at recovering the true divergence, compared to other heuristic divergence estimators or event-based optic flow methods. With GPU acceleration, our method also achieves competitive runtimes.

Keywords: Event cameras · Time-to-contact · Divergence · Optic flow

1 Introduction

Flying insects employ seemingly simple sensing and processing mechanisms for flight guidance and control. Notable examples include the usage of observed optic flow (OF) by honeybees for flight control and odometry [48], and observed OF asymmetries for guiding free flight behaviours in fruit flies [54]. Taking inspiration from insects to build more capable flying machines [17,49,50], a branch of biomimetic engineering, is an active research topic[1].

---

[1] Please see supplementary material for a demo program.

Supplementary Information The online version contains supplementary material available at https://doi.org/10.1007/978-3-031-25056-9_1.

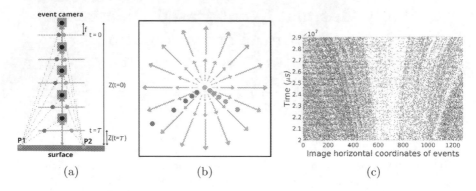

Fig. 1. (a) An event camera with focal length f undergoing ventral descent to a surface over time 0 to τ seconds. (b) Two example event trajectories triggered by observation of scene points $P1$ and $P2$ on the surface. The event trajectories are asynchronous observations of the radial optic flows, which emanate from the focus of expansion (FOE). (c) Actual event data produced under ventral descent (only the image horizontal coordinates and event time stamps are shown here).

The advent of event sensors [10, 28, 42] has helped push the boundaries of biomimetic engineering. Akin to biological eyes, a pixel in an event sensor asynchronously triggers an event when the intensity change at that pixel exceeds a threshold. An (ideal) event sensor outputs a stream of events if it observes continuous changes, *e.g.*, due to motion, else it generates nothing. The asynchronous pixels also enable higher dynamic range and lower power consumption. These qualities make event sensors attractive as flight guidance sensors for aircraft [13, 15, 32, 41, 45, 57] and spacecraft [35, 47, 55, 56].

In previous works, divergence is typically recovered from the OF vectors resulting from event-based OF estimation [41, 47, 55, 56]. However, OF estimation is challenged by noise in the event data (see Fig. 2b), which leads to inaccurate divergence estimates (as we will show in Sect. 6). Fundamentally, OF estimation, which usually includes detecting and tracking features, is more complex than estimating divergence (a scalar quantity).

Our Contributions. We present a *contrast maximisation* [18, 53] formulation for event-based divergence estimation. We examined the geometry of ventral landing and derived event-based radial flow, then built a mathematically justified optimisation problem that aims to maximise the contrast of the flow-compensated image; see Fig. 2. To solve the optimisation, we developed an exact algorithm based on branch-and-bound (BnB) [24], which is accelerated using GPU.

Since our aim is divergence estimation, our mathematical formulation differs significantly from the exact methods of [29, 40], who focused on different problems. More crucially, our work establishes the practicality of exact contrast maximisation through GPU acceleration, albeit on a lower dimensional problem.

To test our method, we collected real event data from ventral landing scenarios (Sect. 5), where ground truth values were produced through controlled

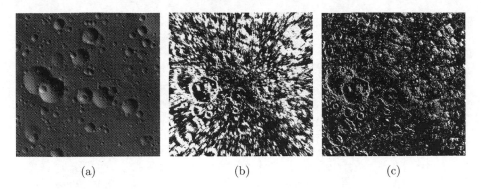

(a)   (b)   (c)

Fig. 2. (a) Intensity image of landing surface. (b) Event image (without motion compensation) for a batch of events generated while observing the surface during vertical landing (image contrast is 0.15). (c) Motion compensated event image produced by our divergence estimation method (image contrast is 0.44).

recording using a robot arm and depth sensing[2]. Results in Sect. 6 show that our method is much more accurate than previous divergence estimators [41,47] that conduct heuristic OF recovery (plane fitting, centroid tracking). Our method also exhibited superior accuracy over state-of-the-art event-based OF methods [20,51,53] that we employed for divergence estimation. The usage of GPU acceleration enabled our global method to achieve competitive runtimes.

2 Related Work

Event-Based Optic Flow. Event-based OF is a rapidly growing area in computer vision. Some event-based OF approaches [7,37] are based on the traditional brightness constancy assumption. Other methods [9,18] expand on the Lucas and Kanade algorithm [31], which assumes constant, localised flow. Alternative approaches include pure event-based time oriented OF estimations [8], computing distance transforms as input to intensity-based OF algorithms [6], and adaptive block matching where OF is computed a points where brightness changes [30]. An extension of event-based OF is event-based motion segmentation [52,53,59], where motion models are fitted to clusters of events that have the same OF.

Learning methods for event-based OF train deep neural networks (DNNs) to output event image OF estimations from input representations of event point clouds. Some of these methods require intensity frames in addition to the event data [14,60,61], whereas others require event data only [19,20,58]. Closely relevant are image and video reconstruction from event methods [38,43,44,46], which employ frame-based and event-based OF techniques. Note that not all event sensors provide intensity frames and event streams.

---

[2] Dataset: https://github.com/s-mcleod/ventral-landing-event-dataset.

Event-Based Divergence Estimation. Using event cameras for flight guidance (*e.g.*, attitude estimation, aerial manoeuvres) is an active research area [11, 12, 21, 35]. The focus of our work is event-based divergence estimation, which is useful for guiding the ventral descent of aircraft [41] and spacecraft [47].

Pijnacker-Hordijk *et al.* [41] employed and extended the plane fitting method of [8] for event-based divergence estimation. The algorithm was also embedded in a closed-loop vertical landing system of an MAV. In the context of planetary landing, Sikorski *et al.* [47] presented a local feature matching and tracking method, which was also embedded in a closed-loop guidance system. However, their system was evaluated only on simulated event data [47].

Since our focus is on the estimation of divergence, we will distinguish against the state-of-the-art methods [41, 47] only in that aspect, *i.e.*, independently of a close-loop landing system. As we will show in Sect. 6, our globally optimal contrast maximisation approach provides much more accurate estimates than [41, 47].

Spiking Neural Networks. Further pushing the boundaries of biomimetic engineering is the usage of spiking neural networks (SNN) [36] for event-based OF estimation [22, 23, 27, 39]. Due to the limited availability of SNN hardware [26], we consider only off-the-shelf computing devices (CPUs and GPUs) in our work, but we note that SNNs are attractive future research.

3   Geometry of Ventral Landing

We first examine the geometry of ventral descent and derive continuous radial OF and divergence, before developing the novel event-based radial flow.

3.1   Continuous-Time Optic Flow

Let $\mathbf{P}(t) = \begin{bmatrix} X(t), Y(t), Z(t) \end{bmatrix}^T$ be a 3D scene point that is moving with velocity $\mathbf{V}(t) = \begin{bmatrix} V_X(t), V_Y(t), V_Z(t) \end{bmatrix}^T$. Assuming a calibrated camera, projecting $\mathbf{P}(t)$ onto the image plane yields a moving image point

$$\mathbf{p}(t) = \begin{bmatrix} x(t), y(t) \end{bmatrix}^T := \begin{bmatrix} fX(t)/Z(t), fY(t)/Z(t) \end{bmatrix}^T, \tag{1}$$

where f is the focal length of the event camera. Differentiating $\mathbf{p}(t)$ yields the *continuous-time OF* of the point, *i.e.*,

$$\mathbf{v}(t) = \begin{bmatrix} \dfrac{fV_X(t) - V_Z(t)x(t)}{Z(t)}, \dfrac{fV_Y(t) - V_Z(t)y(t)}{Z(t)} \end{bmatrix}. \tag{2}$$

Note that the derivations above are standard; see, *e.g.*, [16, Chapter 10].

3.2 Continuous-Time Radial Flow and Divergence

Following [41,47], we consider the ventral landing scenario depicted in Fig. 1a. The setting carries the following assumptions:

- All scene points lie on a fronto-parallel flat surface; and
- The camera has no horizontal motion, i.e., $V_X(t) = V_Y(t) = 0$, which implies that $\mathbf{P}(t) = [X, Y, Z(t)]^T$, i.e., the first two elements of $\mathbf{P}(t)$ are constants.

Sect. 6 will test our ideas using real data which unavoidably deviated from the assumptions. Under ventral landing, (2) is simplified to

$$\mathbf{v}(t) = -\left[V_Z(t)/Z(t)\right]\mathbf{p}(t). \tag{3}$$

Since the camera is approaching the surface, $V_Z(t) \leq 0$, hence $\mathbf{v}(t)$ points away from $\mathbf{0}$ (the FOE) towards the direction of $\mathbf{p}(t)$; see Fig. 1b.

The *continuous-time divergence* is the quantity

$$D(t) = V_Z(t)/Z(t), \tag{4}$$

which is the rate of expansion of the radial OF. The divergence is independent of the scene point $\mathbf{P}(t)$, and the inverse of $D(t)$, i.e., distance to surface over rate of approach to the surface, is the *instantaneous TTC*.

3.3 Event-Based Radial Flow

Let $\mathcal{S}$ be an event stream generated by an event camera during ventral descent. Following previous works, e.g. [18,37,52], we separate $\mathcal{S}$ into small finite-time sequential batches. Let $\mathcal{E} = \{\mathbf{e}_i\}_{i=1}^N \subset \mathcal{S}$ be a batch with N events, where each $\mathbf{e}_i = (x_i, y_i, t_i, p_i)$ contains image coordinates (x_i, y_i), time stamp t_i and polarity p_i of the i-th event. To simplify exposition, we offset the time by conducting $t_i = t_i - \min_i t_i$ such that the period of $\mathcal{E}$ has the form $[0, \tau]$.

Since τ is small relative to the speed of descent, we assume that $V_Z(t)$ is constant in $[0, \tau]$. We can thus rewrite (3) corresponding to $\mathbf{P}(t)$ as

$$\mathbf{v}(t) = \frac{-\nu}{Z_0 + \nu t}\left[\frac{fX}{Z_0 + \nu t}, \frac{fY}{Z_0 + \nu t}\right]^T, \quad \forall t \in [0, \tau], \tag{5}$$

where ν is the constant vertical velocity in $[0, \tau]$, and Z_0 is the surface depth at time 0 (recall also that the horizontal coordinates of $\mathbf{P}(t)$ are constant under ventral descent). Accordingly, divergence is redefined as

$$D(t) = \nu/(Z_0 + \nu t). \tag{6}$$

Assume that $\mathbf{e}_i$ was triggered by scene point $\mathbf{P}(t)$ at time $t_i \in [0, \tau]$, i.e., $\mathbf{e}_i$ is a point observation of the continuous flow $\mathbf{v}(t)$. This means we can recover

$$X = x_i(Z_0 + \nu t_i)/f, \quad Y = y_i(Z_0 + \nu t_i)/f, \tag{7}$$

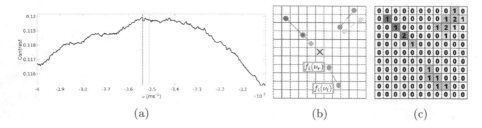

Fig. 3. (a) Plot of the contrast function (14) for an event batch. Blue and green lines represent optimal and ground truth velocities respectively. Figure 1c was produced by the optimal velocity. (b) Event lines $f_i(\nu_\ell) \leftrightarrow f_i(\nu_r)$ corresponding to five events (differentiated by colours) under velocity domain $\mathcal{V} = [\nu_\ell, \nu_r]$. (c) Upper bound image $\bar{H}$ whose computation is GPU accelerated. (Color figure online)

following from (1). Substituting (7) into (5) yields

$$\mathbf{v}(t) = \frac{-\nu}{Z_0 + \nu t} \left[\frac{x_i(Z_0 + \nu t_i)}{Z_0 + \nu t}, \frac{y_i(Z_0 + \nu t_i)}{Z_0 + \nu t} \right]^T, \tag{8}$$

The future event triggered by $\mathbf{P}(t)$ at time t', where $t_i \leq t' \leq \tau$, is

$$\mathbf{p}' = \begin{bmatrix} x_i \\ y_i \end{bmatrix} + \int_{t_i}^{t'} \mathbf{v}(t)dt = \begin{bmatrix} x_i \\ y_i \end{bmatrix} + \Gamma_i(t'), \tag{9}$$

and $\Gamma_i(t')$ is called the *event-based radial flow*

$$\Gamma_i(t') := \left[\frac{x_i(Z_0 + \nu t_i)}{Z_0 + \nu t'} - x_i, \frac{y_i(Z_0 + \nu t_i)}{Z_0 + \nu t'} - y_i \right]^T. \tag{10}$$

While (9) seems impractical since it requires knowing the depth Z_0, we will circumvent *a priori* knowledge of the depth next.

4   Exact Algorithm for Divergence Estimation

Following the batching procedure in Sect. 3.3, our problem is reduced to estimating divergence (6) for each event batch $\mathcal{E}$ with assumed time duration $[0, \tau]$. Here, we describe our novel algorithm to perform the estimation exactly.

4.1   Optimisation Domain and Retrieval of Divergence

There are two unknowns in (6): Z_0 and ν. Since the event camera approaches the surface during descent, we must have $\nu \leq 0$. We also invoke cheirality, *i.e.*, the surface is always in front of the camera, thus $Z_0 + \nu\tau \geq 0$. These imply

$$-Z_0/\tau \leq \nu \leq 0. \tag{11}$$

Note also that $D(t)$ is a ratio, thus the precise values of Z_0 and ν are not essential. We can thus fix, say, $Z_0 = 1$, and scale ν accordingly to achieve the same divergence. The aim is now reduced to estimating ν from $\mathcal{E}$.

Given the estimated ν^*, the divergence can be retrieved as $D(t) = \nu^*/(1 + \nu^* t)$. In our experiments, we take the point estimate $D^* = \nu^*/(1 + \nu^* \tau)$.

4.2 Contrast Maximisation

To estimate ν from $\mathcal{E} = \{e_i\}_{i=1}^N$, we define the *event-based radial warp*

$$f_i(\nu) = \begin{bmatrix} x_i \\ y_i \end{bmatrix} + \begin{bmatrix} \dfrac{x_i(Z_0 + \nu t_i)}{Z_0 + \nu\tau} - x_i, \ \dfrac{y_i(Z_0 + \nu t_i)}{Z_0 + \nu\tau} - y_i \end{bmatrix}^T, \tag{12}$$

which is a specialisation of (9) by grounding $Z_0 = 1$ and warping e_i to the end τ of the duration of $\mathcal{E}$. The motion compensated image

$$H(\mathbf{u}; \nu) = \sum_{i=1}^N \mathbb{I}(f_i(\nu) \text{ lies in pixel } \mathbf{u}), \tag{13}$$

accumulates events that are warped to the same pixel (note that we employ the discrete version of [29]). Intuitively, noisy events triggered by the same scene point will be warped by $f_i(\nu)$ to close by pixels. This motivates contrast [18]

$$C(\nu) = \frac{1}{M} \sum_{\mathbf{u}} (H(\mathbf{u}; \nu) - \mu(\nu))^2, \tag{14}$$

of the motion compensated image as an objective function for estimating ν, where M is the number of pixels in the image, and $\mu(\nu)$ is the mean of $H(\mathbf{u}; \nu)$. Figure 3a illustrates the suitability of (14) for our aim. In particular, the true ν^* is very close to the maximiser of the contrast. Note that, following [18,29] we do not utilise the event polarities, but they can be accommodated if desired.

4.3 Exact Algorithm

As shown in Fig. 3a, contrast (14) is non-convex and non-smooth. Liu *et al.* [29] demonstrated that maximising the contrast using approximate methods (*e.g.*, via smoothing and gradient descent) could lead to bad local optima. Inspired

---

Algorithm 1. Exact contrast maximisation for ventral velocity estimation.

Require: Event batch $\mathcal{E} = \{\mathbf{e}_i\}_i^N$, batch duration $[0, \tau]$, convergence threshold γ.
1: Initialise $\mathcal{V} \leftarrow [-1/\tau, 0]$, $\hat{\nu} \leftarrow$ centre of $\mathcal{V}$.
2: Insert $\mathcal{V}$ into priority queue Q with priority $\bar{C}(\mathcal{V})$.
3: **while** Q is not empty **do**
4:      $\mathcal{V} \leftarrow$ dequeue first item from Q.
5:      **if** $\bar{C}(\mathcal{V}) - C(\hat{\nu}) \leq \gamma$ **then break**.
6:      $\nu_c \leftarrow$ centre of $\mathcal{V}$.
7:      **if** $C(\nu_c) \geq C(\hat{\nu})$ **then** $\hat{\nu} \leftarrow \nu_c$.
8:      Split $\mathcal{V}$ equally into 2 subdomains $\mathcal{V}_1$ and $\mathcal{V}_2$.
9:      **for** $s = 1, 2$ **do**
10:         **if** $\bar{C}(\mathcal{V}_s) \geq C(\hat{\nu})$ **then** insert $\mathcal{V}_s$ into Q with priority $\bar{C}(\mathcal{V}_s)$.
11:     **end for**
12: **end while**
13: **return** $\hat{\nu}$.

---

by [29, 40], we developed a BnB method (Algorithm 1) to exactly maximise (14) over the domain $\mathcal{V} = [-1/\tau, 0]$ of the vertical velocity ν.

Starting with $\mathcal{V}$, Algorithm 1 iteratively splits and prunes $\mathcal{V}$ until the maximum contrast is found. For each subdomain, the upper bound $\bar{C}$ of the contrast over the subdomain (more details below) is computed and compared against the contrast of the current best estimate $\hat{\nu}$ to determine whether the subdomain can be discarded. The process continues until the difference between current best upper bound and the quality of $\hat{\nu}$ is less than a predefined threshold γ. Then, $\hat{\nu}$ is the globally optimal solution, up to γ. Given the velocity estimate from Algorithm 1, Sect. 4.1 shows how to retrieve the divergence.

Upper Bound Function. Following [29, Sec. 3], we use the general form

$$\bar{C}(\mathcal{V}) = \frac{1}{M}\bar{S}(\mathcal{V}) - \underline{\mu}(\mathcal{V})^2, \tag{15}$$

as the upper bound for contrast, where the terms respectively satisfy

$$\bar{S}(\mathcal{V}) \geq \max_{\nu \in \mathcal{V}} \sum_{\mathbf{u}} H(\mathbf{u}; \nu)^2 \quad \text{and} \quad \underline{\mu}(\mathcal{V}) \leq \min_{\nu \in \mathcal{V}} \mu(\nu), \tag{16}$$

as well as equality in (16) when $\mathcal{V}$ is singleton. See [29, Sec. 3] on the validity of (15). For completeness, we also re-derive (15) in the supplementary material.

It remains to specify $\bar{S}$ and $\underline{\mu}$ for our ventral landing case. To this end, we define the upper bound image

$$\bar{H}(\mathbf{u}; \mathcal{V}) = \sum_{i=1}^{N} \mathbb{I}(f_i(\nu_\ell) \leftrightarrow f_i(\nu_r) \text{ intersects with pixel } \mathbf{u}), \tag{17}$$

where $\nu_\ell = \min(\mathcal{V})$ and $\nu_r = \max(\mathcal{V})$, and $f_i(\nu_\ell) \leftrightarrow f_i(\nu_r)$ is the line on the image that connects points $f_i(\nu_\ell)$ and $f_i(\nu_r)$; see Figs. 3b and 3c. We can establish

$$\bar{H}(\mathbf{u}; \mathcal{V}) \geq H(\mathbf{u}; \nu), \quad \forall \nu \in \mathcal{V}, \tag{18}$$

with equality achieved when $\mathcal{V}$ reaches a single point ν in the limit; see supplementary material for the proof. This motivates to construct

$$\bar{S}(\mathcal{V}) = \sum_{\mathbf{u}} \bar{H}(\mathbf{u}; \mathcal{V})^2, \tag{19}$$

which satisfies the 1st condition in (16). For the 2nd term in (15), we use

$$\underline{\mu}(\mathcal{V}) = \frac{1}{M} \sum_{i=1}^{N} \mathbb{I}(f_i(\nu_\ell) \leftrightarrow f_i(\nu_r) \text{ fully lies in the image plane}), \tag{20}$$

which satisfies the second condition in (16); see proof in supplementary material.

GPU Acceleration. The main costs in Algorithm 1 are computing the contrast (14), contrast upper bound (15), and the line-pixel intersections (17). These routines are readily amenable to GPU acceleration via CUDA (this was suggested in [29], but no details or results were reported).

A key step is CUDA memory allocation to reduce the overheads of reallocating memory and copying data between CPU-GPU across the BnB iterations:

```
cudaMallocManaged(&event_image, ...)
cudaMallocManaged(&event_lines, ...)
cudaMalloc(&events, ...)
cudaMemcpy(&events, cpu_events, ...)
```

Summing all pixels (including with squaring the pixels), can be parallelised through the CUDA Thrust library:

```
sum_of_pixels = thrust::reduce(event_image_plane, ...)
sum_of_square_of_pixels = thrust::transform_reduce(
    event_image_plane, square<double>...,plus<double>..., ...)
```

Finally, the computation of all lines and line-pixel intersections can be parallelised through CUDA device functions:

```
get_upper_bound_lines <<< number_of_blocks, block_size >>>
    (events, event_lines, ...)
get_upper_bound_image <<< number_of_blocks, block_size >>>
    (event_lines, event_image_plane, ...)
```

Our CUDA acceleration (on an Nvidia GeForce RTX 2060) resulted in a runtime reduction of 80% for the routines above, relative to pure CPU implementation.

4.4   Divergence Estimation for Event Stream

The algorithm discussed in Sect. 4 can be used to estimate divergence for a single event batch $\mathcal{E}$. To estimate the time-varying divergence for the event stream $\mathcal{S}$, we incrementally take event batches from $\mathcal{S}$, and estimate the divergence of each batch as discrete samples of the continuous divergence value. In Sect. 6, we will report the batch sizes (which affect the sampling frequency) used in our work.

5   Ventral Landing Event Dataset

Due to the lack of event datasets for ventral landing, we collected our own dataset to benchmark the methods. Our dataset consists of

- 1 simulated event sequence (SurfSim) of duration 47 s.
- 7 real event sequences observing 2D prints of landing surfaces (Surf2D-X, where X = 1 to 7) of duration 15 s each.
- 1 real event sequence observing the 3D print of a landing surface (Surf3D) of duration 15 s.

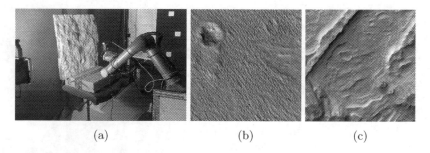

(a)   (b)   (c)

Fig. 4. (a) Our setup for recording real event data under (approximate) ventral landing, shown here with the 3D printed lunar surface. Note the only approximately fronto-parallel surface. (b) (c) [33,34] Two of the planetary surface images that were 2D printed. See supplementary material for more details.

SurfSim was produced using v2e [25] to process a video that was graphically rendered using PANGU [4], of a camera undergoing pure ventral landing on a lunar-like surface (see Fig. 2).

Figure 4 depicts our setup for recording the real event sequences. A UR5 robot arm was employed to execute a controlled linear trajectory towards horizontally facing 2D printed planetary surfaces (Mars, Mercury, *etc.*) and a 3D printed lunar surface (based on the lunar DEM from NASA/JAXA [3]). A Prophesee Gen 4 event camera and Intel RealSense depth sensor were attached to the end effector, while a work lamp provided lighting. Ground truth divergences at 33 Hz were computed from velocities and depths retrieved from the robot arm and depth camera. Note that, despite using a robot arm, the real sequences do not fully obey the ventral landing assumptions, due to the surfaces not being strictly fronto-parallel to the camera, and the 3D surface not being planar.

6   Results

We tested our method, **e**xact **c**ontrast **m**aximisation for event-based **d**ivergence estimation (ECMD), on the dataset described in Sect. 5. We compared against:

- Local plane fitting (PF) method [41];
- 2D centroid tracking (CT) method [47];
- Learning-based dense OF method (E-RAFT) [20]; and
- Tracking and grid search-based OF method (SOFAS) [51,53].

Only PF and CT were aimed at divergence estimation, whereas SOFAS and E-RAFT are general OF methods. While there are other event-based OF methods, *e.g.*, [6,18,19,30,37,58], our list includes the main categories of heuristic, optimisation and learning methods which had publicly available implementations.

Preprocessing. All event sequences were subject to radial undistortion and hot pixel removal. Two versions of the data were also produced:

- *Full resolution*: The original data collected for `Surf2D-X` and `Surf3D` by our Prophesee Gen 4 event camera (1280x760 pixels) and simulated data for `SurfSim` (512x512 pixels), with only the processing above.
- *Resized and resampled*: The effective image resolution of the camera was resized (by scaling the image coordinates of the events) to 160x90 pixels for `Surf2D-X` and `Surf3D`, and 64x64 pixels for `SurfSim`. The size of each event batch was also reduced by 75% by random sampling.

Implementation and Tuning. For ECMD, the convergence threshold γ (see Algorithm 1) was set to 0.025, and an event batch duration of 0.5 s was used. For PF, we used the implementation available at [5]. There were 16 hyperparameters that influence the estimated divergence in nontrivial ways, thus we used the default values. Note that PF is an asynchronous algorithm, thus its temporal sampling frequency is not directly comparable to the other methods. We used our own implementation of CT. The two main parameters were the temporal memory and batch duration, which we tuned to 0.3 s and 0.1 s for best performance. For SOFAS, we used the implementation available at [2] with batch duration of 0.5 s. For E-RAFT, we used the implementation available at [1]. As there was no network training code available, we used the provided DSEC pre-trained network. E-RAFT was configured to output OF estimates every 0.5 s.

All methods were executed on a standard machine with an Core i7 CPU and RTX 2060 GPU. See supplementary material for more implementation details.

Converting of to Divergence. For E-RAFT and SOFAS, the divergence for each event batch was calculated from the OF estimates by finding the rate of perceived expansion of the OF vectors, as defined in [47] as

$$
D = \frac{1}{P\tau} \sum_{k=1}^{P} \left[1 - \frac{\|FOE + \mathbf{p}_k + \tau \mathbf{v}_k\|}{\|FOE + \mathbf{p}_k\||} \right], \tag{21}
$$

where P is the number of OF vectors, $\mathbf{v}_k$ is the k-th OF vector at corresponding pixel location $\mathbf{p}_k$, and τ is the duration of the event batch.

Comparison Metrics. For each method, we recorded:

- *Divergence error*: The absolute error (in %) between the estimated divergence and closest-in-time ground truth divergence for each event batch.
- *Runtime*: The runtime of the method on the event batch.

Since the methods were operated at different temporal resolutions for best performance, to allow comparisons we report results at 0.5 s time resolution, by taking the average divergence estimate of methods that operated at finer time scales (specifically PF at 0.01 s and CT at 0.3 s) over 0.5 s windows.

6.1   Qualitative Results

Figure 5 shows motion compensated event images from the estimated ν across all methods in Sect. 6, including event images produced under ground truth (GT) ν, and when $\nu = 0$ (Zero). In these examples, ECMD returned the highest image contrast (close to the contrast of GT).

Table 1. Divergence estimation results (error and runtime) on event sequences. The final column contains the average error and runtime across all sequences.

Full resolution event sequences											
Dataset		Surf-Sim	Surf 2D-1	Surf 2D-2	Surf 2D-3	Surf 2D-4	Surf 2D-5	Surf 2D-6	Surf 2D-7	Surf-3D	Avg.
Absolute average error per event batch (%)	ECMD	8.34	13.48	7.41	6.11	12.19	10.41	5.12	3.72	12.90	8.85
	PF	37.78	39.11	44.79	44.42	41.19	42.19	43.42	44.03	25.23	40.24
	CT	44.33	43.86	50.27	42.96	50.07	72.80	53.73	60.30	45.85	51.57
	E-RAFT	46.27	50.79	44.81	44.75	41.12	29.90	53.53	49.19	21.82	43.46
	SOFAS	819.33	24.85	24.65	23.63	30.67	34.80	37.05	30.88	71.54	121.93
Average runtime per event batch(s)	ECMD	7.99	6.84	6.95	6.50	5.85	7.62	6.02	3.01	2.36	5.90
	PF	0.21	0.22	0.23	0.30	0.18	0.28	0.22	0.14	0.13	0.21
	CT	35.12	31.58	35.46	41.67	28.48	48.49	42.17	21.43	20.98	33.93
	E-RAFT	0.24	0.25	0.25	0.24	0.25	0.25	0.24	0.25	0.24	0.25
	SOFAS	57.80	61.46	66.43	65.79	55.62	66.23	59.32	13.27	35.83	53.53
Resized and resampled event sequences											
Dataset		Surf-Sim	Surf 2D-1	Surf 2D-2	Surf 2D-3	Surf 2D-4	Surf 2D-5	Surf 2D-6	Surf 2D-7	Surf-3D	Avg.
Absolute average error per event batch (%)	ECMD	18.65	18.86	10.33	7.36	14.45	12.28	7.41	4.25	11.68	11.70
	PF	146.76	58.66	32.79	50.49	52.66	43.80	57.36	30.61	31.64	56.09
	CT	102.20	143.99	51.83	73.35	40.57	81.72	89.83	96.80	43.47	80.42
	E-RAFT	47.40	86.40	72.31	81.67	88.62	71.87	91.26	66.79	21.19	69.72
	SOFAS	1688.21	60.10	50.49	56.29	67.73	47.87	24.42	42.98	63.51	233.51
Average runtime per event batch(s)	ECMD	0.20	0.31	0.29	0.32	0.34	0.34	0.30	0.28	0.23	0.29
	PF	0.12	0.10	0.10	0.10	0.17	0.07	0.12	0.10	0.06	0.10
	CT	7.81	20.65	22.09	30.59	32.79	13.99	27.38	11.63	15.33	20.25
	E-RAFT	0.24	0.24	0.25	0.24	0.24	0.25	0.25	0.25	0.24	0.24
	SOFAS	6.14	36.18	39.06	39.54	38.67	34.63	40.13	36.96	25.27	32.95

6.2   Quantitative Results on Full Resolution Sequences

Figure 6 (top row) plots the estimated divergence for all methods across 3 of the full resolution event sequences; see supplementary material for the rest of the sequences. Observe that ECMD generally tracked the ground truth divergence well and was the most stable method. The estimates of the other methods were either inaccurate or quite erratic. A common failure point was towards the end of SurfSim; this was when the camera was very close to the surface and the limited resolution of the rendered image did not generate informative events.

Table 1 (top) provides average divergence errors and runtimes. Values of the former reflect the trends in Fig. 6; across all sequences, ECMD committed the least error relative to the ground truth. The results suggest that, although ECMD was based upon a pure vertical landing assumption, the method was not greatly affected by violations of the assumption in the real event sequences.

While PF and E-RAFT were more efficient (up to one order of magnitude faster than ECMD), their estimates were rather inaccurate. Despite GPU acceleration, the runtime of ECMD was longer than event batch duration. We will present results on the resized and resampled sequences next.

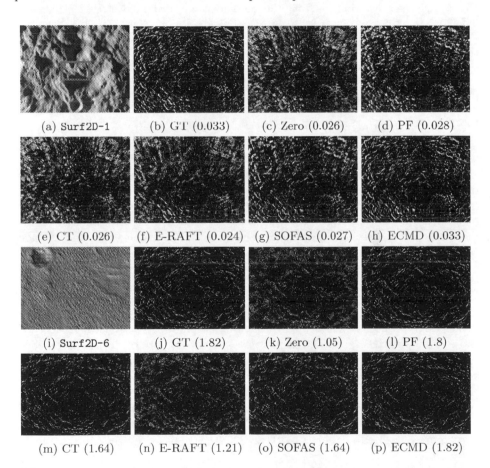

(a) Surf2D-1	(b) GT (0.033)	(c) Zero (0.026)	(d) PF (0.028)
(e) CT (0.026)	(f) E-RAFT (0.024)	(g) SOFAS (0.027)	(h) ECMD (0.033)
(i) Surf2D-6	(j) GT (1.82)	(k) Zero (1.05)	(l) PF (1.8)
(m) CT (1.64)	(n) E-RAFT (1.21)	(o) SOFAS (1.64)	(p) ECMD (1.82)

Fig. 5. (b)–(h) Flow-compensated event images produced for an event batch (full resolution) from Surf2D-1. (j)-(p) Flow-compensated event images produced for an event batch (resized and resampled) from Surf2D-6 [33]. GT = Ground truth. Zero = No warp. Numbers in parentheses are the contrast values (14).

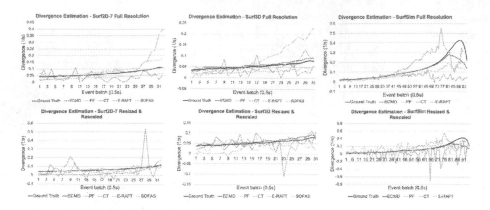

Fig. 6. Divergence estimates from the methods described in Sect. 6 for datasets `Surf2D-7`, `Surf3D`, and `SurfSim`. (Top row) Full resolution sequences. (Bottom row) Resized and resampled sequences. Note that SOFAS was not plotted for `SurfSim` since the divergence estimates were too innacurate (see Table 1).

6.3   Quantitative Results on Resized and Resampled Sequences

Figure 6 (bottom row) plots the estimated divergence for all methods across 3 of the resized and resampled event sequences; see supplementary material for the rest of the sequences. Table 1 (bottom) provides average errors and runtimes.

The results show that the accuracy of all methods generally deteriorated due to the lower image resolution and event subsampling. However, the impact on the accuracy of ECMD has been relatively small, with a difference in average error of within 3% from the full resolution results. Though PF was still the fastest method, the average runtime of ECMD has reduced by about 95%. The average runtime of ECMD was 0.29 s, which was below the batch duration.

7   Conclusions and Future Work

This paper proposes exact contrast maximisation for event-based divergence estimation (ECMD) for guiding ventral descent. Compared to other divergence and OF methods, ECMD demonstrated stable and accurate divergence estimates across our ventral landing dataset. Violations to the ECMD assumption on a pure ventral landing did not greatly affect the divergence estimates. Through GPU acceleration and downsampling, ECMD achieved competitive runtimes.

Future work includes integrating ECMD into a closed loop system [41,47], and comparing GPU-enabled ECMD to SNN event-based OF methods.

Acknowledgement. Sofia Mcleod was supported by the Australian Government RTP Scholarship. Tat-Jun Chin is SmartSat CRC Professorial Chair of Sentient Satellites.

References

1. E-RAFT: dense optical flow from event cameras. https://github.com/uzh-rpg/E-RAFT. Accessed 24 Feb 2022
2. Event contrast maximization library. https://github.com/TimoStoff/events_contrast_maximization. Accessed 27 Feb 2022
3. Moon LRO LOLA - Selene Kaguya TC dem merge 60N60S 59m v1. https://astrogeology.usgs.gov/search/map/Moon/LRO/LOLA/Lunar_LRO_LOLAKaguya_DEMmerge_60N60S_512ppd. Accessed 24 June 2021
4. Planet and asteroid natural scene generation utility product website. https://pangu.software/. Accessed 26 Jan 2022
5. Vertical landing for micro air vehicles using event-based optical flow dataset. https://dataverse.nl/dataset.xhtml?persistentId=hdl:10411/FBKJFH. Accessed 31 Jan 2022
6. Almatrafi, M., Baldwin, R., Aizawa, K., Hirakawa, K.: Distance surface for event-based optical flow. IEEE Trans. Pattern Anal. Mach. Intell. **42**(7), 1547–1556 (2020)
7. Bardow, P., Davison, A.J., Leutenegger, S.: Simultaneous optical flow and intensity estimation from an event camera. In: Proceedings of the IEEE Conference on Computer Vision and Pattern Recognition, pp. 884–892 (2016)
8. Benosman, R., Clercq, C., Lagorce, X., Ieng, S.H., Bartolozzi, C.: Event-based visual flow. IEEE Trans. Neural Netw. Learn. Syst. **25**(2), 407–417 (2014)
9. Benosman, R., Ieng, S.H., Clercq, C., Bartolozzi, C., Srinivasan, M.: Asynchronous frameless event-based optical flow. Neural Netw. **27**, 32–37 (2012)
10. Brandli, C., Berner, R., Yang, M., Liu, S.C., Delbruck, T.: A 240 × 180 130 db 3 μs latency global shutter spatiotemporal vision sensor. IEEE J. Solid-State Circuits **49**(10), 2333–2341 (2014)
11. Chin, T.J., Bagchi, S., Eriksson, A., van Schaik, A.: Star tracking using an event camera. In: IEEE/CVF Conference on Computer Vision and Pattern Recognition Workshops (CVPRW) (2019)
12. Clady, X., et al.: Asynchronous visual event-based time-to-contact. Front. Neurosci. **8**, 9 (2014)
13. Dinaux, R., Wessendorp, N., Dupeyroux, J., Croon, G.C.H.E.D.: FAITH: fast iterative half-Plane focus of expansion estimation using optic flow. IEEE Rob. Autom. Lett. **6**(4), 7627–7634 (2021)
14. Ding, Z., et al.: Spatio-temporal recurrent networks for event-based optical flow estimation. In: AAAI Conference on Artificial Intelligence, pp. 1–13 (2021)
15. Falanga, D., Kleber, K., Scaramuzza, D.: Dynamic obstacle avoidance for quadrotors with event cameras. Sci. Rob. **5**(40), eaaz9712 (2020)
16. Forsyth, D., Ponce, J.: Computer Vision: A Modern Approach. Prentice hall (2011)
17. Fry, S.N.: Experimental approaches toward a functional understanding of insect flight control. In: Floreano, D., Zufferey, J.C., Srinivasan, M.V., Ellington, C. (eds.) Flying Insects and Robots, pp. 1–13. Springer, Heidelberg (2010). https://doi.org/10.1007/978-3-540-89393-6_1
18. Gallego, G., Rebecq, H., Scaramuzza, D.: A unifying contrast maximization framework for event cameras, with applications to motion, depth, and optical flow estimation. In: 2018 IEEE/CVF Conference on Computer Vision and Pattern Recognition, pp. 3867–3876 (2018)
19. Gehrig, D., Loquercio, A., Derpanis, K.G., Scaramuzza, D.: End-to-end learning of representations for asynchronous event-based data. In: Proceedings of the IEEE/CVF International Conference on Computer Vision, pp. 5633–5643 (2019)

20. Gehrig, M., Millhäusler, M., Gehrig, D., Scaramuzza, D.: E-RAFT: dense optical flow from event cameras. In: 2021 International Conference on 3D Vision (3DV), pp. 197–206 (2021)
21. Gómez Eguíluz, A., Rodríguez-Gómez, J.P., Martínez-de Dios, J.R., Ollero, A.: Asynchronous event-based line tracking for Time-to-Contact maneuvers in UAS. In: 2020 IEEE/RSJ International Conference on Intelligent Robots and Systems (IROS), pp. 5978–5985 (2020)
22. Haessig, G., Cassidy, A., Alvarez, R., Benosman, R., Orchard, G.: Spiking optical flow for event-based sensors using IBM's TrueNorth neurosynaptic system. IEEE Trans. Biomed. Circuits Syst. **12**(4), 860–870 (2018)
23. Hagenaars, J.J., Paredes-Vallés, F., de Croon, G.C.H.E.: Self-Supervised learning of Event-Based optical flow with spiking neural networks. In: Neural Information Processing Systems, October 2021
24. Horst, R., Hoang, T.: Global Optimization: Deterministic Approaches. Springer, Heidelberg (1996). https://doi.org/10.1007/978-3-662-03199-5
25. Hu, Y., Liu, S.C., Delbruck, T.: v2e: from video frames to realistic DVS events. In: 2021 IEEE/CVF Conference on Computer Vision and Pattern Recognition Workshops (CVPRW) (2021)
26. Intel: beyond today's AI. https://www.intel.com.au/content/www/au/en/research/neuromorphic-computing.html
27. Lee, C., Kosta, A.K., Zhu, A.Z., Chaney, K., Daniilidis, K., Roy, K.: Spike-FlowNet: event-based optical flow estimation with energy-efficient hybrid neural networks. In: Vedaldi, A., Bischof, H., Brox, T., Frahm, J.-M. (eds.) ECCV 2020. LNCS, vol. 12374, pp. 366–382. Springer, Cham (2020). https://doi.org/10.1007/978-3-030-58526-6_22
28. Lichtsteiner, P., Posch, C., Delbruck, T.: A 128 × 128 120 db 15 µs latency asynchronous temporal contrast vision sensor. IEEE J. Solid-State Circuits **43**(2), 566–576 (2008)
29. Liu, D., Parra, A., Chin, T.J.: Globally optimal contrast maximisation for event-based motion estimation. In: Proceedings of the IEEE/CVF Conference on Computer Vision and Pattern Recognition, pp. 6349–6358 (2020)
30. Liu, M., Delbruck, T.: Adaptive time-slice block-matching optical flow algorithm for dynamic vision sensors. In: British Machine Vision Conference (BMVC) (2018)
31. Lucas, B.D., Kanade, T.: An iterative image registration technique with an application to stereo vision. In: International Joint Conference on Artificial Intelligence, pp. 674–679 (1981)
32. Mueggler, E., Huber, B., Scaramuzza, D.: Event-based, 6-DOF pose tracking for high-speed maneuvers. In: 2014 IEEE/RSJ International Conference on Intelligent Robots and Systems, pp. 2761–2768 (2014)
33. NASA/JPL-Caltech/University of Arizona: Decoding a geological message (2017). https://www.nasa.gov/sites/default/files/thumbnails/image/pia21759.jpg
34. NASA/JPL-Caltech/University of Arizona: Big fans (2018). https://www.nasa.gov/image-feature/jpl/pia22332/big-fans
35. Orchard, G., Bartolozzi, C., Indiveri, G.: Applying neuromorphic vision sensors to planetary landing tasks. In: IEEE Biomedical Circuits and Systems Conference, pp. 201–204 (2009)
36. Orchard, G., Benosman, R., Etienne-Cummings, R., Thakor, N.V.: A spiking neural network architecture for visual motion estimation. In: IEEE Biomedical Circuits and Systems Conference (BioCAS), pp. 298–301 (2013)

37. Pan, L., Liu, M., Hartley, R.: Single image optical flow estimation with an event camera. In: IEEE/CVF Conference on Computer Vision and Pattern Recognition (CVPR), pp. 1669–1678 (2020)
38. Paredes-Vallés, F., de Croon, G.C.H.E.: Back to event basics: self-supervised learning of image reconstruction for event cameras via photometric constancy. In: 2021 IEEE/CVF Conference on Computer Vision and Pattern Recognition (CVPR) (2021)
39. Paredes-Vallés, F., Scheper, K.Y.W., de Croon, G.C.H.E.: Unsupervised learning of a hierarchical spiking neural network for optical flow estimation: from events to global motion perception. IEEE Trans. Pattern Anal. Mach. Intell. **42**(8), 2051–2064 (2020)
40. Peng, X., Wang, Y., Gao, L., Kneip, L.: Globally-optimal event camera motion estimation. In: Vedaldi, A., Bischof, H., Brox, T., Frahm, J.-M. (eds.) ECCV 2020. LNCS, vol. 12371, pp. 51–67. Springer, Cham (2020). https://doi.org/10.1007/978-3-030-58574-7_4
41. Pijnacker Hordijk, B.J., Scheper, K.Y.W., de Croon, G.C.H.E.: Vertical landing for micro air vehicles using event-based optical flow. J. Field Rob. **35**(1), 69 90 (2018)
42. Posch, C., Matolin, D., Wohlgenannt, R.: A QVGA 143 db dynamic range frame-free PWM image sensor with lossless Pixel-Level video compression and time-domain CDS. IEEE J. Solid-State Circuits **46**(1), 259–275 (2011)
43. Rebecq, H., Ranftl, R., Koltun, V., Scaramuzza, D.: Events-to-video: bringing modern computer vision to event cameras. In: 2019 IEEE/CVF Conference on Computer Vision and Pattern Recognition (CVPR) (2019)
44. Rebecq, H., Ranftl, R., Koltun, V., Scaramuzza, D.: High speed and high dynamic range video with an event camera. IEEE Trans. Pattern Anal. Mach. Intell. **43**(6), 1964–1980 (2019)
45. Sanket, N.J., et al.: EVDodgeNet: deep dynamic obstacle dodging with event cameras. In: 2020 IEEE International Conference on Robotics and Automation (ICRA), pp. 10651–10657 (2020)
46. Scheerlinck, C., Rebecq, H., Gehrig, D., Barnes, N., Mahony, R., Scaramuzza, D.: Fast image reconstruction with an event camera. In: Proceedings of the IEEE/CVF Winter Conference on Applications of Computer Vision, pp. 156–163 (2020)
47. Sikorski, O., Izzo, D., Meoni, G.: Event-based spacecraft landing using time-to-contact. In: Proceedings of the IEEE/CVF Conference on Computer Vision and Pattern Recognition, pp. 1941–1950 (2021)
48. Srinivasan, M., Zhang, S., Lehrer, M., Collett, T.: Honeybee navigation EN route to the goal: visual flight control and odometry. J. Exp. Biol. **199**(Pt 1), 237–244 (1996)
49. Srinivasan, M.V.: Honeybees as a model for the study of visually guided flight, navigation, and biologically inspired robotics. Phys. Rev. **91**, 413–406 (2011)
50. Srinivasan, M.V., Thurrowgood, S., Soccol, D.: From visual guidance in flying insects to autonomous aerial vehicles. In: Floreano, D., Zufferey, J.C., Srinivasan, M.V., Ellington, C. (eds.) Flying Insects and Robots, pp. 15–28. Springer, Heidelberg (2010). https://doi.org/10.1007/978-3-540-89393-6_2
51. Stoffregen, T., Kleeman, L.: Simultaneous optical flow and segmentation (SOFAS) using dynamic vision sensor. In: Australasian Conference on Robotics and Automation (2018)
52. Stoffregen, T., Gallego, G., Drummond, T., Kleeman, L., Scaramuzza, D.: Event-based motion segmentation by motion compensation. In: International Conference on Computer Vision, pp. 7243–7252 (2019)

53. Stoffregen, T., Kleeman, L.: Event cameras, contrast maximization and reward functions: an analysis. In: IEEE/CVF Conference on Computer Vision and Pattern Recognition, pp. 12300–12308 (2019)
54. Tammero, L.F., Dickinson, M.H.: The influence of visual landscape on the free flight behavior of the fruit fly drosophila melanogaster. J. Exp. Biol. **205**(Pt 3), 327–343 (2002)
55. Valette, F., Ruffier, F., Viollet, S., Seidl, T.: Biomimetic optic flow sensing applied to a lunar landing scenario. In: 2010 IEEE International Conference on Robotics and Automation, pp. 2253–2260 (2010)
56. Medici, V., Orchard, G., Ammann, S., Indiveri, G., Fry, S.N.: Neuromorphic computation of optic flow data. Technical report, European Space Agency, Advanced Concepts Team (2010)
57. Vidal, A.R., Rebecq, H., Horstschaefer, T., Scaramuzza, D.: Ultimate SLAM? Combining events, images, and IMU for robust visual SLAM in HDR and high-speed scenarios. IEEE Rob. Autom. Lett. **3**(2), 994–1001 (2018)
58. Ye, C., Mitrokhin, A., Fermüller, C., Yorke, J.A., Aloimonos, Y.: Unsupervised learning of dense optical flow, depth and egomotion with Event-Based sensors. In: 2020 IEEE/RSJ International Conference on Intelligent Robots and Systems (IROS), pp. 5831–5838 (2020)
59. Zhou, Y., Gallego, G., Lu, X., Liu, S., Shen, S.: Event-Based motion segmentation with Spatio-Temporal graph cuts. IEEE Trans. Neural Netw. Learn. Syst. (2020)
60. Zhu, A.Z., Yuan, L., Chaney, K., Daniilidis, K.: EV-FlowNet: self-Supervised optical flow estimation for event-based cameras. Rob. Sci. Syst. (2018)
61. Zhu, A.Z., Yuan, L., Chaney, K., Daniilidis, K.: Unsupervised event-based learning of optical flow, depth, and egomotion. In: Proceedings of the IEEE/CVF Conference on Computer Vision and Pattern Recognition, pp. 989–997 (2019). openaccess.thecvf.com

Transfer Learning for On-Orbit Ship Segmentation

Vincenzo Fanizza[1]($\boxtimes$) (ID), David Rijlaarsdam[1] (ID),
Pablo Tomás Toledano González[2], and José Luis Espinosa-Aranda[2] (ID)

[1] Ubotica Technologies, Kluyverweg 1, 2629 HS Delft, The Netherlands
{vincenzo.fanizza,david.rijlaarsdam}@ubotica.com
[2] Ubotica Technologies, Camino de Moledores s/n, Incubadora de Empresas,
13005 Ciudad Real, Spain
{pablo.toledano,josel.espinosa}@ubotica.com

Abstract. With the adoption of edge AI processors for space, on-orbit inference on EO data has become a possibility. This enables a range of new applications for space-based EO systems. Since the development of on-orbit AI applications requires rarely available raw data, training of these AI networks remains a challenge. To address this issue, we investigate the effects of varying two key image parameters between training and testing data on a ship segmentation network: Ground Sampling Distance and band misalignment magnitude. Our results show that for both parameters the network exhibits degraded performance if these parameters differ in testing data with respect to training data. We show that this performance drop can be mitigated with appropriate data augmentation. By preparing models at the training stage for the appropriate feature space, the need for additional computational resources on-board for e.g. image scaling or band-alignment of camera data can be mitigated.

Keywords: Edge AI · On-orbit inference · Data augmentation · Transfer learning · Domain gap for space problems

1 Introduction

With our rapidly changing planet, the need for near real-time EO data has become even more acute [10,26]. The developments in new sensor technology have allowed remote sensing to become an economic and sustainable solution to collect information over large areas of the Earth's surface [4,18]. Compared to the airborne concept, spaceborne sensors achieve greater coverage with a reduced number of flying units, making these sensors an appealing option for EO.

Due to this appeal, satellite images are used in a broad range of applications: from object detection to geographic classification [2,3,46], and from disaster identification to environmental monitoring [21,28,36]. For a number of these tasks the salient features are only a few meters wide, thus small differences between training and testing data may cause features to be represented in a dramatically different way.

L. Karlinsky et al. (Eds.): ECCV 2022 Workshops, LNCS 13801, pp. 21–36, 2023.
https://doi.org/10.1007/978-3-031-25056-9_2

The work presented here consists of an investigation of the robustness of Neural Networks (NNs) in performing a typical EO segmentation task: ship segmentation. An example of this task is shown in Fig. 1.

Fig. 1. Example of a ship segmentation network applied to EO data: original image (left), prediction (center), and ground truth mask (right).

We compare the performance of two segmentation models on both processed and unprocessed EO images applied in different combinations at both the training and testing stages. In particular, this paper focuses on GSD and band misalignment as variable image parameters. These two effects were chosen since both can potentially be mitigated on-board by increasing the computational resources of a spacecraft. GSD can artificially be changed by either resizing images on-board or applying super resolution [12], whereas band misalignment can be mitigated by applying band alignment algorithms. If the need to process data on-board a spacecraft can be mitigated by augmentation techniques during the training stage on-ground, the complexity, cost, and power usage of these systems can be reduced.

The paper is structured as follows: Section 2 offers an overview of the main technologies used in remote sensing, their limitations to the quality of raw images, and an introduction to transfer learning. Section 3 presents an overview of the works most closely related to our research. Section 4 briefly introduces the experiment datasets, while Sect. 5 describes the experimental setup and discusses the main results. Section 6 summarises the content of this paper and proposes areas of further research, intending to bridge the knowledge gap required to deploy on-the-edge AI on a larger scale in space. The main contributions of our work are:

1. We assess the robustness of NNs over different GSDs and levels of band misalignment. We also show that the performance of a model trained on images with GSD different from the testing stage or without band misalignment at the training stage can degrade substantially.
2. We prove that the model performance can be improved by including the perturbing effects in the training dataset and that this is a key step in making on-the-edge AI for space-based EO robust for deployment.

2   Background

While space-based EO data can be acquired with a range of both active and passive sensors, our research here focuses on the application of AI to optical EO data. For the acquisition of optical data, a number of different space cameras are in use. Among these camera systems, hyperspectral imaging has been used over a large number of applications [31]. A hyperspectral imager is a spectral imager capable of capturing a wide spectrum of light, often split up in a number of image bands. Multispectral imaging consists of cameras that rely on faster image acquisition and lighter analysis techniques [31], but collects information on a more limited number of bands [40]. Often, a system-level trade-off is present between these systems: due to the finite detector space available, an increase in spectral resolution reduces the spatial resolution of a system and vice-versa. Among the higher performance optical sensors, pushbrooms cameras are the most widely used solution for both multi- and hyperspectral systems [11]. The working paradigm of these cameras is based on taking advantage of the platform orbital motion to acquire spatial and spectral information in multiple dimensions [11,13].

Despite the advancement in sensor technology, image quality can be compromised by external factors, such as scene-related effects and platform instabilities. In order to mitigate these effects, post-processing techniques are generally applied on-ground [30]. The most thoroughly studied effects are atmospheric scattering and absorption, topography, relative illumination and viewing angle. Several models have been proposed to account for these effects [1,7,25], and are now included in a large number of on-ground image post-processing chains.

Acting disturbances may also vary depending on the type of camera. Despite their wide use on satellites, pushbroom cameras are susceptible to new sources of degradation with respect to staring array cameras [15]. Since, until recently, AI has not been applied to (near) raw space-based EO pushbroom data, the consequences of these effects on the performance of NNs require investigation. Under realistic and non-ideal scenarios, the observing direction of each pixel row will be affected by the pointing instabilities, affecting the spatial and spectral information collected by the instrument. Band misalignment is a first example [14], where the pointing fluctuations between the observation times of a scene from different bands cause the corresponding channels to be geometrically shifted with respect to each other. A representation of band misalignment is shown in Fig. 2. This effect is primarily due to low-frequency pointing disturbances. At higher frequencies additional effects such as jitter introduce contrast loss and defocus [29,41]. These effects are not considered here.

As mentioned above, the second image parameter we investigate is the Ground Sampling Distance (GSD), which is a measure of the ground distance covered by a single pixel. It depends not only on the detector resolution but also on the orbital height: the same camera may capture a scene with finer details if flying in a lower orbit, at the expense of its swath.

In order to achieve satisfactory performance, NNs need to be trained with a representative and sufficiently large dataset [37]. However, in many real-world appli-

Fig. 2. Example of (an extreme case of) band misalignment in EO images. The left image shows an image with band misalignment, the right image shows a post-processed image with band-alignment applied.

cations collecting such data may be unfeasible or resource demanding [24]. To take full advantage of already existing data, transfer learning can be applied [6,43,48]. This technique allows the use for training of additional data outside of the feature space and distribution of the application data [27]. However, large differences between source and target domain may prevent the training data and corresponding labels from being used effectively with no preprocessing [17]. A domain transformation is often implemented, where ideally effects only present in the source domain are removed and effects typical of the target domain are introduced.

3   Related Work

Research into on-orbit AI applications for EO has been mounting recently. In September 2020, ESA launched the 6U-Cubesat Phisat-1 [9], the first AI-enabled satellite performing EO. The mission served to prove the concept of on-orbit computation, running a cloud detection application on the Intel Movidius Myriad 2. In particular, the CloudScout architecture [8] was used to run inference on data collected by the hyperspectral camera Hyperscout-2. The application was trained and tested on Sentinel-2 data augmented to represent Hyperscout-2 [30]. However, the performance difference due to the inclusion of these augmentations was not investigated.

In this work, we present an investigation on networks performing ship segmentation. This choice is dictated by the large interest shown by the scientific community as well as the challenging nature of ship-based tasks, like ship dimension, orientation, and dense aggregation [16,39,45]. Shi et al. [35] developed an on-orbit ship detection and classification algorithm for SAR satellites. The method consists of using a sliding window to identify the suspected target area and perform classification with the MobileNetV2 network [33]. The performance, however, experienced a drop for ships close to the coast, leaving room for improvement in that area. Lin et al. [19] tackles the challenge by designing a sea-land segmentation algorithm that uses accurate coastline data. The method achieves a reduction in computational load compared to traditional algorithms, making it appealing for on-orbit applications. Like the work in [30], neither research investigated the effects that image parameters have on the on-orbit performance of their applications.

Transfer learning techniques have previously been considered for satellite imagery. For the specific case of on-board AI, images coming from a first satellite constitute the source domain, while a second satellite for which little or no data is available is defined to be the target domain. Mateo et al. [22] advocate the importance of considering both the spectral and spatial differences between two imaging systems, improving the performance of the network when the transfer path goes from Landsat-8 to Proba-V. In [23], the authors developed an on-board AI flood mapping algorithm for Phisat-1. The authors utilise Sentinel 2 images as source and augment their dataset to represent more closely images in the target domain. Their image processing includes re-sampling as well as Gaussian noise, jitter, and motion blur. In addition, Wieland et al. [44] achieve similar performance over different satellite sensors with only contrast and brightness augmentations. None of the authors, however, investigated the performance gain of data augmentation for either GSD or band misalignment.

4   Dataset

The Airbus Ship Detection dataset was used for both training and testing. This dataset consists of 192,556 RGB images of 768 × 768 pixels from the SPOT6/7 and Pléiades spacecraft. The spatial resolution of this dataset varies between 0.3 and 6 m GSD [47]. In addition, the dataset contains tabular metadata describing the bounding box of ships in the images. For our experiments, these bounding boxes were used to create segmentation masks and these masks were used as ground truth in both training and testing of the network. Due to the nature of the segmentation subject, i.e. ships, these square segmentation masks are an appropriate approximation of a more precise segmentation mask. As we show in our experiments, the masks enable the NN to achieve satisfactory results.

4.1   Dataset for Ground Sampling Distance Experiments

The dataset used for the GSD experiments is a modified version of the original Airbus dataset. It consists of a subset of the original dataset containing only images with boats divided into two groups: training (34,044 images) and validation (8,512), for a total of 42,556 images.

Some examples of the images contained in the Airbus dataset are shown in Fig. 3.

Fig. 3. Sample images taken from the Airbus dataset

The segmentation masks were created from the provided Run-length encoding [42]. They consist of binary images, in which a pixel can assume a value of 0 or 1 when belonging to the background and ship class, respectively. This transformation was necessary to make the dataset compatible with the MMSegmentation framework [5], chosen for quickly iterating between training pipeline configurations.

The GSD is artificially changed in the images by using the resizing function in the MMSegmentation pipeline. It internally uses OpenCV, which applies a bilinear interpolation to the images and a nearest-neighbour interpolation to the segmentation masks.

4.2   Dataset for Band Misalignment Experiments

For the band misalignment experiments, we split the original Airbus dataset into random training (90%), validation (5%), and test (5%) subsets. This separates them into 173,300 images for training, and 9,628 for both validation and testing.

In order to simulate band misalignment, we augmented our images at both training and testing. This allows us to rapidly iterate and test different configurations of band misalignment without the need for regeneration of the dataset. Firstly, it is assumed that band misalignment is caused by a semi-constant, low-frequency movement in the lateral direction to the flight path of the pushbroom-equipped spacecraft. Secondly, the magnitude of the band misalignment in number of pixels is modeled as a Gaussian distribution with constant mean and variance, and the relative magnitude is cumulative between consecutive bands. Thirdly, the direction of the band misalignment is randomly generated per image. Per image, a probability of having band misalignment can be assigned to ensure a greater variance in the dataset by having some images without misalignment present. The first band of every image is kept in place and the ship segmentation mask is not shifted with respect to this first band. At the testing phase, a random seed is generated to ensure that our experiments are performed on identical and consistent datasets with random band misalignment applied. An example of the output of these augmentations was shown in Fig. 2.

5   Experiments

As stated above, we split our research up into two distinct experiments. The first investigates the robustness of NNs against changing the GSD, the second investigates the robustness of NNs against different levels of band misalignment.

5.1   Ground Sampling Distance Robustness

For ship segmentation, relevant features typically only cover a small number of pixels in the image. As a result, if GSD is not properly characterised during training, it may be challenging for the network to transfer the learned features to the application domain. Our experiments assume that training data with different GSD or lower quality than testing data is available.

Experimental Setup. A Fully Convolutional Network [20] (FCN) with MobileNetV2 [34] as the backbone was used for the GSD experiments. MobileNetV2 is an architecture optimised for mobile and resource-constrained applications, and is therefore suited for space applications. The network is available in the MMSegmentation framework. This model was trained using an Adam optimiser with a learning rate of 0.02.

The metric used to compare results between experiment runs were IoU and accuracy. Intersection over Union (IoU) is the area of overlap between the predicted segmentation and the ground truth. In our case the ground truth is specified in the binary mask, divided by the area of union between the predicted segmentation and the ground truth. The formula for this is presented in Eq. 1.

$$J(y, \hat{y}) = \frac{|y \cap \hat{y}|}{|y \cup \hat{y}|} \tag{1}$$

Here, $\hat{y}$ is the output of the model and y is the ground truth. Accuracy is the percentage of correctly predicted pixels. These calculations are done internally by MMSegmentation using Pytorch methods.

Testing was done using two approaches. Firstly, a sliding window approach was taken. In this approach testing images are first resized to simulate a certain GSD, and subsequently cropped using a sliding windows algorithm where 384×384 pixel crops are fed to the network for testing (Fig. 4).

Secondly, a resized tensor approach was taken. In this approach, the entire input image is resized to a certain input size to simulate the desired GSD. This approach is shown in Fig. 5.

Testing was done by using the two approaches described above with original, 2x (upsampled, simulating 50% GSD decrease), and 0.75x (downsampled, simulating 33% increase) resolution. These parameters were selected to analyse a set of GSD configurations sufficiently close to the original one that (i) could represent a real situation of using images with a new GSD, and (ii) at least some of the main features of the image could be theoretically maintained. The two

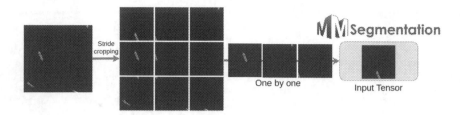

Fig. 4. Sliding window approach [5]

approaches and three testing resolutions were applied to 3 different models: a model trained on unaltered 768 × 768 images, a model trained on downsampled images, and a model trained on images with reduced quality. For the training of the second model, training images were downsampled to a common lower resolution: they were resized by a factor of 0.5 resulting in 384 × 384-pixel images, with GSD ranging from 0.6 to 12 m. The third model was created to simulate a lower-quality camera. To achieve this, the images were again downsampled to a size of 384 × 384 from their 768 × 768 resolution and then upsampled back to the original resolution. This network was tested with the original images without applying any scaling. This way we simulate the use of a worse-quality sensor at training time while we use real images in testing, with the idea of simulating the deployment of the previously trained models in a new satellite with an improved sensor.

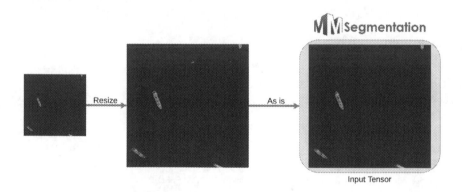

Fig. 5. Resized tensor approach [5]

Results. The parameters and results of the baseline experiment for the sliding window approach is shown in Table 1.

It can be seen that the accuracy and IoU are slightly better for the resized input tensor with respect to the sliding window approach in the baseline experiment. We hypothesise that this difference is due to noise introduced by the scaling algorithms on the testing images, and therefore not relevant for the performance themselves.

Table 1. Results of the baseline experiment for the sliding window approach

Input tensor size	384 × 384		
GSD [m]	0.3–6	0.15–3	0.399–7.98
IoU [%]	70.86	67.49	67.18
Accuracy [%]	83.67	80.3	82.32

Table 2. Results of the baseline experiment for the resized input tensor approach

Input tensor size	768 × 768	1536 × 1536	576 × 576
GSD [m]	0.3–6	0.15–3	0.399–7.98
IoU [%]	71.29	68.71	67.39
Accuracy [%]	84.74	83.13	83.05

Our second experiment, on a model trained on downsampled images simulating increased GSD, shows that testing on images with the original and 33% increased GSD only yields a limited reduction in performance. In fact, a small increase in performance is seen for the resized input tensor approach in the case of 33% increased testing GSD. On the other hand, when testing on the images with the lowest GSD, a large loss in both accuracy and metric occurs. For the sliding window approach, IoU was observed to degrade from 67.49% to 14.25%, while the accuracy dropped from 80.3% to 14.74% (Tables 1 and 3). Similar drops also occurred for the resized input tensor approach (Tables 2 and 4). Images with a better GSD include a greater level of detail, which affects the way ship features are represented. It follows that, between training and testing, a difference in GSD translates into a different representation of ship features. From the results shown in Tables 3 and 4 it can be concluded that, if the difference is not too large, the network is still able to recognise ship features. When the difference becomes significant, however, the model is not able to maintain a sufficient performance.

In the scenario of a training dataset with higher GSD images compared to testing, it is recommended to keep the GSD difference limited to avoid the possible accuracy drop. For our experiment, the combination of a 200% increase of GSD during training and a 33% reduction of GSD during testing achieved the best performance.

In Table 5 the results for our third model, trained on down- and subsequently upsampled data, are presented. Compared to the baseline, a slight improvement in the accuracy can be observed, whereas the IoU is reduced by around 11 for both cases. This indicates that the model is currently detecting more pixels as background, consequently increasing the accuracy as most of the pixels belong to that class. However, the experienced reduction in the metric hints to the fact that boats are not detected correctly in a larger number of cases compared to the baseline.

Table 3. Results of the second experiment for the sliding window approach

GSD [m]	0.3–6	0.15–3	0.399–7.98
IoU [%]	67.16	14.25	63.97
Accuracy [%]	82.02	14.74	75.74
IoU loss compared to baseline [%]	3.7	53.24	3.21
Accuracy loss compared to baseline [%]	1.65	65.56	6.58

Table 4. Results of the second experiment for the resized input tensor approach

GSD [m]	0.3–6	0.15–3	0.399–7.98
IoU [%]	65.74	17.89	67.6
Accuracy [%]	78.8	18.69	83.24
IoU loss compared to baseline [%]	5.55	50.82	−0.29
Accuracy loss compared to baseline [%]	5.94	64.44	0.19

The results prove that using images coming from two different cameras between training and testing generally reduces the segmentation model performance. A reason for the observed poorer behaviour can be found in the artifacts and other degrading effects introduced by the camera, here simulated by the upsampling. The presence of additional perturbations in the testing images challenges the robustness of the trained model and causes a drop in performance. A possible solution to compensate for these unforeseen effects might be including them during training with data augmentation, by implementing an appropriate domain adaptation, which will make the model aware of the additional degradation and eventually more robust over the testing images.

Table 5. Results of the final experiment for the two approaches

	Sliding window	Resized input tensor
GSD [m]	0.3–6	0.3–6
IoU [%]	59.29	60.39
Accuracy [%]	86.96	87.58
IoU loss compared to baseline [%]	11.57	10.9
Accuracy loss compared to baseline [%]	−3.29	−2.84

5.2   Band Misalignment Robustness

The effect of band misalignment in raw data from pushbroom cameras on the performance of a NN applied on-board a spacecraft is significant. The performance of a model on perturbed data with misaligned bands is impacted by the presence of this degrading effect. We hypothesise that this effect can be mitigated by appropriate data augmentation at the training stage. The experiments described here aim to assess the magnitude of this effect for a ship segmentation network, and if it can indeed be mitigated during training.

Experimental Setup. The neural network used in our experiments is U-Net [32] with MobileNetV2 as encoder. U-Net utilises a contracting part to capture context and a decoding part to localise features. The architecture performs well in segmentation tasks. The optimiser for the experiments was Adam with a learning rate of 1e−4.

A combination of Binary Cross Entropy Loss and the Dice score was used as loss function for training. The used loss function can be expressed as follows:

$$Loss = (1 - DSC) - \frac{0.5}{N} \sum_{i=1}^{N} [y_i log(\hat{y}_i) + (1 - y_i) log(1 - \hat{y}_i)] \qquad (2)$$

Here, N is the number of scalar values in the model output (in this case, the number of pixels in the output prediction). $\hat{y}_i$ is the i-th scalar value in the model output, y_i is the corresponding target value. DSC is the Dice score [38] (or F1 score) we implemented as:

$$DSC = \frac{2|y \cap \hat{y}| + 1}{|y| + |\hat{y}| + 1 + \epsilon} \qquad (3)$$

where ϵ is a smoothing factor set at 1e-15. We use the mean Dice score as defined in Eq. 3 as our scoring metric for our band misalignment experiments. For our experiments, two models were trained. The baseline for both models was a pre-trained model trained for 60 epochs on the dataset described in Sect. 4.2. Both models were retrained on the same dataset for 40 epochs, where one model was trained on misaligned data with a mean misalignment of 3.2 pixels and a probability of 70% of presence of misalignment (model 1), and a second model (model 2) was trained on non-misaligned data. During training for both models, random data augmentation was applied in the form of horizontal flipping of images, random brightness adjustments, random contrast adjustments and random shifting, scaling and rotating of the images. Note that these augmentations were omitted during testing.

Results. The effects of different levels of band misalignment on both models have been tested. Both models were subject to increasing levels of band misalignment: 0, 1.6, 3.2 and 10 pixels. The results are shown in Fig. 6. It can be observed how the model not trained on data with band misalignment included (model 2) performs visibly worse when deployed on misaligned EO images. Consequently, the model trained on misaligned data shows greater robustness against the introduced disturbance. Even under very severe conditions, beyond the trained misalignment domain, the performance of model 1 is significantly better than that of model 2. This is not surprising, as the distribution of the training data becomes more representative of the real one when band misalignment is included.

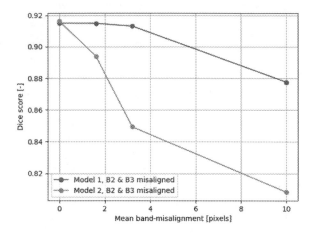

Fig. 6. Comparison of the performance of a ship segmentation under band misalignment. Model 1 has been trained on misaligned images as well as non-misaligned images. Model 2 has been trained only on non-misaligned images.

Since model 1 was trained on misaligned data and a design decision was made to align the ground truth mask of our ship segmentation dataset with the first band, there may be a bias towards such bands in the trained model. If our model would only be using the first band to extract the ship segmentation mask, one could argue that training a model on a single band is a more effective method to achieve robustness, as it intrinsically avoids the band misalignment problem. To this end, an additional experiment was performed on model 1: we performed inference on a dataset with misalignment *only* applied to the first band. This scenario is not representative of real EO data but serves to characterise the level of bias of our model.

As shown in Fig. 7, shifting only the first band reduces the performance of our model. This is expected: our model has never seen this type of noise, and has learned that the ground truth is more related to this band than to the second and third band. However, the performance of model 1 deteriorates less with input of data with severe misalignment (10 pixels) of only band 1 than with input of data containing misalignment of both band 2 and 3. Under these extreme conditions, the model appears more biased towards bands 2 and 3 than towards band 1. We conclude that while some bias towards the first band is present, this bias is not severe.

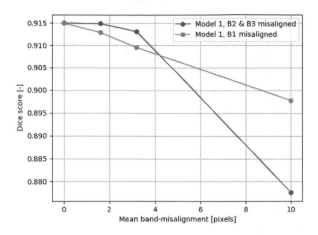

Fig. 7. Comparison of the performance of model 1 under different misalignment variants.

6   Future Work and Conclusion

In this paper, we investigated the effects of variable GSD and band misalignment on the robustness of convolutional neural networks for EO. Two sets of experiments were conducted to assess the impact of each effect separately and were both found to significantly affect the model performance.

With respect to GSD variations, our research proved how large discrepancies between training and testing images translate into a drop in accuracy and IoU up to 65% and 53%, respectively. We also showed how using images coming from different cameras for training and testing induces a loss in performance.

For the investigation of the effect of band misalignment, lower performance was shown by the model trained on non-misaligned images. It was shown that model performance on images containing band misalignment can be made more robust by including this effect at the training stage. In addition to this, an investigation of the bias towards the first band showed this bias to be limited.

Finally, after presenting the different sets of results, some recommendations on possible mitigation strategies were given. In particular, our proposals aim at improving data augmentation and transfer learning techniques for on-board AI applied to EO data. The advantage of this approach goes beyond the reduced loss in performance, as it removes the need to bring better-performing hardware on-orbit. This feature is of clear interest for any space mission, where mass and power consumption requirements are the most stringent. Our work supports the role of training to impact the final performance of a model. Instead of adapting hardware to suit the applied software, the software should be adapted to suit the hardware.

Our research remains a preliminary investigation on the effect of scene- and spacecraft-related disturbances for on-orbit AI applications. Future work may expand the results here presented. Possible focus areas are: (i) how different applications are affected by variations in GSD and band misalignment and (ii) how other image parameters impact the performance of neural networks. We believe the interest in answers to these questions will greatly increase in the upcoming years, with the growth of on-orbit AI for EO.

References

1. Chavez, P.S., et al.: Image-based atmospheric corrections-revisited and improved. Photogramm. Eng. Remote. Sens. **62**(9), 1025–1035 (1996)
2. Cheng, G., Han, J.: A survey on object detection in optical remote sensing images. ISPRS J. Photogramm. Remote. Sens. **117**, 11–28 (2016)
3. Cheng, G., Han, J., Zhou, P., Guo, L.: Multi-class geospatial object detection and geographic image classification based on collection of part detectors. ISPRS J. Photogramm. Remote. Sens. **98**, 119–132 (2014)
4. Cihlar, J.: Land cover mapping of large areas from satellites: status and research priorities. Int. J. Remote Sens. **21**(6–7), 1093–1114 (2000)
5. MMS Contributors: MMSegmentation: Openmmlab semantic segmentation toolbox and benchmark (2020). https://github.com/open-mmlab/mmsegmentation
6. Dai, W., Jin, O., Xue, G.R., Yang, Q., Yu, Y.: EigenTransfer: a unified framework for transfer learning. In: Proceedings of the 26th Annual International Conference on Machine Learning, pp. 193–200 (2009)
7. Franklin, S.E., Giles, P.T.: Radiometric processing of aerial and satellite remote-sensing imagery. Comput. Geosci. **21**(3), 413–423 (1995)
8. Giuffrida, G., et al.: CloudScout: a deep neural network for on-board cloud detection on hyperspectral images. Remote Sens. **12**(14), 2205 (2020)
9. Giuffrida, G., et al.: The ϕ-sat-1 mission: the first on-board deep neural network demonstrator for satellite earth observation. IEEE Trans. Geosci. Remote Sens. **60**, 1–14 (2021)
10. Gong, P.: Remote sensing of environmental change over china: a review. Chin. Sci. Bull. **57**(22), 2793–2801 (2012)
11. Gupta, R., Hartley, R.I.: Linear pushbroom cameras. IEEE Trans. Pattern Anal. Mach. Intell. **19**(9), 963–975 (1997)
12. Guzmán, R., et al.: A compact multispectral imager for the MANTIS mission 12U CubeSat. In: CubeSats and SmallSats for Remote Sensing IV, vol. 11505, p. 1150507. SPIE (2020)
13. Hirschmüller, H., Scholten, F., Hirzinger, G.: Stereo vision based reconstruction of huge urban areas from an airborne Pushbroom camera (HRSC). In: Kropatsch, W.G., Sablatnig, R., Hanbury, A. (eds.) DAGM 2005. LNCS, vol. 3663, pp. 58–66. Springer, Heidelberg (2005). https://doi.org/10.1007/11550518_8
14. Honkavaara, E., Rosnell, T., Oliveira, R., Tommaselli, A.: Band registration of tuneable frame format hyperspectral UAV imagers in complex scenes. ISPRS J. Photogramm. Remote. Sens. **134**, 96–109 (2017)
15. Iwasaki, A.: Detection and estimation satellite attitude jitter using remote sensing imagery. Adv. Spacecraft Technol. **13**, 257–272 (2011)

16. Kang, M., Ji, K., Leng, X., Lin, Z.: Contextual region-based convolutional neural network with multilayer fusion for SAR ship detection. Remote Sens. **9**(8), 860 (2017)

17. Kouw, W.M., Loog, M.: An introduction to domain adaptation and transfer learning. arXiv preprint arXiv:1812.11806 (2018)

18. Kramer, H.J., et al.: Observation of the Earth and its Environment: Survey of Missions and Sensors, vol. 1982. Springer, Heidelberg (2002). https://doi.org/10.1007/978-3-642-56294-5

19. Lin, X., Xu, Q., Han, C.: Shoreline data based sea-land segmentation method for on-orbit ship detection from panchromatic images. In: 2018 Fifth International Workshop on Earth Observation and Remote Sensing Applications (EORSA), pp. 1–5. IEEE (2018)

20. Long, J., Shelhamer, E., Darrell, T.: Fully convolutional networks for semantic segmentation. CoRR abs/1411.4038 (2014). https://arxiv.org/abs/1411.4038

21. Manfreda, S., et al.: On the use of unmanned aerial systems for environmental monitoring. Remote Sens. **10**(4), 641 (2018)

22. Mateo-García, G., Laparra, V., López-Puigdollers, D., Gómez-Chova, L.: Transferring deep learning models for cloud detection between Landsat-8 and Proba-V. ISPRS J. Photogramm. Remote. Sens. **160**, 1–17 (2020)

23. Mateo-Garcia, G., et al.: Towards global flood mapping onboard low cost satellites with machine learning. Sci. Rep. **11**(1), 1–12 (2021)

24. Mitchell, T.M.: Machine learning and data mining. Commun. ACM **42**(11), 30–36 (1999)

25. Moran, M.S., Jackson, R.D., Slater, P.N., Teillet, P.M.: Evaluation of simplified procedures for retrieval of land surface reflectance factors from satellite sensor output. Remote Sens. Environ. **41**(2–3), 169–184 (1992)

26. Navalgund, R.R., Jayaraman, V., Roy, P.: Remote sensing applications: an overview. Current Sci. **93**, 1747–1766 (2007)

27. Pan, S.J., Yang, Q.: A survey on transfer learning. IEEE Trans. Knowl. Data Eng. **22**(10), 1345–1359 (2009)

28. Peng, D., Zhang, Y., Guan, H.: End-to-end change detection for high resolution satellite images using improved UNet++. Remote Sens. **11**(11), 1382 (2019)

29. Perrier, R., Arnaud, E., Sturm, P., Ortner, M.: Estimation of an observation satellite's attitude using multimodal Pushbroom cameras. IEEE Trans. Pattern Anal. Mach. Intell. **37**(5), 987–1000 (2014)

30. Phiri, D., Simwanda, M., Salekin, S., Nyirenda, V.R., Murayama, Y., Ranagalage, M.: Sentinel-2 data for land cover/use mapping: a review. Remote Sens. **12**(14), 2291 (2020)

31. Qin, J., Chao, K., Kim, M.S., Lu, R., Burks, T.F.: Hyperspectral and multispectral imaging for evaluating food safety and quality. J. Food Eng. **118**(2), 157–171 (2013)

32. Ronneberger, O., Fischer, P., Brox, T.: U-net: convolutional networks for biomedical image segmentation (2015). https://doi.org/10.48550/ARXIV.1505.04597. https://arxiv.org/abs/1505.04597

33. Sandler, M., Howard, A., Zhu, M., Zhmoginov, A., Chen, L.C.: MobileNetV2: inverted residuals and linear bottlenecks. In: Proceedings of the IEEE Conference on Computer Vision and Pattern Recognition, pp. 4510–4520 (2018)

34. Sandler, M., Howard, A., Zhu, M., Zhmoginov, A., Chen, L.C.: MobileNetV2: inverted residuals and linear bottlenecks (2018). https://doi.org/10.48550/ARXIV.1801.04381. https://arxiv.org/abs/1801.04381

35. Shi, H., He, G., Feng, P., Wang, J.: An on-orbit ship detection and classification algorithm for SAR satellite. In: IGARSS 2019–2019 IEEE International Geoscience and Remote Sensing Symposium, pp. 1284–1287. IEEE (2019)

36. Shi, W., Zhang, M., Zhang, R., Chen, S., Zhan, Z.: Change detection based on artificial intelligence: state-of-the-art and challenges. Remote Sens. **12**(10), 1688 (2020)

37. Simard, P.Y., Steinkraus, D., Platt, J.C., et al.: Best practices for convolutional neural networks applied to visual document analysis. In: ICDAR, vol. 3. Edinburgh (2003)

38. Sudre, C.H., Li, W., Vercauteren, T., Ourselin, S., Jorge Cardoso, M.: Generalised dice overlap as a deep learning loss function for highly unbalanced segmentations. In: Cardoso, M.J., et al. (eds.) DLMIA/ML-CDS -2017. LNCS, vol. 10553, pp. 240–248. Springer, Cham (2017). https://doi.org/10.1007/978-3-319-67558-9_28

39. Tang, J., Deng, C., Huang, G.B., Zhao, B.: Compressed-domain ship detection on spaceborne optical image using deep neural network and extreme learning machine. IEEE Trans. Geosci. Remote Sens. **53**(3), 1174–1185 (2014)

40. Teke, M., Deveci, H.S., Haliloğlu, O., Gürbüz, S.Z., Sakarya, U.: A short survey of hyperspectral remote sensing applications in agriculture. In: 2013 6th International Conference on Recent Advances in Space Technologies (RAST), pp. 171–176. IEEE (2013)

41. Teshima, Y., Iwasaki, A.: Correction of attitude fluctuation of terra spacecraft using ASTER/SWIR imagery with parallax observation. IEEE Trans. Geosci. Remote Sens. **46**(1), 222–227 (2007)

42. Tsukiyama, T., Kondo, Y., Kakuse, K., Saba, S., Ozaki, S., Itoh, K.: Method and system for data compression and restoration (Apr 29 1986), uS Patent 4,586,027

43. Weiss, K., Khoshgoftaar, T.M., Wang, D.D.: A survey of transfer learning. J. Big Data **3**(1), 1–40 (2016). https://doi.org/10.1186/s40537-016-0043-6

44. Wieland, M., Li, Y., Martinis, S.: Multi-sensor cloud and cloud shadow segmentation with a convolutional neural network. Remote Sens. Environ. **230**, 111203 (2019)

45. Yang, X., et al.: Automatic ship detection in remote sensing images from google earth of complex scenes based on multiscale rotation dense feature pyramid networks. Remote Sens. **10**(1), 132 (2018)

46. Yang, Y., Newsam, S.: Geographic image retrieval using local invariant features. IEEE Trans. Geosci. Remote Sens. **51**(2), 818–832 (2012)

47. Zhang, Z., Zhang, L., Wang, Y., Feng, P., He, R.: ShipRSImageNet: a large-scale fine-grained dataset for ship detection in high-resolution optical remote sensing images. IEEE J. Sel. Top. Appl. Earth Observations Remote Sens. **14**, 8458–8472 (2021)

48. Zhuang, F., et al.: A comprehensive survey on transfer learning. Proc. IEEE **109**(1), 43–76 (2020)

Spacecraft Pose Estimation Based on Unsupervised Domain Adaptation and on a 3D-Guided Loss Combination

Juan Ignacio Bravo Pérez-Villar[1,2](✉) (iD), Álvaro García-Martín[2] (iD),
and Jesús Bescós[2] (iD)

[1] Deimos Space, 28760 Madrid, Spain
juan-ignacio.bravo@deimos-space.com
[2] Video Processing and Understanding Lab, Univ. Autónoma de Madrid,
28049 Madrid, Spain
juanignacio.bravo@estudiante.uam.es,
{alvaro.garcia,j.bescos}@uam.es

Abstract. Spacecraft pose estimation is a key task to enable space missions in which two spacecrafts must navigate around each other. Current state-of-the-art algorithms for pose estimation employ data-driven techniques. However, there is an absence of real training data for spacecraft imaged in space conditions due to the costs and difficulties associated with the space environment. This has motivated the introduction of 3D data simulators, solving the issue of data availability but introducing a large gap between the training (source) and test (target) domains. We explore a method that incorporates 3D structure into the spacecraft pose estimation pipeline to provide robustness to intensity domain shift and we present an algorithm for unsupervised domain adaptation with robust pseudo-labelling. Our solution has ranked second in the two categories of the 2021 Pose Estimation Challenge organised by the European Space Agency and the Stanford University, achieving the lowest average error over the two categories (Code is available at: https://github.com/JotaBravo/spacecraft-uda).

Keywords: Spacecraft · Pose estimation · Unsupervised domain adaptation · 3D loss

1 Introduction

Relative navigation between a chaser and a target spacecraft is a key capability for future space missions, due to the relevance of close-proximity operations within the realm of active debris removal or in-orbit servicing. Camera sensors are a growing suitable choice to aid in the relative navigation around non-cooperative targets due to their reduced cost, low-mass and low-power requirements compared to active sensors. However, data acquired in real operational scenarios is often scarce or not available, limiting the deployment of data-driven

L. Karlinsky et al. (Eds.): ECCV 2022 Workshops, LNCS 13801, pp. 37–52, 2023.
https://doi.org/10.1007/978-3-031-25056-9_3

Fig. 1. Sample images of the SPEED+ Dataset. From left to right: Synthetic (training & validation), Sunlamp (test), Lightbox (test).

algorithms. This limitation is nowadays overcome by using data simulators, which solve the issue of data scarcity but introduce the problem of domain shift between the training and test datasets.

The 2021 Spacecraft Pose Estimation Challenge (SPEC2021) [7], organised by the European Space Agency and the Stanford University, was designed to bridge the performance gap between computer-simulated and operational data for the task of non-cooperative spacecraft pose estimation. SPEC2021 builds around the SPEED+ dataset [25] containing 60,000 computer-generated images of the Tango spacecraft with associated pose labels and 9531 unlabelled hardware-in-the-loop test images of a half-scale mock-up model. Test images are grouped into two subsets: Sunlamp, consisting of 2791 images featuring strong illumination and reflections over a black background; and Lightbox, consisting of 6740 images with softer illumination but increased noise levels and the presence of the Earth in the background (see sample images in Fig. 1). SPEC2021 evaluates the accuracy in pose estimation for each test subset.

Under this Challenge, we have developed a single algorithm that has ranked second in both Sunlamp and Lightbox categories, with the best total average error over the two datasets. Our main contributions are:

– A spacecraft pose estimation algorithm that incorporates 3D structure information during training, providing robustness to intensity based domain-shift.
– An unsupervised domain adaptation scheme based on robust pseudo-label generation and self-training.

2   Related Work

This section briefly presents the related work on pose estimation and unsupervised domain adaptation.

2.1   Pose Estimation

Pose estimation is the process of estimating the rigid transform that aligns a model frame with a sensor, body, or world reference frame [13]. Similarly to [27] we categorise these methods into: end-to-end, two-stage, and dense.

End-to-end methods map the two-dimensional images or their feature representations into the six-dimensional output space without employing geometrical solvers. Handcrafted approaches compare image regions with a database of templates to return the 6 Degree-of-freedom (DoF) pose associated with the template [10,32]. Learning-based approaches employ CNNs to either learn the direct regression of the 6DoF output space [14,15,20,28] or classify the input into a discrete pose [34]. End-to-end methods have been explored in [30], where an AlexNet was used to classify input images with labels associated to a discrete pose. The authors in [28] employed a network to regress the position and do continuous orientation estimation of the spacecraft with soft assignment coding.

Two-stage pose estimation methods first estimate the 2D image projection of the 3D object key-points and then apply a Perspective-n-Point (PnP) algorithm over the set of 2D-3D correspondences to retrieve the pose. Traditionally, this has been solved via key-point detection, description and matching between the images and a 3D model representation based on handcrafted [1,2,19] or learnt [21,36] key-point detectors, descriptors and matchers. Other approaches directly employ CNNs designed for 2D key-point localisation that detect the vertices of the 3D object bounding-box or regress the key-point position by employing heatmaps, hence unifying the detection, description and matching into a single network [22,39]. These approaches are the current state-of-the-art in satellite pose estimation. The authors from [4] employed an HRNet and non-linear pose refinement which obtained the first position in the Spacecraft Pose Estimation Challenge of 2019. In [11] a multi-scale fusion scheme with a 3D loss independent of the depth was employed to improve the pose estimation accuracy under different scale distributions of the spacecraft.

Dense pose estimation methods establish dense correspondences, at pixel or patch level, between a 2D image and a 3D surface object representation. These approaches can be based on depth data, RGB images or image pairs [6,12,27].

Discussion: Non-cooperative spacecraft pose estimation presents specific conditions that can be incorporated as prior knowledge into the algorithms: 1) the target will not appear occluded in the image; 2) and the targets have highly reflective materials that coupled with small orbital periods frequently lead to heterogeneous illumination conditions during rendezvous; and 3) the estimation should provide some degree of interpretability so the navigation filter can handle failure cases. End-to-end pose estimation methods, that confront the problem as a classification task, are limited by the error associated with the discretisation of the pose-space and require a correct sampling of the attitude space to obtain uniform representations [16]. In addition, the methods that directly regress the pose fail to generalise, as the formulation might be closer to the image retrieval problem rather than to pose estimation one [29]. Two-stage pose estimation methods do not provide robustness to occlusions and can be affected by key-points falling outside the image space. Finally, dense methods provide robustness to occlusions by producing a prediction for every pixel or patch in the image, but increase the difficulty of the learning task due to the larger output space [27]. These also impose additional constraints on the input data as methods may need depth or surface normals, which were not available for the Challenge.

These considerations motivated our choice of a two-stage approach. Keypoints provide a certain degree of interpretability, as the output is easily understandable and can be directly related with a pose. These methods also allow for continuous output estimation without compromising generalisation and do not require the additional ground-truth data, usually depth, that dense methods do. In addition, two-stage methods have shown better performance compared to end-to-end methods in 6DoF pose estimation [31] and are commonly used into the state-of-the-art methods for the spacecraft domain [4, 11].

2.2 Domain Adaptation

In visual-based navigation for space applications the distribution of the data to be acquired during the mission is often unknown, as no or little data is available prior the mission. This entails the introduction of computer graphic simulators [26] to generate the data required to design and test the algorithms. Although computer-graphic simulated images have greatly improved their quality and fidelity, there still exists a noticeable domain gap between the simulated (source) data and the real (target) data. Methods to overcome the reduction in algorithms performance for the same task under data with different distributions fall under the category of Domain Adaptation [23]. More specifically, when labels are only provided in the source domain is referred as Unsupervised Domain Adaptation (UDA). UDA methods can be categorised into: 1) domain-invariant feature learning; 2) aligning input data; and 3) self-training via pseudo-labelling.

Domain-invariant feature learning aims to align the source and target domain at feature level. It is based on the assumption that exists a feature representation that does not vary across domains, i.e. the information captured by the feature extractor is the one that does not depend on the domain. This family of methods try to learn the representation either by minimising the distribution shift under some divergence metric [35], enforcing that both feature representations can be used to reconstruct the target or source domain data [9], or employing adversarial training [8].

Input Data Alignment. Instead of aligning the domains at the feature level, these approaches aim to align the source domain to the target domain at input level, a mapping usually confronted via generative adversarial networks [33, 42].

Self-training approaches employ generated pseudo-labels to iteratively improve the model. However, these labels generation process is inevitably noisy. Some works have tried to mitigate the effect of the label noise by designing robust losses [41] but they often fail to handle real-world noisy data [40]. Other approaches have used self-label correction that relies on the agreement between different learners [17] or stages of the pipeline [40] to correct the noisy labels. Online domain adaptation is employed in [24] for satellite pose estimation across different domains. A segmentation-based loss is employed to update the feature extractor of a network, adapting the features to the new input-domain.

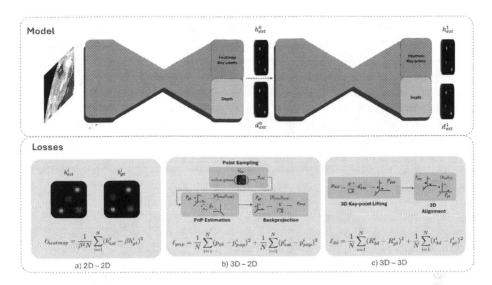

Fig. 2. Proposed pipeline. Model (top) - A two-stack hourglass network with two heads for key-point extraction and depth prediction. Losses (bottom) - A combination of: a) 2D-2D comparison between the predicted key-point heatmaps and the ground-truth heatmaps; b) 2D-3D comparison between the 3D key-point positions estimated via PnP and the ground-truth positions; c) 3D-3D comparison between the pose that aligns the lifted 2D estimated key-points with the 3D model and the ground-truth pose.

Discussion: Feature learning methods rely on the existence of feature representations that do not vary across domains, and assume that similar semantic objects are shared between domains during training. However, in pose estimation the semantic objects (key-points) are pose dependent and thus, the same semantic objects at training might not appear in both domains, hence affecting the alignment performance. Input alignment methods are able to transfer general illumination and contrast properties. However, they might introduce artefacts, and reflectances due to the object materials and sensor specific noise between train and test are often not captured. Our proposal is based on pseudo-labelling, which allows to learn from the target domain providing directly interpretable results. In addition, pseudo-labelling provides state-of-the-art results in UDA for the semantic segmentation task [40].

3   Pose Estimation

This section introduces the proposed method for pose estimation. In our setting, the target frame is the geometrical centre of the target spacecraft, represented by a set of 11 3D key-points. The sensor frame is expressed at the centre of the image. At test time, we use a Perspective-n-Point solver over the 3D spacecraft key-points and their estimated 2D image projections to retrieve the pose.

In this challenging scenario where, apart from a noticeable domain gap, surfaces reflectivity and illumination conditions further generate great change, objects in the source and target domains keep the same 3D structure. This motivates the use of an architecture and learning losses that not only rely on image textures, but make explicit use of the 3D key-point structure.

At architecture level (see Fig. 2 - Model), we propose the use of a stacked hourglass network with two heads to jointly regress the key-point position and their associated depth, enabling 3D-aware feature learning. At loss level (see Fig. 2 - Losses), we formulate a triple learning objective: a) in the 2D-2D loss we compare the mean squared error between the key-point head output with a 2D Gaussian blob centred on the ground-truth key-point location; b) in the 3D-2D loss we incorporate pose information by comparing the coordinates of the 3D key-points projected onto the image using the estimated pose via PnP with the corresponding ground-truth coordinates in the image; c) in the 3D-3D loss we incorporate 3D key-point structure: the estimated depth along the camera intriniscs is used to lift the estimated key-point positions, and these lifted key-points are aligned in 3D with the ground-truth key-points using a least square estimator. Then, the retrieved pose from the alignment is compared to the ground-truth pose.

3.1   2D-2D: Key-Point Heatmap Loss

The ground-truth heatmaps h_{gt} are obtained by representing a circular Gaussian centred at the ground-truth image key-point coordinates $p_{gt} = [u_{gt}, v_{gt}]^T$. Given an image of the target spacecraft, represented by the 3D key-points $P_{gt} = [x, y, z]^T$ and acquired with camera intrinsics K in a ground-truth pose with rotation matrix R_{gt} and translation vector t_{gt}, the ground-truth key-point coordinates can obtained via the projection expression:

$$p_{gt} = K R_{gt}^T (P_{gt} + R_{gt} t_{gt}). \tag{1}$$

The heatmaps h_{est}^i predicted by each head i are learnt by comparing the output of the key-point prediction head vs the ground-truth heat-map h_{gt} via mean squared error. To keep a fixed range for the key-point heatmap loss, we normalise each heatmap by its maximum value and weight them by a parameter $\beta = 1e3$, ensuring a large difference between the minimum and maximum value of the heatmap. The key-point heatmap loss is then computed via:

$$\ell_{heatmap} = \frac{1}{\beta^2 N} \sum_{i=1}^{N} (h_{est}^i - \beta h_{gt}^i)^2. \tag{2}$$

3.2   3D-2D: PnP Loss

Pose information is here considered by comparing the ground-truth 2D key-point image coordinates p_{gt} with the projection of the 3D key-points P_{gt} onto the image, obtained using the rigid transform estimated via PnP (see Fig. 2 - Model b)). The predicted key-point locations p_{est}^i used to estimate such rigid transform are originally expressed as heatmaps. The application of the arg-max function over h_{est}^i to retrieve the predicted key-point pixel positions p_{est}^i is a non-differentiable operation which cannot be used for training. We instead use the integral pose regression method [37] and apply an expectation operation over the soft-max normalized heatmaps h_{est}^i to retrieve the pixel coordinates p_{est}^i. Then, we employ the backpropagatable PnP algorithm from [5] to retrieve the estimated rotation matrix R_{pnp}^i and translation vector t_{pnp}^i.

The PnP loss evaluates the mean squared error between the ground-truth key-point pixel locations p_{gt} and the key-point pixel locations p_{pnp}^i obtained by projecting P_{gt} onto the image with R_{pnp}^i and t_{pnp}^i as in Eq. 1. The loss has an additional regularisation term, comparing p_{est} with p_{pnp}, to ensure convergence of the estimated key-point coordinates the desired positions [5]. The final PnP loss term ℓ_{pnp} is computed via:

$$\ell_{pnp} = \frac{1}{N} \sum_{i=1}^{N} (p_{gt} - p_{pnp}^i)^2 + \frac{1}{N} \sum_{i=1}^{N} (p_{est}^i - p_{pnp}^i)^2. \tag{3}$$

3.3   3D-3D: Structure Loss

We incorporate 3D structure by comparing the rotation R_{3d}^i and translation t_{3d}^i that align the predicted key-points lifted to the 3D space P_{3d}^i with the key-points representing the spacecraft P_{gt} with the ground-truth pose (see Fig. 2 - Model c)). Following the approach described in Sect. 3.1, we retrieve the estimated key-point coordinates p_{est}^i. Then we lift them to the 3D space by applying the camera intrinsics K and the estimated depth d_{est}^i. Instead of providing direct supervision to the depth, we choose to sample the depth prediction head at the position of the ground-truth key-point positions p_{gt}^i, and supervise the output with the 3D alignment loss. Sampling at the ground-truth key-point positions instead of the estimated positions helps convergence, as the number of combinations of key-point locations and depth values that satisfy a given rotation and translation is not unique. R_{3d}^i and t_{3d}^i are found by minimising the sum of square distances between the target points P_{gt} and the lifted 3d points P_{3d}^i:

$$R_{3d}^i, t_{3d}^i = \mathtt{argmin}_{R,t} \sum \|RP_{3d}^i + t - P_{gt}\|^2 \tag{4}$$

The algorithm that solves this expression is referred as the Umeyama [38] (or Kabsch) algorithm and has a closed form solution. We employ the Procrustes solver from [3] to obtain the arguments that minimise the above expression. The loss is then computed by:

$$\ell_{3d} = \frac{1}{N} \sum_{i=1}^{N} (R_{3d}^i - R_{gt}^i)^2 + \frac{1}{N} \sum_{i=1}^{N} (t_{3d}^i - t_{gt}^i)^2 \tag{5}$$

The final learning objective is then defined by a weighted sum of the three losses. The values of $\gamma_1 = \gamma_2 = 0.1$ are selected so the contributions of the 2D-3D and 3D-3D losses represent a 10% of the total learning objective.

$$\ell_{final} = \ell_{heatmap} + \gamma_1 \ell_{pnp} + \gamma_2 \ell_{3d}. \tag{6}$$

4   Domain Adaptation

The losses defined in Sect. 3 to consider texture, pose and structure improve the performance in the test set for the task of spacecraft pose estimation (see Sect. 5.3). However, we still observe a large performance gap between the source domain (validation) and the two target domains (test). This gap motivates the introduction of unsupervised domain adaptation techniques. Given a dataset of images in the source domain I_s and the associated ground-truth rotation matrix R_{gt} and translation vector t_{gt} that define the pose, we aim to train a model M such that when evaluated on a target dataset I_t produces correct ground-truth key-point positions, without using the ground-truth pose. Our proposal relies on self-training with robust pseudo-labelling generation. As discussed before, pseudo-labels provide a strong supervision signal on the target domain but are intrinsically noisy. If not handled properly, this noise can degrade the overall performance. In our setting, we reduce the noise during label generation by enforcing consistent pose estimates across the network head.

The proposed approach is summarised in Algorithm 1. We first pre-train a model M (Fig. 2 - Model) on the source dataset. Then, we evaluate the images from a given target dataset I_t on the model M. The model M will generate a prediction on the key-point locations, expressed as heatmaps h_{est}^i at each head $i = 1, 2$ of the network. We consider the pixel coordinates of the maximum value of the heatmap to be the predicted key-point pixel coordinates p_{est}^i. Differently to the optimisation of the losses ℓ_{pnp} and ℓ_{3d}, here we do not require this step to be differentiable, and thus a standard argmax operation suffices: $p_{est}^i = argmax(h_{est}^i)$. Next, we perform key-point filtering based on the heatmap response. First, we remove the key-points falling close to the edges of the image. The remaining key-points p_{est}^i are sorted based on the total sum of the heatmap h_{est}^i, and the seven key-points with strongest response are kept. The remaining key-points are noted as p_{filt}^i. We employ the EPnP (Efficient Perspective-n-Point) algorithm [18] under a RANSAC loop to estimate the pose with the filtered key-points p_{filt}^i at each head i.

The pose at each level is estimated with the prediction of each head p_{filt}^i by applying the EPnP (Efficient Perspective-n-Point) algorithm [18] under a RANSAC loop. A new pose pseudo-label is introduced into the test set if the RANSAC-EPnP routine converges for all the heads i. The solution corresponding to the largest total key-point response is chosen as the resulting pose. The convergence of the RANSAC routine is controlled by two parameters: the confidence parameter c that controls the probability of the method to provide a useful result, and the reprojection error r that defines the maximum allowed distance between a point and its reprojection with the estimated pose to be considered an inlier. Modifying the convergence criteria (i.e. c and r) allows to control the amount of label noise. In our case, we choose $c = 0.999$ and $r = 2.0$.

Algorithm 1. Pseudo-Labelling for Spacecraft Pose Estimation.

1: **while** convergence criteria **do**
2:     **if** first iteration **then**
3:         Initialise M with weights trained on source domain
4:     **else**
5:         Initialise M with weights from previous iteration
6:     **end if**
7:     **for** $j = 1$ to J **do**
8:         $h_{est}^i = M(I_t^j)$
9:         **for** $i = 1 : 2$ **do**
10:             $p_{est}^i = \text{argmax}(h_{est}^i)$
11:             $p_{filt}^i = \text{FilterKeypoints}(p_{est}^i)$
12:             $R_{est}^i, t_{est}^i, \text{flag}^i = \text{PoseEstimation}(p_{filt}^i, P_{gt}, c = 0.999, r = 2.0)$
13:         **end for**
14:         **if** $\text{flag}^i ==$ True for all i **then**
15:             Choose R_{est}, t_{est} from key-point response
16:             Add R_{est}, t_{est} pseudo-label to I_t^j
17:         **end if**
18:     **end for**
19:     Retrain M for a number of epochs
20: **end while**

---

5 Evaluation

5.1 Metrics

The evaluation metric used in the Challenge was the pose score S_{total}. The pose score is based on the combination of the position score S_{pos} and the orientation score S_{ori}. The position score for an image j is defined as the absolute difference between the estimated and ground-truth translation vectors, normalised by the module of the ground-truth position vector. The position score accounts for the precision in position of the robotic platform, which is of 2.173 millimetre per metre of ground truth distance. Values of S_{pos} lower than 0.002173 are zeroed.

$$S_{pos}^j = \frac{\left| t_{est}^j - t_{gt}^j \right|_2}{\left| t_{gt}^j \right|_2} \tag{7}$$

The orientation score S_{ori} is defined as the angle that aligns the estimated and ground-truth quaternion orientations. This metric also accounts for the precision of the robotic arm which is $0.169^{\underline{o}}$, zeroing S_{ori} than the precision.

$$S_{ori}^j = 2 \cdot arccos(\left|\langle q_{est}^j, q_{gt}^j \rangle\right|) \tag{8}$$

The total score is the average of the sum of the orientation and position scores corresponding the N images of the test set:

$$S_{total}^j = \frac{1}{J} \sum_{j=1}^{J} (S_{pos}^j + S_{ori}^j) \tag{9}$$

5.2   Challenge Results

The ground-truth labels of the two test sets (Lightbox and Sunlamp) from the SPEED+ dataset have not been made publicly available. Instead, the Challenge evaluation was performed by submitting the estimated poses to a evaluation server; the results were then published on public leaderboards only if the obtained score improved a previous result from the same participant. We provide a summary of the results by showing the error achieved by the 10 first teams on the Lightbox and Sunlamp datasets on Table 1 and Table 2 respectively. Teams are represented in decreasing order based on the achieved rank in the leaderbord, corresponding the first row to the winner of the category. The winning teams were *TangoUnchained* for the Lightbox dataset and *lava1302* for the Sunlamp dataset. Our solution *VPU* ranked 2nd on both datasets, and achieved the best total average error as shown in Table 3, where we represent the averaged scores of the two datasets. This shows that our solution is robust to different domain changes and can be applied without fine-tuning on different settings with different illumination and noise conditions.

Table 1. Summary of the SPEC2021 results for the lightbox dataset.

Team name	S_{pos}^j	S_{ori}^j	S_{total}^j
TangoUnchained	**0.0179**	**0.0556**	**0.0734**
VPU (Ours)	0.0215	0.0799	0.1014
lava1302	0.0483	0.1163	0.1646
haoranhuang_njust	0.0315	0.1419	0.1734
u3s_lab	0.0548	0.1692	0.2240
chusunhao	0.0328	0.2859	0.3186
for graduate	0.0753	0.4130	0.4883
Pivot SDA AI & Autonomy Sandbox	0.0721	0.4175	0.4895
bbnc	0.0940	0.4344	0.5312
ItTakesTwoToTango	0.0822	0.5427	0.6248

For the Challenge, we trained the model described in Sect. 3 during 35 epochs on the Synthetic dataset at a resolution of 640×640 pixels. Then, we trained individual models following Algorithm 1 for the Sunlamp and Lightbox target datasets during 50 iterations. Figure 3 and Fig. 4 provide visual results of the estimated poses for the Sunlamp and Lightbox datasets. The estimated poses

Table 2. Summary of the SPEC2021 results for the Sunlamp dataset.

Team name	S^j_{pos}	S^j_{ori}	S^j_{total}
lava1302	**0.0113**	**0.0476**	**0.0588**
VPU (Ours)	0.0118	0.0493	0.0611
TangoUnchained	0.0150	0.0750	0.0899
u3s_lab	0.0320	0.1089	0.1409
haoranhuang_njust	0.0284	0.1467	0.1751
bbnc	0.0819	0.3832	0.4650
for graduate	0.0858	0.4009	0.4866
Pivot SDA AI & Autonomy Sandbox	0.1299	0.6361	0.7659
ItTakesTwoToTango	0.0800	0.6922	0.7721
chusunhao	0.0584	0.0584	0.8151

Table 3. Summary of the SPEC2021 results averaged over both Datasets.

Team name	S^j_{pos}	S^j_{ori}	S^j_{total}
VPU (Ours)	0.0166	**0.0646**	**0.0812**
TangoUnchained	**0.0164**	0.0653	0.0816
lava1302	0.0298	0.0820	0.1117
haoranhuang_njust	0.0456	0.1443	0.1742
u3s_lab	0.0434	0.1391	0.1824
for graduate	0.0805	0.4070	0.4874
bbnc	0.0879	0.4088	0.4981
chusunhao	0.0456	0.17215	0.56685
Pivot SDA AI & Autonomy Sandbox	0.0999	0.5268	0.6266
ItTakesTwoToTango	0.0811	0.61745	0.69845

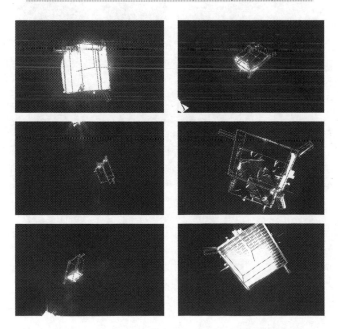

Fig. 3. Example results of the proposed algorithm on the Sunlamp test dataset. The estimated pose is represented with a wire-frame model over the image.

are represented with the 3-axis plot and the wireframe model projected over the image. They show that the proposed method can handle strong reflections, presence of other elements on the image, complex background and close views of the object.

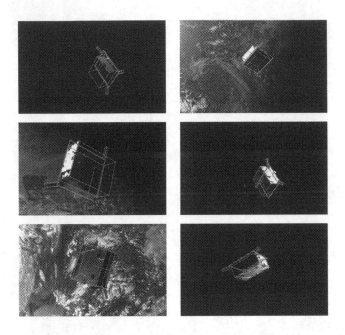

Fig. 4. Example results of the proposed algorithm on the Lightbox test dataset. The estimated pose is represented with a wire-frame model over the image.

5.3   Additional Experiments

We show in Fig. 5 the effect of iterating over Algorithm 1 on the number of generated pseudo-labels. The left plot corresponds to the strategy followed over the Sunlamp dataset, where the reprojection error that controls the convergence criteria of RANSAC-EPnP remains fixed during all iterations with a value of 2.0. It can be observed that the percentage of pseudo-labels reaches 90% after 10 iterations and slowly increases from there, suggesting a possible overfit to wrongly estimated labels. The right plot corresponds to the Ligthbox dataset, where a different approach was followed. Motivated by the fact that the Lightbox test set is more visually similar to the training set than the Sunlamp test set, we hypothesize that restricting the label generation by reducing the reprojection error would lead to better pseudo-labels and hence better performance. To test this, the reprojection error was decreased after 10 iterations (red dashed line) to 1.0. It can be observed that the percentage of pseudo-labels quickly drops and then grows at a slower rate. However, based on the final results we cannot conclude that this strategy improved the results compared to using a fixed reprojection error, as the achieved errors for both test sets are in a similar range.

To show the effectiveness of the proposed method on the target test sets, we evaluated the proposed pose estimation algorithm against a baseline considering only the 2D heatmap loss. The models were trained on the source synthetic dataset during 10 epochs and tested over a pseudo-test dataset. The pseudo-test set is obtained by running Algorithm 1 until approximately 30 % of the target dataset is labelled. This number is chosen as a compromise between the necessary number of pseudo-labels to obtain a representative analysis and the quality of the labels obtained. The pseudo-test set approach is followed as the real labels of the test set are not made publicy available in the challenge. The results are shown in Table 4 with the validation set from the source dataset as reference. It can be observed that the proposed method, despite obtaining similar values on the validation (Source set) is able to improve the results on the target (test) scts.

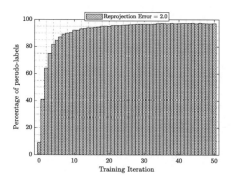 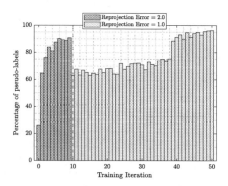

Fig. 5. Percentage of pseudo-labels created in the dataset as a function of the training iterations over Algorithm 1. Left plot represents the pseudo-labels over the Sunlamp dataset for a constant reprojection error constraint of 2.0 for the RANSAC EPnP algorithm. Right plot represents the pseudo-labels over the Lightbox dataset for a variable reprojection error that starts at 2.0 and drops to 1.0 after 10 iterations.

Table 4. Influence of the proposed pose estimation method on the position, orientation and total score on a pseudo-test set and on the validation set.

Method	Sunlamp			Lightbox			Validation		
	S_{pos}^j	S_{ori}^j	S_{total}^j	S_{pos}^j	S_{ori}^j	S_{total}^j	S_{pos}^j	S_{ori}^j	S_{total}^j
Baseline	0.3095	0.9229	1.2324	0.2888	0.8935	1.1823	**0.0086**	0.02627	**0.03490**
Proposed	**0.2138**	**0.8554**	**1.0692**	**0.2837**	**0.7709**	**1.0546**	0.0088	**0.0256**	0.0355

6   Conclusions

We have presented a method for spacecraft pose estimation that integrates structure information in the learning process by incorporating pose and 3D alignment information in a combined loss function. We argue that incorporating such information provides robustness to domain shifts when the imaged objects keep the same structure although they are captured under different illumination conditions. In addition we have introduced a method for pseudo-labelling in pose estimation that exploits the consensus within a network and provides robustness to label noise. Our solution has ranked second on both datasets for the SPEC2021 Challenge, achieving the best average score.

Acknowledgements. This work is supported by Comunidad Autónoma de Madrid (Spain) under the Grant IND2020/TIC-17515.

References

1. Alcantarilla, P.F., Bartoli, A., Davison, A.J.: KAZE features. In: Fitzgibbon, A., Lazebnik, S., Perona, P., Sato, Y., Schmid, C. (eds.) ECCV 2012. LNCS, vol. 7577, pp. 214–227. Springer, Heidelberg (2012). https://doi.org/10.1007/978-3-642-33783-3_16
2. Bay, H., Ess, A., Tuytelaars, T., Van Gool, L.: Speeded-up robust features (SURF). Comput. Vis. Image Underst. **110**(3), 346–359 (2008)
3. Brégier, R.: Deep regression on manifolds: a 3D rotation case study (2021)
4. Chen, B., Cao, J., Parra, A., Chin, T.J.: Satellite pose estimation with deep landmark regression and nonlinear pose refinement. In: Proceedings of the IEEE/CVF International Conference on Computer Vision Workshops, pp. 0–0 (2019)
5. Chen, B., Parra, A., Cao, J., Li, N., Chin, T.J.: End-to-end learnable geometric vision by backpropagating PNP optimization. In: Proceedings of the IEEE/CVF Conference on Computer Vision and Pattern Recognition, pp. 8100–8109 (2020)
6. Doumanoglou, A., Kouskouridas, R., Malassiotis, S., Kim, T.K.: Recovering 6D object pose and predicting next-best-view in the crowd. In: Proceedings of the IEEE Conference on Computer Vision and Pattern Recognition, pp. 3583–3592 (2016)
7. European Space Agency: Spacecraft pose estimation challenge 2021 (2021). https://kelvins.esa.int/pose-estimation-2021/
8. Ganin, Y., Lempitsky, V.: Unsupervised domain adaptation by backpropagation. In: International Conference on Machine Learning, pp. 1180–1189. PMLR (2015)
9. Ghifary, M., Kleijn, W.B., Zhang, M., Balduzzi, D., Li, W.: Deep reconstruction-classification networks for unsupervised domain adaptation. In: Leibe, B., Matas, J., Sebe, N., Welling, M. (eds.) ECCV 2016. LNCS, vol. 9908, pp. 597–613. Springer, Cham (2016). https://doi.org/10.1007/978-3-319-46493-0_36
10. Hinterstoisser, S., Lepetit, V., Ilic, S., Fua, P., Navab, N.: Dominant orientation templates for real-time detection of texture-less objects. In: 2010 IEEE Computer Society Conference on Computer Vision and Pattern Recognition, pp. 2257–2264. IEEE (2010)
11. Hu, Y., Speierer, S., Jakob, W., Fua, P., Salzmann, M.: Wide-depth-range 6d object pose estimation in space. In: Proceedings of the IEEE/CVF Conference on Computer Vision and Pattern Recognition, pp. 15870–15879 (2021)

12. Kehl, W., Milletari, F., Tombari, F., Ilic, S., Navab, N.: Deep learning of local RGB-D patches for 3D object detection and 6D pose estimation. In: Leibe, B., Matas, J., Sebe, N., Welling, M. (eds.) ECCV 2016. LNCS, vol. 9907, pp. 205–220. Springer, Cham (2016). https://doi.org/10.1007/978-3-319-46487-9_13

13. Kelsey, J.M., Byrne, J., Cosgrove, M., Seereeram, S., Mehra, R.K.: Vision-based relative pose estimation for autonomous rendezvous and docking. In: 2006 IEEE Aerospace Conference, p. 20. IEEE (2006)

14. Kendall, A., Cipolla, R.: Geometric loss functions for camera pose regression with deep learning. In: Proceedings of the IEEE Conference on Computer Vision and Pattern Recognition, pp. 5974–5983 (2017)

15. Kendall, A., Grimes, M., Cipolla, R.: PoseNet: a convolutional network for real-time 6-DoF camera relocalization. In: Proceedings of the IEEE International Conference on Computer Vision, pp. 2938–2946 (2015)

16. Kuffner, J.J.: Effective sampling and distance metrics for 3D rigid body path planning. In: IEEE International Conference on Robotics and Automation, 2004. Proceedings, ICRA 2004, vol. 4, pp. 3993–3998. IEEE (2004)

17. Lee, K.H., He, X., Zhang, L., Yang, L.: CleanNet: transfer learning for scalable image classifier training with label noise. In: Proceedings of the IEEE Conference on Computer Vision and Pattern Recognition, pp. 5447–5456 (2018)

18. Lepetit, V., Moreno-Noguer, F., Fua, P.: EPnP: an accurate O(N) solution to the PnP problem. Int. J. Comput. Vision $81(2)$, 155–166 (2009)

19. Lowe, D.G.: Distinctive image features from scale-invariant keypoints. Int. J. Comput. Vision $60(2)$, 91–110 (2004)

20. Mahendran, S., Ali, H., Vidal, R.: 3D pose regression using convolutional neural networks. In: Proceedings of the IEEE International Conference on Computer Vision Workshops, pp. 2174–2182 (2017)

21. Mishchuk, A., Mishkin, D., Radenovic, F., Matas, J.: Working hard to know your neighbor's margins: local descriptor learning loss. In: Advances in Neural Information Processing Systems, vol. 30 (2017)

22. Newell, A., Yang, K., Deng, J.: Stacked hourglass networks for human pose estimation. In: Leibe, B., Matas, J., Sebe, N., Welling, M. (eds.) ECCV 2016. LNCS, vol. 9912, pp. 483–499. Springer, Cham (2016). https://doi.org/10.1007/978-3-319-46484-8_29

23. Pan, S.J., Yang, Q.: A survey on transfer learning. IEEE Trans. Knowl. Data Eng. $22(10)$, 1345–1359 (2009)

24. Park, T.H., D'Amico, S.: Robust multi-task learning and online refinement for spacecraft pose estimation across domain gap. arXiv preprint arXiv:2203.04275 (2022)

25. Park, T.H., Märtens, M., Lecuyer, G., Izzo, D., D'Amico, S.: SPEED+: Next-generation dataset for spacecraft pose estimation across domain gap. arXiv preprint arXiv:2110.03101 (2021)

26. Parkes, S., Martin, I., Dunstan, M., Matthews, D.: Planet surface simulation with PANGU. In: Space ops 2004 Conference, p. 389 (2004)

27. Peng, S., Liu, Y., Huang, Q., Zhou, X., Bao, H.: PVNET: pixel-wise voting network for 6DoF pose estimation. In: Proceedings of the IEEE/CVF Conference on Computer Vision and Pattern Recognition, pp. 4561–4570 (2019)

28. Proença, P.F., Gao, Y.: Deep learning for spacecraft pose estimation from photorealistic rendering. In: 2020 IEEE International Conference on Robotics and Automation (ICRA), pp. 6007–6013. IEEE (2020)

29. Sattler, T., Zhou, Q., Pollefeys, M., Leal-Taixe, L.: Understanding the limitations of CNN-based absolute camera pose regression. In: Proceedings of the IEEE/CVF Conference on Computer Vision and Pattern Recognition, pp. 3302–3312 (2019)
30. Sharma, S., Beierle, C., D'Amico, S.: Pose estimation for non-cooperative spacecraft rendezvous using convolutional neural networks. In: 2018 IEEE Aerospace Conference, pp. 1–12. IEEE (2018)
31. Shavit, Y., Ferens, R.: Introduction to camera pose estimation with deep learning. arXiv preprint arXiv:1907.05272 (2019)
32. Shi, J.F., Ulrich, S., Ruel, S.: Spacecraft pose estimation using principal component analysis and a monocular camera. In: AIAA Guidance, Navigation, and Control Conference, p. 1034 (2017)
33. Shrivastava, A., Pfister, T., Tuzel, O., Susskind, J., Wang, W., Webb, R.: Learning from simulated and unsupervised images through adversarial training. In: Proceedings of the IEEE Conference on Computer Vision and Pattern Recognition, pp. 2107–2116 (2017)
34. Su, H., Qi, C.R., Li, Y., Guibas, L.J.: Render for CNN: viewpoint estimation in images using CNNs trained with rendered 3D model views. In: Proceedings of the IEEE International Conference on Computer Vision, pp. 2686–2694 (2015)
35. Sun, B., Saenko, K.: Deep CORAL: correlation alignment for deep domain adaptation. In: Hua, G., Jégou, H. (eds.) ECCV 2016. LNCS, vol. 9915, pp. 443–450. Springer, Cham (2016). https://doi.org/10.1007/978-3-319-49409-8_35
36. Sun, J., Shen, Z., Wang, Y., Bao, H., Zhou, X.: LoFTR: detector-free local feature matching with transformers. In: Proceedings of the IEEE/CVF Conference on Computer Vision and Pattern Recognition, pp. 8922–8931 (2021)
37. Sun, X., Xiao, B., Wei, F., Liang, S., Wei, Y.: Integral human pose regression. In: Ferrari, V., Hebert, M., Sminchisescu, C., Weiss, Y. (eds.) ECCV 2018. LNCS, vol. 11210, pp. 536–553. Springer, Cham (2018). https://doi.org/10.1007/978-3-030-01231-1_33
38. Umeyama, S.: Least-squares estimation of transformation parameters between two point patterns. IEEE Trans. Pattern Anal. Mach. Intell. **13**(04), 376–380 (1991)
39. Wang, J., et al.: Deep high-resolution representation learning for visual recognition. IEEE Trans. Pattern Anal. Mach. Intell. **43**(10), 3349–3364 (2020)
40. Zhang, P., Zhang, B., Zhang, T., Chen, D., Wang, Y., Wen, F.: Prototypical pseudo label denoising and target structure learning for domain adaptive semantic segmentation. In: Proceedings of the IEEE/CVF Conference on Computer Vision and Pattern Recognition, pp. 12414–12424 (2021)
41. Zhang, Z., Sabuncu, M.: Generalized cross entropy loss for training deep neural networks with noisy labels. In: Advances in Neural Information Processing Systems, vol. 31 (2018)
42. Zhu, J.Y., Park, T., Isola, P., Efros, A.A.: Unpaired image-to-image translation using cycle-consistent adversarial networks. In: Proceedings of the IEEE International Conference on Computer Vision, pp. 2223–2232 (2017)

MaRF: Representing Mars as Neural Radiance Fields

Lorenzo Giusti[1,2(✉)], Josue Garcia[3], Steven Cozine[3], Darrick Suen[3], Christina Nguyen[3], and Ryan Alimo[2,4]

[1] Sapienza University of Rome, Rome, Italy
lorenzo.giusti@uniroma1.it
[2] Jet Propulsion Laboratory, California Institute of Technology, Pasadena, CA 91109, USA
[3] University of California, San Diego, San Diego, CA, USA
[4] OPAL AI Inc., Los Angeles, CA, USA

Abstract. The aim of this work is to introduce MaRF, a novel framework able to synthesize the Martian environment using several collections of images from rover cameras. The idea is to generate a 3D scene of Mars' surface to address key challenges in planetary surface exploration such as: planetary geology, simulated navigation and shape analysis. Although there exist different methods to enable a 3D reconstruction of Mars' surface, they rely on classical computer graphics techniques that incur high amounts of computational resources during the reconstruction process, and have limitations with generalizing reconstructions to unseen scenes and adapting to new images coming from rover cameras. The proposed framework solves the aforementioned limitations by exploiting Neural Radiance Fields (NeRFs), a method that synthesize complex scenes by optimizing a continuous volumetric scene function using a sparse set of images. To speed up the learning process, we replaced the sparse set of rover images with their neural graphics primitives (NGPs), a set of vectors of fixed length that are learned to preserve the information of the original images in a significantly smaller size. In the experimental section, we demonstrate the environments created from actual Mars datasets captured by Curiosity rover, Perseverance rover and Ingenuity helicopter, all of which are available on the Planetary Data System (PDS).

1 Introduction

A critical aspect of space exploration is understanding remote environments based on the limited data returned by spacecrafts. These environments are often too challenging or expensive to visit in person and can only be explored through images. While spacecraft imaging is an essential data source for learning about faraway locations, learning through two-dimensional images is drastically different from how geologists study Earth's environments [1]. To address this challenge, mixed reality (MR), a human-computer interface, combines interactive computer graphics environments with real-world elements, recreating a whole sensory world entirely through computer-generated signals of an environment

Fig. 1. Illustration of collaborative space exploration with scientists being virtually present as avatars onto the Mars' surface through mixed reality technology (Image credit: NASA/JPL-Caltech).

[13,21]. MR overlays the computer-generated sensory signals onto the actual environment, enabling the user to experience a combination of virtual and real worlds concurrently. This technology allows scientists and engineers to interpret and visualize science data in new ways and experience environments otherwise impossible to explore. For instance, researchers at the NASA Goddard Space Flight Center (GSFC) have developed immersive tools for exploring regions from the depths of our oceans to distant stars and galaxies [16]. Another example is OnSight [2,5], a framework developed through a collaboration between NASA JPL's Opérations Lab and Microsoft that creates an immersive 3-D terrain model of sites that Curiosity rover has visited and allows scientists to collaboratively study the geology of Mars as they virtually meet at those sites as shown in Fig. 1. All these applications provide scientists and engineers an immersive experience of virtual presence in the field. Our contribution is an end-to-end framework that takes spacecrafts' image collections from JPL's Planetary Data System (PDS) [15,20], filters them and learn the radiance field of Mars surface. Our framework is capable of recreating a synthetic view of the Martian scene in order to have a better sense of both the geometry and appearance of an environment that cannot be physically explored yet. Since the proposed mixed-reality framework captures both the scene's geometry and appearance information, enabling the synthesis of previously unknown viewpoints, it allows scientists and engineers to have a higher scientific return with respect to planetary geology, robotic navigation and many more space applications. Furthermore, the framework can also be used for compressing the images from the Mars Exploratory Rovers (MER) since the *entire* scene is encoded in a Multi-Layer Perceptron (MLP) with a storage size of several orders of magnitude less than a single PDS image, all while preserving as much informative content as the original image collection.

This paper is structured as follows: Sect. 2 provides a review of the related works and technologies that use mixed reality for space applications, followed by

(a)   (b)

Fig. 2. Rendered view synthesis of the Perseverance rover at Jet Propulsion Laboratory: (a) frontal view; (b) side view.

the required mathematical background for understanding how a Neural Radiance Fields model for view synthesis is trained, and lastly the landscape of the most important techniques that exploit the original NeRF idea to solve the problems most closely related to ours. Section 3 contains the core of this work. It provides a detailed description of the experimental setup and a review of the results obtained. Section 4 contains the methodology used to account for reconstruction instabilities using a bootstrap technique to measure the uncertainty of the reconstructions.

2   Related Works

In this section, we review mixed reality technologies for space applications, provide the required background for training a NeRF model, discuss related methods for photo-realistic view synthesis and explain the possibility of reducing the computational time required for achieving optimal results.

2.1   Mixed Reality for Space Applications

Mixed reality has been successfully applied in many fields of *aerospace engineering*. For example, it can be used to guide operators through a given task with the highest possible efficiency, achieved by superimposing 2-D and 3-D instructions and animations atop the working, real-world area [23]. Consider also that MR allows technicians to communicate with engineers while inspecting or repairing aircraft engines. Using this technology, a technician can be on any work site, wearing an immersive display with a camera, microphone, and speaker, while the engineer guides the technician remotely. For *navigation and guidance* environments, immersive displays were initially used to provide pilots with flight information [9]. The use of mixed reality to display flight paths and obstacles can decrease pilot workload during landing operations in environments with scarcity of visibility. Augmented Reality (AR) systems have shown high potential for application in unmanned flight control and training. In *simulations*, there have been numerous AR projects have investigate such technology and its application for pilot or astronaut training. For astronaut training and support, European Space Agency (ESA) developed the Computer Aided Medical Diagnostics and Surgery System (CADMASS) project [19], which was designed to help astronauts perform mild to high-risk medical examinations and surgeries in space. The CADMASS project was designed to help, through mixed reality, those without medical expertise perform surgeries and similar procedures.

2.2 Neural Radiance Fields (NeRF)

In this section, we provide a brief background for synthesizing novel views of large static scenes by representing the complex scenes as Neural Radiance Fields [17]. In NeRF, a scene is represented with a Multi-Layer Perceptron (MLP). Given the 3D spatial coordinate location $\mathbf{r} = (x, y, z)$ of a voxel and the viewing direction $d = (\theta, \phi)$ in a scene, the MLP predicts the volume density σ and the view-dependent emitted radiance or color $\mathbf{c} = [R, G, B]$ at that spatial location for volumetric rendering. In the concrete implementation, $\mathbf{r}$ is first fed into the MLP, and outputs σ and intermediate features, which are then fed into an additional fully connected layer to predict the color c. The volume density is hence decided by the spatial location $\mathbf{r}$, while the color $\mathbf{c}$ is decided by both X and viewing direction $\mathbf{r}$. The network can thus yield different illumination effects for different viewing directions. This procedure can be formulated as:

$$[R, G, B, \sigma] = F_\Theta(x, y, z, \theta, \phi), \tag{1}$$

where Θ is the set of learnable parameters of the MLP. The network is trained by fitting the rendered (synthesized) views with the reference (ground-truth) views via the minimization of the total squared error between the rendered and true pixel colors in the different views. NeRF uses a rendering function based on classical volume rendering. Rendering a view from NeRF requires calculating the expected color $C(\mathbf{r})$ by integrating the accumulated transmittance T along the camera ray $\mathbf{r}(t) = \mathbf{o} + t\mathbf{d}$, for $\mathbf{o}$ being the origin of cast ray from the desired virtual camera, with near and far bounds t_n and t_f. This integral can be expressed as

$$C(\mathbf{r}) = \int_{t_n}^{t_f} T(t)\sigma(\mathbf{r}(t))c(\mathbf{r}(t), \mathbf{d})dt, \tag{2}$$

where $T(t) = \exp\left(-\int_{t_n}^{t} \sigma(\mathbf{r}(s)ds\right)$ reflects the cumulative transmittance along the ray from t_n to t, or the possibility that the beam will traverse the path without interacting with any other particle. Estimating this integral $C(\mathbf{r})$ for a camera ray traced across each pixel of the desired virtual camera is essential when producing a view from our continuous neural radiance field. To solve the

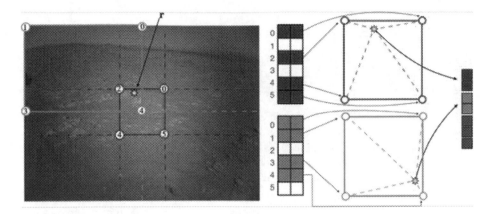

Fig. 3. Example of multi-resolution hash encoding proposed in [18] for the 2D case on a picture collected from the Ingenuity helicopter. In this case, the number of resolution levels L is set to be equal to 2. The green boxes represent auxiliary inputs concatenated to the embedding vector before going into the neural network. (Color figure online)

equation (2), the integral is approximated using the quadrature rule [14] by sampling a finite number of points t_i uniformly along the ray $\mathbf{r}(t)$. Formally:

$$\hat{C}(\mathbf{r}) = \sum_{i=1}^{N} T_i(1 - exp(-\sigma_i\underbrace{(t_{i+1} - t_i)}_{\delta_i}))\mathbf{c}_i$$

where $\mathbf{c}_i$ is the color of the $i - th$ sample, σ_i its density and $T_i = exp\left(-\sum_{j=1}^{i-1}\sigma_j\delta_j\right)$ estimates the transmittance. An example of NeRF results is shown in Fig. 2.

2.3 NeRF Landscape

Due to its computational time requirements, learning a view synthesis using Neural Radiance Fields might be intractable for huge natural scenes like the Martian environment. Despite the most successful breakthroughs in the field [22, 26, 30, 32], the issue was partially solved with the introduction of Plenoxels [29], where they proved how a simple grid structure with learned spherical harmonics coefficients may converge orders of magnitudes quicker than the fastest NeRF alternative without compromising the quality of the results. Recently, this structure was enhanced in [18], in which they defined a multi-scale grid structure that looks up embeddings by querying coordinates in $\mathcal{O}(1)$ time using hash tables [28]. Essentially, their work focuses on the concept of leveraging a grid-like structure to store learning embeddings that may be interpolated to obtain correct values for each position. Multiple hash tables of differing resolutions are grouped in geometric progressions from the coarsest to the finer in the encoding structure. At each resolution level, they find the corners of the voxel containing the query coordinate, acquire their encodings, then interpolate them to create a

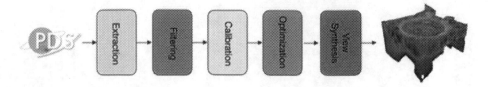

Fig. 4. Sequence of operations required to get from the spacecrafts images to the neural view synthesis of the Mars' surface. The steps are, from left to right: (i) extract data from the Planetary Data System; (ii) filter image that do not meet the requirements needed for the reconstruction; (iii) calibrate the intrinsic and extrinsic camera parameters; (iv) training the algorithm to produce the view.

collection of feature vectors for the provided coordinate, which is concatenated into a single input embedding; pictorially represented in Fig. 3. There is a unique coordinate for each vertex in the coarse levels, but at the fine levels, the number of embeddings is less than the number of vertices, and a spatial hash function is used to extract the required feature vector. At further resolutions, it is feasible to trade off time requirements and quality of the results by increasing the size of the hash tables. These taught input embeddings allows to shrink the dimension of the network reducing drastically the overall computational time required for synthesize a scene, solving a critical bottleneck in past works. Implementing the entire process in a highly optimized low-level GPU code implementation, allows this approach to converge in a time that is several orders of magnitude less than the original NeRF implementation. Also, consider that the hash table may be queried in parallel on all resolution levels for all pixels, without requiring any form of control flow. Moreover, the cascading nature of the multi-resolution grid enables the use of a coarse one to one mapping between entries and spatial locations, and a fine "hash-table-like" structure with a spatial function that performs collision resolution via a gradient based method by averaging the colliding gradients, resulting in a sparse grid by design. If the distribution of points changes to a concentrated location anywhere in the picture, the collisions will become uncommon, and a better hashing function may be trained, thereby automatically adapting to the changing data on the fly without any task-specific structural adjustments.

3   Mars Radiance Fields (MaRF)

In this section we provide the experimental setup used as well as the results obtained.

In particular we implemented a pipeline[1] (Fig. 4) that is able to train a NeRF view synthesis from spacecraft images, available on the planetary data system (PDS), consisting of the following phases:

---

[1] Code available at: https://github.com/lrnzgiusti/MaRF.

Extraction: To retrieve the spacecrafts' images, we queried the Planetary Data System by specifying the exact location of the images (i.e. `/mars2020/mars2020_mastcamz_ops_raw/`) and extracted *all* the images present at that path across every *sol*. Each image comes with an associated label files, which includes metrics such as the spacecraft's systems, the time of the image being taken and camera parameters (both intrinsics and extrinsics) in CAHVOR format [27].

Filtering: Neural Radiance Fields provide a view synthesis' resolution proportional to the one of the images used for learning the scene. However, raw data provided by the *extraction* phase might include images of varying resolution, including smaller thumbnail images; noisy images and duplicates with different color filters applied. Thus, before training the view synthesis, the datasets are first forwarded through a bank of filters that keep images according to: (a) file dimension; (b) image shape; (c) duplicates up to a color filter; (d) grayscale; (e) color histogram; (f) blur detection. The first two filters (a-b) are parameterized using threshold values, i.e. if an image has a storage size less than a value β or the shape is too small we remove since almost surely it does not meet the minimum quality requirements. The duplicate filter (c) utilizes a perceptual hash function [31] to identify photos with a near-identical hashes and remove images in which the environment is the same, but the color filters are different. The grayscale filter (d) has two means of action: it identifies any photos with only one color channel, and filters those photos out. However, some grayscale photos have all three color channels. These photos can be identified because for each pixel, their RGB values will be equal. The color histogram filter (e) analyzes the image channels in all photos that pass the previous filters and builds a histogram of the average color intensities. Then, a photo is filtered if more than half of the count of saturation values in a photo are more than a standard deviation away from the average saturation value. This methodology was adopted because Mars' images in general have consistent color channel values (Fig. 5 (b)) while images that does not contain the Mars' environment will have far different color histograms (i.e. images containing parts of the spacecraft or glitched images Fig. 5 (a)). The blur filter works by taking the variance of the Laplacian of the image, which numerically determines the sharpness of an image [4]. By setting a threshold value τ on the sharpness we filter out all the images that have a sharpness value *less* than τ.

Calibration: For each image, the camera intrinsics and extrinsics models are provided in CAHVOR format [27]. However, training a Neural Radiance Fields model requires these parameters to be expressed in a pinhole camera model [12], which can be represented via the camera projection matrix as:

$$\mathbf{P} = \mathbf{KR}\,[\mathbf{I}| - \mathbf{t}] \tag{3}$$

where $\mathbf{R}$ is a 3×3 rotation matrix, $\mathbf{I}$ is a 3×3 identity matrix, $\mathbf{t}$ is a 3×1 translation vector and $\mathbf{K}$ is the camera calibration matrix defined as:

$$K = \begin{pmatrix} f_x & 0 & c_x \\ 0 & f_y & c_y \\ 0 & 0 & 1 \end{pmatrix} \tag{4}$$

(a)   (b)

Fig. 5. Different images retrieved from the planetary data system (PDS). In the first case (a) the image is removed during the filtering stage while in (b) the preprocessing pipeline does not detect any defection and marks it as ready for the reconstruction.

Here, **K** contains the intrinsic camera parameters where f_x and f_y are the focal lengths of the camera, which for square pixels assume the same value, and c_x and c_y are the offsets of the principal point from the top-left corner of the image.

Although there exists techniques that provide a conversion between CAHVOR and pinhole camera models [10] we rely the calibration phase on Structure from Motion (SfM) algorithms [24, 25] since the conversion defined in [10] requires parameters not readily available on the PDS. An illustration of the results from the calibration process is shown in Fig. 6.

Optimization: To properly tune the model's hyper-parameters, we rely on random search [8]. The metric used to quantitatively measure the reconstruction quality is the *peak signal-to-noise ratio* between the underlying scene represented by S described by a set of images $\{I_j\}_{j=1}^N$, where each image I_j is captured by a viewpoint $\mathbf{r}_j$ and the view synthesis $\hat{S}$, composed by a set of rendered images $\{\hat{I}_j\}_{j=1}^N$ where the elements $\hat{I}_j$ are rendered from the same viewpoint of the ground truth $\mathbf{r}_j$. Then, for a single image, the *PSNR* between the ground truth and the synthesized one is computed as:

$$PSNR(I, \hat{I}) = 10 \cdot \log_{10}\left(\frac{max(I^2)}{MSE(I, \hat{I})}\right), \tag{5}$$

where $MSE(I, \hat{I}) = \frac{(I-\hat{I})^2}{W \cdot H \cdot C}$. Then, to get the peak signal-to-noise ratio between S and $\hat{S}$, we compute the average on the peak signal-to-noise ratios of the images that compose the scene.

View Synthesis: For the scope of this work, we used different datasets coming from the following sources: **(a)** Curiosity rover's science cameras [7]; **(b)** Ingenuity helicopter's color camera [3]; **(c)** Perseverance rover's Mastcam-Z [6]. In Table 1, we shown a quantitative assessment of the synthesis process by comparing the peak signal to noise ratio (PSNR) of the proposed framework for the

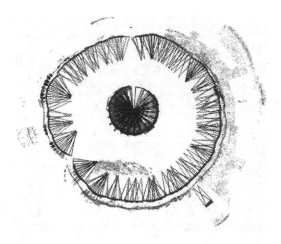

Fig. 6. Top-view of the camera calibration process for the Perseverance dataset.

aforementioned datasets and listing the results of our method after training for
10 s to 15 min. For the Perseverance dataset, the PSNR decreases with respect
to the training time. This fact is due to a slight misalignment of the camera
models involved during the training process leading to sparse black spots in the
view synthesis as shown in Fig. 7d. Although this misalignment influences the
quantitative results of the scene, the overall quality of the reconstruction is not
compromised (Fig. 8).

Table 1. Peak signal to noise ratio (PSNR) of MaRF across the selected spacecraft
datasets for different execution times of the reconstruction.

Quantitative results			
	Curiosity	Perseverance	Ingenuity
PSNR (**10 s**)	23.02	18.09	22.07
PSNR (**60 s**)	26.22	18.08	23.00
PSNR (**5 m**)	31.01	17.82	23.86
PSNR (**10 m**)	34.71	17.71	24.25
PSNR (**15 m**)	36.82	17.63	24.24

4   Bootstrapping the Uncertainty

In this section we propose a technique to account for fluctuations during the
training of the Mars' radiance field, as well as computing a view synthesis' confidence in terms of an uncertainty map of the reconstruction. The aim of these

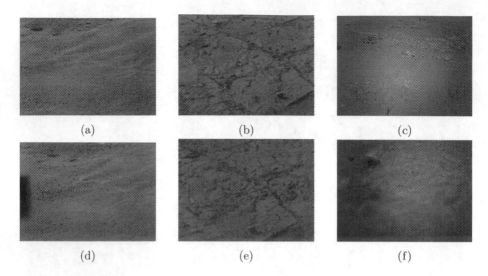

(a)   (b)   (c)

(d)   (e)   (f)

Fig. 7. A demonstration of the view synthesis quality for spacecrafts' images. Top row: Reference test images of (a) Perseverance; (b) Curiosity; (c) Ingenuity datasets for which we evaluated the proposed framework. Bottom row: rendered view synthesis of the reference images using MaRF. The rendering of the reconstruction for the Perseverance dataset (d) contains a black rectangle due to a misalignment during the camera calibration.

techniques is to provide useful insights and allow scientist and engineering teams to stay in the loop when it comes to decision making for the Mars' missions, maximizing the scientific return. For example, in a simulated environment, the uncertainty map provides a support to the navigation team to determine the best sequencing plan, while being aware of potential pitfalls. To compute the uncertainty map of the synthesized scene, we follow the bootstrap method [11]. In particular, once the model's hyper-parameters have been optimized, we repeat the training process B times using the same setup and then the spatial locations in which the model is unstable is computed as the pixel-wise standard deviation of the rendered frames. Formally:

1. Optimize the models' hyper-parameters according to the method defined in the Sect. 3.
2. Let $\{\hat{\mathcal{S}}_b\}_{b=1}^{B}$ be the set of bootstrapped view synthesis of the same scene.
3. Render each scene to produce a multi-set $\{\{\hat{I}_{j,b}\}_{j=1}^{N}\}_{b=1}^{B}$ of bootstrapped rendered images from the viewpoints $\mathbf{r}_j$.
4. Let $\tilde{I}_{j,b}$ be the gray-scale version of $\hat{I}_{j,b}$ and stack them into $\tilde{\mathbf{I}}_j$, a 3D tensor in which the bootstrap results are placed along the third dimension.
5. Let $\tilde{\mathbf{I}}_j = \frac{1}{B}\sum_b \tilde{\mathbf{I}}_{j,b}$ be the expected rendering at viewpoint $\mathbf{r}_j$.
6. Let $\Sigma_j = \sum_b \sigma\left(\tilde{\mathbf{I}}_{j,b}\right)$ be the uncertainty map at viewpoint $\mathbf{r}_j$, computed as the pixel-wise standard deviation along its third dimension.

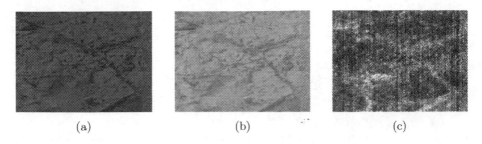

(a)   (b)   (c)

Fig. 8. Uncertainty estimation. **(a)** Pixel-wise mean out of the bootstrapped render-ings. **(b)** Grayscale version of **(a)**, used to compute the uncertainty maps. **(c)** Pixel-wise uncertainty map; here uncertainty is proportional to the pixel intensity.

By this point, we have the uncertainty maps for all the viewpoints of the scene. With this, we are able to put them in a sequence to provide a fly-through video of the geometric and texture uncertainties.

5   Conclusion and Discussion

In this work we presented *MaRF*, an end-to-end framework able to convert space-crafts images available on the Planetary Data System (PDS) into an synthetic view of the Mars surface by exploiting Neural Radiance Fields. It consists of an extract, transform, load (ETL) pipeline that (i) retrieves images from the PDS; (ii) filters them according to a minimum quality requirements; (iii) calibrates the cameras using structure from motion (SfM) algorithms, (iv) Optimizes the model's hyper-parameters and (v) obtains the view synthesis of the Martian environment by learning its radiance field using a neural network that optimizes a continuous volumetric scene function using the sparse set of images that have been calibrated during phase (iii). We have evaluated the proposed technique on three different spacecraft datasets: Curiosity rover, Perseverance rover and Ingenuity helicopter. We also equip our framework with a method that is able to account for reconstruction instabilities and provide an estimation of the uncer-tainty of the reconstruction. We believe that equipping future space missions with tools like MaRF, at the intersection of mixed reality and artificial intel-ligence, will make a significant impact in terms of scientific returns, allowing collaborative space exploration, enhancing planetary geology and having a bet-ter sense of remote environments that cannot yet be explored.

References

1. Abercrombie, S., et al.: Multi-platform immersive visualization of planetary, aster-oid, and terrestrial analog terrain. In: 50th Annual Lunar and Planetary Science Conference (2019)
2. Abercrombie, S.P., et al.: Onsight: multi-platform visualization of the surface of mars. In: AGU Fall Meeting Abstracts, vol. 2017, pp. ED11C-0134 (2017)

3. Balaram, J., Aung, M., Golombek, M.P.: The ingenuity helicopter on the perseverance rover. Space Sci. Rev. **217**(4), 1–11 (2021)
4. Bansal, R., Raj, G., Choudhury, T.: Blur image detection using laplacian operator and open-cv. In: 2016 International Conference System Modeling & Advancement in Research Trends (SMART), pp. 63–67. IEEE (2016)
5. Beaton, K.H., et al.: Mission enhancing capabilities for science-driven exploration extravehicular activity derived from the NASA BASALT research program. Planet. Space Sci. **193**, 105003 (2020)
6. Bell, J., et al.: The mars 2020 perseverance rover mast camera zoom (Mastcam-Z) multispectral, stereoscopic imaging investigation. Space Sci. Rev. **217**(1), 1–40 (2021)
7. Bell, J.F., III., et al.: The mars science laboratory curiosity rover mastcam instruments: preflight and in-flight calibration, validation, and data archiving. Earth Space Sci. **4**(7), 396–452 (2017)
8. Bergstra, J., Bengio, Y.: Random search for hyper-parameter optimization. J. Mach. Learn. Res. **13**(2) (2012)
9. Collinson, R.P.: Introduction to Avionics Systems. Springer, Dordrecht (2013). https://doi.org/10.1007/978-94-007-0708-5
10. Di, K., Li, R.: Cahvor camera model and its photogrammetric conversion for planetary applications. J. Geophys. Res. Planets **109**(E4) (2004)
11. Efron, B., Tibshirani, R.J.: An Introduction to the Bootstrap. CRC Press, Boca Raton (1994)
12. Hartley, R., Zisserman, A.: Multiple View Geometry in Computer Vision. Cambridge University Press, Cambridge (2003)
13. Mahmood, T., Fulmer, W., Mungoli, N., Huang, J., Lu, A.: Improving information sharing and collaborative analysis for remote geospatial visualization using mixed reality. In: 2019 IEEE International Symposium on Mixed and Augmented Reality (ISMAR), pp. 236–247. IEEE (2019)
14. Max, N.: Optical models for direct volume rendering. IEEE Trans. Vis. Comput. Graph. **1**(2), 99–108 (1995). https://doi.org/10.1109/2945.468400
15. McMahon, S.K.: Overview of the planetary data system. Planet. Space Sci. **44**(1), 3–12 (1996)
16. Memarsadeghi, N., Varshney, A.: Virtual and augmented reality applications in science and engineering. Comput. Sci. Eng. **22**(3), 4–6 (2020)
17. Mildenhall, B., Srinivasan, P.P., Tancik, M., Barron, J.T., Ramamoorthi, R., Ng, R.: Nerf: representing scenes as neural radiance fields for view synthesis. In: ECCV (2020)
18. Müller, T., Evans, A., Schied, C., Keller, A.: Instant neural graphics primitives with a multiresolution hash encoding. arXiv preprint arXiv:2201.05989 (2022)
19. Nevatia, Y., Chintamani, K., Meyer, T., Blum, T., Runge, A., Fritz, N.: Computer aided medical diagnosis and surgery system: towards automated medical diagnosis for long term space missions. In: 11th Symposium on Advanced Space Technologies in Robotics and Automation (ASTRA). ESA (2011)
20. Padams, J., Raugh, A., Hughes, S., Joyner, R., Crichton, D., Loubrieu, T.: Fair and the planetary data system. In: AGU Fall Meeting Abstracts, vol. 2021, pp. U43E-11 (2021)
21. Paelke, V., Nebe, K.: Integrating agile methods for mixed reality design space exploration. In: Proceedings of the 7th ACM Conference on Designing Interactive Systems, pp. 240–249 (2008)
22. Park, K., et al.: Nerfies: deformable neural radiance fields. In: Proceedings of the IEEE/CVF International Conference on Computer Vision, pp. 5865–5874 (2021)

23. Safi, M., Chung, J., Pradhan, P.: Review of augmented reality in aerospace industry. Aircr. Eng. Aerosp. Technol. (2019)
24. Schönberger, J.L., Frahm, J.M.: Structure-from-motion revisited. In: Conference on Computer Vision and Pattern Recognition (CVPR) (2016)
25. Schönberger, J.L., Zheng, E., Frahm, J.-M., Pollefeys, M.: Pixelwise view selection for unstructured multi-view stereo. In: Leibe, B., Matas, J., Sebe, N., Welling, M. (eds.) ECCV 2016. LNCS, vol. 9907, pp. 501–518. Springer, Cham (2016). https://doi.org/10.1007/978-3-319-46487-9_31
26. Sitzmann, V., Martel, J., Bergman, A., Lindell, D., Wetzstein, G.: Implicit neural representations with periodic activation functions. Adv. Neural. Inf. Process. Syst. **33**, 7462–7473 (2020)
27. Tate, C., Hayes, A., Perucha, M.C., Paar, G., Deen, R.: Mastcam-z geometric calibration: an alternative approach based on photogrammetric and affine solutions specific to filter, focus, and zoom. In: Lunar and Planetary Science Conference, p. 2414 (2020)
28. Teschner, M., Heidelberger, B., Müller, M., Pomerantes, D., Gross, M.H.: Optimized spatial hashing for collision detection of deformable objects. In: VMV, vol. 3, pp. 47–54 (2003)
29. Yu, A., Fridovich-Keil, S., Tancik, M., Chen, Q., Recht, B., Kanazawa, A.: Plenoxels: radiance fields without neural networks (2021)
30. Yu, A., Ye, V., Tancik, M., Kanazawa, A.: pixelnerf: neural radiance fields from one or few images. In: Proceedings of the IEEE/CVF Conference on Computer Vision and Pattern Recognition, pp. 4578–4587 (2021)
31. Zauner, C.: Implementation and benchmarking of perceptual image hash functions (2010)
32. Zhang, K., Riegler, G., Snavely, N., Koltun, V.: NeRF++: analyzing and improving neural radiance fields. arXiv preprint arXiv:2010.07492 (2020)

Asynchronous Kalman Filter
for Event-Based Star Tracking

Yonhon Ng[1] , Yasir Latif[2(✉)] , Tat-Jun Chin[2] , and Robert Mahony[1]

[1] Australian National University, Canberra ACT, Australia
{yonhon.ng,robert.mahony}@anu.edu.au
[2] University of Adelaide, Adelaide, SA, Australia
{yasir.latif,tat-jun.chin}@adelaide.edu.au

Abstract. High precision tracking of stars both from ground and from orbit is a vital capability that enables autonomous alignment of both satellite and ground-based telescopes. Event cameras provide high-dynamic range, high temporal resolution, low latency asynchronous "event" data that captures illumination changes in a scene. Such data is ideally suited for estimating star motion since it has minimal image blur and can capture low-intensity changes in irradiation typical of astronomical observations. In this work, we propose a novel Asynchronous Event-based Star Tracker that processes each event asynchronously to update a Kalman filter that estimates star position and velocity in an image. The proposed tracking method is validated on real and simulated data and shows state-of-that-art tracking performance against existing approach.

Keywords: Star tracking · Event camera · Asynchronous · Kalman filter

1 Introduction

The night sky is full of stars that are visible to the naked eye and a lot can be seen with more sensitive optics and longer exposure times. Traditionally, optical sensors [12] have been used for astronomical applications such as star identification, tracking, and subsequently attitude estimation. Event sensors offer several unique advantages such as microsecond temporal resolution and asynchronous operation. What this means is that pixels can report relative change in intensity millions of time a second and they can do so independently of each other. This unique combination makes event cameras ideal for application in high-speed, low latency tracking scenarios. For example, when tracking fast moving space debris from ground based telescopes or tracking fast moving object such as meteors as they rapidly move through the earth's atmosphere, both high temporal resolution and asynchronous operation are highly desirable. Additionally, in contrast to conventional optical sensors, event sensors only fire when intensity change

Supplementary Information The online version contains supplementary material available at https://doi.org/10.1007/978-3-031-25056-9_5.

is detected at a particular pixel. An event camera looking at a mostly black sky with star sparingly sprinkled over it, therefore, generates a minute amount of data and in the process, consumes minimal power. Power is often a mission critical resource in robotics application.

To take advantage of the asynchronous nature of the event sensor, algorithms needs to be developed that can cater for and process asynchronous data effectively. In this work, we present the design of an asynchronous Kalman filter that takes advantage of this asynchronous high temporal resolution data. Kalman filters are the de facto standard for tracking in robotic and aeronautic domains. Additionally, we look towards the sky for inspiration and show the performance of the proposed Kalman filter for a star tracking application. We restrict the scope of this tracking problem to the task of recovering the motion of the night sky in two-dimensions. This is a slightly different problem compared to typical star-tracker literature, where the attitude of the vehicle is also estimated (e.g. [4]). We note that incorporating the attitude estimation is a logical next step, and leave this for future work. More concretely, this paper makes the following contributions:

- We propose an Asynchronous Event-based Kalman filter for Star Tracking that processes each event individually to estimate the $2D$ motion of a star field.
- With rigorous experiment on both simulated and real datasets, we show that our method is significantly more accurate at tracking fast star motion at very fine temporal resolution.
- We show that the computational overhead of the process is negligible and introduces very little latency in the overall pipeline, allowing near event-rate updates.

2   Related Work

The task of star tracking and identification has traditionally been addressed with frame-based camera with long exposure time to capture fainter stars. In the field of astronomy these are often referred to as "star sensors" [10] and are use in attitude estimation of satellites and alignment of ground-based telescopes. Such sensors suffer from image blur if the telescope is not correctly compensated for the earth's motion and are not well suited to the task of tracking. In contrast, event sensors offer the advantage of high dynamic range and robustness to image blur and appears ideal for star tracking.

Based on the processing paradigm followed, *Event based methods* can be divided into two categories. The first approach, which we term **frame-based**, is to convert the event stream into a sequence of "images" by integration over time (creating a frame) so they can be used as input to a traditional image based pipeline. Instead of utilizing each event as it becomes available, techniques in this category accumulate events together before processing. This can be done, for example, to accumulate enough data to rise above the noise floor. One of the first event based approaches to star tracking problem [4] converts star tracking into a blob tracking problem by integrating events to generate an "event image". Similarly, Cohen *et al.* [5] were able to track Low Earth Orbit satellites during the night as well as, more surprisingly, during the day using time surfaces, another event integration mechanism. More recently, [2] showed that various timescales can be taken into account to compute satellite orientation using a bank of Hough Transforms over various time windows. The main disadvantage of frame based methods is that they fail to utilize the high temporal resolution offered by an event sensor, limiting their ability to respond to sudden motions or high speed objects. This has motivated the need for **asynchronous methods** – methods that utilize individual events for estimation– which have been proposed for application such as intensity aided image reconstruction [13,15], corner detection [9], among others. This asynchronous paradigm allows algorithms to take advantage of the unique characteristics of the event sensor. This work aims to take advantage of the full potential of an event sensor by designing a computationally inexpensive Kalman filter to process each event individually.

When an optical sensor looks at the sky, distant objects should appear as point sources of light. However, limitations of lens design and fabrication as well as atmospheric effects leads to the light from distant objects being spread in a distribution, termed a "point spread function" (PSF) [7], over the receptor array. The PSF function maps a point source of light into the observed blob in the captured image. For reliable imaging, an estimate of the PSF of the system needs to be carried out [12,17]. We will use the widely accepted Gaussian approximation of the PSF in this work (see Sect. 4.2).

3   Problem Formulation

In this section, we provide an overview of the event generation process, followed by an outline of the proposed Kalman filter.

The output of an event camera is usually encoded using Address Event Representation (AER) [3] whereby each event is represented with a tuple (t_i, p_i, ρ_i), with timestamp t_i, pixel coordinates $p_i = (p_i^x, p_i^y)$, and polarity ρ_i representing direction of intensity change.

The events at each pixel can be treated independently, such that the continuous time, biased event stream $E(p, t)$ can be written as

$$E(p, t) = \sum_{i=1}^{\infty} c^p \left(\rho_i - b^p \right) \delta(t - t_i) \delta_p(p_i), \tag{1}$$

where p is the pixel of interest, t is the time-variable, c^p and b^p are the contrast threshold and bias of pixel p, (t_i, p_i, ρ_i) are the event AER data stream, $\delta(t - t_i)$ is the Dirac delta function (a unit impulse at time $t = t_i$) and $\delta_p(p_i)$ is the Kroneker delta (unity for $p = p_i$ and zero otherwise).

An event is triggered when the log intensity change exceeds the biased contrast threshold. Let L^p be the log intensity reconstructed from direct integration of events between time t_0 and t_1 at pixel p, then

$$L^p = \int_{t=t_0}^{t_1} E(p, t) dt = \log \left(\frac{I^p(t_1) + I_o}{I^p(t_0) + I_o} \right), \tag{2}$$

where log is the natural logarithmic function, $I^p(t)$ is the intensity value at time t and pixel p, and I_o is a log intensity offset [13] introduced to avoid negative log intensities. We use $I_o = 1$.

For the problem of star tracking, we assume that the stars viewed through the telescope can be locally approximated to lie on a planar star field image. This is due to the zoom of the telescope and the very small effective field-of-view. Thus, the star tracking problem can then be formulated using a Kalman filter, with state

$$\xi = \begin{bmatrix} v \\ \mu \end{bmatrix} \in \mathbb{R}^4, \tag{3}$$

where $v = (v_x, v_y)$ is the linear velocity and $\mu = (\mu_x, \mu_y)$ is the position (2D coordinates) of the camera frame $\{\mathcal{C}\}$ with respect to the star field image frame $\{\mathcal{S}\}$. The kinematics of the state is then

$$\dot{\xi} = \begin{bmatrix} 0 \\ v \end{bmatrix}. \tag{4}$$

Note that in this work, we assume the roll of the camera is negligible compared to the yaw and pitch such that the two frames $\{\mathcal{C}\}$ and $\{\mathcal{S}\}$ are assumed to be of the same orientation in the image plane. On the time scales of the problems considered (tens of seconds), this is certainly the case.

The time varying coordinate of a particular j star $s^j = (s_x^j, s_y^j)$ is

$$s^j(t) = \mu(t) + \kappa^j, \tag{5}$$

where (t) is used to highlight the time-varying nature of the variables, and κ^j is the constant coordinate of the jth star with respect to the star field image frame $\{\mathcal{S}\}$. Given a known star map, κ^j can be easily retrieved.

4   Method

The data received from an event sensor is far from ideal as it contains noise. Several aspect of (1) need to be estimated via various calibration steps for robust estimation. These are presented as follows.

4.1   Event Camera Calibration

Due to the sparse nature of the star field, noise and error in events can have a significant effect on the quality of image reconstruction and tracking. Inspired by Wang et al. [14], the event camera that is used in the experiment is first calibrated for the per-pixel contrast threshold and bias.

In this work, we do not assume to have access to event-frame registered intensity images. A $30s$ calibration dataset is collected, where an event camera is moved randomly at a fixed height parallel to an unstructured, equally textured surface (e.g. a carpeted floor). A method similar to the event-only offline calibration proposed in [14] can then be used. More robust calibration parameters are obtained by splitting the $30s$ dataset into 5 sections, each individually calibrated, and the per-pixel median of the contrast threshold and bias is computed.

The number of events for different pixels in a sample calibration dataset is shown in Fig. 1. Assuming equal stimulation of all pixels, the number of events for each pixel should follow a Poisson distribution. However, a very long tail can be observed from the figure, where pixels that get triggered much more often than the other pixels are termed the "hot pixels". Hot pixels are particularly detrimental when the signal of interest is very sparse (*e.g.* star field). Thus, these pixels are identified as outliers and their corresponding contrast threshold is set to zero.

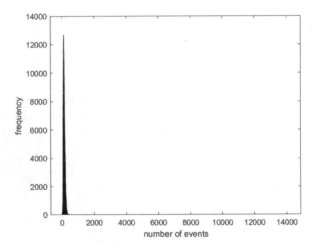

Fig. 1. Histogram showing number of events for different pixels in a calibration dataset. Some pixels with much larger number of triggered events (> 3500) are treated as hot pixels, and have their contrast threshold set to zero, meaning that their output is never considered.

4.2   Stars' Log Intensity Model

As previously mentioned, due to the imperfection of the optical system and other non-local disturbances, a point source of light is observed as a blob in the captured image, which can be described by the point spread function (PSF). Gaussian point spread function is usually assumed for conventional star images taken with a typical frame-based camera [6,12]. Assuming each star can be fitted with a two-dimensional Gaussian function (for a particular optical system with calibrated PSF) such that

$$\phi^j(p,t) = A^j \exp\left(-\frac{(p_x - s_x^j)^2 + (p_y - s_y^j)^2}{2(\sigma^j)^2}\right), \tag{6}$$

where A^j is the intensity magnitude of the star, $p = (p_x, p_y)$ is the coordinates of the pixel, $s^j(t) = (s_x^j, s_y^j)$ is the coordinates of the jth star pixel location at time t, and σ^j is the standard deviation of the Gaussian function.

Let a pixel at time t_0 have an intensity $I^p(t_0) = 0$ representing an area with no visible star, following (2) and using Taylor expansion,

$$
\begin{aligned}
L^p &= \log(\phi^j(p, t_1) + 1) \\
&= \phi^j(p, t_1) - \frac{\phi^j(p, t_1)^2}{2} + \frac{\phi^j(p, t_1)^3}{3} - \frac{\phi^j(p, t_1)^4}{4} + \dots \\
&\approx \phi^j(p, t_1).
\end{aligned}
\tag{7}
$$

The final equation is true due to the nature of an event camera, where it has an effective exposure time between 0.3–3 ms [11], such that the value of A^j is close to zero when viewing most stars. Hence, stars viewed through an event camera will have an image reconstruction that can be approximated using the same Gaussian function ϕ in (6).

It was found that the event camera is noisy, even after calibration due to various effects such as transistor switch leakage [8], refractory period [16] and other noise sources. Thus, we employ a high-pass filter [13] with a low cut-off frequency to reconstruct the log intensity of the visible stars asynchronously. The use of a high-pass filter also ensures that the assumption $I^p(t_0) = 0$ is valid, and (7) is a good approximation. The asynchronous update of intensity values using high-pass filter with biased events is then

$$
L^p(t) = L^p(t_0^p) \exp(-\alpha(t - t_0^p)) + c^p \left(\rho_i^p - b^p \right),
\tag{8}
$$

where α is the cutoff frequency, and t_0^p is the previous event timestamp at pixel p.

Then, we propose to augment the star map with calibrated star magnitude A^j and size σ^j such that we can assume these to be known when running the star tracker. Alternatively, they can be estimated in real-time using the moment method [1].

4.3 Star Tracking

The star tracking algorithm is designed to be asynchronous, such that each event is processed as it arrives. Due to the nature of event camera, which captures the rate of change of intensity, the likelihood of detecting a log intensity described by a Gaussian function can be computed as follows.

Without the loss of generality, assuming a star is centred around $(0, 0)$, and dropping the subscript j for ease of notation, the cross section of the 2D Gaussian across the "$p_x - \phi$" plane can be described by the Gaussian

$$
\phi'(p_x) = A \exp \left(-\frac{p_x^2}{2\sigma^2} \right).
\tag{9}
$$

The likelihood is then

$$l(p_x) = \left| \frac{d\phi'}{dp_x} \right| = \left| -A \frac{p_x}{\sigma^2} \exp\left(-\frac{p_x{}^2}{2\sigma^2} \right) \right|,$$ (10)

which is illustrated in Fig. 2.

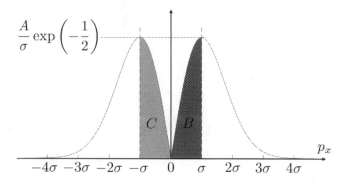

Fig. 2. Likelihood of event generation at different values of p_x for a star centered around (0,0).

An event camera is noisy, thus, we propose to use the pixel coordinate p_i of an event that results in a reconstructed log intensity above a threshold of $A \exp(-0.5)$ as a direct measurement of the star location s. This corresponds to using only log intensity within 1σ away from the mean star pixel location, and treating the other events as outliers, such that

$$y = p_i + \nu, \quad \nu \sim N(\lambda, \Sigma),$$ (11)

where ν is the uncertainty of the measurement, approximated with a normal distribution with mean λ, and covariance matrix Σ. The mean and covariance can be computed using the expected value method. For a one-dimensional Gaussian, the mean is

$$\lambda = \mathbb{E}[y] = \frac{\int_{p_x=-\sigma}^{\sigma}(p_x l)\ dp_x}{\int_{p_x=-\sigma}^{\sigma} l\ dp_x} = 0,$$

and the variance is

$$\Sigma = \mathbb{E}[(y - \lambda)^2] = \frac{\int_{p_x=-\sigma}^{\sigma}((p_x)^2 l)\ dp_x}{\int_{p_x=-\sigma}^{\sigma} l\ dp_x}$$

$$= 0.4585\sigma^2,$$

such that the denominator is the normalisation of the likelihood, which corresponds to dividing by the area of $(B+C)$ in Fig. 2 for 1D Gaussian, or the volume obtained when rotating the area B around the vertical axis by 2π radians for

2D Gaussian. Then, for a two-dimensional Gaussian, the mean and covariance matrix are

$$\lambda = \mathbb{E}[y] = 0, \tag{12}$$

$$\Sigma = \mathbb{E}[(y - \lambda)(y - \lambda)^\top] = 0.2825\sigma^2 I_2. \tag{13}$$

Finally, a Kalman filter is designed to track the motion of the stars, where the complete algorithm is summarised in Algorithm 1.

Algorithm 1. Asynchronous Event-based Star Tracker

Require: Star map parameters: κ^j, A^j, σ^j. Event camera calibration: c_p, b_p
   while New event (t_i, p_i, σ_i) **do**
      Update log intensity $L^p(t)$ using high-pass shown in (8) at $p = p_i$
      Predict the stars coordinates at time $t = t_i$ using (4) (5)
      Find the ID k of the star closest to event pixel p_i
      if $L^p(t) > A^k \exp(-0.5)$ **then**
         Update the state (3) using Kalman filter update step and measurement (11)
      end if
   end while

The proposed asynchronous event-based star tracker is computationally very lightweight. The computational complexity of the proposed asynchronous event-based star tracker is $\mathcal{O}(n)$, where n is the number of events. Thus, it is suitable to be implemented in resource constrained robotics platforms and low-powered devices.

5   Experiments

In this section, we first present the results for the pre-processing steps described in Sect. 4. We then show quantitative tracking results for high speed tracking in simulation, with comparison against a frame-based method. Finally, qualitative results are presented for real data where the event camera was mounted to a telescope and recorded real stars.

5.1   Event Camera Calibration

The calibration scheme described in Sect. 4.1 is performed to compute the per-pixel contrast threshold and bias. The effect of calibration is demonstrated in Fig. 3 which shows an example log intensity reconstruction before and after cal-

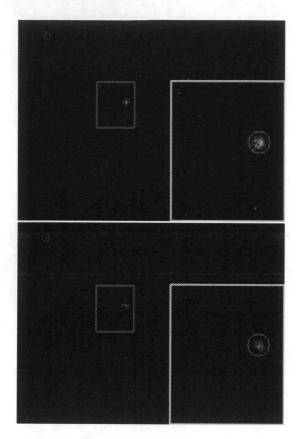

Fig. 3. High pass filter log intensity reconstruction (**top**) before and (**bottom**) after calibration. The red circle plots the estimated stars location. The inset image on the bottom right of each image shows a magnified region within the blue box, showing two hot pixels being removed due to calibration and better reconstruction of the star (Figure best seen on screen). (Color figure online)

ibration. It can be observed that the calibration successfully removed spurious white dots (hot pixels) from the reconstruction, and help make the reconstructed log intensity of the star more Gaussian-like.

5.2   Gaussian Model for Star's Log Intensity

The high pass reconstruction of star's log intensity is verified to be well approximated by a Gaussian function shown in (6). An example Gaussian fitting of the log intensity around a star using real event data is shown in Fig. 4. The red star represent the median reconstructed log intensity while the approximated Gaussian is represented by the coloured underlying mesh.

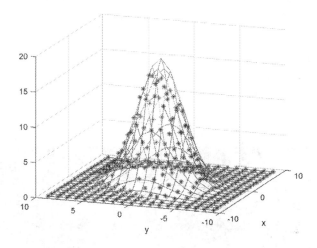

Fig. 4. Log intensity of the brightest star in Fig. 3. The coloured mesh is the fitted Gaussian, while the red * is the median reconstructed log intensity of the star from the real dataset. (Color figure online)

5.3   Simulated Star Tracking

To verify the efficacy of the proposed method using ground truth motion, we designed an event simulator for the task of star tracking. The simulator generates the night sky with specified number of stars moving with known velocities. Event are generated using the method described in Sect. 3. The simulator includes realistic additive noise in the event timestamp, as well as 50% outlier events randomly generated that do not correspond to real intensity changes. The simulator also allows us to test our proposed algorithm in tracking very fast moving objects.

The events are generated from simulated star field images at the rate of $0.1MHz$ (time resolution $1e^{-5}$) for a duration of $0.2s$. The simulated star field contains 5 visible stars, with position κ^j randomly generated, A^j between [0.045, 0.345], and σ^j between [1.2, 9.2]. The contrast threshold is assumed to be 0.01. The simulated motion of the star field image is set to a circular arc with initial velocity $v(0) = (2000, 0)$ pixel/s and an angular velocity of $1000°/s$. The Kalman filter state is initialised with an error of 10 pixels in both axis for position μ, and 500 pixel/s for velocity v.

An example simulation result is shown in Fig. 6, where the tracking performance is compared to the ground truth and a baseline algorithm. The baseline algorithm is our re-implementation of the tracking part in the method proposed in [4], where an "event image" is first generated by accumulating events over a short time interval (best result obtained using $0.01s$, with example event images shown in Fig. 5). Then, Iterative Closet Point (ICP) algorithm is used to compute the inter-frame star-field motion. Due to the frame-based nature of the baseline algorithm, it can only obtain estimates sparsely in time as seen from the sparse green dots in Fig. 6. The baseline algorithm also does not estimate the velocity, thus, we only compare our estimated velocity with the ground truth in Fig. 6(e)(f).

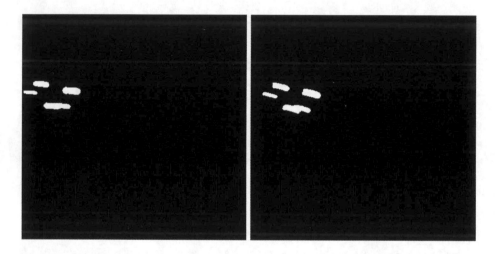

Fig. 5. First two event images generated following the baseline algorithm [4], which shows the blobs where events occurred during a short time interval (10 ms). Median filter is also applied to reduce noise in the event images. Accumulating events leads to the streaks from fast moving stars, introducing lag in estimation and difficulty in localizing. Full trajectory can be seen in Fig. 6

It can be observed that our proposed algorithm successfully tracked the motion of the star field accurately, with fast convergence from the wrong initial estimate. It can also be noted that the estimated velocity is noisier, due to the fact that the measurement does not directly measure this part of the state. A slowdown ($0.03\times$ real time) video of the tracking is available in the supplementary material. It can also be seen that the baseline method fails to track reliably in the fast motion setting as the motion leads to significant blur (streaks observed in Fig. 5 which should have been points), which degrades the performance of ICP. When using smaller intervals for generating frames, the signal to noise ratio drops significantly, and the noise suppression method (median filter) suppressing the true signal, leading to worse estimates than the ones shows for the 0.01 s frame duration in Fig. 6.

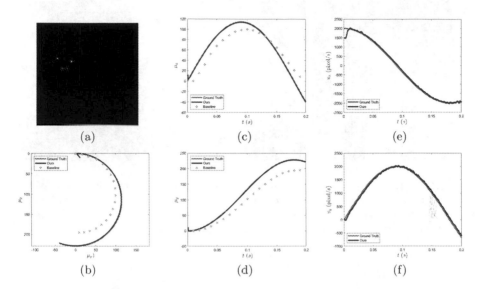

Fig. 6. Simulation results showing comparison of our proposed method (red) with the ground truth (blue) and a baseline algorithm [4] (green). (**a**) Star field image with tracked star position highlighted in red circle, (**b**) trajectory of the star field motion, (**c**) x position, (**d**) y position, (**e**) x velocity, and (**f**) y velocity of star field image. (Color figure online)

5.4   Real Data Star Tracking

The real data experiment is performed using a Prophesee third generation event camera with VGA resolution, mounted on a telescope with focal length of 2.5 m. The telescope is maintained mostly stationary with respect to Earth, pointed at some stars under a clear, dry winter night, such that most of the star motion is due to the Earth's rotation.

However, the true motion of the star field is not known. Thus, the tracking performance can only be verified visually. Two different dataset were collected. The first dataset looks at multiple similarly bright stars moving at approximately constant velocity to the left of the image. The second dataset only has one bright star visible most of the time, and we artificially introduce a very fast oscillatory motion by bumping the telescope near the start of the sequence. The result is shown in Fig. 7, which shows a good tracking performance. The videos with log intensity reconstruction and tracking are available in the supplementary material.

Fig. 7. Star tracking using real event data, showing the reconstructed log intensity and tracking performance (red circles). **Top**: Dataset 1 with multiple stars moving linearly from right to left. **Bottom** Dataset 2 with only one star and large oscillation at the start. Different columns show tracking results at different times.

6   Conclusion

In this paper, we proposed an asynchronous star tracking algorithm for an event camera, where the star field motion is updated with the arrival of each event. The simulated and real data experiment shows that the proposed algorithm can produce very high speed, low latency, robust and accurate tracking of the star field. In addition, the simplicity of the algorithm makes it ideal for resource constrained robotics platform. This work can be extended to also estimate the full 3D attitude of the camera in future work.

Acknowledgements. We would like to thank reviewers for the excellent feedback. We would also like to thank SmartSat CRC for funding the research, and Professor John Richards for his generous assistance in collecting the real event dataset using his telescope.

References

1. Anthony, S.M., Granick, S.: Image analysis with rapid and accurate two-dimensional gaussian fitting. Langmuir **25**(14), 8152–8160 (2009)
2. Bagchi, S., Chin, T.J.: Event-based star tracking via multiresolution progressive hough transforms. In: Proceedings of the IEEE/CVF Winter Conference on Applications of Computer Vision, pp. 2143–2152 (2020)
3. Chan, V., Liu, S.C., van Schaik, A.: Aer ear: a matched silicon cochlea pair with address event representation interface. IEEE Trans. Circuits Syst. I Regul. Pap. **54**(1), 48–59 (2007). https://doi.org/10.1109/TCSI.2006.887979

4. Chin, T.J., Bagchi, S., Eriksson, A., Van Schaik, A.: Star tracking using an event camera. In: Proceedings of the IEEE/CVF Conference on Computer Vision and Pattern Recognition Workshops (2019)
5. Cohen, G., et al.: Event-based sensing for space situational awareness. J. Astronaut. Sci. **66**(2), 125–141 (2019)
6. Darling, J., Houtz, N., Frueh, C., DeMars, K.J.: Recursive filtering of star tracker data. In: AIAA/AAS Astrodynamics Specialist Conference, p. 5672 (2016)
7. Delbracio, M., Musé, P., Almansa, A., Morel, J.M.: The non-parametric sub-pixel local point spread function estimation is a well posed problem. Int. J. Comput. Vision **96**(2), 175–194 (2012)
8. Delbruck, T.: Frame-free dynamic digital vision. In: Proceedings of International Symposium on Secure-Life Electronics, Advanced Electronics for Quality Life and Society, vol. 1, pp. 21–26. Citeseer (2008)
9. Duo, J., Zhao, L.: An asynchronous real-time corner extraction and tracking algorithm for event camera. Sensors **21**(4), 1475 (2021)
10. Gai, E., Daly, K., Harrison, J., Lemos, L.: Star-sensor-based satellite attitude/attitude rate estimator. J. Guid. Control. Dyn. **8**(5), 560–565 (1985)
11. Gallego, G., et al.: Event-based vision: a survey. IEEE Trans. Pattern Anal. Mach. Intell. **44**(1), 154–180 (2020)
12. Liebe, C.C.: Accuracy performance of star trackers - a tutorial. IEEE Trans. Aerosp. Electron. Syst. **38**(2), 587–599 (2002)
13. Scheerlinck, C., Barnes, N., Mahony, R.: Continuous-time intensity estimation using event cameras. In: Jawahar, C.V., Li, H., Mori, G., Schindler, K. (eds.) ACCV 2018. LNCS, vol. 11365, pp. 308–324. Springer, Cham (2019). https://doi.org/10.1007/978-3-030-20873-8_20
14. Wang, Z., Ng, Y., van Goor, P., Mahony, R.: Event camera calibration of per-pixel biased contrast threshold. In: Australasian Conference on Robotics and Automation (2019)
15. Wang, Z., Ng, Y., Scheerlinck, C., Mahony, R.: An asynchronous kalman filter for hybrid event cameras. In: International Conference on Computer Vision (ICCV) (2021)
16. Yang, M., Liu, S.C., Delbruck, T.: A dynamic vision sensor with 1% temporal contrast sensitivity and in-pixel asynchronous delta modulator for event encoding. IEEE J. Solid-State Circuits **50**(9), 2149–2160 (2015)
17. Zhang, B., Zerubia, J., Olivo-Marin, J.C.: Gaussian approximations of fluorescence microscope point-spread function models. Appl. Opt. **46**(10), 1819–1829 (2007)

Using Moffat Profiles to Register Astronomical Images

Mason Schuckman[1], Roy Prouty[1], David Chapman[1,2], and Don Engel[1(✉)]

[1] University of Maryland, Baltimore County, Baltimore, MD 21250, USA
{mschuck1,proutyr1,dchapm2,donengel}@umbc.edu
[2] University of Miami, Coral Gables, FL 33146, USA

Abstract. The accurate registration of astronomical images without a world coordinate system or authoritative catalog is useful for visually enhancing the spatial resolution of multiple images containing the same target. Increasing the resolution of images through super-resolution (SR) techniques can improve the performance of commodity optical hardware, allowing more science to be done with cheaper equipment. Many SR techniques rely on the accurate registration of input images, which is why this work is focused on accurate star finding and registration. In this work, synthetic star field frames are used to explore techniques involving star detection, matching, and transform-fitting. Using Moffat stellar profiles for stars, non-maximal suppression for control-point finding, and gradient descent for point finding optimization, we are able to obtain more accurate transformation parameters than that provided other modern algorithms, e.g., AstroAlign. To validate that we do not over-fit our method to our synthetic images, we use real telescope images and attempt to recover the transformation parameters.

Keywords: Image registration · Feature detection · Feature matching · Astronomy · Starfield · Super-resolution

1 Introduction

Precise high-resolution image registration enables difference image analysis and image stacking. In astronomy, difference image analysis enables the study of a wide variety of phenomena from the detection and parameterization of objects within our solar system [7,8] to the dynamics of astrophysical jets [5] to detecting changes in starlight that might indicate the existence of exoplanets [16]. Image stacking in astronomy enables the observer to circumvent constraints imposed by object tracking, detector sensitivity, atmospheric conditions, and pixel saturation to generate high signal-to-noise ratio (SNR) imagery [14].

Super-resolution is also enabled by image registration and, similar to image stacking, can generate higher SNR images. Super-resolution methods exploit relative motions between the detector and object to extract a more complete observation from distinct and incomplete observations. In this way, multiple low-resolution observations of an object can be used to generate a higher-resolution

L. Karlinsky et al. (Eds.): ECCV 2022 Workshops, LNCS 13801, pp. 80–95, 2023.
https://doi.org/10.1007/978-3-031-25056-9_6

"super-resolved" image that constitutes more information of the object [2]. Therefore, precise image registration must be done since any mis-alignment could result in erroneous super-resolved frames.

Super-resolving astronomical imagery represents a software solution to a fundamental problem whose solution has otherwise been hardware-driven. By this, the answer to diffraction-limited optical systems has been to develop optical systems with larger apertures. As examples of this paradigm, we note the recent launch and successful deployment of the James Webb Space Telescope [18] and Thirty Meter Telescope [21]. The cost of such hardware-driven endeavors quickly approach prohibitive levels [20]. The increasing cost and engineering complexity of these hardware-driven endeavors invite the development of more cost-effective algorithmic alternatives.

The remainder of this section discusses some assumptions and background to provide context for the image registration problem we aim to solve. Additionally, we address the choice of Moffat PSF as a tool for generating representative stellar profiles for synthetic astronomical images. The remaining sections are the related work, methodology, experimental design, results, conclusions, and future work. After an exploration of related work aimed at the problem of image registration, we discuss our novel algorithms for image registration. That is, we discuss the actual synthesis of synthetic images used for development and testing as well as the algorithms we developed to detect and match features as well as recover the parameters governing the transformation relating two successive frames. Next, we present the set of experiments we ran as tests of our algorithms and discuss their results. Finally, we draw conclusions from these results and discuss next steps for our work.

1.1   Image Registration

Image registration as mentioned above has many applications across many domains, such as medical imaging; agriculture; emergency response; national defense; oil and gas exploration; microscopy; and astronomy. Image registration can be used to register images captured at different times or captured by different sensors [24]. Frames captured can also be registered to models of the scene to enable enhanced digital representations for data assimilation studies [15,24].

Given the variety of both image domains and operational needs, there is no universal image registration algorithm. Therefore, every image registration task can be optimized for the particular task, ideally exploiting the information present in the frames to arrive at more performant algorithms.

Some of the earliest work in image registration took place in the early 1970s [1,9], driven by the need to process data from the U.S. Geological Survey's LandSat 1 satellite, originally known as Earth Resources Technology Satellite (ERTS). As the first satellite with a multispectral sensor, ERTS images needed to be registered across both time and light frequency. Initial approaches were based on Fast Fourier Transforms (FFT), with more recent approaches applying a broad array of alternative techniques [24]. These approaches can be considered to fall into two families: intensity-based (aka area-based) approaches and feature-based approaches.

Feature-based image registration pipelines, such as the approach to be described in this paper, consist of four main steps [24], (1) feature detection; (2) feature matching (3) recovery of transformation parameters; and (4) image resampling and transformation. Here, we are focused on the first three.

In this work, we are principally concerned with advancing image registration algorithms which would be used to enhance commodity institutional observatory telescopes (e.g. those present on the roofs of some university physics buildings), enabling functionality (super-resolution, change detection, etc.) otherwise unachievable from telescopes with that level of budget, optics, and environmental considerations (altitude, air pollution, light pollution, etc.). As the output of such telescopes is not particularly data-intensive, we focus our benchmarks on the quality of output rather than the time required. Similarly, the datasets we explore herein (both synthesized and natural) are chosen with this family of applications in mind. Notably, such images are likely to be quite sparse compared to satellite images of earth, medical images, and other domains for which image registration research has most frequently been tailored. Instead, as discussed in the following section, these sparse images consist of empty space, stars, and much rarer non-stellar objects. We note that there exist close-form solutions to recovering the properties of affine transformations when at least three control points are known perfectly. The noise inherent in these datasets makes such closed-form approaches intractable due to difficulty in identifying control points precisely. We anticipate this work may be of interest for additional domains, besides astronomy, where images are similarly sparse.

1.2   Moffat PSF

In this paper, we develop an algorithm for precise image registration of star fields. The aim is to exploit the sparsity of these star field scenes to achieve higher image registration metrics than leading software packages. While these scenes may include a variety of astrophysical phenomena (galaxies, nebulae, planetary bodies, etc.) which are relatively diffuse, the vast majority of the light comes from stars. 9,096 stars are visible to the naked eye [10], which dwarfs the quantity of visible planets and visible galaxies and other deep-sky objects, of which there are about twelve [11]. Stars which are best modelled as point sources which have been modified by their interaction with the atmosphere and the interstellar medium to take the form of a Moffat point spread function (PSF) [17]. The analytical form of the normalized Moffat PSF is given below:

$$f(x, y; \alpha, \beta) = \frac{\beta - 1}{\pi \alpha^2} \left[1 + \left(\frac{x^2 + y^2}{\alpha^2} \right) \right]^{-\beta} \tag{1}$$

where x and y are the distance in perpendicular directions along the surface sensor, and α and β are "seeing" parameters which are dependent on the medium through which the observation is taking place. That is, α and β capture the differences between a roof-based telescope at a university campus, a high-altitude telescope at a major observatory, and a space-based telescope. Moffat PSFs are more numerically stable than other analytic PSFs (namely a Gaussian PSF), and Moffat PSFs capture the behavior of the "wings" of stellar profiles [23]. Furthermore, the Moffat PSF performs well for datasets collected from seeing-limited instruments endemic to backyard astronomy or polluted university-town rooftops.

2 Related Work

2.1 AstroAlign and Astrometry.net

AstroAlign [3] is the closest recent effort to this paper, taking a shared interest in the problem of aligning astronomical images without using the Word Coordinate System (WCS) [6] metadata which is frequently provided in the FITS image files. AstroAlign focuses on registering pairs images to each other without determining the WCS position of those images in the sky. By contrast, Astrometry.net [12] provides a web-based service for adding WCS data to a FITS image based on registration between an uploaded image and an internal database of the known sky.

AstroAlign differs from our work in several key respects. Our motivating use case is super-resolution (SR), so we are more interested in high-precision alignment (allowing for the delineation of sub-pixels) than is necessary for other use cases of astronomical image alignment, such as change detection, non-SR image stacking, and stitching images together to cover a larger area. Moreover, AstroAlign is built with multimodal imaging in mind. In our algorithm development, we presume that the images being aligned are taken by a single instrument with consistent settings. Thus, AstroAlign anticipates the need for image scaling as part of the alignment process, while our algorithm requires that image pairs were captured at a consistent scale. Even so, some of the methodologies in our approach mirror those in AstroAlign, such as the use of brighter stars in the image to define triangles for matching. Differences include our algorithm using a point spread function to model detected features, and the refinement of our image alignment transformations through gradient descent across a high-dimensional parameter space, including parameters of the point spread function which distributes the light of individual stars over multiple pixels.

2.2 Use of Image Registration in Astronomical Super Resolution

Li et al. (2018) [13] develop a novel approach to astronomical super resolution which requires high-precision, sub-pixel image registration of the sort proposed herein. In their work, they build upon the image registration approach of Tabor et al. (2007) [22], which uses a triangle-based approach distinct from those of AstroAlign and Astrometry.net. Similar to Astrometry.net, Tabor is principally concerned with matching a single image to a whole-sky database in order to determine the global coordinates of the image, and is therefore focused primarily with optimizing for speed against a large catalog, rather than high-precision, sub-pixel mapping between image pairs.

2.3   Sequential Image Registration for Astronomical Images

In "Sequential Image Registration for Astronomical Images," Shahhosseini et al. (2012) [19] are concerned with image stacking to enhance the visibility of very dim, distant objects. In this use case, the telescope's movement is a significant enough factor that objects (stars, etc.) appear as streaks, so an algorithm is developed which uses the Hough transform to detect the linear equations of these streaks and calculate the camera's motion. This is then used to constrain the search space for the transformations necessary to stack the images.

3   Methodology

This study focuses on the generation of synthetic starfields and the recovery of transformation parameters that transform an original starfield frame to a transformed frame. The remainder of this section focuses on the generation of synthetic images, the extraction of features from these images, matching these features between the frames, and then the recovery of the transformation parameters as described above.

3.1   Image Synthesis

For our experiments, two synthetic starfield frames are generated. These will be referred to as F and $\mathcal{T}_{\mathbf{t}}(F)$. They are related by the transform, $\mathcal{T}_{\mathbf{t}}$ as described below. Where the subscript $\mathbf{t} = (\theta, \delta_x, \delta_y)$ is the parameter vector that defines the operation of $\mathcal{T}_{\mathbf{t}}$. Here, θ is the angle in degrees measured positive by the right hand rule, and the two δ_x and δ_y are the pixel-center translations in the standard Cartesian x and y directions, measured positive from the origin. The origin is located at the bottom-left of the frames represented in this work.

The first frame (F) is represented by a table of integers with indexed rows $i \in [0, I]$ and indexed columns from $j \in [0, J]$. Integers are used here to reflect the analog-to-digital quantization errors endemic in CCD or CMOS technology. As a result, each pixel value (noted $F_{i,j}$) will contain the following additive terms.

1. Signal from Background Sky Illumination
2. Signal from Targeted Objects
3. Randomly-Sampled Shot Noise

Each $F_{i,j}$ is attributed signal from the background sky illumination (S). This is generated as an integer randomly drawn from a discrete uniform distribution with minimum and maximum values of s_{min} and s_{max}. An independent sampling of this distribution is made for each pixel in the frame.

Then, the signal from the targeted objects, i.e., those appearing in the starfield or equivalently the frame field of view (a.k.a. scene), is generated for each pixel and added to $F_{i,j}$. This signal is computed via the continuous Moffat PSF given in Eq. 1 and down-sampled to an integer. Note that for direct comparison to Eq. 1, $x = i_c - i$ and $y = j_c - j$. We assume that given shape parameters α and β that define a Moffat Distribution are a constant for the frame. For each object, we scale the maximum extent of the distribution to a randomly generated object intensity. The α and β parameters for each frame are randomly sampled from two different uniform distributions. The uniform distribution for the α parameter ranged from α_{min} to α_{max}. The uniform distribution for the β parameter ranged from β_{min} to β_{max}.

The total number of objects (e.g., stars) admitted to the frame is N. For each nth object, a star center location (i_n, j_n) is drawn from a pair of continuous uniform distributions with support on $[0, I]$ and $[0, J]$, respectively. Also for each object the scaling factor (o_n), which accounts for a variety of star brightnesses, is randomly sampled from a continuous uniform distribution on the range $[o_{min}, o_{max}]$. Note that future sections of this work refer to the set of objects as H. Each $h \in H$ is associated with the object location and scaling factor as defined above.

After each pixel has been attributed the signal from the background sky illumination and the targeted object illumination, each pixel value is resampled according to a random distribution. This random resampling of the pixel value is ideally Poisson-distributed, however this distribution approaches the Normal distribution $(\mathcal{N}(\mu, \sigma))$ in the limit of large values, which we assume outright.

The integer values range from 0 to $2^{16} - 1$ which is consistent with uint16, one of the native FITS data types. The integer values are clipped to this range after the previous resampling step. So each pixel value, $F_{i,j}^{\text{orig}}$ (for original) is replaced with $F_{i,j}^{\text{new}}$ as follows.

$$F_{i,j}^{\text{new}} \sim \mathcal{N}\left(\mu = F_{i,j}^{\text{orig}}, \sigma = \sqrt{F_{i,j}^{\text{orig}}}\right) \tag{2}$$

Following the generation of F, we generate $\mathcal{T}_t(F)$ by transforming the set of object locations $\{(i_n, j_n)\}$ according to $\mathcal{T}_t$. We refer to this operation on the set of objects as $\mathcal{T}_t(H)$. We adopt the form of a rigid rotation and translation transformation matrix as given below. Note that we abuse this notation with, e.g., $\mathcal{T}_t(F)$ and $\mathcal{T}_t(H)$, to imply that the transformation with the transformation parameters $\mathbf{t}$ has acted on the objects given parenthetically.

$$\mathcal{T}_t = \begin{pmatrix} \cos\theta & -\sin\theta & \delta_x \\ \sin\theta & \cos\theta & \delta_y \\ 0 & 0 & 1 \end{pmatrix} \tag{3}$$

Of note here is that there is no scaling factor, and we admit this is due to our application domain being successive frames from the same instrument and detector. Furthermore, we note that this assumes that the two images or sets of control points are related only by a global affine transform such as that shown above. This is generally not a safe assumption. One alternative to this is to seek to fit non-linear transformations. Another alternative is to seek to fit linear transformation on subimages and aggregate the results to the larger images. Neither of these alternatives are explored in this work, but represent clear opportunities for future work.

By taking each location as $(i, j, 1)^T$, we can generate the transformed pixel location, (i'_n, j'_n), and associate with it the object scaling factor, O_n. These, along with a new background sky illumination (S') and a new set of Moffat parameters (α', β'), allow for the generation of a new frame. The new background sky illumination and new set of Moffat parameters mimics the change in atmosphere and other environmental conditions at the time of a subsequent (hypothetical) starfield frame observation. To mimic the changes in instrument-level noise on such a new observation, we resample each pixel according to Eq. (2).

The resulting set of pixel values with their locations $(i', j', 1)^T$ comprise $\mathcal{T}_t(F)$. For those transformed pixel locations that lie outside the extent of the image, i.e., $i' > I$ or $j' > J$, the pixels and their values are discarded. Additionally, the integer values are clipped to the uint16 range as discussed above (Fig. 1).

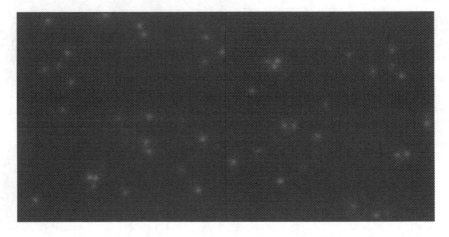

Fig. 1. A sample pair of frames, F *(left)* and $F' = \mathcal{T}_t(F)$ *(right)*. The image extents are $I, J = 400\text{pix}, 400\text{pix}$. The background sky illumination levels are $S, S' = 1000, 1000$. The Moffat parameters are $(\alpha, \beta), (\alpha', \beta') = (6, 1.5), (6, 1.5)$. In the original frame, there are $N = 32$ objects. Each object illumination is scaled to between $o_{min} = 0$ and $o_{max} = 200$. The transformation parameter vector is $\mathbf{p} = (107°, -8\text{pix}, 14\text{pix})$.

3.2   Feature Detection

Non-maximum Suppression for Candidate Control Point Finding. First, we normalize the image to the nominal FITS uint16 bounds as mentioned in the previous subsection. To do this, we find the minimum and maximum pixel values for the frame and then linearly scale all pixel values to the range $[0, 2^{16} - 1]$.

Second, this normalized image is convolved with a Gaussian kernel of size $k = 31$. This kernel size was determined experimentally to reduce false-positives by suppressing hot pixels and other erroneous local maxima without destroying the profiles of brighter stars. This approach results in some dimmer star profiles not being used as control points, with a stronger dependency on brighter stars. The variance of the Gaussian used to determine the kernel parameters is determined by the kernel size. The OpenCV [4] documentation provides more on this. The Gaussian kernel convolution (a.k.a. Gaussian blurring) is done with the OpenCV function `Imgproc.GaussianBlur`.

Third, we build a set P of those pixels p which are the brightest within their 13×13 square window. We define W_p as the window centered on and excluding p. For each $p \in P$, we define the "average pixel difference" by Eq. 4:

$$\sum_{q \in W_p} \frac{p-q}{168} = p - \overline{W_p} \tag{4}$$

We determine the mean and standard deviation of all average pixel differences and denote them as μ_{avg} and σ_{avg}. We define a subset $C \subseteq P$ that is the set of p for which the associated average pixel difference (see Eq. 4) is greater than $\mu_{\mathrm{avg}} + 0.8 \cdot \sigma_{\mathrm{avg}}$.

$$C = \left\{ p : \left(p - \overline{W_p} \right) > (\mu_{\mathrm{avg}} + 0.8 \cdot \sigma_{\mathrm{avg}}) \right\} \tag{5}$$

Together with the convolution of the Gaussian kernel, this filtering of locally brightest pixels has the effect of selecting points of interest in a sparse image of objects.

Moffat Fitting to Candidate Control Points. For each candidate control point in $p \in C$, we extract a subimage ($\mathcal{A}$) from around the candidate. This subimage size is experimentally determined to be 15×15 to allow for appropriate fitting of the Moffat PSF while avoiding the inclusion of a neighboring star. These candidates which are less than 8 pixels away from the edge of the image are ignored. We sort these subimages by the total sum of their pixel values and then take up to 50 of the brightest subimages. We then attempt to fit the Moffat PSF (Eq. 1) using gradient descent.

$$M(i, j, \alpha, \beta, o_n, i_c, j_c, s) = o_n \cdot f\left((i - i_c), (j - j_c); \alpha, \beta\right) + s \tag{6}$$

Using Eq. 6 and a set of initial parameters to synthesize a subimage, we find the sum of squared differences between $\mathcal{A}$ and the synthesized subimage. This sum of squared differences is then used to update the initial parameters and make subsequent guesses that approach the best set of parameters that fit $\mathcal{A}$. The parameters used in this implementation of gradient descent are found in Table 1. Note that the step size decreases in size according to a decay rate. The next step size (Δ_{next}) is related to the decay rate (γ) and the current step size ($\Delta_{\mathrm{current}}$) via the following schematic: $\Delta_{\mathrm{next}} = \Delta_{\mathrm{current}}/\gamma$.

Table 1. Gradient Descent fitting parameters with initial values, initial step sizes, step size decay rates, and required precision.

Parameter	Initial values	Initial step size (Δ)	Decay rate (γ)	Stopping precision
α	8	0.6	1.005	0.1
β	2	0.6	1.005	0.1
s	$\mathrm{avg}(\mathcal{A})$	$\min\left\{\frac{\mathrm{avg}(\mathcal{A})}{2}, 500\right\}$	1.05	1
o_n	$\max(\mathcal{A}) - \mathrm{avg}(\mathcal{A})$	$\min\left\{\frac{\max(\mathcal{A})-\mathrm{avg}(\mathcal{A})}{2}, 50\right\}$	1.005	1
i_c	7.5	0.6	1.005	0.05
j_c	7.5	0.6	1.005	0.05

After the appropriate stopping precision has been reached for each parameter or 5,000 iterations have been passed, the gradient descent algorithm terminates. The final set of parameters is then filtered to ensure (1) o_n and s are both bound to the `uint16` range, (2) $\alpha \in [0.5, 2000.0]$, and (3) $\beta \in [0.5, 2000.0]$.

This process is repeated for each candidate control point in C. Upon completion, the best fit α, β, o_n, i_c, j_c, and s are recorded. While the α, β, and s are largely frame-wide parameters, the control point location (i_c, j_c) and peak brightness o_n constitute a recovered h_n, called h_n^*. The set of these compose H^* for the original frame and $\mathcal{T}_t(H)^*$ for the transformed frame. These will be the topic of our experiments in Sect. 4.

3.3   Feature Matching

Now, we attempt to find the transformation between each of the sets of control points. These are the recovered control points from frame F, i.e., H^*, and the recovered control points from frame $\mathcal{T}_t(F)$, i.e., $\mathcal{T}_t(H)^*$.

To accomplish this, we sort the control points by their brightness (o_n). Taking the top ten brightest control points, we loop through each of the $\binom{10}{3} = 120$ possible choices of three control points. For each set of three control points in each frame, we build a triangle. This gives us 120 possible triangles from each of the two frames. We then look for matching pairs of triangles between the original and transformed frames.

Because the images are required to have been captured at the same scale, we score how well pairs of triangles match based on their perimeters and edge lengths. Pairs of triangles are deemed to be a match if both of the following two conditions are met:

1. The perimeters of the triangles must be within two pixel lengths of each other.
2. For each triangle, the length of each pair of edges is summed to give three numbers (AB+BC, AC+CB, CA+AB), which are then sorted into ascending order (x, y, z). The absolute differences between the corresponding positions in this sorted list for the two triangles are summed $(|x_1 - x_2| + |y_1 - y_2| + |z_1 - z_2|)$. This sum must be two or fewer pixel lengths.

After the above conditions are met the triangle vertices are sorted by the dot product of their connected edges. The sorted vertices are then thought to represent a one-to-one correspondence to each other between the frames. This assumption allows us to directly estimate the transformation parameters that are used as an initial guess to the more precise estimate discussed in Sect. 3.4. We find the initial guess of the transform rigid rotation by analyzing the angular displacement between the leg connecting the top two vertices in the sorted list. We find this initial guess of the translation parameters by inverting the rotation of the vertices from the control points in the transformed frame $(\mathcal{T}_t(H)^*)$ and then finding the vertical and horizontal displacements.

Finally, we take this rough transformation estimate and apply its inverse to the remaining control points in the transformed frame. If at least six other control points are found to be within two pixel widths of their original frame control point counterparts, this rough estimate is used as the initial position of the transformation gradient descent algorithm described below in Sect. 3.4.

3.4   Recovery of Transformation Parameters

Using Eq. 3 and the set of initial parameters estimated in the previous section (t^*), we find the sum of squared differences between $\mathcal{T}_{t^*}(H^*)$ and $\mathcal{T}_t(H)^*$. This sum of squared differences is then used to update the initial parameters and make subsequent guesses that approach the best set of parameters that fit $\mathcal{T}_t$. As in Sect. 3.2, the step size decreases each step according to a decay rate as given schematically as before. For all of the transformation parameters, the step size is 0.1, the decay rate is 1.0001, and the precision sought is 0.000001.

4   Experimental Design

4.1   Synthetic Images

With images generated as described in Sect. 3.1, we can explore the performance of our feature detection, feature matching, and transformation parameter recovery. In the following, we are interested in the recovery of various values and relationships between the objects belonging to H and $\mathcal{T}_t(H)$. Recall that H is the set of objects generated for frame F and that $\mathcal{T}_t(H)$ are those same objects used in the generation of $\mathcal{T}_t(F)$.

Comparing Recovered Control Points to Known Control Points. In this experiment we seek to validate the Moffat fitting algorithm as described in Sect. 3.2. To do this, we focus only on the generated frame F and the "true" object locations used to generate F, i.e., $h \in H$. For the frame we generate, we execute the algorithm we describe in Sects. 3.2–3.3 to recover the set of objects called H^*. This recovered set of objects is compared to the known set of objects, H, used to generate the frame. The comparison is done by analyzing the average Euclidean distance between the control points.

For our comparison to AstroAlign, we used the same generated frames and passed the frame to `AstroAlign.find_transform(...)`.

Comparing Recovered Transformation Parameters to Generated Transformation Parameters. In this experiment we seek to validate the transformation fitting algorithm as described in Sect. 3.4. To do this, we generate a pair of frames as described in Sect. 3.1. Then, we execute the algorithm we describe in Sects. 3.2–3.4. The recovered transformation parameters are compared to the generated transformation parameters with the absolute difference in the case of the rigid rotation comparison and average Euclidean distance between the translation parameters.

4.2   Comparing Recovered Transformation Parameters to Generated Transformation Parameters

To validate portions of this pipeline with actual telescope observations, we used the SDSS archive to download three FITS files corresponding to three different scenes. One was in the region of Orion, one centered on the Andromeda Galaxy, and one in the region of Camelopardalis. All scenes represent a high density of stellar objects, but also gradients in background sky illumination. Additionally, the objects have some artifacts introduced by the optical systems that performed the observation.

For this experiment, we test the feature detection, feature matching, and transformation parameter recovery algorithms as described in Sects. 3.2–3.4. Notably, we have no knowledge of the "true" object locations with these real-world images.

5   Results

5.1   Synthetic Images

Comparing Recovered Control Points to Generated Control Points. In Fig. 2, we see that our algorithm out-performs the AstroAlign implementation. Our implementation demonstrates little change in the reported error as a function of increasing background sky illumination. While the AstroAlign implementation reports higher error (by our metric), we see that this error decreases as a function of increasing background sky illumination. While counterintuitive, one reason performance increases with added noise could be that fewer stars are selected by the AstroAlign algorithm to generate their transformation, and this downselection provides a better result.

In Fig. 3, we see that our algorithm out-performs the AstroAlign implementation. Our implementation demonstrates an increase in reported error as a function of increasing number of objects. While the AstroAlign implementation reports higher error (by our metric), we see that this error increases slightly faster than our implementation as a function of number of increasing objects.

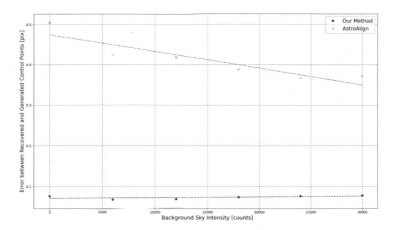

Fig. 2. Experimental results comparing recovered control points to generated control points. For this figure, five trials were run using different random seeds for the distribution sampling. The error reported on the vertical axis is the average Euclidean distance between the recovered and generated control points for a synthetic frame, measured in pixels. The horizontal axis is the Background Sky Intensity (S) as discussed in Sect. 3.1. For these frames, $N = 40$ objects were generated with fixed Moffat PSF parameters $\alpha = 6$ and $\beta = 1.5$. The plotted markers represent the actual reported values according to our metric, and the dashed lines are the lines of best fit.

Comparing Recovered Transformation Parameters to Generated Transformation Parameters. In Fig. 4, the average errors are reported on both axes. So the marks closer to the origin (bottom-left) of both the left and right graphs are best. In this figure, we see that our algorithm out-performs the AstroAlign implementation. While both implementations see an increase in their reported average errors when moving from left to right ($N = 100$ objects to $N = 300$ objects), the change in average error is larger for the AstroAlign implementation. This may be due to the way AstroAlign determined positive detection of control points, and so the larger number of objects can generate higher transformation parameter error in some trials (Table 2).

5.2   Comparing Recovered Transformation Parameters to Generated Transformation Parameters with SDSS Imagery

We see that our implementation of the translation transformation parameters outperforms AstroAlign's algorithm in each case. For the scene around Orion, the five trials yielded an AstroAlign average error 24× our average error. For the scene around Andromeda, the five trials yielded an AstroAlign average error 31× our average error. For the scene around Camelopardalis, the five trials yielded an AstroAlign average error 16× our average error.

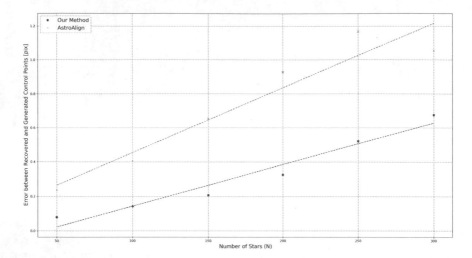

Fig. 3. Experimental results comparing recovered control points to generated control points. For this figure, five trials were run using different random seeds for the distribution sampling. The error reported on the vertical axis is the average Euclidean distance between the recovered and generated control points for a synthetic frame, measured in pixels. The horizontal axis is the number of objects in the frame (N) as discussed in Sect. 3.1. For these frames, a variable number of objects were generated with a background sky illumination of $S = 24000$[counts] with fixed Moffat PSF parameters $\alpha = 6$ and $\beta = 1.5$. The plotted markers represent the actual reported values according to our metric, and the dashed lines are the lines of best fit.

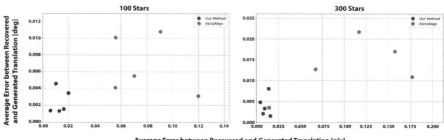

Fig. 4. Experimental results comparing recovered transformation parameters to generated transformation parameters when using $N = 100$ objects (left) and $N = 300$ objects (right). For these figures, five trials for each N were run using different random seeds for the distribution sampling. The vertical axes represent the absolute error between the generated and recovered transformation rotation. The horizontal axes represent the average Euclidean distance between the recovered and generated transformation translations. For these frames, the Moffat PSF parameters $\alpha = 6$ and $\beta = 1.5$ were used with a Background Sky Illumination of $S = 24000$[counts].

Table 2. Reported error in average Euclidean distance [pix] between the recovered and generated transformation translations between our implementation and that of AstroAlign. Each reported error represents the average of five trials where random transformation parameters were drawn and used to generate F and $\mathcal{T}_t(F)$.

Scene	Errors	
	Our approach	AstroAlign
Orion	0.000109	0.002642
Andromeda	0.000822	0.025552
Camelopardalis	0.000149	0.002454

6   Conclusion

We find that our algorithm for the detection and matching of features as control points in astronomical images out performs a recently published library of similar algorithms. Also, we find that our recovery of the rigid rotation and translation parameters operating between two frames outperforms the same library. We arrive at these conclusions based on the use of synthetic images we believe to be representative of real-world astronomical scenes as well as based on validation of these algorithms with actual astronomical images.

7   Future Work

Our development of the IR pipeline as discussed and tested above would be bolstered by further improvements to image synthesis and a ongoing exploration of the parameters used in the image synthesis steps. Additionally, future work could entail searching for additional transformation parameters to allow for any scaling or warping between image pairs.

Since any misalignment, even at the sub-pixel level, can result in erroneous super-resolved frames, we note that the significant increase in precision we demonstrate is of particular importance to our application domain of super-resolution. While we have utilized metrics to directly evaluate the registration of aligned image pairs, the ultimate test of this method for this particular use case is to conduct future work which would allow super-resolved images themselves to be evaluated relative to higher-resolution ground truths (e.g. super-resolved images from a land-based commodity telescope evaluated against images from a space telescope).

Acknowledgements. The material is based upon work supported by NASA under award number 80GSFC21M0002 and by a UMBC Undergraduate Research Award.

References

1. Anuta, P.E.: Spatial registration of multispectral and multitemporal digital imagery using fast fourier transform techniques. IEEE Trans. Geosci. Electron. **8**(4), 353–368 (1970). https://doi.org/10.1109/TGE.1970.271435

2. Bannore, V.: Iterative-Interpolation Super-Resolution Image Reconstruction, A Computationally Efficient Technique. Studies in Computational Intelligence, pp. 77–91 (2009). https://doi.org/10.1007/978-3-642-00385-1_5

3. Beroiz, M., Cabral, J., Sanchez, B.: Astroalign: a python module for astronomical image registration. Astron. Comput. **32**, 100384 (2020). https://doi.org/10.1016/j.ascom.2020.100384

4. Bradski, G.: The OpenCV library. Dr. Dobb's J. Softw. Tools (2000)

5. Gómez, J.L., et al.: Probing the innermost regions of AGN jets and their magnetic fields with RadioAstron. I. Imaging BL Lacertae at 21 microarcsecond resolution. arXiv (2015)

6. Greisen, E.W., Calabretta, M.R.: Representations of world coordinates in fits. Astron. Astrophys. **395**(3), 1061–1075 (2002). https://doi.org/10.1051/0004-6361:20021326

7. Gural, P.S., Larsen, J.A., Gleason, A.E.: Matched filter processing for asteroid detection. Astron. J. **130**(4), 1951–1960 (2005). https://doi.org/10.1086/444415

8. Gural, P.S., Otto, P.R., Tedesco, E.F.: Moving object detection using a parallax shift vector algorithm. Publ. Astron. Soc. Pac. **130**(989), 074504 (2018). https://doi.org/10.1088/1538-3873/aac1ff

9. Keating, T.J., Wolf, P.R., Scarpace, F.: An improved method of digital image correlation. Photogramm. Eng. Remote Sens. **41**, 993–1002 (1975)

10. King, B.: 9,096 stars in the sky - is that all? (2014). https://skyandtelescope.org/astronomy-blogs/how-many-stars-night-sky-09172014/

11. King, B.: The eyes have it - deep-sky observing without equipment (2018). https://skyandtelescope.org/observing/deep-sky-naked-eye/

12. Lang, D., Hogg, D.W., Mierle, K., Blanton, M., Roweis, S.: Astrometry.net: blind astrometric calibration of arbitrary astronomical images. Astron. J. **139**(5), 1782–1800 (2010). https://doi.org/10.1088/0004-6256/139/5/1782

13. Li, Z., Peng, Q., Bhanu, B., Zhang, Q., He, H.: Super resolution for astronomical observations. Astrophys. Space Sci. **363**(5), 1–15 (2018). https://doi.org/10.1007/s10509-018-3315-0

14. Loke, S.C.: Astronomical image acquisition using an improved track and accumulate method. IEEE Access **5**, 9691–9698 (2017). https://doi.org/10.1109/access.2017.2700162

15. Mandel, J., Beezley, J., Coen, J., Kim, M.: Data assimilation for wildland fires. IEEE Control. Syst. **29**(3), 47–65 (2009). https://doi.org/10.1109/mcs.2009.932224

16. Marois, C., Macintosh, B., Véran, J.P.: Exoplanet imaging with LOCI processing: photometry and astrometry with the new SOSIE pipeline. In: Adaptive Optics Systems II, pp. 77361J–77361J-12 (2010). https://doi.org/10.1117/12.857225

17. Moffat, A.F.J.: A theoretical investigation of focal stellar images in the photographic emulsion and application to photographic photometry. Astron. Astrophys. (1969). https://ui.adsabs.harvard.edu/abs/1969A&A....3.455M

18. Rieke, G.H., et al.: The mid-infrared instrument for the James Webb space telescope, VII: the MIRI detectors. Publ. Astron. Soc. Pac. **127**(953), 665–674 (2015). https://doi.org/10.1086/682257

19. Shahhosseini, S., Rezaie, B., Emamian, V.: Sequential image registration for astronomical images. In: 2012 IEEE International Symposium on Multimedia, pp. 314–317 (2012). https://doi.org/10.1109/ISM.2012.65
20. Stahl, H.P.: Survey of cost models for space telescopes. Opt. Eng. **49**(5), 053005–053005-8 (2010). https://doi.org/10.1117/1.3430603
21. Stone, E., Bolte, M.: Development of the thirty-meter telescope project. Curr. Sci. **113**(04), 628–630 (2017). https://doi.org/10.18520/cs/v113/i04/628-630
22. Tabur, V.: Fast algorithms for matching CCD images to a stellar catalogue. Publ. Astron. Soc. Aust. **24**(4), 189–198 (2007). https://doi.org/10.1071/AS07028
23. Trujillo, I., Aguerri, J., Cepa, J., Gutiérrez, C.: The effects of seeing on Sérsic profiles - II. The Moffat PSF. Mon. Not. R. Astron. Soc. **328**(3), 977–985 (2001). https://doi.org/10.1046/j.1365-8711.2001.04937.x
24. Zitová, B., Flusser, J.: Image registration methods: a survey. Image Vis. Comput. **21**(11), 977–1000 (2003). https://doi.org/10.1016/s0262-8856(03)00137-9

Mixed-Domain Training Improves Multi-mission Terrain Segmentation

Grace Vincent[1,2(✉)], Alice Yepremyan[2], Jingdao Chen[3], and Edwin Goh[2]

[1] Electrical and Computer Engineering, North Carolina State University,
Raleigh, USA
gmvincen@ncsu.edu

[2] Jet Propulsion Laboratory, California Institute of Technology, Pasadena, USA
{alice.r.yepremyan,edwin.y.goh}@jpl.nasa.gov

[3] Computer Science and Engineering, Mississippi State University, Starkville, USA
chenjingdao@cse.msstate.edu

Abstract. Planetary rover missions must utilize machine learning-based perception to continue extra-terrestrial exploration, with little to no human presence. Martian terrain segmentation has been critical for rover navigation and hazard avoidance to perform further exploratory tasks, e.g. soil sample collection and searching for organic compounds. Current Martian terrain segmentation models require a large amount of labeled data to achieve acceptable performance, and also require retraining for deployment across different domains, i.e. different rover missions, or different tasks, i.e. geological identification and navigation. This research proposes a semi-supervised learning approach that leverages unsupervised contrastive pretraining of a backbone for a multi-mission semantic segmentation for Martian surfaces. This model will expand upon the current Martian segmentation capabilities by being able to deploy across different Martian rover missions for terrain navigation, by utilizing a mixed-domain training set that ensures feature diversity. Evaluation results of using average pixel accuracy show that a semi-supervised mixed-domain approach improves accuracy compared to single domain training and supervised training by reaching an accuracy of 97% for the Mars Science Laboratory's Curiosity Rover and 79.6% for the Mars 2020 Perseverance Rover. Further, providing different weighting methods to loss functions improved the models correct predictions for minority or rare classes by over 30% using the recall metric compared to standard cross-entropy loss. These results can inform future multi-mission and multi-task semantic segmentation for rover missions in a data-efficient manner.

Keywords: Semantic segmentation · Semi-supervised learning · Planetary rover missions

1 Introduction

Planetary rover missions such as Mars 2020 (M2020) [25] and the Mars Science Laboratory (MSL) [10] have been critical for the exploration of the Martian surfaces without a human presence. Terrain segmentation has been necessary for a

L. Karlinsky et al. (Eds.): ECCV 2022 Workshops, LNCS 13801, pp. 96–111, 2023.
https://doi.org/10.1007/978-3-031-25056-9_7

rovers ability to navigate and maneuver autonomously, and it involves identifying features across the surface's terrain e.g., soil, sand, rocks, etc. Automated pipelines that utilize deep learning (DL) can enable the efficient (pixel-wise) classification of large volumes of extra-terrestrial images [23] for use in downstream tasks and analyses. DL is notoriously sample inefficient [13], often requiring on the order of 10,000 examples to achieve good performance or retraining when moving across different domains (e.g., different missions, planets, seasons lighting conditions, etc.) [21]. As such, significant annotation campaigns are required to generate a large number of labeled images for model training. Furthermore, specialized annotations in the planetary science domain require the knowledge of experts such as geologists and rover drivers, which incurs additional cost. Inconsistencies with the taxonomy and lighting further complicate this issue.

To address this circular problem of requiring annotated examples to develop an automated annotation pipeline, this research aims to leverage semi-supervised learning for semantic segmentation (i.e., pixel-wise classification), with limited annotated examples across multiple missions. This framework will utilize a large number of unlabeled images through contrastive pretraining [3,4] for a data-efficient way to extract general features and has been shown to outperform supervised learning [8]. This backbone model is then finetuned using a mixed-domain dataset that captures the diverse feature information across the different planetary missions. The ability to achieve state-of-the-art semantic segmentation performance with minimal labels will be integral for conducting downstream science (geological identification) and engineering (navigation) tasks across the M2020 and MSL missions, as well as future Lunar and Ocean Worlds missions.

In summary, the contributions of this work are 1) analyzing a mixed-domain training set for multi-mission segmentation that isolates the change in domain and the change in number of samples to understand the effects of different training set compositions, 2) introducing a contrastive self-supervised pretraining network to improve performance on limited-label multi-mission terrain segmentation, and 3) addressing the class imbalance through reweighting loss functions.

The rest of the paper is organized as follows. Section 2 describes current computer vision applications in planetary science as well as the transfer learning methods, while Sect. 3 details the data collection of the AI4Mars datasets and summarizes the technical approach. Section 4 presents in detail the results of the contributions listed above.

2   Related Work

2.1   Planetary Computer Vision

Planetary computer vision applications have increased in the recent decades to increase capability and automation of planetary exploration and data analysis. The Planetary Data System (PDS) allows for the archival of image data collected from different NASA missions, e.g. MER [5], MSL [10], HiRISE [16], etc. These datasets have been used to train classifiers to detect classes of interest [18,24]. Development of annotated datasets allowed for deployment of multi-label Convolutional Neural Networks (CNNs) for classification of both science and engineering tasks for individual missions [15,23] which showed performance improvements over Support Vector Machine (SVM) classifiers. Additional planetary applications include the Science Captioning of Terrain Images (SCOTI) [19] which generates text captions for terrain images.

Terrain Segmentation. Semantic segmentation for planetary rover missions on Mars has primarily focused on engineering-based tasks such as terrain classification for downstream navigation [6,12,17]. The development of the annotated AI4Mars dataset [21] has enabled the segmentation of Martian surfaces. SPOC (Soil Property and Object Classification) [20] leverages the AI4Mars dataset to segment the Martian terrain by utilizing a fully-convolutional neural network.

Self-supervised Learning. Self-supervised and semi-supervised learning can potentially alleviate the significant effort used to annotate rover images in the AI4Mars dataset. Self-supervised networks have aided in the process of annotating images by developing clusters of relevant terrain classes utilizing unlabeled images [18]. Contrastive learning has emerged as an important self-supervised learning technique where a network is trained on *unlabeled* data to maximize agreement between randomly augmented views of the same image and minimize agreement between those of different images [3,4]. By using these contrastive-pretrained weights (as opposed to supervised pretrained weights) as a starting point for supervised *finetuning*, it has been shown that contrastive pretraining improves performance on Mars terrain segmentation when only limited annotated images are available [8]. This work extends prior work by finetuning the generalized representations obtained through contrastive pretraining on mixed-domain datasets to improve performance across multiple missions (Fig. 1).

3   Methodology

3.1   Dataset Composition

This section presents the two primary datasets used in this work, as well as the mixed-domain dataset resulting from their combination. Note that the expert-labeled test sets for MSL and M2020 (322 and 49 images, respectively) are held constant throughout this paper for evaluation. We only manipulate and vary the training set size (number of images) and composition.

MSL AI4Mars Dataset. The AI4Mars dataset [21] is comprised of more than 50k images from the PDS captured by the MSL's grey-scale navigation camera (NAVCAM) and color mast camera (Mastcam) as well as MER. Citizen scientists around the world annotated MSL images at the pixel-level with one of four class labels, i.e., soil, bedrock, sand, and big rock, resulting in more than 16,000 annotated training images. A testing set of 322 images was composed by three experts with unanimous pixel-level agreement.

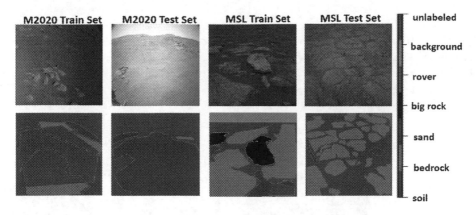

Fig. 1. Sample images and segmentation masks for M2020 and MSL from the AI4Mars dataset [21]. Top row shows the rover captured images and the bottom row shows the corresponding annotations by citizen scientists (train set) or experts (test set).

Additionally, the MSL dataset provides two additional masks—one for rover portions present in the image and another for pixels beyond 30m from the rover. While [21] treats pixels that fall under these two masks as null or unlabeled, this paper investigates the feasibility and utility of treating them as additional classes, namely "rover" and "background", respectively. Unlabeled pixels, or pixels with insufficient consensus in the citizen scientist annotations, are removed from the metrics calculations in training and evaluation.

M2020 AI4Mars Dataset. To evaluate multi-mission segmentation on the M2020 mission, this paper leverages images from M2020's color NAVCAM that were annotated using the same citizen scientist workflow for the MSL AI4Mars dataset. The M2020 dataset consists of a training set of 1,321 color images that span Sols 0–157 with four classes, i.e., soil, bedrock, sand, big rock. The testing set was comprised of 49 grey-scale images from Sols 200–203 manually annotated by an expert. The M2020 and MSL datasets differ primarily in the lack of rover and range masks for M2020 in addition to the image dimensions (1280×960 on M2020 compared to 1024×1024 on MSL). In addition, M2020 training images are full color whereas MSL is grey-scale. The test sets on both MSL and M2020 are grey-scale.

Deployment for MSL and M2020. To deploy a single trained model across different missions, we merge the two datasets together, creating a mixed-domain dataset. In doing so, we ensure feature diversity across the domains such that the model does not overfit to one mission. To investigate the dependence of our mixed-domain approach on the composition the combined dataset, we performed two different experiments with a different maximum number of images set to match the total number of images in the M2020 training set, 1,321 images (Fig. 2a), and the total number of images in the MSL training set, 16,064 images (Fig. 2b). This maximum number of images allows for the number of images from both the MSL AI4Mars and M2020 datasets to be varied to see the importance of one dataset over the other, since the M2020 dataset is roughly 1/10th the size of the MSL dataset.

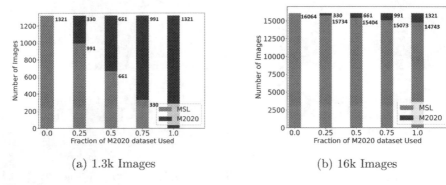

(a) 1.3k Images   (b) 16k Images

Fig. 2. The compositions of the MSL + M2020 mixed datasets across the two training set size experiments, based on different proportions of the M2020 dataset that are present (0.0, 0.25, 0.5, 0.75, 1.0). The number of present images is shown to the right of each bar.

3.2 Semi-supervised Finetuning

In this work, we train a segmentation network based on the DeepLab architecture [2] to perform semantic segmentation on Mars terrain images. We finetune pretrained backbones on a combined MSL and M2020 dataset for 50 iterations with a batch size of 16 and a learning rate of 10^{-5}. To investigate model performance under limited labels, we also vary the number of labeled training images provided from 1%-100% of the available 17,385 training images.

The DeepLab architecture combines a ResNet encoder [11] with a segmentation module based on Atrous Spatial Pyramid Pooling (ASPP) [2]. Due to the relatively small number of labeled images, on the order of 10k, we leverage transfer learning rather than training the segmentation model from scratch. In other words, we initialize the encoder with pretrained weights and randomly initialize the decoder weights before finetuning for 50 additional epochs on the labeled Mars images.

The choice of pretrained weights used to initialize the encoder has been shown to play an important role in downstream segmentation performance. In particular, transferring ResNet weights that were pretrained using the SimCLR framework on unlabeled ImageNet images [4] enables better Mars terrain segmentation performance under limited labels compared to ResNet weights that were pretrained in a supervised fashion (i.e., with labels) on ImageNet [8]. In this work, we investigate whether the contrastive pretraining framework is advantageous and generalizes weights for different missions (M2020 vs. MSL) on a mixed domain dataset.

4   Results

4.1   Analysis of Dataset Composition

Mixed-Domain Training Improves Multi-mission Segmentation Accuracy. Table 1 compares segmentation results between supervised and contrastive pretraining methods for the MSL and M2020 evaluation sets. The supervised pretraining tests are referred to as SPOC because they follow [1, 20, 21] in initializing the encoder using weights that were pretrained in a supervised manner to perform classification on the ImageNet dataset. The contrastive pretraining results are referred to as SimCLR as they leverage the weights [3, 4] that were pretrained in an *unsupervised* manner on unlabeled ImageNet images. The MSL, M2020, and MSL combined with M2020 were the training sets used to evaluate performance across MSL and M2020 test sets for both pretraining methods.

Table 1. Baseline supervised transfer learning (SPOC) compared with semi-supervised fine-tuning results (SimCLR) across MSL and M2020. Combining MSL and M2020 training images provides significant performance improvement on M2020 and a modest improvement on MSL.

Evaluated On:	MSL				M2020			
Trained On:	In-Domain (16,064 Images)		Mixed Domain (17,385 Images)		In-Domain (1,321 Images)		Mixed Domain (17,385 Images)	
	Acc	F1	Acc	F1	Acc	F1	Acc	F1
SPOC	0.959	0.840	0.965	0.860	0.662	0.495	0.745	0.544
SimCLR	0.065	0.000	**0.070**	0.885	0.732	0.539	**0.796**	0.610

When finetuned and evaluated on MSL, a model using either SPOC or SimCLR achieves an average pixel accuracy of 96%. However, we see a large performance decrease when the model is finetuned and evaluated on M2020, with an average pixel accuracy 66.2% and 73.2% for SPOC and SimCLR, respectively. This decrease in performance could be attributed to the large discrepancy in dataset sizes. As mentioned in Sect. 3.1, the number of images in the M2020 dataset is only 8.2% of MSL (i.e., 1,321 vs. 16,064 images). Furthermore, M2020 only has 49 expert labeled images in the test set, compared to 322 test images in the MSL dataset.

However, finetuning with a combined dataset of MSL and M2020 images (MSL + M2020 in Table 1) improves performance on both MSL and M2020 test sets across both pretraining methods. With SimCLR pretraining, finetuning on the combined dataset provides a 6% accuracy and 7% F-1 improvement on M2020 compared to finetuning only on M2020 data. On the other hand, SPOC experiences a higher accuracy improvement (8.3%) and a smaller F-1 improvement (4.9%) on the M2020 test set when using the combined dataset. The improvement on MSL with a combined dataset is marginal, with a 0.5% increase in accuracy across SPOC and SimCLR when comparing combined dataset finetuning to MSL-only finetuning. For M2020, this performance improvement is driven mainly by the increased number of training examples, which prevents the model from overfitting to only a few M2020 images.

Despite the limited M2020 examples, SimCLR demonstrates higher sample efficiency compared to SPOC, enabling a higher overall pixel accuracy. Yet, SPOC's larger accuracy improvement could indicate a tendency for the *task-specific* supervised pretrained weights to overfit to classes that are more frequent or more closely related to the ImageNet pretraining task. In contrast, SimCLR's larger F-1 improvement at the cost of accuracy could indicate that the *task-agnostic* contrastive pretrained weights tend to be updated in a more uniform fashion across all classes. In other words, contrastive pretraining acts as a form of implicit regularization in limited-labels regimes such as that encountered by M2020 and enables the model to favor variance in the bias-variance trade-off.

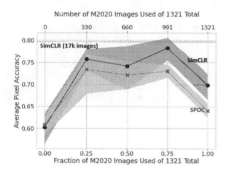 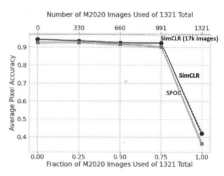

(a) Evaluation on M2020 using 1.3k Imgs   (b) Evaluation on MSL using 1.3k Imgs

Fig. 3. Sensitivity of the mixed-domain approach to the combination of MSL images with a specified proportion of M2020 images when the dataset size is capped at 1,321. Solid lines represent the average performance across the different seeds while the bands indicate a 95% confidence interval.

Model Performance is Sensitive to Images Used in Dataset Composition. The initial results (Table 1) establish that using a combined, mixed-domain, dataset for finetuning improves performance on M2020 and MSL across SPOC (ImageNet supervised) and SimCLR (ImageNet contrastive) pretraining. This improvement can be attributed to the increased number of training examples (using the combined 17k training set vs. the 16k MSL or the 1.3k M2020 training sets) and the increased *variety* or variance in the domain of training images. To decouple these potentially-synergistic factors and isolate the mixed-domain effect, we performed two experiments where we fixed the total number of examples at 1,321 (i.e., the original number of M2020-only labeled images) and 16,064 (i.e., the original number of MSL-only labeled images), then experimented with the number of M2020 images used to replace MSL images in our mixed-domain dataset as illustrated in Fig. 2. For each dataset split, we ran multiple tests across different seeds to investigate how specific combinations of M2020 images present in the dataset (and conversely, specific combinations of MSL images *not* present) affect model performance.

Figure 3 shows the average pixel accuracy across the different sensitivity tests for each fraction of M2020 images used in a total dataset size of 1,321 images. Unsurprisingly, finetuning the SimCLR-pretrained weights with the full 17,385-image MSL+M2020 dataset (79.6%; black dotted line in Fig. 3a) outperforms the average runs across all fractions of M2020 images present. However, the performance is improved when there is more *variety* in the training set when compared to an in-domain or complete domain shift training set. This variety from the MSL images could be attributed to the increased number of sites, 77, that MSL explored compared to the 5 sites M2020 explored, thus contributing to performance improvements over the in-domain training.

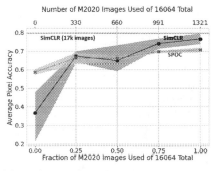 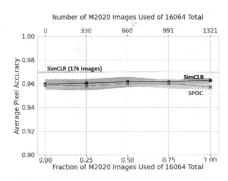

(a) Evaluation on M2020 using 16k Imgs   (b) Evaluation on MSL using 16k Imgs

Fig. 4. Sensitivity of the mixed-domain approach to the combination of MSL images with a specified proportion of M2020 images when the dataset size is capped at 16,064. Solid lines represent the average performance across the different seeds while the bands indicate a 95% confidence interval.

We see the same improvement when the number of training images is increased to 16k (Fig. 4a) since the increased fraction of M2020 images corresponds to a more diverse training set. This indicates that, independent of the total number of images, the fortuitous replacement of a specific combination of MSL images with M2020 images increases the dataset diversity and is important to the mixed-domain multi-mission terrain segmentation task. A larger dataset simply increases the probability that these important images will be seen by the model during training.

The sensitivity analysis of different pretraining methods and image samples had a marginal affect on the performance of the MSL test set, so long as there was MSL data present in training. Figure 3b shows that when a model is trained solely on M2020 images then evaluated on MSL, overall performance is decreased by more than 50% from the 17,385-image baseline. This significant decrease in performance results from the lack of MSL data present in training as well as the inherent decreased variety from the M2020 dataset that only explored 5 sites. Therefore when MSL images are present in training, there is minimal behavior shift is expected as the MSL image provide more variety in images and make up the bulk of the training set when the data is capped at 16,064 images. Still, Figs. 3b and 4b shows across the sampling methods SimCLR outperforms SPOC's supervised pretraining.

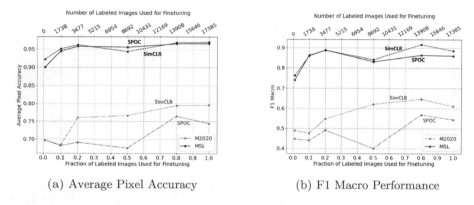

(a) Average Pixel Accuracy   (b) F1 Macro Performance

Fig. 5. Average Pixel Accuracy and F1-Macro performance vs fraction of labeled images from the mixed-domain dataset used in finetuning the ResNet model. Red lines represent evaluation on the M2020 test set and blue lines represent evaluation on MSL. (Color figure online)

4.2   Segmentation with Fewer Labels

The generation of annotated datasets for the purpose of training deep learning models, specifically from such specific domains as planetary missions, requires significant analysis by experts such as planetary scientists, rover operators, and mission planners. This incurs additional time and cost, which can prohibit the application of computer vision-based algorithms on shorter or smaller (i.e., lower-budget) missions. Thus, utilizing fewer labeled images decreases the need for such efforts. Using the combined training sets from Sect. 3.1, we vary the number of labeled images used to finetune our network. The training datasets will vary in size from 173 (1% of the available labeled data) to 17,385 (100% of the available labeled data), note that the expert-labeled test sets are not varied. We stratify the sampled images across the two datasets; the ratio of M2020 to MSL images stays constant across all dataset sizes. For example, a label fraction of 0.1 would contain 1,739 images, of which 8.2% (143) would be M2020 images while the remaining 91.2% (1,586) would be MSL images.

Figure 5 compares the average pixel accuracy and F1 Macro score achieved by the contrastive (SimCLR) and supervised (SPOC) pretraining methods on both MSL and M2020 test sets when finetuned on varying numbers of labels.

SimCLR and SPOC perform similarly on the MSL evaluation set, achieving >90% accuracy across the range of label fractions tested. However, SimCLR exhibits a slight accuracy advantage over SPOC below 20% labels as well as a slight F1 score advantage above 50% labels.

SimCLR exhibits a larger performance advantage over SPOC on the M2020 test set. At 20% labels, SimCLR is able to maintain 92.5% of its full accuracy at 100% labels (i.e., 3.5% performance degradation), whereas SPOC experiences a 5.4% performance decrease.

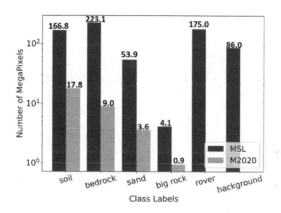

Fig. 6. Class distribution (in megapixels) of the training sets for M2020 and MSL. Note the absence of rover and background masks in the M2020 dataset.

4.3 Ablation Studies

Addressing Class Imbalance with Weighted Loss Functions. Class imbalance is an important problem in computer vision tasks such as semantic segmentation because the minority classes are often relevant, e.g., identifying big rocks when navigating rovers. Given that the Martian landscape is relatively monotonous and not as label diverse as other standard computer vision datasets [7,14], we expect to encounter class imbalance in most planetary datasets.

Figure 6 highlights the class imbalance on the MSL and M2020 datasets. Both datasets exhibit different class distributions—bedrock is the most frequent terrain element encountered by MSL, whereas soil is the most frequent class encountered by M2020—but have the same minority classes of sand and big rock. This section presents investigations into different loss functions aimed at improving performance on minority classes such as big rock and sand.

Cross-Entropy Loss. Most of the existing work in semantic segmentation utilize the cross-entropy loss function. This loss function is widely used in classification tasks not only because it encourages the model to output class probability distributions that match the ground truth, but also because of its mathematical properties that enable efficient backpropagation during the training of neural networks. Following the derivation in [22], the cross-entropy loss can be written in the form given in Eq. 1,

$$CE = -\sum_{c=1}^{C} N_c \log P_c \tag{1}$$

where $c \in 1, ..., C$ is the class among a set of C classes, N_c is the number of examples that belong to class c, and $P_c = \left(\prod_{n:y_n=c}^{N_c} P_n^c\right)^{1/N_c}$ represents the model's geometric mean confidence across all examples belonging to class c. Equation 1 highlights that a "vanilla" cross-entropy loss implementation biases towards classes with larger N_c for imbalanced datasets, and could reduce model performance as it does not consider the probability mass distribution across all classes [9]. Thus, we experimented with several loss functions to improve overall model performance as well as performance on individual classes.

$$InvCE = -\sum_{c=1}^{C} \frac{1}{freq(c)} N_c \log(P_c) \tag{2}$$

$$= -\sum_{c=1}^{C} \frac{1}{N_c} N_c \log(P_c) = -\sum_{c=1}^{C} \log(P_c) \tag{3}$$

Inverse Frequency Loss. A common method of handling class imbalance is to weight the loss function so as to prioritize specific classes. The most common weighting method is to use inverse frequency, which assigns more importance to minority/rare classes (i.e., classes with few examples) [22]. The frequency of a class is found by dividing the number of pixels in a class c, N_c, by the total number of pixels, N. Inverse frequency weighting can over-weight the minority classes, so to reduce this effect and the addition of false positives, we normalize the weights.

Recall Loss. The recall metric measures a model's predictive accuracy—it is given by $\frac{TP}{TP+FN}$, which is the number of true positives (i.e., pixels that the model *correctly predicted* as belonging to the positive class) divided by the sum of true positives and false negatives (i.e., the total number of pixels that actually belong to the positive class). Recall loss [22] is a novel method that weights a class based on the training recall metric performance as seen in Eq. 4 and even extends the concept of inverse frequency weighting. The true positive and false negative predictions are used with respect to class $c \in 1, ..., C$ and the time dependent optimization step, t, to optimize the geometric mean overall classes at t, $p_{n,t}$.

$$RecallCE = -\sum_{c=1}^{C}(1 - \frac{TP_{c,t}}{FN_{c,t} + TP_{c,t}})N_c \log(p^{c,t})$$

$$= -\sum_{c=1}^{C} \sum_{n:y_i=c} (1 - R_{c,t}) \log(p_{n,t}) \tag{4}$$

Recall loss is weighted by the class-wise false negative rate. We expect the model to produce more false negatives (i.e., exhibit low recall) on rare classes that are more challenging to classify, whereas more prevalent classes will produce fewer false negatives. This allows the recall loss to dynamically favor classes with lower per-class recall performance ($R_{c,t}$) during training. Recall loss has been shown to improve accuracy while maintaining a competitive mean Intersection-over-Union (IoU).

Table 2. Performance metrics across different loss functions on M2020 evaluation.

Loss function	Accuracy	F1 Macro	mIoU	Big rock recall
Cross entropy	0.795	0.610	0.480	0.120
Inverse frequency	0.757	0.582	0.449	0.350
Recall CE	0.781	0.607	0.474	0.164
Inverse frequency + Recall CE	0.772	0.591	0.458	**0.480**

When evaluating model performance on an imbalanced dataset, overall pixel accuracy does not provide adequate insight into the model's performance on individual classes. As such, we use the F1 score, which is the harmonic mean of precision and recall. The F1 score takes into account both the false positives and false negatives, whereas recall only accounts for the false positives along with the true positives. Furthermore, we show the mean intersection over union (mIoU), which is the percent of overlap in the prediction and ground truth.

Table 2 compares the performance utilizing different loss functions when fine-tuning a contrastive-pretrained ResNet-101 model on the 17,385-image mixed-domain training set. Results show that weighting the loss function improves the minority class positive predictions, pixel accuracy, and the F1 Macro score. In particular, Recall Cross Entropy (CE) alone improves Big Rock recall but decreases mIoU value when compared to the baseline CE loss. The combined Recall CE and Inverse Frequency loss have the highest recall on the Big Rock class with a value of 0.48. The confusion matrices in Fig. 7 better visualize the "rebalancing" of classes when using the combined loss functions over the CE loss. Figure 7b shows that the combined weighted loss function was able to improve correct predictions of big rock over the cross-entropy loss. Additionally, Cross-entropy loss model tends to over predict soil since it is the most frequent class (Fig. 7a), but Inv. Freq and Recall is able to address all classes more uniformly at a small expense of lower Soil recall.

Bigger Models Increase Performance. To evaluate the importance of model size, specifically the number of parameters, we simplified our experiments. Four different models, MobileNet v2, ResNet-50, ResNet-101, and ResNet-101 with 2x width, were used to perform finetuning with a training set of the MSL dataset (16,064 images). Figure 8 shows that across a varied number of labeled images

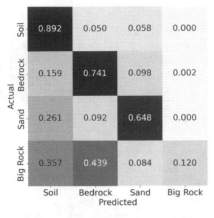 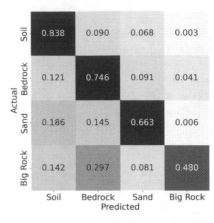

(a) Cross Entropy – Acc: 79.5%   (b) Inv Freq + Recall CE – Acc: 77.2%

Fig. 7. Confusion matrices for Cross Entropy loss and Inverse Frequency + Recall Cross Entropy loss as a result of the ablation studies. The confusion matrix shows a 36% increase in ability to positively predict big rock using Inv Freq + Recall CE loss compared to Cross Entropy loss.

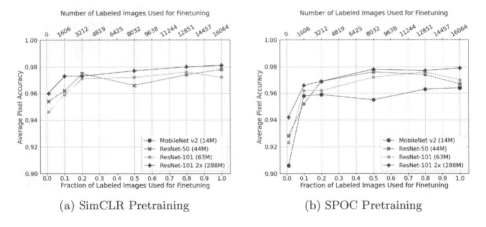

(a) SimCLR Pretraining   (b) SPOC Pretraining

Fig. 8. Comparison of SPOC and SimCLR across different sized models when finetuned on a varying number of labeled images from the MSL dataset and evaluated on the MSL test set. Note that contrastive pretraining weights are not available for the MobileNet v2 model.

used in finetuning, that larger models with an increased number of parameters have the highest pixel accuracy. In comparison to the MobileNet v2 (14 M parameters), the ResNet-101 2x (288 M parameters) improves accuracy at 1% of the available labeled images by 5% and when using 100% of the available labeled images there is 2% increase. Additionally, it is shown that when using these larger model sizes (e.g. ResNet-101 and ResNet-101 2x) with fewer annotated images improves accuracy over smaller models (e.g. MobileNet v2, ResNet-50), but the increase in accuracy with the increase of available labels starts to diminish. These diminishing returns in pixel accuracy question the trade-off in compute expense and accuracy specifically on a constrained edge-device like a planetary rover.

5   Conclusion

This research proposes a state-of-the-art semi-supervised mixed-domain training approach for multi-mission Mars terrain segmentation by combining annotated MSL images from the novel AI4Mars dataset with newly-annotated M2020 images to finetune a contrastive pretrained model, thereby improving performance across both MSL and M2020 test sets.

The idea of combined datasets is not novel, yet decoupling the effects of change in domain training and change in number of samples revealed their relative importance within the context of multi-mission deployment. It was found that while the number of images plays a significant role in enabling the performance gain from a combined dataset, the additional variance in a mixed-domain dataset prevents the model from overfitting to unhelpful examples. We then show that a self-supervised contrastive pretraining method outperforms the published baseline [20] by 5%, using only 20% of the available labels on the combined dataset. Finally, adding different class weighting methods aided in addressing the class imbalance issue that is inherently present within the planetary datasets. By improving performance on the minority classes (i.e. big rock and sand), we can improve a rover's ability to detect and avoid these potentially hazardous areas.

Areas of future work include expanding this approach to handle multi-task learning (e.g. scientific vs engineering class sets). The model would become valuable as a mission- and task-agnostic method for semantic segmentation in planetary explorations. Additionally, evaluating the performance of the larger model sizes for a multi-mission effort will aid in the development of a distillation pipeline for the deployment of a smaller, computationally less expensive model on a planetary rover.

Acknowledgments. This research was carried out at the Jet Propulsion Laboratory, California Institute of Technology, under a contract with the National Aeronautics and Space Administration (80NM0018D0004), and was funded by the Data Science Working Group (DSWG). The authors also acknowledge the Extreme Science and Engineering Discovery Environment (XSEDE) Bridges at Pittsburgh Supercomputing Center for providing GPU resources through allocation TG-CIS220027. The authors would like to thank Hiro Ono and Michael Swan for facilitating access to AI4Mars. U.S. Government sponsorship acknowledged.

References

1. Atha, D., Swan, R.M., Didier, A., Hasnain, Z., Ono, M.: Multi-mission terrain classifier for safe rover navigation and automated science. In: IEEE Aerospace Conference (2022)

2. Chen, L.C., Papandreou, G., Kokkinos, I., Murphy, K., Yuille, A.L.: DeepLab: semantic image segmentation with deep convolutional nets, atrous convolution, and fully connected CRFs. IEEE Trans. Pattern Anal. Mach. Intell. **40**(4), 834–848 (2017)

3. Chen, T., Kornblith, S., Norouzi, M., Hinton, G.: A simple framework for contrastive learning of visual representations. In: International Conference on Machine Learning, pp. 1597–1607. PMLR (2020)

4. Chen, T., Kornblith, S., Swersky, K., Norouzi, M., Hinton, G.E.: Big self-supervised models are strong semi-supervised learners. Adv. Neural. Inf. Process. Syst. **33**, 22243–22255 (2020)

5. Crisp, J.A., Adler, M., Matijevic, J.R., Squyres, S.W., Arvidson, R.E., Kass, D.M.: Mars exploration rover mission. J. Geophys. Res. Planets **108**(E12) (2003)

6. Cunningham, C., Ono, M., Nesnas, I., Yen, J., Whittaker, W.L.: Locally-adaptive slip prediction for planetary rovers using gaussian processes. In: 2017 IEEE International Conference on Robotics and Automation (ICRA), pp. 5487–5494. IEEE (2017)

7. Deng, J., Dong, W., Socher, R., Li, L.J., Li, K., Fei-Fei, L.: ImageNet: a large-scale hierarchical image database. In: CVPR 2009 (2009)

8. Goh, E., Chen, J., Wilson, B.: Mars terrain segmentation with less labels. arXiv preprint arXiv:2202.00791 (2022)

9. Grandini, M., Bagli, E., Visani, G.: Metrics for multi-class classification: an overview. arXiv preprint arXiv:2008.05756 (2020)

10. Grotzinger, J.P., et al.: Mars science laboratory mission and science investigation. Space Sci. Rev. **170**(1), 5–56 (2012)

11. He, K., Zhang, X., Ren, S., Sun, J.: Deep residual learning for image recognition. In: Proceedings of the IEEE Conference on Computer Vision and Pattern Recognition, pp. 770–778 (2016)

12. Higa, S., et al.: Vision-based estimation of driving energy for planetary rovers using deep learning and terramechanics. IEEE Robot. Autom. Lett. **4**(4), 3876–3883 (2019)

13. Li, C., et al.: Efficient self-supervised vision transformers for representation learning. arXiv preprint arXiv:2106.09785 (2021)

14. Lin, T.-Y., et al.: Microsoft COCO: common objects in context. In: Fleet, D., Pajdla, T., Schiele, B., Tuytelaars, T. (eds.) ECCV 2014. LNCS, vol. 8693, pp. 740–755. Springer, Cham (2014). https://doi.org/10.1007/978-3-319-10602-1_48

15. Lu, S., Zhao, B., Wagstaff, K., Cole, S., Grimes, K.: Content-based classification of mars exploration rover pancam images. In: 52nd Lunar and Planetary Science Conference, p. 1779. No. 2548 in Lunar and Planetary Science Conference (2021)

16. McEwen, A.S., et al.: Mars reconnaissance orbiter's high resolution imaging science experiment (HiRISE). J. Geophys. Res. Planets **112**(E5) (2007)

17. Ono, M., et al.: Data-driven surface traversability analysis for mars 2020 landing site selection. In: 2016 IEEE Aerospace Conference, pp. 1–12. IEEE (2016)

18. Panambur, T., Chakraborty, D., Meyer, M., Milliken, R., Learned-Miller, E., Parente, M.: Self-supervised learning to guide scientifically relevant categorization of martian terrain images. In: Proceedings of the IEEE/CVF Conference on Computer Vision and Pattern Recognition, pp. 1322–1332 (2022)

19. Qiu, D., et al.: SCOTI: science captioning of terrain images for data prioritization and local image search. Planet. Space Sci. **188**, 104943 (2020)

20. Rothrock, B., Kennedy, R., Cunningham, C., Papon, J., Heverly, M., Ono, M.: SPOC: deep learning-based terrain classification for mars rover missions. In: AIAA SPACE 2016, p. 5539 (2016)

21. Swan, R.M., et al.: AI4MARS: a dataset for terrain-aware autonomous driving on mars. In: Proceedings of the IEEE/CVF Conference on Computer Vision and Pattern Recognition, pp. 1982–1991 (2021)
22. Tian, J., Mithun, N., Seymour, Z., Chiu, H.P., Kira, Z.: Striking the right balance: recall loss for semantic segmentation. arXiv preprint arXiv:2106.14917 (2021)
23. Wagstaff, K., et al.: Mars image content classification: three years of NASA deployment and recent advances. arXiv preprint arXiv:2102.05011 (2021)
24. Wagstaff, K., Lu, Y., Stanboli, A., Grimes, K., Gowda, T., Padams, J.: Deep mars: CNN classification of mars imagery for the PDS imaging atlas. In: Proceedings of the AAAI Conference on Artificial Intelligence, vol. 32 (2018)
25. Williford, K.H., et al.: The NASA mars 2020 rover mission and the search for extraterrestrial life. In: From Habitability to Life on Mars, pp. 275–308. Elsevier (2018)

CubeSat-CDT: A Cross-Domain Dataset for 6-DoF Trajectory Estimation of a Symmetric Spacecraft

Mohamed Adel Musallam$^{(\boxtimes)}$ ⓘ, Arunkumar Rathinam ⓘ,
Vincent Gaudillière ⓘ, Miguel Ortiz del Castillo ⓘ, and Djamila Aouada ⓘ

SnT, University of Luxembourg, Luxembourg 1359, Luxembourg
{mohamed.ali,arunkumar.rathinam,vincent.gaudilliere,
djamila.aouada}@uni.lu
https://cvi2.uni.lu/

Abstract. This paper introduces a new cross-domain dataset, *CubeSat-CDT*, that includes 21 trajectories of a real CubeSat acquired in a laboratory setup, combined with 65 trajectories generated using two rendering engines – *i.e.* Unity and Blender. The three data sources incorporate the same 1U CubeSat and share the same camera intrinsic parameters. In addition, we conduct experiments to show the characteristics of the dataset using a novel and efficient spacecraft trajectory estimation method, that leverages the information provided from the three data domains. Given a video input of a target spacecraft, the proposed end-to-end approach relies on a Temporal Convolutional Network that enforces the inter-frame coherence of the estimated 6-Degree-of-Freedom spacecraft poses. The pipeline is decomposed into two stages; first, spatial features are extracted from each frame in parallel; second, these features are lifted to the space of camera poses while preserving temporal information. Our results highlight the importance of addressing the domain gap problem to propose reliable solutions for close-range autonomous relative navigation between spacecrafts. Since the nature of the data used during training impacts directly the performance of the final solution, the *CubeSat-CDT* dataset is provided to advance research into this direction.

1 Introduction

With the increase in the number of space missions and debris [3,9,17], the need for Space Situational Awareness (SSA) – referring to the key ability of inferring reliable information about surrounding space objects from embedded sensors – is growing rapidly. Moreover, the highest level of autonomy is required to meet the need for reactivity and adaptation during on-orbit operations. Due to their low cost and power consumption combined with their high frame rate, cameras

This work was funded by the Luxembourg National Research Fund (FNR), under the project reference BRIDGES2020/IS/14755859/MEET-A/Aouada, and by LMO (https://www.lmo.space).

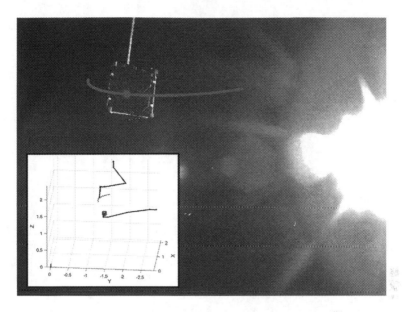

Fig. 1. Illustration of the SnT Zero-G Laboratory data. Top: chaser camera field of view featuring a CubeSat target with its projected trajectory (in red). Bottom left: instantaneous 3D positions of the robotic arms simulating the chaser (camera in blue) and target (trajectory in red) spacecrafts. (Color figure online)

represent suitable sensors for SSA. Consequently, vision-based navigation is the preferred route for performing autonomous in-orbit operations around a target spacecraft [24]. To do so, the core task consists in estimating both the position and attitude — referred to as pose — of the target over time. Furthermore, Deep Learning (DL) techniques have been proven to be successful in a wide variety of visual applications such as image classification, object detection or semantic segmentation [7]. Therefore, their use in monocular spacecraft pose estimation has gained interest accordingly. Moreover, the results from the first edition of the Satellite Pose Estimation Challenge (SPEC) [15] – organized by the Advanced Concepts Team (ACT) of the European Space Agency and the Space Rendezvous Laboratory (SLAB) of Stanford University – have shown promising outcomes in that direction. The scope of the SPEC was limited to single-frame pose estimation from synthetic images only (Fig. 1).

Due to the appearance gap between images from synthetic and real domains, DL-based algorithms trained on synthetic data typically suffer from significant performance drop when tested on real images [15]. As a consequence, existing spaceborne images captured from previous missions are sometimes combined with synthetic data. In particular, the Cygnus dataset [4] contains 540 pictures of the Cygnus spacecraft in orbit in conjunction with 20k synthetic images generated with Blender [1]. However, the main limitation of spaceborne images is

Table 1. Overview of existing SSA datasets. In the table, *sim* refers to simulated images and *lab* to laboratory data.

	SPARK [20]	SPEED [15]	SPEED+ [21]	URSO [26]	SwissCube [13]	Cygnus [4]	Prisma12K [22]	Prisma25 [6]	ch8CubeSatspsCDT
Synthetic images	150k	15k	60k	15k	50k	20k	12k	-	**14k**
Non-synthetic images	-	305	10k	-	-	540	-	25	**8k**
Object classes	15	1	1	2	1	1	1	1	**1**
Image Resolution	1024 1024	1920 1200	1920 1200	1080 960	1024 1024	1024 1024	752 580	-NA-	**1440 1080**
Visible	✓	✓	✓	✓	✓	✓	✓	✓	✓
Color	✓	✗	✗	✓	✓	✓	✗	✗	✓
Depth	✓	✗	✗	✗	✗	✗	✗	✗	✗
Mask	✓	✗	✗	✗	✓	✗	✗	✗	✗
6D Pose	✓	✓	✓	✓	✓	✓	✓	✓	✓
Rendering	sim	sim + lab	sim + lab	sim	sim	sim + real	sim	real	**sim + lab**
Trajectories	✗	✗	✗	✗	✓	✗	✗	✗	✓
Public dataset	✓	✓	✓	✓	✓	✗	✗	✗	✓

the lack of accurate pose labels and their limited diversity in terms of pose distribution. To overcome these difficulties, laboratory setups trying to mimic space conditions currently represent the de facto target domain for spacecraft pose estimation algorithms [21]. Moreover, laboratories offer monitoring mockup poses and environmental conditions to ensure a higher quality of the data [23,30]. Therefore, the ongoing second edition of the SPEC will rank the pose estimation algorithms based on their performance on two laboratory datasets, while training images were generated synthetically. That so-called SPEED+ dataset [21] is the first of its kind for vision-only spacecraft pose estimation that combines the information from 60k synthetic images with 10k others acquired from a robotic laboratory setup. However, laboratory pose labels are not publicly available.

The SPEED+ dataset offers no temporal consistency between the images. While single-frame spacecraft pose estimation is needed to initialize any multi-frame tracking algorithm [16,28], leveraging the temporal information provided through consecutive image acquisitions is likely to improve the robustness and reliability of any deployable method. However, to the best of our knowledge, no existing dataset features temporally consistent image sequences for 6-DoF spacecraft trajectory estimation.

In this paper, we propose a novel dataset with the aim to foster research on both the domain gap reduction and temporal information processing for spacecraft pose estimation. To the best of our knowledge, this is the first dataset featuring multiple (synthetic-2x and real-1x) domains with temporal information for spacecraft 6-DoF trajectory estimation (86 trajectories). We also propose a baseline algorithm that leverages the data from the three different modalities contained in the dataset.

In addition, man-made objects and specially spacecrafts often present a high degree of symmetry by construction. However, most existing datasets rely on non-symmetrical target spacecrafts [4,6,8,13,15,20–22,26]. Indeed, estimating the pose – especially orientation – of a symmetric object is a difficult task that has been receiving interest from the research community [5,12,25]. To develop it further, our dataset features a highly-symmetrical spacecraft – a 1U CubeSat. Moreover, instead of a mockup made with inadequate materials, a real space-compliant spacecraft is used.

Our contributions are summarized below:

- A comparative analysis of the existing SSA datasets.
- A cross-domain spacecraft trajectory dataset, that will be made publicly available and referred to as *CubeSat-CDT*, where "CDT" stands for Cross-Domain Trajectories.
- A novel algorithm for spacecraft trajectory estimation, built upon previous work [19], whose training has been made possible by the creation of the CubeSat-CDT dataset.
- A detailed analysis of the proposed dataset based on the training and testing of the aforementioned algorithm.

The paper is organised as follows. Section 2 provides a comparative review of existing SSA datasets. Section 3 describes the proposed CubeSat-CDT dataset. Section 4 presents the proposed method to leverage the temporal information for trajectory estimation and cross domain data validation. Section 5 presents an analysis of the performance of our proposed approach on the CubeSat-CDT dataset.

2   Related Datasets

Multiple datasets are provided to address the SSA challenge. Some are freely available for research [13,15,20,21,26], others are proprietary from some space companies and are not publicly available [4,6,22]. These datasets are discussed below, highlighting their contributions, limitations and difference with CubeSat-CDT. Table 1 summarizes this study.

SPEED [15] consists of 15k grayscale images of the Tango spacecraft, with a resolution of 1920×1200 pixels. Two different sets of images are available: a) 300 generated in a robotic laboratory environment with a custom illumination set up to recreate the space environment; b) 15k images of the exact mock-up generated using Open-GL to generate photo-realistic images. The main limitation of this study is the unbalanced number of samples from the laboratory and the synthetic data, as well as the absence of pose labels for most of the laboratory data.

SPEED+ [21] introduces the first dataset designed to minimize the gap between synthetic simulated and real data. In addition to the 60k synthetic generated images of the Tango satellite using Open-GL, it incorporates 10k laboratory images acquired with two different sources of illumination: *lightbox*, with diffuser plates for albedo simulation, and *sunlamp* to mimic direct high-intensity homogeneous light from the Sun.

SPARK [20] is a synthetic dataset generated using the Unity3D game engine as a simulation environment. It has 150k images of 10 different satellite models along with 5 debris objects combined in one class. It contains a large diversity in sensing conditions for all the targets, and provides RGB and depth information. Nonetheless, SPARK, intended to be used for spacecraft recognition and pose estimation, does not include any trajectory.

URSO [26] provides 15k images of two different targets, the 'Dragon' spacecraft and the 'Soyuz' one, with different operating ranges and at a resolution

Table 2. Minimum and maximum distances between the CubeSat and the camera for each data domain of the proposed CubeSat CDT dataset.

Data Domain $\diagdown$ Distance	Minimum	Maximum
Zero-G lab	0.65 m	1.2 m
Synthetic (SPARK)	0.85 m	3.8 m
Synthetic (Blender)	0.40 m	2.2 m

of 1080 × 960 pixels. The images were randomly generated and sampled around the day side of the Earth from low Earth orbit altitude with an operating range between 10m and 40m. All images are labelled with the corresponding target pose with respect to the virtual vision sensor.

SwissCube [13] is made of 500 scenes consisting of 100 frame sequences for a total of 50k images. It is the first public dataset that includes spacecraft trajectories, in particular a 1U CubeSat. Nonetheless, the main limitation of the dataset is the domain, as it only contains synthetically generated images using Mitsuba 2 render.

Cygnus [4] includes 20k synthetic images generated with Blender in addition to 540 real images of the Northrop Grumman Enhanced Cygnus spacecraft. They perform several augmentation techniques on the synthetic data including various types of randomized glare, lens flares, blur, and background images. However, the main limitation of the dataset is that it is not publicly accessible.

PRISMA12K [22] is created using the same rendering software used in SPEED. However, PRISMA12K replicates the camera parameters used during the PRISMA mission targeting the Mango satellite. It comprises 12k grayscale images of the Tango spacecraft using the same pose distribution presented in SPEED.

PRISMA25 [6] contains 25 spaceborne images captured during the rendezvous phase of the PRISMA mission. This real dataset is used to evaluate the performance of the algorithms developed using PRIMSA12K. The main limitation of this dataset is the number of real case examples and the lack of diversity in the target's pose.

The CubeSat-CDT dataset, presented in Sect. 3, contains multiple trajectories in three different domains with real spacecraft (see Table 1). We believe such a dataset opens new possibilities for studying trajectory estimation of spacecrafts.

3   Proposed CubeSat-CDT Dataset

The proposed CubeSat Cross-Domain Trajectory (CDT) dataset contains 21 trajectories of a real CubeSat acquired in a laboratory setup, combined with 50 trajectories generated using Unity [10], and 15 trajectories generated using Blender [1]. Combining a total of 22k high-quality and high-resolution images of a 1U CubeSat moving in predefined trajectories. Table 2 shows the minimum

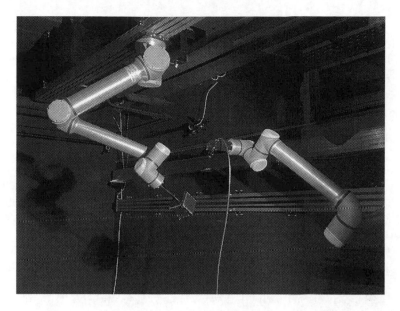

Fig. 2. Illustration of the Zero-Gravity Laboratory, located at the Interdisciplinary Centre for Security, Reliability and Trust (SnT) of the University of Luxembourg [23].

and maximum distances between the CubeSat and the camera according to the data domain.

3.1   Zero-G Laboratory Data

For this work we utilized the Zero-Gravity facility (Zero-G Lab) at the Inter-disciplinary Center for Security, Reliability and Trust (SnT) of the University of Luxembourg [23]. In this facility we reconstruct the space environment repli-cating multiple environmental conditions from the illumination to the motion in space. The Zero-G Lab, is a 3×5 meters robotic based laboratory used for real-time simulation of on-orbit servicing missions. It consists of two industrial, highly accurate robotic arms with six degrees of freedom each from Universal Robots (model UR10) [27]. They can simulate the 6D dynamic motion of two satellites during orbital rendezvous. Both arms are mounted on top of a Cobo-track rail system with a maximum translational velocity of 1.45 m/s. One robot is mounted on the wall at a linear track of 5m length and the other one is mounted on the ceiling at a linear track of 4.6m length. On both tracks the robotic manip-ulators can be moved and thus the final approach can be completely simulated (see Fig. 2).

The camera mounted as a payload of the wall robotic arm is the FLIR Black-fly S BFS-U3-16S2C. This camera is a lightweight (<50 g) and cost effective solution for space-sensitive imaging applications. This camera has a variety of features, including precise control over exposure, gain, white balance, and color correction. A fixed focal lens of 12 mm, designed for pixels that are $\leq 2.2\,\mu$m, was added to the camera. This provides a high level resolution ($>$200 lp/mm) across the sensor.

3.2 Unity-Based SPARK Synthetic Data

The first subset of synthetic data, referred to in the following as *SPARK Synthetic Data*, was generated using the SPARK space simulation environment based on Unity3D game engine [20]. It can generate close to realistic visual data of the target model under multiple configurable lighting conditions and with or without Earth background. The virtual target was programmed to describe predefined trajectories while the camera was fixed. The camera intrinsic parameters were fixed to match those of the Zero-G Lab data.

3.3 Blender-Based Synthetic Data

To increase the diversity of the dataset, we used a second rendering engine for generating data. The Blender-based Synthetic Data was generated using the Blender open-source 3D computer graphics software [1]. The Computer Aided Design (CAD) model of the CubeSat was imported into Blender. The intrinsic camera parameters were adapted to match the physical properties of the camera used in the Zero-G Lab. The pose information from the Zero-G Lab generated datasets were used as groundtruth labels to render the individual images. In Blender, the camera is fixed in origin, and the target is moved relative to the camera based on the pose information. This enables easier image acquisition and verification of the groundtruth labels after rendering.

3.4 Discussion

The proposed CubeSat-CDT dataset will foster new research on leveraging the temporal information while dealing with the symmetries of a 1U CubeSat, therefore addressing the challenges of estimating the 6-DoF poses of this common platform used in a wide range of new space missions. As presented in Fig. 3, there is a difference in the high-frequency components of the spectrum of the real and synthetic images. Therefore, it is crucial to take into consideration the data domain used for training a DL-based solution. Indeed, the domain gap can lead to a considerable generalization error if it is not addressed.

The trajectories defined for the CubeSat were designed to emulate close-range operations when a 1U CubeSat is deployed in orbit. Figure 4 presents the position and orientation of the trajectories performed in the three datasets. As summarized in Table 2, the 1U CubeSat relative distance to the camera frame varies from 0.40 m to 3.8 m depending on the dataset. Different illumination conditions were taken into account to emulate solar flares and reflections in the solar panels to better generalize the satellite detection and subsequent pose estimation.

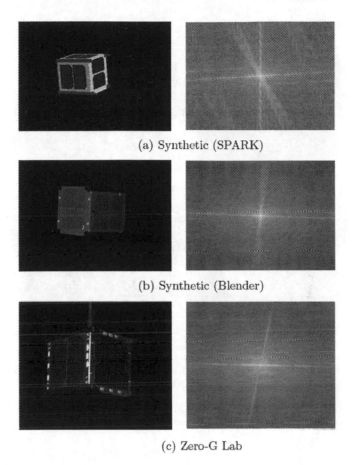

(a) Synthetic (SPARK)

(b) Synthetic (Blender)

(c) Zero-G Lab

Fig. 3. Qualitative comparison of example images in the spatial (left) and frequency (right) domains for (a) synthetic data generated with SPARK, (b) synthetic data generated with Blender, (c) real data acquired in the Zero-G Laboratory.

4   Proposed Baseline for Spacecraft Trajectory Estimation

In this section, we introduce a baseline to evaluate the effects of both the domain gap and temporal information on the CubeSat trajectory estimation.

4.1   Problem Formulation

The problem of trajectory estimation consists in estimating the 3D positions and attitudes – *i.e.* orientations – of a target spacecraft in the camera reference frame along the recorded sequence.

Following the notations introduced in [19], let $V_I = \{I_1, \cdots, I_N\}$ be a sequence of RGB images featuring the observed spacecraft, N being the total number of frames. Acquisition is made using a camera with known intrinsics

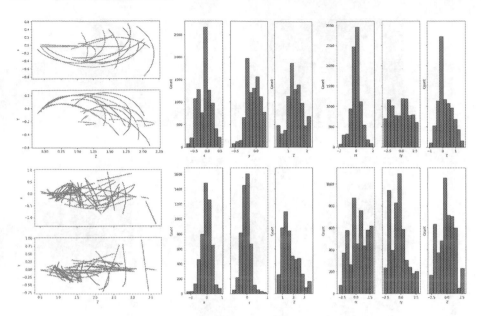

Fig. 4. Trajectory analysis for the Blender & Zero-G Lab (First row) and SPARK (Second row) subsets. From left to right: trajectory analysis in x and y vs z axis, range distribution for all the axes, angle distribution for all the axes.

$K \in \mathbb{R}^{3 \times 3}$. The goal is to estimate the trajectory $\mathcal{Y} = \{(t_1, \mathsf{R}_1), \cdots, (t_N, \mathsf{R}_N)\}$, where $(t_i, \mathsf{R}_i) \in \mathbb{R}^3 \times \mathrm{SO}_3(\mathbb{R})$ is the 3D location t and rotation R of the spacecraft captured at frame i.

4.2   Proposed Approach

The proposed model is composed of an EfficientNet B2 [29] backbone that takes a sequence of images and processes each frame in parallel then passes the learned features to a TCN model to compute the 6-DoF poses over the full sequence.

Spatial Feature Extraction For a given video sequence V_I, the EfficientNet B2 [29] feature extractor,

$$f : I \in \mathbb{R}^{M \times N} \mapsto Z \in \mathbb{R}^{\Psi}, \tag{1}$$

is applied frame by frame on V_I resulting in a sequence of estimated learned features $\hat{\mathcal{X}} = \{f(I_1), \cdots, f(I_N)\}$. In Eq. (1), $\Psi = 128$ is the dimension of the extracted CNN features, M is the image dimension, and N is the number of frames.

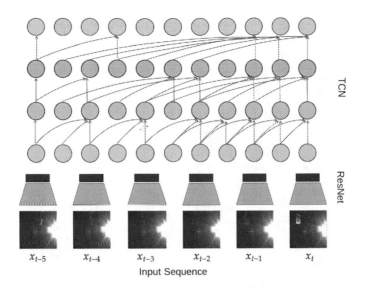

Fig. 5. Our TCN-based trajectory estimation model. At time t, we leverage the spatial features of each frame to estimate the target poses using a TCN, leading to smooth and accurate trajectory estimation.

3D Trajectory Estimation. Given a sequence of spatial features $\mathcal{X} \in \mathbb{R}^{\Psi \times N}$, the goal is to lift into the 6D space of poses. To that end, we need to estimate a function $g(\cdot)$ that maps $\mathcal{X}$ to its corresponding 6D sequence, such that:

$$g : \mathcal{X} \in \mathbb{R}^{\Psi \times N} \mapsto \mathcal{Y} \in \left(\mathbb{R}^3 \times \mathrm{SO}_3(\mathbb{R})\right)^N. \tag{2}$$

Independently estimating the poses from each frame would lead to temporally inconsistent results. Therefore, the function $g(\cdot)$ is approximated by a Sequence-to-Sequence Temporal Convolutional Network (*Seq2Seq TCN*) model using 1D temporal convolution. Such a model is illustrated in Fig. 5.

Finally, the sequence of poses is obtained using the composition of the functions $f(\cdot)$ and $g(\cdot)$, such that

$$\hat{y} = g \circ f(V_I). \tag{3}$$

4.3   Justification of the Proposed Approach

We note that TCNs represent a variation of convolutional neural network for sequence modelling tasks. Compared to traditional Recurrent Neural Networks (RNNs), TCNs offer more direct high-bandwidth access to past and future information. This allows TCN to be more efficient to model the temporal information of the input data with fixed size [18]. TCN can be causal; meaning that there is no information "leakage" from future to past, or non-causal where past and future information is considered. The main critical component of the TCN is the

Table 3. Pose MSE when regressed frame per frame independently, for the three data domains.

Train set \ Test set	Lab (Zero-G)	Synthetic (SPARK)	Synthetic (Blender)
Lab (Zero-G)	0.05 m/12.98°	0.40 m/115.27°	0.50 m/86.77°
Synthetic (SPARK)	0.26 m/92.25°	0.15 m/72.38°	0.60 m/126.31°
Synthetic (Blender)	0.30 m/102.14°	0.47 m/127.59°	0.14 m/88.30°

Table 4. Pose MSE when regressed by the Temporal Convolutional Network for the three data domains.

Data domain \ Model	Temporal	Single frame
Lab (Zero-G)	0.02 m/11.27°	0.05 m/12.98°
Synthetic (SPARK)	0.10 m/105.7°	0.15m/72.38°
Synthetic (Blender)	0.08 m/54.27°	0.14 m/ 88.30°

dilated convolution layer [11], which allows to properly treat temporal order and handle long-term dependencies without an explosion in model complexity. For simple convolution, the size of the receptive field of each unit - block of input which can influence its activation - can only grow linearly with the number of layers. In the dilated convolution, the dilation factor d increases exponentially at each layer. Therefore, even though the number of parameters grows only linearly with the number of layers, the effective receptive field of units grows exponentially with the layer depth.

Convolutional models enable parallelization over both the batch and time dimension while RNNs cannot be parallelized over time [2]. Moreover, the path of the gradient between output and input has a fixed length regardless of the sequence length, which mitigates the vanishing and exploding gradients. This has a direct impact on the performance of RNNs [2].

5   Experiments

To analyse further the features of the proposed CubeSat-CDT dataset, we conducted two sets of experiments. First, we analyse the gaps between the three different domains by focusing only on single-frame CubeSat pose estimation. Second, we demonstrate the importance of leveraging the temporal information for more accurate predictions.

As presented in Table 4, our proposed method reduces the pose prediction error, on average by a factor of 2. Furthermore, we note that our approach provides a smoother and a more temporally coherent trajectory as highlighted in Fig. 6 and 7.

We use a PoseNet model [14] with an EfficientNet [29] backbone for feature extraction, followed by fully connected layers for pose regression.

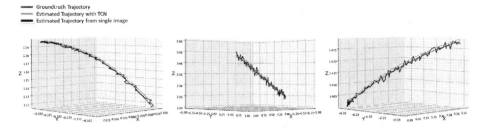

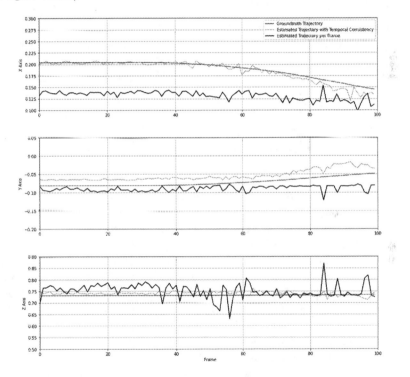

Fig. 6. Groundtruth trajectories of the CubeSat center (in green), estimated trajectories using TCN (in red) and estimated positions using single-frame regressions (in blue). From left to right: example sequences from Zero-G Lab, SPARK and Blender. (Color figure online)

Fig. 7. Per Axis position estimations for *Sequence*7 from Zero-G Laboratory data.

5.1   Domain Gap Analysis

In the first experiment, we trained our pose estimation model in a cross-validation manner – *i.e.*, training on one domain subset and testing on another one – in order to assess the gaps between the different domains. When training and testing on the same subset, we used a 80%-20% split of the data. The temporal information is not used to evaluate the impact of the domain in this setup, so that the pose of the 1U CubeSat is regressed by single frame processing. To eliminate the temporal dependency in the datasets, training data was randomly shuffled.

The quantitative results, presented in Table 3, confirm the relevance of the cross-domain data and the need for applying techniques to minimize the gap between domains. The best inter-domain results were obtained when training the model on the SPARK synthetic data then testing on the Zero-G Lab data (0.26 m/92.25°), with an average position error increment of 0.11 m (0.15 m/72.38° when testing on SPARK). Due to the symmetric nature of the satellite under test, the orientation error is considerably high. Indeed, there is not enough information to discriminate between the different CubeSat faces.

5.2   Impact of Temporal Information

The second experiment was designed to assess the impact of temporal information on pose estimations. The model proposed in Sect. 4 was trained on batches of randomly selected images from the training dataset as an input. The images were then split into 80% for training and 20% for testing.

Given the sequence of estimated learned features computed during the first stage, the second part of the model applies a Seq2Seq TCN model using 1D temporal convolution, to produce the 6-DoF poses of the full sequence.

6   Conclusions

In this paper, we investigated the problem of trajectory estimation of a symmetric spacecraft using only RGB information. We collected 21 real CubeSats trajectories in a laboratory environment along with two different synthetic datasets generated in Unity (50) and Blender (15). We proposed a model composed of an EfficientNet B2 backbone to process a sequence of frames in parallel and then pass the learned features to a Temporal Convolutional Network to compute the final 6-DoF poses. Our experimental results show the importance of leveraging the temporal information to estimate the pose of an object in space and increase accuracy compared to a direct pose regression per frame. Furthermore, the results demonstrate the relevance of the data domain used to train the proposed model on the final performance. The CubeSat Cross-Domain Trajectory dataset will be publicly shared with the research community in order to enable further research on minimizing the domain gap between synthetic and real data, leveraging temporal information for pose estimation and computing the pose of highly-symmetric objects.

References

1. Blender 3.0 reference manual - blender manual. https://docs.blender.org/manual/en/latest/index.html
2. Bai, S., Kolter, J.Z., Koltun, V.: An empirical evaluation of generic convolutional and recurrent networks for sequence modeling. arXiv:1803.01271 (2018)

3. Biesbroek, R., Aziz, S., Wolahan, A., Cipolla, S., Richard-Noca, M., Piguet, L.: The clearspace-1 mission: ESA and clearspace team up to remove debris. In: Proceedings of 8th European Conference on Space Debris, pp. 1–3 (2021)
4. Black, K., Shankar, S., Fonseka, D., Deutsch, J., Dhir, A., Akella, M.R.: Real-time, flight-ready, non-cooperative spacecraft pose estimation using monocular imagery. arXiv preprint arXiv:2101.09553 (2021)
5. Corona, E., Kundu, K., Fidler, S.: Pose estimation for objects with rotational symmetry. In: IEEE International Conference on Intelligent Robots and Systems, pp. 7215–7222, December 2018. https://doi.org/10.1109/IROS.2018.8594282
6. D'Amico, S., Bodin, P., Delpech, M., Noteborn, R.: Prisma. In: D'Errico, M. (eds) Distributed Space Missions for Earth System Monitoring. Space Technology Library, vol. 31, pp. pp. 599–637. Springer, New York, NY (2013). https://doi.org/10.1007/978-1-4614-4541-8_21
7. Dung, H.A., Chen, B., Chin, T.J.: A spacecraft dataset for detection, segmentation and parts recognition. In: Proceedings of the IEEE/CVF Conference on Computer Vision and Pattern Recognition (CVPR) Workshops, pp. 2012–2019, June 2021
8. Garcia, A., Musallam, M.A., Gaudilliere, V., Ghorbel, E., Al Ismaeil, K., Perez, M., Aouada, D.: LSPNet: a 2d localization-oriented spacecraft pose estimation neural network. In: Proceedings of the IEEE/CVF Conference on Computer Vision and Pattern Recognition (CVPR) Workshops, pp. 2048–2056, June 2021
9. GSFC: NASA's Exploration & In-space Service. NASA. https://nexis.gsfc.nasa.gov/
10. Haas, J.K.: A history of the unity game engine (2014)
11. Holschneider, M., Kronland-Martinet, R., Morlet, J., Tchamitchian, P.: A real-time algorithm for signal analysis with the help of the wavelet transform. In: Combes, J.M., Grossmann, A., Tchamitchian, P. (eds) Wavelets. IPTI, pp. 286–297. Springer, Heidelberg (1990). https://doi.org/10.1007/978-3-642-97177-8_28
12. Hu, J., Ling, H., Parashar, P., Naik, A., Christensen, H.I.: Pose estimation of specular and symmetrical objects. CoRR abs/2011.00372 (2020)
13. Hu, Y., Speierer, S., Jakob, W., Fua, P., Salzmann, M.: Wide-depth-range 6d object pose estimation in space. In: Proceedings of the IEEE/CVF Conference on Computer Vision and Pattern Recognition, pp. 15870–15879 (2021)
14. Kendall, A., Grimes, M., Cipolla, R.: Posenet: a convolutional network for real-time 6-DOF camera relocalization. In: Proceedings of the IEEE International Conference on Computer Vision, pp. 2938–2946 (2015)
15. Kisantal, M., Sharma, S., Park, T.H., Izzo, D., Märtens, M., D'Amico, S.: Satellite pose estimation challenge: dataset, competition design, and results. IEEE Trans. Aerosp. Electron. Syst. **56**(5), 4083–4098 (2020)
16. Lepetit, V., Fua, P.: Monocular model-based 3d tracking of rigid objects: a survey. Found. Trends Comput. Graph. Vis. **1**(1), 1–89 (2005)
17. Marchand, E., Chaumette, F., Chabot, T., Kanani, K., Pollini, A.: Removedebris vision-based navigation preliminary results. In: IAC 2019–70th International Astronautical Congress, pp. 1–10 (2019)
18. Mishra, N., Rohaninejad, M., Chen, X., Abbeel, P.: A simple neural attentive meta-learner (2017)
19. Musallam, M.A., del Castillo, M.O., Al Ismaeil, K., Perez, M.D., Aouada, D.: Leveraging temporal information for 3d trajectory estimation of space objects. In: Proceedings of the IEEE/CVF International Conference on Computer Vision (ICCV) Workshops, pp. 3816–3822, October 2021

20. Musallam, M.A., et al.: Spacecraft recognition leveraging knowledge of space environment: simulator, dataset, competition design and analysis. In: 2021 IEEE International Conference on Image Processing Challenges (ICIPC), pp. 11–15. IEEE (2021)

21. Park, T.H., Märtens, M., Lecuyer, G., Izzo, D., D'Amico, S.: Speed+: next generation dataset for spacecraft pose estimation across domain gap. arXiv preprint arXiv:2110.03101 (2021)

22. Park, T.H., Sharma, S., D'Amico, S.: Towards robust learning-based pose estimation of noncooperative spacecraft. arXiv preprint arXiv:1909.00392 (2019)

23. Pauly, L., et al.: Lessons from a space lab - an image acquisition perspective (2022). https://doi.org/10.48550/ARXIV.2208.08865, arxiv.org:2208.08865

24. Pellacani, A., Graziano, M., Fittock, M., Gil, J., Carnelli, I.: Hera vision based GNC and autonomy. In: 8th European Conference For Aeronautics and SP (2019). https://doi.org/10.13009/EUCASS2019-39

25. Pitteri, G., Ramamonjisoa, M., Ilic, S., Lepetit, V.: On object symmetries and 6d pose estimation from images. In: 2019 International Conference on 3D Vision, 3DV 2019, Québec City, QC, Canada, 16–19 September 2019. pp. 614–622. IEEE (2019)

26. Proença, P.F., Gao, Y.: Deep learning for spacecraft pose estimation from photorealistic rendering. In: 2020 IEEE International Conference on Robotics and Automation (ICRA), pp. 6007–6013. IEEE (2020)

27. Robots, U.: Ur10. https://www.universal-robots.com/products/ur10-robot/. Accessed on 11 March 2022

28. Taketomi, T., Uchiyama, H., Ikeda, S.: Visual SLAM algorithms: a survey from 2010 to 2016. IPSJ Trans. Comput. Vision App. $9(1)$, 1–11 (2017). https://doi.org/10.1186/s41074-017-0027-2

29. Tan, M., Le, Q.V.: Efficientnet: Rethinking model scaling for convolutional neural networks. In: Chaudhuri, K., Salakhutdinov, R. (eds.) Proceedings of the 36th International Conference on Machine Learning, ICML 2019, 9–15 June 2019, Long Beach, California, USA. Proceedings of Machine Learning Research, vol. 97, pp. 6105–6114. PMLR (2019)

30. University, S.: Space rendezvous laboratory. https://damicos.people.stanford.edu/. Accessed 11 March 2022

Data Lifecycle Management in Evolving Input Distributions for Learning-based Aerospace Applications

Somrita Banerjee[1]([✉])(iD), Apoorva Sharma[1], Edward Schmerling[1](iD),
Max Spolaor[2], Michael Nemerouf[2], and Marco Pavone[1](iD)

[1] Stanford University, Stanford, CA 94305, USA
{somrita,apoorva,schmrlng,pavone}@stanford.edu
[2] The Aerospace Corporation, El Segundo, CA 90245, USA
{max.spolaor,michael.k.nemerouf}@aero.org

Abstract. As input distributions evolve over a mission lifetime, maintaining performance of learning-based models becomes challenging. This paper presents a framework to incrementally retrain a model by selecting a subset of test inputs to label, which allows the model to adapt to changing input distributions. Algorithms within this framework are evaluated based on (1) model performance throughout mission lifetime and (2) cumulative costs associated with labeling and model retraining. We provide an open-source benchmark of a satellite pose estimation model trained on images of a satellite in space and deployed in novel scenarios (e.g., different backgrounds or misbehaving pixels), where algorithms are evaluated on their ability to maintain high performance by retraining on a subset of inputs. We also propose a novel algorithm to select a *diverse* subset of inputs for labeling, by characterizing the information gain from an input using Bayesian uncertainty quantification and choosing a subset that maximizes collective information gain using concepts from batch active learning. We show that our algorithm outperforms others on the benchmark, e.g., achieves comparable performance to an algorithm that labels 100% of inputs, while only labeling 50% of inputs, resulting in low costs and high performance over the mission lifetime.

Keywords: Data lifecycle management · Out-of-distribution detection · Batch active learning · Learning-based aerospace components

1 Introduction

Motivation. Learning-based models are being increasingly used in aerospace applications because of their ability to deal with high-dimensional observations and learn difficult-to-model functions from data. However, the performance of these models depends strongly on the similarity of the test input to the training data. During a mission lifetime, anomalous inputs and changing input distributions are very common. For example, a model that uses camera images to estimate the pose of a satellite could encounter out-of-distribution (OOD) inputs

L. Karlinsky et al. (Eds.): ECCV 2022 Workshops, LNCS 13801, pp. 127–142, 2023.
https://doi.org/10.1007/978-3-031-25056-9_9

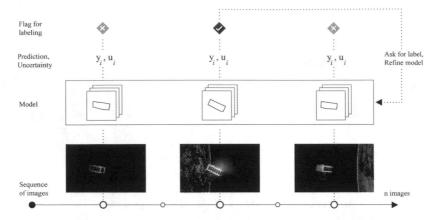

Fig. 1. Given a sequence of inputs, a model outputs a prediction and, potentially, an uncertainty. Data lifecycle management for a model requires adapting to changing input distributions, e.g., novel backgrounds, by periodically refining the model with inputs that were flagged, i.e., inputs that the model is uncertain about.

such as different backgrounds, lighting conditions, lens flares, or misbehaving pixels [9, 12]. For rover dynamics models that are adaptive and learning-enabled, the OOD inputs could result from encountering novel terrain or sensor degradation [23]. Without any intervention, these OOD inputs would then lead the model to produce incorrect predictions and potentially unsafe results.

For safety-critical and high consequence applications, such as those in aerospace, it is crucial that the model remains trustworthy throughout the mission lifetime. In order to support long deployment lifetimes, during which the input distribution may evolve, it is critical to design models that are resilient to such changes. This highlights the need for frameworks like Aerospace's Framework for Trusted AI in High Consequence Environments [20] and methods of retraining the model during the mission lifetime, so that it can adapt to evolving conditions. We refer to this challenge as the problem of "data lifecycle management," entailing how to detect when a learning-based model may be uncertain or performing poorly, and how to design strategies to mitigate this poor performance, e.g., by referring to a human operator or fallback system, or obtaining labels for new inputs to retrain and adapt the model for continued high performance.

Closing the data lifecycle loop is challenging because it requires storing data, downlinking data, obtaining oracle labels, and uplinking labels for retraining, all of which are expensive operations, and must usually be done in a batched episodic manner to ensure safety and computational feasibility. These challenges are discussed in more detail in Sect. 2.1. In order to practically retrain models during their lifetimes, we need onboard real-time decision making about which inputs are most informative to store and label. In this work, we focus on managing the data lifecycle within the context of the framework shown in Fig. 1, which consists of the following steps: select a subset of test inputs for labeling, obtain labels for these inputs (either by expert humans or an oracle fallback system), and use these labels to retrain/adapt the model over time, so that the model can adapt to changing distributions.

Related work. The data lifecycle framework, as presented, can be broadly divided into two algorithmic components: (1) selecting a subset of inputs for labeling and (2) retraining the model using these inputs. For the first task, techniques such as uncertainty quantification [1] and OOD detection [24] are useful for detecting inputs the model is most uncertain about, has not encountered previously, and is likely to perform poorly on. However, these techniques only provide information about a single test input and do not capture the mutual information correlation between multiple test inputs. A practical solution to selecting inputs requires augmenting these techniques with a notion of group uncertainty and choosing subsets that are collectively maximally informative. Choosing subsets of inputs that maximize information gain has been the focus of active learning [18], especially batch mode active learning [6,7,14,17]. Some approaches for selecting inputs include labeling the inputs that lead to greatest expected error reduction [16], that would change the model the most [4], for which the model has the highest predictive uncertainty [25], or that maximize diversity and information gain [14]. However, these works do not address how to quantify the information gain from an input. Our proposed algorithm bridges the gap by using Bayesian uncertainty to express the information gain from an input and techniques from batch active learning to choose a diverse subset.

The literature also lacks a benchmark to evaluate these algorithms in a full data lifecycle context. For example, the WILDS benchmark [10] is useful to evaluate the robustness of models to distribution shifts, but does not close the loop on evaluating the effect of labeling and retraining the model on the lifelong performance of the model. Similarly, the second algorithmic component, i.e., retraining the model, has been the focus of continual learning algorithms [13], for which benchmarks such as CLEAR have been developed [8,11] but these benchmarks do not consider the impact of judiciously selecting which inputs to label on the downstream retraining task. This paper addresses this gap by presenting a benchmark for the full data lifecycle.

Contributions. As our first contribution, we propose a benchmark that evaluates algorithms for subselecting and retraining along two macro-level metrics (1) model performance throughout mission lifetime and (2) cumulative costs associated with labeling and retraining. While this benchmark is generalizable to any application, we provide open-source interfaces and test sequences from the ExoRomper dataset, developed by The Aerospace Corporation. The ExoRomper simulation dataset was assembled in support of Aerospace's Slingshot program [21], which recently launched a commercial 12U smallsat bus, Slingshot 1, with 19 "plug and play" payloads, including ExoRomper. The ExoRomper payload is designed with thermal and visible light cameras facing a maneuverable model spacecraft and leverages an AI accelerator payload also on Slingshot 1 to experiment with machine learning-based pose estimation solutions. The ExoRomper dataset contains simulated camera images of the maneuverable model spacecraft as viewed on-orbit, along with ground truth labels of the satellite model's pose. Test sequences constructed from this dataset include different types of OOD images which can be used to evaluate a pose estimation model and compare different algorithms in the context of a data lifecycle.

In order to balance labeling costs and maintain model performance, we need onboard real-time decision making about which inputs are most informative to store and label. This leads us to our second contribution—a novel algorithm, Diverse Subsampling using Sketching Curvature for OOD Detection (DS-SCOD)—which selects a diverse subset of OOD inputs for labeling and retraining. This algorithm leverages Jacobians from a state-of-the-art OOD detection and uncertainty quantification algorithm, SCOD, [19] and concepts from Bayesian active learning [14] to select a diverse subset of inputs for labeling, which are then used to adapt the model to changing distributions. We show that this algorithm outperforms other algorithms (naive, random, and uncertainty threshold-based) on the benchmark, i.e., results in high model performance throughout the mission lifetime, while also incurring low labeling costs.

Organization. The paper is organized as follows. Section 2 discusses the challenges associated with data lifecycle management and introduces two tools that are instrumental to our approach: OOD detection algorithms and batch active learning methods. Section 3 describes the problem formulation for data lifecycle management. Section 4 presents our open-source benchmark for evaluating data lifecycle algorithms. Section 5 introduces our novel data subselection algorithm, DS-SCOD. Section 6 presents results of evaluating our algorithm on the benchmark using the ExoRomper dataset where it outperforms other methods, e.g. achieves performance comparable to an algorithm that labels 100% of inputs while only labeling 50% of inputs. Finally, Sect. 7 provides conclusions and future research directions.

2   Background

2.1   Challenges for Data Lifecycle Management

Data lifecycle management, especially for aerospace applications, has a number of challenges. We use the example of pose estimation for CubeSats [15], such as the Slingshot satellite, to contextualize these challenges, which include:

1. Data collection and storage may be limited by onboard storage capacity, especially for large input sizes, constant streams of data, and having to store inputs for future labeling.
2. Data transmission may be limited by cost and practicality, e.g., CubeSats often have intermittent connectivity to ground stations, narrow downlink windows, and limited bandwidth [15].
3. Operations associated with labeling are expensive, e.g., obtaining a label from a human expert.
4. Retraining the model is expensive and limited by onboard computation power [3,15], especially for space hardware on CubeSats. Rather than retraining on each individual flagged input, it is more computationally feasible to retrain on batches of flagged inputs.

5. Lastly, building safe and trusted AI systems often requires model updates to be episodic and versioned in nature, in order to ensure adequate testing. This results in a batched, rather than streaming, source of ground truth labels as well as delays between flagging inputs and receiving labels.

These challenges highlight the need for a judicious, real-time, and onboard decision-making algorithm that selects inputs for storage, downlinking, labeling, and retraining. In this work, we model some of these challenges by designing our algorithms to operate on batches and explicitly optimizing for low labeling cost. Next, we discuss two key tools that enable our algorithmic design.

2.2   Out-of-Distribution Detection

A key requirement of data lifecycle management is to be able to detect when inputs to the model are no longer similar to those seen during training, i.e., detect when inputs are OOD and the model is likely to perform poorly. Detecting when inputs are OOD is useful in many ways, e.g., as a runtime monitor or anomaly/outlier detector so that the corresponding outputs can be discarded or a fallback system can be used. OOD detection is also useful in deciding which inputs to use to adapt the model, so model performance does not degrade as input distributions evolve. A principled approach to OOD is to quantify the functional uncertainty of the model for each test input, i.e., how well does the training data determine the predictions at a new test input. This work employs the Bayesian approach to uncertainty quantification developed by Sharma et al., Sketching Curvature for Out-of-Distribution Detection (SCOD) [19], which is summarized below.

Consider a pre-trained model f with weights $\boldsymbol{w} \in \mathbb{R}^N$ that maps from input $\boldsymbol{x} \in \mathcal{X}$ to the parameters $\boldsymbol{\theta} \in \Theta$ of a probability distribution $p_\theta(y)$ on targets $\boldsymbol{y} \in \mathcal{Y}$, where $\boldsymbol{\theta} = f(\boldsymbol{x}, \boldsymbol{w})$. The model is trained on a dataset of examples $\mathcal{D} = \{\boldsymbol{x}_i, \boldsymbol{y}_i\}_{i=1}^M$. The prior over the weights of the model is $p(\boldsymbol{w})$ and the posterior distribution on these weights given the dataset is $p(\boldsymbol{w} \mid \mathcal{D})$. Characterizing the posterior and marginalizing over it produces the posterior predictive distribution

$$p(\boldsymbol{y} \mid \boldsymbol{x}) = \int p(\boldsymbol{w} \mid \mathcal{D}) p_w(\boldsymbol{y} \mid \boldsymbol{x}) d\boldsymbol{w}. \tag{1}$$

Estimating Posterior Offline. To tractably compute this posterior for Deep Neural Networks, SCOD uses the Laplace second order approximation of the log posterior $\log p(\boldsymbol{w} \mid \mathcal{D})$ about a point estimate $\boldsymbol{w}^*$. If the prior on the weights is the Gaussian $p(\boldsymbol{w}) = \mathcal{N}(0, \epsilon^2 I_N)$, the Laplace posterior is also a Gaussian, with mean equal to the point estimate $\boldsymbol{w}^*$, and covariance given by

$$\Sigma^* = \left(H_L + \epsilon^{-2} I_N\right)^{-1}, \tag{2}$$

where H_L is the Hessian with respect to $\boldsymbol{w}$ of the log likelihood of the training data, namely, $L(\boldsymbol{w}) = \sum_{i=1}^M \log p_{w^*}(\boldsymbol{y}^{(i)} \mid \boldsymbol{x}^{(i)})$. The Hessian can in turn be well approximated by the dataset Fisher information matrix, i.e.,

$$H_L = \left.\frac{\partial^2 L}{\partial \boldsymbol{w}^2}\right|_{\boldsymbol{w}=\boldsymbol{w}^*} \approx M F_{\boldsymbol{w}^*}^{\mathcal{D}}. \tag{3}$$

The dataset Fisher information matrix $F_{\boldsymbol{w}^*}^{\mathcal{D}}$ describes the impact of perturbing the weights $\boldsymbol{w}$ on the predicted output distribution for each output in the dataset $\mathcal{D}$, i.e.,

$$F_{\boldsymbol{w}^*}^{\mathcal{D}} = \frac{1}{M} \sum_{i=1}^{M} F_{\boldsymbol{w}^*}^{(i)} \tag{4}$$

$$F_{\boldsymbol{w}^*}^{(i)}(\boldsymbol{x}_i, \boldsymbol{w}^*) = \boldsymbol{J}_{f,\boldsymbol{w}^*}^T F_{\boldsymbol{\theta}}^{(i)}(\boldsymbol{\theta}) \boldsymbol{J}_{f,\boldsymbol{w}^*}, \tag{5}$$

where $\boldsymbol{J}_{f,\boldsymbol{w}^*}$ is the Jacobian matrix $\frac{\partial f}{\partial \boldsymbol{w}}$ evaluated at $(\boldsymbol{x}_i, \boldsymbol{w}^*)$ and $F_{\boldsymbol{\theta}}^{(i)}(\boldsymbol{\theta})$ is the Fisher information matrix of the distribution $p_{\boldsymbol{\theta}}(y)$. In the case where $p_{\boldsymbol{\theta}}(y)$ is a Gaussian with mean $\boldsymbol{\theta}$ and covariance Σ, the Fisher information matrix is given by

$$F_{\boldsymbol{\theta}}^{(i)}(\boldsymbol{\theta}) = \Sigma(\boldsymbol{x}^{(t)}, \boldsymbol{w}^*)^{-1}. \tag{6}$$

SCOD uses a matrix sketching-based approach to efficiently compute a low-rank approximation of the dataset Fisher $F_{\boldsymbol{w}^*}^{\mathcal{D}}$, which in turn enables the computation of the Laplace posterior Σ^* using Eqs. 2, 3, and 4. This is done offline.

Online Computation of Predictive Uncertainty. For a test input $\boldsymbol{x}^{(t)}$, the predictive uncertainty is computed as the expected change in the output distribution (measured by a KL divergence δ_{KL}) when weights are perturbed according to the Laplace posterior Σ^*. Sharma et al. show that this expectation can be related back to the Fisher information matrix for this test input $F_{\boldsymbol{w}^*}^{(t)}$, i.e.,

$$\mathrm{Unc}(\boldsymbol{x}^{(t)}) = \mathop{\mathbb{E}}_{d\boldsymbol{w} \sim \mathcal{N}(0,\Sigma^*)} \left[\delta_{\mathsf{KL}}(\boldsymbol{x}^{(t)}) \right]$$
$$\approx \mathrm{Tr}\left(F_{\boldsymbol{w}^*}^{(t)} \Sigma^* \right). \tag{7}$$

Therefore, for a test input $\boldsymbol{x}^{(t)}$, SCOD uses a forward pass to calculate the output $p(\boldsymbol{y}) = f(\boldsymbol{x}^{(t)}, \boldsymbol{w})$, a backward pass to calculate the Jacobian $\boldsymbol{J}_{f,\boldsymbol{w}^*}$ at $\boldsymbol{x}^{(t)}$, and the output covariance $\Sigma(\boldsymbol{x}^{(t)}, \boldsymbol{w}^*)$. Then, using Eqs. 5, 6, and 7, the predictive uncertainty $\mathrm{Unc}(\boldsymbol{x}^{(t)})$ can be efficiently computed. This predictive uncertainty can be thought of as a measure of epistemic uncertainty: if it is high, then the prediction for input $\boldsymbol{x}^{(t)}$ is *not* well supported by the training data, while if it is low, then the training data supports a confident prediction. Therefore, this estimate is useful for detecting OOD inputs.

2.3   Bayesian Batch Active Learning

Detecting inputs that have high uncertainty allows those inputs to be flagged for labeling and retraining. However, in practice, a limited budget for retraining precludes selecting every uncertain input. While a naive solution might be to select the subset with the *highest* uncertainties, this can lead to subsets with redundancy: including many similar inputs while leaving out other inputs that are also informative. Instead, it is important that we not select inputs greedily, but rather sets of *diverse* inputs for which labels would jointly provide more information. To do so, we take inspiration from active learning, where the goal is to select the most informative inputs, and, in particular, batch active learning, which selects a batch of informative inputs instead of a single input, allowing retraining to be done episodically rather than sequentially, further lowering labeling costs [6,7,14,17]. This work closely utilizes the Bayesian batch active learning approach developed by Pinsler et al. [14], which is summarized below.

Consider a probabilistic predictive model $p(y \mid x, \theta)$ parameterized by $\theta \in \Theta$ mapping from inputs $x \in \mathcal{X}$ to a probabilistic output $y \in \mathcal{Y}$. The model is trained on the dataset $\mathcal{D}_0 = \{x_i, y_i\}_{i=1}^n$, resulting in the prior parameter distribution $p(\theta|\mathcal{D}_0)$. During deployment, the model encounters a set of test inputs $\mathcal{X}_p = \{x_i\}_{i=1}^m$. It is assumed that an oracle labeling mechanism exists which can provide labels $\mathcal{Y}_p = \{y_i\}_{i=1}^m$ for the corresponding inputs. From a Bayesian perspective, it is optimal to label every point and use the full dataset $\mathcal{D}_p = (\mathcal{X}_p, \mathcal{Y}_p)$ to update the model. The resultant data posterior is given by Bayes' rule

$$p(\theta|\mathcal{D}_0 \cup (\mathcal{X}_p, \mathcal{Y}_p)) = \frac{p(\theta|\mathcal{D}_0)\, p(\mathcal{Y}_p|\mathcal{X}_p, \theta)}{p(\mathcal{Y}_p|\mathcal{X}_p, \mathcal{D}_0)}. \tag{8}$$

In practice, only a subset of points $\mathcal{D}' = (\mathcal{X}', \mathcal{Y}') \subseteq \mathcal{D}_p$ can be selected. From an information theoretic perspective, it is desired that the query points $\mathcal{X}' = \{x_i\}_{i=1}^k$ (where $k \leq m$) are maximally informative, which is equivalent to minimizing the expected entropy, $\mathbb{H}$, of the posterior, i.e.,

$$\mathcal{X}^* = \underset{\mathcal{X}' \subset \mathcal{X}_p, |\mathcal{X}'| \leq k}{\arg \min} \ \mathbb{E}_{\mathcal{Y}' \sim p(\mathcal{Y}'|\mathcal{X}', \mathcal{D}_0)} \left[\mathbb{H} \left[\theta | \mathcal{D}_0 \cup (\mathcal{X}', \mathcal{Y}') \right] \right]. \tag{9}$$

Solving Eq. 9 requires considering all possible subsets of the pool set, making it computationally intractable. Instead, Pinsler et al. choose the batch $\mathcal{D}'$ such that the updated log posterior $\log p(\theta|\mathcal{D}_0 \cup \mathcal{D}')$ best approximates the complete data log posterior $\log p(\theta|\mathcal{D}_0 \cup \mathcal{D}_p)$. The log posterior, following Eq. 8 and taking the expectation over unknown labels $\mathcal{Y}_p$, is given by

$$\underset{\mathcal{Y}_p}{\mathbb{E}} \left[\log p(\theta|\mathcal{D}_0 \cup (\mathcal{X}_p, \mathcal{Y}_p)) \right] = \underset{\mathcal{Y}_p}{\mathbb{E}} \left[\log p(\theta|\mathcal{D}_0) + \log p(\mathcal{Y}_p|\mathcal{X}_p, \theta) - \log p(\mathcal{Y}_p|\mathcal{X}_p, \mathcal{D}_0) \right]$$

$$= \log p(\theta|\mathcal{D}_0) + \sum_{i=1}^m \left(\underbrace{\underset{y_i}{\mathbb{E}} \left[\log p(y_i|x_i, \theta) \right] + \mathbb{H} \left[y_i|x_i, \mathcal{D}_0 \right]}_{\mathcal{L}_i(\theta)} \right), \tag{10}$$

where $\mathcal{L}_i(\theta)$ is the belief update as a result of choosing input $x_i \in \mathcal{X}_p$. The subset $\mathcal{X}'$ can now be chosen to best approximate the complete belief update, i.e., by choosing

$$\boldsymbol{c}^* = \underset{\boldsymbol{c}}{\text{minimize}} \; \left\| \sum_i \mathcal{L}_i - \sum_i c_i \mathcal{L}_i \right\|^2$$
$$\text{subject to} \quad c_i \in \{0, 1\} \quad \forall i,$$
$$\sum_i c_i \leq k. \tag{11}$$

Here $\boldsymbol{c} \in \{0, 1\}^m$ is a weight vector indicating which of the m points to include in the subset of maximum size k. Pinsler et al. show that a relaxation,

$$\boldsymbol{c}^* = \underset{\boldsymbol{c}}{\text{minimize}} \; (\mathbf{1} - \boldsymbol{c})^T \boldsymbol{K} (\mathbf{1} - \boldsymbol{c})$$
$$\text{subject to} \quad c_i \geq 0 \quad \forall i,$$
$$\sum_i c_i \|\mathcal{L}_i\| = \sum_i \|\mathcal{L}_i\|, \tag{12}$$

where K is a kernel matrix $K_{mn} = \langle \mathcal{L}_m, \mathcal{L}_n \rangle$, can be solved efficiently in real-time using Frank-Wolfe optimization, yielding the optimal weights $\boldsymbol{c}^*$ after k iterations. The number of non-zero entries in $\boldsymbol{c}^*$ will be less than or equal to k, denoting which inputs (up to a maximum of k) are to be selected from the batch of m inputs. Empirically, this property leads to smaller subselections as more data points are acquired. We leverage this Bayesian approach to selecting maximally informative subsets in our algorithm design, which enables maintaining high model performance while reducing labeling costs. In the next section, we discuss the framework and problem formulation guiding our approach.

3   Problem Formulation and Framework

In this paper, we consider the problem of data lifecycle management as applied to model adaptation via retraining using the specific framework shown in Fig. 1. This framework assumes that the model receives inputs in a sequence of batches and that the model outputs a prediction for each input. Additionally, the model may also output an estimate of uncertainty for each input, such as the epistemic uncertainty estimate provided by SCOD. Based on the outputs for each batch, a decision must be made about which (if any) inputs to flag for labeling. If flagged, the input is labeled by an oracle and then used to retrain the model.

In the context of this framework, we seek to (1) develop a benchmark that evaluates algorithms not just for their accuracy and sample efficiency on a single batch, but over lifecycle metrics such as average improvement in lifetime model performance and cumulative lifetime labeling cost, and (2) develop an algorithm to select inputs for labeling, that are ideally diverse and informative, in order to minimize costs associated with labeling and retraining. In support of the first goal, we propose an evaluation benchmark that is described in the next section. In support of the second goal, we present a novel algorithm for diverse subset selection in Sect. 5.

4   Evaluation Benchmark

In this section, we present a benchmark to evaluate algorithms that decide which inputs to flag for labeling. A flagging algorithm is evaluated based on how well it manages the performance of a pre-trained model on a sequence of test input batches. The two metrics used for this evaluation are: the cumulative labeling cost, i.e., how many inputs were flagged for labeling, and the rolling model performance, i.e., the prediction loss of the model during the deployment lifetime. The ideal flagging algorithm keeps model performance high while incurring a low labeling cost. This benchmark procedure is described in Procedure 1.

Procedure 1: Evaluation benchmark for "flagging" algorithms

Input: *model, test_seq, test_labels, flagger*
Output: *labeling_cost, rolling_performance*

1 Initialize *flagged* vector and *outputs*
2 **foreach** *i, batch* ∈ *test_seq* **do**
3   *outputs[i]* ←*model(batch)*   // store model predictions
4   *flags = flagger(batch)*   // flag to request oracle labels
5   **if** *any(flags)* **then**
6   *flagged_inputs = batch[flags]*   // store flagged inputs
7   Update *flagged[i]*
8   *true_labels = test_labels[i][flags]*   // get oracle labels
9   *model* ←refine_model(*flagged_inputs, true_labels*)   // refine model

   /* calculate benchmark metrics   */
10 *labeling_cost* = cumsum(*flagged*)
11 *rolling_performance* = -[loss(*outputs[i]*, *test_labels[i]*) for *i, batch* ∈ *test_seq*]
12 **return** *labeling_cost, rolling_performance*

We provide an open-source benchmark to test flagging algorithms[1]. This interface includes a pre-trained pose estimation model and test sequences drawn from the ExoRomper dataset, as well as a categorization of different "types" of out-of-distribution images, such as those with different backgrounds (Earth/space), lens flares, or with added sensor noise. These input sequences can be passed in sequentially to the model, or as batches. Additionally, the benchmark also contains implementations of some simple flagging algorithms to compare against, such as flagging every input (**naive_true**), flagging none of the inputs (**naive_false**), and randomly flagging k inputs from each batch (**random_k**). The benchmark also contains implementations of two proposed flagging algorithms: (1) **scod_k_highest** (or **scod** for short), which wraps the pre-trained model with a SCOD wrapper that provides an estimate of epistemic uncertainty for each input. From each batch, the k inputs with the highest SCOD uncertainties are flagged for labeling, and (2) **diverse**, Diverse Subsampling using SCOD (DS-SCOD), which is a novel algorithm that uses the Jacobians from SCOD and Bayesian active learning to select a subset of inputs that are not only out-of-distribution but also *diverse*. This algorithm is further described in the next section.

---

[1] Code for the benchmark and our experiments is available at https://github.com/StanfordASL/data-lifecycle-management.

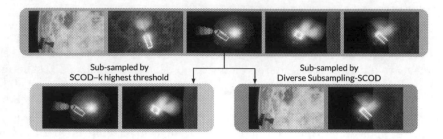

Fig. 2. Given a set of 5 random images drawn from the ExoRomper dataset and flagged as OOD by SCOD, a subset of 2 images is generated using SCOD-k highest threshold and Diverse Subsampling using SCOD. Although SCOD-k selects the 2 images with the highest predictive uncertainty, it appears, at least to a human eye, that the images selected by DS-SCOD are more semantically diverse.

5 Diverse Subsampling Using SCOD (DS-SCOD)

To obtain performance improvement without high cost, it is necessary that the inputs chosen for labeling and retraining should provide information to the model that improves downstream performance, e.g., OOD inputs for which the model has high predictive uncertainty. However, an approach that greedily selects the k most uncertain images from a batch may be suboptimal. Consider a batch of inputs containing multiple "types" of OOD, e.g., images with an Earth background or with a lens flare, where flagging the k highest uncertainty images could lead to subsets containing only one type, while neglecting the information available from other types of OOD images. Downstream, this would lead to improved model performance for only the OOD type selected. In contrast, DS-SCOD seeks to capture the mutual correlation between multiple inputs, leading to a subset that contains diverse images and is maximally informative for the model. As discussed later in Sect. 6, empirical testing shows that diverse subselecting leads to higher downstream performance, while allowing for low labeling cost. Figure 2 illustrates the different results from completing the task of subselecting 2 images from a set of 5 when using scod only, i.e., selecting the 2 images with the highest uncertainty, vs. using diverse, i.e., optimizing for diversity.

DS-SCOD is based on two governing principles: (1) that the inputs selected for labeling should provide information that improves downstream performance, e.g., inputs where the model has high uncertainty or OOD inputs and (2) that the subselected inputs from each batch be chosen to maximize *collective* information gain. Therefore, the representation of information gain due to an input must lend itself to expressing the total information gain after selecting a subset of multiple inputs. An ideal choice for representing this information gain is the Bayesian posterior for each input that is provided by Bayesian OOD algorithms such as SCOD, as described in Sect. 2.2. Specifically, the Jacobian of the output with respect to the weights, $\boldsymbol{J}^{(i)}_{f,w^\star}$, encodes the uncertainty for a single input, i.e., the impact of small weight changes on the output, and also encodes a correlation in uncertainty where $\boldsymbol{J}^{(i)}$ and $\boldsymbol{J}^{(j)}$ determine whether changes in output for inputs i and j are driven by the same set of weights. The key insight driving

DS-SCOD is that OOD algorithms, which quantify uncertainty to detect differences between test inputs and a training distribution, are also naturally a powerful tool to detect differences between samples and quantify the uncertainty reduction resulting from selecting a batch for retraining.

The core methodology employed by DS-SCOD is to use batch active learning to select the maximally informative batch, as described in Sect. 2.3, where the information value of an input is expressed using a Bayesian belief update that we obtain from SCOD, as described in Sect. 2.2. In DS-SCOD, we express the belief update, $\mathcal{L}_i(\boldsymbol{\theta})$ as a result of choosing input $\boldsymbol{x}_i$, from Eq. 10, as

$$\mathcal{L}_i(\boldsymbol{\theta}) = \left(F_{\boldsymbol{\theta}}^{(i)}(\boldsymbol{\theta}) \right)^{1/2} \boldsymbol{J}_{f,\boldsymbol{w}^*}^{(i)}. \tag{13}$$

where $\boldsymbol{J}_{f,\boldsymbol{w}^*}^{(i)}$ is the Jacobian of the output with respect to the weights evaluated at $\boldsymbol{x}_i$ and $F_{\boldsymbol{\theta}}^{(i)}(\boldsymbol{\theta})$ is the Fisher information matrix of the output probability distribution $p_{\boldsymbol{\theta}}(y)$ from Eq. 6. Both of these expressions are calculated efficiently online by SCOD. Next, from each batch of size m, we select a maximum of k inputs to flag for labeling by solving the optimization problem in Eq. 12. Specifically, the optimization takes the form

DS-SCOD:

$$\boldsymbol{c}^* = \underset{\boldsymbol{c}}{\text{minimize}} \ (\mathbf{1} - \boldsymbol{c})^T \boldsymbol{K} \, (\mathbf{1} - \boldsymbol{c})$$

$$\text{where} \qquad K_{ij} = \langle \mathcal{L}_i, \mathcal{L}_j \rangle = \text{Tr} \left(\left(F_{\boldsymbol{\theta}}^{(i)} \right)^{1/2} \boldsymbol{J}_{f,\boldsymbol{w}^*}^{(i)} \boldsymbol{J}_{f,\boldsymbol{w}^*}^{(j)T} \left(F_{\boldsymbol{\theta}}^{(j)} \right)^{1/2} \right)$$

$$\text{subject to} \quad c_i \geq 0 \quad \forall i,$$

$$\sum_i c_i \left\| \left(F_{\boldsymbol{\theta}}^{(i)} \right)^{1/2} \boldsymbol{J}_{f,\boldsymbol{w}^*}^{(i)} \right\| = \sum_i \left\| \left(F_{\boldsymbol{\theta}}^{(i)} \right)^{1/2} \boldsymbol{J}_{f,\boldsymbol{w}^*}^{(i)} \right\|. \tag{14}$$

This optimization is solved using the Frank-Wolfe algorithm in real-time. By minimizing the difference between the belief update after choosing all m points and after choosing inputs according to $\boldsymbol{c}$, the loss of information is minimized. This results in a maximally informative subsample of inputs, without exceeding the given labeling budget k. In the next section, we discuss the results of testing the DS-SCOD algorithm on the benchmark.

6   Experimental Results

The benchmark described in Sect. 4 was used to test DS-SCOD and other data lifecycle algorithms. The task for the benchmark was to estimate pose of a satellite given a camera image. The ExoRomper dataset was filtered into three categories: images with a space background, with Earth in the background, and with a lens flare (examples in Fig. 4). A pose estimation model was trained on 300 images drawn *only* from the space background set (considered "in-distribution")

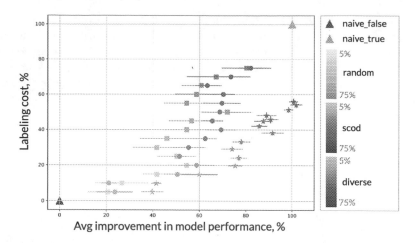

Fig. 3. Comparing performance of the naive, random, scod, and diverse (DS-SCOD) algorithms in terms of average improvement in model performance over the lifecycle and the labeling cost. The latter three algorithms were evaluated with labeling budgets between 5% and 75% at intervals of 5%. The metrics are reported in percentages relative to the naive_false (0%) and naive_true (100%) algorithms.

using a deep CNN architecture inspired by [2], i.e., the winner of the Kelvins satellite pose estimation challenge organized by ESA and Stanford University [9]. The benchmark supports constructing test sequences by randomly sampling images from a configurable mix of categories. Our evaluation test sequence consisted of 100 batches of 20 images sampled from all three categories, with increasing numbers of degraded pixels ($\leq 10\%$), to simulate degrading sensors. From each batch, inputs could be flagged by the subselection algorithm. If any inputs were flagged, the corresponding ground truth labels were looked up, and the model was retrained on those labels for 200 epochs at a low learning rate of 0.001, before the start of the next batch. Thus, the evaluation was episodic but without delays between batch flagging and retraining.

As shown in Fig. 3, the algorithms were evaluated on two metrics: the improvement in model performance as a result of retraining, averaged over the full test sequence, and the labeling cost, which is the number of images flagged for labeling expressed as a percentage of the total number of images. For some algorithms, we set a value k that determines what percentage of inputs the algorithm is allowed to select for labeling, henceforth referred to as the *budget*. The following algorithms were tested:

- **naive_false**: This simple algorithm never flags any image. Thus, the labeling cost is zero. The performance of this algorithm is used as a baseline for other algorithms, therefore the percentage improvement in model performance is also zero.
- **naive_true**: This simple algorithm flags every image. Thus, the labeling cost is 100%. The performance of this algorithm improves due to the retraining. Since no algorithm can do better than having every input labeled (barring

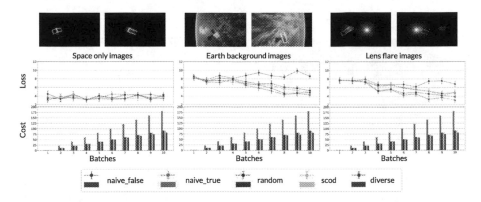

Fig. 4. Comparing algorithm performance on images of a single category. For each successive batch, the average loss of the model and the cumulative cost are plotted for each of the algorithms: naive_false, naive_true, random, scod, and diverse. For the latter three algorithms, the relabeling budget is set to 50%.

stochasticity, as we discuss later), we use the performance of this algorithm as the 100% benchmark.

- random k%: This algorithm randomly selects k% of inputs from each batch for labeling. This budget is varied between 5% and 75%, at intervals of 5%.
- scod k%: This algorithm wraps the pre-trained model with a SCOD wrapper which computes the uncertainty for each input. From each batch, the top k% of inputs with the highest SCOD uncertainties are flagged for labeling.
- diverse k%: This algorithm uses DS-SCOD, as presented in Sect. 5, to select not only uncertain inputs, but also a diverse subset of inputs. For each batch, the optimization problem from Eq. 14 is solved efficiently. This results in flagging *up to* k% of inputs, in contrast to the previous algorithms that select a fixed, i.e. exactly k%, number of inputs.

We observe from Fig. 3 that random sampling leads to an improvement in performance; periodically retraining the model allows it to adapt to distribution shifts. SCOD-k performs better than random sampling, on average, for a given budget since it picks the most uncertain inputs which results in higher information gain and downstream model performance. Diverse subsampling, DS-SCOD, outperforms random and SCOD-k by achieving better performance for a given budget. We observe that when the labeling budget is low, e.g., 10% to 25%, DS-SCOD expends the full budget, i.e., chooses exactly k% of inputs to label. However, due to the relaxed optimization process, when given a higher budget, e.g., 40% to 75%, DS-SCOD uses only a fraction of the budget, not exceeding 60% in Fig. 3. Remarkably, DS-SCOD achieves performance comparable to naive_true while incurring only 50% of the labeling cost. DS-SCOD's performance sometimes slightly exceeds 100% on our benchmark, i.e., makes slightly better predictions than naive_true, due to the stochastic nature of the stochastic gradient descent we use to retrain the model. Overall, DS-SCOD results in the highest average lifetime model performance, while reducing labeling cost.

To further understand the performance of DS-SCOD, the algorithms were benchmarked on test sequences drawn only from *one* of the three categories, as shown in Fig. 4. The labeling budgets were set to 50%. For the space-only images, all the algorithms resulted in relatively constant test loss over time. This result is unsurprising because the model was trained on space background images, and likely reached convergence on these "in-distribution" images. For both the Earth background and lens flare images, all of the labeling algorithms, naive_true, random, scod, and diverse, led to a decrease in loss over time, as compared to naive_false. This indicates that retraining is effective in increasing model performance when encountering OOD inputs. We observe that naive_true achieves the greatest reduction in loss by labeling every input in each batch. Further, we note that the diverse (DS-SCOD) algorithm achieves a reduction in loss most comparable to naive_true while accruing a labeling cost of less than 50% that of naive_true. Therefore, we show that DS-SCOD offers a cost-effective solution to adapting learning-based models to changing input distributions.

7   Conclusion

In this work, we discussed the challenges associated with data lifecycle management for learning-based components. We presented a framework for online adaptation by labeling a subset of test inputs and using the labels to retrain the model. We provided a benchmark for evaluating algorithms, that choose which subset of test inputs should be labeled, in the context of two macro-level metrics: the average model performance over the lifecycle and the cost of labeling. Lastly, we presented a novel subsampling algorithm, Diverse Subsampling using SCOD (DS-SCOD), inspired by ideas from Bayesian uncertainty quantification and Bayesian batch active learning. We showed that this algorithm, when evaluated on the benchmark, achieves performance comparable to labeling every input while incurring 50% of the labeling costs.

Interesting future directions include adding a delay between flagging inputs for labels and receiving labels for retraining, to better simulate the data lifecycle [5]. Another direction would be to expand the benchmark to other applications beyond satellite pose estimation, as well as different tasks and types of OOD inputs, since algorithm performance may be better suited to some tasks than others. Another promising research direction is to evaluate the effect of retraining algorithms on the improvement in model performance and design a retraining algorithm that makes maximal use of the information gain from diverse labeled inputs while avoiding catastrophic forgetting, e.g., by borrowing concepts from transfer learning and continual learning [13,22].

Acknowledgments. This work is supported by The Aerospace Corporation's University Partnership Program, and by the Stanford Graduate Fellowship (SGF). The NASA University Leadership initiative (grant #80NSSC20M0163) provided funds to assist the authors with their research, but this article solely reflects the opinions and conclusions of its authors and not any NASA entity. The authors would like to thank Rohan Sinha and the reviewers for their helpful comments.

References

1. Abdar, M., et al.: A review of uncertainty quantification in deep learning: techniques, applications and challenges. Inf. Fusion **76**, 243–297 (2021)
2. Chen, B., Cao, J., Parra, A., Chin, T.J.: Satellite pose estimation with deep landmark regression and nonlinear pose refinement. In: IEEE International Conference on Computer Vision (2019)
3. Doran, G., Daigavane, A., Wagstaff, K.: Resource consumption and radiation tolerance assessment for data analysis algorithms onboard spacecraft. IEEE Trans. Aerosp. Electron. Syst. **58**, 5180–5189 (2022)
4. Freytag, A., Rodner, E., Denzler, J.: Selecting influential examples: active learning with expected model output changes. In: Fleet, D., Pajdla, T., Schiele, B., Tuytelaars, T. (eds.) ECCV 2014. LNCS, vol. 8692, pp. 562–577. Springer, Cham (2014). https://doi.org/10.1007/978-3-319-10593-2_37
5. Grzenda, M., Gomes, H.M., Bifet, A.: Delayed labelling evaluation for data streams. Data Min. Knowl. Disc. **34**(5), 1237–1266 (2020)
6. Guo, Y., Schuurmans, D.: Discriminative batch mode active learning. In: Conference on Neural Information Processing Systems (2007)
7. Hoi, S.C., Jin, R., Zhu, J., Lyu, M.R.: Batch mode active learning and its application to medical image classification. In: International Conference on Machine Learning (2006)
8. Hsu, Y.C., Liu, Y.C., Kira, Z.: Re-evaluating continual learning scenarios: a categorization and case for strong baselines. In: Conference on Neural Information Processing Systems (2018)
9. Kisantal, M., Sharma, S., Park, T.H., Izzo, D., Märtens, M., D'Amico, S.: Satellite pose estimation challenge: dataset, competition design, and results. IEEE Trans. Aerosp. Electron. Syst. **56**(5), 4083–4098 (2020)
10. Koh, P.W., et al.: WILDS: a benchmark of in-the-wild distribution shifts. In: International Conference on Machine Learning (2021)
11. Lin, Z., Shi, J., Pathak, D., Ramanan, D.: The CLEAR benchmark: continual learning on real-world imagery. In: Conference on Neural Information Processing Systems - Datasets and Benchmarks Track (2021)
12. Murray, G., Bourlai, T., Spolaor, M.: Mask R-CNN: detection performance on SPEED spacecraft with image degradation. In: IEEE International Conference on Big Data (2021)
13. Parisi, G.I., Kemker, R., Part, J.L., Kanan, C., Wermter, S.: Continual lifelong learning with neural networks: A review. Neural Netw. **113**, 54–71 (2019)
14. Pinsler, R., Gordon, J., Nalisnick, E., Hernández-Lobato, J.M.: Bayesian batch active learning as sparse subset approximation. In: Conference on Neural Information Processing Systems (2019)
15. Poghosyan, A., Golkar, A.: CubeSat evolution: analyzing CubeSat capabilities for conducting science missions. Prog. Aerosp. Sci. **88**, 59–83 (2017)
16. Roy, N., McCallum, A.: Toward optimal active learning through sampling estimation of error reduction. In: International Conference on Machine Learning (2001)
17. Sener, O., Savarese, S.: Active learning for convolutional neural networks: a core-set approach. In: International Conference on Learning Representations (2018)
18. Settles, B.: Active learning. Synth. Lect. Artif. Intell. Mach. Learn. **6**(1), 1–114 (2012)
19. Sharma, A., Azizan, N., Pavone, M.: Sketching curvature for efficient out-of-distribution detection for deep neural networks. In: Proceedings of Conference on Uncertainty in Artificial Intelligence (2021)

20. Slingerland, P., et al.: Adapting a trusted AI framework to space mission autonomy. In: IEEE Aerospace Conference (2022)
21. Weiher, H., Mabry, D.J., Utter, A.C.: Slingshot: in-space modularity test platform. In: AIAA Aerospace Sciences Meeting (2022)
22. Weiss, K., Khoshgoftaar, T.M., Wang, D.D.: A survey of transfer learning. J. Big Data **3**(1), 1–40 (2016). https://doi.org/10.1186/s40537-016-0043-6
23. Wellhausen, L., Ranftl, R., Hutter, M.: Safe robot navigation via multi-modal anomaly detection. IEEE Robot. Autom. Lett. **5**(2), 1326–1333 (2020)
24. Yang, J., Zhou, K., Li, Y., Liu, Z.: Generalized out-of-distribution detection: a survey (2021). arxiv.org:2110.11334
25. Yang, Y., Loog, M.: Active learning using uncertainty information. In: IEEE International Conference on Pattern Recognition (2016)

Strong Gravitational Lensing Parameter Estimation with Vision Transformer

Kuan-Wei Huang[1], Geoff Chih-Fan Chen[2], Po-Wen Chang[3], Sheng-Chieh Lin[4], ChiaJung Hsu[5], Vishal Thengane[6], and Joshua Yao-Yu Lin[7(✉)]

[1] Carnegie Mellon University, Pittsburgh, USA
[2] University of California, Los Angeles, USA
[3] Ohio State University, Columbus, USA
[4] University of Kentucky, Lexington, USA
[5] Chalmers University of Technology, Gothenburg, Sweden
[6] Mohamed bin Zayed University of Artificial Intelligence, Abu Dhabi, UAE
[7] University of Illinois at Urbana-Champaign, Champaign, USA
yaoyuyl2@illinois.edu

Abstract. Quantifying the parameters and corresponding uncertainties of hundreds of strongly lensed quasar systems holds the key to resolving one of the most important scientific questions: the Hubble constant (H_0) tension. The commonly used Markov chain Monte Carlo (MCMC) method has been too time-consuming to achieve this goal, yet recent work has shown that convolution neural networks (CNNs) can be an alternative with seven orders of magnitude improvement in speed. With 31,200 simulated strongly lensed quasar images, we explore the usage of Vision Transformer (ViT) for simulated strong gravitational lensing for the first time. We show that ViT could reach competitive results compared with CNNs, and is specifically good at some lensing parameters, including the most important mass-related parameters such as the center of lens θ_1 and θ_2, the ellipticities e_1 and e_2, and the radial power-law slope γ'. With this promising preliminary result, we believe the ViT (or attention-based) network architecture can be an important tool for strong lensing science for the next generation of surveys. The open source of our code and data is in https://github.com/kuanweih/strong_lensing_vit_resnet.

1 Introduction

The discovery of the accelerated expansion of the Universe [1,2] and observations of the Cosmic Microwave Background (CMB; e.g., [3]) established the standard cosmological paradigm: the so-called Λ cold dark matter (CDM) model, where Λ represents a constant dark energy density. Intriguingly the recent direct 1.7% H_0 measurements from Type Ia supernovae (SNe), calibrated by the traditional Cepheid distance ladder ($H_0 = 73.2 \pm 1.3$ km s^{-1} Mpc^{-1}; SH0ES collaboration [4]), show a 4.2σ tension with the Planck results ($H_0 = 67.4 \pm 0.5$ km s^{-1} Mpc^{-1}

K.-W. Huang, G.C.-F. Chen and Y.-Y. Lin—Equal contribution

[5]). However, a recent measurement of H_0 from SNe Ia calibrated by the Tip of the Red Giant Branch ($H_0 = 69.8 \pm 0.8(\text{stat}) \pm 1.7(\text{sys})$ km s^{-1} Mpc^{-1}; CCHP collaboration [6]) agrees with both the Planck and SH0ES results. The spread in these results, whether due to systematic effects or not, clearly demonstrates that it is crucial to reveal unknown systematics through different methodology.

Strongly lensed quasar system provides such a technique to constrain H_0 at low redshift that is completely independent of the traditional distance ladder approach (e.g., [7–9]). When a quasar is strongly lensed by a foreground galaxy, its multiple images have light curves that are offset by a well-defined time delay, which depends on the mass profile of the lens and cosmological distances to the galaxy and the quasar [10]. However the bottleneck of using strongly lensed quasar systems is the expensive cost of computational resources and man power. With commonly used Markov chain Monte Carlo (MCMC) procedure, modeling single strongly lensed quasar system requires experienced modelers with a few months effort in order to obtain robust uncertainty estimations and up to years to check the systematics (e.g., [11–16]). This is infeasible as ~ 2600 of such systems with well-measured time delays are expected to be discovered in the upcoming survey with the Large Synoptic Survey Telescope [17,18].

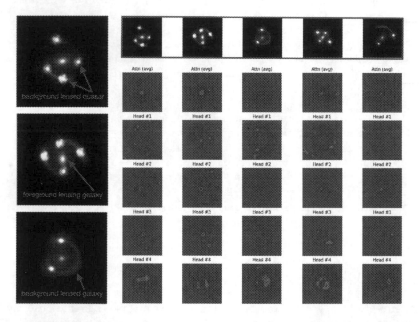

Fig. 1. Left panel: simulated strong lensing imaging with real point spread functions (top two: space-based telescope images; bottom: ground-based adaptive-optics images). Each image contains the lensing galaxy in the middle, the multiple-lensed quasar images, and the lensed background host galaxies (arc). **Right panel**: Vision Transformer attention map: the overall average attentions are focusing on the strong lens system. Each individual head is paying attention to different subjects such as attention heads #2 are focusing the center of lens, heads #1 and #3 are looking into particular lensed quasars, and heads #4 are dealing with the arc.

Deep learning provides a workaround for the time-consuming lens modeling task by directly mapping the underlying relationships between the input lensing images and the corresponding lensing parameters and their uncertainties. Hezaveh et al. [19] and Perreault Levasseur et al. [20] first demonstrated that convolution neural networks (CNNs) can be an alternative to the maximum likelihood procedures with seven orders of magnitude improvement in speed. Since then, other works adopt CNN for strong lensing science related inference [21–30].

In this work, instead of using traditional CNN-based models, we explore the attention-based Vision Transformer (ViT, [31,32]) that has been shown to be more robust compared with CNN-based models [33]. Furthermore, ViT retains more spatial information than ResNet [34] and hence is perfectly suitable for the strong lensing imaging as the quasar configuration and the spatially extended background lensed galaxy provide rich information on the foreground mass distribution (see Fig. 1).

2   Data and Models

In Sect. 2.1, we describe the strong lensing simulation for generating the datasets in this work. In Sect. 2.2, we describe the deep learning models we use to train on the simulated dataset for strong lensing parameters and uncertainty estimations.

2.1   Simulation and Datasets

Simulating strong lensing imaging requires four major components: the mass distribution of the lensing galaxy, the source light distribution, the lens light distribution, and the point spread function (PSF), which convolves images depending on the atmosphere distortion and telescope structures. We use the LENSTRONOMY package [35,36] to generate 31,200 strong lensing images with the corresponding lensing parameters for our imaging multi-regression task. For the mass distribution, we adapt commonly used (e.g., [15,37]) elliptically symmetric power-law distributions [38] to model the dimensionless surface mass density of lens galaxies,

$$\kappa_{\rm pl}(\theta_1, \theta_2) = \frac{3 - \gamma'}{1 + q} \left(\frac{\theta_{\rm E}}{\sqrt{\theta_1^2 + \theta_2^2/q^2}} \right)^{\gamma' - 1}, \tag{1}$$

where γ' is the radial power-law slope ($\gamma' = 2$ corresponding to isothermal), $\theta_{\rm E}$ is the Einstein radius, and q is the axis ratio of the elliptical isodensity contour. The light distribution of the lens galaxy and source galaxy are described by elliptical Sérsic profile,

$$I_{\rm S}(\theta_1, \theta_2) = I_{\rm s} \exp \left[-k \left(\left(\frac{\sqrt{\theta_1^2 + \theta_2^2/q_{\rm L}^2}}{R_{\rm eff}} \right)^{1/n_{\rm sérsic}} - 1 \right) \right], \tag{2}$$

where I_s is the amplitude, k is a constant such that R_eff is the effective radius, q_L is the minor-to-major axis ratio, and $n_\mathrm{sérsic}$ is the Sérsic index [39]. For the PSFs, we use six different PSF structures including three real Hubble space telescope PSFs generated by Tinytim [40] and corrected by the real HST imaging [15], and three adaptive-optics (AO) PSFs reconstructed from ground-based Keck AO imaging [41–43]. Three example images are shown in Fig. 1.

We split the whole simulated dataset of 31,200 images into a training set of 27,000 images, a validation set of 3,000 images, and a test set of 1,200 images. We rescale each image as $3 \times 224 \times 224$ and normalize pixel values in each color channel by the mean [0.485, 0.456, 0.406] and the standard deviation [0.229, 0.224, 0.225] of the datasets. Each image has eight target variables to be predicted in this task: the Einstein radius θ_E, the ellipticities e_1 and e_2, the radial power-law slope γ', the coordinates of mass center θ_1 and θ_2, the effective radius R_eff, and the Sérsic index $n_\mathrm{sérsic}$.

2.2   Models

We use the Vision Transformer (ViT) as the main model for our image multi-regression task of strong lensing parameter estimations. Inspired by the original Transformer models [31] for natural language processing tasks, Google Research proposed the ViT models [32] for computer vision tasks. In this paper, we leverage the base-sized ViT model (ViT-Base), which was pre-trained on the ImageNet-21k dataset and fine-tuned on the ImageNet 2012 dataset [44].

Taking advantage of the transfer learning concept, we start with the pre-trained ViT-Base model downloaded from the module of HUGGINGFACE'S TRANSFORMERS [45], and replace the last layer with a fully connected layer whose number of outputs matches the number of target variables in our regression tasks. The ViT model we use thus has 85,814,036 trainable parameters, patch size of 16, depth of 12, and 12 attention heads.

Alongside the ViT model, we also train a ResNet152 model [46] for the same task as a comparison between ViT and the classic benchmark CNN-based model. We leverage the pre-trained ResNet152 model from the TORCHVISION package [47] and modify the last layer accordingly for our multi-regression purpose.

For regression tasks, the log-likelihood can be written as a Gaussian log-likelihood [48]. Thus for our task of K targets, we use the negative log likelihood as the loss function [20]:

$$
\begin{aligned}
\mathrm{Loss}_n &= -\mathcal{L}\left(\boldsymbol{y}_n, \hat{\boldsymbol{y}}_n, \hat{\boldsymbol{s}}_n\right) \\
&= \frac{1}{2}\left(\sum_{k=1}^{K} e^{-\hat{s}_{n,k}} \left\| y_{n,k} - \hat{y}_{n,k}\right\|^2 + \hat{s}_{n,k} + \ln 2\pi\right)
\end{aligned}
\tag{3}
$$

where $(\boldsymbol{y}_n, \hat{\boldsymbol{y}}_n, \hat{\boldsymbol{s}}_n)$ are the (target, parameter estimation, uncertainty estimation) for the nth sample, and $(y_{n,k}, \hat{y}_{n,k}, \hat{s}_{n,k})$ are the (target, parameter estimation, uncertainty estimation) for the n-th sample of the k-th target. We note that in practice, working with the log-variance $\hat{s}_n = \ln \hat{\sigma}_n^2$ instead of the variance $\hat{\sigma}_n^2$ improves numerical stability and avoids potential division by zero during the training process [49]. Choosing this loss function instead of the commonly used mean squared error results in the uncertainty prediction as well as the parameter prediction, which provides more statistical information than point-estimation-only predictions.

It is worth noting that we apply dropout before every hidden layers for both models with dropout rate of 0.1 to approximate Bayesian networks for the uncertainty estimate, but not for the attention layers in the ViT model. This is to include the "epistemic" uncertainties in neural networks by leaving dropout on when making predictions, together with the "aleatoric" uncertainties described by $\hat{\sigma}_n^2$ to account for intrinsic noise from the data. We refer readers to [20,48] for detailed discussion and derivation of the uncertainties.

Using the training set of 27,000 images, we train our ViT-Base and ResNet152 models with the loss function in Equation (3), the Adam optimizer [50] with 0.001 for the initial learning rate, the batch size of 20. Based on the validation set of 3,000 images, we evaluate the model predictions by the mean squared error across all 8 target variables to determine the best models. We then report the performance of the best ViT and ResNet models according to the test set of 1,200 images in Sect. 3.

3   Results

In this section, we present the performance of the best ViT and ResNet models on the test set of 1,200 images regards of our image multi-regression task of the strong lensing parameter and uncertainty estimation. Following the procedure in [20], for each model, we execute the prediction on the test set for 1000 times with dropout on to catch the epistemic uncertainty of the model. For each parameter prediction $\hat{y}_n$ and uncertainty prediction $\hat{\sigma}_n$ amongst the 1000 predictions, we draw a random number from a Gaussian distribution $N(\hat{y}_n, \hat{\sigma}_n)$ as the prediction of the parameter. Therefore for each test sample, we have 1000 predicted parameters so that we take the mean and standard deviation as the final parameter and uncertainty predictions respectively.

The overall root mean square errors (RMSEs) for the lensing parameter estimation are 0.1232 for our best ViT-Base model and 0.1476 for our best ResNet152 model. The individual RMSEs for each target variable are summarized in Table 1. Our models indicate that except for the Einstein radius θ_E, the attention-based model ViT-Base outperforms the CNN-based model ResNet152 for all the other parameters in this image multi-regression task. Despite a higher RMSE of the Einstein radius for our best ViT than that of our best ResNet, it still reaches the benchmark precision of about 0.03 arcsec [19].

Table 1. Comparison of RMSE of the parameter predictions between ViT and ResNet.

Target	ViT	ResNet
Overall	**0.1232**	0.1476
θ_E [arcsec]	0.0302	**0.0221**
γ'	**0.0789**	0.0816
θ_1 [arcsec]	**0.0033**	0.0165
θ_2 [arcsec]	**0.0036**	0.0169
e_1	**0.0278**	0.0364
e_2	**0.0206**	0.0347
R_{eff} [arcsec]	**0.0241**	0.0487
$n_{\mathrm{sérsic}}$	**0.0790**	0.0959

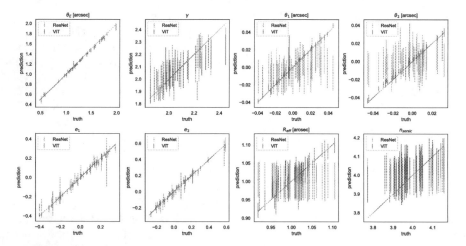

Fig. 2. Comparison of predicted parameters and 1-σ uncertainties to the ground truths for all eight targets between our best Vision Transformer (ViT) model and our best ResNet model for 50 random chosen test samples. For each target, the ground truth and the model prediction are shown on the x-axis and y-axis respectively. The ViT model outperforms the ResNet model for θ_1, θ_2, e_1, e_2, and R_{eff} while maintaining a competitive performance for the other parameters.

Using the prediction of the mean values and the corresponding uncertainties on the strong lensing parameters, we randomly select 50 test samples to illustrate the predictions from our ViT and ResNet models in Fig. 2. Overall, the ViT model outperforms the ResNet model for θ_1, θ_2, e_1, e_2, and R_{eff} while maintaining a competitive performance for the other parameters. We note that both models cannot well capture the features of $n_{\mathrm{sérsic}}$. In Fig. 3, we show the percentage error between the predictions and the ground truths for all 1200 test samples, supporting the statements above.

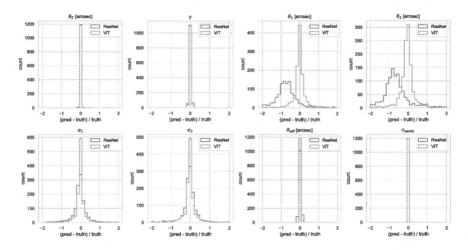

Fig. 3. Comparison of predicted parameters to ground truths for all eight targets between our best Vision Transformer (ViT) model and our best ResNet model. Each panel shows the histogram of the percentage errors between the prediction means and the ground truths of the 1200 test samples for each target. The ViT model outperforms the ResNet model for θ_1, θ_2, e_1, e_2, and R_{eff} while maintaining a competitive performance for the other parameters.

4   Conclusion

Strongly lensed quasar systems provide unique tools to resolve the recent 4-σ tension between the direct measurements of the H_0 and the prediction from the standard cosmological model (ΛCDM model) [4,6,7,42,51]. One of the key requirements is the mass parameter estimations of hundreds of strong lensing systems in order to achieve statistically significant results [52]. While this challenge cannot be achieved by the traditional and time-consuming MCMC method, deep neural network models can be a perfect alternative technique to efficiently achieve this goal. For example, Hezaveh et al. [19] and Perreault Levasseur et al. [20] showed that CNN-based models could be used to estimate the values and the corresponding uncertainties of the parameters given the strong lensing images.

In this work, we explored the recent state-of-the-art ViT as it has the advantage of capturing long-range interaction of pixels compared to CNN-based models. As a supervised multi-regression task, we trained ViT-Base as well as ResNet152 for the parameter and uncertainty estimations, using the dataset of 31,200 strong lensing images.

We show that ViT could reach competitive results compared with CNNs, and is specifically good at some lensing parameters, including the most important mass-related parameters such as the center of lens θ_1 and θ_2, the ellipticities e_1 and e_2, and the radial power-law slope γ'. With this promising preliminary result, we believe the ViT (or attention-based) network architecture can be an important tool for strong lensing science for the next generation of surveys.

Note that the mass distribution of real lensing galaxies are much more complicated than the simple power-law model and hence can potentially affect the H_0 measurement [53–63]. This effect has been illustrated with cosmological hydrodynamic simulations [56,64]. In the future work, we plan to train the neural network to directly learn from realistic hydrodynamic simulations without the need of mass profile assumptions. This cannot be achieved by traditional MCMC method, while neural network is the only way to directly test this long-standing debate about the possible systematics regarding the degeneracy between lensing H_0 results and the mass profile assumptions. We plan to open-source our code and datasets.

References

1. Perlmutter, S., et al.: Measurements of omega and lambda from 42 high-redshift supernovae. Astrophys. J. **517**, 565–586 (1999)
2. Riess, A.G., et al.: Observational evidence from supernovae for an accelerating universe and a cosmological constant. Astron. J. **116**, 1009–1038 (1998)
3. Hinshaw, G., et al.: Nine-year Wilkinson microwave anisotropy probe (WMAP) observations: cosmological parameter results. Astrophys. J. Suppl. Ser. **208**, 19 (2013)
4. Riess, A.G., et al.: Cosmic distances calibrated to 1% precision with Gaia EDR3 parallaxes and Hubble space telescope photometry of 75 Milky Way Cepheids confirm tension with ΛCDM. Astrophys. J. Lett. **908**(1), L6 (2021)
5. Planck Collaboration, N., et al.: Planck 2018 results. VI. Cosmological parameters. Astron. Astrophys. **641**, A6 (2020)
6. Freedman, W.L., et al.: Calibration of the tip of the red giant branch. Astrophys. J. **891**(1), 57 (2020)
7. Wong, K.C., et al.: H0LiCOW – XIII. A 2.4 per cent measurement of H_0 from lensed quasars: 5.3σ tension between early- and late-Universe probes. Monthly Notices R. Astron. Soc. **498**(1), 1420–1439 (2020)
8. Treu, T., Marshall, P.J.: Time delay cosmography. Astron. Astrophys. Rev. **24**, 11 (2016)
9. Suyu, S.H., Chang, T.-C., Courbin, F., Okumura, T.: Cosmological distance indicators. Space Sci. Rev. **214**(5), 91 (2018)
10. Refsdal, S.: On the possibility of determining Hubble's parameter and the masses of galaxies from the gravitational lens effect. Mon. Not. R. Astron. Soc. **128**, 307 (1964)
11. Wong, K.C., et al.: H0LiCOW - IV. Lens mass model of HE 0435–1223 and blind measurement of its time-delay distance for cosmology. Mon. Not. R. Astron. Soc. **465**, 4895–4913 (2017)
12. S. Birrer, et al. H0LiCOW - IX. Cosmographic analysis of the doubly imaged quasar SDSS 1206+4332 and a new measurement of the Hubble constant. Mon. Not. R. Astron. Soc. **484**, 4726–4753 (2019)
13. Rusu, C.E., et al.: H0LiCOW XII. Lens mass model of WFI2033-4723 and blind measurement of its time-delay distance and H_0. Mon. Not. R. Astron. Soc. **498**(1), 1440–1468 (2020)

14. Chen, G.C.-F., et al.: Constraining the microlensing effect on time delays with a new time-delay prediction model in H_0 measurements. Mon. Not. R. Astron. Soc. **481**(1), 1115–1125 (2018)
15. Chen, G.C.F., et al.: SHARP - VIII. J 0924+0219 lens mass distribution and time-delay prediction through adaptive-optics imaging. Mon. Not. R. Astron. Soc. **513**, 2349–2359 (2022)
16. Shajib, A.J., et al.: STRIDES: a 3.9 per cent measurement of the Hubble constant from the strong lens system DES J0408–5354. Mon. Not. R. Astron. Soc. **494**(4), 6072–6102 (2020)
17. LSST Science Collaboration, et al.: LSST Science Book, Version 2.0. arXiv e-prints, page arXiv:0912.0201, December 2009
18. Oguri, M., Marshall, P.J.: Gravitationally lensed quasars and supernovae in future wide-field optical imaging surveys. Mon. Not. R. Astron. Soc. **405**, 2579–2593 (2010)
19. Hezaveh, Y.D., Levasseur, L.P., Marshall, P.J.: Fast automated analysis of strong gravitational lenses with convolutional neural networks. Nature. **548**(7669), 555–557 (2017)
20. Levasseur, L.P., Hezaveh, Y.D., Wechsler, R.H.: Uncertainties in parameters estimated with neural networks: application to strong gravitational lensing. Astrophys. J. Lett. **850**(1), L7 (2017)
21. Brehmer, J., Mishra-Sharma, S., Hermans, J., Louppe, G., Cranmer, K.: Mining for dark matter substructure: inferring sub halo population properties from strong lenses with machine learning. Astrophys. J. **886**(1), 49 (2019)
22. Wagner-Carena, S., et al.: Hierarchical inference with Bayesian neural networks: an application to strong gravitational lensing. Astrophys. J. **909**(2), 187 (2021)
23. Lin, J.Y.-Y., Yu, H., Morningstar, W., Peng, J., Holder, G.: Hunting for dark matter Subhalos in strong gravitational lensing with neural networks. In: 34th Conference on Neural Information Processing Systems, October 2020
24. Park, J.W., et al.: Large-scale gravitational lens modeling with Bayesian neural networks for accurate and precise inference of the Hubble constant. Astrophys. J. **910**(1), 39 (2021)
25. Morgan, R., Nord, B., Birrer, S., Lin, J.Y.-Y., Poh, J.: Deeplenstronomy: a dataset simulation package for strong gravitational lensing. J. Open Source Softw. **6**(58), 2854 (2021)
26. Morningstar, W.R., et al.: Analyzing interferometric observations of strong gravitational lenses with recurrent and convolutional neural networks. arXiv preprint arXiv:1808.00011 (2018)
27. Coogan, A., Karchev, K., Weniger, C.: Targeted likelihood-free inference of dark matter substructure in strongly-lensed galaxies. In 34th Conference on Neural Information Processing Systems, October 2020
28. Ostdiek, B., Rivero, A.D., Dvorkin, C.: Extracting the subhalo mass function from strong lens images with image segmentation. Astrophys. J. **927**(1), 3 (2022)
29. Ostdiek, B., Rivero, A.D., Dvorkin, C.: Image segmentation for analyzing galaxy-galaxy strong lensing systems. Astron. Astrophys. **657**, L14 (2022)
30. Thuruthipilly, H., Zadrozny, A., Pollo, A.: Finding strong gravitational lenses through self-attention. arXiv preprint arXiv:2110.09202 (2021)
31. Vaswani, A., et al.: Attention is all you need. In: Advances in Neural Information Processing Systems, vol. 30 (2017)
32. Dosovitskiy, A., et al.: An image is worth 16x16 words: Transformers for image recognition at scale. In: 9th International Conference on Learning Representations, ICLR 2021, Virtual Event, Austria, 3–7 May 2021. OpenReview.net, 2021

33. Paul, S., Chen, P.-Y.: Vision transformers are robust learners. In: AAAI (2022)
34. Raghu, M., Unterthiner, T., Kornblith, S., Zhang, C., Dosovitskiy, A.: Do vision transformers see like convolutional neural networks? In: Beygelzimer, A., Dauphin, Y., Liang, P., Wortman Vaughan, J. (eds.) Advances in Neural Information Processing Systems (2021)
35. Birrer, S., Amara, A.: lenstronomy: multi-purpose gravitational lens modelling software package. Phys. Dark Univ. **22**, 189–201 (2018)
36. Birrer, S., et al.: Lenstronomy ii: a gravitational lensing software ecosystem. J. Open Sourc. Softw. 6(62), 3283 (2021)
37. Suyu, S.H., et al.: Two accurate time-delay distances from strong lensing: implications for cosmology. Astrophys. J. **766**, 70 (2013)
38. Barkana, R.: Fast calculation of a family of elliptical mass gravitational lens models. Astrophys. J. **502**, 531 (1998)
39. Sérsic, J.L.: Atlas de galaxias Australes. Observatorio Astronomico, Cordoba, Argentina (1968)
40. Krist, J.E., Hook, R.N.: NICMOS PSF variations and tiny Tim simulations. In: Casertano, S., Jedrzejewski, R., Keyes, T., Stevens, M. (eds.) The 1997 HST Calibration Workshop with a New Generation of Instruments, p. 192, January 1997
41. Chen, G.C.-F., et al.: SHARP - III. First use of adaptive-optics imaging to constrain cosmology with gravitational lens time delays. Mon. Not. R. Astron. Soc. **462**, 3457–3475 (2016)
42. Chen, G.C.-F., et al.: A SHARP view of H0LiCOW: H_0 from three time-delay gravitational lens systems with adaptive optics imaging. Mon. Not. R. Astron. Soc. **490**(2), 1743–1773 (2019)
43. Geoff C.-F. Chen, Treu, T., Fassnacht, C.D., Ragland, S., Schmidt, T., Suyu, S.H.: Point spread function reconstruction of adaptive-optics imaging: meeting the astrometric requirements for time-delay cosmography. Mon. Not. R. Astron. Soc. **508**(1), 755–761 (2021)
44. Deng, J., Dong, W., Socher, R., Li, L.-J., Li, K., Li, F.-F.: Imagenet: a large-scale hierarchical image database. In: 2009 IEEE Conference on Computer Vision and Pattern Recognition, pp. 248–255 (2009)
45. Wolf, T., et al.: Transformers: state-of-the-art natural language processing. In: Proceedings of the 2020 Conference on Empirical Methods in Natural Language Processing: System Demonstrations, pp. 38–45. Association for Computational Linguistics, October 2020
46. He, K., Zhang, X., Ren, S., Sun, J.: Deep Residual Learning for Image Recognition. arXiv e-prints, arXiv:1512.03385, December 2015
47. Paszke, A., et al.: Pytorch: an imperative style, high-performance deep learning library. In: Wallach, H., Larochelle, H., Beygelzimer, A., d'Alché-Buc, F., Fox, E., Garnett, R., (eds.), Advances in Neural Information Processing Systems, vol. 32, pp. 8024–8035. Curran Associates Inc (2019)
48. Gal, Y., Ghahramani, Z.: Dropout as a Bayesian approximation: representing model uncertainty in deep learning. In: Balcan, M.F., Weinberger, K.Q. (eds.), Proceedings of The 33rd International Conference on Machine Learning, volume 48 of Proceedings of Machine Learning Research, pp. 1050–1059. PMLR, New York, New York, USA, 20–22 June 2016
49. Kendall, A., Gal, Y.: What Uncertainties Do We Need in Bayesian Deep Learning for Computer Vision? arXiv e-prints. arXiv:1703.04977, March 2017
50. Kingma, D.P., Ba, J.: Adam: A Method for Stochastic Optimization. arXiv e-prints. arXiv:1412.6980, December 2014

51. Abdalla, E., et al.: Cosmology intertwined: a review of the particle physics, astrophysics, and cosmology associated with the cosmological tensions and anomalies. J. High Energy Astrophys. **34**, 49–211 (2022)

52. Suyu, S.H., et al.: The Hubble constant and new discoveries in cosmology. ArXiv e-prints. arxiv:1202.4459, February 2012

53. Falco, E.E., Gorenstein, M.V., Shapiro, I.I.: On model-dependent bounds on H(0) from gravitational images Application of Q0957 + 561A.B. Astrophys. J. Lett. **289**, L1–L4 (1985)

54. Gorenstein, M.V., Falco, E.E., Shapiro, I.I.: Degeneracies in parameter estimates for models of gravitational lens systems. Astrophys. J. **327**, 693 (1988)

55. Schneider, P., Sluse, D.: Mass-sheet degeneracy, power-law models and external convergence: impact on the determination of the Hubble constant from gravitational lensing. Astron. Astrophys. **559**, A37 (2013)

56. Xu, D., et al.: Lens galaxies in the Illustris simulation: power-law models and the bias of the Hubble constant from time delays. Mon. Not. R. Astron. Soc. **456**, 739–755 (2016)

57. Gomer, M., Williams, L.L.R.: Galaxy-lens determination of H_0: constraining density slope in the context of the mass sheet degeneracy. J. Cosmol. Astropart. Phys. **2020**(11), 045 (2020)

58. Kochanek, C.S.: Over constrained gravitational lens models and the Hubble constant. Mon. Not. R. Astron. Soc. **493**(2), 1725–1735 (2020)

59. Blum, K., Castorina, E., Simonović, M.: Could quasar lensing time delays hint to a core component in Halos, instead of H_0 tension? Astrophys. J. Lett. **892**(2), L27 (2020)

60. Millon, M., et al.: TDCOSMO. I. An exploration of systematic uncertainties in the inference of H_0 from time-delay cosmography. Astron. Astrophys. **639**, A101 (2020)

61. Ding, X., et al.: Time delay lens modelling challenge. Mon. Not. R. Astron. Soc. **503**(1), 1096–1123 (2021)

62. Birrer, S., et al.: TDCOSMO. IV. Hierarchical time-delay cosmography – joint inference of the Hubble constant and galaxy density profiles. Astron. Astrophys. **643**, A165 (2020)

63. Chen, G.C.-F., Fassnacht, C.D., Suyu, S.H., Yıldırım, A., Komatsu, E., Bernal, J.L.: TDCOSMO. VI. Distance measurements in time-delay cosmography under the mass-sheet transformation. Astron. Astrophys. **652**, A7 (2021)

64. Tagore, A.S., et al.: Reducing biases on H_0 measurements using strong lensing and galaxy dynamics: results from the EAGLE simulation. Mon. Not. R. Astron. Soc. **474**(3), 3403–3422 (2018)

End-to-end Neural Estimation of Spacecraft Pose with Intermediate Detection of Keypoints

Antoine Legrand[1,2,3]($\boxtimes$) (ID), Renaud Detry[2] (ID),
and Christophe De Vleeschouwer[1] (ID)

[1] UniversitÉ Catholique de Louvain, Louvain-la-Neuve, Belgium
{antoine.legrand,christophe.devleeschouwer}@uclouvain.be
[2] Katholieke Universiteit Leuven, Leuven, Belgium
renaud.detry@kuleuven.be
[3] Aerospacelab, Mont-Saint-Guibert, Belgium

Abstract. State-of-the-art methods for estimating the pose of space-crafts in Earth-orbit images rely on a convolutional neural network either to directly regress the spacecraft's 6D pose parameters, or to localize pre-defined keypoints that are then used to compute pose through a Perspective-n-Point solver. We study an alternative solution that uses a convolutional network to predict keypoint locations, which are in turn used by a second network to infer the spacecraft's 6D pose. This formulation retains the performance advantages of keypoint-based methods, while affording end-to-end training and faster processing. Our paper is the first to evaluate the applicability of such a method to the space domain. On the SPEED dataset, our approach achieves a mean rotation error of 4.69° and a mean translation error of 1.59% with a throughput of 31 fps. We show that computational complexity can be reduced at the cost of a minor loss in accuracy.

Keywords: Spacecraft pose estimation · End-to-end · Keypoints

1 Introduction

Space agencies and the private sector are showing a growing interest and demand for missions that involve proximity operations, such as on-orbit servicing (repairing, refuelling, inspection) or space debris mitigation. While some of those missions conduct proximity operations via tele-operation, the consensus is that autonomous operations are safer and cheaper, which in turn motivates the development of guidance systems that allow a spacecraft to navigate by itself at close range of its target. A key component of this capability is to estimate on-board the 6D pose, i.e. position and orientation, of the target spacecraft.

In terrestrial applications, 6D pose estimation is often achieved with the help of a Lidar or depth camera. Unfortunately, the cost, mass and power requirements of space-grade Lidars are obstacles to their integration in an orbital probe.

© The Author(s), under exclusive license to Springer Nature Switzerland AG 2023
L. Karlinsky et al. (Eds.): ECCV 2022 Workshops, LNCS 13801, pp. 154–169, 2023.
https://doi.org/10.1007/978-3-031-25056-9_11

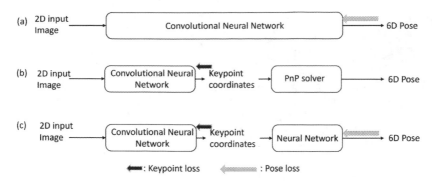

Fig. 1. Overview of the existing methods to estimate a spacecraft's pose. **(a)** Direct methods use a convolutional neural network to directly predict the target 6D pose. **(b)** Keypoints-based methods exploit a convolutional neural network to predict pre-defined keypoint coordinates which are then used to recover the pose through a Perspective-n-Point (PnP) solver. **(c)** Our solution combines both methods as it relies on a first, convolutional, neural network to extract the coordinates of pre-defined keypoints and a neural network to predict pose from those coordinates. The solution is therefore both keypoints-based and end-to-end trainable.

As a result, there is a strong interest in solutions that rely solely on cameras, and in particular on a single monocular camera.

To deploy monocular pose estimation on orbit, multiple challenges must be overcome. Orbital lighting conditions are very different from those encountered on Earth: the lack of atmospheric diffusion causes extreme contrast between exposed and shadowed surfaces. This problem is further exacerbated by specularity: the absence of dust contamination maintains the specularity of clean metal surfaces. In addition, spacecraft design often includes near symmetrical shapes, which requires us to resolve symmetry ambiguities from relatively subtle features. Finally, pose estimation must be carried out using the limited resources offered by space-grade hardware [5, 8, 11].

Traditional pose estimators have struggled to overcome the challenges listed above [30]. Recently, as for terrestrial applications, significant improvements have been brought by the use of deep convolutional neural networks [28]. In space robotics, CNN pose estimators generally adopt one of the two following strategies. The first strategy, depicted in Fig. 1(a), trains a model that directly regresses the 6D pose of the target from a camera image [22, 28]. The mapping from image data to 6D poses is subject to no constraints other than the model's loss function, and the inductive bias of classical CNN layers. By contrast, the second type of model, depicted in Fig. 1(b) uses CNNs to predict image keypoints, often set in pre-defined locations on the object [2, 3, 9, 16, 21], and adopts an iterative solver such as RANSAC-PnP to turn the keypoints into pose parameters. Despite its attractive end-to-end form, the direct-regression approach (a) generally under-performs compared to the second, keypoint-based, solution (b) [17].

In this paper, we consider a third strategy, depicted in Fig. 1(c). Here, a neural network replaces the iterative solver. This network estimates the target's

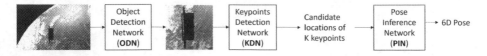

Fig. 2. Framework overview. The high-resolution input image is processed by an Object Detection Network to detect a Region-Of-Interest. The region is fed into a Keypoint Detection Network that regresses multiple candidate locations of K pre-defined keypoints. Finally, a Pose Inference Network computes the spacecraft 6D pose from those candidate locations.

pose from the keypoints predicted by the upstream CNN. This approach enables the end-to-end training of both networks from a loss of direct interest for the spacecraft pose estimation task instead of a surrogate loss. Furthermore, as the iterative solving required by RANSAC-PnP is replaced by an inference through a network that can be parallelized, the proposed method is faster than its PnP-based counterpart. A similar strategy has been tested successfully on terrestrial images of natural objects in [14]. Our work is the first to explore its applicability to images of satellites and spacecrafts orbiting the Earth. We study the relationship between model complexity and pose accuracy, and we validate our work on the standard SPEED [29] dataset.

The rest of the paper is organized as follows. Section 2 presents our method. Section 3 introduces our validation set-up, and presents the results of our experiments. Section 4 concludes.

2 Method

Figure 2 depicts our spacecraft pose estimation framework. The input image goes through an Object Detection Network (ODN) that outputs a square region forming a minimal bounding box around the object. Then, this region is processed by a Keypoint Detection Network (KDN) that predicts multiple candidate pixel coordinates for K pre-defined keypoints. Finally, we map keypoint coordinates to the spacecraft's 6D pose with a neural network that we refer to as the Pose Inference Network (PIN).

This section discusses the KDN (Sect. 2.1) and PIN (Sect. 2.2), as well as the approach we followed to train these two networks jointly (Sect. 2.3). We assume that object detection is carried out with one of the many solutions previously discussed in the literature [1,3,9,21,23] and discuss it no further. In Sect. 3, we limit the scope of our evaluation to the KDN and PIN, and use ground truth bounding boxes.

2.1 Keypoint Detection Network

The Keypoint Detection Network used in this paper was originally proposed by Hu *et al.* [15] to predict the locations of K pre-defined keypoints in natural and terrestrial images. As depicted in Fig. 3, this network consists of a backbone and two heads. The first head performs a segmentation task that separates the

Segmentation head

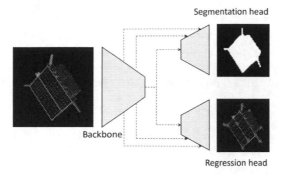

Backbone

Regression head

Fig. 3. Architecture of the keypoint detection network [15]. It is made of a backbone and two decoding heads. The first head performs a segmentation of the spacecraft while the second head regresses candidate locations of some pre-defined keypoints. Dashed lines represent skip connections [25].

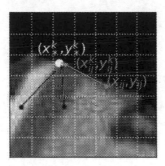

Fig. 4. Illustration of some offsets predicted by the KDN. (x_*^k, y_*^k) are the ground-truth coordinates of the keypoint k. (x_{ij}, y_{ij}) are the coordinates of the center of a patch. This patch predicts an offset (x_{ij}^k, y_{ij}^k) pointing toward keypoint k.

spacecraft from the background (empty space or Earth's disk) while the second head predicts one composite field, f, for each 3D keypoint. Each cell of these composite fields encodes a 2D vector that points to the location of the associated keypoint, and a parameter that models the confidence associated to this offset.

The width/height of the maps produced by either head of the network are an eighth of the width/height of the input image. The segmentation stream outputs a map that predicts, for each cell, whether the corresponding 8×8 patch in the input image belongs to the foreground or the background. The fields $f \in R^{K \times 3 \times W \times H}$ regressed by the second head can be represented over the output map as:

$$f_{ij}^k = [x_{ij}^k, y_{ij}^k, c_{ij}^k] \tag{1}$$

where i and j index the output feature cell over the feature map height H and width W respectively, and $1 \leq k \leq K$ denotes the keypoint index. As depicted in Fig. 4, x and y denotes the coordinates of the vector pointing from the center of the cell to the keypoint location while c is the confidence of cell in its prediction.

In summary, the KDN fulfills three tasks: (a) Segmentation of the spacecraft, (b) Offsets prediction on predefined keypoint locations, (c) Confidence estimation on the accuracy of these predictions. Following Hu *et al.* [15], we model these three tasks with three loss functions respectively denoted by $\mathcal{L}_{seg}$, $\mathcal{L}_{kpts}$ and $\mathcal{L}_{conf}$, and we train the KDN with a loss written as

$$\mathcal{L}_{KDN} = \beta_{seg}\mathcal{L}_{seg} + \beta_{kpts}\mathcal{L}_{kpts} + \beta_{conf}\mathcal{L}_{conf}, \tag{2}$$

where β_{seg}, β_{kpts} and β_{conf} are numerical parameters that weight the contribution of each task in the KDN.

Fig. 5. Computation of pseudo ground-truth segmentation masks. **left**: Keypoint coordinates may be computed by projecting the spacecraft model on the image plane according to its relative pose. **middle**: Body Edges are directly computed from the body keypoints while antennas may be approximated as a straight line segment between their corresponding keypoint and their basis on the spacecraft body. **right**: The approximate segmentation mask can be recovered by filling in the body parts.

We define the segmentation loss $\mathcal{L}_{seg}$ as the Binary Cross Entropy between the predicted segmentation map and the ground-truth segmentation mask [15]. As SPEED [29] does not contain segmentation masks, we define approximate segmentation masks from the known keypoint coordinates as depicted in Fig. 5.

The keypoint loss $\mathcal{L}_{kpts}$ is the term associated to the mean error between the predicted keypoint coordinates and their ground truth, computed only in offset map locations that belong to the foreground, $\mathcal{M}$. Let Δ_{ij}^k denote the prediction error of the kth keypoint from cell (i,j)'s location, with

$$\Delta_{ij}^k = \sqrt{(x_{ij} + x_{ij}^k - x_*^k)^2 + (y_{ij} + y_i^k - y_*^k)^2}, \tag{3}$$

where (x_{ij}, y_{ij}) are coordinates of the center of the patch of the input image that corresponds to cell (i,j) in the output map, and (x_*^k, y_*^k) are the coordinates of keypoint k in the input image. We define the keypoint loss as the sum of all prediction errors [15], with

$$\mathcal{L}_{kpts} = \frac{1}{K}\frac{1}{|\mathcal{M}|}\sum_{(i,j)\in\mathcal{M}}\sum_{k=1}^{K}\Delta_{ij}^k. \tag{4}$$

Finally, we design the confidence loss $\mathcal{L}_{conf}$ to control the training of the confidence map prediction. As in previous work [15], this is done by adopting a

MSE loss to ensure that the confidence map fits a decreasing exponential of the offset approximation error. Formally,

$$\mathcal{L}_{conf} = \frac{1}{K} \frac{1}{|\mathcal{M}|} \sum_{(i,j)\in\mathcal{M}} \sum_{k=1}^{K} (c_{ij}^k - e^{-\tau \Delta_{i,j}^k})^2. \tag{5}$$

This network was evaluated on SPEED [10], in conjunction with a RANSAC-PnP pose solver, during SPEC2019 [17] where it achieved the second best performance. Unlike the SPEC2019 winner [3], the network is inherently resilient to occlusions because it can predict the location of keypoints hidden or located outside the image, via the predictions made by patches that are in the frame. This relaxes significantly the constrains on the rest of the system and therefore motivates its selection as a baseline for this paper.

2.2   Pose Inference Network

This section presents how the 2D keypoint location candidates are turned into a 3D pose, using a neural network trained to convert 2D cues into 3D pose. This gives the opportunity to make the pose prediction system end-to-end, but requires a specific training for each target satellite and each camera intrinsic parameters.

The Keypoint Detection Network computes $W \times H$ predictions of the location of each keypoint. Among those predictions, those made by background cells are highly uncertain. Hence, the second network only processes the predictions made by foreground cells. Furthermore, since the input resolution of the second network is fixed, we sample m predictions per keypoint out of the foreground predictions.

Because each prediction associates a location on the 2D image with a 3D reference keypoint, those predictions are referred as 2D-3D correspondences. The 2D locations are predicted by the Keypoint Detection Network while the 3D keypoints are learned by the Pose Inference Network and implicitly encoded in its weights. In the text below, we refer to the set of correspondences associated to a single keypoint as a *cluster*. This term alludes to the typical clustered nature of predictions around the keypoint's ground truth position.

To infer the spacecraft pose from those clusters, we use the network proposed by Hu *et al.* [14], as depicted in Fig. 6, where each prediction, under the form of a 4 dimensional vector containing the patch center coordinates and the predicted offsets ($[x_{ij}, y_{ij}, \hat{x}_{ij}^k, \hat{y}_{ij}^k]$), is supplied as input to a MLP unit, in charge of extracting a high-dimensional representation of the prediction. The representations corresponding to a same keypoint are then pooled to obtain a representation of the whole cluster. The spacecraft pose is computed through a MLP fed by the representations of all clusters.

In order to train the network, we make use of the 3D reconstruction loss [14,18,32] that captures the 3D-error made on all the keypoints between the estimated spacecraft pose and the ground-truth:

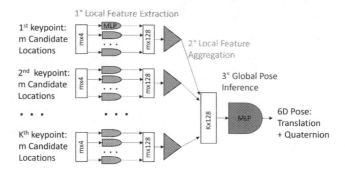

Fig. 6. Architecture of the Pose Inference Network [14]. Local features are extracted from every candidate location through a MLP which weights are shared across all candidates and all keypoints. Those local features are then aggregated per keypoint into a single representation. Finally, the final pose is inferred through a MLP fed by those keypoint representations.

$$\mathcal{L}_{PIN} = \frac{1}{K} \sum_{i=1}^{K} ||(\hat{\mathbf{R}}\mathbf{p}_i + \hat{\mathbf{t}}) - (\mathbf{R}\mathbf{p}_i + \mathbf{t})||_2, \tag{6}$$

where $\hat{\mathbf{R}}$ and $\mathbf{R}$ are the estimated and true rotation matrices, $\hat{\mathbf{t}}$ and $\mathbf{t}$ are the estimated and true translation vectors, $\mathbf{p}_i$ is the 3D coordinates of the ith keypoint in the spacecraft 3D model and K is the number of keypoints.

2.3   Joint Training of KDN and PIN

Training a concatenation of the KDN and PIN with the PIN loss only is unlikely to converge, because the output of the PIN can only make sense if its input consists of keypoint predictions clustered around their ground truth. We therefore train the model as follows: We first train the KDN alone, using the combined loss described in Sect. 2.1. Next, we train a concatenation of the KDN and PIN with all KDN weights frozen, using the 3D reconstruction loss described in Sect. 2.2. Finally, we unfreeze the KDN weights and fine-tune both networks jointly. Since both aim at complementary but different goals, the total loss must represent both of them. We opt for a weighted sum of $\mathcal{L}_{KDN}$ and $\mathcal{L}_{PIN}$:

$$\mathcal{L}_{Tot} = \mathcal{L}_{KDN} + \beta_{pose}\mathcal{L}_{PIN} \tag{7}$$

3   Experiments

This section considers the on-orbit pose estimation task. Section 3.1 discusses the dataset and implementation details considered in the following. Section 3.2 assesses the proposed framework performance on the SPEED [29] dataset and compares its performance with previous works. Section 3.3 studies how complexity reduction affects the pose estimation accuracy. Finally, Sect. 3.4 presents some ablation studies of the Keypoint Detection Network and the Pose Inference Network.

3.1   Dataset and Implementation Details

In this section, we evaluate our framework's performance on SPEED [29] and compare it to previous works. SPEED was used in the Satellite Pose Estimation Challenge (SPEC) [17] that was organized in 2019 by the Advanced Concepts Team (ACT) of the European Space Agency (ESA) and the Space Rendezvous Laboratory (SLAB) of Stanford University. It consists of 15,000 synthetic images and 305 real images. An OpenGL-based rendering pipeline was used to produce the 1920×1200, grayscale, synthetic images depicting the target spacecraft at distances between 3 and 40.5 m. On half of them, random Earth images were added. The synthetic images were post-processed with Gaussian blur and Gaussian noise. The dataset is split into 12,000 images for training and 3,000 for testing. For each training image, the ground-truth pose label is provided under the from of a translation vector and a unit quaternion that represents the relative rotation between the camera and the target spacecraft [17]. The real images were produced in the Testbed for Rendezvous and Optical Navigation (TRON) facility of SLAB [20].

80% of the training set was used to train the model while the model accuracy was evaluated on the remaining 20%. Training was carried on a NVIDIA Tesla A100 while inference was performed on a NVIDIA GeForce RTX 2080 Ti to provide a fair comparison with previous works.

Two assumptions are adopted through the whole section.

- We have at our disposal a perfect Object Detection Network to crop the input image from its initial resolution of 1920×1200 pixels to the KDN input resolution, i.e. 608×608 pixels. In practice, we follow the process used in [21] where the crop is taken as the smallest square containing all the keypoints enlarged by 20% in inference or by a random percentage (from 0 up to 50%) during training. In addition, both horizontal and vertical shifts of at most 20% of the crop size are applied during training.
- A wireframe model of the spacecraft is provided. Here, as SPEED does not contain the spacecraft model, we used the one recovered by Chen *et al.* [3]. In this model, the keypoints are defined as the 8 corners of the spacecraft body combined with the top of the 3 antennas as depicted in Fig. 5 (a).

The KDN backbone and decoding streams were pre-trained on Linemod [12, 15] and then trained for 200 epochs using SGD with a momentum of 0.9 and a weight decay of $1e^{-4}$. After a grid search, the initial learning rate was set to $5e^{-3}$ and divided by 10 at [50, 75, 90] % of the training. We used brightness, contrast and noise data augmentation as explained in Sect. 3.4. After a grid search, the loss hyperparameters were fixed to $\beta_{seg} = 1$, $\beta_{kpts} = 4$, $\beta_{conf} = 1$, $\tau = 5$ and $\beta_{pose} = 1$. Regarding the strategy used to select the $K \times m$ correspondences

Table 1. Our framework performance compared to previous works. Our two-network solution achieves slightly better performance than the solution combining the first network with a PnP strategy, at a higher throughput. It is also close to, but not yet on par with, state-of-the art solutions. Values from [4, 15, 16, 21] are copied from their original paper, or from [16] (indicated by the 1 in superscript).

Metric		Ours		Previous works			
		DarkNet-PIN	DarkNet-PnP	Segdriven [15]	KRN [21]	DLR [3]	WDR [16]
Complexity [M]	↓	72.1	71.2	89.2[1]	5.64	176.2[1]	51.5
Throughput [fps]	↑	31	23	12[1]	∼70	0.7[1,*]	18–35*
$score_{ESA}$ [/]	↓	0.098	0.112	0.02 [17]	0.073	0.012[1]	0.016–0.018
mean $E_{T,N}$ [%]	↓	1.589	2.561	–	1.9	–	–
median $E_{T,N}$ [%]	↓	1.048	0.820	–	–	–	–
mean E_T [m]	↓	0.201	0.267	–	0.211	0.0359	–
median E_T [m]	↓	0.089	0.078	–	0.124	0.0147	–
mean E_R [°]	↓	4.687	4.957	–	3.097	0.728	–
median E_R [°]	↓	3.272	1.402	–	2.568	0.521	–

out of the KDN predictions, we observed that selecting the correspondences as the m most confident predictions per keypoint did not provide any improvement compared to randomly sampling them from the predicted foreground mask. We therefore adopted this second, simpler, strategy.

3.2   Comparisons with a PnP-based Solution and Prior Works

Table 1 ranks competing solutions according to three metrics: complexity, frames per second, and pose accuracy. We define complexity as the number of parameters in a model. We opted for this definition because of its generality and hardware-independence, by contrast to floating point operations which depend on software implementation and hardware configuration. In addition to this definition of complexity, we provide an empirical measure of complexity given by the processing rate (FPS) of each model on a GeForce RTX 2080 Ti. The third metric provided in Table 1 reflects pose accuracy. We follow the definition proposed by ESA in SPEC2019, which is defined as a sum of normalized translation and rotation errors [17]:

$$score_{ESA} = \frac{1}{N} \sum_{i=1}^{N} \left[E_{T,N}^{(i)} + E_R^{(i)} \right] \tag{8}$$

The normalized translation error on image i, $E_{T,N}^{(i)}$ is defined by the following equations where $t_*^{(i)}$ and $t^{(i)}$ are the ground-truth and estimated translations, respectively.

$$E_{T,N}^{(i)} = \frac{E_T^{(i)}}{\left| t_*^{(i)} \right|_2} \quad \text{where} \quad E_T^{(i)} = \left| t_*^{(i)} - t^{(i)} \right|_2 \tag{9}$$

For each image i, the rotation error, in radians, between the predicted and ground-truth quaternions is computed as:

$$E_R^{(i)} = 2 \arccos \left(\left| \left\langle q_*^{(i)}, q^{(i)} \right\rangle \right| \right) \tag{10}$$

where $q_*^{(i)}$ and $q^{(i)}$ are the ground-truth and estimated quaternions, respectively.

As highlighted by Table 1, our solution, which combines a Keypoint Detection Network with a Pose Inference Network, achieves a pose estimation accuracy similar to the one obtained when using a more classical RANSAC-PnP solver applied on the keypoints detected by the same KDN. Furthermore, our solution runs 34% faster than the RANSAC-PnP approach. However, even if both solutions lead to a decent accuracy, they remain beyond the state-of-the-art methods. Those preliminary results are promising, and demonstrate the relevance of studying architectures composed of two networks when dealing with space images[1]. Paths for improvements include joint end-to-end training of both networks, but also end-to-end learning of keypoints, for example using an approach similar to D2-Net [6].

3.3   Complexity Reduction

Although SPEC2019 showed that it was possible to accurately estimate the pose of a spacecraft using a conventional camera and neural networks, it also highlighted the fact that those methods are often too complicated to run on space-grade hardware [17]. This section therefore evaluates the complexity-accuracy trade-off of our framework. Because the PIN accounts for only 1.3% of the total size of our model, we limit this section to a discussion of the complexity-accuracy trade-off of the KDN.

Backbone. Figure 7 summarizes the complexity-accuracy trade-off obtained by 6 different backbones (DarkNet-53 [24], ShuffleNet v2 [19], EfficientNet [31], MobileNet v2 [27], MobileNet v3 small [13] and MobileNet v3 large [13]). All

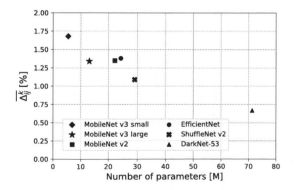

Fig. 7. Complexity/accuracy trade-off for 6 different KDN backbones. The common trend is that larger networks achieve lower errors on keypoint locations.

---

[1] The accuracy of our solution must still be measured on spaceborne images.

of them were pre-trained on ImageNet [26], except DarkNet-53 which was pre-trained on Linemod [12], and then trained on SPEED using the training strategy described in Sect. 3.1. Figure 7 reveals that the large version of MobileNet v3 offers the best trade-off among the selected architectures.

MobileNet v3 small, large and ShuffleNet offer interesting trade-offs. MobileNet v3 large offers a trade that contrast sharply with DarkNet, and it positions itself as a compromise between MobileNet v3 small and ShuffleNet. We therefore chose to use it as an additional baseline.

Segmentation and Regression Heads. To further reduce the KDN complexity, we may simplify the 2 decoding heads as it becomes relevant when using most simple backbones. Each head is made of 3 stages that process the features map at different spatial resolutions. Each stage is made of 3 blocks that implement depth-wise convolutions. To simplify the network, we simply decrease the number of blocks in each stage.

Figure 8 summarizes the complexity-accuracy trade-off that occurs in a KDN made of a large MobileNet v3 with a regression stream made of 3 stages composed of 1,2 or 3 blocks and a segmentation stream made of 3 stages composed of 1,2 or 3 blocks.

We draw the following conclusions:

- Adopting a single block per stage in the segmentation head reduces by 30% the number of total parameters without impacting accuracy.
- Reducing the number of parameters in the regression head by a factor 0.65 inflates the keypoint estimation error, $\overline{\Delta_{ij}^k}$, by a factor 1.3, a trade that many end-users are likely to consider worthy.

As a result, we decided to use a large MobileNet v3 with regression and segmentation streams made of 3 stages composed of 2 and 1 blocks, respectively.

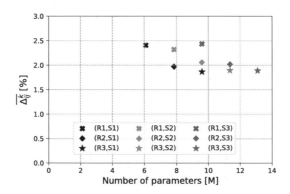

Fig. 8. Complexity/accuracy trade-off for a KDN based on MobileNet v3 large with 9 different heads architectures. Each setting is labelled (RX,SY) where X is the number of blocks per stage in the regression head and Y is the number of blocks per stage in the segmentation one. Reducing the complexity of the regression head increases the error on the keypoints. The complexity of the segmentation head has a negligible impact on the predictions accuracy.

Table 2. Average error on the keypoint locations ($\mathbb{E}\left[\Delta_{ij}^k\right]$) in percent of the image size for different data augmentation techniques

Bright./Contr	Noise	Rotation	Background	$\mathbb{E}\left[\Delta_{ij}^k\right]$ [%]
				1.33
✓				0.62
	✓			1.18
		✓		1.34
			✓	1.66
✓	✓			0.67

Such a network achieves a mean error of 2.05% on the keypoint locations while using only 7.8 millions parameters. Compared to our first baseline, based on DarkNet, which uses 71.2 millions parameters to reach a 0.67% accuracy, it represents a solution that may fit on space-grade hardware while achieving a sufficient accuracy on the keypoint locations. This results in an ESA-score twice larger than the one obtained with DarkNet in Table 1.

3.4   Ablation Study

This section studies the impact of the data augmentation techniques used to train the KDN and evaluates the impact of the PIN architecture on the complexity-accuracy trade-off.

KDN Data Augmentation. Table 2 summarizes the mean error on the keypoint predictions normalized by the image size obtained by the KDN when trained with different data augmentation techniques. Each data augmentation is applied randomly on each image with a 50% probability.

Table 3. Pose estimation accuracy for 2 global pose inference MLPs of different complexities.

MLP layers	mean E_T [%]	mean E_R [°]	score$_{ESA}$ [/]	parameters [M]
$1408 \rightarrow 512 \rightarrow 256 \rightarrow 7$	1.59	4.69	0.098	**0.86**
$1408 \rightarrow 1024 \rightarrow 512 \rightarrow 7$	**1.37**	**3.87**	**0.081**	1.97

Table 4. Pose Estimation Accuracy achieved by the Pose Inference Network when fed with the predictions emitted by the ground-truth foreground cells or with the predictions made by the foreground cells predicted by the KDN.

Candidates sampling strategy	mean E_T [%]	mean E_R [°]	score$_{ESA}$ [/]
PIN fed with foreground ground-truth	2.30	6.60	0.138
PIN fed with predicted mask	**1.59**	**4.69**	**0.098**

– **Brightness/Contrast.** We make use of brightness/contrast data augmentation to account for the wide diversity of illumination scenarios encountered in orbit. In practice, we follow the method proposed by Park *et al.* [21]. The contrast is modified by a random factor in between 0.5 and 2 while the brightness is randomly increased by at most 10%. Such a data augmentation provides a reduction of 0.7% on the mean error on the keypoint locations, $\overline{\Delta}_{ij}^k$.
– **Gaussian Noise.** Adding Gaussian noise which variance is randomly chosen below 10% of the image range decreases by 0.14% the average error.
– **Rotation.** Random rotations of $[90,180,270]°$ did not provide any improvements compared to unaugmented training.
– **Background.** We tested a random background data augmentation technique. Using the segmentation mask recovered in 2.1, we paste the masked original image on an image randomly taken from the PASCAL-VOC dataset [7]. Unfortunately, this data augmentation did not provide any improvement, even worse, it decreased the keypoint detection accuracy. The reason of such a decrease probably comes from the raw cut-and-paste strategy used in the data augmentation. In particular, since the background and spacecraft brightness's are not similar, the network may have learned to detect changes of brightness which do not exist on the original image.
– As both **Brightness/Contrast and Noise** data augmentation improve the KDN performance, both were combined, leading to an error reduction of 0.66%.

PIN Architecture. In this section, we discuss some elements of the Pose Inference Network that impact the accuracy of our system.

Despite its appealing performance, the Pose Inference Network is not yet on par with existing solutions. However, as summarized in Table 3, adding more parameters in the global pose inference MLP improves its ability to predict accurate pose labels. We may therefore expect that a smarter architecture may lead to more accurate pose estimates.

Table 4 compares the accuracy of a PIN network fed with candidate locations extracted from either the segmentation map predicted by the KDN or the foreground mask. Using the ground-truth foreground mask leads to an ESA-score increased by 50% compared to the KDN segmentation output. Since the misclassifications happen mainly in some specific parts of the image such as the spacecraft edges, antennas or darker areas, we believe that there is a link between the nature of the different patches and the accuracy of their predictions.

The PIN could therefore be enhanced by associating a keypoint-specific level of confidence to the prediction of the 2D candidates, e.g. to take into account the visibility of each keypoint in the 2D image. Furthermore, the Pose Inference Network could be modified to learn by itself the 3D keypoints according to its ability to accurately predict them.

4 Conclusions

This paper has investigated a spacecraft pose estimation framework adopting a first, convolutional, neural network to predict candidate locations of some pre-defined keypoints, from which a fixed number of candidates are randomly sampled from a predicted foreground mask to feed a second neural network that infers the spacecraft 6D pose.

The method was tested on SPEED [29] where it achieves a mean rotation error of 4.69°, a mean translation error of 1.59% and an ESA-score of 0.098 for a throughput of 31 fps. This is promising in terms of accuracy, although behind the performance of state-of-the-art solutions. Furthermore, our work reveals that the complexity of the network initially recommended in [14] can be significantly reduced to run on space-grade hardware while preserving a reasonable accuracy. Finally, we highlighted that the Pose Inference Network can be improved either by exploring smarter architectures or by feeding it with the confidence associated to the keypoint candidates.

Acknowledgments. Special thanks go to Mikko Viitala and Jonathan Denies for the supervision of this work within Aerospacelab. The research was funded by Aerospacelab and the Walloon Region through the Win4Doc program. Christophe De Vleeschouwer is a Research Director of the Fonds de la Recherche Scientifique - FNRS. Computational resources have been provided by the supercomputing facilities of the Université catholique de Louvain (CISM/UCL) and the Consortium des Équipements de Calcul Intensif en Fédération Wallonie Bruxelles (CÉCI) funded by the Fond de la Recherche Scientifique de Belgique (F.R.S.-FNRS) under convention 2.5020.11 and by the Walloon Region.

References

1. Black, K., Shankar, S., Fonseka, D., Deutsch, J., Dhir, A., Akella, M.R.: Real-time, flight-ready, non-cooperative spacecraft pose estimation using monocular imagery. arXiv preprint arXiv:2101.09553 (2021)
2. Carcagnì, P., Leo, M., Spagnolo, P., Mazzeo, P.L., Distante, C.: A lightweight model for satellite pose estimation. In: Sclaroff, S., Distante, C., Leo, M., Farinella, G.M., Tombari, F. (eds) Image Analysis and Processing. ICIAP 2022. LNCS, vol. 13231, pp. 3–14. Springer, Cham (2022). https://doi.org/10.1007/978-3-031-06427-2_1
3. Chen, B., Cao, J., Parra, A., Chin, T.J.: Satellite pose estimation with deep landmark regression and nonlinear pose refinement. In: Proceedings of the IEEE/CVF International Conference on Computer Vision Workshops (2019)
4. Chen, B., Parra, A., Cao, J., Li, N., Chin, T.J.: End-to-end learnable geometric vision by backpropagating PNP optimization. In: Proceedings of the IEEE/CVF Conference on Computer Vision and Pattern Recognition, pp. 8100–8109 (2020)
5. Cosmas, K., Kenichi, A.: Utilization of FPGA for onboard inference of landmark localization in CNN-based spacecraft pose estimation. Aerospace 7(11), 159 (2020)
6. Dusmanu, M., Rocco, I., Pajdla, T., Pollefeys, M., Sivic, J., Torii, A., Sattler, T.: D2-net: a trainable CNN for joint description and detection of local features. In: Proceedings of the IEEE/CVF Conference on Computer Vision and Pattern Recognition, pp. 8092–8101 (2019)

7. Everingham, M., Van Gool, L., Williams, C.K., Winn, J., Zisserman, A.: The pascal visual object classes (VOC) challenge. Int. J. Comput. Vision **88**(2), 303–338 (2010)
8. Furano, G., et al.: Towards the use of artificial intelligence on the edge in space systems: challenges and opportunities. IEEE Aerosp. Electron. Syst. Mag. **35**(12), 44–56 (2020). https://doi.org/10.1109/MAES.2020.3008468
9. Garcia, A., et al.: LSPNet: a 2d localization-oriented spacecraft pose estimation neural network. In: Proceedings of the IEEE/CVF Conference on Computer Vision and Pattern Recognition, pp. 2048–2056 (2021)
10. Gerard, K.: Segmentation-driven satellite pose estimation. Kelvins Day Presentation (2019). https://indico.esa.int/event/319/attachments/3561/4754/pose_gerard_segmentation.pdf
11. Goodwill, J., et al.: Nasa spacecube edge TPU smallsat card for autonomous operations and onboard science-data analysis. In: Proceedings of the Small Satellite Conference. No. SSC21-VII-08, AIAA (2021)
12. Hinterstoisser, S., Lepetit, V., Ilic, S., Holzer, S., Bradski, G., Konolige, K., Navab, N.: Model based training, detection and pose estimation of texture-less 3D objects in heavily cluttered scenes. In: Lee, K.M., Matsushita, Y., Rehg, J.M., Hu, Z. (eds.) ACCV 2012. LNCS, vol. 7724, pp. 548–562. Springer, Heidelberg (2013). https://doi.org/10.1007/978-3-642-37331-2_42
13. Howard, A., et al.: Searching for mobilenetv3. In: Proceedings of the IEEE/CVF International Conference on Computer Vision, pp. 1314–1324 (2019)
14. Hu, Y., Fua, P., Wang, W., Salzmann, M.: Single-stage 6d object pose estimation. In: Proceedings of the IEEE/CVF Conference on Computer Vision and Pattern Recognition, pp. 2930–2939 (2020)
15. Hu, Y., Hugonot, J., Fua, P., Salzmann, M.: Segmentation-driven 6d object pose estimation. In: Proceedings of the IEEE/CVF Conference on Computer Vision and Pattern Recognition, pp. 3385–3394 (2019)
16. Hu, Y., Speierer, S., Jakob, W., Fua, P., Salzmann, M.: Wide-depth-range 6d object pose estimation in space. In: Proceedings of the IEEE/CVF Conference on Computer Vision and Pattern Recognition, pp. 15870–15879 (2021)
17. Kisantal, M., Sharma, S., Park, T.H., Izzo, D., Märtens, M., D'Amico, S.: Satellite pose estimation challenge: Dataset, competition design, and results. IEEE Trans. Aerosp. Electron. Syst. **56**(5), 4083–4098 (2020)
18. Li, Y., Wang, G., Ji, X., Xiang, Yu., Fox, D.: DeepIM: deep iterative matching for 6D pose estimation. In: Ferrari, V., Hebert, M., Sminchisescu, C., Weiss, Y. (eds.) ECCV 2018. LNCS, vol. 11210, pp. 695–711. Springer, Cham (2018). https://doi.org/10.1007/978-3-030-01231-1_42
19. Ma, N., Zhang, X., Zheng, H.-T., Sun, J.: ShuffleNet V2: practical guidelines for efficient CNN architecture design. In: Ferrari, V., Hebert, M., Sminchisescu, C., Weiss, Y. (eds.) Computer Vision – ECCV 2018. LNCS, vol. 11218, pp. 122–138. Springer, Cham (2018). https://doi.org/10.1007/978-3-030-01264-9_8
20. Park, T.H., Bosse, J., D'Amico, S.: Robotic testbed for rendezvous and optical navigation: Multi-source calibration and machine learning use cases. arXiv preprint arXiv:2108.05529 (2021)
21. Park, T.H., Sharma, S., D'Amico, S.: Towards robust learning-based pose estimation of noncooperative spacecraft. In: 2019 AAS/AIAA Astrodynamics Specialist Conference (2019)
22. Proença, P.F., Gao, Y.: Deep learning for spacecraft pose estimation from photorealistic rendering. In: 2020 IEEE International Conference on Robotics and Automation (ICRA), pp. 6007–6013. IEEE (2020)

23. Rathinam, A., Gao, Y.: On-orbit relative navigation near a known target using monocular vision and convolutional neural networks for pose estimation. In: i-SAIRAS (2020)
24. Redmon, J., Farhadi, A.: Yolov3: An incremental improvement. arXiv preprint arXiv:1804.02767 (2018)
25. Ronneberger, O., Fischer, P., Brox, T.: U-net: convolutional networks for biomedical image segmentation. In: Navab, N., Hornegger, J., Wells, W.M., Frangi, A.F. (eds.) MICCAI 2015. LNCS, vol. 9351, pp. 234–241. Springer, Cham (2015). https://doi.org/10.1007/978-3-319-24574-4_28
26. Russakovsky, O., et al.: Imagenet large scale visual recognition challenge. Int. J. Comput. Vision **115**(3), 211–252 (2015)
27. Sandler, M., Howard, A., Zhu, M., Zhmoginov, A., Chen, L.C.: Mobilenetv 2: inverted residuals and linear bottlenecks. In: Proceedings of the IEEE Conference on Computer Vision and Pattern Recognition, pp. 4510–4520 (2018)
28. Sharma, S., Beierle, C., D'Amico, S.: Pose estimation for non-cooperative spacecraft rendezvous using convolutional neural networks. In: 2018 IEEE Aerospace Conference, pp. 1–12. IEEE (2018)
29. Sharma, S., D'Amico, S.: Pose estimation for non-cooperative rendezvous using neural networks. arXiv preprint arXiv:1906.09868 (2019)
30. Sharma, S., Ventura, J., D'Amico, S.: Robust model-based monocular pose initialization for noncooperative spacecraft rendezvous. J. Spacecr. Rocket. **55**(6), 1414–1429 (2018)
31. Tan, M., Le, Q.: Efficientnet: rethinking model scaling for convolutional neural networks. In: International Conference on Machine Learning, pp. 6105–6114. PMLR (2019)
32. Xiang, Y., Schmidt, T., Narayanan, V., Fox, D.: PoseCNN: A convolutional neural network for 6d object pose estimation in cluttered scenes. arXiv preprint arXiv:1711.00199 (2017)

Improving Contrastive Learning on Visually Homogeneous Mars Rover Images

Isaac Ronald Ward[1,3]($\boxtimes$) , Charles Moore[2], Kai Pak[1] , Jingdao Chen[2], and Edwin Goh[1]

[1] Jet Propulsion Laboratory, California Institute of Technology, Pasadena, CA, USA
{isaac.r.ward,kai.pak,edwin.y.goh}@jpl.nasa.gov
[2] Computer Science and Engineering, Mississippi State University, Starkville, MS, USA
cam1271@msstate.edu, chenjingdao@cse.msstate.edu
[3] Department of Computer Science , University of Southern California, Los Angeles, CA, USA

Abstract. Contrastive learning has recently demonstrated superior performance to supervised learning, despite requiring no training labels. We explore how contrastive learning can be applied to hundreds of thousands of unlabeled Mars terrain images, collected from the Mars rovers Curiosity and Perseverance, and from the Mars Reconnaissance Orbiter. Such methods are appealing since the vast majority of Mars images are unlabeled as manual annotation is labor intensive and requires extensive domain knowledge. Contrastive learning, however, assumes that any given pair of distinct images contain distinct semantic content. This is an issue for Mars image datasets, as any two pairs of Mars images are far more likely to be semantically similar due to the lack of visual diversity on the planet's surface. Making the assumption that pairs of images will be in visual contrast — when they are in fact not — results in pairs that are falsely considered as negatives, impacting training performance. In this study, we propose two approaches to resolve this: 1) an unsupervised deep clustering step on the Mars datasets, which identifies clusters of images containing similar semantic content and corrects false negative errors during training, and 2) a simple approach which mixes data from different domains to increase visual diversity of the total training dataset. Both cases reduce the rate of false negative pairs, thus minimizing the rate in which the model is incorrectly penalized during contrastive training. These modified approaches remain fully unsupervised end-to-end. To evaluate their performance, we add a single linear layer trained to generate class predictions based on these contrastively-learned features and demonstrate increased performance compared to supervised models; observing an improvement in classification accuracy of 3.06% using only 10% of the labeled data.

Keywords: Self-supervised learning · Contrastive learning · Unsupervised learning · Unlabeled images · Multi-task learning · Planetary science · Astrogeology · Space exploration · Representation learning · Robotic perception · Mars rovers

© The Author(s), under exclusive license to Springer Nature Switzerland AG 2023
L. Karlinsky et al. (Eds.): ECCV 2022 Workshops, LNCS 13801, pp. 170–185, 2023.
https://doi.org/10.1007/978-3-031-25056-9_12

1   Introduction

The primary goal for exploring Mars is to collect data pertaining to the planet's geology and climate, identify potential biological markers to find evidence of past or present life, and study the planet in preparation for eventual human exploration [15]. As it is currently infeasible for humans to do this work, autonomous rovers have emerged as the primary means to explore the Martian surface and collect images and other data. The deluge of images produced from these rovers — on the order of hundreds of thousands of images — has provided the opportunity to apply deep learning (DL) based computer vision methods to tackle a variety of science and engineering challenges.

Existing DL-based approaches have tackled the aforementioned challenge using supervised (transfer) learning and typically require thousands of annotated images to achieve reasonable performance [20, 21]. Although a manual labeling approach may seem viable, Mars images contain subtle class differences and fine-grained geological features that require highly specialized scientific knowledge and expertise, meaning that scaling manual efforts to the volume of required data is difficult, if not impractical. Several efforts at training citizen scientists to annotate such datasets have been met with some success [18]. However, these logistically complex efforts can introduce inconsistencies that must be resolved by experts [19] to prevent ambiguities/bias.

Self-supervised learning (SSL) can help to circumvent the need for large-scale labeling efforts, since these techniques do not require labeled data to train. Contrastive learning (CL) — a type of SSL — has demonstrated much success in this domain in the past few years, and continues to gain research momentum thanks to the promise of leveraging unlabeled images to achieve state-of-the-art performance on large-scale vision benchmarks [3]. These techniques generally create psuedo-labels by leveraging the intrinsic properties of images, such as prior knowledge that two views (augmented crops) originating from the same source image must belong to the same semantic class. These pseudo-labels are then used in conjunction with a contrastive loss function to help the model learn representations that attract similar (positive) samples and repel different (negative) samples, and in doing so, identify an image's defining visual features.

The contrastive loss function implicitly assumes that an image and it's views define a unique semantic classes. While this implicit assumption (herein referred to as the *contrastive assumption*) may be practical in diverse vision datasets with a large number of balanced classes, it is problematic for planetary science applications due to the homogeneity of planetary images. This, in combination with small batch sizes, increases the probability that multiple source images within the same batch are semantically similar, which further invalidates this assumption. Mars rover images, for example, are regularly taken immediately after one another, and 90% of their content may be overlapping. This results in the incorrect assignment of pseudo-labels, diminishing performance overall. **In short, any two images from a Mars rover dataset may not necessarily provide the contrast required for effective contrastive learning.**

In this work, we demonstrate how contrastive learning can be used to extract useful features from Mars rover images without the need for labels. We also explore how simple modifications to the contrastive learning process can improve performance when the contrastive assumption is violated, without introducing the need for any human supervision. Our key contributions are thus:

1. An improved method of contrastive training with a cluster-aware approach that improves the contrastive loss formulation.
2. Evidence that mixed-domain datasets can markedly improve the performance of downstream vision tasks by increasing semantic diversity.
3. Demonstrations of how SSL followed by supervised fine-tuning/linear evaluation with limited labels can exceed the performance of published supervised baselines on Mars-related vision tasks.

2   Related Work

Supervised Learning for Mars Images. Wagstaff et al. proposed a fully supervised approach to training AlexNet-based [13] Mars classification models [20,21]. In these works, benchmark datasets were created to validate the classification performance of the trained models: the Mars Science Laboratory (MSL) dataset, and the High Resolution Imaging Science Experiment (HiRISE) dataset. In this work, we use these datasets to benchmark and compare the downstream task performance of our trained feature extractors (more details in Sect. 3). Other supervised approaches have since improved on the initial results reported by Wagstaff et al. by using attention-based models [2].

Semi-supervised Learning for Mars Images. Wang et al. engineered a semi-supervised learning approach tailored for the semantic content of Mars rover images [22]. Their approach ignores problematic (redundant) training samples encountered during contrastive learning by making use of labels. Their proposed multiterm loss function contains both supervised and unsupervised terms, thus creating a semi-supervised approach.

Self-supervised Learning for Mars Images. Panambur et al. extract granular geological terrain information without the use of labels for the purpose of clustering sedimentary textures encountered in 30,000 of Curiosity's Mast camera (Mastcam) images [16]. They modify a neural network architecture that was originally designed for texture classification so that it can support self-supervised training, and use a metric learning objective that leverages triplet loss to generate a learning signal. The K-nearest neighbors (KNN) algorithm is then used on the embeddings to cluster, and thus support the querying of the data. The results of this deep clustering are validated by planetary scientists, and a new taxonomy for geological classification is presented.

Self-supervised Learning for Earth Observations. Learning based approaches for Earth observation predictions have long been of interest to the space community [11,14]. Wang et al. introduced the 'Self-Supervised Learning for Earth Observation, Sentinel-1/2' (SSL4EO-S12) dataset and illustrate how

SSL can be used to achieve comparable or superior performance to fully supervised counterparts [23]. This work uses techniques such as MoCo [6,9] and DINO [1], the former of which considers contrastive learning as a dictionary lookup problem, and builds a dynamic dictionary solution that leverages a queue with a moving-averaged encoder, and the latter of which leverages the properties of vision transformers [7] trained with large batch sizes.

Contrastive Learning for Mars Images. A Simple Framework for Contrastive Learning (SimCLR, proposed in [3] and improved upon in [4]) has been used to train discriminant and performant feature extractors in a self-supervised manner across many domains. In [8], a deep segmentation network is pretrained on unlabeled images using SimCLR and trained further in a supervised manner on a limited set of labeled segmentation data (only 161 images). This approach outperforms fully supervised learning approaches by 2–10%.

Relevant Improvements to Contrastive Learning. A number of approaches to detect and counteract issues relating to violating the contrastive assumption are outlined in [5,12]. Such techniques generate a support set of views for any given image, and use cosine similarities between the support set images and an incoming view to detect if the potential pair will be a false negative. Techniques that use this manner of false negative detection have demonstrably outperformed standard contrastive learning techniques on the ImageNet dataset [17].

3   Datasets

Table 1. Statistics and descriptions of the datasets used in this work.

Dataset name	Description	Images (or train/test/val.)	Labels
Perseverance	Mars terrain images captured from the Perseverance rover across 19 cameras	112,535	No
Curiosity	Mars terrain images captured from the Curiosity rover across 17 cameras	100,000	No
MSL v2.1 [20,21]	Mars terrain images captured from the Curiosity rover, partitioned into 19 classes	5920/300/600	Yes
HiRISE v3.2 [20,21]	Orbital images of Mars captured from the Mars Reconnaissance Orbiter, partitioned into 8 classes	6997/2025/1793	Yes

3.1   Perseverance Rover Images

The primary source of training data in this work are unlabeled images from the Perseverance rover. We downloaded $112,535$ Perseverance rover images, captured

(a) Unlabeled Curiosity images (left), and (b) unlabeled Perseverance images (right)

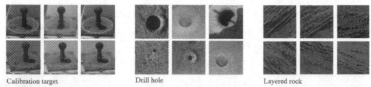

Calibration target   Drill hole   Layered rock

(c) Labeled MSL v2.1 images [21,20], with 3 of the 19 classes shown.

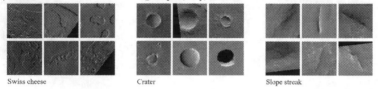

Swiss cheese   Crater   Slope streak

(d) Labeled HiRISE v3.2 images [21,20], with 3 of the 8 classes shown.

Fig. 1. The various datasets used in this work. The two unlabeled datasets (a and b) are entirely unstructured, but can be leveraged by contrastive learning techniques for training feature extractors. The two labeled datasets (c and d) are used for linear evaluation and benchmarking. We note that three of the four datasets are in the domain of Mars rover images (a, b, and c), whereas the HiRISE v3.2 dataset (d) is in the domain of orbital images. Further outlined in Table 1.

during its traversal of the planet Mars from Sols 10–400. These images can be accessed through NASA's Planetary Data System (PDS)[1]. Each image is accompanied by metadata that outlines which of the rover's 19 different cameras was used to take the image, the time of capture, and more. This results in a set of images that vary in resolution, colour profile, and semantic content. Image contents include the Mars surface, landscapes, terrain features, geologic features, the night sky, astronomical targets, calibration targets, the Perseverance rover itself, and more (as seen in Fig. 1).

---

[1] https://pds.nasa.gov/.

3.2   Curiosity Rover Images

We consider two sets of Curiosity rover images in this work; an unlabeled set, and a labeled set. The unlabeled set consists of 100,000 raw rover images, and is similar in semantic content to the Perseverance rover images. This dataset is used for contrastive learning in the same domain as the Perseverance rover images. The labeled dataset (herein referred to as MSL v2.1) was compiled and annotated by Wagstaff et al.'s [20,21] and consists of 6820 images divided into 19 classes of interest.

The MSL v2.1 dataset is further divided by Sol into train, validation, and test subsets (see Table 1). We use MSL v2.1's testing subset to benchmark the performance of the feature extraction networks that we train with contrastive learning, thus permitting the comparison of our results with the fully supervised approaches outlined in Wagstaff et al.'s work. We note that the MSL v2.1 dataset is imbalanced: one of the 19 classes (entitled 'nearby surface') accounts for 34.76% of the images, whereas the 'arm cover' class accounts for only 0.34% of the images.

3.3   Mars Reconnaissance Orbiter Images

We additionally use a labeled dataset of Mars images taken from the Mars Reconnaissance Orbiter's HiRISE camera, again compiled and annotated by Wagstaff et al. and split into train, validation, and test subsets (see Table 1). This dataset (herein referred to as HiRISE v3.2) is again used entirely for benchmarking, but instead focuses on a separate visual domain (orbital images of Mars rather than rover based images of Mars). Similarly to MSL v2.1, the HiRISE v3.2 dataset is imbalanced: one of the 8 classes (entitled 'other') accounts for 81.39% of the images, whereas the 'impact ejecta' class accounts for only 0.68% of the images.

4   Methods

4.1   Contrastive Learning

Our baseline method in this work is based on SimCLR (v2) [3,4], which is a contrastive learning framework that enables deep learning models to learn efficient representations without the need for labeled images. We introduce two approaches to address violations of the contrastive assumption (illustrated in Fig. 2). The first approach modifies the SimCLR framework to use the results from an unsupervised clustering process to identify violations, and the second modification addresses the same issue by altering the distribution of the data that the model ingests.

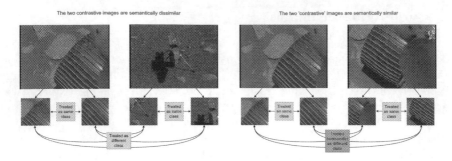

Fig. 2. The contrastive assumption is being held (left), as the two sampled images *are* in visual contrast. The contrastive assumption is being violated (right) as the two sampled images are not in contrast, *but they are assumed to be by the baseline contrastive learning framework*. The violating pairs are thus given an incorrect pseudo-label during training, causing reduced performance. This issue is exacerbated as more semantically similar images are sampled into a training batch.

Whereas traditional supervised learning techniques attach a manually generated annotation or label to an input image, SimCLR operates by automatically annotating pairs of images. Two views from each source image in a batch are taken and augmented. All pairs of views are analysed and labeled as a *positive* pair if they came from the same source image, otherwise they are labeled as a *negative* pair. This work follows SimCLR in generating augmented views of each training image by taking a random crop of the source image and applying the following transformations at random: horizontal flipping, color jittering, and grayscaling [4].

A neural network then completes a forward pass over each of the views in the batch, creating a batch of embedding vectors. For any given positive pair of views, we know that their embeddings should be similar, and for any given negative pair of views, their embeddings should be dissimilar. In practice, cosine similarity is used as the similarity measure for the embedding vectors, and this defines an objective which can be optimised using gradient descent (Eq. 1), thus allowing the network to be trained.

$$\mathcal{L}_{\text{contrastive}} = -\log \frac{\exp\left(\text{sim}\left(z_i, z_j\right)/\tau\right)}{\sum_{k=1}^{M} \mathbb{1}_{[k \neq i]} \exp\left(\text{sim}\left(z_i, z_k\right)/\tau\right)} \tag{1}$$

where z_x refers to the xth view's representation, $\text{sim}(u, v)$ is the cosine similarity of some embedding vectors u and v, B is the batch size, M is the number of views in the batch ($M = 2 \cdot B$), τ is a temperature parameter, and i, j, and k are view indices. Note how identical views ($k = i$) are ignored when calculating the loss.

4.2   Cluster-Aware Contrastive Learning

The assumptions of standard SimCLR fail when two views from different images actually contain the same or highly similar visual content. In this case, they will be labeled incorrectly as a negative pair; they are a *false negative* (see Fig. 2, right). Our first modification to SimCLR addresses this using prior information relating to if two images actually contain similar visual content.

We gather this information using the unsupervised clustering technique outlined in [16], based on a ResNet-18 backbone. We input the unlabeled training dataset, and the output is a partition of this dataset into some manually defined number of clusters K. Clusters are defined by the similarity of their semantic content, with a specific focus on geological texture as a result of the deep texture encoding module that is incorporated into the architecture [25]. Since this clustering approach is unsupervised, the end-to-end contrastive training process still requires no supervision.

During training, we define positive view pairs as views which came from the same source image *or cluster*. This has the effect of turning false negatives into true positives; taking view pairs that would have otherwise been incorrectly labeled during training, and converting them into correctly labeled instances.

$$\mathcal{L}_{\text{cluster-aware}} = -\log \frac{\exp\left(\text{sim}\left(z_i, z_j\right)/\tau\right)}{\sum_{k=1}^{M} \mathbb{1}_{[c_k \neq c_i]} \exp\left(\text{sim}\left(z_i, z_k\right)/\tau\right)} \tag{2}$$

The modified loss function for the cluster-aware contrasive learning method is outlined in Eq. 2, where c_x refers to the cluster index of the xth view in the batch. The key difference being that view pairs are considered negative based on their cluster index, rather than their view index.

4.3   Mixed-Domain Contrastive Learning

We observe that the rate of false negative view pairs is related to the semantic homogeneity of a training dataset, and as such, we hypothesize that increasing the visual variance of the dataset should decrease the rate of false negatives encountered during contrastive learning. We achieve this by simply concatenating and shuffling two visually different datasets and then performing baseline SimCLR contrastive training on the resultant dataset.

However, mixed-domain training may also have negative side-effects in that injecting too many out-of-domain images may ultimately reduce performance on Mars-related vision tasks. In this study, to achieve a reduced rate of false negative view pairs while preserving a sufficient amount of in-domain images, we perform mixed-domain contrastive learning by combining unlabeled Curiosity images and ImageNet images during pretraining.

4.4   Benchmarking Learned Feature Extraction with Linear Evaluation

Both SimCLR and our modified variants take a set of unlabeled training images and train a neural network to extract discriminant features from such images. Naturally, it is our desire to quantify the success of this training paradigm, and the resultant feature extractor. To do so, we apply the contrastively-trained backbone to supervised Mars-related vision tasks, namely MSL v2.1 and HiRISE v3.2, which represent in-domain and out-of-domain challenges respectively (relative to the unlabeled training dataset which is comprised of Mars rover images).

When benchmarking, we precompute the set of features for the benchmark images using the contrastively-trained model, and then use a single dense/linear layer of 128 neurons to learn a mapping from the features to the images' class. We train this evaluation layer using the features we extracted from the MSL v2.1 and HiRISE v3.2 training subsets and their corresponding labels, before benchmarking the performance on the datasets' corresponding testing subsets. Importantly, the parameters of the contrastively-trained model remain untouched in this process; we are only testing how useful the extracted features are when completing downstream Mars-related vision tasks. This form of linear evaluation protocol is consistent with other contrastive-learning works [3,4,6,9,23].

5   Results and Discussion

5.1   Adhering to the Contrastive Assumption Strengthens Performance

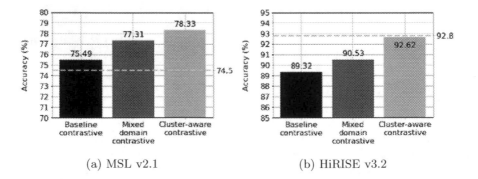

(a) MSL v2.1   (b) HiRISE v3.2

Fig. 3.   Performance improvements with mixed-domain contrastive learning and cluster-aware contrastive learning. With a pretraining process that requires no labels, contrastive pretraining generally outperforms the supervised transfer learning baseline (marked with a dotted line) with a model one third the size.

We began by investigating how addressing violations of the contrastive assumption improve the quality of learned features. We trained a feature extractor using the baseline SimCLR and the two modified SimCLR approaches on the unlabeled Perseverance and Curiosity datasets outlined in Sect. 3, and then compared their performance on the MSL v2.1 classification benchmark, with the results outlined in Fig. 3.

Figure 3 shows that each of the contrastive learning approaches (baseline, mixed-domain, cluster-aware) exceeds Wagstaff et al.'s published baseline results (obtained using supervised transfer learning) on the MSL v2.1 benchmark [20].

We note that other published techniques do outperform our model on the MSL v2.1 classification benchmark, but these approaches incorporate labels into a semi-supervised learning approach (95.86% test accuracy on MSL v2.1) [22], or take a fully supervised approach with attention-based models (81.53% test accuracy on MSL v2.1) [2].

Importantly, no labels were required to train our contrastive feature extractors, whereas Wagstaff et al.'s model required a supervised dataset in addition to 1M+ labeled ImageNet images (for pretraining). Moreover, we note that Wagstaff et al.'s fully supervised approach used a neural network backbone with approximately 60M parameters (AlexNet [13]), whereas our approaches use a backbone with approximately 21M parameters (ResNet-50 [10,24]).

We see that the baseline contrastive approach is outperformed by the mixed-domain approach, which is then outperformed by the cluster-aware approach. A potential explanation for this is that the mixed-domain approach *minimises* contrastive assumption violations, and the cluster-aware approach *corrects* said violations, hence the performance difference.

We note that the performance gains associated with our proposed modified contrastive learning approaches are comparable to those reported in literature using similar approaches designed to mitigate incorrectly labeled view pairs during contrastive training. These works report accuracy increases ranging from approximately 3–4% [5,12].

5.2   Learned Features Generalise to Out-of-domain Tasks

We extended our testing to the out-of-domain HiRISE v3.2 benchmark — a classification task comprised entirely of orbital images of Mars, reporting the results in Fig. 3b. Wagstaff et al.'s fully supervised model was trained on HiRISE data (orbital images), whereas our contrastive models were trained on Perseverance data (rover images) — the HiRISE v3.2 task is thus out-of-domain for our contrastive approaches, but in-domain for the fully supervised technique.

We observe that the fully supervised technique outperforms our contrastive approaches, likely due to the in-domain/out-of-domain training difference. Regardless, the contrastive methods still demonstrate comparable performance to a fully supervised approach with only a linear classification head, thus illustrating the general nature of the learned feature extraction; performance is maintained even under a domain shift and without any fine tuning of the feature extractor's parameters. Put simply: our model has never encountered an orbital image of Mars during training, yet still performs competitively with a supervised model that was trained specifically on Martian orbital images.

As in Sect. 5.1, we observe a performance increase from the baseline contrastive method, to the mixed-domain method, to the cluster-aware method.

5.3   Mixed-Domain Approaches Increase Dataset Variability and Performance

The 'mixed domain contrastive' results in Fig. 3 indicate that increasing the semantic heterogeneity of the training data by mixing datasets from different visual domains increases the performance of the trained feature extractors with respect to downstream classification tasks.

Table 2. Analysis of mixed-domain contrastive learning

Domain(s)/training datasets	FN per batch (%)	MSL Acc. (%)	HiRISE Acc. (%)
100K Curiosity	15.0	75.60	89.70
100K ImageNet	0.08	76.54	91.89
50K Curiosity + 50K ImageNet	0.30	77.31	90.53

Table 2 quantifies this effect; each row represents a different training set (or combination of training sets) that were used during training. The first two rows represent baseline contrastive learning, and the last row represents *mixed-domain* contrastive learning — as the training dataset is a mix of two domains. The *number* of training images remains constant in all cases. More details on this simulation are provided in this work's supplementary file. We expect that performance will be worst when the model is trained on pure datasets that are highly semantically homogeneous (e.g. 100,000 Curiosity images). We also expect the inverse to be true — diverse datasets (e.g. ImageNet) will result in a more performant model, as the contrastive assumption will be violated less often during training.

Moreover, we expect to find a trade off in performance; training on data that is similar to the benchmark data will result in increased performance with respect to that benchmark, but performance will still be hampered due to semantic homogeneity. A mix of two domains, one which shares the domain of the benchmark dataset, and one which provides semantic heterogeneity, should then result in increased performance.

Column two of Table 2 shows the average percentage of false negatives view pairs out of all the view pairs encountered in a given training batch. These results are derived from a simulation which samples multiple batches and checks for FNs using labels. Note that the more diverse dataset (ImageNet) has a far lower false negative view pairs rate than the less diverse Curiosity dataset, and mixing these datasets 50–50 results in a FN rate equal to a 'diversity-weighted' average of the two constituent datasets (ImageNet is far more diverse than the Curiosity images, with 1000 classes and 19 classes respectively).

Column three shows our results on the MSL v2.1 benchmark. We note that the model trained on pure Curiosity data (i.e., the same domain as the benchmark dataset) is the *least* performant model. Training on pure ImageNet increases the performance further, and training on an equal mix of both domains (i.e., mixed-domain contrastive learning) provides the best performance.

In column four, we see equivalent results, but on a benchmark domain which is *not* represented in any training dataset (HiRISE v3.2's orbital images). In this case, the benefit of having in-domain training images (i.e. Curiosity images) is removed; the most semantically heterogenous training dataset (ImageNet) provides the best performance and out-of-domain generalisation. In fact, using any Curiosity images in this case *reduces performance*. This raises a useful conclusion: if no data from the target testing domain is available, use the most semantically heterogeneous dataset available (when using contrastive learning). If data from the target testing domain is available, then using a mix of training data may provide increased performance.

6   Ablation Studies

6.1   10% of the Labels Is Sufficient to Transfer to Other Tasks

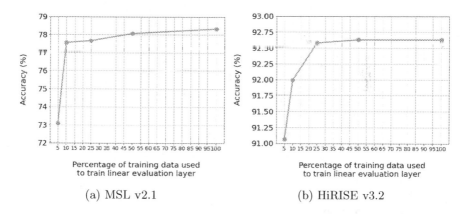

(a) MSL v2.1   (b) HiRISE v3.2

Fig. 4. The task performance as a function of different percentages of training data used during linear evaluation. In all cases, we pretrain on the full unlabeled Perseverance dataset using our proposed cluster-aware contrastive learning, then perform linear evaluation with limited task-specific training data, then benchmark and report the results on the task's testing set. We note that the accuracy return for increasing amounts of training data quickly diminishes.

Expanding on the results outlined in Sect. 6.2, we observe in Fig. 4 that performance remains relatively unchanged even when a fractional amount of the benchmark training data is used. On the MSL v2.1 benchmark, accuracy is within 1% of the best score even when only 10% of the task-specific training data is being used, though a distinct performance drop is noted when this proportion is reduced to 5%. On the HiRISE v3.2 benchmark, accuracy is within 0.5% of the best score even at 25% of the task-specific training data. The relatively higher proportion of training data required on the out-of-domain HiRISE v3.2 classification task may suggest that more training data is required to specialise to an out-of-domain task.

Overall, these results suggest that the majority of the learning has occurred during contrastive training and without labels, necessitating only a small number of task-specific labels to leverage the features extracted by the contrastively-trained neural network. This demonstrates key benefits of contrastive learning; the resulting models are relatively general and can be optimised for specific tasks with only a limited number of training samples (especially when compared to a fully supervised training pipeline).

6.2 Cluster-Aware Contrastive Learning Is Sensitive to the Number of Clusters Chosen

We altered the number of clusters (K) during the unsupervised clustering stage of our cluster-aware contrastive approach and benchmarked the results.

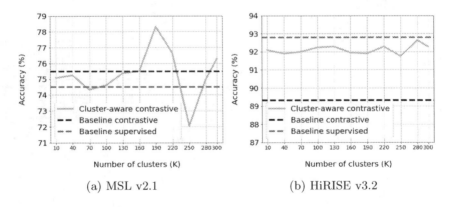

(a) MSL v2.1   (b) HiRISE v3.2

Fig. 5. The performance of the contrastively-trained feature extractors, as measured with respect to an in-domain (left) and out-of-domain (right) downstream task, is dependent on the number of clusters computed prior to training. Too few clusters and images that are dissimilar in content may be assigned to the same cluster, too many clusters and images that are similar in content may be assigned to different clusters — both will cause incorrect pseudo-labels during contrastive training.

We note that the hyperparameter K requires careful optimisation, otherwise the cluster-aware method may not improve (and may decrease) performance. We posit that too few clusters may fail to adequately address violations of the contrastive assumption; images with different semantic content may be placed into the same clusters due to the coarseness of the clustering, thus failing to solve the issue at hand. With too many clusters, the effectiveness of the method may diminish, with only weakly contrasting images being labeled as negatives.

We observe that for the out-of-domain task HiRISE v3.2, the tuning of the hyperparameter K is less important, with any choice of K producing meaningful performance gains when compared to the baseline contrastive case. We hypothesize that our modified contrastive methods only attend to features inherent to the types of images that were encountered during contrastive training, thus resulting in nuances of orbital images being ignored.

7   Conclusion

In this work, we have explored and quantified a common issue native to contrastive learning on semantically homogeneous datasets; violations of the contrastive assumption leading to false negative view pairs. We analysed this issue within the purview of Mars images taken from the rovers Curiosity and Perseverance, and from the Mars Reconnaissance Orbiter, and we proposed two modifications to contrastive learning tailored for this domain.

Our experiments demonstrate how these modifications — cluster-aware and mixed-domain contrastive learning — improve upon baseline contrastive learning by training more discriminant and powerful feature extractors. In each case, we compare our results to a fully supervised baseline, and note that for in-domain classification tasks our methods result in feature extractors that exceed the performance of their fully supervised counterparts, even when using only 10% of the available labels. It is our hope that this work illustrates how the benefits of contrastive learning can be applied to the large scale space datasets — even when the underlying visual data is semantically homogeneous or not sufficiently contrastive — thus increasing the scope for contrastive learning in the domain of space images.

Acknowledgments. This research was carried out at the Jet Propulsion Laboratory, California Institute of Technology, under a contract with the National Aeronautics and Space Administration (80NM0018D0004), and was funded by the Data Science Working Group (DSWG). The authors also acknowledge the Extreme Science and Engineering Discovery Environment (XSEDE) Bridges at Pittsburgh Supercomputing Center for providing GPU resources through allocation TG-CIS220027. U.S. Government sponsorship acknowledged.

References

1. Caron, M., et al.: Emerging properties in self-supervised vision transformers. In: Proceedings of the International Conference on Computer Vision (ICCV) (2021)
2. Chakravarthy, A.S., Roy, R., Ravirathinam, P.:MRSCAtt: a spatio-channel attention-guided network for mars rover image classification. In: Proceedings of the IEEE/CVF Conference on Computer Vision and Pattern Recognition, pp. 1961–1970 (2021)

3. Chen, T., Kornblith, S., Norouzi, M., Hinton, G.: A simple framework for contrastive learning of visual representations. In: International Conference on Machine Learning, pp. 1597–1607. PMLR (2020)
4. Chen, T., Kornblith, S., Swersky, K., Norouzi, M., Hinton, G.: Big self-supervised models are strong semi-supervised learners. arXiv preprint arXiv:2006.10029 (2020)
5. Chen, T.S., Hung, W.C., Tseng, H.Y., Chien, S.Y., Yang, M.H.: Incremental false negative detection for contrastive learning. arXiv preprint arXiv:2106.03719 (2021)
6. Chen, X., Fan, H., Girshick, R., He, K.: Improved baselines with momentum contrastive learning. arXiv preprint arXiv:2003.04297 (2020)
7. Dosovitskiy, A., et al.: An image is worth 16x16 words: transformers for image recognition at scale. arXiv preprint arXiv:2010.11929 (2020)
8. Goh, E., Chen, J., Wilson, B.: Mars terrain segmentation with less labels. arXiv preprint arXiv:2202.00791 (2022)
9. He, K., Fan, H., Wu, Y., Xie, S., Girshick, R.: Momentum contrast for unsupervised visual representation learning. arXiv preprint arXiv:1911.05722 (2019)
10. He, K., Zhang, X., Ren, S., Sun, J.: Deep residual learning for image recognition. In: Proceedings of the IEEE Conference on Computer Vision and Pattern Recognition, pp. 770–778 (2016)
11. Helber, P., Bischke, B., Dengel, A., Borth, D.: Eurosat: A novel dataset and deep learning benchmark for land use and land cover classification. IEEE J. Select. Top. Appl. Earth Obser Remote Sens. 12(7), 2217–2226 (2019)
12. Huynh, T., Kornblith, S., Walter, M.R., Maire, M., Khademi, M.: Boosting contrastive self-supervised learning with false negative cancellation. In: Proceedings of the IEEE/CVF Winter Conference on Applications of Computer Vision, pp. 2785–2795 (2022)
13. Krizhevsky, A., Sutskever, I., Hinton, G.E.: ImageNet classification with deep convolutional neural networks. In: 25th Proceedings of Advances in Neural Information Processing Systems (2012)
14. Kucik, A.S., Meoni, G.: Investigating spiking neural networks for energy-efficient on-board AI applications. a case study in land cover and land use classification. In: Proceedings of the IEEE/CVF Conference on Computer Vision and Pattern Recognition, pp. 2020–2030 (2021)
15. NASA: Mars exploration rover mission goals, http://mars.nasa.gov/mer/mission/science/goals
16. Panambur, T., Chakraborty, D., Meyer, M., Milliken, R., Learned-Miller, E., Parente, M.: Self-supervised learning to guide scientifically relevant categorization of Martian terrain images. arXiv preprint arXiv:2204.09854 (2022)
17. Russakovsky, O., et al.: ImageNet large scale visual recognition challenge. Int. J. Comput. Vision 115(3), 211–252 (2015)
18. Sprinks, J.C., Wardlaw, J., Houghton, R., Bamford, S., Marsh, S.: Mars in motion: an online citizen science platform looking for changes on the surface of mars. In: AAS/Division for Planetary Sciences Meeting Abstracts# 48. vol. 48, pp. 426–401 (2016)
19. Swan, R.M., et al.: Ai4mars: a dataset for terrain-aware autonomous driving on mars. In: Proceedings of the IEEE/CVF Conference on Computer Vision and Pattern Recognition, pp. 1982–1991 (2021)
20. Wagstaff, K., et al.: Mars image content classification: three years of NASA deployment and recent advances. arXiv preprint arXiv:2102.05011 (2021)
21. Wagstaff, K.L., Lu, Y., Stanboli, A., Grimes, K., Gowda, T., Padams, J.: Deep mars: CNN classification of mars imagery for the PDS imaging atlas. In: Thirty-Second AAAI Conference on Artificial Intelligence (2018)

22. Wang, W., Lin, L., Fan, Z., Liu, J.: Semi-supervised learning for mars imagery classification. In: 2021 IEEE International Conference on Image Processing (ICIP), pp. 499–503. IEEE (2021)
23. Wang, Y., Braham, N.A.A.A., Albrecht, C.M., Xiong, Z., Liu, C., Zhu, X.X.: Ssl4eo-s12: a large-scale multimodal multitemporal dataset for self-supervised learning in earth observation (2022)
24. Wightman, R.: Pytorch image models (2019). http://github.com/rwightman/pytorch-image-models. https://doi.org/10.5281/zenodo.4414861
25. Xue, J., Zhang, H., Dana, K.: Deep texture manifold for ground terrain recognition. In: Proceedings of the IEEE Conference on Computer Vision and Pattern Recognition, pp. 558–567 (2018)

Monocular 6-DoF Pose Estimation for Non-cooperative Spacecrafts Using Riemannian Regression Network

Sunhao Chu[1]([envelope]) [iD], Yuxiao Duan[1] [iD], Klaus Schilling[2] [iD], and Shufan Wu[1] [iD]

[1] Shanghai Jiao Tong University, Shanghai 200240, China
{Chu1220,duanyx,shufan.wu}@sjtu.edu.cn
[2] Julius-Maximilians-University Würzburg, Würzburg, Germany
schi@informatik.uni-wuerzburg.de

Abstract. As it is closely related to spacecraft in-orbit servicing, space debris removal, and other proximity operations, on-board 6-DoF pose estimation of non-cooperative spacecraft is an essential task in on-going and planned space mission design. Spaceborne navigation cameras, on the other hand, face the challenges of rapidly changing light conditions, low signal-to-noise ratio in space imagery, and real-time demand, which are not present in terrestrial applications. To address this issue, we propose an EfficientNet-based method that regresses position and orientation. The rotation regression loss function is converted into the Riemannian geodesic distance between the predicted values and ground-truth labels, which speeds up the rotation regression and limits the error to a desirable range. Moreover, several data augmentation techniques were proposed to address the overfitting issue caused by the small scale of the spacecraft dataset in this paper. In the SPARK2022 challenge, our method achieves state-of-the-art pose estimation accuracy.

Keywords: Non-cooperative spacecraft · 6-DoF pose estimation · Monocular system · Riemannian geometry

1 Introduction

The number of spacecraft launched into orbit has increased exponentially since the launch of the first artificial satellite, Sputnik, in 1957. As a result, gathering information and knowledge about objects orbiting the Earth, also known as space situational awareness (SSA), has emerged as an important research topic. On-board pose estimation for non-cooperative spacecraft, a subtopic of SSA, is a crucial technology for ongoing and upcoming space surveillance, autonomous rendezvous, satellite formation, debris removal and in-orbit servicing missions [18], such as Surrey Space Center's RemoveDEBRIS mission [3], DARPA's Phoenix program [4], and NASA's Restore-L mission [16]. However, image processing based on handcrafted feature descriptions and prior knowledge, on the other hand, is inapplicable to spacecraft with varying structural and physical

© The Author(s), under exclusive license to Springer Nature Switzerland AG 2023
L. Karlinsky et al. (Eds.): ECCV 2022 Workshops, LNCS 13801, pp. 186–198, 2023.
https://doi.org/10.1007/978-3-031-25056-9_13

characteristics. Second, due to mission operational constraints, a priori informa-tion of the target spacecraft is rarely available, nor is it desirable in cases where full autonomy is required. Furthermore, onboard cameras must contend with rapidly changing lighting conditions, low signal-to-noise ratio space scenarios, and real-time demand [26].

Since 2015, pose estimation algorithms have shifted from handcrafted features to CNNs, owing to the boom in deep learning techniques. The milestone work is the PoseNet proposed in [8]. An end-to-end convolution neural network is trained to regress the 6-DOF camera pose from a single RGB image with no additional post-processing required. In 2018, deep neural networks were first attempted in the application of spacecraft pose estimation [17]. Sharma et al. first went through images collected in the PRISMA task and used OpenGL to simulate and generate spacecraft images at different viewing angles and distances owing to limited annotations of space imagery data, AlexNet is then used to solve a viewpoint classification problem, demonstrating that CNNs have great potential for use in estimating the attitude of spacecraft.

In this paper, we further investigate the spacecraft pose estimation method. We develop a novel Spacecraft Pose Estimation Network (SPENet) which intro-duces a new attitude representation – a 9D vector as an orientation regression output. The 9D vector can be converted onto $SO(3)$ space via reshaping and SVD special orthogonalization. The orientation loss is treated as the Rieman-nian geodesic distance between the predicted and ground-truth rotation matrix. The network's weights are updated by the backpropagation of orientation loss jointly with location loss during the training process. Several data augmenta-tion techniques for spacecraft pose estimation are also discussed. Experimental results show that these data augmentation techniques effectively improve the generalisation of the network in the context of a limited dataset scale, and our proposed method performs well on the SPARK2022 dataset [15].

The following is the paper's outline: Sect. 2 reviews previous research on spacecraft pose estimate techniques, Sect. 3 details our suggested architecture, SPENet, Sect. 4 contains the comparative analysis and ablation study, and Sect. 5 summarizes the whole paper.

2   Related Work

2.1   Template-Matching Monocular Pose Estimation

By using manual feature descriptors, templates match a 2-Dimensional RGB image with the target spacecraft's wireframe 3D model in order to extract and detect target features such as keypoints, corners, and edges. The Perspective-n-Points (PnP) Problem is then solved to get the relative pose.

A representative work is that D'Amico et al. [2] utilized the Canny + HT method in conjunction with a Low-Pass Filter. The PRISMA image dataset was used to test this approach, which showed flexibility in terms of spacecraft shape, but lacked robustness to lighting variations and background textures when com-pared to other methods. As an extension of this approach, Sharma et al. [19]

introduced Weak Gradient Elimination to remove the imagery's background. During the next stage, features were extracted and detected using the Sobel + Hough Transform algorithm. They synthesized the detected features using geometric constraints, which greatly reduces the search space of the EPnP method for solving the template-matching problem.

2.2   Appearance-Based Monocular Pose Estimation

In contrast to template-based methods that employ image processing to find feature points, appearance-based methods only consider the spacecraft's appearance. Shi et al. [21] developed the only appearance-based method for estimating the pose of a spacecraft using a monocular camera. The proposed Euler-PCA algorithm consists of two components: offline training and in-flight testing. The disadvantage of this method is that the test image must be compared to each image in the training dataset for each posture solution, which is computationally intensive if the number of training frames is substantial. In addition, it was demonstrated that the performance of the PCA method degrades with image noise, which contradicts true space imagery. These drawbacks restrict the growth of appearance-based techniques.

2.3   CNN-Based Monocular Pose Estimation

Pose estimation methods have shifted from traditional feature-based algorithms to CNN-based methods as deep learning techniques have grown in popularity in recent years. A CNN-based pose estimation approach was introduced first by Kendall et al. [8] using a multilayer perceptron (MLP) regressor to recover the camera's location and rotation from an input image by simply affixing a MLP regressor to the Inception backbone. This approach, known as PoseNet, provided a lightweight and quick option for positioning and orientation even though it was far less accurate than 3D-based methods. Furthermore, they extended their framework to a Bayesian model capable of determining the uncertainty of camera relocalization via Monte Carlo sampling with dropout [6]. Extensions of PoseNet, also include the following works: GPoseNet [1], LSTM-PN [24], GeoPoseNet [7], etc.

The preceding work demonstrates that CNNs outperform feature-based algorithms in terms of robustness to poor lighting conditions and computational complexity. As a result, CNN-based methods were transferred from terrestrial to space-borne applications, with the difference being that the target pose was regressed relative to the camera frame rather than the camera pose relative to the world frame.

In Sharma et al. 's study [17], the spacecraft pose estimation problem was viewed as a viewpoint classification problem, and then the AlexNet algorithm was used to solve it. In [18], they suggested an SPN network made up of three separative branches and five convolutional layers. The Region Proposal Network detects a bounding box surrounding the target in the first branch. Three fully-connected layers are used to solve the categorization and regression problems, resulting in the relative attitude of the target spacecraft. Combining the bounding box information with the attitude information, we can solve for the relative position by reducing the distance between the corners of the bounding box and the extremal points.

It is proposed by Proenca and Gao that a neural network architecture for pose estimation can be developed based on orientation soft classifications, which allows modeling orientation ambiguity using a combination of Gaussian functions [14]. URSO datasets and the Spacecraft Pose Estimation Dataset (SPEED) [9] were utilized to validate their framework [19]. In their study, they concluded that estimating orientation using soft classification yielded better results than direct regression, as well as providing the means to model uncertainty.

Hu et al. [5] pointed out that the method in [18] was flawed in several ways. In order to address these shortcomings, they introduced a hierarchical end-to-end trainable 6D pose estimation architecture that produced robust and scale-insensitive 6D poses. A RANSAC-based PnP method uses information across scales to infer a single reliable pose from these 3D-to-2D correspondences for each level of the pyramid. This differs from most networks, which obtain pose prediction only from the top layer.

3   Proposed Method

Our novel SPENet framework is depicted in Fig. 1, where we use the EfficientNet-B0 [22] architecture with pre-trained weights as the backbone network owing to the good accuracy-complexity trade-off. Two MLP branches are added after the global average pooling layer in order to anticipate the relative position and attitude of the target spacecraft.

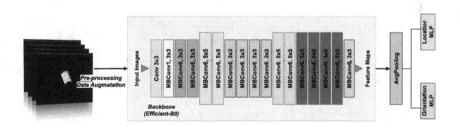

Fig. 1. Network architecture for spacecraft pose estimation

In the following, we will present our approach in terms of backbone networks, prediction head branches, and loss functions, data augmentation, etc.

3.1   Backbone

Many experiments have been conducted in the search for the most efficient CNN architecture. Many bottlenecks, such as accuracy, parameters, FLOPs, or inference time, must be perfectly balanced in an efficient architecture. Efficientnet is an excellent solution to this issue. Despite having fewer parameters, the model transfers well and achieves SOTA accuracy on CIFAR-100 (91.7%), Flowers (98.8%), and three other datasets for transfer learning [22].

By using a compound coefficient, EfficientNet uniformly scales all depth, width, and resolution dimensions in a CNN architecture (Fig. 2a). The EfficientNet scaling method uniformly scales network width, depth, and resolution, unlike previous work that indiscriminately scales them. Mobile inverted bottleneck MBConv with squeeze-and-excitation optimization (Figs. 2b and 2c) serves as its main building block.

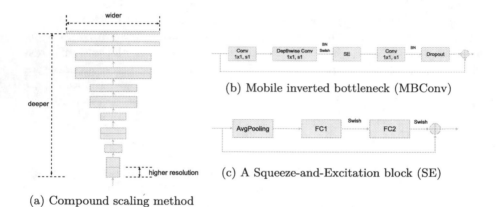

(b) Mobile inverted bottleneck (MBConv)

(c) A Squeeze-and-Excitation block (SE)

(a) Compound scaling method

Fig. 2. EfficientNet's compound scaling method and its main building block

3.2   Pose Prediction Heads and Loss Function

Pose is an element of the Special Euclidean group in three dimensions $SE(3)$, which represents a rigid transformation in three dimensions. A pose consists of two parts: a translation component of the Euclidean space $\mathbb{R}^3$ and a rotation component of the special orthogonal group $SO(3)$. Since the objective of the SPENet is to regress the pose of the non-cooperative spacecraft w.r.t. the camera coordinate system, the translation and rotation components can be decoupled into two regression branches. Here, we use the loss function proposed by SPEED+ challenge [13] as reference:

$$loss = L_{\mathrm{loc}} + L_{\mathrm{ori}}, \tag{1}$$

L_{loc} and L_{ori} denote the losses of the two branches, respectively. In the validation process, the equation (1) is called the *ESA SCORE*.

Location Estimation Branch. The spatial location regression branch is a straightforward two-layer MLP, because the translation component is in the

Euclidean space $\mathbb{R}^3$; however, like in the SPEED+ chanllenge, we minimize the relative error rather than the absolute Euclidean distance:

$$L_{\text{loc}} = \frac{\left\| t^{(i)} - t_{gt}^{(i)} \right\|_2}{\left\| t_{gt}^{(i)} \right\|_2},\tag{2}$$

where $t^{(i)}$ and $t_{gt}^{(i)}$ are the estimated and ground-truth translation vectors, respectively.

Orientation Estimation. Optimization on the special orthogonal group $SO(3)$, and more generally on Riemannian manifolds, has been studied extensively in academia. Since $SO(3)$ is not topologically homeomorphic from any subset of 4D Euclidean spaces, for 3D rotations, all representations are discontinuous in four or less dimensional real Euclidean spaces with Euclidean topology, which applies to all classical representations, i.e., Euler angles, rotation axes/angles, and unit quaternions [25].

Theoretical results show that for a given number of neurons, functions with higher smoothness/continuity have lower approximation errors. Therefore, in the context of neural networks, manifold-based gradient optimization, discontinuity and singularity is a pressing problem. To solve this problem, rotational representations based on SVD special orthogonalization are proposed [10].

Given a orientation prediction head raw 9D output, it can be reshaped to 3×3 matrix M with SVD decomposition $U\Sigma V^T$. The special orthogonalization

$$R_{est} = \text{SVDO}^+(M) := U\Sigma'V^T, \text{ where } \Sigma' = \text{diag}\left(1, \ldots, 1, \det\left(UV^T\right)\right)\tag{3}$$

will map the raw 9D output onto $SO(3)$, see Fig 3.

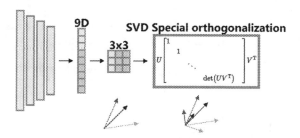

Fig. 3. SVD Special orthogonalization based 9D Rotation representation

The geodesic distance between rotations represented by rotation matrices R_{est} and R_{gt} is the angle of the difference rotation represented by the rotation matrix $\Delta R = R_{est}^{-1}R_{gt}$ [12].

We can retrieve the angle of the difference rotation from the trace of ΔR,

$$\text{tr}\,\Delta R = 1 + 2\cos\theta.\tag{4}$$

Thus, we can obtain the orientation estimation head's loss function

$$L_{\mathrm{ori}} = \theta = \arccos \frac{\mathrm{tr}\,\Delta R - 1}{2}. \tag{5}$$

When inference, R_{est} can be converted to quaternion q.

3.3   Data Pre-processing and Augmentation

Contrary to image classification, data pre-processing and augmentation for pose estimation includes both labels and images; as a result, common image transformations (such as cropping and flipping) must be carefully considered because they will unavoidably ruin the dataset by unpredictable changing the target pose and camera intrinsic matrix.

After several trials, we used the following data pre-processing augmentation tricks, including:

1. To balance the computational cost and loss of image detail, we first scaled the image from 1440×1080 to 640×480. In this case, the camera intrinsic matrix is changed and the pose label remains unchanged.
2. Image random shift, scaling, and rotation are used to simulate the in-plane rotation and the distance variation of the satellite. The camera's intrinsic matrix remains unchanged in this case. After image augmentation, we can easily obtain the 3D coordinates of the satellite's bounding box corners and their 2D projection coordinates, and the updated pose label can be obtained by solving the PnP problem [5].
3. Warping the image to simulate the camera orientation perturbations $R_{perturb}$, the updated position label becomes $R^T_{perturb}t$ and the orientation label becomes $q(R^T_{perturb}R(q))$ [14].
4. Changing the contrast and exposure of the image at random, adding Gaussian noise, blurring the image, and dropping out patches. In this instance, the camera intrinsic matrix and pose label remain unchanged.

Figure 4 shows the change in image and label before and after data augmentation.

4   Experiments

4.1   Experimental Setup

Datasets. We evaluate our approach using the SPARK2022 dataset [15], which is proposed by the SnT Lab at the University of Luxembourg. The SPARK2022 dataset was generated in a photorealistic space simulation environment in Unity3D, aiming at object detection and pose estimation of spacecraft leveraging knowledge of the space environment.

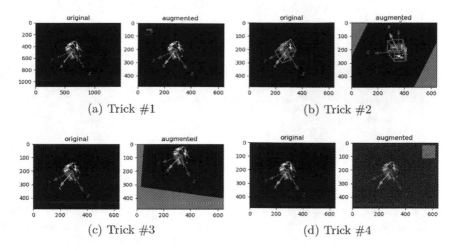

(a) Trick #1

(b) Trick #2

(c) Trick #3

(d) Trick #4

Fig. 4. Comparison of data before and after augmentation tricks

Implementation Details. Our model is implemented in PyTorch. We use a pre-trained EfficientNet-B0 as features extractor. The two MLP heads, regressing the position and orientation vectors, respectively, expand the backbone channel dimension to 1024 with a single hidden layer. We optimize our model to minimize the loss in Eq. (1) using AdamW [11], with $\beta_1 = 0.9$, $\beta_2 = 0.999$ and $\epsilon = 10^{-8}$ as default. Throughout all experiments we use a batch size of 16 and ReduceLROnPlateau learning rate scheduler with an initial value of $lr = 10^{-4}$. All experiments reported in this paper were carried out on a computer with an Intel Core i9-10900X CPU @ 3.70GHz, 64 GB of RAM and an NVIDIA GeForce RTX 3090 GPU.

4.2   Comparative Analysis

We compare results to two CNN-based approaches – PoseNet [8] and Trans-PoseNet [20] – on SPARK2022 dataset. Although these two networks are structurally consistent with our proposed network SPENet, they are used to regress the pose of the camera system relative to the world system as discussed in Sects. 2.3, 3.3, and 3.2, and the data augmentation method (random crop) involved in their implementation ruins the spacecraft pose labels. Therefore, in the experiments, for both these network implementations the random crop augmentation was removed and only the resizing of the images (trick #1) and the image enhancement operations (trick #4) were included. As for our SPENet implementation, we compared the use of the full data augment pipeline (trick # 1,2,3 and 4) with the above-mentioned image-only augment pipeline (trick # 1 and 4). The experimental results are shown in Table 1.

It is clear that the networks used for camera relocalization fail to address the problem. Possible reasons for this failure are that the L_2 loss they used is not suitable for pose regression and the lack of random crop augmentation

Table 1. Results of pose estimation algorithms for the SPARK2022 dataset. We present the average position/orientation errors. Bold is used to highlight the best results.

Method	Position Error [m]	Orientation error [deg]	ESA SCORE
PoseNet	0.4047	28.64	0.5601
TransposeNet	0.3810	48.15	0.9016
SPENet (no pose augment)	0.2729	19.45	0.3795
SPENet	**0.1098**	**4.398**	**0.09263**

results in insufficient samples in the training set. Similarly, our method that performs only image enhancement without pose label augmentation also faces the overfitting problem. In contrast, our proposed network SPENet with the full data augment pipeline solves the non-cooperative spacecraft six-degree-of-freedom pose-estimation task well, limiting the estimated position error to within 0.15 m and the pointing error to within 5°C.

4.3 Ablation Study

We conducted two ablation experiments on the SPARK2022 dataset to investigate the effects of different architecture design choices. Our ablation study focuses on two main aspects of our approach: (a) rotation parameters predicted by the orientation head, and (b) the backbone architecture.

Rotation Parameters. We consider four rotational representations for our choice: Euler angles, quaternions, r6d [25] and svd parameter [10]. The implementation details are all the same except for the difference in the rotation regression head. We convert the rotation parameters of the raw output to a rotation matrix on the SO(3) space, and then calculate the geodesic distance loss by Eq. (5). For Euler angles and quaternions, the conversion to rotation matrices is explicitly defined as

$$R_{est}(\varphi, \vartheta, \psi) = \begin{bmatrix} \cos\psi & \sin\psi & 0 \\ -\sin\psi & \cos\psi & 0 \\ 0 & 0 & 1 \end{bmatrix} \begin{bmatrix} 1 & 0 & 0 \\ 0 & \cos\vartheta & \sin\vartheta \\ 0 & -\sin\vartheta & \cos\vartheta \end{bmatrix} \begin{bmatrix} \cos\varphi & \sin\varphi & 0 \\ -\sin\varphi & \cos\varphi & 0 \\ 0 & 0 & 1 \end{bmatrix} \quad \text{(Euler angles)}$$

$$R_{est}(q) = \begin{bmatrix} 2\left(q_0^2 + q_1^2\right) - 1 & 2\left(q_1 q_2 - q_0 q_3\right) & 2\left(q_1 q_3 + q_0 q_2\right) \\ 2\left(q_1 q_2 + q_0 q_3\right) & 2\left(q_0^2 + q_2^2\right) - 1 & 2\left(q_2 q_3 - q_0 q_1\right) \\ 2\left(q_1 q_3 - q_0 q_2\right) & 2\left(q_2 q_3 + q_0 q_1\right) & 2\left(q_0^2 + q_3^2\right) - 1 \end{bmatrix} \quad \text{(quaternions)}$$

$$(6)$$

For the svd parameter, the rotation matrix is defined in Eq. (3), and the conversion of the r6d parameter to a rotation matrix is given in [25]

$$R_{est}\left(\begin{bmatrix} a_1 & a_2 \end{bmatrix}\right) = \begin{bmatrix} b_1 & b_2 & b_3 \end{bmatrix}, \quad b_i = \begin{cases} N\left(a_1\right) & i = 1 \\ N\left(a_2 - \left(b_1 \cdot a_2\right) b_1\right) & i = 2 \\ b_1 \times b_2 & i = 3. \end{cases} \quad \text{(r6d)} \quad (7)$$

where a_1, a_2 are three-dimensional vectors consisting of the first three and the last three of the r6d parameters, $N(\cdot)$ denotes a normalization function.

Table 2. A study of the ablation of the rotation parameter using the SPARK2022 dataset.

Rotation parameter	Position error [m]	Orientation error [deg]	ESA SCORE
Euler angles	0.245	14.56	0.2899
Quaternions	0.2394	12.12	0.2432
r6d	0.1222	4.903	0.1031
svd	**0.1098**	**4.398**	**0.09263**

The results obtained with these parameters are shown in Table 2 and Fig. 5. The r6d and svd parameter achieve a better performance compared to the traditional ones. As mentioned in Sect. 3.2, parameters less than four dimensions are not suitable for rotation representations in neural networks.

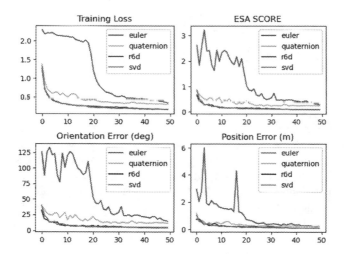

Fig. 5. Training loss and performance on the validation set for the rotation parameters ablation study

Backbone Network. We consider two CNN networks for our backbone architecture choice: EfficientNet-B0 and EfficientNetv2-M [23]. The results obtained with these backbones are shown in Table 3 and Fig. 6. With appropriate deeper models, EfficientNetv2-M is able to achieve better performance than EfficientNet-B0, at the cost of memory and runtime. Considering the cost of on-board computing and real-time requirements, we still choose the lighter one, EfficientNet-B0, as our feature extractor.

Table 3. Ablations of the backbone network, evaluated on the SPARK2022 dataset.

Backbone	Position error [m]	Orientation error [deg]	ESA SCORE
EfficientNetv2-M	**0.09521**	**3.808**	**0.08025**
EfficientNet-B0	0.1098	4.398	0.09263

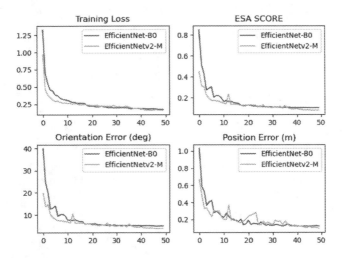

Fig. 6. Training loss and performance on the validation set for the backbone networks ablation study

5   Conclusions

In this paper, we propose a novel CNN-based approach for non-cooperative spacecraft pose regression. Different rotational representation forms on the $SO(3)$ space are investigated, and the rotation regression loss function is converted into the Riemannian geodesic distance between the predicted and ground-truth values, which speeds up the rotation regression and limits the error to a desirable range. Furthermore, our proposed data augment and preprocessing techniques effectively solve the overfitting problem caused by the small scale of the spacecraft dataset, while also balancing inference speed and prediction error. On the SPARK2022 dataset, our approach provides state-of-the-art pose estimation accuracy.

Acknowledgements. This work is supported by National Natural Science Foundation of China (Grant No: U20B2054).

References

1. Cai, M., Shen, C., Reid, I.D.: A hybrid probabilistic model for camera relocalization. In: BMVC (2018)

2. D'Amico, S., Benn, M., Jørgensen, J.L.: Pose estimation of an uncooperative spacecraft from actual space imagery. Int. J. Space Sci. Eng. **2**(2), 171–189 (2014)
3. Forshaw, J.I., et al.: RemoveDEBRIS: an in-orbit active debris removal demonstration mission. Acta Astronaut. **127**, 448–463 (2016). https://doi.org/10.1016/j.actaastro.2016.06.018
4. Henshaw, C.G.: The DARPA phoenix spacecraft servicing program: overview and plans for risk reduction. In: International Symposium on Artificial Intelligence, Robotics and Automation in Space (i-SAIRAS). European Space Agency (2014)
5. Hu, Y., Speierer, S., Jakob, W., Fua, P., Salzmann, M.: Wide-depth-range 6d object pose estimation in space. In: Proceedings of the IEEE/CVF Conference on Computer Vision and Pattern Recognition, pp. 15870–15879 (2021)
6. Kendall, A., Cipolla, R.: Modelling uncertainty in deep learning for camera relocalization. In: 2016 IEEE iInternational Conference on Robotics and Automation (ICRA), pp. 4762–4769. IEEE (2016)
7. Kendall, A., Cipolla, R.: Geometric loss functions for camera pose regression with deep learning. In: Proceedings of the IEEE Conference on Computer Vision and Pattern Recognition, pp. 5974–5983 (2017)
8. Kendall, A., Grimes, M., Cipolla, R.: PoseNet: a convolutional network for Real-Time 6-DOF Camera Relocalization. In: 2015 IEEE International Conference on Computer Vision (ICCV), pp. 2938–2946. IEEE, Santiago, Chile (December 2015). DOIurl10.1109/ICCV.2015.336
9. Kisantal, M., Sharma, S., Park, T.H., Izzo, D., Märtens, M., D'Amico, S.: Satellite pose estimation challenge: dataset, competition design, and results. IEEE Trans. Aerosp. Electron. Syst. **56**(5), 4083–4098 (2020)
10. Levinson, J., et al.: An analysis of SVD for deep rotation estimation. Adv. Neural. Inf. Process. Syst. **33**, 22554–22565 (2020)
11. Loshchilov, I., Hutter, F.: Decoupled weight decay regularization. arXiv preprint arXiv:1711.05101 (2017)
12. Mahendran, S., Lu, M.Y., Ali, H., Vidal, R.: Monocular object orientation estimation using Riemannian regression and classification networks (2018). 10.48550/ARXIV.1807.07226
13. Park, T.H., Märtens, M., Lecuyer, G., Izzo, D., D'Amico, S.: SPEED+: next-generation dataset for spacecraft pose estimation across domain gap. arXiv preprint arXiv:2110.03101 (2021)
14. Proença, P.F., Gao, Y.: Deep learning for spacecraft pose estimation from photorealistic rendering. In: 2020 IEEE International Conference on Robotics and Automation (ICRA), pp. 6007–6013. IEEE (2020)
15. Rathinam, A., et al.: SPARK 2022 dataset : spacecraft detection and trajectory estimation (June 2022). https://doi.org/10.5281/ZENODO.6599762, type: dataset
16. Reed, B.B., Smith, R.C., Naasz, B.J., Pellegrino, J.F., Bacon, C.E.: The Restore-L Servicing Mission. In: AIAA SPACE 2016. American Institute of Aeronautics and Astronautics, Long Beach, California (September 2016). https://doi.org/10.2514/6.2016-5478
17. Sharma, S., Beierle, C., D'Amico, S.: Pose estimation for non-cooperative spacecraft rendezvous using convolutional neural networks. In: 2018 IEEE Aerospace Conference, pp. 1–12. IEEE (2018)
18. Sharma, S., D'Amico, S.: Neural network-based pose estimation for noncooperative spacecraft rendezvous. IEEE Trans. Aerosp. Electron. Syst. **56**(6), 4638–4658 (2020). https://doi.org/10.1109/TAES.2020.2999148

19. Sharma, S., Ventura, J., D'Amico, S.: Robust model-based monocular pose initialization for noncooperative spacecraft rendezvous. J. Spacecr. Rocket. **55**(6), 1414–1429 (2018)
20. Shavit, Y., Ferens, R., Keller, Y.: Learning multi-scene absolute pose regression with transformers. In: Proceedings of the IEEE/CVF International Conference on Computer Vision, pp. 2733–2742 (2021)
21. Shi, J.F., Ulrich, S., Ruel, S.: Spacecraft pose estimation using principal component analysis and a monocular camera. In: AIAA Guidance, Navigation, and Control Conference, p. 1034 (2017)
22. Tan, M., Le, Q.: Efficientnet: Rethinking model scaling for convolutional neural networks. In: International Conference on Machine Learning, pp. 6105–6114. PMLR (2019)
23. Tan, M., Le, Q.: Efficientnetv2: Smaller models and faster training. In: International Conference on Machine Learning. pp. 10096–10106. PMLR (2021)
24. Walch, F., Hazirbas, C., Leal-Taixe, L., Sattler, T., Hilsenbeck, S., Cremers, D.: Image-based localization using LSTMS for structured feature correlation. In: Proceedings of the IEEE International Conference on Computer Vision, pp. 627–637 (2017)
25. Zhou, Y., Barnes, C., Lu, J., Yang, J., Li, H.: On the continuity of rotation representations in neural networks. In: Proceedings of the IEEE/CVF Conference on Computer Vision and Pattern Recognition, pp. 5745–5753 (2019)
26. Zhu, X., Song, X., Chen, X., Lu, H.: Flying spacecraft detection with the earth as the background based on superpixels clustering. In: 2015 IEEE International Conference on Information and Automation, pp. 518–523. IEEE, Lijiang, China (August 2015). https://doi.org/10.1109/ICInfA.2015.7279342

W02 - Vision for Art

W02 - Vision for Art

The Workshop on Computer VISion for ART (VISART) is a forum for the presentation, discussion, and publication of Computer Vision (CV) techniques for the understanding of art. This workshop brings together leading researchers in the fields of CV, ML, IR with Art History, Visual Studies, and Digital Humanities along with museum curators to focus on art and Cultural Heritage problems. The potential uses of Computer Vision for cultural history and cultural analytics have created great interest in the Humanities, with large projects on applying Computer Vision in galleries and museums, including the Getty and MoMA (in collaboration with Google). A key feature of this workshop is the close collaboration between scholars of Computer Vision and the Arts and Humanities, thus both exposing new technical possibilities to the Arts and Humanities, as well as offering new artistic and humanistic perspectives on computer vision.

October 2022

Alessio Del Bue
Peter Bell
Leonardo L. Impett
Noa Garcia
Stuart James

HyperNST: Hyper-Networks for Neural Style Transfer

Dan Ruta[1(⊠)], Andrew Gilbert[1], Saeid Motiian[2], Baldo Faieta[2], Zhe Lin[2], and John Collomosse[1,2]

[1] University of Surrey, Guildford, UK
dan.sebastian.ruta@gmail.com
[2] Adobe Research, San Jose, CA, USA

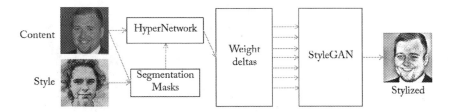

Abstract. We present HyperNST; a neural style transfer (NST) technique for the artistic stylization of images, based on Hyper-networks and the StyleGAN2 architecture. Our contribution is a novel method for inducing style transfer parameterized by a metric space, pre-trained for style-based visual search (SBVS). We show for the first time that such space may be used to drive NST, enabling the application and interpolation of styles from an SBVS system. The technical contribution is a hyper-network that predicts weight updates to a StyleGAN2 pre-trained over a diverse gamut of artistic content (portraits), tailoring the style parameterization on a per-region basis using a semantic map of the facial regions. We show HyperNST to exceed state of the art in content preservation for our stylized content while retaining good style transfer performance.

Keywords: Datasets and evaluation · Image and video retrieval · Vision + language · Vision applications · and systems

1 Introduction

Neural style transfer (NST) methods seek to transform an image to emulate a given appearance or 'style' while holding the content or structure unchanged. Efficient, stylistically diverse NST remains an open challenge. Feed-forward NST methods are fast yet typically fail to represent a rich gamut of styles. At the same time, optimization based approaches can take several seconds or minutes at inference time, lacking the speed for practical, creative use. Moreover, NST algorithms are often driven by one or more exemplar 'style' images, rather than an intuitive parameter space, impacting their controllability as a creative tool.

L. Karlinsky et al. (Eds.): ECCV 2022 Workshops, LNCS 13801, pp. 201–217, 2023.
https://doi.org/10.1007/978-3-031-25056-9_14

In this work, we propose a fast feed-forward method for driving neural stylization (NST) parameterized by a metric embedding for style representation. Our approach is based upon a hyper-network trained to emit weight updates to a StyleGAN2 [13] model, trained on a large dataset of artistic portraits (e.g. AAHQ [16]) in order to specialize it to the depiction of the given target style. Our work is inspired by the recent StyleGAN-NADA [5], in which a CLIP [20] objective is optimized by fine-tuning a pre-trained StyleGAN2 model to induce NST. We extend this concept to a feed-forward framework using a hyper-network to generate the weight updates. Furthermore, we introduce the use of a metric parameter space (ALADIN [22]) originally proposed for style driven visual search to condition the hyper-network prediction (vs. CLIP in [5]) and adaptively drive this parameterization using a semantic map derived from the source and target image. Without loss of generality, our experiments focus on the challenging domain of facial portraits driven using a semantic segmentation algorithm for this content class. We show our method improves target image content retention versus the state of the art while retaining comparable accuracy for diverse style transfer from a single model – and despite using a hyper-network exhibiting comparable inference speed to leading feed-forward NST [19]. Moreover, our method exhibits good controllability; using a metric space for our style code enables intuitive interpolation between diverse styles and region-level controllability of those parameters. We adopt the recent ALADIN style code for this purpose, raising the novel direction of unifying style based visual search and stylization from a single representation.

2   Related Work

Neural Style Transfer (NST). The seminal work of Gatys et al. [6] enabled artistic style transfer through neural models. This work demonstrates the correlation between artistic style and features extracted from specific layers in a pre-trained vision model. The AdaIN work [8] introduced parameterized style transfer through first and second moment matching, via mean and standard deviation values extracted from random target style images. MUNIT [9] explores domain translation in images through the de-construction of images into semantic content maps and global appearance codes. ALADIN [22] explored the creation of a metric space modeling artistic style, across numerous areas of style. The embedding space was trained in conjunction with AdaIN and a modified version of the MUNIT model and weakly supervised contrastive learning over the BAM-FG dataset. A follow-up work [21] studying multi-modal artistic style representation expanded upon ALADIN, pushing the representation quality further through a vision transformer model. CycleGAN also explored domain transfer in images, but through learned model weight spaces, encoding pairwise image translation functions into separate generators for each image domain. Enforced by cyclic consistency, the translation quality between a pair of image domains was high, at the cost of requiring bespoke models for each domain translation to be trained. Using StyleGAN as a generator model, Swapping Autoencoders [19] directly learn the embedding of images into a StyleGAN generator's weight space while simultaneously encoding a vector representation of the visual appearance of an image externally. These models separately focus on landscapes, buildings, bedrooms, or faces.

StyleGAN Inversion. The evolution of the StyleGAN models [10, 12, 13] explore generation of extremely realistic portrait images. They use weight modulation based editing of visual appearance in the generated images. They also undertake preliminary investigations into the inversion of existing images into the GANs' weight spaces for reconstruction. The work in e4e [26] includes an undertaking of a deeper analysis of real image embedding into the StyleGAN weight space, including the quality/editability tradeoff this imposes. Their work enables multiple vectors of editability for images generated through StyleGAN across several domains. Restyle [1] improves the reconstruction quality of images embedded into the weight space by executing three fine-tuning optimization steps at run-time. Also, using StyleGAN as target generators, HyperStyle [2] embeds real portrait images into the weight space with high fidelity. Similar to our approach, they use a hyper-network to generate weight updates for StyleGAN, trained to infer weights to bridge the gap between quick rough inversions and the fully detailed reference portrait images. Strengths of this approach are the high reconstruction fidelity and strong photorealistic editing control for portrait photos. They further undertake some early explorations at domain adaptation for images by changing StyleGAN checkpoints. However, the HyperStyle work focuses on photorealism and does not enable region-based control or style space parameterization as we propose in HyperNST.

Our approach is inspired by the StyleGAN-NADA [5] work, which explores style transfer in the StyleGAN weight space through CLIP-based optimization. Though effective, this incurs long-running optimization passes, which are impractical for wide use. Moreover, this method has no built-in methods to effectively embed real portrait images into the weight space for reconstruction and editing. Recently, FaRL [28] undertakes representation learning of facial images, through multi-modal vision-text data, with face segmentation models that cope well with various visual appearances for portraits.

3   Methodology

Domain adaptation models like CycleGAN perform well at image translation, where style features from a style image are correctly mapped to the matching semantic features in a content image. A limitation of CycleGAN is that transfer only between a single pair of image domains (styles) is possible. Hypernetwork [7] models are used to predict the weights of a target model. Such a hypernetwork can be conditioned on some input before inferring the target model weights. Early experiments with a CycleGAN backbone yielded promising results, especially when we re-framed the training process into a more manageable task by learning weight offsets to a pre-trained checkpoint rather than a from-scratch model. Motivated by these findings, we use a more extensive, modern generator such as StyleGAN2 with weight updates predicted using a hyper-network. This echoes the recent optimization based approach StyleGAN-NADA which updates weights via fine-tuning for stylization rather than hyper-network prediction. HyperStyle [2] uses a channel-wise mean shift in the target generator, which significantly reduces the target number of trainable weights into a practical range.

3.1 HyperNST Architecture

We compose the stylization pipeline of our model with a hyper-network set up to generate weight deltas for a target frozen, pre-trained StyleGAN2 model. Figure 1 shows an architecture diagram of HyperNST, showing the losses and the conditioning process of the hyper-network upon the content image and semantically arranged ALADIN style codes of the style image.

We direct the training to find a set of weight deltas for StyleGAN2, which can generate the same image as the content image for reconstruction. We begin with the GAN inversion, using e4e to reconstruct a given content portrait image roughly. An encoder encodes this image into a tensor used by the hyper-network weight generator modules along with style information to predict weight deltas to apply to the frozen, pre-trained StyleGAN2. Stylization then occurs by changing the style embedding in the hypernetwork conditioning.

3.2 Conditioning on Style, and Facial Semantic Regions

We further introduce conditioning on a target style image by injecting a style representation embedding. We project the initial $16 \times 16 \times 512$ encoding into $16 \times 16 \times 256$, using ALADIN [22] codes to compose the other half of the $16 \times 16 \times 512$ tensor. Each ALADIN code is a 1×256 vector representing the artistic style of an image. We project semantically arranged ALADIN embeddings into a second $16 \times 16 \times 256$, which we then concatenate together with the first to form the $16 \times 16 \times 512$ tensor upon which the hypernetwork is conditioned.

Facial portraits are very heavily grounded in their semantic structure. We therefore condition the stylization process on these semantic regions to ensure that the content is maintained. We execute this using segmentation masks extracted via FaRL [28].

To condition on semantic regions, we aim to use the ALADIN style code representing only the style contained within a given semantic region. Given that ALADIN codes can only be extracted from square or rectangular inputs, we use the average ALADIN code extracted across several patches per semantic region.

We use FaRL to compute the semantic regions, extract patches from the image randomly, and use mean intersection over union (IoU) to ensure that patches mainly cover pixels attributed to these respective semantic classes for each region. The ALADIN codes from these patches are averaged to form the average ALADIN code for a given semantic class. These average style codes are tiled and re-arranged on a per-pixel basis to match the semantic segmentation maps of the *content* image before they are projected by an encoder into the second $16 \times 16 \times 256$ tensor, as above.

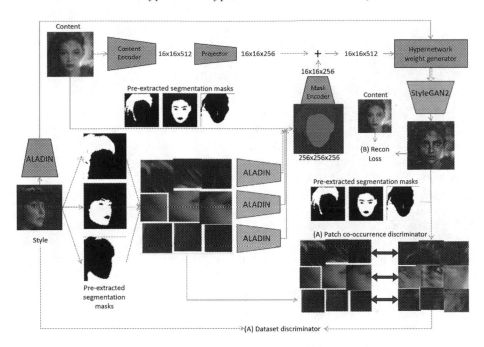

Fig. 1. Architecture diagram of our approach. Facial semantic segmentation regions used in conditioning via ALADIN and guiding via patch co-occurence discriminator a HyperStyle model into embedding a content portrait image into an AAHQ+FFHQ trained StyleGAN2 model, and using ALADIN style codes to stylize it towards the style of a style image. Blue modules represent frozen modules, and orange modules represent modules included in the training. + represents concatenation in the channels dimension. (A) represents the stylization pass losses, and (B) represents the reconstruction pass losses. Not pictured for clarity: the reconstruction pass uses the same semantic regions as the content image for the $256 \times 256 \times 256$ semantically arranged ALADIN conditioning. (Color figure online)

This projected $16 \times 16 \times 256$ semantically arranged ALADIN tensor and the original $16 \times 16 \times 256$ tensor encoded from the content image, are concatenated in the channel dimension to $16 \times 16 \times 512$, making up the final tensor which the hypernetwork weight delta generating modules operate over.

3.3   HyperNST Training Process

We use three iterations per batch refining methodology during training. We include a patch co-occurrence discriminator D_{patch} to introduce a style loss and an image-level discriminator D. The hypernetwork weight deltas generator H is conditioned on the encoded content images $E_c(c)$, and ALADIN style codes arranged by the semantic segmentation mask of the content image $M(x, E_{mask})$, seen in Eq. 1.

$$M(x, E_{mask}) = E_{mask} \left(rearrange \left(\frac{\sum_{p=1}^{P} A(crop(x, SM_x)_p)}{P}, SM_c \right) \right) \quad (1)$$

$$y(c, s) = G(\theta + H(M(s) \frown E_c(c), A(s)))) \quad (2)$$

During training, two forward passes are executed, one for stylization (A) (Eq. 2), and one for reconstruction (B). The reconstruction pass (B) loads content images and their respective ALADIN style code arranged by the content image semantic map and performs the original reconstruction code (Eq. 3). The stylization pass (A) performs style transfer by loading the content images c, and ALADIN style codes of other style images s. During stylization, the generation is performed with the mixed features combined to create stylized images, with the patch co-occurrence discriminator providing a learning signal to train the stylization (Eq. 4). In this pass, the discriminator is also trained, with the stylized images as *fake*, and the target style images as *real* (Eq. 5).

$$\mathcal{L}_{rec}(c) = (c, y(c, c))) \quad (3)$$

$$\mathcal{L}_{pcd}(D_{patch}) = \mathbb{E}_{s \sim S, c \sim C} \left[-log(D_{patch}(crops(s), crops(y(c, s)))) \right] \quad (4)$$

$$\mathcal{L}_{disc}(c, s, D) = \mathbb{E}_{s \sim S, c \sim C} \left[-log(D(s, y(c, s))) \right] \quad (5)$$

where: c is content image, s is style image, C are all content images, S are all style images, θ are the original StyleGAN2 weights, H is the hypernetwork weight delta generator, G is the StyleGAN2 generator, A is ALADIN, P is number of patches per semantic region, SM are pre-extracted semantic segmentation masks, and $\frown$ represents concatenation.

$$\mathcal{L} = \lambda_1 \mathcal{L}_{rec} + \lambda_2 (\mathcal{L}_{disc} + \mathcal{L}_{pcd}) \quad (6)$$

The final loss is shown in Eq. 6, with Sect. 4.3 describing ablations for the λ values.

3.4   Stylized Target Generator

The hypernetwork model is trained to infer weight delta values for weights in a StyleGAN2 model, which acts as the generator. Given that we work in the artistic style domain, we thus need the target generator to be able to model the weight space of a generator already able to produce high quality highly stylized images. We thus first fine-tune an FFHQ [11] StyleGAN2 model on the Artstation Artistic Face (AAHQ) dataset [16]. Largely popular in research centered around the facial images domain, the FFHQ dataset has been used by the majority of reference papers targeting this domain. Given our exploration into artistically manipulating the visual features of portrait images, we further use this AAHQ dataset, which encompasses portraits from across a large and varied corpus of artistic media. We continue to include FFHQ images, to ensure we keep high quality modeling capabilities for features more often found in photographic images than artistic renderings (like glasses and beards) to ensure real world images can still be encoded well.

We first train our model simply for ALADIN conditioned reconstruction of the AAHQ and FFHQ datasets for all layers, including the toRGB layers, which have been shown to target textures and colors [2]. We then fine-tune this pretrained checkpoint with the goal of stylization, where we include the patch co-occurrence training for guidance and ALADIN code swapping in conditioning for style transfer.

During the stylization fine-tuning step, we freeze the content encoder and train only the hypernetwork weight delta prediction modules. This is to prevent the stylization signal from negatively affecting the model's embedding abilities to reconstruct real images with high accuracy.

We run the stylization fine-tuning only on the part of the target StyleGAN layers. We find that the further into the StyleGAN2 model we apply stylization fine-tuning, the more the stylization affects colorization rather than textures and adjustments to the structure. We include the toRGB layers originally omitted in HyperStyle for their texture adjustments, and we train the weight deltas generation modules for layers 13 onward (out of 25). The weight delta generation for layers before this are frozen after their initial training, therefore still allowing reconstruction, but no more extended training during the stylization stage. This ensures that the overall facial structure of the images is not too greatly affected during stylization training. Layer 13 is a sweet middle spot with the best balance between stylization and retaining good face structure reconstruction. A visual example of this phenomenon can be seen in Fig. 3.

We further make changes to the patch co-occurrence discriminator. The Swapping Autoencoders model generates images in 1024×1024 resolution, with a discriminator operating over 8 patches of 128×128 dimensions. Our hypernetwork model generates images with a resolution of 256×256 pixels, therefore, we adjust our discriminator's patch sizes to a lower size of 32×32.

3.5 Region Mask Driven Patch Discriminator

We also use facial semantic segmentation masks in the patch co-occurrence discriminator, to provide a style signal separated by semantic region.

In the original discriminator, patches are extracted from across the entire image at random. Instead, our process is repeated for each semantic class' map. Patches are extracted at random, ensuring that the mean intersection over union (IoU) mostly covers pixels attributed to the respective semantic class. As in the original case, the losses from the patches are averaged to form the style learning signal. However, instead of comparing the stylized patches with patches randomly selected from the reference style image, we select patches bound by the respective semantic region - e.g. style hair patches are matched with stylized hair patches, background with background, face with face, etc.

Implementation. The semantic regions predicted by FaRL contain several regions which individually cover only small areas in the pixel space. Given our use of the regions as patches we extract for ALADIN and the patch co-occurrence discriminator, we could not accurately use these small regions without the patches mainly containing pixel data from the surrounding regions. Instead, we group up the regions into 3 larger classes which typically cover larger areas of a portrait: (1) Hair, (2) Face, and (3) Background. Due to overlap that cannot be avoided, the *Face* region also contains eyes, noses, and lips. Furthermore, some images do not contain hair, for which this semantic class is not used. In the semantically driven patch co-occurrence loss, we use 3 patches for each region, totaling 9 patches.

4   Evaluation

Experiments generally required around 24 h to converge. We ran our experiments on a single NVIDIA RTX 3090 with 24 GB of VRAM. The HyperNST experiments were executed with a batch size of 1, which required around 22 GB of VRAM, due to the high number of weights needed in the hypernetwork configuration.

4.1   Datasets

We create a test dataset for use in evaluation, using images from FFHQ [11], and AAHQ [16]. We extract 100 random content images from FFHQ and 100 random style images from AAHQ. Together, these result in 10'000 evaluation images, when models are used to stylize all combinations.

We measure the content similarity using the two LPIPS variants (Alexnet [15], and VGG [25]), and we measure the style similarity using SIFID.

LPIPS [27] evaluates the variance between two images based on perceptual information better aligned to human perception compared to more traditional statistics based methods. We use this LPIPS variation to compute the average variation between each stylized image and its original content image. A lower average variation value would indicate a more similar semantic structure, therefore, a lower value is better.

Like Swapping Autoencoders, we employ the Single Image Fréchet Inception Distance (**SIFID**) introduced in SinGAN [23]. This metric evaluates FID, but uses only a single image at a time, which is most appropriate when evaluating style transfer computed using a single source sample. A lower value indicates better style transfer here.

Finally, we measure the time it takes to synthesize the stylized image in seconds per image. A low inference time is essential for the practicality of a model in real use cases. All timings were computed on an RTX 3090 GPU.

Table 1. Overall results, comparing metrics for our model's content, style, and timing and the closest most similar model, Swapping AutoEncoders (top), and some further, more traditional methods (bottom). All methods were trained using AAHQ as the style dataset and FFHQ as the content dataset.

Model	LPIPS (Alexnet)	LPIPS (VGG)	SIFID	Time (s/img)	Interpolation	Editing
SAE [19]	0.334500	0.4857	1.948	0.10	✓	✗
HyperNST (ours)	0.000042	0.0017	2.279	0.35	✓	✓
Gatys [6]	0.000164	0.0030	1.369	14.43	✗	✗
ArtFlow [3]	0.000080	0.0022	1.347	0.32	✗	✗
PAMA [17]	0.000109	0.0029	0.522	0.14	✓	✗
SANet [18]	0.000180	0.0068	0.486	0.11	✓	✗
NNST [14]	0.000149	0.0030	0.871	55.40	✗	✗
ContraAST [4]	0.000133	0.0035	0.666	0.10	✗	✗

4.2 Baselines

To test the performance of our approach, we train 7 other methods on the FFHQ and AAHQ datasets: Swapping Autoencoders [19], Gatys [6], ArtFlow [3], PAMA [17], SANet [18], NNST [14], and ContraAST [4]. From these, Swapping Autoencoders is the most closely related, as it also operates over a StyleGAN2 space, thus, we separate it from the other more traditional methods which do not.

Table 1 shows the LPIPS, SIFID, and timing results computed for our own HyperNST method, and the other methods we evaluate against. The results show the superiority of HyperNST in retaining the most semantic structure in the facial reconstructions, indicated by the lowest LPIPS value. Meanwhile, we achieve comparable SIFID values (2.279) to Swapping AutoEncoders (1.948), indicating a small gap in stylization quality. Only our HyperNST model can embed a portrait into a model with semantic editing capabilities. Finally, the inference time of our model is similar to other methods and low enough for practical use of the model in real applications.

One disadvantage of the hypernetwork based approach is the current limitation with how strongly the StyleGAN2 model can be fine-tuned. As per our findings and the HyperStyle authors', hyper-learning weight deltas for every individual weight would be infeasible on current hardware (requiring over 3 billion parameters for a StyleGAN2 model) - we are limited by the technology of our time. So at best, we can aim to at most match a method where the model is fully trained, the closest comparison being Swapping AutoEncoders. The Swapping AutoEncoders model also fine-tunes a StyleGAN2 model. However, the more straightforward training approach affords the luxury to fine-tune individual weights in the generator, compared to the channel-wise mean shifts, currently possible with a hyper-network on today's hardware.

Nevertheless, hyper-network based NST is interesting to study, as such limitations are temporary. Both model and hardware improvements with subsequent works will reduce such limitations in time. We can compare hyper-network approaches directly with fully-trained counterparts for the current state of the art.

Table 2. Ablation results for varying the style loss strength

Style loss strength	LPIPS (Alexnet)	LPIPS (VGG)	SIFID
0.5	0.001480	0.002657	3.312
1.0	0.001310	0.002626	2.590
2.0	0.000042	0.001688	2.279
5.0	0.000097	0.000191	2.463
10.0	0.000091	0.000222	2.571

Fig. 2. Qualitative HyperNST visualization of portrait style transfer.

Figure 4 visualizes some style transfer results obtained with a Swapping Autoencoders (SAE) model, re-trained with the AAHQ dataset. Figure 2 shows the same visualization, obtained with HyperStyle. It is worth noting that the SAE model generates 1024 × 1024 images, whereas the target StyleGAN2 model in the hypernetwork setting uses a smaller, 256 × 256 version. Quality of stylization is somewhat equivocal for these approaches; in most cases, the SAE results

Content   L3   L7   L13   L17   L22   Style

Fig. 3. Representative visualization of the effect of choosing where in the target Style-GAN2 layers to start fine-tuning hyper-weight generators for. The left-most image is the reference content image, and the right-most is the reference style image. In the middle, from left to right, images represent visualizations for layers 3, 7, 13, 17, and 22.

Fig. 4. Baseline comparison to Swapping Autoencoders (re-trained with AAHQ), the closest work - a fully per-weight-trained model.

are slightly more faithful in their style transfer accuracy but are less faithful at retaining the semantic structure of the portraits vs. HyperNST. The SAE model also has difficulty stylizing backgrounds.

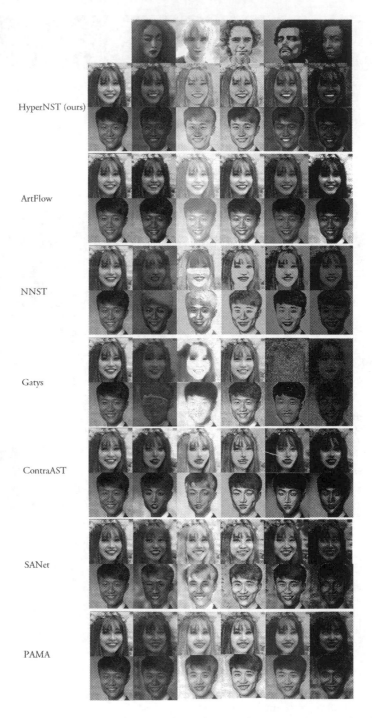

Fig. 5. Stylization comparison of HyperNST (1st pair), against Artflow (2nd), NNST (3rd), Gatys et al. (4th), ContraAST (4th), SANet (5th), and PAMA (6th)

Table 3. Ablation results for varying the starting StyleGAN2 layer past which layers are included in the stylization fine-tuning.

Layer	LPIPS (Alexnet)	LPIPS (VGG)	SIFID
3	0.000140	0.00399	3.123
7	0.000190	0.00302	2.792
13	0.000042	0.00169	2.279
17	0.000153	0.00268	3.337
22	0.000042	0.00104	3.506

4.3 Ablations

We perform the stylization fine-tuning on only part of the StyleGAN2 layers' hyper-weight generators. Out of 25 layers, we find layer 13 to be a good balance between inducing style transfer and retaining semantic structure. The earlier the target StyleGAN2 layers that hyper-weight generators get fine-tuned for stylization, the stronger the emphasis on style is, therefore losing structure reconstruction quality. Conversely, the later the starting layer is, the less pronounced the stylization is, thereby focusing the training more on the semantic reconstruction quality and instead just performing a color transfer. Figure 3 shows a representative visualization of this effect, and Table 3 shows quantitative metrics. The images further to the left represent a deeper fine-tuning of layers of style and losing structure. The images to the right show a more shallow fine-tuning for style, retaining more structure from the original content image. We targeted these specific layers, as these are toRGB layers in the target StyleGAN2 layer, which are known to more significantly affect color and texture.

Finally, we ablate the use of facial semantic region information in the model and training pipeline. We explore the effect of including the semantic regions as a conditioning signal in the pipeline (cond) and as a way to match region types in the patch co-occurrence discriminator (loss). We populate the results in Table 4, showing the usefulness of these components.

Table 4. Ablation results for varying the facial semantic information used in the architecture.

Experiment	LPIPS (Alexnet)	LPIPS (VGG)	SIFID
Baseline (with both cond and loss)	0.00004	0.0017	2.279
No mask cond and no mask loss	0.00017	0.0024	2.892
No mask cond	0.00012	0.0025	2.848
No mask loss	0.00016	0.0026	3.438

4.4 Downstream Experiments

Style Interpolation. In Fig. 6, we visualize a controllability aspect of the stylization process based on interpolations done in ALADIN style space. Basing the stylization upon a metric style embedding space, it is possible to perform

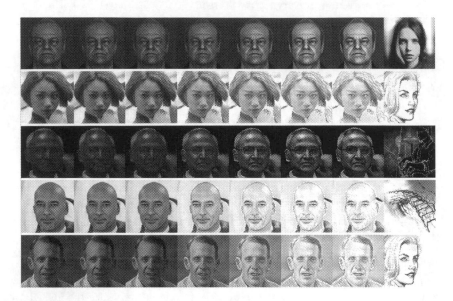

Fig. 6. Visualization of style interpolation between a content image, and a target style image. Note in this example, we show that a non-portrait style image can also be used.

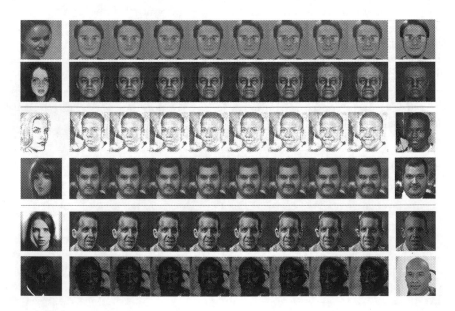

Fig. 7. Semantic face/portrait editing capabilities. The top two example rows range across the *age* vector, the second two example rows show the *smile* vector, and the third shows the *pose* rotation vector.

smooth transitions between the style of a reference content image and a target style image. We also show in this example that it is possible to perform stylization using style images not containing portraits. In this case, the image-wide ALADIN code is used instead of semantically arranged per-feature ALADIN codes.

Semantic Controlability. One advantage over the SAE model is the editable method of GAN inversion that the HyperNST model uses. Coupled with Inter-FaceGAN [24], latent directions in the StyleGAN2 weight space can be used to expose semantic controls over portrait aspects such as age, pose, and facial expression. On top of the artistic style aspects of the style transfer capabilities of HyperNST, these semantic abilities offer a deeper level of artistic control over the editing and generation of portraits. Figure 7 visualizes some style transfer examples, where these latent directions are also used in tandem for more advanced control. We show the *smile*, *pose*, and *age* vectors.

5 Conclusion

We explored a novel application of a class of models, hypernetworks, to the task of style transfer. We explored the merits and drawbacks of such an approach and what is currently possible. The hypernetwork approach is excellent at maintaining high quality semantic structure and allows for artistic controls, such as semantic editing and style-code-based stylization interpolation.

The main limitation of the current implementation of hyper-weight prediction is the very high VRAM consumption of such a model (22GB for a batch size of 1) during training, forcing weight deltas to affect the target generator only on a per-channel mean shifting basis. It is possible that the style transfer quality could be pushed further with a larger batch size. For example, the batch size used by Swapping Autoencoders is 16. This could be mitigated by exploring smaller alternatives to StyleGAN2, or by implementing GPU optimization tricks such as memory/compute tradeoff training approaches based on gradient checkpointing. Such reductions in VRAM use could also afford an increase in batch size, bringing the training more in line with previous works, primarily affecting discriminators. A more promising direction could be to more deeply analyse which layers can be omitted from training (i.e. need not be updated by the hyper network). In this paper, we focused our attention on the domain of human portraits. However, other domains would be equally important to study, further generalizing this novel approach to neural artistic stylization using more general semantic segmentation techniques or larger semantically diverse models such as StyleGAN XL, once hardware evolves to support hyper-network inference of larger models' weights.

References

1. Alaluf, Y., Patashnik, O., Cohen-Or, D.: Restyle: a residual-based stylegan encoder via iterative refinement. CoRR, abs/2104.02699 (2021)

2. Alaluf, Y., Tov, O., Mokady, R., Gal, R., Bermano, A.H.: Hyperstyle: Stylegan inversion with hypernetworks for real image editing. CoRR, abs/2111.15666 (2021)
3. An, J., Huang, S., Song, Y., Dou, D., Liu, W., Luo, J.: Artflow: unbiased image style transfer via reversible neural flows. CoRR, abs/2103.16877 (2021)
4. Chen, H., et al.: Artistic style transfer with internal-external learning and contrastive learning. In: Beygelzimer, A., Dauphin, Y., Liang, P., Wortman Vaughan, J. (eds.) Advances in Neural Information Processing Systems (2021)
5. Gal, R., Patashnik, O., Maron, H., Chechik, G., Cohen-Or, D.: Stylegan-nada: Clip-guided domain adaptation of image generators. CoRR, abs/2108.00946 (2021)
6. Gatys, L.A., Ecker, A.S., Bethge, M.: Image style transfer using convolutional neural networks. In: Proceedings of CVPR, pp. 2414–2423 (2016)
7. Ha, D., Dai, A.M., Le, Q.N.: Hypernetworks. CoRR, abs/1609.09106 (2016)
8. Huang, X., Belongie, S.: Arbitrary style transfer in real-time with adaptive instance normalization. In: Proceedings of ICCV (2017)
9. Huang, X., Liu, M.-Y., Belongie, S., Kautz, J.: Multimodal unsupervised image-to-image translation. In: Ferrari, V., Hebert, M., Sminchisescu, C., Weiss, Y. (eds.) ECCV 2018. LNCS, vol. 11207, pp. 179–196. Springer, Cham (2018). https://doi.org/10.1007/978-3-030-01219-9_11
10. Karras, T., et al.: Alias-free generative adversarial networks. In: Proceedings of NeurIPS (2021)
11. Karras, T., Laine, S., Aila, T.: A style-based generator architecture for generative adversarial networks. CoRR, abs/1812.04948 (2018)
12. Karras, T., Laine, S., Aila, T.: A style-based generator architecture for generative adversarial networks. In: Proceedings of CVPR (2019)
13. Karras, T., Laine, S., Aittala, M., Hellsten, J., Lehtinen, J., Aila, T.: Analyzing and improving the image quality of stylegan. CoRR, abs/1912.04958 (2019)
14. Kolkin, N., Kucera, M., Paris, S., Shechtman, E., Shakhnarovich, G.: Neural neighbor style transfer (2022)
15. Krizhevsky, A., Sutskever, I., Hinton, G.E.: ImageNet classification with deep convolutional neural networks. In: Proceedings of the 25th International Conference on Neural Information Processing Systems, , NIPS 2012, Vol. 1, pp. 1097–1105, Red Hook, NY, USA, 2012. Curran Associates Inc. (2012
16. Liu, M., Li, Q., Qin, Z., Zhang, G., Wan, P., Zheng, W.: Blendgan: implicitly GAN blending for arbitrary stylized face generation. CoRR, abs/2110.11728 (2021)
17. Luo, X., Han, Z., Yang, L., Zhang, L.: Consistent style transfer. CoRR, abs/2201.02233 (2022)
18. Park, D.Y., Lee, K.H.: Arbitrary style transfer with style-attentional networks. CoRR, abs/1812.02342 (2018)
19. Park, T., et al.: Swapping autoencoder for deep image manipulation. CoRR, abs/2007.00653 (2020)
20. Radford, A., et al.: Learning transferable visual models from natural language supervision. arXiv preprint arXiv:2103.00020(2021)
21. Ruta, D., et al.: Artistic style tagging and captioning (2022)
22. Dan Ruta, S., et al.: All layer adaptive instance normalization for fine-grained style similarity. arXiv preprint arXiv:2103.09776 (2021)
23. Shaham, T.R., Dekel, T., Michaeli, T.: SinGan: learning a generative model from a single natural image. CoRR, abs/1905.01164 (2019)
24. Shen, Y., Yang, C., Tang, X., Bolei Zhou, X.:InterFaceGAN: interpreting the disentangled face representation learned by gans. CoRR, abs/2005.09635 (2020)
25. Simonyan, K., Zisserman, A.: Very deep convolutional networks for large-scale image recognition (2014)

26. Tov, O., Alaluf, Y., Nitzan, Y., Patashnik, O., Cohen-Or, D.: Designing an encoder for stylegan image manipulation. CoRR, abs/2102.02766 (2021)
27. Zhang, R., Isola, P., Efros, A.A., Shechtman, E., Wang, O.: The unreasonable effectiveness of deep features as a perceptual metric. CoRR, abs/1801.03924 (2018)
28. Zheng, Y., et al.: General facial representation learning in a visual-linguistic manner. CoRR, abs/2112.03109 (2021)

DEArt: Dataset of European Art

Artem Reshetnikov$^{(\boxtimes)}$, Maria-Cristina Marinescu ,
and Joaquim More Lopez

Barcelona Supercomputing Center, Barcelona, Spain
{artem.reshetnikov,maria.marinescu,joaquim.morelopez}@bsc.es

Abstract. Large datasets that were made publicly available to the research community over the last 20 years have been a key enabling factor for the advances in deep learning algorithms for NLP or computer vision. These datasets are generally pairs of aligned image/manually annotated metadata, where images are photographs of everyday life. Scholarly and historical content, on the other hand, treat subjects that are not necessarily popular to a general audience, they may not always contain a large number of data points, and new data may be difficult or impossible to collect. Some exceptions do exist, for instance, scientific or health data, but this is not the case for cultural heritage (CH). The poor performance of the best models in computer vision - when tested over artworks - coupled with the lack of extensively annotated datasets for CH, and the fact that artwork images depict objects and actions not captured by photographs, indicate that a CH-specific dataset would be highly valuable for this community. We propose DEArt, at this point primarily an object detection and pose classification dataset meant to be a reference for paintings between the XIIth and the XVIIIth centuries. It contains more than 15000 images, about 80% non-iconic, aligned with manual annotations for the bounding boxes identifying all instances of 69 classes as well as 12 possible poses for boxes identifying human-like objects. Of these, more than 50 classes are CH-specific and thus do not appear in other datasets; these reflect imaginary beings, symbolic entities and other categories related to art. Additionally, existing datasets do not include pose annotations. Our results show that object detectors for the cultural heritage domain can achieve a level of precision comparable to state-of-art models for generic images via transfer learning.

Keywords: Deep learning · Computer vision · Cultural heritage · Object detection

1 Introduction

Cultural heritage (CH) is important not only for historians or cultural and art institutions. This is an area that permeates society and lends itself to research

Supplementary Information The online version contains supplementary material available at https://doi.org/10.1007/978-3-031-25056-9_15.

education and cultural or social projects. Tapping deeper into the richness and potential of CH rests on making it explainable and accessible, and requires efficient indexing and search capabilities. These are built on the premise that quality metadata exists and is available. GLAM (Gallery, Library, Archive, Museum) institutions have been annotating CH artefacts with rich metadata for a long time, but their approach has two shortcomings: (1) the annotations are generated manually as part of a slow and laborious process, and (2) the annotations are usually about the context and making of the artifact or its author. The assumption is that one sees the object and there is little need to describe it visually. This isn't always true for people (e.g. in the case of the visually impaired) and it is certainly false when information is needed for automatic consumption by computers. Enabling visual content annotations via an automatic recommendation process could address both these problems. A good way to approach this challenge is to develop deep learning models for image classification, object detection, or scene understanding. The precision of such models rests on the existence of a large quality dataset, annotated for the target task. Datasets such as MS COCO [18] or Pascal VOC [7] exist and many deep learning algorithms have been developed that return very good results for photographs of everyday life. When testing these models on artworks, the precision of object detection drops significantly (Sect. 4). This is due to reasons of both form and content: paintings are executed in many styles, they have been created to transmit meaning, they capture symbols, depict imaginary beings or artifacts, and reflect objects not in use anymore or actions that we don't often see in photographs. Additionally, there exists an inherent limitation of the number of existing CH artifacts as opposed to the conceptually unbounded number of photographs that can be taken. All these factors make the existing approaches inadequate for CH. This paper introduces the largest and most comprehensive object detection dataset for CH with annotated poses. To the extent of our knowledge, the most extensive CH dataset to date has about 6000 images and annotates 7 classes [9, 11, 27]. Our dataset has 15K images of European paintings between the XIIth and the XVIIIth century. We manually annotate all objects (labels and bounding boxes) corresponding to instances of our 69 classes. Our object detection model has a precision of 31.2%, which is within 87% of the state-of-the-art models for photographs. To reach this precision we use transfer learning based on the Faster RCNN [23] model trained on the MS COCO dataset [18]. In addition to class labels, bounding boxes representing human-like objects also include pose labels, for all the images in our dataset. This is a type of information that identifies an important subset of human-related action verbs. Our paper makes the following main contributions: (1) a dataset of 15K images of paintings between the XIIth-XVIIIth century, annotated for object detection and pose classification; (2) a proof-of-concept that an object detection model can achieve an average precision that is close to that of state-of-the-art models for other domains, and (3) an extensive evaluation, including comparisons with SotA models in computer vision and existing object detections models for cultural heritage. We will make the dataset and model available in August 2022 (DOI:10.5281/zenodo.6984525).

2 Related Work

The publishing of large datasets such as Pascal VOC [7], MS COCO [18], or Open Images [17] enabled researchers in computer vision to improve the performance of the detection of basic object categories in real-life photographs. MS COCO [18] contains images of objects in their natural environments, which makes this dataset amenable for application in different domains such as video surveillance, security, mobile applications, face and people recognition, etc; neither of these datasets performs well when applied to CH images. The problem of object detection for paintings has not been studied extensively and is inherently more difficult due to issues such as a big variation in style, the lack of annotated data, and a small dataset by deep learning standards. Artworks often contain objects that are not present in (recent) everyday life; additionally, CH needs more precise annotations than just broad classes. For instance, we need to know that a person is a monk, a king, or a knight to even start understanding the content or meaning of a scene; a person would not serve the purpose. In reality, what we primarily find in CH is research that is based on restricting the object detection problem to a single class, namely a person. Westlake et al. (2016) [27] proposed to perform people detection in a wide variety of artworks (through the newly introduced PeopleArt dataset) by fine-tuning a network in a supervised way. PeopleArt includes 4821 annotated images. Ginosar et al. (2014) [10] used CNNs for the detection of people in cubist artworks; the authors used 218 paintings of Picasso (a subset of PeopleArt) to fine-tune a pre-trained model. Yarlagadda et al. (2010) [28] collected a dataset for the detection of different types of crowns in medieval manuscripts. More recent approaches for object detection in CH are based on weakly supervised learning or use other ways to generate synthetic data. Gonthier et al. (2018) [11] introduce IconArt, a new dataset with 5955 paintings extracted from WikiCommons (of which 4458 are annotated for object detection) that annotate 7 classes of objects; their approach is based on a (weakly supervised) multiple-instances learning method. Kadish et al. (2021) [15] propose to use style transfer (using AdaIn [14]) to generate a synthetic dataset that recognizes the class person starting with 61360 images from the COCO dataset. Both IconArt [11] and PeopleArt [27] have important weaknesses: they don't represent significant variations in the representation of class instances, and the list of classes is not representative of the objects represented in cultural heritage. Lastly, Marco Fioorucci et al. (2020)'s [8] survey identifies PrintArt, introduced by Carniero et al. (2012) [3], which contains 988 images and identifies 8 classes. Several other computer vision datasets for cultural heritage exist, such as SemArt (Garcia et al.(2018)) [9], OmniArt (Strezoski et al.(2017)) [26], ArtDL (Milani et al. (2020)) [20], or The Met Dataset (Garcia et al.(2021)) [29]. However, all these datasets focus on image classification or caption generation and they don't have bounding box annotations that are necessary for object detection. For the task of pose classification from 2D images, most leading approaches are using part detectors to learn action-specific classifiers. Maji et al. [19] train action-specific pose lets and for each instance create a pose let activation vector that is being classified using SVMs. Hoai et al. [7]

use body-part detectors to localize the human and align feature descriptors and achieve state-of-the-art results on the PASCAL VOC 2012 action classification challenge. Khosla et al. [16] identify the image regions that distinguish different classes by sampling regions from a dense sampling space and using a random forest algorithm with discriminative classifiers. However, we didn't find any work related to CH. Our approach is based on using CNN for pose classification.

3   Dataset

3.1   Object Categories

Selecting the set of CH classes for object detection is a non-trivial exercise; they must be representative for the time period and they should represent a wide range of the possible iconographic meanings. We are exclusively interested in visual classes; this excludes concepts such as thought, intention, mother, etc, given that they are based on assumptions or knowledge not directly apparent from an image. Starting from the MS COCO Dataset [18], we perform a first step of chronological filtering to keep only the classes whose first known use (per the Merriam Webster dictionary [6]) is found before the XIXth century; Table 1 shows these results. Some of them may look incorrect but they are polysemic words with different meanings in the past, e.g. (1) TV: "A sort of annotation in manuscripts" (First known use: 1526) or (2) Keyboard: "A bank of keys on a musical instrument" (First known use: 1776).

Table 1. MS COCO Dataset classes after chronological filtering

'person','car', 'bus', 'train', 'truck', 'boat', 'bench', 'bird', 'cat', 'dog', 'horse', 'sheep', 'cow', 'elephant', 'bear','zebra','giraffe', 'umbrella', 'tie', 'skis', 'kite', 'tennis racket', 'bottle', 'wine glass', 'cup', 'fork', 'knife', 'spoon', 'bowl', 'banana', 'apple', 'sandwich', 'orange', 'broccoli', 'carrot', 'hot dog', 'cake', 'chair', 'couch', 'bed', 'dining table', 'toilet', 'tv', 'mouse', 'keyboard', 'oven', 'sink', 'refrigerator', 'book', 'clock', 'vase', 'scissors', 'toothbrush'

In a second step, objects not depicted in the present or not real - but present in paintings - must be added to complete the class list. We collect artistic topics which correspond to categories represented in paintings by using Wikimedia Commons as the point of entry. This includes categories that match the regular expression "Paintings of *". For each class C in the filtered MS COCO classes (Table 1), we query the Wikimedia API for the category "Paintings_of_C" to find out whether it qualifies as a painting class. Table 2 shows the MS COCO classes that qualify as painting classes. Most of these classes represent generic individuals and objects devoid of any symbolic or iconographic meaning. To avoid missing a whole range of possible objects which are significant in paintings, we need to include a set of important additional classes. Consider a painting of the crucifixion; Jesus Christ should not be referred to as just a person if we don't want to miss his historical significance and his role in what forms the basis of western iconography.

Table 2. MS COCO Dataset classes representing painting classes

'people', 'trains', 'boats', 'birds', 'cats', 'dogs', 'horses', 'sheep', 'cows', 'elephants', 'bears', 'zebras', 'knives', 'bananas', 'apples', 'oranges', 'chairs', 'beds', 'mice', 'books', 'clocks', 'vases'

The Wikimedia Commons categories and subcategories are very useful to identify new and richer painting classes by querying for Paintings of_COCO class_or its hyponyms; "Paintings of humans" or "Paintings of women" as stand-ins for the MS COCO class person returns categories such as "Paintings_of_ angels_with_humans". As a result, angels is added as a new painting class. The pseudocode below illustrates the algorithm of adding new classes starting from an initial class C (in the class list of Table 2). To avoid generating a huge list of classes in T, we go through a maximum of three nested levels of recursion; this already results in a significant number of classes (963). For this reason, we refine the pseudocode to further filter the classes added to T based on scope and the Wikimedia category hierarchy, as explained below; these cases, including exceptions, are captured in the pseudocode as the predicate filter_out. The scope extends to those classes that are found in vocabularies that are commonly used by cultural heritage institutions. Classes were maintained when it was possible to find them in the vocabularies used by one of the following collections:

Table 3. Classes selection pseudocode

$T = \{C\}$

For each painting class $C \in T$

  Obtain set S of WCo subcategories containing C (sg/pl)

   For each S_c in S

    $S = S/\{S_c\}$

    If S_c is represented by a string with no '_' character // no further relationships exist

      If $S_c \notin T$ and not *filter_out*(S_c) // S_c not already inferred from WCo categories

        $T = T \cup S_c$

      Otherwise

        L_S = list of all words preceded by '_' in string representing S_c

        For each $K_S \in L_S$

          If $K_S \notin T$ and not *filter_out*(K_S) // K_S not already inferred from WCo categories

            $T = T \cup K_c$

Wikidata, Brill Iconclass AI Test Set, Netherlands Institute for Art History, and the Europeana Entity Collection. Class names that include geographical or positional qualifiers are maintained in form of the unqualified class, e.g. the class kings of Portugal is included as kings (if it isn't already part of the set); likewise, sitting person is retained as person. Categories with less than 10 child subcategories in Wikimedia taxonomy are filtered out[1]. This number is the result of experiments to reach a tradeoff between the final number of classes and covering a sufficiently representative set. Exceptions apply to maintain (a) all categories that are subclasses of MS COCO in Table 2 and (b) categories that have less than 10 subcategories but were identified as being of specific interest. The categories of specific interest are those which iconographically characterize a category already included as a class, according to the criteria explained in Sect. 3.1, or those that are related to a class that is already included because they are linked or redirected in a linked data database such as Dbpedia.

A last case is also the most interesting: we do not include classes that refer to complex concepts that may be important in CH due to their symbolic value, for instance recognizable events such as Annunciation, Adoration of the Magi, the Passion of Jesus Christ, etc. The reason for this choice is that there usually exists a big variation in the way that they are represented. We believe that a better way to solve this challenge is at a semantic level, by reasoning about sets of simple objects and their relationships to infer higher-level knowledge.

As a result of this process we obtain 69 classes that are either subclasses of a MS COCO superclass (e.g: fisherman - Person; Judith - Person; swan - animal), they are identified with iconographic features, they bear symbolic meaning (e.g: dragon, Pegasus, angel).

3.2   Pose Categories

The DEArt dataset also contains a classification of poses for human-like creatures. We believe pose classification is useful for several reasons. Being able to automatically detect poses and associate them with bonding boxes is an effective way to create rich data structures - e.g. Knowledge graphs - that can be used for search, browse, inference, or better Query and Answer systems. While descriptions are mostly meant to be consumed by end-users, triples of objects connected by relationships are ideal for machine consumption; poses provide such a subset of action verbs. NLP methods to extract triples (i.e. entities and relationships) from descriptions works best when parsing and analyzing phrases that are simple, otherwise it may wrongly identify objects and/or the relationships between them. At the same time, good image descriptions in natural language tend to be complex; see for instance the descriptions in SemiArt [9] or The Prado Museum's website. This makes triple extraction challenging and underlines the importance of automatic pose classification for bounding-boxes. SemiArt [9] contains descriptions that add up to more than 3000 statements which include pose verbs in our

---

[1] Other filters may be implemented, e.g. testing the number of artworks in a category rather than the number of subcategories.

list. Some of these refer to word meanings which are not pose-related, e.g. "fall in love", "move to [a city]", "the scene is moving", "run a business", or "stand out". We set up an experiment and ran SpaCy [13]with 500 of these statements. Of these, 104 have a subject that is a pronoun or an anaphoric expression (e.g. "the other"), in 49 the subject is wrongly detected (e.g. "Sheltered by this construction stands the throne, with the Madonna and Child and four saints.": subject of stands - sheltered), and 26 detect an incomplete subject. Given that SpaCy [13] parses phrase by phrase, rather than taking into consideration the larger context, it isn't able to resolve co-references. Other more advanced methods such as NeuralCoref [5] may be used, although in our experience these don't work fully well for complex language. Regardless of this, the problem of associating poses with bounding boxes stands when there exists more than one bounding box with the same label.

In addition to this challenge, at a higher level of abstraction, the pose of a human-like creature may have symbolic meaning, especially in iconographic images. Different poses may thus give the artwork a very different meaning.

To take the first step towards these goals, we generate a list of relevant poses based on a mix of two approaches: one manual, based on the opinion of cultural heritage specialists, and the second automatic, based on unsupervised learning. By analyzing about 200 random images from our dataset, a cultural heritage expert suggested 12 main pose categories which appear relevant to our dataset and can be interesting for the future enrichment of metadata with actions (see Table 4). Independently, we extracted all the verbs present in the descriptions of the 19245 images in the SemiArt [9] dataset. All descriptions were tokenized and, using part-of-speech tagging, we created a pool of verbs from which we chose the 2000 most common. This approach is based on verbs clustering and the intuition is that verbs appear in the same cluster if they are related to the similar actions. Iterative filtering allows us to keep clusters which are related to poses of human-like creatures in artworks. For this purpose, we use pre-trained GloVe [22] vectors to represent verbs in an embedding space so that we can maintain

Table 4. List of classes suggested by specialists

'walk/move/run/flee', 'bend', 'sit', 'stand', 'fall', 'push/pull', 'throw', 'kneel', 'lift/squat', 'beg/pray', 'bow', 'step on', 'drink/eat', 'lay down'

Table 5. List of classes extracted by suggested approach

'lift/squat', 'move', 'watch', 'falls', 'kneel', 'ride', 'stand', 'eat', 'sit', 'pull/push', 'fly', 'sleep', 'bend', 'beg/pray'

Table 6. Final list of classes

'bend',' fall',' kneel','lay down', 'partial',' pray',' push/pull',' ride',' sit/eat', 'squats','stand','walk/move/run', 'unrecognizable'

their semantic meaning. In each iteration we define the best value for the number of clusters based on computing a score for a number clusters between 1 to 500 and choosing the number with the highest score. We then delete clusters which are not suitable for our purpose, i.e. are not about poses and represent no visual image. The best (but also more computationally expensive) approach to defining the optimal number of clusters during each iteration is to use the silhouette score [1]. The silhouette coefficient for an instance is computed as $(b - a)/max(a, b)$, where a is the mean distance to the other instances in the same cluster (i.e., the mean intra-cluster distance) and b is the mean nearest-cluster distance (i.e., the mean distance to the instances of the next closest cluster, defined as the one that minimizes b, excluding the instance's own cluster). The silhouette coefficient can vary between -1 and +1. A coefficient close to +1 means that the instance is well inside its own cluster and far from other clusters, 0 means that it is close to a cluster boundary, and a coefficient close to -1 means that the instance may have been assigned to the wrong cluster. After 42 iterations we were able to define 14 main clusters that are related to poses of human-like creatures (See Table 5)

Our final list of categories of poses is the intersection of the two approaches, to which we add the class "ride", while we include class "eat" into class "sit" based on the fact that someone who eats is also sitting and almost never actually eating but rather having food before them. We also add the class "partial" and "unrecognizable" for those pictures where the human-like creature is only partially represented (e.g. a man from the torso up) and for those images where action can't be recognized (see Table 6)

3.3   Image Collection Process

Everyday language defines something to be "iconic" to mean that it is well-established, widely known, and acknowledged especially for distinctive excellence. Research in computer vision (CV) uses its own concept of "iconic" to follow the definition introduced by Berg et al. (2009) [2]. Typical iconic-object images include a single object in a canonical perspective [21]. Iconic-scene images are shot from canonical viewpoints [21] and commonly lack people. Non-iconic images contain several objects with a scene in the background. Figure 1 gives some typical examples. For the rest of this paper, we use the CV terminology when we refer to "iconic".

Iconic images are easier to find using search engines or tags on data aggregation platforms, and they are more straightforward to annotate manually. At the same time, they tend to lack the contextual information present in non-iconic images and thus generally represent concepts with no greater complexity or symbolism. In non-iconic images, additional objects play a key role in scene understanding because they bring in an additional layer of symbolic information. To be able to perform inference and iconographic understanding of scenes in the future, our goal is to collect a dataset with a majority of non-iconic images. Our data comes from the following sources: the Europeana Collections, Wikimedia Commons, British Museum, Rijksmuseum, Wikipedia, The Clark Museum, The Cleveland Museum of Art, Harvard Art Museum, Los Angeles

Fig. 1. Examples of Iconic-object, iconic-scene and non-iconic images

County Museum of Art, The Leiden Collection, Paul Mellon Centre, Philadelphia Museum of Art, National Gallery of Scotland, Museo Nacional Thyssen-Bornemisza, Victoria and Albert Museum, National Museum in Warsaw, Yale University Art Gallery, The Metropolitan Museum of Art, National Gallery of Denmark, The National Gallery of Art, The Art Institute of Chicago, WikiArt, PHAROS: The International Consortium of Photo Archives. When possible, in addition to checking the time and place of the creation of the artwork, we filter by license type and only select images open for reuse. Figure 2 illustrates the number of images downloaded from each source.

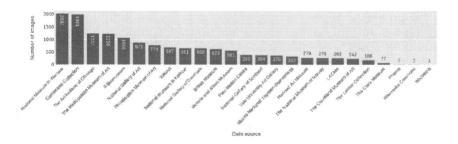

Fig. 2. Distribution of images by data source

3.4 Image Annotation

The annotations we are interested in are of two types: bounding boxes with their object labels and pose labels associated with bounding boxes.

We annotate both objects and poses of human-like objects in images following the rules described in "The PASCAL Visual Object Classes (VOC) Challenge" [7] and set three objectives: (1) Consistency: all annotators should follow the same rules with regard to the list of classes, the position of bounding boxes, and truncated objects. We achieve consistency by exclusively using the project team members (from two institutions) as annotators and following guidelines we drew after extensive discussions. We use LabelMe as the user interface [25] for object annotations. (2) Accuracy: the bounding boxes should be tight, and object labels must be from the list of classes. Poses of human-being creatures were classified according to the list of classes. We achieve accuracy by manual checking. (3) Exhaustiveness: all object instances must be labeled. We achieve exhaustiveness by manual checking. The object labels and bounding boxes are generated as follows:

The object labels and bounding boxes are generated as follows:

1. An original image selection process chose paintings after the XIIth century until the XVIIIth century. The result is a dataset of 15K images, of which about 80% are non-iconic. Our research team members annotated 10K images manually;
2. The rest of 5K was annotated in a semi-supervised manner by using the currently trained model and correcting the annotations manually, in 3 ingestion phases of 2K each, followed by retraining. The training was done over 70% of the dataset. Semi-supervised learning helps us reduce the time for the annotation process, given that the human annotator does not start from zero, but rather accepts or fixes annotations recommended by the current model. Each data ingestion phase consists of several steps: (1) Collecting data, (2) Detecting objects using the model trained on the previous version of the dataset, (3) Manually correcting bounding boxes to resize them around the objects, eliminate falsely detected objects, rename falsely labeled boxes, and manually annotate undetected objects, and (4) Include corrected data into the dataset and retrain the model for the next ingestion iteration. During the entire annotation process, the dataset is periodically observed to double-check the quality of the annotations. This step is performed approximately every time 2000 new images are added, and it evaluates the 10 classes with the most number of instances: person, tree, boat, angel, halo, nude, bird, horse, book, and cow. The evaluation subset is created by randomly extracting 100 images per category.

The same process was applied for the annotation of poses of human-like creatures. Using semi-supervised learning, the research team (we) periodically trained a classification model based on annotated data. In later steps, the research team (we) corrected the generated classes rather than annotating them manually.

3.5 Dataset Statistics

Figure 3a illustrates the total number of instances per class for each of the 69 categories. A class person is by far the most frequent in the dataset, currently with around 46990 instances; the bar for a class person is truncated to make minority classes such as "saturno", "stole" or "holy shroud" visible. Given that instances of minority classes are rarely the subject of iconic images but they can appear in non-iconic images, adding complex scene images may help increase the total number of minority instances.

Figures 3b and 3c illustrate the amount of contextual information [21] in our dataset, which we define as the average number of object classes and instances per image. The figures confirm that during the process of data collection we pay special attention to choosing non-iconic images to be able to obtain a rich dataset. During the annotation process, we made a periodic health check analysis of the dataset to understand which classes were over or under-represented (minority classes). The results directed our search to prioritize images that include minority classes. Differently from photographs, sometimes we have inherent limitations because we cannot produce additional old paintings if they don't exist. The entire annotation process resulted in 105799 bounding boxes. Additionally, we classified 56230 human-like creatures.

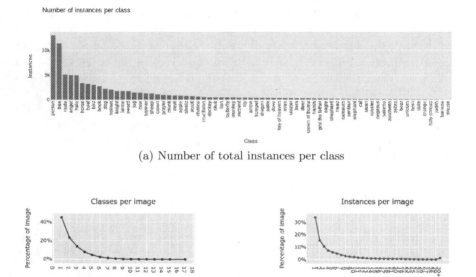

(a) Number of total instances per class

(b) Proportion of the dataset with specific number of classes

(c) Proportion of the dataset with specific number of instances

Fig. 3. Dataset statistics

4   Experiments

4.1   Object Detection

We implement a deep learning architecture based on Faster R-CNN [23] to train object detection models over the DEArt dataset. We split our dataset into 70% training, 15% validation, and 15% test sets (Training set: 10500, Validation/Test sets: 2250). The choice of images is random within each class, and we use the annotated-images[2] Python library to select images such that these percentages

---

[2] https://github.com/SaberD/annotated-images.

are as closely as possibly met for each of the 69 classes. After comparing several possibilities among the different ImageNet models [24], we concluded that residual networks (ResNet [12]) are the best architecture for feature extraction - which coincides with [13]. Taking into account that photograph datasets such as MS COCO [18] contain about 20 times more images, and to increase the final precision of object detection, we decided to use a transfer learning approach. Concretely, we use the Resnet-152 V1 [12] Object detection model, pretrained on the MS COCO 2017 [18] dataset. Given that our dataset is in Pascal VOC format, we chose AP@0.5 per class and mAP@0.5 [7] as evaluation metrics.

Our Faster R-CNN [23] model results in an mAP@0.5 = 31.2; the average precisions by class are presented in Fig. 4.

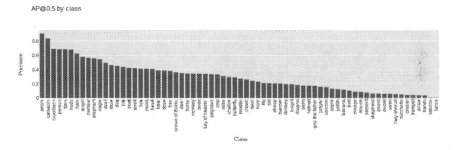

Fig. 4. The average precision by class for Faster RCNN trained on DEArt dataset

The results of our experiments show that including many complex (non-iconic) images in the training set may not always help when evaluating class by class. Such examples may act as noise and contaminate the model if it isn't rich enough to capture the variability in a visual representation of the objects, especially if we consider the sizeable number of classes we cover relative to the number of images of the dataset. Furthermore, we compared the results of the evaluation with object detection datasets. Table 7 shows the results of applying three pretrained models (trained on MS COCO, PASCAL VOC, Open Images) over test set of DEArt; the last rows show the results of our the model tained on our dataset. We present AP@0.5 results for the 16 classes that are included both in MS COCO and our dataset. We can clearly see that our model significantly improves the precision of detection of every-day objects as they are depicted in cultural heritage artifacts, when compared to the photograph-based models. At the same time, MS COCO, PASCAL VOC, and Open Images do not cover - and therefore could not recognize - the other 53 cultural heritage-specific classes. On a different note we want to mention some examples of the fall in precision of the CV models when applied to our dataset: for class person, MS COCO goes from 0.36 (when tested over their data) to 0.25 (over our data), while PASCAL VOC goes from 0.22 down to 0.05. For class sheep, MS COCO goes from 0.28 to 0.15 and PASCAL VOC from 0.17 to 0.004. For class horse, 0.44 to 0.2 (MS COCO) and 0.33 to 0.03 (PASCAL VOC).

4.2   Pose Classification

We built a small version of the Xception [4] network for pose classification. Taking into account that we have a large amount of data, we decided to train the model from scratch using a customized architecture. We used KerasTunner to tune the hyperparameters and architecture, and we split our dataset into 70% training, 15% validation, and 15% test sets. The dataset is highly unbalanced, which made us decide to use the F1 score for the evaluation of the model. Evaluation over all 12 classes shows F1 = 0.471, with weighted F1 = 0.89. The minority pose classes are ride, squat, fall, and push/pull (See Table 8).

5   Discussion

5.1   Poses

Rich metadata in machine-readable form is fundamental to enable better or novel functionalities related to search, browse, recommendation, or question answering. While detection of entities is a relatively common task, learning relationship labels is not. On the other hand, having information about the actions that objects perform or the relationships between them allows the creation of structures such as knowledge graphs, which enable new functionality e.g. inference. A subset of such actions is the set of poses. One may think of these as attributes of the object, while others may consider them as relationships with other objects, e.g. person lies on the bench, angel falls from a cloud, priest prays to an icon, etc. In either case, they qualify objects in new ways that can improve both user experience and machine-exploitable knowledge.

Table 7. Examples for precision of different models (reported in [18]) when tested over DEArt

	apple	banana	bear	bird	boat	book	cat	cow	dog	elephant	horse	mouse	orange	person	sheep	zebra
MS COCO	0.04	0.008	0.03	0.12	0.24	0.05	0.04	0.23	0.12	0.21	0.2	0	0.4	0.25	0.15	0.89
Open Images	0.008	0.005	0.12	0.01	0.07	0.00007	0.04	-	0.08	0.38	0.09	0	0	0.07	0.04	0.5
Pascal VOC	-	-	-	0.02	0.05	-	0.09	0.13	0.02	-	0.03	-	-	0.05	0.004	-
DEArt	0.13	0.12	0.39	0.15	0.42	0.34	0.21	0.33	0.45	0.56	0.34	0.09	0.09	0.68	0.2	0.91

Table 8. F1 score for pose classification by classes

Class	bend	fall	kneel	lie down	partial	pray	push/pull	ride	sit/eat	squats	stand	unrecogn.	walk
F1	0.33	0.08	0.32	0.80	0.90	0.33	0.09	0.10	0.83	0.12	0.84	0.89	0.50

It may be the case that some poses one expects to be detected are not present in our list. This is due to filtering out verbs that are below the 2000 occurrences; while we could play with this number, training a good pose classifier requires a reasonable minimum number of instances. It is conceivable that some of the pose words we eliminated can be in fact very important for the symbolic meaning of artwork despite not being frequent. In this case, we may want to allow the cultural heritage specialist to add pose words to the list and find/produce more aligned image - caption pairs for those minority poses. Lastly, poses further the possibility of drawing symbolic meaning from paintings, especially the iconographic ones.

5.2 Generating New Object Labels Without Annotation

While we consider our class set to be representative, it is by no means complete. The list of classes could be extended both in breadth (e.g. including body parts, fish, broom, etc.) as well as in depth (e.g. adding refinements of existing classes such as carpenter, beggar, archangel, fisherman, etc.). Some of these classes may be very useful in subsequent inference steps to extract symbolic meaning from a painting. Given the relatively small number of CH images, some class set extensions could be useful while others would only make the model less accurate. A way forward to deal with this challenge is to use additional technology to complement deep learning approaches, for instance approaches based on common knowledge. These can allow further refinement of class labels (and thus generate object labels not in the class list without the need of manual annotation) or even the inference of probable relationships between objects in an image.

5.3 Limitations

Our dataset targets European cultural heritage imagery, of which we focus in particular on the period between the XIIth and the XVIIIth centuries. We also collected a 10K dataset of European artwork between the XIXth century and now, and we tested our (XII–XVIII) model over this dataset. Section 4.2 shows that it is possible to receive acceptable results. On the other hand, it is hard to know how well or poorly a European art model will generalize when applied to visual arts of other cultures; we expect that not very well. More annotations than only poses would be useful to fully inter-relate the objects detected in a painting, and therefore interconnect multiple bounding boxes. This is part of future work.

Acknowledgement. This research has been supported by the Saint George on a Bike project 2018-EU-IA-0104, co-financed by the Connecting Europe Facility of the European Union.

References

1. Aranganayagi, S., Thangavel, K.: Clustering categorical data using silhouette coefficient as a relocating measure. In: International Conference on Computational Intelligence and Multimedia Applications (ICCIMA 2007). vol. 2, pp. 13–17. IEEE (2007)

2. Berg, T.L., Berg, A.C.: Finding iconic images. 2009 IEEE Computer Society Conference on Computer Vision and Pattern Recognition Workshops, pp. 1–8 (2009)
3. Carneiro, G., da Silva, N.P., Del Bue, A., Costeira, J.P.: Artistic image classification: an analysis on the PRINTART database. In: Fitzgibbon, A., Lazebnik, S., Perona, P., Sato, Y., Schmid, C. (eds.) ECCV 2012. LNCS, vol. 7575, pp. 143–157. Springer, Heidelberg (2012). https://doi.org/10.1007/978-3-642-33765-9_11
4. Chollet, F.: Xception: Deep learning with depthwise separable convolutions. In: Proceedings of the IEEE Conference on Computer Vision and Pattern Recognition, pp. 1251–1258 (2017)
5. Clark, K., Manning, C.D.: Deep reinforcement learning for mention-ranking coreference models. arXiv preprint arXiv:1609.08667 (2016)
6. Dictionary.PDF, W.: The merriam webster dictionary. In: The Merriam Webster Dictionary (2016)
7. Everingham, M., Gool, L.V., Williams, C.K.I., Winn, J.M., Zisserman, A.: The pascal visual object classes (VOC) challenge. Int. J. Comput. Vision **88**, 303–338 (2009)
8. Fiorucci, M., Khoroshiltseva, M., Pontil, M., Traviglia, A., Bue, A.D., James, S.: Machine learning for cultural heritage: a survey. Pattern Recognit. Lett. **133**, 102–108 (2020)
9. Garcia, N., Vogiatzis, G.: How to read paintings: semantic art understanding with multi-modal retrieval. In: Leal-Taixé, L., Roth, S. (eds.) ECCV 2018. LNCS, vol. 11130, pp. 676–691. Springer, Cham (2019). https://doi.org/10.1007/978-3-030-11012-3_52
10. Ginosar, S., Haas, D., Brown, T., Malik, J.: Detecting people in cubist art. In: Agapito, L., Bronstein, M.M., Rother, C. (eds.) ECCV 2014. LNCS, vol. 8925, pp. 101–116. Springer, Cham (2015). https://doi.org/10.1007/978-3-319-16178-5_7
11. Gonthier, N., Gousseau, Y., Ladjal, S., Bonfait, O.: Weakly supervised object detection in artworks. ArXiv abs/1810.02569 (2018)
12. He, K., Zhang, X., Ren, S., Sun, J.: Deep residual learning for image recognition. In: 2016 IEEE Conference on Computer Vision and Pattern Recognition (CVPR), pp. 770–778 (2016)
13. Honnibal, M., Montani, I.: spaCy 2: Natural language understanding with Bloom embeddings, convolutional neural networks and incremental parsing (2017, to appear)
14. Huang, X., Belongie, S.J.: Arbitrary style transfer in real-time with adaptive instance normalization. In: 2017 IEEE International Conference on Computer Vision (ICCV), pp. 1510–1519 (2017)
15. Kadish, D., Risi, S., Løvlie, A.S.: Improving object detection in art images using only style transfer. In: 2021 International Joint Conference on Neural Networks (IJCNN), pp. 1–8 (2021)
16. Khosla, A., Yao, B., Fei-Fei, L.: Integrating randomization and discrimination for classifying human-object interaction activities. In: Fu, Y. (ed.) Human-Centered Social Media Analytics, pp. 95–114. Springer, Cham (2014). https://doi.org/10.1007/978-3-319-05491-9_5
17. Kuznetsova, A., et al.: The open images dataset v4. Int. J. Comput. Vision **128**, 1956–1981 (2020)
18. Lin, T.-Y., Maire, M., Belongie, S., Hays, J., Perona, P., Ramanan, D., Dollár, P., Zitnick, C.L.: Microsoft COCO: common objects in context. In: Fleet, D., Pajdla, T., Schiele, B., Tuytelaars, T. (eds.) ECCV 2014. LNCS, vol. 8693, pp. 740–755. Springer, Cham (2014). https://doi.org/10.1007/978-3-319-10602-1_48

19. Maji, S., Bourdev, L., Malik, J.: Action recognition from a distributed representation of pose and appearance. In: CVPR 2011, pp. 3177–3184. IEEE (2011)
20. Milani, F., Fraternali, P.: A dataset and a convolutional model for iconography classification in paintings. J. Comput. Cultural Herit. **14**, 1–18 (2021)
21. Palmer, S., Rosch, E., Chase, P.: Canonical perspective and the perception of objects. Attention and performance IX, pp. 135–151 (1981)
22. Pennington, J., Socher, R., Manning, C.D.: Glove: Global vectors for word representation. In: Proceedings of the 2014 Conference on Empirical Methods in Natural Language Processing(EMNLP), pp. 1532–1543 (2014)
23. Ren, S., He, K., Girshick, R.B., Sun, J.: Faster r-CNN: towards real-time object detection with region proposal networks. IEEE Trans. Pattern Anal. Mach. Intell. **39**, 1137–1149 (2015)
24. Russakovsky, O., et al.: Imagenet large scale visual recognition challenge. Int. J. Comput. Vision **115**, 211–252 (2015)
25. Russell, B.C., Torralba, A., Murphy, K.P., Freeman, W.T.: Labelme: a database and web-based tool for image annotation. Int. J. Comput. Vision **77**, 157–173 (2007)
26. Strezoski, G., Worring, M.: Omniart: a large-scale artistic benchmark. ACM Trans. Multim. Comput. Commun. Appl. **14**, 88:1–88:21 (2018)
27. Westlake, N., Cai, H., Hall, P.: Detecting people in artwork with cnns. ArXiv abs/1610.08871 (2016)
28. Yarlagadda, P., Monroy, A., Carqué, B., Ommer, B.: Recognition and analysis of objects in medieval images. In: ACCV Workshops (2010)
29. Ypsilantis, N.A., García, N., Han, G., Ibrahimi, S., van Noord, N., Tolias, G.: The met dataset: Instance-level recognition for artworks. ArXiv abs/2202.01747 (2021)

How Well Do Vision Transformers (VTs) Transfer to the Non-natural Image Domain? An Empirical Study Involving Art Classification

Vincent Tonkes and Matthia Sabatelli[✉]

Department of Artificial Intelligence and Cognitive Engineering,
University of Groningen, 9712 CP Groningen, The Netherlands
m.sabatelli@rug.nl

Abstract. Vision Transformers (VTs) are becoming a valuable alternative to Convolutional Neural Networks (CNNs) when it comes to problems involving high-dimensional and spatially organized inputs such as images. However, their Transfer Learning (TL) properties are not yet well studied, and it is not fully known whether these neural architectures can transfer across different domains as well as CNNs. In this paper we study whether VTs that are pre-trained on the popular ImageNet dataset learn representations that are transferable to the non-natural image domain. To do so we consider three well-studied art classification problems and use them as a surrogate for studying the TL potential of four popular VTs. Their performance is extensively compared against that of four common CNNs across several TL experiments. Our results show that VTs exhibit strong generalization properties and that these networks are more powerful feature extractors than CNNs.

Keywords: Vision Transformers · Convolutional Neural Networks · Transfer learning · Art classification

1 Introduction

Since the introduction of AlexNet, roughly a decade ago, Convolutional Neural Networks (CNNs) have played a significant role in Computer Vision (CV) [15]. Such neural networks are particularly well-tailored for vision-related tasks, given that they incorporate several inductive biases that help them deal with high dimensional, rich input representations. As a result, CNNs have found applications across a large variety of domains that are not *per-se* restricted to the realm of natural images. Among such domains, the Digital Humanities (DH) field is of particular interest. Thanks to a long tradition of works that aimed to integrate advances stemming from technical disciplines into the Humanities, they have been serving as a challenging real-world test-bed regarding the applicability of CV algorithms. It naturally follows that over the last years, several works have studied the potential of CNNs within the DH (see [11] for a survey about the topic), resulting in a significant number of successful applications

L. Karlinsky et al. (Eds.): ECCV 2022 Workshops, LNCS 13801, pp. 234–250, 2023.
https://doi.org/10.1007/978-3-031-25056-9_16

that range from the classification of artworks [23,30,36,43] to the detection of objects within paintings [12,29], automatic style classification [7] and even art understanding [2].

A major breakthrough within the CV community has recently been achieved by the Vision Transformer (VT) [9], a novel neural architecture that has gained state-of-the-art performance on many standard learning benchmarks including the popular ImageNet dataset [8]. Exciting as this may be, we believe that a plain VT is not likely to become as valuable and powerful as a CNN as long as it does not exhibit the strong generalization properties that, over the years, have allowed CNNs to be applied across almost all domains of science [1,25,39]. Therefore, this paper investigates what VTs offer within the DH by studying this family of neural networks from a Transfer Learning (TL) perspective. Building on top of the significant efforts that the DH have been putting into digitizing artistic collections from all over the world [33], we define a set of art classification problems that allow us to study whether pre-trained VTs can be used outside the domain of natural images. We compare their performance to that of CNNs, which are well-known to perform well in this setting, and present to the best of our knowledge the very first thorough empirical analysis that describes the performance of VTs outside the domain of natural images and within the domain of art specifically.

2 Preliminaries

We start by introducing some preliminary background that will be used throughout the rest of this work. We give an introduction about supervised learning and transfer learning (Sect. 2.1), and then move towards presenting some works that have studied Convolutional Neural Networks and Vision Transformers from a transfer learning perspective.

2.1 Transfer Learning

With Transfer Learning (TL), we typically denote the ability that machine learning models have to retain and reuse already learned knowledge when facing new, possibly related tasks [26,46]. While TL can present itself within the entire machine learning realm [4,37,42,45], in this paper we consider the supervised learning setting only, a learning paradigm that is typically defined by an input space $\mathcal{X}$, an output space $\mathcal{Y}$, a joint probability distribution $P(X,Y)$ and a loss function $\ell : \mathcal{Y} \times \mathcal{Y} \to \mathbb{R}$. The goal of a supervised learning algorithm is to learn a function $f : \mathcal{X} \to \mathcal{Y}$ that minimizes the expectation over $P(X,Y)$ of ℓ known as the expected risk. Classically, the only information that is available for minimizing the expected risk is a learning set $\mathcal{L}$ that provides the learning algorithm with N pairs of input vectors and output values $(\mathbf{x}_1, y_1), ..., (\mathbf{x}_N, y_N)$ where $\mathbf{x}_i \in \mathcal{X}$ and $y_i \in \mathcal{Y}$ are i.i.d. drawn from $P(X,Y)$. Such learning set can then be used for computing the empirical risk, an estimate of the expected risk, that can be used for finding a good approximation of the optimal function f^* that minimizes the aforementioned expectation. When it comes to TL, however, we assume that next to the information contained within $\mathcal{L}$, the learning algorithm also has access to an additional learning set defined as $\mathcal{L}'$. Such a learning set can then be used alongside $\mathcal{L}$ for finding a function that better minimizes the loss function ℓ. In this work, we consider $\mathcal{L}'$ to be the ImageNet dataset, and f to come in the form of either a pre-trained Convolutional Neural Networks or a pre-trained Vision Transformers, two types of neural networks that over the years have demonstrated exceptional abilities in tackling problems modeled by high dimensional and spatially organized inputs such as images, videos, and text.

2.2   Related Works

While to this date, countless examples have studied the TL potential of CNNs [24, 25, 34, 40], the same cannot yet be said for VTs. In fact, papers that have so far investigated their generalization properties are much rarer. Yet, some works have attempted to compare the TL potential of VTs to that of CNNs, albeit strictly outside the artistic domain. E.g., in [21] the authors consider the domain of medical imaging and show that if CNNs are trained from scratch, then these models outperform VTs; however, if either an off-the-shelf feature extraction approach is used, or a self-supervised learning training strategy is followed, then VTs significantly outperform their CNNs counterparts. Similar results were also observed in [44] where the authors show that both in a single-task learning setting and in a multi-task learning one, transformer-based backbones outperformed regular CNNs on 13 tasks out of 15. Positive TL results were also observed in [16], where the TL potential of VTs is studied for object detection tasks, in [41] where the task of facial expression recognition is considered, and in [10], where similarly to [21], the authors successfully transfer VTs to the medical imaging domain. The generalization properties of VTs have also been studied outside of the supervised learning framework: E.g. in [5] and [6], a self-supervised learning setting is considered. In [5] the authors show that VTs can learn features that are particularly general and well suited for TL, which is a result that is confirmed in [6], where the authors show that VTs pre-trained in a self-supervised learning manner transfer better than the ones trained in a pure supervised learning fashion. While all these works are certainly valuable, it is worth noting that, except for [21] and [10], all other studies have performed TL strictly within the domain of natural images. Therefore, the following research question

> *"How transferable are the representations of pre-trained VTs when it comes to the non-natural image domain?"*

remains open. We will now present our work that helps us answer this question by considering **image datasets from the artistic domain.** Three art classification tasks are described, which serve as a surrogate for identifying which family of pre-trained networks, among CNNs and VTs, transfers better to the non-natural realm.

Fig. 1. Samples from the three distinct classification tasks used throughout this paper.

3   Methods

Our experimental setup is primarily inspired by the work presented in Sabatelli et al. [30], where the authors report a thorough empirical analysis that studies the TL properties of pre-trained CNNs that are transferred to different art collections. More specifically, the authors investigate whether popular neural architectures such as VGG19 [32] and ResNet50 [13], which come as pre-trained on the ImageNet1k [8] dataset, can get successfully transferred to the artistic domain. The authors consider three different classification problems and two different TL approaches: an off-the-shelf (OTS) feature extraction approach where pre-trained models are used as pure feature extractors and a fine-tuning approach (FT) where the pre-trained networks are allowed to adapt all of their pre-trained parameters throughout the training process. Their study suggests that all the considered CNNs can successfully get transferred to the artistic domain and that a fine-tuning training strategy yields substantially better performance than the OTS one. In this work we investigate whether these conclusions also hold for transformer-based architectures, and if so, whether this family of models outperforms that of CNNs.

3.1   Data

Similar to Sabatelli et al. [30] we also use data stemming from the *Rijksmuseum Challenge* dataset [22] (see Fig. 1 for an illustration). This dataset consists of a large collection of digitized artworks that come together with xml-formatted metadata which can be used for defining a set of supervised learning problems [22,33]. Following [22] and [30] we focus on three, well-known, classification problems namely: (1) *Type classification*, where the goal is to train a model such that it is able to distinguish classes such as 'painting', 'sculpture', 'drawing', etc.; (2) *Material classification*, where the classification problem is defined by labels such as 'paper', 'porcelain', 'silver', etc.; and finally, (3) *Artist classification*, where, naturally, the model has to predict who the creator of a specific artwork is. While

Table 1. Overview of the used datasets. Values between brackets show the situation before balancing operations were performed. 'Sample overlap' gives the average overlap between 2 of the 5 randomly generated sets per task (i and j where $i \neq j$).

Task	# Samples	# Classes	Sample overlap
Type classification	9607 (77628)	30 (801)	0.686
Material classification	7788 (96583)	30 (136)	0.798
Artist classification	6530 (38296)	30 (8592)	1

the full dataset contains 112,039 images, due to computational reasons in the present study we only use a fraction of it as this allows us to run shorter yet more thorough experiments. A smaller dataset also allows us to study an additional research question which we find worth exploring and that [30] did not consider in their study, namely: *"How well do CNNs (and VTs) transfer when the size of the artistic collection is small?"* To this end we decided to select the 30 most occurring classes within their dataset and to set a cap of 1000 randomly sampled instances per class. Table 1 summarizes the datasets used in the present study. Between brackets we report values as they were before balancing operations were performed. For all of our experiments we use 5 fold cross-validation and use 80% of the dataset as training-set, 10% as validation-set and 10% as testing-set.

3.2 Neural Architectures

In total, we consider eight different neural architectures, of which four are CNN-based networks while the remaining four are VT-based models. All models are pre-trained on the ImageNet1K dataset. When it comes to CNNs we consider ResNet50 [13] and VGG19 [32], as the first one has widely been adopted by researchers working at the intersection of computer vision and digital heritage [2,43], whereas the latter was among the best performing models considered by Sabatelli et al. [30]. Next to these two architectures we also consider two additional, arguably more recent networks, namely ConvNext [19], which is a purely CNN-based model that is inspired by VTs' recent successes, and that therefore fits well within the scope of this work, and EfficientNetV2 [35], which is a network that is well known for its computational efficiency and potentially faster training times. Regarding the VTs, we use models that have 16×16 patch sizes. As a result, we start by considering the first original VT model presented in [9] which we refer to as ViT. We then consider the Swin architecture [18] as it showed promising TL performance in [44] and the BeiT [3] and the DeiT [38] transformers. The usage of DeiT is motivated by Matsoukas et al. work [21] reviewed in Sect. 2.2 where the authors compared it to ResNet50. Yet, note that in this work, we use the version of the model known as the 'base' version that does not take advantage of distillation learning in the FT stage. All VT-based models have ≈ 86 million trainable parameters, and so does ConvNext. ResNet50 and EfficientNetV2 are, however much smaller networks as they come with ≈ 25.6

and ≈ 13.6 million trainable parameters, respectively. Lastly, `VGG19` is by far the largest model of all with its 143.7 million trainable parameters.

3.3   Training Procedure

For all experiments, images are resized to a 224×224 resolution by first scaling them to the desired size along the shortest axis (retaining aspect ratio), and then taking a center crop along the longer axis. In addition, all images are normalized to the RGB mean and standard deviation of the images presented within the ImageNet1K dataset ([0.485, 0.456, 0.406] and [0.229, 0.224, 0.225]). For all models we replace the final linear classification layer with a new layer with as many output nodes as there are classes to classify (C) within the dataset and optimize the parameters of the model θ such that the categorical cross-entropy loss function

$$\mathcal{L}(\theta) = -\mathbb{E}_{(\mathbf{x},y) \sim P(X,Y)} \sum_{i=1}^{C} 1(y = i) \log p_{\text{model}} f_i(\mathbf{x}; \theta), \qquad (1)$$

is minimized.

Training is regularized through the early stopping method, which interrupts training if for 10 epochs in a row no improvement on the validation loss is observed. The model with the lowest validation loss is then benchmarked on the final testing set. For our OTS experiments, we use the Adam optimizer [14] with standard `PyTorch` [27] parameters ($\text{lr} = 1\text{e}{-}3$, $\beta_1 = 0.9$, $\beta_2 = 0.999$), and use a batch size of 256. For the FT experiments, hyperparameters are partially inspired by [21,44]. We again use the Adam optimizer, this time initialized with $\text{lr} = 1\text{e}{-}4$ which gets reduced by a factor of 10 after three epochs without improvements on the validation loss; inspired by [20], the batch size is now reduced to 32; whereas label smoothing (0.1) and dropout (p = 0.2) are used for regularizing training even further. Finally, input images are augmented with random horizontal flips and rotations in a $\pm 10°$ range.

3.4   Hardware and Software

All experiments are conducted on a single compute node containing one 32 GB `Nvidia V100` GPU. The FT experiments take advantage of the V100's mixed precision acceleration. An exception is made for the *Type Classification* experiment, as this one is also used to compare OTS TL and FT in terms of time/accuracy trade-offs (see Sect. 6 for further details). The `PyTorch` machine learning framework [27] is used for all experiments, and many pre-trained models are taken from its `Torchvision` library. Exceptions are made for `EfficientNetV2`, `Swin`, `DeiT` and `Beit`, which are taken from the `Timm` library[1]. We release all source code and data on the following GitHub link[2].

---

[1] https://timm.fast.ai/.

[2] https://github.com/IndoorAdventurer/ViTTransferLearningForArtClassification.

4 Results

We now present the main findings of our study and report results for the three aforementioned classification problems and for all previously introduced architectures. All networks are either trained with an OTS training scheme (Sect. 4.1) or with a FT one (Sect. 4.2). Results come in the form of line plots and tables: the former visualize the performance of all models in terms of accuracy on the validation set, whereas the latter report the final performance that the best validation models obtained on the separate testing-set. All line plots report the average accuracy obtained across five different experiments, while the shaded areas correspond to the standard error of the mean ($\pm s \div \sqrt{N}$, where $N = 5$). The dashed lines represent CNNs, whereas continuous lines depict VTs. Plots end when an early stop occurred for the first of the five trials. Regarding the tables, a green-shaded cell marks the best overall performing model, while yellow and red cells depict the second best and worst performing networks, respectively. We quantitatively assess the performance of the models with two separate metrics: the accuracy and the balanced accuracy, where the latter is defined as the average recall over all classes. Note that compared to the plain accuracy metric, the balanced accuracy allows us to penalize type-II errors more when it comes to the less occurring classes within the dataset.

4.1 Off-The-Shelf Learning

For this set of experiments, we start by noting that overall, both the CNNs and the VTs can perform relatively well on all three different classification tasks. When it comes to (a) *Type Classification* we can see that all models achieve a final accuracy between $\approx 80\%$ and $\approx 90\%$, whereas on (b) *Material Classification* the performance deviates between $\approx 75\%$ and 85%, and on (c) *Artist Classification* we report accuracies between 75% and 90%. Overall, however, as highlighted by the green cells in Table 2, the best performing model on all classification tasks is the VT `Swin`, which confirms the good TL potential that this architecture has and that was already observed in [44]. Yet, we can also observe that the second-best performing model is not a VT, but the `ConvNext` CNN. Despite performing almost equally well, it is important to mention though that `ConvNext` required more training epochs to converge when compared to `Swin`, as can clearly be seen in all three plots reported in Fig. 2. We also note that on the *Type Classification* task the worst performing model is the `Beit` transformer (82.26%), but when it comes to the classification of the materials and artists the worst performing model becomes `EfficientNetV2` with final accuracies of 75.96% and 73.92% respectively. Among the different VTs the `Beit` transformer also appears to be the architecture that requires the longest training. In fact, as can be seen in Fig. 3, this network does not exhibit substantial "Jumpstart-Improvements" as the other VTs (learning starts much lower in all plots as is depicted by the green lines).

Several other interesting conclusions can be made from this experiment: we observed that the `VGG19` network yielded worse performance than `ResNet50`, a result which is not in line with what was observed by Sabatelli et al. [30] where VGG19 was the best performing network when used with an OTS training scheme. Also, differently from [30], the most challenging classification task in our experiments appeared to be that of *Material Classification* as it resulted in the overall lowest accuracies. These results seem to suggest that even though both studies considered images stemming from the same artistic collection, the training dataset size can significantly affect the TL performance of the different models, a result which is in line with [28].

In general, VTs seem to be very well suited for an OTS transfer learning approach, especially regarding the Swin and DeiT architectures which performed well across all tasks. Equally interesting is the performance obtained by ConvNext which is by far the most promising CNN-based architecture. Yet, on average, the performance of the VTs is higher than the one obtained by the CNNs: on *Type Classification* the first ones perform on average $\approx 86.5\%$ while the CNNs reach an average classification rate of $\approx 85.51\%$, whereas on *Material* and *Artist Classification* VTs achieve average accuracies of $\approx 81.8\%$ and $\approx 85.8\%$ respectively, whereas CNNs $\approx 79.5\%$ $\approx 81.2\%$. These results suggest that this family of methods is better suited for OTS TL when it comes to art classification problems.

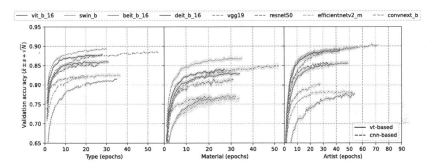

Fig. 2. The validation accuracy obtained by all architectures when trained with an off-the-shelf (OTS) feature extraction approach. We can see that the best performing models are the VT Swin and the CNN ConvNext.

4.2   Fine-Tuning

When it comes to the fine-tuning experiments, we observe, in part, consistent results with what we have presented in the previous section. We can again note that the lower classification rates have been obtained when classifying the material of the different artworks, while the best performance is again achieved when classifying the artists of the heritage objects. The Swin VT remains the network that overall performs best, while the ConvNext CNN remains the overall second

Table 2. The performance of all models trained with an OTS approach on the final testing sets. We can see that the best overall model is the `Swin` transformer (green cells) followed by the `ConvNext` CNN architecture (yellow cells). The worst performing models are `Beit` when it comes to type classification and `EfficientNetV2` when it comes to material and artist classification.

Model	Type		Material		Artist	
	Accuracy	Bal. accuracy	Accuracy	Bal. accuracy	Accuracy	Bal. accuracy
vit_b_16	86.06% $\pm 1.06\%$	84.13% $\pm 1.57\%$	81.78% $\pm 0.48\%$	67.38% $\pm 1.37\%$	84.89% $\pm 0.46\%$	81.42% $\pm 0.42\%$
swin_b	89.43% $\pm 0.93\%$	87.47% $\pm 1.02\%$	85.87% $\pm 0.35\%$	71.19% $\pm 1.60\%$	90.40% $\pm 0.65\%$	88.64% $\pm 0.78\%$
beit_b_16	82.26% $\pm 0.72\%$	77.75% $\pm 0.27\%$	76.87% $\pm 0.96\%$	60.16% $\pm 1.56\%$	79.70% $\pm 0.69\%$	75.35% $\pm 1.09\%$
deit_b_16	88.18% $\pm 0.66\%$	85.36% $\pm 0.54\%$	82.80% $\pm 1.12\%$	66.46% $\pm 1.03\%$	88.13% $\pm 0.76\%$	85.62% $\pm 0.87\%$
vgg19	83.93% $\pm 0.72\%$	83.35% $\pm 0.81\%$	76.87% $\pm 0.44\%$	61.39% $\pm 1.47\%$	82.01% $\pm 0.66\%$	78.10% $\pm 0.77\%$
resnet50	85.51% $\pm 0.64\%$	82.33% $\pm 1.85\%$	80.99% $\pm 0.82\%$	65.51% $\pm 0.93\%$	87.71% $\pm 1.06\%$	85.12% $\pm 1.34\%$
eff. netv2_m	83.41% $\pm 0.76\%$	82.05% $\pm 1.25\%$	75.96% $\pm 1.24\%$	59.15% $\pm 1.24\%$	78.62% $\pm 1.87\%$	73.92% $\pm 0.96\%$
convnext_b	89.19% $\pm 0.64\%$	86.95% $\pm 1.38\%$	84.14% $\pm 0.92\%$	69.10% $\pm 1.05\%$	90.13% $\pm 0.94\%$	87.84% $\pm 1.07\%$

best performing model, even though on *Artist Classification* it is slightly outperformed by `ResNet50`. Unlike our OTS experiments, however, this time, we note that VTs are on average the worst-performing networks. When it comes to the *Type Classification* problem, the `ViT` model achieves the lowest accuracy (although note that if the balance accuracy metric is used, then the worst performing network becomes `VGG19`). Further, when considering the classification of materials and artists, the lowest accuracies are instead achieved by the `Beit` model. While it is true that overall the performance of both CNNs and VTs significantly improves through fine-tuning, it is also true that, perhaps surprisingly, such a training approach reduces the differences in terms of performance between these two families of networks. This result was not observed in [44], where VTs outperformed CNNs even in a fine-tuning training regime. While these results indicate that an FT training strategy does not seem to favor either VTs or CNNs, it is still worth noting that this training approach is beneficial. In fact, no matter whether VTs or CNNs are considered, the worst-performing fine-tuned model still performs better than the best performing OTS network. The only exception to this is *Material Classification*, where `BeiT` obtained a testing accuracy of 85.75% after fine-tuning which is slightly lower than the one obtained by the `Swin` model trained with an OTS training scheme (85.87%).

5   Discussion

Our results show that VTs possess strong transfer learning capabilities and that this family of models learns representations that can generalize to the artistic realm. Specifically, as demonstrated by our OTS experiments, when these architectures are used as pure feature extractors, their performance is, on average, substantially better than the one of CNNs. To the best of our knowledge, this is the first work that shows that this is the case for non-natural and artistic images. Next to be attractive to the computer vision community, as some light is shed on the generalization properties of VTs, we believe that these results are

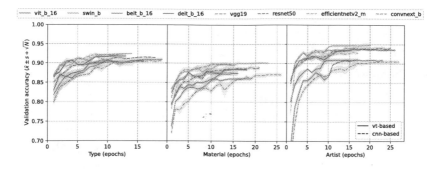

Fig. 3. The validation accuracy learning curves obtained by all models when a fine-tuning FT training approach is followed. We can again see that the `Swin` and `ConvNext` architectures are among the best performing networks.

Table 3. The testing set performance of all models trained with a FT approach. We can see that compared to the results presented in Table 2, all models perform substantially better, yet differences in terms of performance between the VTs and the CNNs now seem to be smaller.

Model	Type		Material		Artist	
	Accuracy	Bal. accuracy	Accuracy	Bal. accuracy	Accuracy	Bal. accuracy
vit_b_16	90.11% ±0.35%	87.40% ±0.29%	87.42% ±0.45%	73.46% ±1.44%	92.05% ±0.44%	89.77% ±0.38%
swin_b	92.17% ±0.98%	89.71% ±1.03%	89.35% ±0.68%	77.16% ±2.98%	95.05% ±0.47%	93.94% ±0.81%
beit_b_16	90.81% ±0.41%	87.95% ±0.69%	85.74% ±0.37%	72.12% ±1.38%	91.27% ±1.13%	88.83% ±1.63%
deit_b_16	91.78% ±0.64%	89.22% ±0.90%	87.85% ±1.12%	74.42% ±1.99%	93.37% ±1.15%	91.67% ±1.55%
vgg19	90.54% ±0.37%	87.05% ±1.03%	85.74% ±1.40%	72.43% ±3.03%	92.20% ±0.49%	90.18% ±0.72%
resnet50	91.78% ±0.44%	88.24% ±0.59%	88.69% ±0.99%	77.97% ±2.25%	94.72% ±0.74%	93.41% ±1.05%
eff. netv2_m	90.87% ±0.67%	88.34% ±1.37%	87.55% ±1.15%	75.31% ±1.60%	92.65% ±0.54%	90.84% ±0.51%
convnext_b	92.15% ±0.40%	89.82% ±1.18%	88.79% ±1.07%	78.40% ±1.26%	94.60% ±0.54%	93.13% ±0.61%

also of particular interest for practitioners working in the digital humanities with limited access to computing power. As the resources for pursuing an FT training approach might not always be available, it is interesting to know that between CNNs and VTs, the latter models are the ones that yield the best results in the OTS training regime.

Equally interesting and novel are the results obtained through fine-tuning, where the performance gap between VTs and CNNs gets greatly reduced, with the latter models performing only slightly worse than the former. In line with the work presented in [30], which considered CNNs exclusively, we clearly show that this TL approach also substantially improves the performance of VTs.

6 Additional Studies

We now present four additional studies that we hope can help practitioners that work at the intersection of computer vision and the digital humanities. Specifically we aim to shed some further light into the classification performance of both CNNs and VTs, while also providing some practical insights that consider the training times of both families of models.

6.1 Saliency Maps

We start by performing a qualitative analysis that is based on the visual investigation of saliency maps. For this set of studies we consider the `ConvNext` CNN and the `Swin` and `DeiT` VTs. When it comes to the former, saliency maps are computed through the popular GradCam method [31], whereas attention-rollout is used when it comes to the `DeiT` architecture. `Swin`'s saliency maps are also computed with a method similar to GradCam, with the main difference being that instead of taking an average of the gradients per channel as weights, we directly multiply gradients by their activation and visualize patch means of this product. Motivated by the nice performance of VTs in the OTS transfer learning setting, we start by investigating how the representation of an image changes with respect to the depth of the network. While it is well known that CNNs build up a hierarchical representation of the input throughout the network, similar studies involving transformer-based architectures are rarer. In Fig. 4 we present some examples that show that also within VTs, the deeper the network becomes, the more the network starts focusing on lower level information within the image.

In Fig. 5 we show how different network architectures and TL approaches result in different saliency maps. In the first image of Fig. 5 we can observe that the `ConvNext` architecture miss-classifies a 'Dish' as a 'Plate' when an OTS approach is used. Note that this is a mistake not being made by the transformer-based architectures. However, we observe that after the fine-tuning stage, the CNN can classify the image's *Type* correctly after having shifted its attention toward the bottom and center of the dish rather than its top. Interestingly, none of the transformer-based architecture focuses on the same image regions, independently of whether an OTS or a FT approach is used. We can see that most saliency maps are clustered within the center of the image, both when an OTS training strategy is used and when an FT approach is adopted. When looking at the second image of Fig. 5, we see that all networks, independently from the adopted TL approach, correctly classify the image as a *'Picture'*. Yet the saliency maps of the `ConvNext` change much more across TL approaches, and we again see that the transformer-based architectures focus more on the central regions of the image rather than on the borders. Similar behavior can also be observed in the last image of Fig. 5.

6.2 Dealing with Small Artistic Collections

It is not uncommon for heritage institutions to deal with datasets far smaller than those typically used by the computer vision community. While it is true that the number of samples used throughout this study is far smaller than the one used within the naturalistic domain, we still wondered whether the results reported in Sect. 4 would generalize to even smaller artistic collections. Inspired by [17] we have designed the following experiment where we consider the *Type Classification*

Fig. 4. Saliency maps for different attention layers of successively deeper transformer blocks. The deeper the network, the lower level the representations learned by the Vision Transformer become.

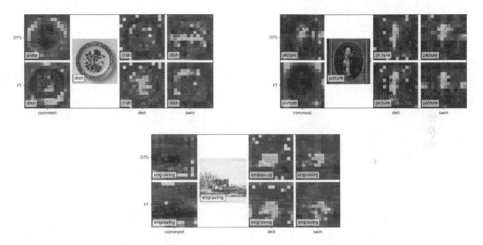

Fig. 5. Saliency maps computed for ConvNext, DeiT and Swin when classifying the *"Type"* of three heritage objects. Saliency maps are computed both for the off the shelf experiments as well as for the fine-tuning ones.

task. We consider the top 15 most occurring classes within the original dataset presented in Table 1 and scale the number of samples four times by a factor $\sqrt[4]{\frac{1}{10}} \approx 0.56$. This results in five datasets which are respectively 100%, 56%, 32%, 18% and 10% the size of the original dataset shown in Table 1. Note that we ensure that the distribution of samples per class remains the same across all datasets. We then trained all models following the exact experimental protocol described in Sect. 3. Our results are reported in Fig. 6 where we can observe how the testing accuracy decreases when smaller portions of the full dataset are taken. Note that the x-axes show a logarithmic scale, with the rightmost value being roughly 10% the size of the leftmost one. For both an OTS approach (left plot of Fig. 6), as well as a FT one (right plot of Fig. 6), we show that the findings discussed in Sects. 4.1 and 4.2 generalize to smaller datasets.

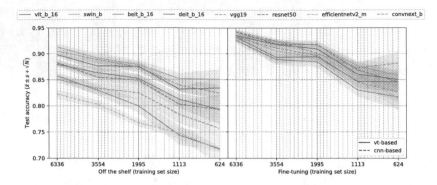

Fig. 6. The testing accuracy obtained on the *Type Classification* problem when gradually reducing the size of the dataset. We can observe that different dataset sizes do not affect the best performing models, which as presented in Sect. 4 remain the `Swin` VT and the `ConvNext` CNN.

6.3 Training Times

We now report some final results that describe the training times that are required by CNNs and VTs when trained with the aforementioned TL strategies. These results have also been obtained on the *Type* classification task but with the V100s mixed precision capabilities disabled. In Fig. 7 we show the classification accuracy as reported in Tables 2 and 3 and plot it against the average duration of one training epoch. This allows us to understand the time/accuracy trade-offs that characterize all neural networks. As shown by the blue dots, we can observe that when it comes to VTs trained in an OTS setting, all transformer-based architectures require approximately the same number of seconds to successfully go through one training epoch (≈ 40). This is, however, not true for the CNNs, which on average, require less time and for which there is a larger difference between the fastest OTS model (`ResNet50` requiring ≈ 20 s) and the slowest model (`ConvNext`). While, as discussed in Sect. 4, VTs are more powerful feature extractors than CNNs, it is worth noting that the gain in performance these models have to offer comes at a computational cost. Note, however, that this is not true anymore when it comes to an FT training regime, as all models (red and purple dots) now perform almost equally well. Yet it is interesting to point out that the computational costs required by the VTs stay approximately the same across all architectures, which is not the case for the CNN-based networks as there is a clear difference between the fastest fine-tuned network (`ResNet50`) and the slowest (`ConvNext`). We believe that the design of a transformer-based architecture that, if fine-tuned, results in the same training times as `ResNet50` provides an interesting avenue for future work.

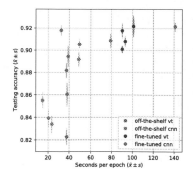

Fig. 7. Time/accuracy trade-offs for VTs and CNNs trained either in an off-the-shelf setting or with a fine-tuning strategy.

7 Conclusion

This work examined how well VTs can transfer knowledge to the non-natural image domain. To this end, we compared four popular VT architectures with common CNNs, in terms of how well they transfer from ImageNet1k to classification tasks presented in the *Rijksmuseum Challenge* dataset. We have shown that when fine-tuned VTs performed on par with CNNs and that they performed even better than their CNN counterparts when used as feature extractors. We believe that our study proves that VTs can become a valuable alternative to CNN based architectures in the non-natural image domain and in the realm of artistic images specifically. Especially `Swin` and `DeiT` showed promising results throughout this study and, therefore, we aim to investigate their potential within the Digital Humanities in the future. To this end, inspired by [33] we plan on using them in a multi-task learning setting; as backbone feature extractors when it comes to the object detection within artworks [12,29], and finally, in a self and semi-supervised learning setting. We also plan on performing a similar analysis for the `ConvNext` CNN, as this architecture showed very good performance as well.

To conclude, we believe that the study presented in this work opens the door for a more fundamental CV question that deserves attention: *"What makes transformer-based architectures such powerful feature extractors?"*. We plan on investigating whether the results obtained within the domain of DH will also generalize to other non-natural image datasets with the hope of partially answering this question.

References

1. Ackermann, S., Schawinski, K., Zhang, C., Weigel, A.K., Turp, M.D.: Using transfer learning to detect galaxy mergers. Mon. Not. R. Astron. Soc. **479**(1), 415–425 (2018)

2. Bai, Z., Nakashima, Y., Garcia, N.: Explain me the painting: multi-topic knowledgeable art description generation. In: Proceedings of the IEEE/CVF International Conference on Computer Vision, pp. 5422–5432 (2021)
3. Bao, H., Dong, L., Piao, S., Wei, F.: BEiT: BERT pre-training of image transformers. In: ICLR 2022 (2022)
4. Bengio, Y.: Deep learning of representations for unsupervised and transfer learning. In: Proceedings of ICML Workshop on Unsupervised and Transfer Learning, pp. 17–36. JMLR Workshop and Conference Proceedings (2012)
5. Caron, M., Touvron, H., Misra, I., Jégou, H., Mairal, J., Bojanowski, P., Joulin, A.: Emerging properties in self-supervised vision transformers. In: Proceedings of the IEEE/CVF International Conference on Computer Vision, pp. 9650–9660 (2021)
6. Chen, X., Xie, S., He, K.: An empirical study of training self-supervised vision transformers. In: Proceedings of the IEEE/CVF International Conference on Computer Vision, pp. 9640–9649 (2021)
7. Chu, W.T., Wu, Y.L.: Image style classification based on learnt deep correlation features. IEEE Trans. Multimedia **20**(9), 2491–2502 (2018)
8. Deng, J., Dong, W., Socher, R., Li, L.J., Li, K., Fei-Fei, L.: ImageNet: a large-scale hierarchical image database. In: 2009 IEEE Conference on Computer Vision and Pattern Recognition, pp. 248–255. IEEE (2009)
9. Dosovitskiy, A., et al.: An image is worth 16×16 words: transformers for image recognition at scale. arXiv preprint arXiv:2010.11929 (2020)
10. Duong, L.T., Le, N.H., Tran, T.B., Ngo, V.M., Nguyen, P.T.: Detection of tuberculosis from chest x-ray images: boosting the performance with vision transformer and transfer learning. Expert Syst. Appl. **184**, 115519 (2021)
11. Fiorucci, M., Khoroshiltseva, M., Pontil, M., Traviglia, A., Del Bue, A., James, S.: Machine learning for cultural heritage: a survey. Pattern Recogn. Lett. **133**, 102–108 (2020)
12. Gonthier, N., Gousseau, Y., Ladjal, S., Bonfait, O.: Weakly supervised object detection in artworks. In: Proceedings of the European Conference on Computer Vision (ECCV) Workshops (2018)
13. He, K., Zhang, X., Ren, S., Sun, J.: Deep residual learning for image recognition. In: Proceedings of the IEEE Conference on Computer Vision and Pattern Recognition, pp. 770–778 (2016)
14. Kingma, D.P., Ba, J.: Adam: a method for stochastic optimization. arXiv preprint arXiv:1412.6980 (2014)
15. Krizhevsky, A., Sutskever, I., Hinton, G.E.: ImageNet classification with deep convolutional neural networks. In: Advances in Neural Information Processing Systems, vol. 25. Curran Associates, Inc. (2012)
16. Li, Y., Xie, S., Chen, X., Dollar, P., He, K., Girshick, R.: Benchmarking detection transfer learning with vision transformers. arXiv preprint arXiv:2111.11429 (2021)
17. Liu, Y., Sangineto, E., Bi, W., Sebe, N., Lepri, B., Nadai, M.: Efficient training of visual transformers with small datasets. In: Ranzato, M., Beygelzimer, A., Dauphin, Y., Liang, P., Vaughan, J.W. (eds.) Advances in Neural Information Processing Systems, vol. 34, pp. 23818–23830. Curran Associates, Inc. (2021). proceedings.neurips.cc/paper/2021/file/c81e155d85dae5430a8cee6f2242e82c-Paper.pdf
18. Liu, Z., et al.: Swin transformer: hierarchical vision transformer using shifted windows. In: Proceedings of the IEEE/CVF International Conference on Computer Vision, pp. 10012–10022 (2021)
19. Liu, Z., Mao, H., Wu, C.Y., Feichtenhofer, C., Darrell, T., Xie, S.: A convnet for the 2020s. arXiv preprint arXiv:2201.03545 (2022)

20. Masters, D., Luschi, C.: Revisiting small batch training for deep neural networks. arXiv preprint arXiv:1804.07612 (2018)
21. Matsoukas, C., Haslum, J.F., Söderberg, M., Smith, K.: Is it time to replace CNNs with transformers for medical images? arXiv preprint arXiv:2108.09038 (2021)
22. Mensink, T., van Gemert, J.: The Rijksmuseum challenge: museum-centered visual recognition. In: ACM International Conference on Multimedia Retrieval (ICMR) (2014)
23. Milani, F., Fraternali, P.: A dataset and a convolutional model for iconography classification in paintings. J. Comput. Cult. Heritage (JOCCH) **14**(4), 1–18 (2021)
24. Minaee, S., Kafieh, R., Sonka, M., Yazdani, S., Soufi, G.J.: Deep-Covid: Predicting Covid-19 from chest x-ray images using deep transfer learning. Med. Image Anal. **65**, 101794 (2020)
25. Mormont, R., Geurts, P., Maree, R.: Comparison of deep transfer learning strategies for digital pathology. In: Proceedings of the IEEE Conference on Computer Vision and Pattern Recognition (CVPR) Workshops, June 2018
26. Pan, S.J., Yang, Q.: A survey on transfer learning. IEEE Trans. Knowl. Data Eng. **22**(10), 1345–1359 (2009)
27. Paszke, A., et al.: Automatic differentiation in pytorch (2017)
28. Sabatelli, M.: Contributions to deep transfer learning: from supervised to reinforcement learning. Ph.D. thesis, Universitè de Liegè, Liegè, Belgique (2022)
29. Sabatelli, M., et al.: Advances in digital music iconography: benchmarking the detection of musical instruments in unrestricted, non-photorealistic images from the artistic domain. Digital Humanit. Q. **15**(1) (2021)
30. Sabatelli, M., Kestemont, M., Daelemans, W., Geurts, P.: Deep transfer learning for art classification problems. In: Leal-Taixé, L., Roth, S. (eds.) ECCV 2018. LNCS, vol. 11130, pp. 631–646. Springer, Cham (2019). https://doi.org/10.1007/978-3-030-11012-3_48
31. Selvaraju, R.R., Cogswell, M., Das, A., Vedantam, R., Parikh, D., Batra, D.: Grad-CAM: visual explanations from deep networks via gradient-based localization. In: Proceedings of the IEEE International Conference On Computer Vision, pp. 618–626 (2017)
32. Simonyan, K., Zisserman, A.: Very deep convolutional networks for large-scale image recognition. arXiv preprint arXiv:1409.1556 (2014)
33. Strezoski, G., Worring, M.: OmniArt: multi-task deep learning for artistic data analysis. arXiv preprint arXiv:1708.00684 (2017)
34. Talo, M., Baloglu, U.B., Yıldırım, Ö., Acharya, U.R.: Application of deep transfer learning for automated brain abnormality classification using MR images. Cogn. Syst. Res. **54**, 176–188 (2019)
35. Tan, M., Le, Q.: EfficientNetv2: smaller models and faster training. In: International Conference on Machine Learning, pp. 10096–10106. PMLR (2021)
36. Tan, W.R., Chan, C.S., Aguirre, H.E., Tanaka, K.: Ceci n'est pas une pipe: a deep convolutional network for fine-art paintings classification. In: 2016 IEEE International Conference on Image Processing (ICIP), pp. 3703–3707. IEEE (2016)
37. Taylor, M.E., Stone, P.: Transfer learning for reinforcement learning domains: a survey. J. Mach. Learn. Res. **10**(7), 1633–1685 (2009)
38. Touvron, H., Cord, M., Douze, M., Massa, F., Sablayrolles, A., Jégou, H.: Training data-efficient image transformers & distillation through attention. In: International Conference on Machine Learning, vol. 139, pp. 10347–10357 (2021)
39. Van Den Oord, A., Dieleman, S., Schrauwen, B.: Transfer learning by supervised pre-training for audio-based music classification. In: Conference of the International Society for Music Information Retrieval (ISMIR 2014) (2014)

40. Vandaele, R., Dance, S.L., Ojha, V.: Deep learning for automated river-level monitoring through river-camera images: an approach based on water segmentation and transfer learning. Hydrol. Earth Syst. Sci. **25**(8), 4435–4453 (2021)
41. Xue, F., Wang, Q., Guo, G.: Transfer: learning relation-aware facial expression representations with transformers. In: Proceedings of the IEEE/CVF International Conference on Computer Vision, pp. 3601–3610 (2021)
42. Ying, W., Zhang, Y., Huang, J., Yang, Q.: Transfer learning via learning to transfer. In: International Conference on Machine Learning, pp. 5085–5094. PMLR (2018)
43. Zhong, S.H., Huang, X., Xiao, Z.: Fine-art painting classification via two-channel dual path networks. Int. J. Mach. Learn. Cybern. **11**(1), 137–152 (2020)
44. Zhou, H.Y., Lu, C., Yang, S., Yu, Y.: Convnets vs. transformers: whose visual representations are more transferable? In: Proceedings of the IEEE/CVF International Conference on Computer Vision, pp. 2230–2238 (2021)
45. Zhu, Z., Lin, K., Zhou, J.: Transfer learning in deep reinforcement learning: a survey. arXiv preprint arXiv:2009.07888 (2020)
46. Zhuang, F., et al.: A comprehensive survey on transfer learning. Proc. IEEE **109**(1), 43–76 (2020)

On-the-Go Reflectance Transformation Imaging with Ordinary Smartphones

Mara Pistellato$^{(\boxtimes)}$ ⓘ and Filippo Bergamasco ⓘ

DAIS, Universitá Ca'Foscari Venezia, 155, via Torino, Venezia, Italy
{mara.pistellato,filippo.bergamasco}@unive.it

Abstract. Reflectance Transformation Imaging (RTI) is a popular technique that allows the recovery of per-pixel reflectance information by capturing an object under different light conditions. This can be later used to reveal surface details and interactively relight the subject. Such process, however, typically requires dedicated hardware setups to recover the light direction from multiple locations, making the process tedious when performed outside the lab.

We propose a novel RTI method that can be carried out by recording videos with two ordinary smartphones. The flash led-light of one device is used to illuminate the subject while the other captures the reflectance. Since the led is mounted close to the camera lenses, we can infer the light direction for thousands of images by freely moving the illuminating device while observing a fiducial marker surrounding the subject. To deal with such amount of data, we propose a neural relighting model that reconstructs object appearance for arbitrary light directions from extremely compact reflectance distribution data compressed via Principal Components Analysis (PCA). Experiments shows that the proposed technique can be easily performed on the field with a resulting RTI model that can outperform state-of-the-art approaches involving dedicated hardware setups.

Keywords: Reflectance Transformation Imaging · Neural network · Camera pose estimation · Interactive relighting

1 Introduction

In Reflectance Transformation Imaging (RTI) an object is acquired with different known light conditions to approximate the per-pixel Bi-directional Reflectance Distribution Function (BRDF) from a static viewpoint. Such process is commonly used to produce relightable images for Cultural Heritage applications [6,19] or perform material quality analysis [4] and surface normal reconstruction. The flexibility of such method makes it suitable for several materials,

---

Supplementary Information The online version contains supplementary material available at https://doi.org/10.1007/978-3-031-25056-9_17.

and the resulting images can unravel novel information about the object under study such as manufacturing techniques, surface conditions or conservation treatments. Among the variety of practical applications in Cultural Heritage field, we can mention enhanced visualisation [6,21], documentation and preservation [13,15,16] as well as surface analysis [3]. Moreover, RTI techniques can be effectively paired with other tools as 3D reconstruction [23–25,36] or multispectral imaging [8] to further improve the results.

In the majority of the cases, the acquisition of RTI data is carried out with specialised hardware involving a light dome and other custom devices that need complex initial calibration. Since the amount of processed data is significant, several compression methods have been proposed for RTI data representation to obtain efficient storage and interactive rendering [9,27]. In addition to that, part of the proposals focus on the need of low-cost portable solutions [12,28,38], including mobile devices [31] to perform the computation on the field.

In this paper we first propose a low-cost acquisition pipeline that requires a couple of ordinary smartphones and a simple marker printed on a flat surface. During the process, both smartphones acquire two videos simultaneously: one device acting as a static camera observing the object from a fixed viewpoint, while the other provides a trackable moving light source. The two videos are synchronised and then the marker is used to recover the light position with respect to a common reference frame, originating a sequence of intensity images paired with light directions. The second contribution of our work is an efficient and accurate neural-network model to describe per-pixel reflectance based on PCA-compressed intensity data. We tested the proposed relighting approach both on a synthetic RTI dataset, involving different surfaces and materials, and on several real-world objects acquired on the field.

2   Related Work

The literature counts a huge number of different methods for both acquisition and processing of RTI data for relighting. In [22] the authors give a comprehensive survey on Multi-Light Image Collections (MLICs) for surface analysis. Many approaches employ the classical polynomial texture maps [14] to (i) define the per-pixel light function, (ii) store a representation of the acquire data, and (iii) dynamically render the image under new lights. Similar techniques are the so-called Hemispherical Harmonics coefficients [17] and Discrete Modal Decomposition [26]. In [9] the authors propose a new method based on Radial Basis Function (RBF) interpolation, while in [27] a compact representation for web visualisation employing PCA is presented. The authors in [18] present the High-light Reflectance Transformation Imaging (H-RTI) framework, where the light direction is estimated by detecting its specular reflection on one or more spherical objects captured in the scene. However, such setup involves several assumptions such as constant light intensity and orthographic camera model, that in practice make the model unstable. Other techniques that have been proposed to estimate light directly from some scene features are [1,2], while in the authors [9] propose a novel framework to expand the H-RTI technique.

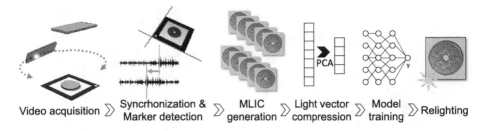

Video acquisition ≫ Syncrhonization & Marker detection ≫ MLIC generation ≫ Light vector compression ≫ Model training ≫ Relighting

Fig. 1. Complete mobile-based RTI acquisition and relighting pipeline.

Recently, neural networks have been employed successfully in several Computer Vision tasks, including RTI. In particular, the encoder-decoder architecture is used in several applications for effective data compression [33]. The work in [30] presents a NN-based method to model light transport as a non-linear function of light position and pixel coordinates to perform image relighting. Other related work using neural networks are [39], in which a subset of optimal light directions is selected, and [29] where a convolutional approach is adopted. Authors in [5] propose an autoencoder architecture to perform relighting of RTI data: the architecture is composed by an encoder part where pixel-wise acquired values are compressed, then the decoder part uses the light information to output the expected pixel value. They also propose two benchmark datasets for evaluation.

3   Proposed Method

Our method follows the classical procedure employed in the vast majority of existing RTI applications: the whole pipeline is presented in Fig. 1. First, several images of the object under study are acquired varying the lighting conditions. In our case, the operation uses the on-board cameras and flash light of a pair of ordinary smartphones while taking two videos. The two videos are then synchronised and the smartphones positions with respect to the scene are recovered using a fiducial marker: in this way we obtain light position and reflectance image for each frame. Such data is processed to create a model that maps each pair (*pixel, light direction*) to an observed reflectance value. Section 3.1 gives a detailed description of this process. This results in a Multi-Light Image Collection (MLIC), that is efficiently compressed by projecting light vectors to a lower-dimensional space via PCA. Then, we designed a neural model defined as a small Multi-Layer Perceptron (MLP) to decode the compressed light vectors and extrapolate the expected intensity of a pixel given a light direction. In Sect. 3.2 the neural reflectance model and data compression are illustrated in detail. Finally, the trained model is used to dynamically relight the object by setting the light direction to any (possibly unseen) value.

3.1   Data Acquisition

Data acquisition is performed using two smartphones and a custom fiducial marker as shown in Fig. 2 (left). The object to acquire is placed at the centre of a marker composed by a thick black square with a white dot at one corner.

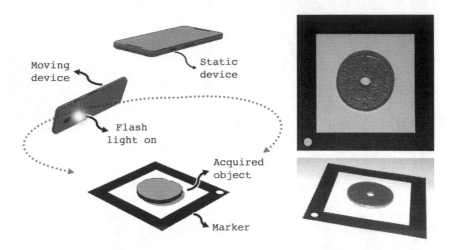

Fig. 2. Left: Proposed RTI acquisition setup. Right: Example frames acquired by the static and moving devices.

One device is located above the object, with the camera facing it frontally so that it produces images as depicted in Fig. 2 (top-right). This device, called *static*, must not move throughout the acquisition, so we suggest to attach it to a tripod. The second device, called *moving*, is manually moved around the object with an orbiting trajectory. The flash led-light located close to the backward-facing camera must be kept on all the time to illuminate the object from different locations. This will allow the static device to observe how the reflectance of each pixel changes while moving the light source.

Both the devices record a video during the acquisition. For now, let's consider those videos as just sequences of images perfectly synchronised in time. In other words, the acquisition consists in a sequence of M images $(I_0^s, I_1^s, \ldots, I_M^s)$ acquired from the static device paired with a sequence $(I_0^m, I_1^m, \ldots, I_M^m)$ acquired from the moving device at times $t_0, t_1, \ldots, t_M$.

After video acquisition, each image is processed to detect the fiducial marker. For the static camera, this operation is needed to locate the 4 corners (c_0, c_1, c_2, c_3) of the inner white square (i.e. the internal part of the marker inside the thick black border). This region is then cropped to create a sequence of $(\mathcal{I}_0, \ldots, \mathcal{I}_N)$ images composed by $W \times H$ pixels commonly referred as Multi-light Image Collection (MLIC). Note that N can be lower than M because the fiducial marker must be detected in both I_i^s and I_i^m to be added to the MLIC.

Each $\mathcal{I}_i$ is a single-channel grayscale image containing only the luminance of the original I_i^s image. We decided to model only the reflectance intensity (and not the wavelength) as a function of the light's angle of incidence for two reasons. First, we cannot change the colour of the light source and, second, it is uncommon to have iridescent materials where the incindent angle affects the reflectance spectrum [11]. Therefore, we convert all the images to the YUV colour space to store only the Y channel in the MLIC. To deal with the colour, we store the pixel-wise averages $\bar{U} = \frac{1}{N} \sum_N U_i$ and $\bar{V} = \frac{1}{N} \sum_N V_i$ for further processing.

The marker is also detected in the moving camera image sequence, but for a different purpose. We assume that the flash light is so close to the camera optical centre that can be considered almost at the same point. So, by finding the pose of the camera (R, t) in the marker reference frame, we can estimate the location of the light source (i.e. the moving camera optical center) with respect to the object. This operation is simply performed by computing the Homography H mapping $c_0 \ldots c_3$ to the marker model points $\begin{pmatrix} 0 \\ 0 \\ 1 \end{pmatrix}, \begin{pmatrix} W \\ 0 \\ 1 \end{pmatrix}, \begin{pmatrix} W \\ H \\ 1 \end{pmatrix}, \begin{pmatrix} 0 \\ H \\ 1 \end{pmatrix}$ and then factorising it as:

$$K^{-1}H = \alpha \begin{pmatrix} | & | & | \\ r_1 & r_2 & t \\ | & | & | \end{pmatrix} \tag{1}$$

where K is the intrinsic camera matrix, r_1, r_2 are the first two columns of the rotation matrix R, and α is a non-zero unknown scale factor [10]. Since R must be orthonormal, α can be approximated as $2/(\|r_1\| + \|r_2\|)$ and r_3 as $r_1 \times r_2$. Since (R, t) maps points from the camera reference frame to the marker (i.e. object) reference frame, the vector t represents the light position and $\mathcal{L} = t/\|t\|$ the light direction. Since the light is not at the infinity, each object point actually observes a slightly different light direction vector $\mathcal{L}$. However, as usually done in other RTI applications, we consider this difference negligible so that we collect a single light direction vector for each image.

After the data acquisition process, we end up with the MLIC $(\mathcal{I}_0 \ldots \mathcal{I}_N)$ and $(\mathcal{L}_0, \ldots, \mathcal{L}_N)$ vectors together with $\bar{U}, \bar{V}$. This is all the data we need to generate our reflectance model and proceed with dynamic relighting. At this point, we need to do some considerations regarding the acquisition procedure:

- Pixel reflectance data is collected only from the static camera, while the moving camera is used just to estimate the light direction. This implies that the final result quality is directly affected by the quality of the static camera (i.e. resolution, noise, etc.). Therefore, we suggest to use a good smartphone for that. Conversely, the moving device can be cheap as long as images are sufficiently well exposed to reliably detect the marker.
- The moving camera must be calibrated a-priori to factorise H. In practice, the calibration is not critical and can simply be inferred from the lens information provided in the EXIF metadata. We also used this approach in all our experiments.
- The ambient should be illuminated uniformly and constantly over time. Ideally, the moving device flash light should be the only one observed by the object. Since that is typically impractical, it is at least sufficient that the contribution from ambient illumination is negligible with respect to the provided moving light.

– The orbital motion should uniformly span the top hemisphere above the object with a certain constant radius. Indeed, we only consider light direction so changes in the reflectance due to light proximity with respect to the object will not be properly accounted by the model.

Video Synchronisation. Video recording is manually started (roughly at the same time) in the two devices. So, it is clear that the two frame sequences are not synchronised out of the box (i.e. the ith frame of the static device will not be taken at the same time as the ith frame of the moving one). That will never be the case without an external electronic triggering but we can still obtain a reasonable synchronisation exploiting the audio signal of the two videos [37].

We first extract the two audio tracks and then compute the time offset in seconds that maximises the Time-lagged Cross-correlation [32]. Once the offset is known, initial frames from the video starting first are dropped to match the two sequences. Note also that, if the framerates are different, frames must be dropped from time to time from the fastest video to keep it in sync with the other. In the worst case, the time skewness is 1/FPS where FPS is the framerate of the slowest video. Nevertheless, since the moving device is orbited around the object very slowly, such time skewness will have a negligible effect in the estimation of $(\mathcal{L}_0, \ldots, \mathcal{L}_N)$.

Fiducial Marker Detection. Detecting the four internal corners of the proposed fiducial marker can be simply performed with classical image processing. We start with Otsu's image thresholding [20] followed by hierarchical closed contour extraction [34]. Each contour is then simplified with the Ramer-Douglas-Peucker algorithm and filtered out if resulting in a number of points different than 4. All the black-to-white 4-sides polygons contained into a white-to-black 4-side polygon are good candidates for a marker. So, we check the midpoint of each closest corresponding vertex pairs searching for the white dot. If exactly one white dot is found among corresponding pairs, than the four vertexes of the internal polygon can be arranged in clockwise order starting from the one closest to the dot. This results in the four corners $c_0, \ldots, c_3$.

Since we expect to see exactly one instance of the marker in every frame, this simple approach is sufficient in practice. We decided not use popular alternative markers (see for instance [7]) because they typically reserve the internal payload area to encode the marker id. Of course, any method will work as long as it results in a reasonably accurate localisation of the camera while providing free space to place the object under study.

3.2   The Reflectance Model

To perform interactive relighting, we need first to model how the reflectance changes when varying the light direction. Our goal is to define a function:

$$f(\mathbf{p}, \vec{l}) \rightarrow (y, u, v) \tag{2}$$

producing the intensity y and colour u, v (in the YUV space) of a pixel $\mathbf{p} = (x, y) \in \mathbb{N} \times \mathbb{N}$ when illuminated from a light source with direction $\vec{l} = (l_u, l_v)$[1]. Once the model is known, relighting can be done by choosing a light $\vec{l}$ and evaluating f for every pixel $\mathbf{p}$ of the target image.

The data acquisition process described before produces a sampling of f for some discrete values of $\mathbf{p}$ and $\vec{l}$. This sampling is dense on the pixels, since we acquire an entire image for every light, but typically sparse in the amount of the observed light directions, especially if using a light dome where this number is limited to a few dozens. In our case, the sampling of $\vec{l}$ is a lot denser since we acquire an entire video composed by thousands of frames. However, directions are highly correlated in space as we follow a continuous circular trajectory (See Fig. 3, Left).

The challenge is to: (i) provide a realistic approximation of f for previously unseen light directions while (ii) using a very compact representation so that it can be easily transferred, stored and evaluated even on a mobile phone. The two problems are related because the selection of what family of functions to use for f affects how many parameters are needed to describe the chosen one. For example, in [14] each pixel is independently modelled as a 6-coefficients biquadratic function of the light direction, requiring the storage of $6 \times M \times N$ values.

Inspired by *NeuralRTI* proposed by Dulecha et al. [5], we also represent f as a Multi-Layer Perceptron trained from the data acquired with the smartphones. However, we have two substantial differences with respect to their approach. First, we avoid the auto-encoder architecture for data compression. Since we do not use a light dome, the number of light samples changes in each acquisition and is at least an order of magnitude greater. NeuralRTI would not be feasible in our case, as it results in a network taking in input vectors of thousands of elements. Also, the network architecture itself depends on N (variable in our case) producing a different layout for each acquired object. Instead, we compress such vectors with classical PCA to feed the MLP acting as a decoder. Interestingly, this tend to produce better results not only on our data but also on images acquired with a classical light dome. Second, the light vector $\vec{l}$ is not concatenated as-is to the network input but projected to a higher-dimensional Fourier space with random frequencies as discussed in [35]. This has a positive effect on the ability of the network to reconstruct the correct pixel intensity.

3.3   Neural Model

Our proposed neural model $\mathcal{Z}(\mathbf{k}_p, \vec{l}) \rightarrow y$ works independently for each pixel (i.e. it does not consider the spatial relationship among those) and recovers the intensity information y without the colour. It takes as input a compressed light vector $\mathbf{k}_p = (k_0, k_1, \ldots, k_B) \in \mathbb{R}^B$ of any pixel p and a light direction $\vec{l}$ to produce the intensity for pixel p.

---

[1] l_u and l_v range between $[-1 \ldots 1]$ respectively as they are the first two components of a (unitary-norm) 3D light direction vector pointing toward the light source.

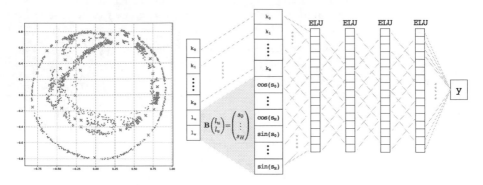

Fig. 3. Left: 2D plot of the first two components of the light direction vectors ($\mathcal{L}_0 \ldots \mathcal{L}_n$). Each point is associated to an image in the MLIC. Note the circular trajectory. Right: Network architecture composed by 5 fully-connected layers with ELU activation. (Color figure online)

The model $\mathcal{Z}$ is composed by an initial (non-trainable) projection of the light vector followed by a MLP arranged in 5 layers consisting in 16 neurons each, all using the ELU activation function except for the output realised with a single neuron with linear activation (Fig. 3, right). The network input I is a $(B + 2H)$-dimensional vector created by concatenating the B values of $\mathbf{k}_p$ with the projection of $\vec{l}$ to an H-dimensional Fourier space with random frequencies.

Specifically, let $\mathbf{B}$ be a $H \times 2$ matrix where each element is sampled from a Gaussian distribution $\mathcal{N}(0, \sigma^2)$. This matrix is generated once for each acquired object, it is not trained, and is common to all the pixels. The network input is then obtained as:

$$I = \left(k_0, \ldots k_B, \cos(s_0), \ldots \cos(s_H), \sin(s_0) \ldots \sin(s_H)\right) \tag{3}$$

$$\text{where} \quad \left(s_0 \ldots s_H \right)^T = \mathbf{B} \begin{pmatrix} l_u \\ l_v \end{pmatrix}. \tag{4}$$

Once $\mathcal{Z}$ is trained, it can be used for relighting as follows:

$$f(\mathbf{p}, \vec{l}) = \left(\mathcal{Z}(\mathbf{k_p}, \vec{l}), \ \bar{U}(\mathbf{p}), \ \bar{V}(\mathbf{p}) \right). \tag{5}$$

Creating the Compressed Light Vectors $\mathbf{k_p}$. The size of the neural model $\mathcal{Z}$ depends by the $2H$ values of the matrix $\mathbf{B}$, its internal weights, and the light vectors $\mathbf{k}_p$ (one for each pixel, for a total of $W \cdot H \cdot B$ values). It is obvious that most of the storage is spent for the light vectors since the number of image pixels is far greater than the other variables. Considering that we acquire roughly 1 minute of video at 30 FPS, our MLIC is composed by $\approx$2000 images cropped to a size of 400×400 pixels for a total of 320 MB. So, using all the acquired data as-is to define f jeopardises the idea of doing interactive relighting directly on a mobile app or in a web browser. Since we assume that all the pixels observe the same light vector, the acquired MLIC $(\mathcal{I}_0 \dots \mathcal{I}_N)$ can be represented as a $W \times H \times N$ N-channel image in which, for each pixel, a vector of N values (corresponding to light directions $\mathcal{L}_0 \dots \mathcal{L}_N$) have been observed. In [5] the authors use an auto-encoder to produce an intermediate encoded representation of the observed light vectors of each pixel, and then just the decoder for relighting. This works well for the light dome in which N is typically less than 50. We tried their approach with $N = 1500$ lights and realised that the network struggles to converge to an effective encoded representation.

We propose a more classical approach in which the encoding of the light vectors is not based on Deep Learning. Let $\mathbf{K}_p$ be the N-dimensional light vector of the pixel p. We propose to use Principal Component Analysis on all the light vectors acquired $\mathbf{K}_{p_0} \dots \mathbf{K}_{p_{W \times H}}$ to find a lower-dimensional space of B orthogonal bases. Then, the encoded $\mathbf{k}_p$ is obtained by projecting $\mathbf{K}_p$ into that space. In the experimental section we show how the number of bases B can be very small compared to N while still producing high-quality results.

Network Training and Implementation Details. The network model $\mathcal{Z}$ is trained by associating each input I with the expected reflectance intensity. Specifically, we combined the encoded vector $\mathbf{k}_p$ of each pixel, with all the possible light directions $\mathcal{L}_0 \dots \mathcal{L}_N$ to produce the input $I_{x,y,n}$ ($0 \leq y < H$, $0 \leq x < W$, $0 \leq n < N$). The output associated to $I_{x,y,n}$ is simply the value of $\mathcal{I}_n(x,y)$, that is the intensity value observed for light n at pixel (x, y). This results in a total of $W \cdot H \cdot N$ data samples to be used for training. Note that, regardless the amount of pixels (i.e. the image resolution) and light directions involved (i.e. number of frames in the video), the network architecture remains unchanged. Therefore, it is easy to compute how much storage is needed for the model, depending only on the number of PCA bases B and image resolution. Considering the acquired data size discussed before, and supposing to use $B = 8$ bases and $H = 10$ frequencies, we have to store $400 \times 400 \times 8$ compressed light vectors (≈ 5 MB with single precision), and 1252 values for network weights and $\mathbf{B}$.

Finally, we adopted a classical Mean Absolute Error (MAE) loss function:

$$\text{MAE} = \frac{1}{W \cdot H \cdot N} \sum_{x,y,n} |\mathcal{Z}(\mathbf{k}_{\mathbf{p}=(x,y)}, \vec{l}_n) - \mathcal{I}_n(x, y)|. \tag{6}$$

4   Experimental Results

We started by analysing the behaviour of our proposed MLP model with respect to two relevant parameters, namely the number of PCA bases B and the parameter σ used to sample the frequencies of the light projection matrix $\mathbf{B}$. Then, we quantitatively and qualitatively validated our method with respect to fully-synthetic data as well as with real-world smartphones acquisition.

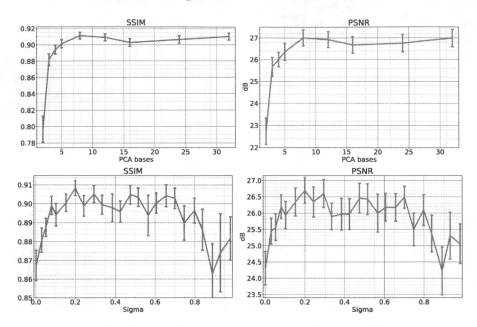

Fig. 4. First row: SSIM and PSNR values increasing the number of PCA bases for data compression. Second row: SSIM and PSNR values increasing the sigma.

In all our tests we fixed $H = 10$ so that the matrix $\mathbf{B}$ has size 10×2 always projecting the input light vector $\vec{l}$ into a 20-dimensional space. Note that the values of the matrix $\mathbf{B}$ are randomly sampled before starting the training and never optimised. During the training we used Adam optimiser with a learning rate of 10^{-3} for the first 20 epochs and then reduced to 10^{-4} for another 20.

Real-World Datasets. For the real data we used some coins as test objects, acquired using an Apple iPhone 11 acting as static device and a Samsung Galaxy A40 as the moving one. Videos have been processed as described in Sect. 3.1 resulting in a MLIC with roughly $2K$ images for each coin. Then, for each MLIC we randomly selected 25 lights for the test set and discarded the closest light directions (within a radius) so that the learning process is not trained on similar conditions. An example of acquired light directions for the dataset *Coin1* is shown in Fig. 3 (left), where the blue dots are the ∼1920 lights used for training and the red crosses the ones extracted for test.

4.1 Parameter Study

We first studied the effect of the number B (i.e. PCA bases) on the final relighting quality. Therefore, in the first test we projected the acquired MLIC data into an increasing number of PCA bases, and proceeded with the training process as described. The plots in Fig. 4 (first row) show the resulting PSNR (Peak Signal-to-Noise Ratio) and SSIM (Structural Similarity Index) for the test set while

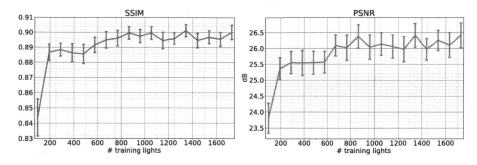

Fig. 5. Average SSIM and PSNR on test set increasing the number of lights used during training procedure for our method.

raising the number of bases B, from 2 to 32. Note that such results have been computed by taking the average among all the acquired datasets and repeating the training 10 times due to the random nature of the process. The error bars denote the standard error. We can notice that with $B = 2$ the relighting quality is quite low, and increasing the number of bases from 3 to 8 corresponds to an increasing reconstruction quality. Both PSNR and SSIM value stabilise for $B > 8$, meaning that a higher number of bases would not further improve the output quality. In all our tests we observed that a PCA projection with 8 bases offers a good compression for our smartphone-acquired data. Moreover, we tested the same 8-bases compression for classical RTI datasets where a dome with equispaced lights is used: interestingly, results are numerically and qualitatively better with respect to the autoencoder compression technique as shown in the first row of Table 1.

The next experiment results are shown in the second row of Fig. 4: we analysed the relighting quality against the value of σ (on x-axis) used to generate the random values in the matrix **B**. The test was repeated 50 times for each different dataset. We can observe that values around $\sigma = 0.3$ offer good results in terms of average PSNR and SSIM on the test set, exhibiting also a smaller standard error. As stated in [35], σ is a free parameter that has to be tuned for a particular problem. However, our light directions have unitary norm so, once the optimal σ is defined, it will remain the same regardless the object to reconstruct. Therefore, we used 0.3 in all our real-world tests.

We also tested the effect of the number of acquired light directions (the size of N) against the final reconstruction accuracy while keeping the same network layout (Fig. 5). This increases the size of the training set but not the storage space required for the model. Both SSIM and PSNR increase with N, probably because the network can be trained better if a large variety of light conditions can be used. Nevertheless, this increase is almost negligible when the number of light samples exceeds 700–800. So, assuming an acquisition in which a carefully planned circular motion around the object is performed, an average video duration of 40 s at 25 FPS would be sufficient.

Table 1. Relight comparison for different methods.

Dataset	Polynomial		RBF		NeuralRTI		Our	
	PSNR	SSIM	PSNR	SSIM	PSNR	SSIM	PSNR	SSIM
SynthRTI	22.7451	0.7932	22.6828	0.8353	26.3658	0.8540	**26.4075**	**0.8553**
Coin1	24.0562	0.8643	25.6791	**0.9152**	25.6846	0.8940	**27.0019**	0.9118
Coin2	25.8627	0.8798	26.8939	**0.9197**	26.9147	0.8937	**28.0361**	0.9105
Coin3	24.2319	0.8642	25.4360	**0.9009**	25.0304	0.8808	**25.7479**	0.8975
Coin4	27.6388	0.9155	28.1845	0.9369	27.8950	0.9176	**29.6954**	**0.9398**

4.2 Comparisons

We compared our relighting approach with two classical light interpolation methods, namely polynomial texture maps [14] (from now on, identified as *polynomial*) and RBF. Moreover, we tested against the already discussed learning-based method *NeuralRTI* [5]. We recall that NeuralRTI architecture changes with the number of input lights (the length of the encoder input is $3N$ because the network works in the RGB space), so we trained it on our acquired data by randomly selecting 100 lights among the training set, since the training process becomes unfeasible for highest input dimensions.

In addition to our data acquired with smartphones, we also validated the method on classical RTI configurations represented by the synthetic dataset presented in [5]. Such data is generated simulating a dome with 69 lights, divided into separate train and test sets of 49 and 20 lights respectively. In all comparisons we set $B = 8$ (PCA bases), $\sigma = 0.3$ and $H = 10$. Table 1 shows the comparison results. The values in the first row represent the average SSIM and PSNR for the whole SynthRTI dataset. To better evaluate our method comprising not only the reflectance model but also the smartphone-based data acquisition, we show the results for all the objects of the real-world dataset acquired as proposed. Overall, our method exhibits the higher PSNR value, while in some cases relighted data interpolated with RBF give a slightly higher SSIM, but with a significantly smaller PSNR with respect to our method. Note however that RBF is significantly slower in the relighting phase. Also, our values are slightly better with respect to NeuralRTI for the synthetic dataset, where the training lights are sampled uniformly on a dome setup. This indicates that our proposed PCA compression and decoder network still improves the encoder-decoder architecture of [5]. Note that we did not tune any parameter for our results, concluding that the number of PCA bases does not depend on the specific dataset.

Qualitative examples for our acquired dataset are shown in Fig. 6, where we display the relighting of three coins with two different test lights (last column shows the ground truth, GT). We can notice that our method is able to recover the object reflectance with high accuracy, especially for the shadows projected near the coins, while the other methods tend to generate light blooms or blurry shadows. Moreover, we notice that NeuralRTI slightly alters the output tint

Polynomial   RBF   NeuralRTI   Our   GT

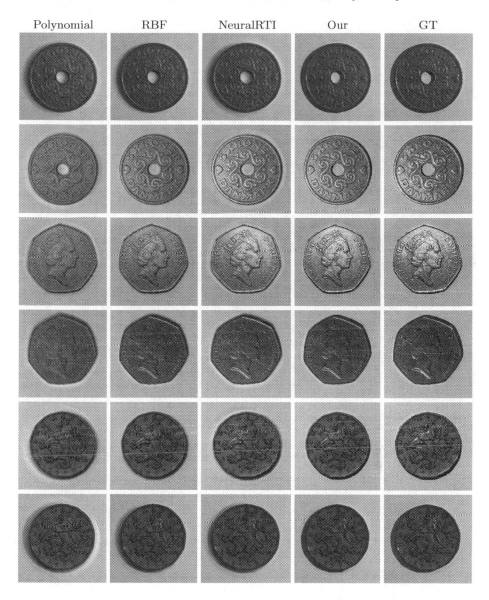

Fig. 6. Relighting comparison of real-world data acquired with two smartphones. The last column (ground truth, GT) shows the actual pictures from the test set.

Polynomial   RBF   NeuralRTI   Our   GT

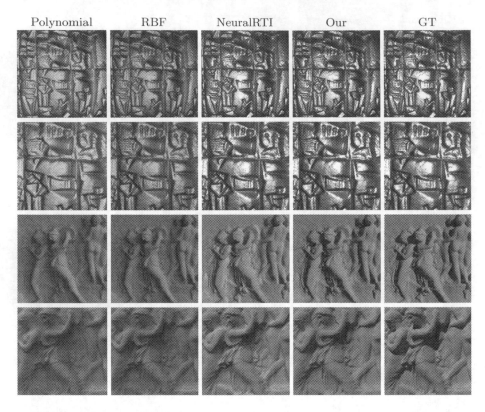

Fig. 7. Qualitative comparison on synthetic data generated with a dome configuration.

with respect to the original: this can be seen in particular in the first two rows. Probably, directly modelling each pixel intensity and colour is more difficult to handle for the network than just the intensity. Using the average UV-value is easier and produces more stable results for non-iridescent objects. Finally, in Fig. 7 we show a couple of outputs for the synthetic dataset. Our results are quite similar with respect to NeuralRTI but also in this case our shadow areas are sharper and the images exhibit a higher contrast.

5   Conclusions

In this paper we proposed a low-cost technique to perform image relighting on the go using two smartphones for data acquisition. A practical video processing pipeline extracts the MLIC that is compressed and used to train a neural relighting model. Extensive tests in both synthetic and real-world settings show that our network effectively hallucinates images from unseen light directions with high quality. The presented setup can be easily operated directly on the field, with no need of expensive and specialised hardware, allowing researchers to carry out part of their work in an effective and fast way.

References

1. Ackermann, J., Fuhrmann, S., Goesele, M.: Geometric point light source calibration. In: VMV, pp. 161–168 (2013)
2. Ahmad, J., Sun, J., Smith, L., Smith, M.: An improved photometric stereo through distance estimation and light vector optimization from diffused maxima region. Pattern Recogn. Lett. **50**, 15–22 (2014)
3. Ciortan, I., Pintus, R., Marchioro, G., Daffara, C., Giachetti, A., Gobbetti, E., et al.: A practical reflectance transformation imaging pipeline for surface characterization in cultural heritage (2016)
4. Coules, H., Orrock, P., Seow, C.E.: Reflectance transformation imaging as a tool for engineering failure analysis. Eng. Fail. Anal. **105**, 1006–1017 (2019)
5. Dulecha, T.G., Fanni, F.A., Ponchio, F., Pellacini, F., Giachetti, A.: Neural reflectance transformation imaging. Visual Comput. **36**, 2161–2174 (2020). https://doi.org/10.1007/s00371-020-01910-9
6. Earl, G., et al.: Reflectance transformation imaging systems for ancient documentary artefacts. In: Electronic Visualisation and the Arts (EVA 2011), pp. 147–154 (2011)
7. Garrido-Jurado, S., Muñoz-Salinas, R., Madrid-Cuevas, F.J., Marín-Jiménez, M.J.: Automatic generation and detection of highly reliable fiducial markers under occlusion. Pattern Recogn. **47**(6), 2280–2292 (2014)
8. Giachetti, A., Ciortan, I., Daffara, C., Pintus, R., Gobbetti, E., et al.: Multispectral RTI analysis of heterogeneous artworks (2017)
9. Giachetti, A., Ciortan, I.M., Daffara, C., Marchioro, G., Pintus, R., Gobbetti, E.: A novel framework for highlight reflectance transformation imaging. Comput. Vis. Image Underst. **168**, 118–131 (2018)
10. Hartley, R., Zisserman, A.: Multiple View Geometry in Computer Vision, 2nd edn. Cambridge University Press, New York (2003)
11. Kinoshita, S., Yoshioka, S., Miyazaki, J.: Physics of structural colors. Rep. Progress Phys. **71**(7), 076401 (2008). https://doi.org/10.1088/0034-4885/71/7/076401
12. Kinsman, T.: An easy to build reflectance transformation imaging (RTI) system. J. Biocommun. **40**(1), 10–14 (2016)
13. Kotoula, E., Kyranoudi, M.: Study of ancient Greek and Roman coins using reflectance transformation imaging. E-conservation Mag. **25**, 74–88 (2013)
14. Malzbender, T., Gelb, D., Wolters, H.: Polynomial texture maps. In: Proceedings of the 28th Annual Conference on Computer Graphics and Interactive Techniques, pp. 519–528 (2001)
15. Manfredi, M., et al.: Measuring changes in cultural heritage objects with reflectance transformation imaging. In: 2013 Digital Heritage International Congress (Digital-Heritage), vol. 1, pp. 189–192. IEEE (2013)
16. Manrique Tamayo, S.N., Valcárcel Andrés, J.C., Osca Pons, M.: Applications of reflectance transformation imaging for documentation and surface analysis in conservation. Int. J. Conserv. Sci. **4**, 535–548 (2013)
17. Mudge, M., et al.: Image-based empirical information acquisition, scientific reliability, and long-term digital preservation for the natural sciences and cultural heritage. In: Eurographics (Tutorials), vol. 2(4) (2008)
18. Mudge, M., Malzbender, T., Schroer, C., Lum, M.: New reflection transformation imaging methods for rock art and multiple-viewpoint display. In: Ioannides, M., Arnold, D., Niccolucci, F., Mania, K. (eds.) The 7th International Symposium on Virtual Reality, Archaeology and Cultural Heritage, vol. 6, pp. 195–202. Vast (2006)

19. Mytum, H., Peterson, J.: The application of reflectance transformation imaging (RTI) in historical archaeology. Hist. Archaeol. **52**(2), 489–503 (2018)
20. Otsu, N.: A threshold selection method from gray-level histograms. IEEE Trans. Syst. Man Cybern. **9**(1), 62–66 (1979). https://doi.org/10.1109/TSMC.1979.4310076
21. Palma, G., Corsini, M., Cignoni, P., Scopigno, R., Mudge, M.: Dynamic shading enhancement for reflectance transformation imaging. J. Comput. Cult. Heritage (JOCCH) **3**(2), 1–20 (2010)
22. Pintus, R., Dulecha, T.G., Ciortan, I., Gobbetti, E., Giachetti, A.: State-of-the-art in multi-light image collections for surface visualization and analysis. In: Computer Graphics Forum, vol. 38, pp. 909–934. Wiley Online Library (2019)
23. Pistellato, M., Albarelli, A., Bergamasco, F., Torsello, A.: Robust joint selection of camera orientations and feature projections over multiple views, pp. 3703–3708 (2016). https://doi.org/10.1109/ICPR.2016.7900210
24. Pistellato, M., Bergamasco, F., Albarelli, A., Torsello, A.: Dynamic optimal path selection for 3D triangulation with multiple cameras, vol. 9279, pp. 468–479 (2015)
25. Pistellato, M., Bergamasco, F., Albarelli, A., Torsello, A.: Robust cylinder estimation in point clouds from pairwise axes similarities, pp. 640–647 (2019). https://doi.org/10.5220/0007401706400647
26. Pitard, G., et al.: Discrete modal decomposition: a new approach for the reflectance modeling and rendering of real surfaces. Mach. Vis. Appl. **28**(5), 607–621 (2017)
27. Ponchio, F., Corsini, M., Scopigno, R.: Relight: a compact and accurate RTI representation for the web. Graph. Models **105**, 101040 (2019)
28. Porter, S.T., Huber, N., Hoyer, C., Floss, H.: Portable and low-cost solutions to the imaging of paleolithic art objects: a comparison of photogrammetry and reflectance transformation imaging. J. Archaeol. Sci. Rep. **10**, 859–863 (2016)
29. Rainer, G., Jakob, W., Ghosh, A., Weyrich, T.: Neural BTF compression and interpolation. In: Computer Graphics Forum, vol. 38, pp. 235–244. Wiley Online Library (2019)
30. Ren, P., Dong, Y., Lin, S., Tong, X., Guo, B.: Image based relighting using neural networks. ACM Trans. Graph. (ToG) **34**(4), 1–12 (2015)
31. Schuster, C., Zhang, B., Vaish, R., Gomes, P., Thomas, J., Davis, J.: RTI compression for mobile devices. In: Proceedings of the 6th International Conference on Information Technology and Multimedia, pp. 368–373. IEEE (2014)
32. Shen, C.: Analysis of detrended time-lagged cross-correlation between two nonstationary time series. Phys. Lett. A **379**(7), 680–687 (2015)
33. Smys, S., Chen, J.I.Z., Shakya, S.: Survey on neural network architectures with deep learning. J. Soft Comput. Paradigm (JSCP) **2**(03), 186–194 (2020)
34. Suzuki, S., Be, K.: Topological structural analysis of digitized binary images by border following. Comput. Vis. Graph. Image Process. **30**(1), 32–46 (1985). https://doi.org/10.1016/0734-189X(85)90016-7, https://www.sciencedirect.com/science/article/pii/0734189X85900167
35. Tancik, M., et al.: Fourier features let networks learn high frequency functions in low dimensional domains. Adv. Neural. Inf. Process. Syst. **33**, 7537–7547 (2020)
36. Uribe, M.D.G., Wheatley, D.W.: Rock art an digital technologies: the application of reflectance transformation imaging (RTI) and 3D laser scanning to the study of late bronze age Iberian stelae. Menga: Revista de prehistoria de Andalucía (4), 187–203 (2013)
37. Vieira, M., Guimarães, P.V., Violante-Carvalho, N., Benetazzo, A., Bergamasco, F., Pereira, H.: A low-cost stereo video system for measuring directional wind waves. J. Marine Sci. Eng. **8**(11), 831 (2020)

38. Watteeuw, L., et al.: Light, shadows and surface characteristics: the multispectral portable light dome. Appl. Phys. A **122**(11), 1–7 (2016)
39. Xu, Z., Sunkavalli, K., Hadap, S., Ramamoorthi, R.: Deep image-based relighting from optimal sparse samples. ACM Trans. Graph. (ToG) **37**(4), 1–13 (2018)

Is GPT-3 All You Need for Visual Question Answering in Cultural Heritage?

Pietro Bongini[ID], Federico Becattini[✉][ID], and Alberto Del Bimbo[ID]

University of Florence, MICC, Florence, Italy
{Pietro.Bongini,Federico.Becattini,AlbertoDel.Bimbo}@unifi.it

Abstract. The use of Deep Learning and Computer Vision in the Cultural Heritage domain is becoming highly relevant in the last few years with lots of applications about audio smart guides, interactive museums and augmented reality. All these technologies require lots of data to work effectively and be useful for the user. In the context of artworks, such data is annotated by experts in an expensive and time consuming process. In particular, for each artwork, an image of the artwork and a description sheet have to be collected in order to perform common tasks like Visual Question Answering. In this paper we propose a method for Visual Question Answering that allows to generate at runtime a description sheet that can be used for answering both visual and contextual questions about the artwork, avoiding completely the image and the annotation process. For this purpose, we investigate on the use of GPT-3 for generating descriptions for artworks analyzing the quality of generated descriptions through captioning metrics. Finally we evaluate the performance for Visual Question Answering and captioning tasks.

Keywords: Visual Question Answering · GPT-3 · Image captioning · Natural language processing · Computer vision

1 Introduction

Cultural Heritage often relies on digital resources to engage and attract visitors. From audio-guides to smartphone applications, museum visits are becoming increasingly more interactive, allowing users to deepen concepts without the need of a human assistant or after the visit is concluded. Forms of gamification are also important, favoring engagement especially for young visitors and instructional purposes. Artificial Intelligence and Computer Vision are playing a large part in the development of such smart visits and applications [5,6,8,13]. A notable machine learning application that has recently found usage in cultural heritage is Visual Question Answering (VQA), which exploits both Computer Vision and Natural Language Processing to allow users to ask questions on the content of an image [6]. The advantage of VQA is that it allows museums to develop smart guides and interactive gamification approaches. However, for pictorial art, most questions posed by users concern contextual information rather than what is actually depicted in a painting.

L. Karlinsky et al. (Eds.): ECCV 2022 Workshops, LNCS 13801, pp. 268–281, 2023.
https://doi.org/10.1007/978-3-031-25056-9_18

To address this limitation, an evolution of VQA known as Contextual Question Answering (CQA) was proposed [6]. The authors explicitly focused on cultural heritage applications, combining visual and contextual cues to answer questions. The contextual information is derived from a textual meta-data, which is fed to the model along with the question and the image. In this way the VQA/CQA model has to learn to attend either relevant parts of an image or relevant sections of the text to provide an adequate answer. The need of a textual data nonetheless opens a new issue, namely where to obtain such description. Information sheets for artworks may already be available to museum curators yet extending this kind of application to new data becomes time-consuming and requires a domain expert.

In this paper we explore the usage of a generative natural language processing model to automatically create contextual information to be fed to a CQA model. In fact, recently, generative text models have been finding large diffusion with groundbreaking results. Among these we find GPT-3, a generative model trained on a massive corpus of textual data regarding several domains, including art [7]. GPT-3 is capable of generating a description starting from a textual query and it has been demonstrated that the model includes knowledge of the entities described in the training data, for example paintings and artworks. We therefore investigate the possibilities and the limitations of GPT-3 in applications for cultural heritage, with a specific focus on question answering. In particular, we explore the quality of the textual description of artworks that the model is able to generate and we evaluate their applicability for visual and contextual question answering.

The main contributions of our work are the following:

We propose an automatic approach to generate textual information sheets of artworks exploiting GPT-3. We find that the model has excellent knowledge of art concepts and event details of specific paintings.

We propose a method to answer both visual and contextual questions which is artwork agnostic, i.e. it does not require any additional data or training to be adapted to a new set of images.

- We explore the applicability of GPT-3 in cultural heritage applications. To the best of our knowledge we are the first to apply GPT-3 to the art domain.

2 Related Work

Natural Language Processing (NLP) in recent years has evolved at an extremely fast pace, converging to a set of well defined application paradigms [33]. Such paradigms include text classification, matching, machine reading comprehension, sequence to sequence translation, sequence tagging and language modeling. Despite the wide variety of tasks [1, 9, 28], some recent noticeable approaches have been shown to perform well as generic pre-training for NLP models [7, 11]. In particular, this can be attributed to the introduction of attention models, based on the transformer architecture [36]. The effectiveness of models such as BERT [11] stems from the capability of processing text bidirectionally exploiting the self-attention mechanism of transformers to obtain word level representations that are informed of their surrounding context within the sentence. Whereas BERT is built exploiting the encoder part of the transformers, another state of the art approach for NLP, Generative Pre-trained Transformer (GPT) [22], is built stacking transformer decoder blocks and is trained to predict the next word in a sentence. The model has then been improved in subsequent versions, GPT-2 [23] and GPT-3 [7], yielding larger and more effective models.

Interestingly, GPT-3 has been trained using a large quantity of internet data, meaning that the training process has distilled into the model common sense knowledge making it able to generate essays and even poetry [10]. In this paper we exploit GPT-3 as a generator of textual content describing artworks, showing that it can be used for interactive applications for cultural heritage such as captioning [19] and Visual Question Answering (VQA) [2]. VQA is a recent trend in machine learning that bridges the Natural Language Processing and Computer Vision domains [4]. The goal is to answer questions regarding the content of an image through artificial intelligence. This involves several sub-tasks such as object detection [15] and recognition [16], question reasoning [20]. Typical VQA approaches use Convolutional Neural Networks (CNNs) to interpret images and Recurrent Neural Networks (RNNs) to process questions. The authors of [1] proposed a bottom-up attention mechanism looking at salient objects in images. Differently from previous approaches that considered regularly spaced image portions [30], they use object Faster R-CNN [25] features as attention candidates. In the past few years multiple Transformer-based approaches reached impressive performances on this task [17,32,34,38].

Recently, a few approaches [3,6,14,35] have addressed VQA in the cultural heritage domain. A dataset of questions and answers for art related questions has been recently proposed [3], exploiting an ontology based framework to extract data with question templates. The authors of [6] and [14] found that to make the best out of VQA for museum applications, a model must be able to integrate some source of external knowledge in order to address contextual questions, i.e. questions concerning non-visual cues such as name of the author, year and artistic style. In particular, [6] used a question classifier to understand if visual of contextual knowledge is required. Depending on the output of the classifier a VQA model is used, otherwise a purely textual based question answering model is used discarding the image content. In this work we explore the effectiveness of using GPT-3 to generate artwork captions, suitable for such a visual and contextual question answering model.

Other approaches have been used to answer questions relying on captions, yet only regarding visual content [29]. The most similar approach to ours is instead [39], which used GPT-3 for VQA. However, differently from us, the authors feed GPT-3 with questions and descriptions generated by an image captioner directly to obtain an answer. We, instead, aim at extracting the domain specific knowledge from GPT-3 which is requested to correctly answer a question.

3 GPT-3

To provide to the reader a better understanding of our work, here we present a brief background context about GPT-3, the third version of Generative Pre-Trained Transformer [7]. This is an autoregressive language model with 175 billion parameters that can be used for different tasks without any finetuning, achieving strong performances.

The architecture of the GPT-3 Transfomer model is made of 96 attention layers. While language models like BERT [11] use the Encoder to generate embeddings from the raw text which can be used in other machine learning applications, GPT-3 use the Decoder half, so it takes embeddings as inputs and produces text. In particular the GPT-3 language model has the ability to generate natural language text that can be hard to distinguish from human-written text, to the point that research has been carried out to asses whether GPT-3 could pass a written Turing test [12].

Concretely, during inference, the target of the new task y is directly predicted conditioned on the given context C and the new task's input x, as a text sequence generation task. Note that all C, x and y are text sequences. For example, $y = (y^1, ..., y^T)$. Therefore, at each decoding step t we have

$$y^t = \arg\max_{y^t} p_W(y^t|C, x, y < t) \tag{1}$$

where W are the weights of the pretrained language model, which are frozen for all new tasks. The context $C = h, x_1, y_1, ..., x_n, y_n$ consists of an optional prompt head h and n in-context examples ($\{x_i, y_i\}^n_{i=1}$) from the new task.

4 Method

In a Cultural Heritage context, the information useful to answer questions about a specific artwork is contained in the artwork image and in its contextual description. Finding such a description might not be trivial, since it might require a domain expert to write it down. At the same time, it is quite costly to train a Visual Question Answering model that takes in input both the image and the description. This is also not straightforward, since the two modalities need to be blended and matched together. Consequently, the main idea of this work is to generate new descriptions for artworks based on a specific prompt or a specific question and directly use these descriptions to answer visual and contextual questions. The overall pipeline of our proposed work is as follows:

1. **GPT-3 caption generation.** We use GPT-3 to generate descriptions of artworks, leveraging its memorization capabilities that allowed the model retain relevant information about training instances. An important aspect in this phase in to feed the correct prompt in input to GPT-3 in order to obtain realistic and correct descriptions. We consider two different types of input prompt:

General description generation for question answering

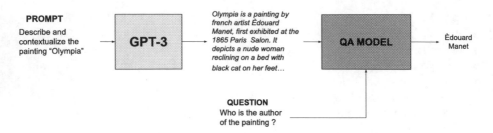

Fig. 1. Scheme of our method for answering questions using a general generated description. A prompt with a specific structure is given in input to GPT-3. Subsequently the generated text is fed together with the question to a Question Answering model that outputs the answer.

- **General** - A general prompt where the expected output is a general description of the artwork. The input text follows the structure:
  "Describe and Contextualize the painting < painting_name >"

- **Question-based** - A specific question based prompt. The input text follows the structure:
  "Painting < painting_name > < question >".
  The expected generated text by GPT-3 is a small text snippet that consists in a couple of sentences, focused on the topic of the question.

2. **Question answering.** Once the description has been generated in the previous step, we can exploit it to answer both visual and contextual questions through a Question Answering language model. For this purpose we use a pretrained version of DistilBert [27] fine-tuned on the SQUAD [24] dataset. We feed in input to the DistilBert model the generated text from the previous step together with the question. The answer given as output will be the final answer of our method.

Figures 1 and 2 show a scheme of the two variants of our method. More precisely, in Fig. 1 the general input prompt for GPT-3 yields the generation of a long description of the artwork (similar to a museum information sheet). On the other hand, the question-based prompt in Fig. 2 yields only the generation of a brief output text, which we find suitable for answering the question. In conclusion, these two schemes follow roughly the same structure. The difference is in the input prompt that in the case of Fig. 1 is more general and in Fig. 2 is more task oriented.

Question-based description generation for question answering

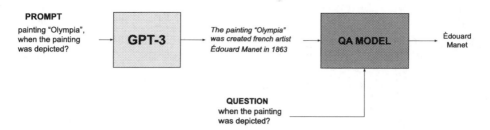

Fig. 2. Scheme of our method for answering questions using a question-based generated description. A prompt containing the name of the painting and the question is given in input to GPT-3. Subsequently the generated text is fed together with the question to a Question Answering model that outputs the answer.

5 Experiments

In this section we first outline the experimental setting for the experiments carried out in this paper, presenting dataset and experimental protocol and we then move on to a discussion of the results.

5.1 Dataset

For our experiments, we use the Artpedia dataset [31]. Artpedia contains a collection of 2,930 artworks, associated to a variable number of textual descriptions gathered from WikiPedia. Sentences are labelled as a visual descriptions or as a contextual descriptions. Contextual descriptions regard information about the artwork that does not directly describe its visual content. For instance, contextual descriptions can describe the historical context of the artwork, its author, the artistic influence or the museum where a painting is exhibited. The dataset contains 28,212 descriptions, 9,173 of which are labelled as visual and the remaining 19,039 as contextual. The Artpedia dataset has been extended with Question-Answer annotations in [6]. In fact, a subset of the images have been associated with visual and contextual questions, derived from the corresponding captions. In this work we follow the dataset split of [6].

5.2 Experimental Protocol

Following prior work such as [6], we evaluate visual questions and contextual questions with different metrics. In fact, visual question answering and traditional text-based question answering are often treated in two different ways. Visual Question Answering is considered as a classification problem, meaning that a model has to pick an answer from a predefined dictionary of possible candidates containing a few words each. This stems from the fact that questions in most datasets are a way of guiding attention towards specific objects or attributes in the image, without requiring any complex form of language reasoning. Question Answering on the other hand is based on a set of sentences, which may contain rare or out-of-dictionary words. The task is in fact defined as identifying a subset of the textual description that contains the answer.

In light of this, to evaluate visual questions we rely on accuracy:

$$Accuracy = \frac{N_c}{N_a} \tag{2}$$

where N_c is the number of correct answers and N_a the number of total answers.

For text-based question answering, instead, we use both accuracy and F1-measure, a metric that takes into account the global correctness of the answer:

$$F1 = 2 \times \frac{Precision \times Recall}{Precision + Recall} \tag{3}$$

Where *Precision* is defined as:

$$Precision = \frac{N_{Cw}}{|ans|} \tag{4}$$

with N_{Cw} is the number of common words between the output answer and the ground truth answer and *ans* the number of words in the generated answer.

Recall instead is defined as:

$$Recall = \frac{N_{Cw}}{|gt|} \tag{5}$$

where $|gt|$ is the number of words in the ground truth.

We also evaluate the quality of the descriptions generated by GPT-3, considering it as a standalone image captioning model. We use the following standard metrics for captioning:

- *BLEU1* [21]: BiLingual Evaluation Understudy (BLEU) is the most commonly used metric for machine translation and image captioning. BLEU scores are based on how similar a generated caption is to a reference caption, computing the precision of the generated words. The downside of BLEU is that it is very sensitive to small changes, such as synonyms or different word order.
- *ROUGE* [18]: differently from BLEU, which measures the precision of the caption, Recall Oriented Understudy of Gisting Evaluation (ROUGE) focuses on quantifying the amount of correct words with respect to the reference. Thus, this metric is recall-based and tends to reward long sentences.
- *CIDEr* [37]: Consensus-based Image Description Evaluation (CIDEr) is an automatic consensus metric that measures the similarity of captions against a set of ground truth sentences written by humans. This metric has been shown to yield a higher agreement with humans generated text since it captures notions of grammar, importance and precision and recall.

Table 1. Image captioning results. We compare our method which generates captions with GPT-3 with the *General* and the *Question-based* approaches. In the *Question-based* approach we concatenate all the outputs of GPT-3 after conditioning it with different questions related to the image. We compare the results against visual captions, contextual captions or both.

Description type	Metric	OFA [38]	Ours General	Ours Question-based
Visual	BLEU1	0.048	**0.181**	0.137
	ROUGE	0.138	**0.188**	0.16
	CIDEr	0.091	0.079	**0.172**
	COSINE	0.113	**0.157**	0.110
Contextual	BLEU1	0.002	**0.168**	0.160
	ROUGE	0.062	0.178	**0.179**
	CIDEr	0.000	**0.248**	0.129
	COSINE	0.082	0.218	**0.324**
All	BLEU1	0.000	0.113	**0.185**
	ROUGE	0.053	0.158	**0.184**
	CIDEr	0.000	0.016	**0.098**
	COSINE	0.122	0.253	**0.341**

– *Cosine Similarity*: we compute the cosine similarity between feature vectors for the generated caption and the reference caption. Features are extracted with the algorithm TF-IDF [26].

5.3 Experimental Results

Captioning Results. We start by assessing the quality of the captions generated by GPT-3. First of all, we ask GPT-3 to generate captions with our *General* approach. In Table 1 we compare the captions using as reference visual captions, contextual captions and both. All reference captions are ground truth captions taken from the Artpedia dataset [31].

Interestingly, the model appears to better results for visual captions using BLEU1 and ROUGE metrics, while using CIDEr and cosine similarity, the model obtains higher results for contextual captions. This may seem counter-intuitive but can be explained looking at the nature of the metrics. BLEU1 and ROUGE in fact respectively check for word-wise precision and recall, while CIDEr and cosine distance perform a sentence level scoring, which is closer to human consensus. We observe that the model is able to obtain good results, especially with the cosine metric, even when using all the captions as reference.

We then evaluate the method by taking a concatenation of the outputs generated by GPT-3 after being conditioned by different questions related to the image. This obviously introduces a strong bias, given also the fact that questions have been generated from information contained in the captions, but at the same time proves the usefulness of such captions for more advanced applications such as visual question answering. As can be seen in Table 1, conditioning GPT-3 with the captions leads to better captions according to most metrics.

Table 2. Experimental results for Visual Question Answering. We compare our approach against VQA-CH [6] to understand whether GPT-3 can replace information sheets for artworks either for visual or contextual questions. We compare two versions of our model, the *General* version, which produces generic descriptions of artworks and the *Question-based* version, where prompts are conditioned with the input question to generate more specific descriptions.

	Visual	Contextual	Accuracy	F1 score
VQA-CH [6]	✗	✓	0.684	0.832
VQA-CH [6]	✓	✗	0.176	0.150
VQA-CH [6]	✓	✓	0.504	0.417
Ours - General	✗	✓	0.557	0.719
Ours - General	✓	✗	0.070	0.055
Ours - General	✓	✓	0.239	0.360
Ours - Question-based	✗	✓	0.473	0.602
Ours - Question-based	✓	✗	0.134	0.202
Ours - Question-based	✓	✓	0.256	0.330

In Table 1 we also provide a baseline as reference, i.e. the output of the state of the art OFA captioning model [38]. We observe that captions generated by OFA do not align well with the ground truth sentences. We attribute this to a domain shift between the datasets commonly used to train captioning models and descriptions of artworks. In fact, the former are sentences written by non-experts while for applications in cultural heritage a domain knowledge is required. This further motivates the usage of GPT-3, which seems to have integrated sufficient knowledge to articulate complex sentences with a domain specific jargon.

VQA Results. To evaluate the Visual Question Answering capabilities of our proposed method, we follow the setting of [6]. However, we do not rely on any vision-based model but rather on a fully textual question answering model based on DistilBert [27], as explained in Sect. 4. In Table 2, we compare our approach to the one of VQA-CH [6]. It has to be noted that, contrary to [6], we do not rely on real textual descriptions, which are known to contain the answer, but we only extract information from GPT-3. This is a strong disadvantage for our method. However, we are not interested in obtaining better results than VQA-CH, but rather our goal is to demonstrate if GPT-3 can act as a substitute of textual descriptions handcrafted by domain experts.

We test our method evaluating the accuracy for visual questions, contextual questions and both together. Quantitative results indicate that captions generated by GPT-3 can yield to high results for contextual questions, yet very low accuracy for visual questions. As for the captioning setting, we impute this behavior to the fact that GPT-3 generates generic descriptions, without including a fine-grained description of the visual content. Thus, on the one hand the question answering model is capable of extracting meaningful information from the generated captions. This means that GPT-3 is indeed capable of integrating domain knowledge during training and is capable of generating a complete

ARTEPEDIA DESCRIPTION	QUESTION	GENERAL GENERATED DESCRIPTION	GENERAL ANSWER	QUESTION-BASED GENERATED DESCRIPTION	QUESTION-BASED ANSWER
'Bathers at Asnières (French: Une Baignade, Asnières) is an oil-on-canvas painting by the French artist Georges Pierre Seurat, the first of his two masterpieces on the monumental scale.', 'The canvas is of a suburban, placid Parisian riverside scene.', 'A combination of complex brushstroke techniques, and a meticulous application of contemporary colour theory bring to the composition a sense of gentle...	Who is the author of this painting?	Bathers at Asnières is a painting by French artist Georges Seurat. The painting depicts a group of men relaxing by the river Seine on a sunny day. The background includes industrial buildings and smokestacks, which contrast with the leisurely scene in the foreground. Seurat used small dots of color to create a sense of light and movement in the painting, which has been called an early example of Pointillism. The painting was completed in 1884, just before Seurat's death at the age of 31. It was exhibited posthumously at the Salon des Indépendants in 1886, where it received mixed reviews from critics. Some praised its technical mastery, while others found it cold	Georges Seurat	The painting Bathers at Asnieres was created by french artist George Seurat in 1884	Georges Seurat
	When was the painting depicted?		1884	Bathers at Asnieres was painted by George Seurat in 1884	1884
	What is the animal in the ground?		X	The animal on the ground is a dog	dog
	Is there a man lying down?		X	There is a man lying down in the painting Bathers at Asnieres, but it is difficult to see. He is on the left side of the painting, near the bottom.	There is a man lying down
Dogs Playing Poker, by Cassius Marcellus Coolidge, refers collectively to an 1894 painting, a 1903 series of sixteen oil paintings commissioned by Brown & Bigelow to advertise cigars, and a 1910 painting., 'Critic Annette Ferrara has described Dogs Playing Poker as "indelibly burned into... All eighteen paintings in the overall series feature anthropomorphized dogs, but the eleven in which dogs are...	Who is the author of this painting?	Dogs Playing Poker is a painting by American artist Cassius Marcellus Coolidge. The painting depicts a group of dogs sitting around a table playing poker. The dogs are all wearing human clothes and have cigars in their mouths. The painting is set in a dimly lit room, with a fireplace in the background. The painting is often seen as an example of kitsch art, due to its subject matter and style. It is also seen as a classic example of Americana, due to its depiction of a group of dogs engaged in a typically American activity.	Cassius Marcellus Coolidge	The painting is anonymous.	X
	Who is sitting on the chair?		dogs	The artist who painted Dogs Playing Poker is named C. M. Coolidge.	X
	What are the dogs doing?		playing poker	The dogs in the painting are playing poker.	dog
	What is the color of the table?		X	table in Dogs Playing Poker is green	green
The Singing Butler is an oil-on-canvas painting made by Scottish artist Jack Vettriano in 1992.'As a contemporary cultural icon, The Singing Butler has been compared to Grant Wood's American Gothic.' It depicts a couple dancing on the damp sand of a beach with a low horizon. To the left and right, a maid and a man hold up umbrellas against the weather...	When was the painting depicted?	The painting "The Singing Butler" was painted by Scottish artist Jack Vettriano in 1992. The painting depicts two people, a man and a woman, standing on a beach with a Butler who is singing and playing the guitar. The background of the painting is a blue sky with white clouds. The painting is set in the early 20th century.	1992	This painting was depicted in 1992.	1992
	What are the two people in the middle doing?		dancing	The two people in the middle are dancing	dancing
	How many umbrellas are there?		X	There are two umbrellas in the painting.	two

Fig. 3. Qualitative results of our method. *Green*: ground truth description from the Artpedia dataset [31] and input question. *Yellow*: general descriptions provided by GPT-3 and answer obtained based on such text. *Blue*: Question-based description and correspondent answer. General descriptions are longer and more detailed than question-based generated descriptions. However, question-based generated descriptions are customized for the specific question. (Color figure online)

information sheet of the artwork. On the other hand, captions appear to be too generic to obtain information about specific details in the image.

To overcome this limitation, we test the model using captions generated without *Question-based* approach. By feeding the answer to GPT-3 along with the title of the artwork, the model is able to generate more specific captions. Such captions, as explained in Sect. 6 are usually shorter but are focused on the prompt. This is particularly interesting since it means that a purely text-based model is capable of addressing a vision-based task. In Table 2 it can be seen that for visual questions alone, our method with question-based captions performs on par or better than the vision-based VQA-CH model.

6 Qualitative Analysis

In this section we provide a qualitative analysis of the captions generated by GPT-3 in order to characterize which kind of information they contain in both the *General* and *Question-based* formulation.

Since the prompts that we feed to GPT-3 are different, with one being more general and the other being question-based, we expect that the corresponding generated text by GPT-3 will be different. In Fig. 3 we can observe these differences. Generated general descriptions are very long and have the aspect of artwork information sheets in which we can find some visual and contextual information. Question-based generated descriptions are instead shorter and contain the knowledge needed to answer to the specific questions. From Fig. 3 we can observe that the general description is very useful to answer to contextual questions but fails on some visual questions. This is likely due to different reasons:

- The generated text does not take into account any specific question and this can lead to the generation of a description without specific information useful to answer to the question.
- Visual questions are very specific since they refer to object relationships, colors, counting, etc. and the GPT-3 model tends to be more shallow in generating its descriptions.

On the other hand, question-based generated descriptions are helpful to answer visual questions but the small generated description useful to answer those specific questions could contain incorrect information leading to wrong answer predictions. In conclusion these two ways of generating text to answer visual and contextual questions have some pros and cons:

- General descriptions are longer and contain several pieces of information about the artwork. However this is fixed and could not contain the information needed to answer some questions.
- Question-based descriptions are generated for specific questions and contain only the information needed to answer the question on which GPT-3 has been conditioned. If the model has not memorized any specific information regarding such questions it may contain mistakes and descriptions will have to be re-computed for each question.

7   Considerations on Complexity and Accessibility of GPT-3

In the previous sections we have demonstrated that GPT-3 could indeed replace the usage of an information sheet handcrafted by a domain expert. However, we need to understand the actual applicability of GPT-3 in a real case application. GPT-3 has 175B parameters, which approximately amounts to 700GB. This means that inference on a single GPU is unfeasible due to current technological limits. The model however has been made available from OpenAI and is accessible through API that have a pricing fee per generated token. These considerations somewhat limit a large-scale usage of the model, especially if a description has to be generated for each question to be answered. On the other hand, generating fixed descriptions offline, one for each artwork, appears a viable solution at least for addressing contextual questions.

8   Conclusions

In this paper we presented a method for Visual Question Answering in the Cultural Heritage domain. In particular we have addressed the problem of data annotation for artworks, generating descriptions with GPT-3. The performances for the VQA task show that the generated descriptions are useful to answer the questions correctly. This technique allows to answer visual and contextual questions focusing only on the generated description and can be used for any artwork. In fact, there is no need to retrain the model to incorporate new knowledge. This is possible thanks to the memorization capabilities of GPT-3, which at training time has observed millions of tokens regarding domain-specific knowledge. Finally the generated description can be integrated as textual input (textual description) in a more complex architecture as [6] in order to address tasks like Visual Question Answering. This is of particular interest for Cultural Heritage due to the domain shift between common VQA and captioning datasets compared to the technical jargon that is needed to properly address questions about art.

Acknowledgement. This work was partially supported by the project ARS01 00421: "PON IDEHA - Innovazioni per l'elaborazione dei dati nel settore del Patrimonio Culturale." This work was partially supported by the European Commission under European Horizon 2020 Programme, grant number 101004545 - ReInHerit.

References

1. Anderson, P., et al.: Bottom-up and top-down attention for image captioning and visual question answering. In: Proceedings of the IEEE Conference on Computer Vision and Pattern Recognition, pp. 6077–6086 (2018)
2. Antol, S., et al.: VQA: visual question answering. In: Proceedings of the IEEE International Conference on Computer Vision, pp. 2425–2433 (2015)
3. Asprino, L., Bulla, L., Marinucci, L., Mongiovì, M., Presutti, V.: A large visual question answering dataset for cultural heritage. In: International Conference on Machine Learning, Optimization, and Data Science, pp. 193–197. Springer, Cham (2021). https://doi.org/10.1007/978-3-030-95470-3_14
4. Barra, S., Bisogni, C., De Marsico, M., Ricciardi, S.: Visual question answering: which investigated applications? Pattern Recogn. Lett. **151**, 325–331 (2021)
5. Becattini, F., Ferracani, A., Landucci, L., Pezzatini, D., Uricchio, T., Del Bimbo, A.: Imaging Novecento. A mobile app for automatic recognition of artworks and transfer of artistic styles. In: Ioannides, M., et al. (eds.) EuroMed 2016. LNCS, vol. 10058, pp. 781–791. Springer, Cham (2016). https://doi.org/10.1007/978-3-319-48496-9_62
6. Bongini, P., Becattini, F., Bagdanov, A.D., Del Bimbo, A.: Visual question answering for cultural heritage. In: IOP Conference Series: Materials Science and Engineering, vol. 949, p. 012074. IOP Publishing (2020)
7. Brown, T.B., et al.: Language models are few-shot learners (2020)
8. Cetinic, E., She, J.: Understanding and creating art with AI: review and outlook. ACM Trans. Multimedia Comput. Commun. Appl. (TOMM) **18**(2), 1–22 (2022)
9. Cornia, M., Stefanini, M., Baraldi, L., Cucchiara, R.: Meshed-memory transformer for image captioning. In: Proceedings of the IEEE/CVF Conference on Computer Vision and Pattern Recognition, pp. 10578–10587 (2020)
10. Dale, R.: GPT-3: what's it good for? Nat. Lang. Eng. **27**(1), 113–118 (2021)

11. Devlin, J., Chang, M.W., Lee, K., Toutanova, K.: BERT: pre-training of deep bidirectional transformers for language understanding. arXiv preprint arXiv:1810.04805 (2018)
12. Elkins, K., Chun, J.: Can GPT-3 pass a writer's Turing test? J. Cult. Analytics **5**(2), 17212 (2020)
13. Fiorucci, M., Khoroshiltseva, M., Pontil, M., Traviglia, A., Del Bue, A., James, S.: Machine learning for cultural heritage: a survey. Pattern Recogn. Lett. **133**, 102–108 (2020)
14. Garcia, N., et al.: A dataset and baselines for visual question answering on art. In: Bartoli, A., Fusiello, A. (eds.) ECCV 2020. LNCS, vol. 12536, pp. 92–108. Springer, Cham (2020). https://doi.org/10.1007/978-3-030-66096-3_8
15. Han, J., Zhang, D., Cheng, G., Liu, N., Xu, D.: Advanced deep-learning techniques for salient and category-specific object detection: a survey. IEEE Signal Process. Mag. **35**(1), 84–100 (2018)
16. Kheradpisheh, S.R., Ganjtabesh, M., Thorpe, S.J., Masquelier, T.: STDP-based spiking deep convolutional neural networks for object recognition. Neural Netw. **99**, 56–67 (2018)
17. Li, X., et al.: Oscar: object-semantics aligned pre-training for vision-language tasks. In: Vedaldi, A., Bischof, H., Brox, T., Frahm, J.-M. (eds.) ECCV 2020. LNCS, vol. 12375, pp. 121–137. Springer, Cham (2020). https://doi.org/10.1007/978-3-030-58577-8_8
18. Lin, C.Y.: ROUGE: a package for automatic evaluation of summaries. In: Text Summarization Branches Out, pp. 74–81 (2004)
19. Liu, S., Zhu, Z., Ye, N., Guadarrama, S., Murphy, K.: Improved image captioning via policy gradient optimization of spider. In: Proceedings of the IEEE International Conference on Computer Vision, pp. 873–881 (2017)
20. Lu, J., Yang, J., Batra, D., Parikh, D.: Hierarchical question-image co-attention for visual question answering. In: Advances in Neural Information Processing Systems, pp. 289–297 (2016)
21. Papineni, K., Roukos, S., Ward, T., Zhu, W.J.: BLEU: a method for automatic evaluation of machine translation. In: Proceedings of the 40th Annual Meeting of the Association for Computational Linguistics, pp. 311–318 (2002)
22. Radford, A., Narasimhan, K., Salimans, T., Sutskever, I., et al.: Improving language understanding by generative pre-training (2018)
23. Radford, A., Wu, J., Child, R., Luan, D., Amodei, D., Sutskever, I., et al.: Language models are unsupervised multitask learners. OpenAI Blog **1**(8), 9 (2019)
24. Rajpurkar, P., Zhang, J., Lopyrev, K., Liang, P.: SQuAD: 100,000+ questions for machine comprehension of text. arXiv preprint arXiv:1606.05250 (2016)
25. Ren, S., He, K., Girshick, R., Sun, J.: Faster R-CNN: towards real-time object detection with region proposal networks. In: Advances in Neural Information Processing Systems, pp. 91–99 (2015)
26. Salton, G., Buckley, C.: Term-weighting approaches in automatic text retrieval. Inf. Process. Manage. **24**(5), 513–523 (1988)
27. Sanh, V., Debut, L., Chaumond, J., Wolf, T.: DistilBERT, a distilled version of BERT: smaller, faster, cheaper and lighter. arXiv preprint arXiv:1910.01108 (2019)
28. Seidenari, L., Galteri, L., Bongini, P., Bertini, M., Del Bimbo, A.: Language based image quality assessment. In: ACM Multimedia Asia, pp. 1–7 (2021)
29. Sheng, S., Laenen, K., Moens, M.-F.: Can image captioning help passage retrieval in multimodal question answering? In: Azzopardi, L., Stein, B., Fuhr, N., Mayr, P., Hauff, C., Hiemstra, D. (eds.) ECIR 2019. LNCS, vol. 11438, pp. 94–101. Springer, Cham (2019). https://doi.org/10.1007/978-3-030-15719-7_12

30. Shih, K.J., Singh, S., Hoiem, D.: Where to look: focus regions for visual question answering. In: Proceedings of the IEEE Conference on Computer Vision and Pattern Recognition, pp. 4613–4621 (2016)
31. Stefanini, M., Cornia, M., Baraldi, L., Corsini, M., Cucchiara, R.: Artpedia: a new visual-semantic dataset with visual and contextual sentences in the artistic domain. In: Ricci, E., Rota Bulò, S., Snoek, C., Lanz, O., Messelodi, S., Sebe, N. (eds.) ICIAP 2019. LNCS, vol. 11752, pp. 729–740. Springer, Cham (2019). https://doi.org/10.1007/978-3-030-30645-8_66
32. Su, W., et al.: VL-BERT: pre-training of generic visual-linguistic representations. arXiv preprint arXiv:1908.08530 (2019)
33. Sun, T.X., Liu, X.Y., Qiu, X.P., Huang, X.J.: Paradigm shift in natural language processing. Mach. Intell. Res. **19**(3), 169–183 (2022)
34. Tan, H., Bansal, M.: LXMERT: learning cross-modality encoder representations from transformers. arXiv preprint arXiv:1908.07490 (2019)
35. Vannoni, F., Bongini, P., Becattini, F., Bagdanov, A.D., Bimbo, A.: Data collection for contextual and visual question answering in the cultural heritage domain (2020)
36. Vaswani, A., et al.: Attention is all you need. In: Advances in Neural Information Processing Systems, pp. 5998–6008 (2017)
37. Vedantam, R., Lawrence Zitnick, C., Parikh, D.: CIDEr: consensus-based image description evaluation. In: Proceedings of the IEEE Conference on Computer Vision and Pattern Recognition, pp. 4566–4575 (2015)
38. Wang, P., et al.: Unifying architectures, tasks, and modalities through a simple sequence-to-sequence learning framework. arXiv preprint arXiv:2202.03052 (2022)
39. Yang, Z., et al.: An empirical study of GPT-3 for few-shot knowledge-based VQA. In: Proceedings of the AAAI Conference on Artificial Intelligence, vol. 36, pp. 3081–3089 (2022)

Automatic Analysis of Human Body Representations in Western Art

Shu Zhao[1] , Almila Akdağ Salah[1] , and Albert Ali Salah[1,2](✉)

[1] Utrecht University, Princetonplein 5, 3584CC Utrecht, The Netherlands
{a.a.akdag,a.a.salah}@uu.nl
[2] Boğaziçi University, Bebek, 34342 Istanbul, Turkey

Abstract. The way the human body is depicted in classical and modern paintings is relevant for art historical analyses. Each artist has certain themes and concerns, resulting in different poses being used more heavily than others. In this paper, we propose a computer vision pipeline to analyse human pose and representations in paintings, which can be used for specific artists or periods. Specifically, we combine two pose estimation approaches (OpenPose and DensePose, respectively) and introduce methods to deal with occlusion and perspective issues. For normalisation, we map the detected poses and contours to Leonardo da Vinci's Vitruvian Man, the classical depiction of body proportions. We propose a visualisation approach for illustrating the articulation of joints in a set of paintings. Combined with a hierarchical clustering of poses, our approach reveals common and uncommon poses used by artists. Our approach improves over purely skeleton based analyses of human body in paintings.

Keywords: Human pose estimation · Hierarchical clustering · Painting analysis

1 Introduction

The human body is expressive of mood and emotions, as well as intentions, and artists have used the expressive possibilities of the body pose to its fullest extent. Body language can reflect embedded societal or gender differences [18], or convey intense emotions, which cannot be discriminated by facial expressions only [1]. Consequently, art historians analyse the poses of subjects in paintings in depth.

The portrayal of emotions via poses of a human body was first documented by the cultural and art historian Aby Warburg with his concept of Pathosformel [9]. The term comes from the combination of "Pathos" (emotion) and "formel" (a formula). Warburg traces the Pathosformel back to ancient Greek vase paintings [15], where a narrative is illustrated in the interactions and compositional relationships between characters. The same visual elements of postures and gestures are used in recurrent narratives. Warburg's *Bilderatlas* contains a rich collection of artifacts through which he studied the adoption of poses by various painters and influences between them. This is one of the reasons why automatic pose estimation in paintings is a useful tool for art historians, as composition transfer can be identified through the similarities between the postures of individual characters [10]. Painters often incorporate stylistic elements by copying human poses depicted by other artists.

The automatic detection and statistical analysis of body shapes and poses in paintings can provide art historians with an overview of artists from a new angle. The statistical analysis in artworks usually focus on genre, style, or artist identification [23,25], whereas pose analysis is relatively rare [9,14,15][1]. It will furthermore be useful in image retrieval for art datasets and archives, as well as for tracking the re-use of visual elements [4] and help with recent efforts in automatic captioning of paintings [2,6,22].

The aim of this paper is to use automatic pose estimation methods to create a tool which will allow pose based analyses of paintings. For this purpose, we start from off-the-shelf pose estimators, and then seek to resolve specific issues pertaining to poses in paintings. More specifically, we combine two pose estimators (to complement each other's shortcomings), followed by an artist-specific normalisation step that corrects occlusions and perspective related distortions based on average body poses depicted by a specific artist. We contribute an artistic pose dataset of Western art that we have semi-automatically annotated for pose ground truth. By applying hierarchical clustering on the detected and corrected poses, we perform a detailed analysis of the joint angles used in paintings in our dataset, and show how the analysis reveals the common and niche pose depictions used and re-used throughout Western art.

This paper is structured as follows. In Sect. 2, we briefly summarise related work on human pose estimation. Section 3 describes our algorithmic pipeline and its various components. Section 4 describes the dataset we have used and annotated in the study. We illustrate our methods via experimental results in Sect. 5, and conclude in Sect. 6.

2   Related Work

Human pose estimation from images can be achieved with 2D or 3D models. There are three different types of approaches, using kinematic models (used for 2D/3D), planar models (used for 2D), and volumetric models (used for 3D). For the kinematic and planar models, body joints are represented with keypoints, and limbs are represented with lines joining those keypoints. While this is suitable for pose analysis in paintings, we will need more than skeleton representations for body depictions, as the same pose can be depicted with different body representations.

We distinguish between top-down and bottom-up approaches for pose estimation. In the top-down pipeline, a human body detector is used to obtain each person's bounding box and a single-person pose estimator is used to predict the

---

[1] For example IconClass is a classification system for image content and includes human body poses (https://iconclass.org/31A23).

locations of keypoints within the bounding box. In the bottom-up pipeline, body joint detectors are used to extract human body joint candidates which are clustered into individual bodies. In general, the bottom-up methods outperform the top-down methods [28]. OpenPose is one of the most commonly used bottom-up models with a real-time performance to estimate the poses for multiple people in one image [3], and we use it to detect keypoints in this work. This is a rapidly growing area, and the keypoint extractor can be updated with more promising approaches. For instance HRNet, while not as widely tested as OpenPose, shows good performance in certain benchmarks [24].

Keypoint estimation is not sufficient for representing body shapes in images. DensePose [7] is a widely used top-down method, which is built on the Faster R-CNN architecture [19]. In addition to keypoints, it provides a segmentation of body parts (i.e. head, torso, arms, hands, legs and feet) for multiple people in the image. We use this approach in this work for detecting body segments, but further enhance the results with keypoints detected by the more accurate OpenPose.

Pose estimation approaches (as well as face detection), are used for automatic analysis of human representations in paintings [5,15,21], but also, the analysis of these key-points and landmarks are further used as a way to statistically analyse artistic datasets [9] and to design higher level analysis and synthesis methods [4,14,26]. Here we detail three related studies further.

Madhu et al. focused on the analysis of figures on Greek vase paintings, which are full of visual narratives, in which the protagonists are depicted by their actions and interactions conveyed through their poses composed against a certain scene [15]. To automatically detect these poses, a styled dataset is generated from the COCO-Persons dataset and a style transfer approach [8] is applied to convert these images to a style similar to Greek vase paintings. Combining a person detector based on Faster R-CNN, and a pose estimator based on HRNet, a model is created and trained on this styled dataset. This model is fine-tuned on a classical archaeology dataset with 2.629 person annotations and 1.728 pose annotations from over 1.000 Greek vase paintings.

In another example by Yaniv et al., facial landmarks are estimated on the portraits [26]. First, a custom artistic portrait dataset is generated with 160 paintings from 10 artists. A multi-task cascaded CNN [27] is used to automatically detect the faces, which are then cropped and resized to images of 256×256 pixels. A landmark detection algorithm is applied to extract 68 facial landmarks using the Dlib-ml toolkit [11]. With the help of a natural-faces dataset with 68 landmark annotations per face [20], the geometric differences between natural faces and artistic faces is documented. Not surprisingly, artistic faces have a larger geometric variation due to artistic exaggeration and deformation.

In the third example, a portion of Aby Warburg's Bilderatlas collection is manually annotated via crowdsourcing for pose key-points [9], and analysed using statistical methods such as hierarchical and two stage clustering to generate an overview of body pose-clusters. To determine the number of clusters automatically, a two-sample Kolmogorov-Smirnov test is run on the distributions of each joints' angle. The results show that certain poses from antiquity indeed resurface in Renaissance, however the context of these poses (and hence their emotional content) is changed [9].

While these studies show the potential of computer vision based analyses in the domain of paintings, annotated datasets are very rare for pose keypoints, body segments or facial landmarks. However, the investment to generate these datasets is worthwhile. One possibility is to apply style transfer to already-annotated datasets of photographic images, and use such an augmented dataset as a training set for pose and landmark estimation in paintings. However, paintings may contain more stylised poses than naturally occurring poses in photographs, and manual annotation of paintings can potentially result in better model training. In the end, both approaches are costly, either in terms of computing or manpower. In this paper, we explore to what extent automatic pose estimation with widely used off-the-shelf tools can be used for body pose and shape analysis in paintings.

3   Methodology

Our general analysis approach is illustrated in Fig. 1. We use OpenPose and DensePose in parallel to obtain a set of keypoints and segments for the depicted bodies in the paintings. For pose analysis of a group of paintings (such as from a single painter or a style), we use the keypoints from OpenPose, prepare visualisations that depict distributions of joint angles, and following [9], dendrograms to find pose groups. For body shape analysis, we use a shape normalisation step, and generate the average contours from all normalised segments to superpose them on original poses.

3.1   Average Contours of Body Segments

In Western art historical literature, the archetype, or the canon is an example that is considered as the highest standard that transcends the aesthetics of a given era [12]. Such an example can be an artwork, but also the depiction of a pose, or more importantly, the human body. The canonical body, i.e. the perfect proportions of a the human body has been a topic of discussion and elaboration since the ancient Greeks [17]. There have been many descriptions of these proportions, as well as drawings. Among these the Vitruvian Man by Leonardo da Vinci is one of the most well known [16].

In our approach, we use the Vitruvian Man as a standard for normalising extracted body contours of artistic poses, as it has been influential in defining the ideal body proportions in Western art. This allows us to compare different poses with each other at the same scale. Using contours, we propose an approach to mitigate the occlusion and perspective issues that are generally present in automatic pose estimation, as well as generate an overview for each artist's preferred style in drawing human bodies: Do they use proportions closer to norm, or do they prefer exaggeration of these by elongating or thickening of body

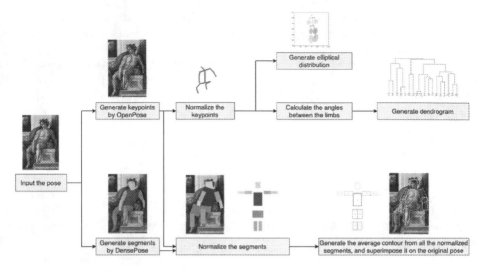

Fig. 1. The pipeline of pose analysis.

segments? Lastly, through the comparison of average contours we illustrate the differences between the contours of natural poses and artistic poses.

To carry out the normalisation process we follow several assumptions: 1) The head size is different for men and women. The (vertical) size of the head is equal to the vertical distance between the nose and the top of the chest. 2) The body proportions are the same for men and women; 3) The body segments are convex, and can be abstracted as rectangles; 4) The body segments are symmetrical between left and right sides. A specific segment is also symmetrical around its centroid.

We use the head size as a normalisation reference, based on which we scale the length of limbs and area of segments. The baseline Vitruvian Man is referenced by a fixed dimension, i.e., (624 × 624) in pixels, with male head size being fixed at **62** pixels, and the female head size by **58** pixels. The median size, i.e. 23.2 cm for men and 21.8 cm for women, is taken as the norm[2]. We transform this proportion to **62** pixels for men, and **58** pixels for women. For an input pose from any image, the scale factor is calculated by dividing the standard head size to the actual head size in pixels. Other limbs are scaled accordingly by the same scale factor.

The output of normalisation is a T-pose figure superimposed on the Vitruvian Man with 10 segments, i.e., head, torso, upper and lower arms, upper and lower legs, respectively. We use the keypoints detected by OpenPose and match these to the DensePose segments, as OpenPose has a higher accuracy for keypoint detection (see Table 1). If a painting has multiple interacting poses, bounding

---

[2] https://en.wikipedia.org/wiki/Human_head.

boxes of segments may overlap. In such cases we will discard the data during matching. We also discard instances if the torso is only partially detected since we use this information for rotating the whole pose to a vertical position. We furthermore rotate each segment separately to a vertical or a horizontal position, as required by a T-pose. Lastly we dilate all the segments to their tightest-fitting rectangles.

3.2 Visualising Joint Distributions

To summarise artists style when it comes to pose geometry, we prepare and visualise distributions for male and female poses by using angle of joints as a reference point. This type of analysis gives on overview of the range of poses for a group of paintings (examples shown in Sect. 5.3).

The procedure starts with the normalisation of keypoints following these three steps: 1) Validate whether a set of indicative keypoints (we use nose, neck and midhip), are detected. 2) Rotate the whole pose to a vertical position, so that the spine is vertical. 3) Use 60 pixels (size of the head) as a scaling factor to normalise all the poses.

After this procedure is applied to all poses in the selected set, the distributions of the keypoints are visualised by fitting Gaussian ellipsoids on the keypoint locations. Each ellipsoid can be thought of a visualisation of the covariance matrix of the spread of the points. Hence, each ellipsoid stands for the standard distribution of one keypoint for a group of poses. However, while traditionally one standard deviation is used for the contours, we use here half a standard deviation for better visibility. The covariance matrix $\sum$ of the vector for one keypoint $X = [X1, X2]$ can be calculated and represented as $\sum = \begin{bmatrix} \sigma_{11} & \sigma_{12} \\ \sigma_{21} & \sigma_{22} \end{bmatrix}$.

The coordinate (x, y) of the ellipsoid for the keypoint with its corresponding radius and scale around it is as follows:

$$x = \frac{\sum_{i=1}^{n} X1_i}{n} \qquad radius_x = \sqrt{1 + \frac{\sigma_{12}}{\sqrt{\sigma_{11} \times \sigma_{22}}}} \qquad scale_x = \sqrt{\sigma_{11}} \times 0.5$$

$$y = \frac{\sum_{i=1}^{n} X2_i}{n} \qquad radius_y = \sqrt{1 - \frac{\sigma_{12}}{\sqrt{\sigma_{11} \times \sigma_{22}}}} \qquad scale_y = \sqrt{\sigma_{22}} \times 0.5$$

3.3 Hierarchical Clustering

The visualisation of joint distributions explains the range of joints depicted in the poses, whereas by looking at joints in relation to each other it is possible to understand the nature of poses as well. To do that, we carry out hierarchical clustering for artistic poses [9]. This process can also illustrate the common and niche poses in detail, which can to some extent explain why a pose detector performs better or worse for some poses.

For hierarchical clustering, the keypoints are normalised in three steps: 1) Validate whether all the 6 torso keypoints are detected, i.e., neck, right and left shoulders, midhip, right and left hips. 2) Rotate the whole pose to a standing-up position, with the spine being vertical. 3) Calculate the inner angles for all the triplets of joints. In total, there are 13 such joint triplets, i.e., 1 triplet of (nose, neck, midhip), 6 triplets of the right body: (shoulder, neck, midhip), (elbow, shoulder, neck), (wrist, elbow, shoulder), (hip, midhip, neck), (knee, hip, midhip), (ankle, knee, hip), and 6 symmetric triplets of the left body. The inner angles of each pose on the 2-dimensional plane are treated as a 13-dimensional vector, representative of each pose. These pose vectors are used for agglomerative hierarchical clustering, in which each pose starts in its own cluster, which is merged with other clusters if their pair-wise Euclidean distance is the smallest. The results are depicted in a dendrogram (see Sect. 5.4), where similar poses are connected with each other by shorter distances, whereas different pose clusters are connected with each other by longer distance.

4   The Artistic Pose (AP) Dataset

To test the inference accuracy of OpenPose and DensePose we decided to generate a dataset by selecting 10 artists from the Painter by Numbers dataset which has 103, 250 paintings from Western art that range from the early 11th century to the 2010s. This is similar in size to earlier studies in the literature [26]. We chose artists with more nude paintings, as these were expected to generate the least problems for pose estimation, but the earlier painters have less nude paintings. We also tried to find a gender balance, as well as a genre balance in choosing our artists. The resulting Artistic Pose dataset (from now on called the AP Dataset[3]) covers a wide range both in time and style: Michelangelo (Renaissance), El Greco (Mannerism), Artemisia Gentileschi (Baroque), Pierre-Paul Prud'hon (Romantism), Pierre-Auguste Renoir (Early Impressionism), Paul Gauguin (Post Impressionism), Felix Vallotton (Magical Realism) and Amedeo Modigliani (Expressionism), Tamara de Lempicka (Art Deco), and Paul Delvaux (Surrealism).

We furthermore make use of a natural-pose dataset to bring forth an understanding of the geometric style of natural poses and to compare them to the artistic poses. COCO Persons dataset is one of these datasets with a wide variety of common activities that are manually annotated with respect to the joints and body segments [13]. More importantly, OpenPose and DensePose are trained with this dataset.

5   Experimental Results

5.1   Comparison of OpenPose and DensePose

We first apply OpenPose and DensePose on the AP dataset and evaluate their performances. To measure accuracy of keypoint detection, we use PCK (Percentage of Correct Keypoints). The keypoints are considered correct if the distance

---

[3] The dataset with manual annotations, as well as all the code are made publicly available at https://github.com/tintinrevient/joints-data.

between the inferred and the true keypoint is within a certain threshold. PCK for each artist is shown in Table 1. For pose analysis, we only use 15 keypoints, namely the nose, neck, midhip, left and right shoulders, elbows, and wrists, left and right hips, knees, and ankles, respectively.

On average for all artists, the inference accuracy of OpenPose is around 80%, and 66% for DensePose. The inference of joints by the paintings of Tamara de Lempicka and Amedeo Modigliani have the least accuracy due to their exaggeration of shapes and proportions. In contrast, Paul Delvaux's paintings have the highest accuracy by both OpenPose and DensePose, because most of the poses are resting or standing with naturally hanging arms, which constitute easy cases for prediction.

Table 1. The measurement of PCK. The columns list the artists, the total number of people depicted (Male and Female), the number of valid keypoints and segments, accuracy of DensePose (Acc.), PCK of OpenPose (PCK-O) and DensePose (PCK-D)

Artist	Subjects	# Keypoints	# Segments	Acc	PCK-O	PCK-D
Michelangelo	15 (14M, 1F)	214	163	79%	87%	63%
El Greco	34 (32M, 2F)	375	259	42%	73%	55%
Artemisia Gentileschi	21 (6M, 15F)	214	182	59%	86%	60%
Pierre-Paul Prud'hon	15 (7M, 8F)	216	196	65%	93%	82%
Pierre-Auguste Renoir	19 (19F)	237	200	49%	80%	63%
Paul Gauguin	31 (4M, 27F)	414	335	64%	82%	74%
Felix Vallotton	20 (1M, 19F)	243	206	66%	86%	67%
Amodeo Modigliani	15 (15F)	180	147	21%	40%	30%
Tamara de Lempicka	18 (1M, 17F)	227	188	54%	69%	52%
Paul Delvaux	35 (4M, 31F)	455	371	82%	89%	89%

In general, OpenPose outperforms DensePose, especially under scenarios such as multiple people, low contrast, and niche poses, as shown in Fig. 2. The universally difficult scenarios are when (1) two people hug or interact closely with each other, (2) the poses have twisted limbs, (3) there are niche perspectives, (4) there are exaggerated shapes, (5) there are exaggerated body proportions. These challenging scenarios are illustrated in the last row of Fig. 2, in the mentioned order.

Next, we test the accuracy of DensePose body segment detection, by the percentage of correctly inferred segments over the total number of visible segments (max. 14), as shown in the last column of Table 1. The factors that impact detection are the number of people in the painting, the occlusions of clothes, interacting people, niche poses and perspectives, artistic effects including unusual colouring and brush usage for body segments, and unusual shapes of body segments.

The paintings of Michelangelo and Paul Delvaux have the highest accuracy. In their paintings, most of the poses are nude, hence without interference of clothes

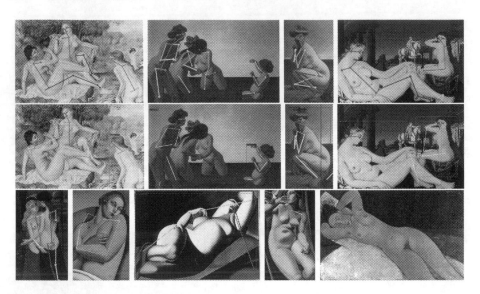

Fig. 2. The joints inferred by OpenPose (first row) and DensePose (second row) under scenarios such as multiple people, low contrast, and niche poses. The last row shows the more difficult scenarios for both OpenPose and DensePose.

and in nude skin colours. For Michelangelo, we have one pose per painting to detect, and for Delvaux, the people are distanced from each other. All the poses from Michelangelo are sitting, and most poses from Delvaux are standing, which form the two most common pose groups, and all of their poses are depicted from a frontal view. Lastly, all of their poses follow natural body proportions. Figure 3 shows successful and failed pose detections including these two artists.

As shown in Table 1, the paintings of Amedeo Modigliani and Pierre-Auguste Renoir have the least accuracy, as most of their poses are very challenging. Lying poses of Modigliani, along with his use of overly slender and elongated body proportions are hard to detect. Similarly, El Greco's elongated bodies are not always easy to segment. If the bodies are painted by blobs of various colours (e.g. Renoir), or if the colour contrast is low (e.g. Gaugin) the inference suffers. The niche perspective poses are also a challenge for inference (e.g. Lempicka). Examples are given in Fig. 3.

The root causes of these failures may include the fact that the COCO dataset used in the training of these models has a lower density of natural poses, such as lying, sitting with back turned to observers, crowds of overlapping people, and such. Instead, many training samples depict single persons, doing sports or with only the upper torso captured. Furthermore, the training data collects mostly common people with common body proportions and height, which are neither too slender, nor too plump. Yet artistic depictions can go to extremes.

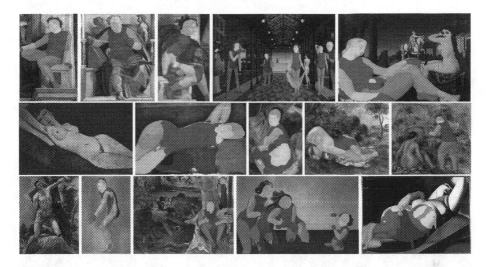

Fig. 3. First row: The detection of body segments for Michelangelo (first three) and Paul Delvaux (last two). Middle row: The inference of segments for Modigliani (first two) and Renoir (last three). Last row, from left to right: The inference of segments for El Greco, Prud'hon, Gauguin, Vallotton and Lempicka.

5.2   Average Contours of Body Segments

We use Michelangelo and Paul Delvaux as examples to demonstrate the normalisation process and the results (see Fig. 4). When we compare these two normalised poses, we see that the body drawn by Michelangelo is more inflated and fills to the brim the under the contours of the Vitruvian Man, whereas Paul Delvaux's woman is slender. Compared with Paul Delvaux's standing pose, Michelangelo's sitting pose exposes three major issues: (1) The right thigh is occluded by the right arm, which leads to the partial segment with a gap in the normalised pose. (2) The thighs are not fully stretched out due to the sitting pose and the viewing perspective. Thus, the corresponding normalised thighs are both shorter. (3) The lower left arm is retracted a little behind the torso, which is further away from the observer. Thus, the corresponding normalised lower left arm is shorter and thinner, compared to the lower right arm.

Figure 4 shows the average contours of the COCO men and women respectively, calculated with the same approach, using 144 men and 150 women from the COCO Person dataset. Superimposing the same pose from Michelangelo and Paul Delvaux on the natural men and women contours shows that Michelangelo tends to exaggerate the muscles of men, as every segment is inflated outside the brim. On the contrary, Paul Delvaux tends to draw the women more slender than their counterparts in the natural poses, as the torso and the limbs shrink a bit width-wise within the borders (see Fig. 4).

In summary, the mean contours can give us an intuitive view of artistic and natural poses, focusing on height and width of body segments. For art historical

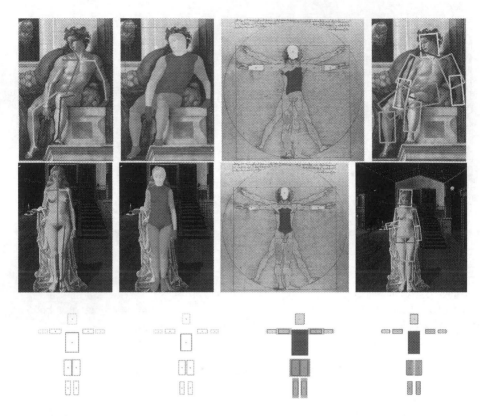

Fig. 4. Michelangelo's man (first row) and Paul Delvaux's woman (middle row). For both rows, left to right: Keypoints by OpenPose, segments by DensePose, normalised DensePose on Vitruvian Man and on the original image. Last row, left to right: 1) Average contour of the COCO men. 2) Average contour of the COCO women. 3) Michelangelo pose imposed on the contour of the COCO men. 4) The Delvaux pose imposed on the contour of the COCO women.

analysis, the segment contours of each artist can be visualised as to how each artist tends to draw human bodies. The contours can be compared with each other to further explore whether there exist significant differences between the drawing styles for each artist, and with that of the COCO men and women contours to analyse whether the artistic contour conforms to or deviates from the natural contours.

5.3   Visualising Joint Distributions

For the AP dataset, after the first step of the validity check, we have 211 poses. Out of these, 67 poses are male, and 144 female (Fig. 5). Different from the distributions of facial landmarks [26], the keypoints are prone to high variance of articulation. For the upper limbs, the wrists occupy a larger ellipsoid than the elbows, as do the elbows compared to the shoulders. Similarly, for the lower limbs, the areas of the ellipsoids are in descending order: ankle > knee > hip. In general, the outer joints swing over a larger circle than the inner joints. We also observe that the joint distributions are similar for both male and female poses, as they are both with arms and legs hanging alongside the torso.

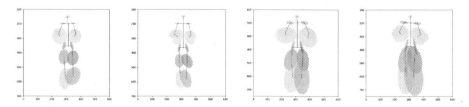

Fig. 5. The mean and 0.5 standard deviation distribution for artistic (AP) and natural poses (COCO). Joints used in the normalisation have smaller variances. From left to right: Males in AP; Females in AP; Males in COCO; Females in COCO.

As a reference, we also analyse the gestures of daily activities captured in photographs. After filtering through the COCO people dataset and a validity check, we use 653 male and 222 female poses for the mean and standard deviation analysis. Compared with artistic poses, natural poses tend to be more varied with wider movement of joints. The natural poses are formed by a wide variety of daily activities and sports, whereas the artistic poses are staged and hence restrained to certain poses (see Fig. 5).

5.4   Hierarchical Clustering

We use the poses from the paintings of Felix Vallotton as an example to demonstrate the hierarchical clustering results. Among the 15 full-body depictions of Felix Vallotton figures, one can find variations such as lying, sitting, and standing poses. The hierarchical clustering, as shown in Fig. 6, cannot distinguish whether a person lies down, sits or stands. But the clusters can differentiate between the stretched and compressed limbs. When the number of clusters is set to 5, cluster 2 and 3 contain the poses with arms thrown up; cluster 4 contains the standing poses with arms relaxed alongside the body; cluster 5 contains the twisted legs and arms.

When the number of clusters is set to 10 in order to generate the hierarchical clustering for all the artists, the outcome shows that there are no clear boundaries between artists. Out of these 10 clusters, there are 2 outstanding clusters with only a small fraction of poses. These are all standing poses with the arms thrown upward. The pose of opening arms appears for example in the religious paintings of El Greco, but similar poses are depicted by Artemisia Gentileschi, Pierre-Paul Prud'hon and is later used by Paul Delvaux as well (see Fig. 7).

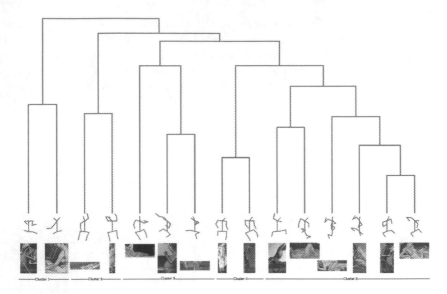

Fig. 6. Dendrogram for all 15 poses drawn by Felix Vallotton from 15 paintings.

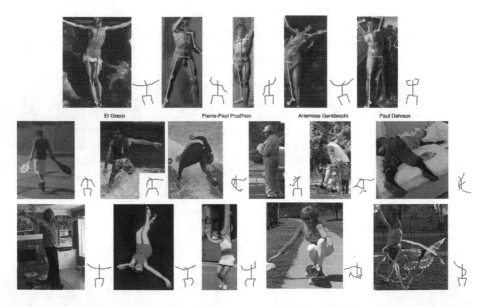

Fig. 7. First row: The standing poses with arms thrown upward into the air, as an example niche pose from the AP dataset. Middle row: The niche natural poses in the COCO dataset for only men. Last row: The niche natural poses in the COCO dataset for only women.

For comparison, we carried out hierarchical clustering in the COCO dataset to find niche poses for men and women. The rare poses for men are usually with more stretched arms, i.e., swinging back and forth, or the ones with more twisted legs. Moreover, the lying poses barely appear for men. The rare poses for women are the ones with twisted or overly stretched legs, or those with arms thrown upward (Fig. 7).

In summary, the hierarchical clustering of the pose vectors can tell the difference of poses in: (1) the orientation, (2) whether the limbs are stretched or compressed, and (3) the completeness of the joints. Via this visualisation method it is also possible to track the transformation and re-use of similar poses by different artists over time. When we compare the artistic and natural poses we see that the poses with arms stretched upward are quite rare for natural poses as they appear only in women's hugging or lying poses, or in sports. It might be assumed that if trained with more niche poses and niche perspectives from natural poses, OpenPose and DensePose might also perform better at inference of artistic poses.

6   Conclusions

In this paper, we evaluated automated human body analysis for paintings in Western artworks. Specifically, we tested two state of the art pose estimation models, namely DensePose and OpenPose, on a curated and manually annotated artistic dataset of 10 Western Artists from various genres and periods, and reported their performances. OpenPose in general performed better than DensePose, especially for paintings with more than one figure where figures are interacting with each other. There is room for improvement in pose segmentation, and fine-tuning on annotated painting datasets may help.

We used a simple, rule-based approach to improve the performance of Dense-Pose by combining it with the keypoints of OpenPose, and proposed a way of visualising a painter's preferred body shapes using a normalisation method using Leonardo's Vitruvian Man. This way, we generate average contours for all poses drawn by a given artist, which can also be used to summarise differences in body shape preferences of artists. For example, a body that looks fat with respect to an average body can be classified as skinny with respect to the artist's specific style. The same holds true for other shape parameters. Our proposed approach improves over purely skeleton-based analysis of human poses.

Finally, we have analysed the distribution of joints in natural photographs and paintings, and shown how our approach can automatically find niche poses preferred by artist. The niche poses we found in the AP dataset via the agglomerative hierarchical clustering are also documented in Impett and Susstrunk's study of Aby Warburg's Bilderatlas [9], which shows that our automatic approach is useful. Our findings also point out to what kind of poses are needed for the training of pose detectors for paintings.

References

1. Aviezer, H., Trope, Y., Todorov, A.: Body cues, not facial expressions, discriminate between intense positive and negative emotions. Science **338**(6111), 1225–1229 (2012)
2. Bai, Z., Nakashima, Y., Garcia, N.: Explain me the painting: multi-topic knowledgeable art description generation. In: Proceedings of the IEEE/CVF International Conference on Computer Vision, pp. 5422–5432 (2021)
3. Cao, Z., Hidalgo, G., Simon, T., Wei, S.E., Sheikh, Y.: OpenPose: realtime multi-person 2D pose estimation using part affinity fields. IEEE Trans. Pattern Anal. Mach. Intell. **43**(1), 172–186 (2019)
4. Castellano, G., Lella, E., Vessio, G.: Visual link retrieval and knowledge discovery in painting datasets. Multimedia Tools Appl. **80**(5), 6599–6616 (2021)
5. Castellano, G., Vessio, G.: Deep learning approaches to pattern extraction and recognition in paintings and drawings: an overview. Neural Comput. Appl. **33**(19), 12263–12282 (2021)
6. Cetinic, E.: Iconographic image captioning for artworks. In: Del Bimbo, A., et al. (eds.) ICPR 2021. LNCS, vol. 12663, pp. 502–516. Springer, Cham (2021). https://doi.org/10.1007/978-3-030-68796-0_36
7. Güler, R.A., Neverova, N., Kokkinos, I.: DensePose: dense human pose estimation in the wild. In: The IEEE Conference on Computer Vision and Pattern Recognition (CVPR) (2018)
8. Huang, X., Belongie, S.: Arbitrary style transfer in real-time with adaptive instance normalization. In: Proceedings of the IEEE International Conference on Computer Vision, pp. 1501–1510 (2017)
9. Impett, L., Süsstrunk, S.: Pose and Pathosformel in Aby Warburg's Bilderatlas. In: Hua, G., Jégou, H. (eds.) ECCV 2016. LNCS, vol. 9913, pp. 888–902. Springer, Cham (2016). https://doi.org/10.1007/978-3-319-46604-0_61
10. Jenicek, T., Chum, O.: Linking art through human poses. In: 2019 International Conference on Document Analysis and Recognition (ICDAR), pp. 1338–1345. IEEE (2019)
11. King, D.E.: Dlib-ml: a machine learning toolkit. J. Mach. Learn. Res. **10**, 1755–1758 (2009)
12. Langfeld, G.: The canon in art history: concepts and approaches. J. Art Historiography **19**, 152–180 (2018)
13. Lin, T.-Y., et al.: Microsoft COCO: common objects in context. In: Fleet, D., Pajdla, T., Schiele, B., Tuytelaars, T. (eds.) ECCV 2014. LNCS, vol. 8693, pp. 740–755. Springer, Cham (2014). https://doi.org/10.1007/978-3-319-10602-1_48
14. Madhu, P., Marquart, T., Kosti, R., Bell, P., Maier, A., Christlein, V.: Understanding compositional structures in art historical images using pose and gaze priors. In: Bartoli, A., Fusiello, A. (eds.) ECCV 2020. LNCS, vol. 12536, pp. 109–125. Springer, Cham (2020). https://doi.org/10.1007/978-3-030-66096-3_9
15. Madhu, P., et al.: Enhancing human pose estimation in ancient vase paintings via perceptually-grounded style transfer learning. arXiv preprint arXiv:2012.05616 (2020)
16. Magazù, S., Coletta, N., Migliardo, F.: The Vitruvian Man of Leonardo da Vinci as a representation of an operational approach to knowledge. Found. Sci. **24**(4), 751–773 (2019)
17. Murtinho, V.: Leonardo's Vitruvian Man drawing: a new interpretation looking at Leonardo's geometric constructions. Nexus Netw. J. **17**(2), 507–524 (2015)

18. Noroozi, F., Kaminska, D., Corneanu, C., Sapinski, T., Escalera, S., Anbarjafari, G.: Survey on emotional body gesture recognition. IEEE Trans. Affect. Comput. **12**, 505–523 (2018)
19. Ren, S., He, K., Girshick, R., Sun, J.: Faster R-CNN: towards real-time object detection with region proposal networks. arXiv preprint arXiv:1506.01497 (2015)
20. Sagonas, C., Tzimiropoulos, G., Zafeiriou, S., Pantic, M.: 300 faces in-the-wild challenge: the first facial landmark localization challenge. In: Proceedings of the IEEE International Conference on Computer Vision Workshops, pp. 397–403 (2013)
21. Sarı, C., Salah, A.A., Akdag Salah, A.A.: Automatic detection and visualization of garment color in Western portrait paintings. Digital Sch. Humanit. **34**(Supplement_1), i156–i171 (2019)
22. Sheng, S., Moens, M.F.: Generating captions for images of ancient artworks. In: Proceedings of the 27th ACM International Conference on Multimedia, pp. 2478–2486 (2019)
23. Silva, J.M., Pratas, D., Antunes, R., Matos, S., Pinho, A.J.: Automatic analysis of artistic paintings using information-based measures. Pattern Recogn. **114**, 107864 (2021)
24. Sun, K., Xiao, B., Liu, D., Wang, J.: Deep high-resolution representation learning for human pose estimation. In: Proceedings of the IEEE/CVF Conference on Computer Vision and Pattern Recognition, pp. 5693–5703 (2019)
25. Wang, J.Z., Kandemir, B., Li, J.: Computerized analysis of paintings. In: The Routledge Companion to Digital Humanities and Art History, pp. 299–312, Routledge (2020)
26. Yaniv, J., Newman, Y., Shamir, A.: The face of art: landmark detection and geometric style in portraits. ACM Trans. Graph. (TOG) **38**(4), 1–15 (2019)
27. Zhang, K., Zhang, Z., Li, Z., Qiao, Y.: Joint face detection and alignment using multitask cascaded convolutional networks. IEEE Signal Process. Lett. **23**(10), 1499–1503 (2016)
28. Zheng, C., et al.: Deep learning-based human pose estimation: a survey. arXiv preprint arXiv:2012.13392 (2020)

ArtFacePoints: High-Resolution Facial Landmark Detection in Paintings and Prints

Aline Sindel[(✉)], Andreas Maier, and Vincent Christlein

Pattern Recognition Lab, FAU Erlangen-Nürnberg, Erlangen, Germany
`aline.sindel@fau.de`

Abstract. Facial landmark detection plays an important role for the similarity analysis in artworks to compare portraits of the same or similar artists. With facial landmarks, portraits of different genres, such as paintings and prints, can be automatically aligned using control-point-based image registration. We propose a deep-learning-based method for facial landmark detection in high-resolution images of paintings and prints. It divides the task into a global network for coarse landmark prediction and multiple region networks for precise landmark refinement in regions of the eyes, nose, and mouth that are automatically determined based on the predicted global landmark coordinates. We created a synthetically augmented facial landmark art dataset including artistic style transfer and geometric landmark shifts. Our method demonstrates an accurate detection of the inner facial landmarks for our high-resolution dataset of artworks while being comparable for a public low-resolution artwork dataset in comparison to competing methods.

Keywords: Facial landmark detection · Convolutional neural networks · Artistic image synthesis · Paintings · Prints

1 Introduction

Facial landmark detection is a key element to analyze face characteristics. In art investigations, portraits of the same artist but also portraits of artists with a similar style are analyzed by comparing facial structures, e. g., by creating hand-drawn tracings of the facial inner lines or the face outline and manually compare the tracings. In this procedure, the art technologists can be supported by automatically detecting facial landmarks in the portrait images and using them to align the images for comparison. Facial landmark detection is a widely explored field for natural images, but is not directly applicable to artworks, since artworks show larger variation in the texture of the images and in the geometry of the landmarks than natural images. In art technology, art examinations are usually based on high-resolution or even macrophotographic images of the artwork. Therefore, there is a demand for a method that can accurately detect facial landmarks in high-resolution images of artworks.

L. Karlinsky et al. (Eds.): ECCV 2022 Workshops, LNCS 13801, pp. 298–313, 2023.
https://doi.org/10.1007/978-3-031-25056-9_20

Fig. 1. Our ArtFacePoints accurately detects facial landmarks in high-resolution paintings and prints. Image sources: Details of (a) Henri de Toulouse-Lautrec, The Streetwalker, The Metropolitan Museum of Art, 2003.20.13; (b) Florentine 16th Century (painter), Portrait of a Young Woman, Widener Collection, National Gallery of Art, 1942.9.51; (c) Paul Gauguin, Madame Alexandre Kohler, Chester Dale Collection, National Gallery of Art, 1963.10.27; (d) Meister des Augustiner-Altars (Hans Traut) und Werkstatt mit Rueland Frueauf d.Ä., Marter der Zehntausend, Germanisches Nationalmuseum Nürnberg, on loan from Museen der Stadt Nürnberg, Kunstsammlungen, Gm 149; (e) North Italian 15th Century (painter), Portrait of a Man, Samuel H. Kress Collection, National Gallery of Art, 1939.1.357; (f) Henri de Toulouse-Lautrec, Albert (René) Grenier, The Metropolitan Museum of Art, 1979.135.14; (g) Cornelis Cort (after Albrecht Dürer), Portret van Barent van Orley, Rijksmuseum, RP-P-2000-140; (h) Lucas Cranach the Elder, Johann I. der Beständige, Kurfürst von Sachsen, Staatliche Kunstsammlungen Dresden SKD, Kupferstich-Kabinett, A 129786, SLUB / Deutsche Fotothek, Loos, Hans, http://www.deutschefotothek.de/documents/obj/ 70243984 (Free access - rights reserved); (i) Hans Brosamer, Sybille von Cleve, Herzog Anton Ulrich-Museum, HBrosamer AB 3.33H; (j) Edgar Degas, Marguerite De Gas the Artist's Sister, The Metropolitan Museum of Art, 2020.10; (k) Thomas Nast, Portrait of the Artist's Wife Sarah Edwards Nast, Reba and Dave Williams Collection, National Gallery of Art, 2008.115.3695; (l) Samuel Amsler, Portret Carl Philip Fohr, Rijksmuseum, RP-P-1954-345

In this paper, we propose ArtFacePoints, a deep-learning-based facial landmark detector for high-resolution images of paintings and prints. It consists of one global and multiple region networks for coarse and fine facial landmark prediction. We employ a ResNet-based encoder-decoder to predict global facial landmarks in downsized images, which are then used to crop regions of the high-resolution image and the global feature maps. With the global feature map regions as prior information, the region networks get an additional impetus to refine the global feature maps using the high-resolution image details. To train and evaluate our method, we created a large facial landmark art dataset by applying style transfer techniques and geometric facial landmark distortion to a public landmark dataset of natural images and by collecting and annotating images of real artworks. In Fig. 1, we show some qualitative examples of our ArtFacePoints for different artistic styles.

2 Related Work

Facial landmark detection is an active research area for natural images [26], hence we will only summarize a selection of regression-based approaches. For instance, one example for a machine-learning-based regression facial landmark method is dlib [11] which employs an ensemble of regression trees. Using convolutional neural networks (CNN) we can broadly differ between methods that directly regress the coordinates using a fully connected layer [7,24,28], or direct heatmap-based approaches [2,13,23,25] that draw the coordinates as Gaussian peaks and formulate the task as the regression between the predicted and target heatmaps, and indirect heatmap-based approaches [3,8,12,18] that use the differentiable spatial softargmax [16] to extract the coordinates from the heatmaps instead of using the argmax and then compute a loss between the predicted and target coordinates.

The high-resolution network (HR-Net) [22,23] was developed for different vision tasks such as for human pose estimation and facial landmark detection. It uses multi-resolution blocks which connect high-to-low resolution convolutions to maintain high-resolution representations. The coordinates of the landmarks are extracted from the heatmap based on the highest response peaks. Recently, the HR-Net was also used for human pose estimation in Greek vase paintings [15], for which style transfer using adaptive instance normalization [9] was applied to a labeled dataset of natural images to train the model.

Chandran et al. [3] proposed a region based facial landmark detector for high-resolution images up to 4K. They detect the coarse landmark positions in a downsized image using the heatmap-based Hourglass network and soft-argmax [16]. By using attention-driven cropping based on the global landmark locations, they extract regions around the eyes, nose, and mouth in the high-resolution image. They predict more accurate landmarks in the regions using for each region specifically trained Hourglass networks. In our approach, we are strongly oriented towards their idea of high-resolution landmarks with a global and regional detection network, however, we make some architectural and conceptual changes, such as using a ResNet-based encoder-decoder or the use of a global feature map as additional input for the regional networks.

For facial landmark detection in artworks, Yaniv et al. [27] proposed Face of Art, which predicts the landmarks in three steps. First, the global landmark location is estimated from a response map, then the landmarks are corrected using a pretrained point distribution model and finally they are tuned using weighted regularized mean shift. They created a synthetic dataset using the style transfer method of Gatys et al. [5] for low-resolution images and geometric landmark augmentations. The latter we adopt for our work. The facial landmark detection method for mangas [21] uses the deep alignment network [13] which uses multiple steps to refine the heatmaps of the previous steps.

(a) AdaIN Synthetic Paintings (Style, content, and generated image)

(b) AdaIN Synthetic Prints (Style, content, and generated image)

(c) CycleGAN Synthetic Paintings   (d) CycleGAN Synthetic Prints

Fig. 2. High-resolution synthetic image generation using AdaIN and CycleGAN. Image sources: (a) Paintings: Vincent van Gogh, Two Children, Musée d'Orsay, Paris, WikiArt; Roy Lichtenstein, Interior with Mirrored Closet, 1991, © Estate of Roy Lichtenstein; Amedeo Modigliani, Seated Young Woman, Private Collection, WikiArt; Georges Seurat, Study for Young Woman Powdering Herself, WikiArt; (b) Prints: Albrecht Dürer, The Virgin and Child Crowned by One Angel (Detail), Rosenwald Collection, National Gallery of Art, 1943.3.3546; Unknown, Adam Zusner (Detail). Wellcome Collection, 9833i; T. Stimmer, Paolo Giovio (Detail), Wellcome Collection, 3587i; Lucas Cranach the Younger, Sigmund (Halle, Erzbischof) (Detail), Staatliche Kunstsammlungen Dresden, Kupferstich-Kabinett, Nr.: B 376, 1, SLUB / Deutsche Fotothek, Kramer, Rudolph, https://www.deutschefotothek.de/documents/obj/70231557 (Free access - rights reserved); (a-d) Photos: Images from the 300-W dataset

3   Artistic Facial Landmarks Dataset Creation

For the training and testing of our method, we collected high-resolution portrait images of different artistic styles from various museums and institutes such as from the Cranach Digital Archive (CDA), Germanisches Nationnalmuseum Nürnberg (GNM), The Metropolitan Museum of Art (MET), National Gallery of Art (NGA), and Rijksmuseum, and additionally also selected images from the WikiArt dataset [17] for training. In order to automatically detect and crop the face region in the images of paintings and prints, we trained and applied the object detector YOLOv4 [1]. The cropped images are either used as surrogates to generate synthetic artworks or are manually labeled.

3.1   Synthetic Image Generation Using Style Transfer and Geometric Augmentations

There exist multiple public datasets for facial landmark detection with ground truth labels such as 300-W [19], however not yet for high-resolution paintings and prints. Thus, we apply style transfer and image-to-image translation techniques to the 300-W dataset to transform the images scaled to $1024 \times 1024 \times 3$ pixels into synthetic paintings and prints.

As style transfer method, we use adaptive instance normalization (AdaIN) [9] which aligns the channel-wise mean and variance of the content image with

Fig. 3. Geometric augmentations of ground truth facial landmarks in synthetic paintings and prints: before and after applying the geometric transformation.

the style image. AdaIN was trained using MS-COCO [14] for the content and WikiArt images [17] for the style images. For our application, we set the alpha parameter, which controls the trade-off between content and style, to 1 to achieve full stylization and apply both content and style image in 1024 × 1024 resolution. Some qualitative results of our high-resolution synthetically stylized portrait images are shown in Figs. 2a and 2b. The synthetic paintings and prints reuse the textures and the color distribution from the style images resulting in a motley set of images.

Further, we use the unpaired image-to-image translation technique Cycle-GAN [29] to learn a mapping between faces in photographs and artworks by exploiting cycle consistency. We train one CycleGAN for each domain pair, i. e., one for photo-to-print and one for photo-to-painting in lower resolution 512 × 512 × 3 and apply them in high-resolution (see Figs. 2c and 2d). Interestingly, the synthetic paintings using CycleGAN clearly show crack structures (craquelure) in the paint typical for old paintings. The synthetic prints express the shading and continuous regions of the photos with tiny lines to mimic the real prints.

Artistic faces not only differ to photos in their textual style but also in their geometric arrangement of the facial landmarks. Faces can be longitudinally or horizontally stretched or unbalanced with, e. g., larger eyes. To account for some degree of artistic variations, we adopt the geometric augmentation strategy of Yaniv et al. [27] to randomly shift or resize single groups of landmarks such as the eyes or mouth, or stretch or squeeze the face. Based on the movements of the landmarks a thin-plate-spline displacement field is computed which is used to warp the synthetic artwork [27]. Some visual examples in Fig. 3 show synthetic images with ground truth annotations in their original pose and after geometric warping of the images.

3.2 Semi-automatic Facial Landmarks Annotation

Additionally to the synthetic dataset, we also include a small number of real artworks for training and validation. For those and also for our test dataset, we annotated the facial landmarks in a semi-automatic manner. We applied the random forest based facial landmark detector dlib [11] to the images and

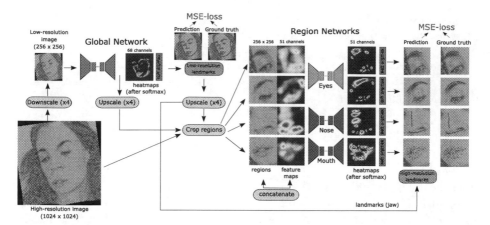

Fig. 4. Our facial landmarks detector ArtFacePoints for high-resolution images of paintings and prints splits the task into a global coarse prediction step and a regional refinement step. The high-resolution image is downsized and fed to the global ResNet encoder-decoder network, which predicts the low-resolution landmarks using softmax on a 68 channels heatmap. Based on the low-resolution landmarks, regions of eyes, nose, and mouth are extracted from the high-resolution image and the upscaled global feature map. For each region, the image patch and feature maps patches are concatenated and are fed to its specific region network to predict the high-resolution landmarks. The global as well as the regional landmarks are compared to the ground truth landmarks using the mean squared error (MSE) loss.

manually corrected the landmarks or in case of a dlib failure, we annotated the landmarks from scratch. For performing the annotations, we have written a small graphical user interface tool in Python that allows annotating from scratch and annotation refinement by enabling the user to move the single landmarks via mouse drag of the items.

4   ArtFacePoints

In this section, our method for high-resolution facial landmark detection in artworks is described, which consists of two main steps, the global and the regional facial landmark detection, as illustrated in Fig. 4.

4.1   Global Facial Landmark Detection

For the global network, we use the encoder-decoder network with stacked ResNet [6] blocks in the bottleneck, which was originally employed for style transfer [10]. We exchanged the transpose convolutional layers in the decoder with a 3×3 convolutional layer and bicubic upsampling to obtain smoother heatmaps. The input to the network is the low-resolution RGB image and the prediction are $N_G = 68$ heatmaps with the same width and height as the input image.

Table 1. Style Art Faces Dataset: Number of images including facial landmarks (68 per image) for the real and the synthetic artworks.

Dataset	Real paintings	Real prints	AdaIN paintings	AdaIN prints	CycleGAN paintings	CycleGAN prints	Total
Train	160	160	511	511	511	511	**2361**
Val	30	30	220	220	220	220	**940**
Test	40	40					**80**

Each heatmap should mark the location of its respective facial landmark as a peak with the highest value. Then, we use the spatial softargmax [16] operator to extract the landmarks from the heatmaps in a differentiable manner.

4.2 Regional Facial Landmarks Refinement

The global landmark predictions in the low-resolution image are upscaled by a factor of 4 to match the high-resolution image. Similar to [3], we automatically extract regions in the high-resolution image around the landmark predictions of the nose, the mouth, and each eye including the eye brow. The region size is padded with a random value between 0.25 to 0.5 of the original region size that is directly estimated from the global landmarks. For inference, the padding is a fixed value of 0.25 of the region size. Analogously, we also extract the same regions from the upscaled feature maps of the global network, which is the direct output of the global network before applying spatial softmax. All regions are scaled to a fixed patch size. The three region networks for eye, nose, and mouth also use the ResNet-based encoder-decoder architecture like the global network, but instead of only feeding the RGB patch, we concatenate the RGB patch (3 channels) with the corresponding regions of the feature maps (N_r channels) as input, where N_r depends on the specific region network (eye: 11, nose: 9, mouth: 20). With the channel fusion, the region network gets the global location of the landmark as prior information, which supports the refinement task. There is no weight sharing between the region networks, such that each network can learn its specific features for the facial sub regions. The high-resolution landmarks are also extracted using spatial softargmax [16].

We use the mean squared error (MSE) loss of predicted landmarks and ground truth landmarks for both the global and regional landmark detection task:

$$\mathcal{L}_{\mathrm{MSE}} = \frac{1}{N_G} \sum_i^{N_G} (\hat{\mathbf{x}}_i - \mathbf{x}_i)^2 + \lambda \sum_r^4 \frac{1}{N_r} \sum_j^{N_r} (\hat{\mathbf{y}}_j - \mathbf{y}_j)^2, \tag{1}$$

where λ is a weighting factor, $\hat{\mathbf{x}}$, $\mathbf{x}$ are the predicted and ground truth global coordinates and $\hat{\mathbf{y}}$, $\mathbf{y}$ are the predicted and ground truth coordinates of one region.

For inference, we need to transfer the high-resolution landmarks from the individual regions back to the global coordinate system. Therefore, we track

Table 2. Quantitative results for our paintings and prints test dataset (1024 × 1024 using the mean error of the 68 landmarks and 51 high-resolution landmarks (without jaw line). * For dlib only 38 out of 40 paintings and 34 out of 40 prints were detected.

Mean Error (ME)	Paintings		Prints	
	68 landmarks	51 landmarks	68 landmarks	51 landmarks
dlib*	20.03±7.77	17.53±7.25	43.18±20.99	34.94±20.02
dlib* (Art)	33.39±24.49	26.84±24.11	109.09±99.81	104.12±108.01
HR-Net	22.29±7.24	19.22±4.36	35.90±16.87	26.28±16.23
HR-Net (Art)	20.01±6.99	17.05±3.50	27.81±9.00	19.67±4.44
Face of Art	**17.93**±7.76	14.33±4.81	27.29±8.44	18.89±5.94
ArtFacePoints (global)	18.87±9.23	13.42±4.80	26.37±8.82	16.74±5.22
ArtFacePoints (w/o FM)	17.97±8.51	12.62±4.62	**25.60**±8.50	15.83±4.79
ArtFacePoints	18.88±8.52	**12.45**±4.44	25.65±8.83	**15.78**±6.20

both, the bounding box coordinates of the extracted regions and the original region sizes. Then, we scale the local coordinates by the region size and add the offset of the bounding box. For the jaw line, we only have the global estimate. Hence, the complete facial landmark prediction is the combination of the global jaw line and the refined regions.

5   Experiments and Results

5.1   Datasets

Our high-resolution facial landmarks dataset of artworks is comprised of real and synthetic paintings and prints as described in Sect. 3. The number of images for training, validation, and test split are summarized in Table 1. All images are of size 1024 × 1024 × 3 and contain each 68 facial landmarks according to the 300-W annotation concept. For the 2924 synthetic artworks, we directly reused the 300-W annotations and for the 460 real artworks, we semi-automatically annotated the landmarks.

Further, we use the public Artistic Faces Dataset [27] for comparison. It consists of a total of 160 images of size 256 × 256 × 3, which are composed of 10 images per artist representing different artistic styles.

5.2   Implementation and Experimental Details

We pretrain the global network for 60 epochs using the Adam optimizer, a learning rate of $\eta = 1 \cdot 10^{-4}$ with linear decay of η to 0 starting at epoch 30, a batch size of 16 and early stopping. Then, we initialize the region networks with the weights of the global network and train both jointly for 30 epochs using $\eta = 1 \cdot 10^{-4}$ with linear decay of η to 0 starting at epoch 10, a batch size of 4 and early stopping. The input image size to crop the regions is 1024 × 1024 and

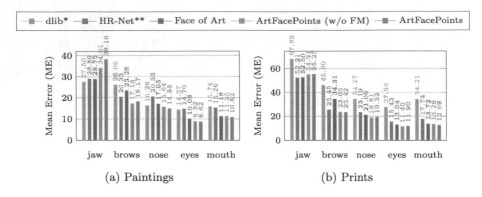

Fig. 5. Quantitative comparison of the facial landmark prediction for individual parts of the face of our high-resolution paintings and prints. * For dlib only 38 out of 40 paintings and 34 out of 40 prints were detected. ** HR-Net is the fine-tuned model on our art dataset.

Table 3. Quantitative comparison for the public Artistic Faces dataset (256 × 256). * For dlib only 134 out of 160 artworks were detected.

Mean Error (ME)	68	51	Jaw	Eye brows	Nose	Eyes	Mouth
dlib*	5.75±3.64	5.42±3.43	**6.74**±5.37	7.01±6.54	4.85±4.10	5.54±3.90	4.82±4.23
HR-Net (Art)	5.28±2.39	4.60±2.45	**7.31**±3.74	5.76±4.08	4.61±3.58	4.19±3.87	4.27±2.11
Face of Art	**4.87**±2.48	**3.88**±2.36	7.83±4.44	5.57±6.02	3.90±4.82	3.23±2.54	**3.41**±3.06
ArtFacePoints (w/o FM)	5.77±3.12	4.27±2.66	10.29±6.37	5.83±5.24	4.51±3.29	**3.19**±3.48	4.02±3.17
ArtFacePoints	5.58±3.19	3.91±2.77	10.60±6.08	**5.55**±6.32	**3.77**±2.57	3.24±3.88	3.55±2.97

the patch size for the global and region networks is 256 × 256. For the loss computation, both global and local landmarks are normalized to $[-0.5, 0.5]$ based on their image or region size and $\lambda = 0.25$ is set to weight each region term equally in the loss.

As comparison methods, we use dlib [11], HR-Net [23], and Face of Art [27]. We retrained dlib and fine-tuned HR-Net using our synthetically augmented art facial landmark dataset including the geometric transformations of the landmarks. To measure the performance, we compute the mean Euclidean error (ME) of the predicted and manually labeled facial landmarks. In contrast to related works [23,27], we do not normalize the error based on the inter-ocular distance, inter-pupil distance, or the diagonal of the bounding box. Especially to assess the accuracy in high-resolution images, we prefer to compare the directly measured pixel distance between the landmarks as the denominator using the eye distance or bounding box distance can become very large and thus the error would become very small.

5.3   Results

The quantitative results for our high-resolution art dataset are summarized in Table 2. Considering all 68 landmarks, all methods are relatively close for the paintings test set, except for the retrained version of dlib on art images that did

Image	Ground truth	dlib	HR-Net	Face of Art	ArtFacePoints

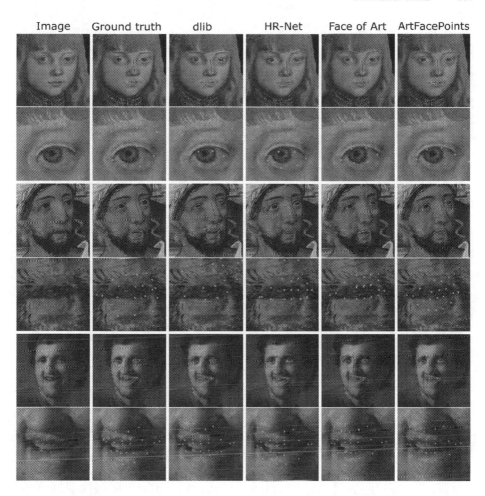

Fig. 6. Qualitative comparison of facial landmark detection for paintings. Image sources: Details of (a) Lucas Cranach the Elder, A Prince of Saxony, Ralph and Mary Booth Collection, National Gallery of Art, 1947.6.1; (b) Meister des Augustiner-Altars (Hans Traut) und Werkstatt mit Rueland Frueauf d.Ä., Marter der Zehntausend, Germanisches Nationalmuseum Nürnberg, on loan from Museen der Stadt Nürnberg, Kunstsammlungen, Gm 149; (c) Rembrandt van Rijn (circle of), Laughing Young Man, Rijksmuseum, SK-A-3934

not work well. By comparing only the inner facial landmarks (51) for which we apply the region refinement, the errors of all methods are considerably reduced, in particular for our ArtFacePoints. The facial landmark detection in prints is more challenging, resulting in overall higher errors. Our ArtFacePoints achieves the lowest error for both the total 68 and also the inner 51 landmarks. For both paintings and prints, fine-tuning of HR-Net shows improvements compared to the pretrained model, hence in the next experiments, we only include dlib (pretrained) and HR-Net (fine-tuned).

Image   Ground truth   dlib   HR-Net   Face of Art   ArtFacePoints

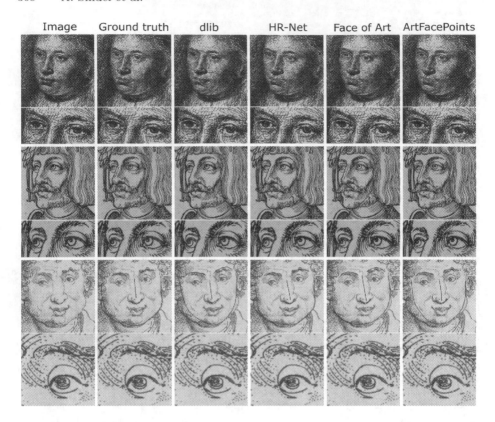

Fig. 7. Qualitative comparison of facial landmark detection for prints. Image sources: Details of (a) Rembrandt van Rijn, The great Jewish bride, Rijksmuseum, RP-P-OB-724; (b) after Hans Baldung Grien, Ulrich von Hutten, British Museum, London, 1911,0708.39, Photo © Thomas Klinke, courtesy of the Trustees of the British Museum; (c) Thomas Rowlandson, Joy with Tranquility, The Elisha Whittelsey Collection, The Metropolitan Museum of Art, 59.533.699

In our ablation study, we compare the global network only (trained for 60 epochs) and two versions of regional refinement: firstly, ArtFacePoints (w/o FM), for which we only use the RGB images as input to the region networks, i. e., without the feature maps and secondly, our proposed ArtFacePoints which uses the concatenation of the RGB channels and the feature maps. We can observe for both paintings and prints that the regional refinement of both variants brings some benefit for the inner facial landmarks. Using the additional input of the feature maps is slightly superior, but ArtFacePoints (w/o FM) works a bit better for including the landmarks of the jaw line.

In Fig. 5, we separately analyze the performance of the individual facial parts. Both variants of ArtFacePoints achieve the lowest errors for eyes, nose, mouth, and brows. The competing method Face of Art is relatively close to ours for the mouth, and it also works good for eyes and nose. HR-Net (fine-tuned) is the third best for the brows, for the eyes and mouth, HR-Net and dlib are comparable for the paintings, but HR-Net is better for the prints and also dlib did not detect landmark estimates for all faces. In general, the largest errors are obtained for the jaw line, which does not show as distinctive features as e. g., the eyes and thus there is more ambiguity in the labeling process where to exactly position the landmarks on the face boundary and sometimes the face boundary is really hard to detect in case of occlusions by beard or hair. This uncertainty is hence also propagated into the models' predictions. The jaw line prediction of dlib is best for the paintings. As we used dlib as initial estimate for the manual labeling and then corrected the landmarks, there might be some bias for dlib regarding the jaw line. For the prints, dlib did not work so well, hence we also had a larger correction effort, which is also visible at the jaw line results. Our method has its limitations for the prediction of jaw lines, which where not specifically refined as only regions of the inner facial landmarks are extracted, but for instance for the registration of portraits only the landmarks of eyes, nose, and mouth are important, for which our method performs best.

Some visual examples for the landmark prediction results are shown in Fig. 6 for the paintings and in Fig. 7 for the prints. For each artwork, we additionally select a zoom in region for a precise comparison of the difference of the competing methods to our ArtFacePoints. For the paintings, in the first zoom region in Fig. 6, the landmarks of the eye are most accurately predicted by ArtFacePoints, for dlib, HR-Net and Face of Art the eye is a bit too small. The second and third zoom regions in Fig. 6 depict the mouth region, for which ArtFacePoints most precisely detects the upper and lower lip. For the prints in Fig. 7, dlib does not achieve an acceptable result for the first two images. In the zoomed regions of the eyes of all three examples, Face of Art and HR-Net miss the eye boundary in some corners and thus are less accurate than our ArtFacePoints.

Further, we tested the facial landmark detection for the low-resolution public Artistic Faces dataset (see Table 3). As our method requires input resolution of 1024×1024, we upscale the 256×256 images to feed them to ArtFacePoints. The mean error is calculated at the low-resolution scale. Regarding the 68 and 51 facial landmarks Face of Art is slightly superior to our method, which is due to our lower performance of the jaw line prediction. However, for the individual regions of the core facial landmarks, ArtFacePoints is on par to Face of Art. HR-Net and dlib numerically also perform quite well for the low-resolution images, except that for dlib only 134 of 160 images could be considered to compute the ME due to dlib's false negative face detections. Thus, we could show that our ArtFacePoints can also be applied to the low-resolutions, but its advantage lies in the accurate prediction of the inner facial features for high-resolution applications.

6   Applications

In this section, we present some examples for the application of our ArtFace-Points to support the visual comparison of similar artworks.

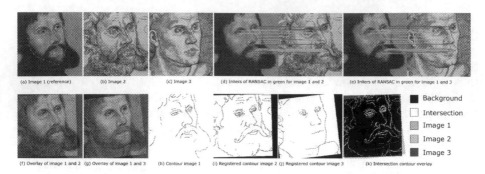

Fig. 8. Registration of paintings and prints using ArtFacePoints: Results for facial landmark detection, matching of inner facial landmarks, image overlays, contours generated using Art2Contour, and intersection contour overlay. Image sources: Details of (a) Lucas Cranach the Elder, Martin Luther as "Junker Jörg" (Detail), Museum der bildenden Künste Leipzig; (b) Lucas Cranach the Elder, Martin Luther as "Junker Jörg" (Detail, mirrored), Germanisches Nationalmuseum Nürnberg, Mp 14637; (c) Lucas Cranach the Elder, Martin Luther as augustinian monk (Detail, mirrored), Klassik Stiftung Weimar, Museen, DK 182/83

6.1   Image Registration Using Facial Landmarks

To be able to visually compare two portraits based on their facial characteristics the images need to be registered. Therefore, the facial landmarks in both images serve as control point pairs for the registration. Since we want to explicitly align the eyes, nose, and mouth in the images, we only take these 41 landmarks as control points. Our aim is to find a global transformation matrix that transforms the source image in such as way that the error between the transformed source control points and target control points is minimal. To robustly compute the transform, we use random sample consensus (RANSAC) [4] that estimates multiple transforms based on random subsets of the 41 control points and then selects the transform with the largest support based on all 41 control points. The aspect ratio of the source image should be kept before and after the registration, thus we estimate a partial affine transform that includes rotation, translation, and scaling but no shearing. This is important for the similarity comparison of the faces to assess e. g., if artists reused some facial structures in two different artworks of a similar motif.

6.2   Facial Image and Contour Comparison

For the visual illustration of facial landmarks based registration in Fig. 8, we picked three artworks by Cranach the Elder that depict Martin Luther as "Junker Jörg" (image 1 and 2) or augustinian monk (image 3). We detected the facial landmarks using ArtFacePoints and then applied RANSAC to predict the transformations between the reference image which is the painting and the two other images, the prints. For both registration pairs, in (d) and (e) the lines connecting the control points in the reference and source image are colored in green and red. The green lines indicate the inliers, which are those points that were selected by RANSAC to compute the transformation and the red lines indicate the outliers that were excluded for the computation. Both examples show a high number of inliers. The first two images of the bottom row of Fig. 8 visualize the registration results as blended image overlays using alpha blending between the target and transformed source image, which indicate high similarity between the facial structures.

Another possibility is to focus only on the facial contours for the comparison. For this task, we generate contour drawings for the three examples using the conditional generative adversarial network Art2Contour [20] and apply the same transformations that we have predicted using the facial landmarks to warp the contour images correspondingly (see (h)-(j) in Fig. 8). Then, we compute an intersection contour overlay in (k). It depicts the contours in white if at least two contours are intersected, and otherwise, i. e., without any intersection, they are drawn in an image-specific color. That allows the comparison of multiple contours in one image, with a direct assignment of contour parts to the images. The intersection overlay depicted in Fig. 8 (k) for the given example images shows that these cross-modal pairs of painting and prints have a very similar shape of the main facial contour with some artistic differences in the contour line of the nose (painting) and in the right part of the jaw line (print with monk).

7   Conclusions

We presented a deep learning method for facial landmark detection in high-resolution artistic images. We employ a heatmap-based global network for coarse coordinates extraction and multiple heatmap-based region networks that operate on high image resolution only for specific regions. To train our method, we created a large synthetically augmented high-resolution dataset by using artistic style transfer and geometric transformations. In the experiments, we showed on our test dataset of paintings and prints, that our method, in comparison to competing methods, more accurately detects the facial landmarks of eyes, nose, and mouth, which are important for facial image registration. Then, we visually demonstrated for some example images the application of facial landmarks for cross-genre registration of paintings and prints and the possibility for facial image and contour comparison. Our method works precisely for the high-resolution landmarks of paintings and prints with moderate artistic shape and texture variations, but its performance of the detection of the jaw line is limited as we do not apply any refinement for this landmark group. Thus, future work will investigate model-based approaches to tune the detection of the facial outline.

Acknowledgement. Thanks to Daniel Hess, Oliver Mack, Daniel Görres, Wibke Ottweiler, GNM, and Gunnar Heydenreich, CDA, and Thomas Klinke, TH Köln, and Amalie Hänsch, FAU Erlangen-Nürnberg for providing image data, and to Leibniz Society for funding the research project "Critical Catalogue of Luther portraits (1519–1530)" with grant agreement No. SAW-2018-GNM-3-KKLB, to the European Union's

Horizon 2020 research and innovation programme within the Odeuropa project under grant agreement No. 101004469 for funding this publication, and to NVIDIA for their GPU hardware donation.

References

1. Bochkovskiy, A., Wang, C.Y., Liao, H.Y.M.: YOLOv4: optimal speed and accuracy of object detection (2020). https://doi.org/10.48550/ARXIV.2004.10934
2. Bulat, A., Tzimiropoulos, G.: How far are we from solving the 2D & 3D face alignment problem? (and a dataset of 230,000 3D facial landmarks). In: 2017 IEEE International Conference on Computer Vision (ICCV), pp. 1021–1030 (2017). https://doi.org/10.1109/ICCV.2017.116
3. Chandran, P., Bradley, D., Gross, M., Beeler, T.: Attention-driven cropping for very high resolution facial landmark detection. In: 2020 IEEE/CVF Conference on Computer Vision and Pattern Recognition (CVPR), pp. 5860–5869 (2020). https://doi.org/10.1109/CVPR42600.2020.00590
4. Fischler, M.A., Bolles, R.C.: Random sample consensus: a paradigm for model fitting with applications to image analysis and automated cartography. Commun. ACM **24**(6), 381–395 (1981). https://doi.org/10.1145/358669.358692
5. Gatys, L.A., Ecker, A.S., Bethge, M.: A neural algorithm of artistic style (2015). https://doi.org/10.48550/ARXIV.1508.06576
6. He, K., Zhang, X., Ren, S., Sun, J.: Deep residual learning for image recognition. In: 2016 IEEE Conference on Computer Vision and Pattern Recognition (CVPR), pp. 770–778 (2016). https://doi.org/10.1109/CVPR.2016.90
7. He, K., Xue, X.: Facial landmark localization by part-aware deep convolutional network. Adv. Multimedia Inf. Process. - PCM **2016**, 22–31 (2016). https://doi.org/10.1007/978-3-319-48890-5_3
8. Honari, S., Molchanov, P., Tyree, S., Vincent, P., Pal, C., Kautz, J.: Improving landmark localization with semi-supervised learning. In: 2018 IEEE Conference on Computer Vision and Pattern Recognition (CVPR), pp. 1546–1555 (2018). https://doi.org/10.1109/CVPR.2018.00167
9. Huang, X., Belongie, S.: Arbitrary style transfer in real-time with adaptive instance normalization. In: 2017 IEEE International Conference on Computer Vision (ICCV), pp. 1501–1510 (2017). https://doi.org/10.1109/ICCV.2017.167
10. Johnson, J., Alahi, A., Fei-Fei, L.: Perceptual losses for real-time style transfer and super-resolution. In: Leibe, B., Matas, J., Sebe, N., Welling, M. (eds.) ECCV 2016. LNCS, vol. 9906, pp. 694–711. Springer, Cham (2016). https://doi.org/10.1007/978-3-319-46475-6_43
11. Kazemi, V., Sullivan, J.: One millisecond face alignment with an ensemble of regression trees. In: 2014 IEEE Conference on Computer Vision and Pattern Recognition (CVPR), pp. 1867–1874 (2014). https://doi.org/10.1109/CVPR.2014.241
12. Kordon, F., Maier, A., Kunze, H.: Latent shape constraint for anatomical landmark detection on spine radiographs. Bildverarbeitung für die Medizin **2021**, 350–355 (2021). https://doi.org/10.1007/978-3-658-33198-6_85
13. Kowalski, M., Naruniec, J., Trzcinski, T.: Deep alignment network: a convolutional neural network for robust face alignment. In: 2017 IEEE Conference on Computer Vision and Pattern Recognition (CVPR) Workshops, pp. 2034–2043 (2017). https://doi.org/10.1109/CVPRW.2017.254

14. Lin, T.-Y., et al.: Microsoft COCO: common objects in context. In: Fleet, D., Pajdla, T., Schiele, B., Tuytelaars, T. (eds.) ECCV 2014. LNCS, vol. 8693, pp. 740–755. Springer, Cham (2014). https://doi.org/10.1007/978-3-319-10602-1_48

15. Madhu, P., et al.: Enhancing Human Pose Estimation in Ancient Vase Paintings via Perceptually-grounded Style Transfer Learning. arXiv:2012.05616 (2020)

16. Nibali, A., He, Z., Morgan, S., Prendergast, L.: Numerical Coordinate Regression with Convolutional Neural Networks. arXiv:1801.07372 (2018)

17. Nichol, K.: Painter by Numbers, WikiArt (2016). https://www.kaggle.com/c/painter-by-numbers

18. Robinson, J.P., Li, Y., Zhang, N., Fu, Y., Tulyakov, S.: Laplace landmark localization. In: 2019 IEEE/CVF International Conference on Computer Vision (ICCV), pp. 10102–10111 (2019). https://doi.org/10.1109/ICCV.2019.01020

19. Sagonas, C., Tzimiropoulos, G., Zafeiriou, S., Pantic, M.: 300 faces in-the-wild challenge: the first facial landmark localization challenge. In: 2013 IEEE International Conference on Computer Vision (ICCV) Workshops, pp. 397–403 (2013). https://doi.org/10.1109/ICCVW.2013.59

20. Sindel, A., Maier, A., Christlein, V.: Art2Contour: salient contour detection in artworks using generative adversarial networks. In: 2020 IEEE International Conference on Image Processing (ICIP), pp. 788–792 (2020). https://doi.org/10.1109/ICIP40778.2020.9191117

21. Stricker, M., Augereau, O., Kise, K., Iwata, M.: Facial Landmark Detection for Manga Images (2018). https://doi.org/10.48550/ARXIV.1811.03214

22. Sun, K., Xiao, B., Liu, D., Wang, J.: Deep high-resolution representation learning for human pose estimation. In: 2019 IEEE/CVF Conference on Computer Vision and Pattern Recognition (CVPR), pp. 5686–5696 (2019). https://doi.org/10.1109/CVPR.2019.00584

23. Sun, K., et al.: High-resolution representations for labeling pixels and regions (2019). https://doi.org/10.48550/ARXIV.1904.04514

24. Sun, Y., Wang, X., Tang, X.: Deep convolutional network cascade for facial point detection. In: 2013 IEEE Conference on Computer Vision and Pattern Recognition (CVPR), pp. 3476–3483 (2013). https://doi.org/10.1109/CVPR.2013.446

25. Wang, X., Bo, L., Fuxin, L.: Adaptive wing loss for robust face alignment via heatmap regression. In: 2019 IEEE/CVF International Conference on Computer Vision (ICCV), pp. 6970–6980 (2019). https://doi.org/10.1109/ICCV.2019.00707

26. Wu, Y., Ji, Q.: Facial landmark detection: a literature survey. Int. J. Comput. Vision **127**(2), 115–142 (2018). https://doi.org/10.1007/s11263-018-1097-z

27. Yaniv, J., Newman, Y., Shamir, A.: The face of art: landmark detection and geometric style in portraits. ACM Trans. Graph. **38**(4) (2019). https://doi.org/10.1145/3306346.3322984

28. Zhang, Z., Luo, P., Loy, C.C., Tang, X.: Facial landmark detection by deep multitask learning. In: Fleet, D., Pajdla, T., Schiele, B., Tuytelaars, T. (eds.) ECCV 2014. LNCS, vol. 8694, pp. 94–108. Springer, Cham (2014). https://doi.org/10.1007/978-3-319-10599-4_7

29. Zhu, J.Y., Park, T., Isola, P., Efros, A.A.: Unpaired image-to-image translation using cycle-consistent adversarial networks. In: 2017 IEEE International Conference on Computer Vision (ICCV), pp. 2242–2251 (2017). https://doi.org/10.1109/ICCV.2017.244

W03 - Adversarial Robustness in the Real World

W03 - Adversarial Robustness in the Real World

Recent deep-learning-based methods achieve great performance on various vision applications. However, insufficient robustness on adversarial cases limits real-world applications of deep-learning-based methods. The AROW workshop aims to explore adversarial examples, as well as to evaluate and improve the adversarial robustness of computer vision systems. In the AROW workshop we discuss topics such as improving model robustness against unrestricted adversarial attacks; improving generalization to out-of-distribution samples or unforeseen adversaries; discovery of real-world adversarial examples; novel architectures with robustness to occlusion, viewpoint, and other real-world domain shifts; domain adaptation techniques for robust vision in the real world; datasets for evaluating model robustness; and structured deep models and explainable AI.

October 2022

Angtian Wang
Yutong Bai
Adam Kortylewski
Cihang Xie
Alan Yuille
Xinyun Chen
Judy Hoffman
Wieland Brendel
Matthias Hein
Hang Su
Dawn Song
Jun Zhu
Philippe Burlina
Rama Chellappa
Yinpeng Dong
Yingwei Li
Ju He
Alexander Robey

TransPatch: A Transformer-based Generator for Accelerating Transferable Patch Generation in Adversarial Attacks Against Object Detection Models

Jinghao Wang[1], Chenling Cui[1], Xuejun Wen[2(✉)], and Jie Shi[2]

[1] School of Computer Science and Engineering, Nanyang Technological University,
Singapore, Singapore
{C190209,CUIC0005}@e.ntu.edu.sg
[2] Singapore Digital Trust Lab, Singapore Research Centre, Huawei,
Singapore, Singapore
{wen.xuejun,shi.jie1}@huawei.com

Abstract. Patch-based adversarial attack shows the possibility to black-box physical attacks on state-of-the-art object detection models through hiding the occurrence of the objects, which causes a high risk in automated security system relying on such model. However, most prior works mainly focus on the attack performance but rarely pay attention to the training speed due to pixel updating and non-smoothing loss function in the training process. To overcome this limitation, we propose a simple but novel training pipeline called **TransPatch**, a transformer-based generator with new loss function, to accelerate the training process. To address the issue of unstable training problem of previous methods, we also compare and visualize the landscape of various loss functions. We conduct comprehensive experiments on two pedestrian and one stop sign datasets on three object detection models, i.e., YOLOv4, DETR and SSD to compare the training speed and patch performance in such adversarial attacks. From our experiments, our method outperforms previous methods within the first few epochs, and achieves absolute 20% ∼ 30% improvements in attack success rate (ASR) using 10% of the training time. We hope our approach can motivate future research on using generator in physical adversarial attack generation on other tasks and models.

Keywords: Adversarial attack · Object detection · Transformer

1 Introduction

Deep learning based object detection models [55] have been widely applied in automated surveillance cameras [19, 21, 43]. However, it has been proven that object detection models are very fragile to the adversarial attack. Digital

---

J. Wang and C. Cui—This work was done during an internship at HUAWEI.

L. Karlinsky et al. (Eds.): ECCV 2022 Workshops, LNCS 13801, pp. 317–331, 2023.
https://doi.org/10.1007/978-3-031-25056-9_21

adversarial attack such as perturbation [36,52], and digital footprint [30] could influence the object detection models in targeted or untargeted manners. Object detection models will mis-classify or neglect some objects on the image that are manipulated by the adversarial pixels. With **digital adversarial attack** being largely unrealistic in automated surveillance camera scenes as it requires the attacker to have direct access to the inner camera system [20,38], **physical adversarial attack** directly on human body is more practical. Such attack requires the generation of adversarial patches held by the target person or printed on the clothes to fool the detector. It is therefore a more challenging task as the patch needs to be robust in various physical circumstances such as different camera angles and distances, distortion of color due to scene light and multiple possible noises.

Most of the previous works have two shortfalls: slow training process and relatively low attack success rate (ASR). Slow training process is common as most of the prior works involve directly updating of the pixels of patch from backpropagation. If we consider the pixel updating as a *zero-layer neural network*, we can see that its optimization is not efficient and its expression power is limited.

In this paper, we propose a novel adversarial attack patch generation pipeline denoted as *TransPatch*. Inspired by the recent progress in natural image synthesis by GAN-based deep generative model, we replace the pixel updating methods with a transformer-based generator and consider our target object detection model as a non-training 'discriminator'. Our generator relies on the loss function given by the target model to improve its ability. The design of loss function is also vital in the generation of the adversarial patches. Instead of using the maximum box objectness or class probability as the loss function, we propose a smoother version of the function and our subsequent visualization shows that our loss function makes gradient decent process easier. Extensive experimental results have shown the advantages of our approach in terms of faster convergence speed and higher attack success rate. Quantitative analysis on different physical scenario also proves the robustness of our patch. Overall, our contributions are summarized as follows:

1. We use a transformer-based generator instead of directly updating the patch to accelerate the process of patch generation.
2. We improve the ASR of adversarial hiding attack on three datasets including person and stop sign class and three target object detection models.

2   Related Work

2.1   Adversarial Attack on Object Detection

Object Detection. Object detection model is an important component in the autonomous driving and security system. It locates the presence of objects with a bounding box and types or classes of the located objects in an image. Starting from 2-staged object detector models, object detection has gained much popularity in computer vision. 2-staged object detectors such as R-CNN [11] and Faster R-CNN [42] generate the region proposals in an image that could belong to a particular object. These particular regions are then classified into specific object classes based on various pre-trained domains. Such models have achieved promising detection precisions but have also shown that the detection speed is relatively slow due to its 2-stage nature. One-stage object detection models, on the other hand, predict the presence and the class scores of the detected objects directly with a single pass. Well-known architectures such as YOLO [1,40,41], SSD [29], and DETR [3] are capable of high precision detection with faster speed as compared to 2-staged object detectors.

2.2   Adversarial Attacks

Adversarial attack examples researches started from white-box based digital attack. Such attacks (i.e. FGSM [12], PGD [32]) assume the adversary has access to both the target model's gradient and pixel information of the input image. However, this assumption may not be applicable in real situations. In black-box based digital attack, attacker constantly inferences the model and uses the predicted results to update their pixel-level perturbations [8,9,36,48,50,52,57]. It has proven in [25,30] that, it is possible to attack the model without manipulating every pixel of the input image. Instead, a 40-by-40 pixel digital footprint is added at some specific location of the image to attack the model. Although such attack has less constraints, the only way to manipulate the pixels of the image captured by security system is through having direct access to the camera, which is largely difficult. Researchers have moved their interest to physical adversarial attacks which only allow direct physical changes to the environment or the target objects. By pasting adversarial patches onto physical stop signs, such attacks can achieve decently high ASR on stop sign detections [4,10,45]. Later works continued their interests to the more challenging person class hiding attacks [16,46,49,51,53]. Although person class has a higher in-class variation such as appearance, clothes and gesture, several successful researches [17,46,49,51,53] have proven the possibility to mislead the object detection by a patch pasted or printed on the t-shirt of the person.

2.3   Generative Models in Adversarial Attack

Generative Adversarial Network (GAN) is widely used in many generative tasks such as image generation [2,22,23,39], text generation [13], music generation [7] and image-to-image transfer [56]. Several works are inspired to use GAN in generating the adversarial patch. In [16], a pre-trained GAN is used and the loss given by the target model is only for updating and searching a specific latent vector input which makes the GAN generate a naturalistic adversarial patch to attack on person class. In [24,28], the GAN in the pipeline is trainable and designed to generate patch to attack road sign class on autonomous driving system. However, the discriminator proposed in these researches is used to make the patch-pasted stop sign as real as possible, which is not significantly applicable to person objects as a pasted patch naturally will not make the person 'unreal'. Therefore, this constraint is erased in our design of pipeline.

3   Methodology

We propose a novel Transformer-based generator pipeline, i.e., TransPatch, to accelerate the training and improve the ASR for the adversarial hiding attack on object detection model. In this section, we will first introduce the overall pipeline, and we will focus on two key components: the generator and the loss function.

3.1   Overall Pipeline

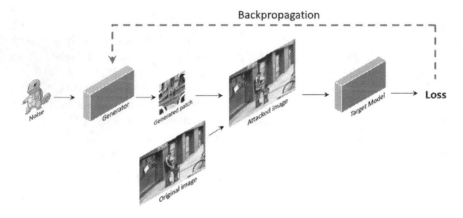

Fig. 1. Overview of the training pipeline.

Components. There are 5 components in the pipeline: input noise, generator, patch applier, target model and loss function. The generator takes a square noise $X \in [0,1]^{C \times L \times L}$ as input and output a same-size adversarial patch $P \in [0,1]^{C \times L \times L}$. We are not using any kinds of random noise such as Gaussian distribution or Bernoulli distribution, instead, we use images from pokemon dataset with random rotation as data augmentation as the input noise. The functionality of patch applier is to paste the generated patch correctly on the pedestrians with various sizes of bounding boxes from the original image. We resize the patch P to $L' = f \times \sqrt{wh}$,where f is a factor to control the size of the pasted patch and w and h represent the width and height respectively of the person bounding box. f is a hyper-parameter and we fix it as 0.3 in all our experiments. The attacked image will be fed into target model. The object detection results (here we only focus on the person class) from the target model is used as the loss function for updating our generator (Fig. 1).

3.2   Generator

Architecture. The generator takes input noise X and output patch P both with dimension $3 \times L \times L$. As shown in the left Stop Sign of Fig. 2, it contains a down-sampling block at the beginning which uses convolutional filters to transform the input noise to the feature map with $192 \times \frac{L}{4} \times \frac{L}{4}$. 5 similar transformer blocks are consequential to the down-sampling block which only change the number of channels innerly as $192 \to 384 \to 768 \to 768 \to 384 \to 192$ but do not change the size of the feature map. Finally, the feature map will be up-sampled by several transpose convolutional filters to the original size $3 \times L \times L$. Both up-sampling and down-sampling blocks are CNN-based and we apply spectral normalization [35] and batch normalization [18] for stable training.

Transformer Block. As shown in the right side of Fig. 2, the number of channels will be adjusted by the point-wise convolutional filters kernel size 1. Following the discussion in [31] that fewer activation functions and fewer normalization layers are better, we directly apply the 25-head attention to the feature map following a layer normalization. We apply layer scale technique [47] in the residual connection where the output of the activation function will be multiplied by a learnable diagonal matrix where its diagonal value λ will be initialized as 10^{-6}. It is proven in our experiments that such technique can improve the stability of training. Following our experiments, we choose Mish [34] as the activation function since it has the best performance among ReLU [37], GELU [14] and Leaky ReLU of slope factor 0.01.

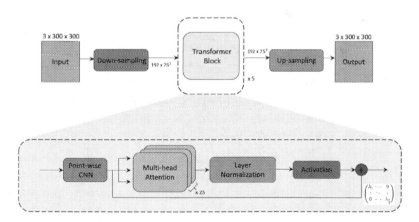

Fig. 2. Generator architecture overview. The top part shows that the generator is constructed by one down-sampling block, 7 transformer blocks and one up-sampling block. The bottom shows the residual connection and the forward flow inStop Sign the transformer block. All transformer blocks are identical to each other excluding the point-wise CNN block for channels adjustment.

3.3   Design of Loss Function

Optimization Objective. We define our generating patch as P, the original image as I and the process of applying patch to the correct location and size of pedestrians BB in the original image as $L(I, P)$. We further define our target model as T, the objectness and class probabilities of an arbitrary bounding box B given by model T as $T_{obj}^B(x), T_{cls}^B(x)$. We should notice that T_{cls}^B is a softmax output with dimension of number of classes (80 in our case).

For hiding person attack, we want our patch to minimize the objectness of bounding boxes whose person class has the highest class probability. Such conditional BB could be written as

$$\{B \mid \underset{person}{\mathbf{argmax}}\ T_{cls}^B\} \tag{1}$$

So the objectness of these conditional BB, which we define it as $\mathbb{O}(x)$, could be written as

$$\mathbb{O}(x) = T_{obj}^{\{B|\underset{person}{argmax}\ T_{cls}^B\}}(x) \tag{2}$$

Hence, our optimization objective is

$$\underset{P}{\mathbf{argmin}}\ \mathbb{O}(L(I, P)) \tag{3}$$

Different Loss Functions. Deriving from Eq. 3, we define our loss function L_{atk} as the mean of top-K of the objectness of these conditional BB, which is represented as $topK$:

$$L_{atk} = \frac{1}{K} \sum^K topK(\mathbb{O}(x)) \tag{4}$$

where the value of K is a hyper-parameter that is subjective to the size of batch, the average number of target object in one image of the dataset and the number of boxes in model's output. For DETR, we use the probability of its additional output class: no-object to calculate objectness. For SSD, we use the probability of background class to get the value of objectness.

Loss functions represented as *MaxObj* and *MaxCls* used by previous methods could be formulated respectively as:

$$L_{atk} = \max(T_{obj}^B(x)) \tag{5}$$

and

$$L_{atk} = \max(T_{cls=person}^B(x)) \tag{6}$$

We briefly summary the differences of the loss function in Table 1.

It is clearly to see that our L_{atk} only focus on a subset of BB since it is intuitively to understand that influencing BB containing other classes with high confidence like the previous loss functions is very hard and not directly responds to our optimization objective Eq. 3. And, we can see that ours is a smoother version to directly get the maximum of all BB. Quantitative analysis is conducted in the following ablation study.

Table 1. Loss functions designs and explanations. The same abbreviation will be used in the following sections

Abbr	Description
topK	Described as above. Formulated in Eq. 4.
MaxObj	The maximum of objectness of all bounding boxes. Formulated in Eq. 5.
MaxCls	The maximum of person class probability of all bounding boxes. Formulated in Eq. 6.

4   Experiments

In this section, we evaluate our proposed TransPatch on pedestrian hiding attack on INRIA Person, CityPersons and Stop Sign datasets. We also show the effect of the transformer blocks and our new design of loss function to outperform the previous methods.

4.1   Datasets

Different Features. The three pedestrians datasets we use have their own strong characteristic and reflect different scenarios of pedestrian detection, which could prove that our method has powerful adaptability. INRIA Person dataset [6] is captured by hand-holding camera with diversity in backgrounds such as square, mountain and grassland. CityPersons dataset [54] is a subset of Cityscapes [5] containing images of pedestrians recording by multiple driving recorders. Stop Sign dataset is generate from all images which is labeled as stop sign in COCO 2017 [27]. Figure 3 shows example from three datasets and Table 2 shows statics to numerically prove the above observation.

Table 2. Quantitative analysis of different dataset predicted by YOLOv4. Under the above nms and confidence setting, for each model (YOLOv4/DETR/SSD), there is a proportion of images with no target object could be detected and are considered as not valid image in our task. Such proportion could reflect the original performance of target models in our datasets

Name	# Total image	# Valid image	# Avg object	Avg size of BB (10^{-2})
INRIA person	902	902/898/823	4.24/6.14/1.92	6.43/4.80/10.40
CityPersons	5000	3363/2644/525	3.63/4.49/1.31	0.82/0.99/3.27
Stop sign	1803	1408/1221/1114	1.12/1.09/1.03	10.47/11.18/13.82

(a)   (b)

(c)

Fig. 3. Example from (a) INRIA person (b) CityPersons (c) Stop sign

Labeling. We randomly split the dataset into training and validation set in the ratio of 9 : 1 and all images are scaled to the same size before labeling. We use the pre-trained target model to extract bounding boxes of all pedestrians in our dataset. We filter through all the boxes of person class using Non-Maximum Suppression (nms) [15] and keep only the boxes with output confidence (objectness × class probability) bigger than 0.5 and IoU threshold of 0.4. The reason why we use such label as the ground true labels instead of using the original annotation provided by the dataset is that our task is not evaluating the performance of the object detection model.

4.2   Adversarial Hiding Attack

Evaluation Metric. In adversarial hiding attack, comparing with AP, ASR is a more suitable and more strict evaluation metric. We define the attack success rate (ASR) as the ratio of the number of hiding persons to the total number of persons in the images. We prove that, under the same circumstance, achieving high ASR is at least as hard as achieving low AP when we measure the adversarial attack result. The Convergence Time (CT) measures the training time cost. We multiply the number of epochs by the time taken of each epoch and use *hour* as the unit of results. We should notice that some methods will converge easily within a shorter time but with a relatively low ASR.

Target Models. Three object detection models are chosen as target models. Their characteristics are summarized in Table 3.

Comparison with Baseline. We compare TransPatch in both ASR and CT of all three datasets under all three target models with the baseline and the results are

Table 3. Comparison between different target models. For fair comparison, the below statistics are based on the models used in our experiments. Release year is based on the year published on *arxir.org*.

Model	Release year	# Parameter
YOLOv4	2020	64M
DETR	2020	41M
SSD	2015	22M

shown in Table 4. The baseline model is the combination of pixel updating and *MaxObj* loss function. It is shown that our method will take a longer time in each epoch due to its complexity in architecture; however, our methods uses significantly fewer number of epochs to converge, which makes our CT shorter than others. Also, our newly design of loss function also helps the training overcome the original local minima.

Table 4. Adversarial attack results and training time consumption of INRIA Person, CityPersons and Stop Sign dataset on YOLOv4, DETR and SSD. Due to the difference of target models and the lack of report of ASR, we train the baseline pixel-updating method [46] with tuning hyper-parameter settings adopting from their paper and official code. The random Gaussian patch and white patch are considered as trivial attack.

Target model	Method	INRIA		CityPersons		Stop sign	
		ASR↑	CT(h)↓	ASR↓	CT(h)↓	ASR↑	CT(h)↓
YOLOv4	Random Gaussian	5.4%	—	22.1%	—	9.7%	—
	White	5.8%	—	23.1%	—	12.0%	—
	Baseline [46]	43.8%	6.4	24.4%	37.0	25.8%	10.8
	TransPatch (ours)	**86.5%**	**0.7**	**68.2%**	**2.6**	**88.2%**	**0.1**
DETR	Random Gaussian	24.7%	—	43.4%	—	33.3%	—
	White	35.9%	—	39.6%	—	31.5%	—
	Baseline [46]	37.9%	6.4	62.5%	16.5	84.6%	8.7
	TransPatch (ours)	**64.5%**	**0.8**	**92.7%**	**0.2**	**94.3%**	**0.1**
SSD	Random Gaussian	10.7%	—	41.2%	—	10.4%	—
	White	20.5%	—	60.5%	—	16.1%	—
	Baseline [46]	64.3%	3.2	20.3%	1.1	27.3%	1.9
	TransPatch (ours)	**80.0%**	**0.2**	**84.4%**	**0.1**	**59.9%**	**0.5**

4.3 Ablation Study

Effect of Transformer Blocks. In Table 5, we prove that the multi-head attention mechanism and the number of transformer blocks both have impact on the performance of adversarial attack by comparing with a CNN-based generator and TransPatch with less number of transformer blocks. We fix *topK* as our loss function in this section.

Effect of Design of Loss Functions. In Table 6, we prove that the design of loss functions has impact on the performance of adversarial attack. In some extreme cases with ill designed loss function, the training even does not converge. The difference of the computational cost is negligible between different loss functions.

Visualization of the Smoothness of Different Loss Functions. To further illustrate the effect, we visualize the landscape of loss function to show that our design of loss function has a smoother gradient. Such method [26] first uniformly samples multiple steps α and β given a closed range and gets two Gaussian random noise δ and η. Representing the weights of the network at epoch k as θ_k, the new value of the loss function $L_{\text{new}} = L(\theta_k + \alpha\delta + \beta\eta)$ will be given by the output of adjusted network (or adjusted patch in baseline model). The points of 3D loss landscape are formed by $(\alpha, \beta, L_{\text{new}})$. Due to the time constraint, we use 200 sample in the training set and 50 steps from $[-1, 1]$ (Fig. 4).

Table 5. Adversarial attack results and number of parameters of different architectures of INRIA Person, CityPersons and Stop Sign on YOLOv4. For *Void*, we only keep the up-sampling and down-sampling block, which could be considered as 0 transformer block. For *CNN*, in each residual-connection block, we connect 3 groups of convolutional filter with kernel size 3 and padding 1, batch normalization and activation function sequentially

Architecture	# Parameters	ASR↑		
		INRIA	CityPersons	Stop sign
Void	2M	28.9%	44.2%	80.6%
9 residual-connection CNN blocks	8M	43.7%	44.8%	**88.9%**
3 transformer blocks	382M	83.8%	65.4%	87.7%
5 transformer blocks (ours)	636M	**86.5%**	**68.2%**	88.2%

Table 6. ASR of of different loss functions on INRIA Person, CityPersons and Stop Sign datasets. Our TransPatch is referred as *transfomer + topK* and the baseline method is referred as *pixel-update + MaxObj*.

Target model	Generation	Loss function	ASR↑		
			INRIA	CityPersons	Stop sign
YOLOv4	Pixel-update	MaxObj	43.8%	24.4%	**25.8%**
	Pixel-update	MaxCls	10.3%	25.6%	15.0%
	Pixel-update	topK	**57.1%**	**65.7%**	6.8%
	Transformer	MaxObj	49.6 %	45.4%	44.4%
	Transformer	MaxCls	14.4 %	48.9%	49.4%
	Transformer	**topK**	**86.5%**	**68.2%**	**88.2%**
DETR	Pixel-update	MaxObj	37.9%	62.5%	84.6%
	Pixel-update	MaxCls	35.1%	76.1%	78.6%
	Pixel-update	topK	**41.8%**	**81.1%**	**85.3%**
	Transformer	MaxObj	49.6 %	79.9%	70.6%
	Transformer	MaxCls	44.7 %	**93.2%**	**94.4%**
	Transformer	**topK**	**64.5%**	92.7%	94.3%
SSD	Pixel-update	MaxObj	64.3%	20.3%	27.3%
	Pixel-update	MaxCls	**87.8%**	18.8%	42.0%
	Pixel-update	topK	57.3%	**26.6%**	**61.0%**
	Transformer	MaxObj	76.1 %	89.6%	**60.0%**
	Transformer	MaxCls	72.0 %	79.7%	57.2%
	Transformer	**topK**	**80.0%**	**84.4%**	50.0%

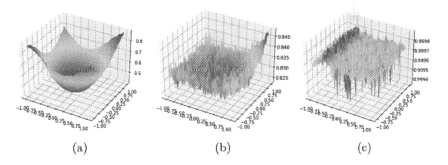

(a)   (b)   (c)

Fig. 4. Loss function landscape of TransPatch at epoch 50 in training INRIA attacking YOLOv4 with (a) *topK* (b) *MaxObj* (c) *MaxCls*. Observing y-axis of the plots, *MaxObj* and *MaxCls* fluctuates intensively and the decrease in loss is insignificant. *topK*, on the other hand, ensures a smooth loss landscape with an obvious minima.

5   Physical Attack

In this section, we further quantitatively investigate our pipeline in the physical attack from two perspectives: additional constrains in the loss function to make our patch after printing become more robust and cycle consistency penalty to make our patch more naturalistic.

5.1   Addition Constrains

In order to perform such attack in a physical manner, we also need to consider the environmental influence to the patch. Two more objectives are introduced: the finite-difference approximation of total variants (**tv**) [33] which constrains the smoothness of the patch and non-printability score (nps) [44] which constrains the printability of the patch in our training.

$$L_{\mathbf{tv}} = \sum_{i,j} \sqrt{(p_{i,j} - p_{i+1,j})^2 + (p_{i,j} - p_{i,j+1})^2} \quad \text{and} \quad L_{\mathbf{nps}} = \sum_{p \in P} \min_{c \in C} \|\hat{p} - p\|$$

During training, we utilize all datasets we have to train the model by sub-sampling 1000 of them in each epoch. Figure 5 shows the qualitative illustration.

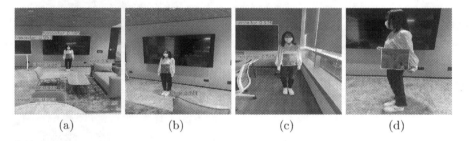

(a)	(b)	(c)	(d)

Fig. 5. Qualitative results of physical attack. The patch from our approach has the capability to attack the model on multiple circumstances such as different (a) distance, (b) view, (c) lighting condition and (d) posture.

6   Conclusion

In this paper, we propose a simple yet efficient pipeline, TransPatch, for physical adversarial attack on state-of-art object detection model. We have shown that the combination of the transformer-based generator and our loss function could significantly accelerate the training and improve the ASR as compared to previous methods. We experimentally validate that our TransPatch remains as an effective method of generating adversarial patches in adversarial attacks on both human and stop signs in all three selected target object detection models.

This particular method comes with challenges as well regarding the need for computational power. We wish that future works can work on such improvements. We also believe that our result can be applied to attacks in automated surveillance camera scenarios and hope that our research could inspire more defense against such attacks in these circumstances.

References

1. Bochkovskiy, A., Wang, C.Y., Liao, H.Y.M.: Yolov4: optimal speed and accuracy of object detection. arXiv preprint arXiv:2004.10934 (2020)
2. Brock, A., Donahue, J., Simonyan, K.: Large scale GAN training for high fidelity natural image synthesis. In: ICLR. OpenReview.net (2019)
3. Carion, N., Massa, F., Synnaeve, G., Usunier, N., Kirillov, A., Zagoruyko, S.: End-to-end object detection with transformers. In: Vedaldi, A., Bischof, H., Brox, T., Frahm, J.-M. (eds.) ECCV 2020. LNCS, vol. 12346, pp. 213–229. Springer, Cham (2020). https://doi.org/10.1007/978-3-030-58452-8_13
4. Chen, S.-T., Cornelius, C., Martin, J., Chau, D.H.P.: ShapeShifter: robust physical adversarial attack on faster R-CNN object detector. In: Berlingerio, M., Bonchi, F., Gärtner, T., Hurley, N., Ifrim, G. (eds.) ECML PKDD 2018. LNCS (LNAI), vol. 11051, pp. 52–68. Springer, Cham (2019). https://doi.org/10.1007/978-3-030-10925-7_4
5. Cordts, M., et al.: The cityscapes dataset for semantic urban scene understanding. In: CVPR, pp. 3213–3223. IEEE Computer Society (2016)
6. Dalal, N., Triggs, B.: Histograms of oriented gradients for human detection. In: CVPR (1), pp. 886–893. IEEE Computer Society (2005)
7. Dong, H., Hsiao, W., Yang, L., Yang, Y.: Musegan: multi-track sequential generative adversarial networks for symbolic music generation and accompaniment. In: AAAI, pp. 34–41. AAAI Press (2018)
8. Dong, Y., et al.: Boosting adversarial attacks with momentum. In: CVPR, pp. 9185–9193. Computer Vision Foundation/IEEE Computer Society (2018)
9. Dong, Y., Pang, T., Su, H., Zhu, J.: Evading defenses to transferable adversarial examples by translation-invariant attacks. In: CVPR, pp. 4312–4321. Computer Vision Foundation/IEEE (2019)
10. Evtimov, I., et al.: Robust physical-world attacks on machine learning models. CoRR abs/1707.08945 (2017)
11. Girshick, R.B., Donahue, J., Darrell, T., Malik, J.: Rich feature hierarchies for accurate object detection and semantic segmentation. In: CVPR, pp. 580–587. IEEE Computer Society (2014)
12. Goodfellow, I.J., Shlens, J., Szegedy, C.: Explaining and harnessing adversarial examples. In: ICLR (Poster) (2015)
13. Guo, J., Lu, S., Cai, H., Zhang, W., Yu, Y., Wang, J.: Long text generation via adversarial training with leaked information. In: AAAI, pp. 5141–5148. AAAI Press (2018)
14. Hendrycks, D., Gimpel, K.: Gaussian error linear units (gelus). arXiv preprint arXiv:1606.08415 (2016)
15. Hosang, J.H., Benenson, R., Schiele, B.: Learning non-maximum suppression. In: CVPR, pp. 6469–6477. IEEE Computer Society (2017)
16. Hu, Y., Chen, J., Kung, B., Hua, K., Tan, D.S.: Naturalistic physical adversarial patch for object detectors. In: ICCV, pp. 7828–7837. IEEE (2021)
17. Huang, L., et al.: Universal physical camouflage attacks on object detectors. In: CVPR, pp. 717–726. Computer Vision Foundation/IEEE (2020)
18. Ioffe, S., Szegedy, C.: Batch normalization: accelerating deep network training by reducing internal covariate shift. In: ICML, JMLR Workshop and Conference Proceedings, vol. 37, pp. 448–456. JMLR.org (2015)
19. Javed, O., Shah, M.: Tracking and object classification for automated surveillance. In: Heyden, A., Sparr, G., Nielsen, M., Johansen, P. (eds.) ECCV 2002. LNCS, vol. 2353, pp. 343–357. Springer, Heidelberg (2002). https://doi.org/10.1007/3-540-47979-1_23

20. Jeong, J., Kwon, S., Hong, M., Kwak, J., Shon, T.: Adversarial attack-based security vulnerability verification using deep learning library for multimedia video surveillance. Multim. Tools Appl. **79**(23–24), 16077–16091 (2020)
21. Madasamy, K., Shanmuganathan, V., Kandasamy, V., Lee, M.Y., Thangadurai, M.: OSDDY: embedded system-based object surveillance detection system with small drone using deep YOLO. EURASIP J. Image Video Process. **2021**(1), 1–14 (2021). https://doi.org/10.1186/s13640-021-00559-1
22. Karras, T., Aila, T., Laine, S., Lehtinen, J.: Progressive growing of gans for improved quality, stability, and variation. In: ICLR. OpenReview.net (2018)
23. Karras, T., Laine, S., Aila, T.: A style-based generator architecture for generative adversarial networks. In: CVPR, pp. 4401–4410. Computer Vision Foundation/IEEE (2019)
24. Kong, Z., Guo, J., Li, A., Liu, C.: Physgan: generating physical-world-resilient adversarial examples for autonomous driving. In: CVPR, pp. 14242–14251. Computer Vision Foundation/IEEE (2020)
25. Lee, M., Kolter, J.Z.: On physical adversarial patches for object detection. CoRR abs/1906.11897 (2019)
26. Li, H., Xu, Z., Taylor, G., Studer, C., Goldstein, T.: Visualizing the loss landscape of neural nets. In: NeurIPS, pp. 6391–6401 (2018)
27. Lin, T.-Y., et al.: Microsoft COCO: common objects in context. In: Fleet, D., Pajdla, T., Schiele, B., Tuytelaars, T. (eds.) ECCV 2014. LNCS, vol. 8693, pp. 740–755. Springer, Cham (2014). https://doi.org/10.1007/978-3-319-10602-1_48
28. Liu, A., et al.: Perceptual-sensitive GAN for generating adversarial patches. In: AAAI, pp. 1028–1035. AAAI Press (2019)
29. Liu, W., et al.: SSD: single shot multibox detector. In: Leibe, B., Matas, J., Sebe, N., Welling, M. (eds.) ECCV 2016. LNCS, vol. 9905, pp. 21–37. Springer, Cham (2016). https://doi.org/10.1007/978-3-319-46448-0_2
30. Liu, X., Yang, H., Liu, Z., Song, L., Chen, Y., Li, H.: DPATCH: an adversarial patch attack on object detectors. In: SafeAI@AAAI. CEUR Workshop Proceedings, vol. 2301. CEUR-WS.org (2019)
31. Liu, Z., Mao, H., Wu, C., Feichtenhofer, C., Darrell, T., Xie, S.: A convnet for the 2020s. CoRR abs/2201.03545 (2022)
32. Madry, A., Makelov, A., Schmidt, L., Tsipras, D., Vladu, A.: Towards deep learning models resistant to adversarial attacks. In: ICLR (Poster). OpenReview.net (2018)
33. Mahendran, A., Vedaldi, A.: Understanding deep image representations by inverting them. In: CVPR, pp. 5188–5196. IEEE Computer Society (2015)
34. Misra, D.: Mish: a self regularized non-monotonic activation function. In: BMVC. BMVA Press (2020)
35. Miyato, T., Kataoka, T., Koyama, M., Yoshida, Y.: Spectral normalization for generative adversarial networks. In: ICLR. OpenReview.net (2018)
36. Moosavi-Dezfooli, S., Fawzi, A., Fawzi, O., Frossard, P.: Universal adversarial perturbations. In: CVPR, pp. 86–94. IEEE Computer Society (2017)
37. Nair, V., Hinton, G.E.: Rectified linear units improve restricted boltzmann machines. In: ICML, pp. 807–814. Omnipress (2010)
38. Payne, B.R.: Car hacking: accessing and exploiting the can bus protocol. J. Cybersecur. Educ. Res. Pract. **2019**(1), 5 (2019)
39. Radford, A., Metz, L., Chintala, S.: Unsupervised representation learning with deep convolutional generative adversarial networks. In: ICLR (Poster) (2016)
40. Redmon, J., Divvala, S., Girshick, R., Farhadi, A.: You only look once: unified, real-time object detection. In: Proceedings of the IEEE Conference on Computer Vision and Pattern Recognition, pp. 779–788 (2016)

41. Redmon, J., Farhadi, A.: Yolo9000: better, faster, stronger. In: Proceedings of the IEEE Conference on Computer Vision and Pattern Recognition, pp. 7263–7271 (2017)

42. Ren, S., He, K., Girshick, R., Sun, J.: Faster r-cnn: towards real-time object detection with region proposal networks. Adv. Neural Inf. Process. Syst. **28**, 1–9 (2015)

43. Baiju, P.S., George, S.N.: An automated unified framework for video deraining and simultaneous moving object detection in surveillance environments. IEEE Access **8**, 128961–128972 (2020)

44. Sharif, M., Bhagavatula, S., Bauer, L., Reiter, M.K.: Accessorize to a crime: real and stealthy attacks on state-of-the-art face recognition. In: CCS, pp. 1528–1540. ACM (2016)

45. Song, D., et al.: Physical adversarial examples for object detectors. In: WOOT @ USENIX Security Symposium. USENIX Association (2018)

46. Thys, S., Ranst, W.V., Goedemé, T.: Fooling automated surveillance cameras: adversarial patches to attack person detection. In: CVPR Workshops, pp. 49–55. Computer Vision Foundation/IEEE (2019)

47. Touvron, H., Cord, M., Sablayrolles, A., Synnaeve, G., Jégou, H.: Going deeper with image transformers. In: ICCV, pp. 32–42. IEEE (2021)

48. Tramèr, F., Kurakin, A., Papernot, N., Goodfellow, I.J., Boneh, D., McDaniel, P.D.: Ensemble adversarial training: attacks and defenses. In: ICLR (Poster). OpenReview.net (2018)

49. Wang, Y., et al.: Towards a physical-world adversarial patch for blinding object detection models. Inf. Sci. **556**, 459–471 (2021)

50. Wu, D., Wang, Y., Xia, S., Bailey, J., Ma, X.: Skip connections matter: on the transferability of adversarial examples generated with resnets. In: ICLR. OpenReview.net (2020)

51. Wu, Z., Lim, S.-N., Davis, L.S., Goldstein, T.: Making an invisibility cloak: real world adversarial attacks on object detectors. In: Vedaldi, A., Bischof, H., Brox, T., Frahm, J.-M. (eds.) ECCV 2020. LNCS, vol. 12349, pp. 1–17. Springer, Cham (2020). https://doi.org/10.1007/978-3-030-58548-8_1

52. Xie, C., Wang, J., Zhang, Z., Zhou, Y., Xie, L., Yuille, A.L.: Adversarial examples for semantic segmentation and object detection. In: ICCV, pp. 1378–1387. IEEE Computer Society (2017)

53. Xu, K., et al.: Adversarial t-shirt! evading person detectors in a physical world. In: Vedaldi, A., Bischof, H., Brox, T., Frahm, J.-M. (eds.) ECCV 2020. LNCS, vol. 12350, pp. 665–681. Springer, Cham (2020). https://doi.org/10.1007/978-3-030-58558-7_39

54. Zhang, S., Benenson, R., Schiele, B.: Citypersons: a diverse dataset for pedestrian detection. In: CVPR, pp. 4457–4465. IEEE Computer Society (2017)

55. Zhao, Z.Q., Zheng, P., Xu, S.t., Wu, X.: Object detection with deep learning: a review. IEEE Trans. Neural Netw. Learn. Syst. **30**(11), 3212–3232 (2019)

56. Zhu, J., Park, T., Isola, P., Efros, A.A.: Unpaired image-to-image translation using cycle-consistent adversarial networks. In: ICCV, pp. 2242–2251. IEEE Computer Society (2017)

57. Zou, J., Pan, Z., Qiu, J., Liu, X., Rui, T., Li, W.: Improving the transferability of adversarial examples with resized-diverse-inputs, diversity-ensemble and region fitting. In: Vedaldi, A., Bischof, H., Brox, T., Frahm, J.-M. (eds.) ECCV 2020. LNCS, vol. 12367, pp. 563–579. Springer, Cham (2020). https://doi.org/10.1007/978-3-030-58542-6_34

Feature-Level Augmentation to Improve Robustness of Deep Neural Networks to Affine Transformations

Adrian Sandru[1,2], Mariana-Iuliana Georgescu[1,2],
and Radu Tudor Ionescu[1,2,3(✉)]

[1] Department of Computer Science, University of Bucharest, 14 Academiei,
Bucharest, Romania
`raducu.ionescu@gmail.com`
[2] SecurifAI, 21D Mircea Voda, Bucharest, Romania
[3] Romanian Young Academy, University of Bucharest, 90 Panduri Street,
Bucharest, Romania

Abstract. Recent studies revealed that convolutional neural networks do not generalize well to small image transformations, e.g. rotations by a few degrees or translations of a few pixels. To improve the robustness to such transformations, we propose to introduce data augmentation at intermediate layers of the neural architecture, in addition to the common data augmentation applied on the input images. By introducing small perturbations to activation maps (features) at various levels, we develop the capacity of the neural network to cope with such transformations. We conduct experiments on three image classification benchmarks (Tiny ImageNet, Caltech-256 and Food-101), considering two different convolutional architectures (ResNet-18 and DenseNet-121). When compared with two state-of-the-art stabilization methods, the empirical results show that our approach consistently attains the best trade-off between accuracy and mean flip rate.

Keywords: Deep learning · Data augmentation · Convolutional neural networks · Robustness to affine transformations

1 Introduction

A series of recent studies [1,3,5,14,15,18,21,22] showed that convolutional neural networks (CNNs) are not properly equipped to deal with small image perturbations. Indeed, it appears that a subtle affine transformation, e.g. a rotation by a few degrees or a translation of a few pixels, can alter the model's decision towards making a wrong prediction. The problem is illustrated by the example shown in Fig. 1, where a deep neural model is no longer able to predict the correct class upon downscaling the input image by a factor of 0.9. To increase the robustness to such small perturbations, researchers [1,3,21,22] proposed various approaches ranging from architectural changes [3,21] and training strategy updates [22] to

© The Author(s), under exclusive license to Springer Nature Switzerland AG 2023
L. Karlinsky et al. (Eds.): ECCV 2022 Workshops, LNCS 13801, pp. 332–341, 2023.
https://doi.org/10.1007/978-3-031-25056-9_22

original image   image scaled by 0.9

hummingbird (0.78)   grasshopper (0.22)

Fig. 1. An image of a hummingbird from Caltech-256 that is wrongly classified as grasshopper (with a probability of 0.22) by a ResNet-18 model, after downscaling it by a factor of 0.9. Best viewed in color.

input data augmentations [13,19]. However, to the best of our knowledge, none of the previous works tried to apply augmentations at the intermediate layers of the neural network. We conjecture that introducing feature-level augmentations improves the robustness of deep CNNs to affine transformations. To this end, we present an augmentation technique that randomly selects a convolutional layer at each mini-batch and applies independent affine transformations (translation, rotation, scaling) on each activation map from the selected layer.

To demonstrate the practical utility of our approach, we conduct experiments with ResNet-18 [7] and DenseNet-121 [9] on three benchmark data sets, namely Tiny ImageNet [17], Caltech-256 [6] and Food-101 [2]. Importantly, we show that feature-level augmentation helps even when the models are trained with standard data augmentation. When compared with two state-of-the-art methods [3,21], the empirical results show that our approach attains the best trade-off between accuracy and mean flip rate. To the best of our knowledge, this is the first time such a comparison is made.

Contribution. In summary, our contribution is twofold:

- We introduce a novel method based on feature-level augmentation to increase the robustness of deep neural networks to affine transformations.
- We conduct an empirical evaluation study to compare state-of-the-art methods addressing the robustness problem among themselves as well as with our approach.

2   Related Work

In literature, there are several approaches towards improving the robustness of deep neural networks to image perturbations. A popular and natural strategy, that proved to work sufficiently well, is to train the network using augmented images, as suggested in [1,13,19]. The intuition behind this approach is to train the model on a wider domain, which can become more similar to the test data.

The experiments conducted by the authors suggest that the robustness and performance improve on all scenarios. Due to its popularity, we apply image augmentation to the baseline models employed in our experiments. An extended solution of using augmented images is mentioned in [22], where the authors employed an additional loss function in order to stabilize the output features of the network in such a manner that a strongly perturbed image should have a similar outcome to the original one. Data augmentation methods are considered to be constrained by the photographers' bias [1]. Thus, the model may only learn to generalize to images from a certain (biased) distribution.

One of the recent architectural changes leading to improvements in the stability of CNNs was introduced by Zhang [21]. The observations of the author centered on the fact that modern CNNs are not shift invariant. Thus, small linear changes applied on the input image may have a negative impact on the final result. The source of this issue is considered to be represented by the downsampling process that usually occurs inside neural networks through pooling operations, which breaks the shift-equivariance. A straightforward solution to solve this issue is to avoid subsampling [1], but this comes with a great computational burden. Therefore, Zhang [21] provides adjustments to conventional operations by including a blur kernel in order to reduce the judder caused by downsampling. We conduct experiments showing that this method can deteriorate the quality of the features and negatively impact the final accuracy of the model, despite improving its stability.

Another new architectural design was proposed in [3], where the authors addressed the downsampling issue by proposing a pooling operation that considers all possible grids and selects the component with the highest norm. Their approach was benchmarked against circular shifts of the input, which do not naturally occur in practical scenarios. Hence, we extend their evaluation to generic (non-circular) affine transformations and compare their approach to our own procedure, showcasing the superiority of our method.

Although researchers explored multiple methods to improve the stability of neural models to image perturbations, it seems that there is no technique that can guarantee a never-failing solution. Hence, we consider that addressing the stability problem with better solutions is of great interest to the computer vision community. To the best of our knowledge, we are the first to propose feature-level augmentation as an enhancement to the stability of neural networks.

3   Method

The proposed method consists of extending the conventional input augmentation procedure by applying it at the feature level. In all our experiments, we include the conventional input augmentation, which is based on randomly shifting the images on both axes with values between -15 and $+15$ pixels, rotating them by -15 to $+15°$, or rescaling them with a factor between 0.4 and 1.15, following [8]. On top of this, we introduce feature-level augmentation (FLA).

We underline that CNNs are usually composed of multiple blocks intercalated with downscaling (pooling) operations. Starting from this observation, we

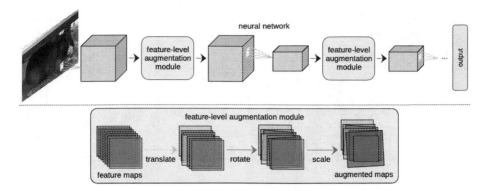

Fig. 2. Our feature-level augmentation module is inserted at different levels in the neural architecture. The proposed feature-level augmentation module individually applies the translation, rotation and scaling operations on each activation (feature) map, with a given probability. Best viewed in color.

capture the features between two randomly selected consecutive blocks, augment the activation maps provided by the first selected block, and train the network from the second block to the output using an augmented version of the activation maps, as shown in Fig. 2.

The augmentation that we propose to employ on the features is based on shifting, rotating and rescaling the activation maps. Due to the fact that one element of a feature map, somewhere deep in the network, is actually the result of processing multiple pixels from the corresponding location in the input, our augmentation procedure should take into account the depth where it is applied, so that the receptive field is not severely affected. Thus, the maximum translation value is scaled accordingly, starting from 15 pixels near the input, gradually going down to 1 pixel for the high-level layers. For rotations, we consider a random value between −15 and +15 degrees, while for the scaling operation, we employ a random resize factor between 0.85 and 1.15. In order to avoid noisy features at the beginning, we activate our augmentation procedure only after two training epochs. Afterwards, we apply feature-level augmentation with a probability of 0.5 on each mini-batch. To increase variety, a feature map has an equal probability of being transformed with any FLA operation. Hence, combining FLA operations is possible.

4   Experiments

4.1   Data Sets

Tiny ImageNet. The Tiny ImageNet data set is a subset of ImageNet [17] containing 120,000 natural images belonging to 200 classes. The resolution of each image is 64×64 pixels. The training set contains 100,000 images, while the validation and test sets contain 10,000 samples each.

Caltech-256. The Caltech-256 [6] data set contains 30,607 images of 256 categories. For each object category, we divide the images into 40 for training, 20 for validation, leaving the remainder (at least 20 samples from each category) for the testing stage.

Food-101. The Food-101 [2] data set is formed of 101,000 images of 101 food types. The original split contains 750 training images and 250 test images for each category. We keep 250 training samples per category to validate the models, leaving us with 500 samples per category for training.

4.2   Evaluation Setup

Evaluation Measures. For the evaluation, we first employ the accuracy between the ground-truth labels and the predictions of the neural models. Following [21], we also use the mean flip rate (mFR) to measure the stability of the models to affine transformations (scaling, rotation and translation). The mFR is measured by how often the predicted label changes, on average, in images with consecutive perturbations. We run each neural model for 5 times, reporting the average scores and the corresponding standard deviations.

Following [8], we create the perturbed version of each test set. We perturb the original test data with rotation, scaling and translation operations. We perform each operation individually on each sample. We rotate the images with angles starting from -15 to $+15$ degrees, using a step of 1 degree. We translate the images by up to 20 pixels in each direction, using a step of 1 pixel. We scale the samples by a scaling factor between 0.4 and 1.15, using a step of 0.025.

In order to quantify the trade-off between the accuracy level of a model and its stability to affine transformations, we define the *trade-off* T as:

$$T = \text{accuracy} - \text{average}_{op}(\text{mFR}_{op}), \tag{1}$$

where $op \in [\text{rotate}, \text{scale}, \text{translate}]$. A higher value for T represents a better trade-off.

Baselines. As neural models, we choose two very widely used architectures, namely ResNet-18 [7] and DenseNet-121 [9]. We train the models using common augmentations applied on the input images, such as random shift, scaling and rotation. This represents our first baseline. In addition, we consider two state-of-the-art methods as baselines for improving the stability of these models, namely anti-aliasing (BlurPool) [21] and adaptive polyphase sampling (APS) [3]. For the BlurPool [21] method, we used the Triangle-3 filter, which obtains the best trade-off between accuracy and stability to affine transformations. We apply the same input augmentations for all baseline models, as well as for our own models based on feature-level augmentation (FLA).

Table 1. Accuracy scores and mFR values (in %) for translation, rotation and scaling operations on the Tiny ImageNet [17], Caltech-256 [6] and Food-101 [2] data sets. Reported results represent the average and the standard deviation over 5 runs. ↑ indicates higher values are better. ↓ indicates lower values are better. Best scores on each data set are highlighted in bold.

Data set	Model	Method	Accuracy ↑	mFR ↓			Trade-Off ↑
				Translate	Rotate	Scale	
Tiny ImageNet	ResNet 18	Baseline	71.50 ± 0.20	12.94 ± 0.14	15.71 ± 0.40	22.29 ± 0.21	54.52
		BP-3 [21]	71.16 ± 0.26	$\mathbf{10.61 \pm 0.33}$	14.98 ± 0.20	21.28 ± 0.25	55.53
		APS [3]	70.30 ± 0.22	15.32 ± 0.38	22.91 ± 0.23	27.86 ± 0.35	48.27
		FLA (ours)	$\mathbf{71.76 \pm 0.16}$	12.00 ± 0.14	$\mathbf{14.42 \pm 0.15}$	$\mathbf{21.11 \pm 0.15}$	$\mathbf{55.91}$
	DenseNet 121	Baseline	76.50 ± 0.37	9.11 ± 0.08	14.19 ± 0.33	18.62 ± 0.19	62.52
		BP-3 [21]	76.57 ± 0.24	$\mathbf{7.70 \pm 0.28}$	13.07 ± 0.39	17.68 ± 0.30	63.75
		APS [3]	76.00 ± 0.21	10.03 ± 0.09	17.77 ± 1.01	21.29 ± 0.26	59.63
		FLA (ours)	$\mathbf{76.60 \pm 0.44}$	8.54 ± 0.18	$\mathbf{12.59 \pm 0.20}$	$\mathbf{17.27 \pm 0.25}$	$\mathbf{63.80}$
Caltech-256	ResNet 18	Baseline	78.96 ± 0.30	5.82 ± 0.15	6.16 ± 0.13	10.07 ± 0.09	71.61
		BP-3 [21]	76.92 ± 0.26	$\mathbf{4.61 \pm 0.14}$	$\mathbf{4.83 \pm 0.16}$	$\mathbf{8.40 \pm 0.19}$	70.97
		APS [3]	78.12 ± 0.40	8.79 ± 0.15	9.86 ± 0.11	13.87 ± 0.10	67.28
		FLA (ours)	78.91 ± 0.06	4.99 ± 0.12	5.45 ± 0.07	9.24 ± 0.05	$\mathbf{72.35}$
	DenseNet 121	Baseline	$\mathbf{83.98 \pm 0.14}$	4.00 ± 0.13	4.28 ± 0.04	7.11 ± 0.03	78.85
		BP-3 [21]	82.91 ± 0.32	$\mathbf{3.26 \pm 0.32}$	$\mathbf{3.52 \pm 0.07}$	$\mathbf{6.17 \pm 0.05}$	78.59
		APS [3]	83.53 ± 0.11	4.57 ± 0.08	5.57 ± 0.05	8.63 ± 0.09	77.27
		FLA (ours)	83.54 ± 0.27	$3.32 + 0.11$	$3.67 + 0.11$	6.31 ± 0.13	$\mathbf{79.10}$
Food-101	ResNet 18	Baseline	$\mathbf{76.29 \pm 0.41}$	6.41 ± 0.03	7.36 ± 0.06	12.39 ± 0.05	67.57
		BP-3 [21]	74.38 ± 0.24	$\mathbf{4.77 \pm 0.11}$	$\mathbf{5.71 \pm 0.09}$	$\mathbf{9.90 \pm 0.09}$	67.58
		APS [3]	76.03 ± 0.20	8.77 ± 0.21	11.41 ± 0.21	16.29 ± 0.08	63.87
		FLA (ours)	76.28 ± 0.33	5.69 ± 0.02	6.63 ± 0.02	11.41 ± 0.04	$\mathbf{68.37}$
	DenseNet 121	Baseline	$\mathbf{83.26 \pm 0.20}$	4.08 ± 0.07	4.85 ± 0.08	8.31 ± 0.08	77.51
		BP-3 [21]	81.94 ± 0.26	$\mathbf{3.26 \pm 0.05}$	$\mathbf{4.00 \pm 0.03}$	$\mathbf{7.05 \pm 0.09}$	77.17
		APS [3]	82.77 ± 0.16	4.60 ± 0.04	6.10 ± 0.08	10.20 ± 0.06	75.71
		FLA (ours)	82.86 ± 0.20	3.46 ± 0.07	4.21 ± 0.10	7.58 ± 0.09	$\mathbf{77.78}$

Hyperparameter Tuning. We train each model for a maximum of 100 epochs, halting the training when the value of the validation loss does not decrease for 10 consecutive epochs. We set the batch size to 16 samples and the learning rate to $5 \cdot 10^{-4}$. We optimize the models using Adam [11], keeping the default values for the parameters of Adam.

4.3   Results

We present the results on the Tiny ImageNet [17], Caltech-256 [6] and Food-101 [2] data sets in Table 1. As mentioned earlier, we compare our approach with a baseline based on standard data augmentation as well as two state-of-the-art methods, namely APS [3] and BlurPool [21].

First, we observe that the APS [3] method does not surpass the accuracy level obtained by the baseline, regardless of the architecture or the data set. It also does not increase the models' stability to affine transformations, even obtaining worse results than the baseline in terms of mFR. The results of APS demonstrate

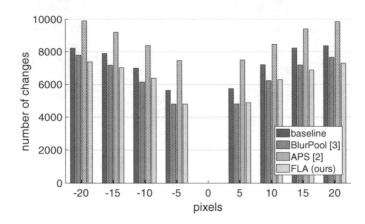

Fig. 3. The number of times the labels predicted by a ResNet-18 model having augmented images as input are different from the labels predicted for the original images (translated with 0 pixels) from Caltech-256. The images are translated from -20 pixels to $+20$ pixels. The ResNet-18 model is trained using various methods addressing the stability problem: input data augmentation (baseline), APS [3], BlurPool [21] and FLA (ours). Best viewed in color.

that the 100% stability to circular shift claimed by Chaman et al. [3] does not increase the network's stability to generic affine transformations.

The BlurPool [21] method increases the models' stability to affine transformations, performing better than the baseline in terms of mFR. However, it also decreases the accuracy of the model in most cases. The only case when BlurPool increases the accuracy level is on Tiny ImageNet for the DenseNet-121 architecture (the baseline reaches an accuracy of 76.50%, while BlurPool reaches a higher accuracy of 76.57%). With one exception (the Tiny ImageNet data set), the models trained with the BlurPool method obtain lower trade-off indices compared to the baseline models.

Different from APS [3] and BlurPool [21], our method consistently attains the best trade-off between accuracy and stability to affine transformations across all models and data sets, as shown in Table 1. On Tiny ImageNet, we obtain the highest accuracy score and the lowest mFR for rotation and scaling, regardless of the network architecture. On Caltech-256, our method obtains the closest accuracy scores to the baselines, while also increasing the models' stability in terms of all mFR scores. Since our method increases the stability of ResNet-18 during multiple runs on Caltech-256, it seems to have an interesting effect on the standard deviation of the reported accuracy, reducing it from 0.30 to 0.06. On the Food-101 benchmark, our approach applied on the ResNet-18 architecture attains an accuracy level on par with the baseline (76.28% $\pm$ 0.33 vs. 76.29% $\pm$ 0.41), but the stability of our approach to affine transformations is significantly higher compared to the stability of the baseline (5.69% vs. 6.41%,

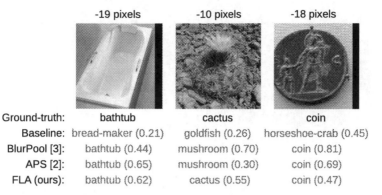

	-19 pixels	-10 pixels	-18 pixels
Ground-truth:	bathtub	cactus	coin
Baseline:	bread-maker (0.21)	goldfish (0.26)	horseshoe-crab (0.45)
BlurPool [3]:	bathtub (0.44)	mushroom (0.70)	coin (0.81)
APS [2]:	bathtub (0.65)	mushroom (0.30)	coin (0.69)
FLA (ours):	bathtub (0.62)	cactus (0.55)	coin (0.47)

Fig. 4. Images perturbed with various translations with ground-truth and predicted labels (by the baseline, BlurPool, APS and FLA methods). Best viewed in color.

6.63% vs. 7.36%, 11.41% vs. 12.39% in terms of the mFR for the translation, rotation and scaling operations, respectively).

In Fig. 3, we illustrate how many times the labels predicted by a ResNet-18 model having translated images as input are different from the labels predicted for the original images (translated with 0 pixels). We observe that our method has a significantly smaller number of instabilities (different predictions) than the baseline and APS [3] methods, while the BlurPool [21] method attains similar results to our method. We also observe that the number of label differences increases with the magnitude of the translation operation for all methods, but the pace seems comparatively slower for our method.

In Fig. 4, we show a couple of images from Caltech-256 that are misclassified by the baseline. We observe that BlurPool and APS induce correct labels for two out of three samples, while FLA is able to correct all labels.

5   Conclusion

In this paper, we have proposed a novel method to address the stability of CNNs against affine perturbations applied on the input images. To improve stability, our method relies on feature-level augmentation. In addition, we are the first who have conducted comparative experiments to assess the performance of state-of-the-art methods [3, 21] addressing the stability problem.

Although there is a recent trend towards focusing on vision transformers [4, 10, 16, 20], our study involves only convolutional architectures. However, the recent work of Liu et al. [12] shows that, upon making proper adjustments, CNNs can obtain comparable results to vision transformers. We thus believe that research related to CNN architectures, such as our own, is still valuable to the computer vision community.

In future work, we aim to study the effect of applying our approach at inference time. This might further improve the stability of neural models, but the stability gains have to be put in balance with the inevitable slowdown in terms of computational time. At the moment, our technique does not affect the inference time at all.

Acknowledgment. This article has benefited from the support of the Romanian Young Academy, which is funded by Stiftung Mercator and the Alexander von Humboldt Foundation for the period 2020–2022.

References

1. Azulay, A., Weiss, Y.: Why do deep convolutional networks generalize so poorly to small image transformations? J. Mach. Learn. Res. **20**, 1–25 (2019)
2. Bossard, L., Guillaumin, M., Van Gool, L.: Food-101 – mining discriminative components with random forests. In: Fleet, D., Pajdla, T., Schiele, B., Tuytelaars, T. (eds.) ECCV 2014. LNCS, vol. 8694, pp. 446–461. Springer, Cham (2014). https://doi.org/10.1007/978-3-319-10599-4_29
3. Chaman, A., Dokmanic, I.: Truly shift-invariant convolutional neural networks. In: Proceedings of CVPR, pp. 3773–3783 (2021)
4. Dosovitskiy, A., et al.: An image is worth 16×16 words: transformers for image recognition at scale. In: Proceedings of ICLR (2021)
5. Goodfellow, I.J., Shlens, J., Szegedy, C.: Explaining and harnessing adversarial examples. In: Proceedings of ICLR (2015)
6. Griffin, G., Holub, A., Perona, P.: Caltech-256 Object Category Dataset. Technical report, California Institute of Technology (2007)
7. He, K., Zhang, X., Ren, S., Sun, J.: Deep residual learning for image recognition. In: Proceedings of CVPR, pp. 770–778 (2016)
8. Hendrycks, D., Dietterich, T.: Benchmarking neural network robustness to common corruptions and perturbations. In: Proceedings of ICLR (2019)
9. Huang, G., Liu, Z., Van Der Maaten, L., Weinberger, K.Q.: Densely connected convolutional networks. In: Proceedings of CVPR, pp. 2261–2269 (2017)
10. Khan, S., Naseer, M., Hayat, M., Zamir, S.W., Khan, F.S., Shah, M.: Transformers in vision: a survey. arXiv preprint arXiv:2101.01169 (2021)
11. Kingma, D.P., Ba, J.: Adam: a method for stochastic optimization. In: Proceedings of ICLR (2015)
12. Liu, Z., Mao, H., Wu, C.Y., Feichtenhofer, C., Darrell, T., Xie, S.: A ConvNet for the 2020s. arXiv preprint arXiv:2201.03545 (2022)
13. Michaelis, C., et al.: Benchmarking robustness in object detection: autonomous driving when winter is coming. arXiv preprint arXiv:1907.07484 (2020)
14. Moosavi-Dezfooli, S.M., Fawzi, A., Frossard, P.: DeepFool: a simple and accurate method to fool deep neural networks. In: Proceedings of CVPR, pp. 2574–2582 (2016)
15. Papernot, N., McDaniel, P., Goodfellow, I., Jha, S., Celik, Z.B., Swami, A.: Practical black-box attacks against machine learning. In: Proceedings of ASIA CCS, pp. 506–519 (2017)
16. Ristea, N.C., et al.: CyTran: cycle-consistent transformers for non-contrast to contrast CT translation. arXiv preprint arXiv:2110.06400 (2021)
17. Russakovsky, O., et al.: ImageNet large scale visual recognition challenge. Int. J. Comput. Vision **115**(3), 211–252 (2015). https://doi.org/10.1007/s11263-015-0816-y

18. Szegedy, C., et al.: Intriguing properties of neural networks. In: Proceedings of ICLR (2014)
19. Volk, G., Müller, S., Bernuth, A.v., Hospach, D., Bringmann, O.: Towards robust CNN-based object detection through augmentation with synthetic rain variations. In: Proceedings of ITSC, pp. 285–292 (2019)
20. Wu, H., et al.: CvT: introducing convolutions to vision transformers. arXiv preprint arXiv:2103.15808 (2021)
21. Zhang, R.: Making convolutional networks shift-invariant again. In: Proceedings of ICML, vol. 97, pp. 7324–7334 (2019)
22. Zheng, S., Song, Y., Leung, T., Goodfellow, I.: Improving the robustness of deep neural networks via stability training. In: Proceedings of CVPR, pp. 4480–4488 (2016)

Benchmarking Robustness Beyond l_p Norm Adversaries

Akshay Agarwal[1,2]([⊠]) [iD], Nalini Ratha[1,2] [iD], Mayank Vatsa[1,2] [iD],
and Richa Singh[1,2] [iD]

[1] University at Buffalo, Buffalo, NY, USA
{aa298,nratha}@buffalo.edu, {mvatsa,richa}@iitj.ac.in
[2] IIT Jodhpur, Karwar, India

Abstract. Recently, a significant boom has been noticed in the generation of a variety of malicious examples ranging from adversarial perturbations to common noises to natural adversaries. These malicious examples are highly effective in fooling almost 'any' deep neural network. Therefore, to protect the integrity of deep networks, research efforts have been started in building the defense against these anomalies of the individual category. The prime reason for such individual handling of noises is the lack of one unique dataset which can be used to benchmark against multiple malicious examples and hence in turn can help in building a true *'universal'* defense algorithm. This research work is an aid towards that goal that created a dataset termed "wide angle anomalies" containing 19 different malicious categories. On top of that, an extensive experimental evaluation has been performed on the proposed dataset using popular deep neural networks to detect these wide-angle anomalies. The experiments help in identifying a possible relationship between different anomalies and how easy or difficult to detect an anomaly if it is seen or unseen during training-testing. We assert that the experiments in seen and unseen category attack training-testing reveals several surprising and interesting outcomes including possible connection among adversaries. We believe it can help in building a universal defense algorithm.

1 Introduction

To protect the integrity of deep neural networks, defense algorithms against modified images are proposed; although, the majority of them deal with a unique category of malicious examples and have shown tremendous success in that [1,5]. It is seen in existing research that the defense algorithms targeting specific attacks fail in identifying the malicious data coming from the same or different malicious examples categories [12,51,54]. This ineffectiveness can be seen as a blind spot that leaves a space for an attacker to attack the system and perform undesired tasks. Therefore, looking at the severeness of this existing limitation, we have studied numerous adversaries intending to develop universal security. In this research, we divide the malicious examples into three broad categories: (i) common corruptions [21,41], (ii) adversarial perturbations [3,4,33], and (iii) natural

L. Karlinsky et al. (Eds.): ECCV 2022 Workshops, LNCS 13801, pp. 342–359, 2023.
https://doi.org/10.1007/978-3-031-25056-9_23

adversary [8, 22, 40, 53]. These different categories follow different rules in crafting the perturbation and hence have significant distribution differences among each other. For instance, in the common corruptions, the noises are uniformly distributed, adversarial perturbations affect the critical regions, and natural examples might occur due to cluttered background or low foreground region [31, 32].

To study such broad robustness, a unique dataset covering such malicious examples is a necessity, and therefore, we have first curated the *"wide angle anomalies"* examples dataset covering the three broad categories mentioned above. In total, the dataset contains approximately 60, 000 images belonging to 20 classes including real and various malicious generation algorithms. Surprisingly, the majority of the malicious examples generation algorithm belonging to the above categories are not explored in existing research, and hence security against them is still a serious concern. Once the wide angle anomalies dataset is prepared we have performed an extensive experimental evaluation using deep convolutional networks to identify these malicious examples. Henceforth, the analysis presented in this paper is an act of benchmarking robustness against such a broad umbrella of malicious examples and in turn helps in building a universal robustness system. We find that there is a connection among the different anomalies coming from the same broad group and can also be used to detect other groups' adversaries. In brief,

- We present a large-scale malicious examples dataset covering 19 different attack generation algorithms. The images consist of a wide distribution shift among the malicious examples due to contradictory ways of generation;
- A benchmark study is presented using a deep convolutional network for the detection of such broad malicious examples categories. The experimental results corresponding to both seen and unseen malicious category in training and testing reveals several interesting and thoughtful insights. We assert the presence of such wide-angle malicious examples dataset and the benchmark study can significantly boost the development of universal robustness.

2   Related Work

In this section, a brief overview of the existing works developed to counter the malicious examples and protect the integrity of the deep neural networks is presented. As mentioned earlier, the majority of the defense work is focused on defending one specific type of malicious example. We first provide a brief overview of the existing defense work tackling artificial adversarial perturbations followed by the defense algorithms countering common corruptions. To the best of our knowledge, no work so far has been presented to build a defense against natural adversarial examples.

The defense algorithms against artificial adversarial perturbations are broadly grouped into (i) detection based, (ii) mitigation based, and (iii) robustness. In the mitigation case, a denoising algorithm is presented to map the noisy data to the clean counterpart. The aim is to reduce the impact of the adversarial perturbation so that the accuracy of the classifier can be restored

[15,18,44]. Robustness-based defenses are one of the most popular defense techniques to make the classifier resilient against noisy test data. The robustness in this category of algorithms is achieved by training the network by utilizing data augmentation techniques including adversarial training [9,46,47]. Adversarial training is one of the powerful defense techniques where the classifiers are either trained or fine-tuned using the adversarial images. However, the probable limitations of the techniques are the computational cost and generalizability against the unseen perturbations [42,50,54]. To address this issue several general-purpose data augmentation techniques are also proposed [6,7,10]. These data augmentation-based defenses also overcome another limitation of adversarial training which is maintaining the performance on the clean images that significantly drops in adversarial training. Another popular and most effective defense against artificial adversarial perturbation is the development of a binary classifier. Recently, several generalized adversarial perturbation detection algorithms are proposed that either utilize the handcrafted features, deep classifiers or a combination of both [1,5,20,30,52].

In contrast to the defense against adversarial perturbations, limited work has been done so far to protect against common corruption. Similar to increasing the robustness against adversarial perturbations, data augmentation is one of the favorite defenses against common corruptions as well [34,36]. In other forms of defense, recently, Schneider et al. [45] have proposed to mitigate the covariate shift by replacing the batch normalization statistics computed over the clean data with the statistics computed over corrupted samples. Another recent work utilizes the mixture of two deep CNN models biased towards low and high-frequency features of an image [43]. The reason might be that noise signals are considered high-frequency information and the author aims to improve the robustness against high-frequency features. The major limitation of the defenses proposed so far is the generalization against the unseen corruptions [14,17,35], computational cost, and degradation performance on the clean or in-distribution images. Interestingly, very limited work has tried to identify/detect the corrupted examples and the majority of them tried to improve the robustness of the model directly which in turn leads to the degradation performance on clean images. The issue of common corruption has recently been explored in other computer vision tasks or models as well such as semantic segmentation [27] and transformers [39]. Therefore, looking at the severity of both common corruptions and artificial perturbations, a resilient defense is critical. Apart from handling these well-known malicious examples, detecting the advanced or recently explored natural adversarial examples [8,22,31] is also important in building a universal defense mechanism. To the best of our knowledge, recent work [2] is the only work which has started an effort in building a unified defense system.

3   Proposed *Wide Angle Anomalies* Dataset

In this research, we have selected different malicious examples generation algorithms which we have broadly grouped into two groups: (i) common corruptions

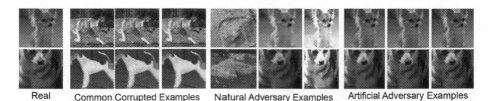

Real   Common Corrupted Examples   Natural Adversary Examples   Artificial Adversary Examples

Fig. 1. Samples from our proposed wide angle anomalies dataset cover a broad spectrum of malicious examples. The covariate shift between the different classes including real is evident which makes the generalizability a tough task in handling a broad spectrum of anomalies.

Table 1. Parameters of the common corruption used.

Noise	GN	UN	SPN	SN	SPKL
Param.	0.08	0.1	0.1	60	0.15

and (ii) adversarial examples. The adversarial examples consist of two broad categories (i) artificial perturbations and (ii) natural adversary.

3.1 Common Corruptions

For the generation of common corruption-induced malicious examples five different popular variants are selected namely Gaussian noise (GN), salt & pepper noise (SPN), uniform noise (UN), shot noise (SN), and speckle noise (SPKN). Each of the selected corruption is applied with low severity with a twofold aim: first is to fool the deep classifiers but at the same time keep the perceptibility of noise pattern minimal. The parameters used with individual common corruption are given in Table 1.

3.2 Adversarial Examples

In contrast to the common corruption, adversarial examples contain the perturbation generated using the classifier itself utilizing its ingredients including image gradient and decision probabilities. For adversarial examples, both artificial perturbation optimization algorithms and natural adversarial examples generation algorithms are selected.

Artificial Perturbations. For artificial adversarial examples, five benchmark algorithms namely fast gradient sign method (FGSM) [19], basic iterative method (BIM) also known as iterative FGSM (IFGSM) [28], projected gradient descent (PGD) [33], DeepFool [38], and Universal perturbation [37] are adopted. FGSM is one of the simplest and most effective adversarial perturbation generation algorithms. It works on the manipulation using the gradient information computed over an image to maximize the loss over the modified image. Mathematically, it can be described as follows: $X^* = X + \eta \cdot sign(\nabla_X J(X, Y_{true}))$. X and X^* represent the clean and FGSM adversarial images, respectively. η controls the strength of added perturbation optimized through the loss function J computed over image X and its associate true label Y_{true}. ∇_X is the gradient concerning X and $sign$ represents the sign function. Even though η tries to control the perturbation visibility it is still highly perceptible with a naked eye and hence to improve that the iterative variants are proposed by Kurakin and Goodfellow [28]. It can be described as follows:

$$X_0^* = X$$
$$X_N^* = Clip_{X,\epsilon}(X_{N-1}^* + \alpha \cdot sign(\nabla_X J(X_{N-1}^*, Y_{true})))$$

where, $Clip_{X,\epsilon}$ represents the scaling of an image in the range of $[X - \epsilon, X + \epsilon]$. Due to its nonlinearity in the gradient direction, several iterations are required to optimize the perturbation. It makes the generated perturbation more effective as compared to FGSM. Compared to the above gradient-based perturbation, the deepfool attack is based on the minimization of the L_2 norm and aims to make sure the adversarial examples jump the decision hyperplane developed by the classifier. The idea is to perturb an image iteratively and with each iteration, the closest decision surface is assumed to be fooled by the updated image. Madry et al. [33] have proposed the PGD attack which is also considered the strongest first-order universal adversary. The optimization iteratively searches for a perturbation vector that minimizes a l_p norm ball around the clean image. The above artificial perturbation generation algorithms whether simple or complex, generate the noise vector individually for each image. To optimize a single perturbation vector that can be applied to multiple images, a universal perturbation is also presented in the literature [37]. The above selected perturbation reflects the wide variety in the generation of adversarial perturbation and hence makes the study of universal defense interesting and a thoughtful step.

Natural Adversary. Recently, several researchers have explored the natural way of crafting adversarial examples. One such way is proposed by Hendrycks et al. [22]. The authors have downloaded the natural images of 200 classes from multiple image hosting websites that can fool the ResNet-50 classifier. In total, the dataset contains $7,500$ adversarial images and is termed Imagenet-A. Later, Li et al. [31] identify several bottlenecks of the natural adversarial examples in ImageNet-A. It is found that the background in the adversarial examples of ImageNet-A is more cluttered due to the presence of multiple objects that might be a possible reason for distribution shift and leads to misclassification. Another possible drawback of the images is that the foreground region in an image occupies a small part as compared to the background. The authors show that removing these limitations can significantly boost the performance of several ImageNet trained models and hence the need for an intelligent way of crafting a natural adversary is highlighted. For that, the authors have presented an ImagNet-A-Plus dataset by minimizing the background information in an image and leaving only one salient object region covering a large portion of an image. The dataset is generated from the images of the ImageNet-A dataset by first filtering out the images containing object proportion 8 times less than the object proportion in the ImageNet images. The filtered images are passed through the ResNet-50 model and the background is clipped from the selected adversarial images so that the target object proportion in images can be increased. Hosseini et al. [24] have proposed another way of generating natural adversarial examples by shifting the color information in images. The assumption of image generation is based on the assertion that the human visual system is biased towards shape while classifying an object as compared to the color information [29]. The authors have utilized the HSV color space due to its closeness to the human visual system as compared to the RGB space. The authors have modified the hue and saturation components of an image keeping the value component intact. The adversarially color shifted images are generated by solving the following optimization problem:

$$min|\delta_S|, \quad s.t.$$
$$\left\{ \begin{array}{c} X_H^* = (X_H + \delta_H) \mod 1 \\ X_S^* = Clip(X_S + \delta_S, 0, 1) \\ X_V^* = X_V \end{array} \right\}$$

where, δ_H and δ_S represent the shift introduced in the hue X_H and saturation X_S components of an image X, respectively. X_V is the value component of an image. Both δ_H and δ_S are scalar values only. The authors have used the 1000 iteration to perturb the color components or till the modified image is misclassified by the VGG classifier, whichever happens first. The authors have reported the success of the attack on the CIFAR10 dataset only. In this research, to keep the adversarial examples of one dataset, we have trained the VGG classifier on the selected ImageNet subset and generated the color shift semantic adversarial examples.

In contrast to the above natural examples which either utilize the limitation of the classifier of not being trained on the kind of images that occur in the adversarial dataset such as ImageNet-A and ImageNet-A-Plus or color shifting the images, Agarwal et al. [8] have utilized the noise inherited at the time image acquisition. When the images are captured from the camera they passed through several intermediate steps, the authors assert that these steps induced some form of manipulation or the environment can itself be noisy. The intuition of the authors is that can this inherited noise be used as an adversarial noise. The inherited noise vector is extracted using several image filtering techniques such as Laplacian and Gaussian. The adversarial examples generation process can be described as follows:

$$Noise = X - \phi(X)$$
$$X^* = Clip(X \circledast (\psi \cdot Noise), 0, 1)$$

where, $Noise$ is generated by subtracting the acquired clean image (X) from its filtered version obtained by applying any image filtering technique ϕ. ψ is a scalar value controlling the strength of the noise. $\circledast$ represents the noise manipulation operator that is either added ($\oplus$) or removed ($\ominus$) from the image with the following assertion: (i) added (referred to as -P) the subtracted component by assuming it as a noise vector and (ii) removed (referred as -S) on the fact that the noise is a high-frequency feature and removing that feature can fool the classifiers which are found highly biased towards shape and texture [16,23,43]. In this research, we have used the Integral, laplacian, and laplacian of Gaussian as three image filtering techniques due to their high effectiveness as compared to the other filtering methods used in the paper [8]. The adversarial examples generated using the above technique are referred to as camera induced perturbation (CIPer) as termed in the original paper. In brief, various classes along with the number of images covered in the proposed research can be summarized as follows:

- Real/Clean Images (3,000)
- Common Corruptions (15,000)
  - Gaussian Noise (GN) (3,000)
  - Uniform Noise (UN) (3,000)
  - Salt & Pepper Noise (SPN) (3,000)
  - Shot Noise (SN) (3,000)
  - Speckle Noise ($SPKN$) (3,000)
- Adversarial Images
  - Artificial Perturbations (15,000)
    * FGSM
    * IFGSM
    * PGD
    * DeepFool (DF)
    * Universal ($Univ.$)
  - Natural Examples (8,763)
    * Subset of ImageNet-A (IN-A)
    * Subset of ImageNet-A-Plus (IN-A-P)
    * Semantic Color-Shift Examples (CS)
  - Camera Induced (18,000)
    * Integral filtering (Int-P and Int-S)
    * Laplace filtering (Lap-P and Lap-S)
    * Laplace of Gaussian filtering (LoG-P and LoG-S)

To generate the malicious images, 3,000 clean images from the validation set of the ImegeNet dataset are first selected [13]. Later, each malicious examples generation algorithm is applied to the selected images except ImageNet-A and ImageNet-A-Plus. The images in this category are directly taken from the images provided by the original contributors. In total, the proposed wide angle anomalies dataset consists of 3,000 clean images and 56,763 malicious images. For the experimental purpose, the first 1500 images of each class are used for training, and the last 1500 images are used for evaluation.

4   Experimental Results and Analysis

In this research, the aim is to study the universal robustness by detecting these wide-angle adversaries, i.e., classifying the images into either real or modified

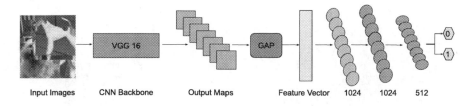

Fig. 2. Malicious examples detection architecture used in this research.

classes. For that, a binary classifier is developed using VGG-16 [48] as a backbone architecture[1]. The robustness performance acts as a benchmark for the study of universal defense by identifying a possible connection between different groups of adversaries. The proposed malicious examples detection architecture is shown in Fig. 2. We have set the gradient of the first 10 layers of VGG equal to zero and finetuned the remaining layers along with training the newly added dense layers from scratch. The architecture is trained for 30 epochs using Adam optimizer where the batch size is set to 32 and initial learning used is $1e^{-4}$.

4.1 Results and Analysis

We have performed an extensive experimental analysis on the collected wide angle anomalies dataset. The experiments are performed in several generalized settings: (i) seen attack generation algorithm such as PGD vs. PGD, (ii) unseen attack generation algorithm such as PGD vs. FGSM, and (iii) unseen attack types such as Natural adversary vs. common corruptions. First, we will present the results and analysis of the seen attack training and testing scenarios followed by the experimental observations on unseen attack settings. In the end, the connection between different malicious examples is established by testing the malicious examples detection algorithm trained on entirely different malicious examples.

Seen Attack Results. In total, in the proposed dataset there are 20 classes belonging to one real and 19 attack class. Therefore, in the seen attack setting, a total of 19 classifiers are trained individually on each attack training data and tested on the same attack type using the testing set. The results of these experiments are shown in Fig. 3. The analysis can be broken down into the following ways: (i) global analysis and (ii) local analysis. In the global analysis, it can be seen that the DF (DeepFool) attack is found highly challenging to detect, i.e., yielding the lowest detection accuracy value of 61.33%. Whereas, the remaining

---

[1] While the results are reported using VGG, similar evaluation analysis (with ± 1–12% as shown in Table 5) is observed across wide range of backbone networks including Xception [11], InceptionV3 [49], DenseNet121 [26], and MobileNet [25]. However, VGG tops each network in the majority of the cases and is hence chosen for detailed study in the paper.

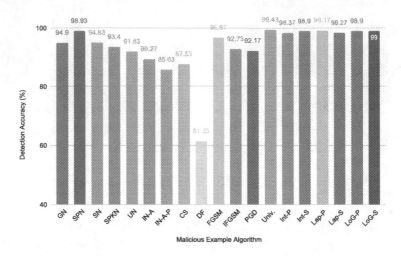

Fig. 3. Malicious examples detection in the seen attack setting, i.e., where training and testing attack generation algorithm is same.

Table 2. Common corruption detection performance in unseen noise training testing conditions. – represents the seen noise training-testing setting and the results are reported in Fig. 3.

Train ↓ Test →	GN	SPN	SN	SPKN	UN	Avg. ± SD
GN	–	**97.90**	94.80	93.00	85.27	**92.74 ± 5.37**
SPN	57.10	–	**58.30**	55.63	51.67	**55.67 ± 2.88**
SN	94.17	**95.83**	–	94.47	87.47	**92.98 ± 3.74**
SPKN	92.80	**95.57**	93.67	–	86.73	**92.19 ± 3.82**
UN	91.87	91.90	**92.00**	91.97	–	**91.93 ± 0.06**

attack detection yields a high accuracy value of at least 85.63% reflecting that defending against adversarial attacks even coming from a variety of algorithms is not difficult even from a simple classification architecture. However, this might gives us a false sense of security as in reality all possible attacks might not be known beforehand.

In the case of local analysis, an observation concerning the different classes of malicious examples can be described. For instance, when the common corruptions are aimed to detect, it is found that *uniform noise* (UN) is one of the toughest corruption to detect. We went ahead to identify the potential reason for such lower performance and found that the perceptibility of the uniform noise is low (last column of common corruption in Fig. 1) as compared to other perturbations and it is approximately similar to artificial adversarial perturbations. It can be seen from another point that the SPN noise is found highly perceptible and hence yield a higher detection performance. In the case of natural

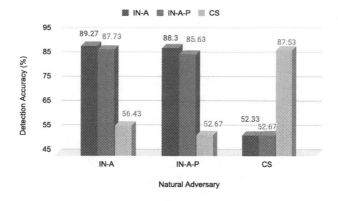

Fig. 4. Unseen natural adversarial examples detection.

adversarial examples, ImageNet-A-Plus images are found less detectable as compared to the other natural adversary. The ImageNet-A-Plus (IN-A-P) can be seen as an advanced version of ImageNet-A (IN-A) where the cluttered background and foreground object region is enhanced. It might be the possible reason for the lower detection performance of these examples. The proposed binary classification algorithm is also found effective in detecting the color shift (CS) semantic natural adversarial examples. The detection performance on each artificial adversarial perturbation except DF is significantly high where the lowest detection accuracy observed is 92.17%. The universal perturbation images are found easiest to detect and demonstrated approximately perfect detection accuracy (99.43%). In comparison to the other malicious attack classes, the detection performance across each variety of CIPer noise is at least 98.27% which makes it less effective in terms of its detection. We want to highlight that such high detection performance observed is in the case where the detection algorithm has seen each malicious class while optimizing the network parameters and therefore, might not be a true indicator of the complexity of the malicious class.

Unseen Attack Detection. From the experimental analysis in seen attack setting, we have seen the impression that it might not be difficult to identify the malicious examples; however, such an impression can be dangerous until the evaluation has been performed under an unseen attack testing setting. Therefore, we have extensively evaluated the generalizability concern and showcased how easy or difficult the detection of malicious examples is in open-world settings. On the common corruption, five-fold cross-validation experiments are performed where at each time one noise type is used for training, and the remaining are used for testing. From the global view, each attack is found similarly generalized in detecting the unseen noise variation except Salt & Pepper noise (SPN). The training on SPN noise is found highly ineffective in identifying other noise variations. While the UN corruption performs similar to other corruptions, the variation in its detection performance is the smallest. Through a multi-fold

Table 3. Artificial adversarial perturbation detection performance in unseen noise training testing conditions. – represents the seen perturbation training-testing setting and reported in Fig. 3.

Train ↓ Test →	DF	FGSM	IFGSM	PGD	Univ.	Avg. ± SD
DF	–	89.60	77.03	77.00	**94.47**	**84.52 ± 8.90**
FGSM	57.63	–	71.20	71.33	**97.80**	**74.49 ± 16.82**
IFGSM	59.10	94.37	–	92.87	**95.30**	**85.41 ± 17.57**
PGD	59.10	94.43	92.17	–	**95.37**	**85.27 ± 17.50**
Univ.	55.00	**62.07**	51.40	51.30	–	**54.94 ± 5.05**

Table 4. Camera induced adversarial perturbation detection performance in unseen noise training testing conditions.

Train ↓ Test →	Int-P	Int-S	Lap-P	Lap-S	LoG-P	LoG-S
Int-P	–	58.77	50.00	56.07	**70.63**	64.87
Int-S	54.33	–	49.27	54.23	52.60	**63.60**
Lap-P	49.63	49.53	–	49.50	**86.47**	49.50
Lap-S	54.70	56.20	49.10	–	49.10	**98.87**
LoG-P	58.40	50.67	**95.23**	49.43	–	49.43
LoG-S	52.43	55.17	49.53	**96.63**	49.53	–

cross-validation experiment it is observed that the generalizability of the unseen corruption detection is somewhat better; however, needs further attention and evaluation against other categories. In an interesting observation, it can also be noticed that the detectors trained on each noise yield the lowest performance on the uniform noise, and the performance on SPN noise is highest. The results of this experimental analysis are reported in Table 2.

In the second unseen attack setting, natural adversary examples are chosen where a three-fold cross-validation experimental evaluation has been performed. Similar to common corruption only one attack is used for training and others are used for testing. This kind of setting represents the worst-case performance where it might be possible that only one type of attack is available for training or in other words limited knowledge about the adversary is known. In contrast to generalizability in common corruption, the robustness in the detection of the natural adversary is poor. The ImageNet-A (IN-A) and ImageNet-A-Plus (IN-A-P) being the same category of natural adversary yields similar and high detection performance; whereas the color shift semantic adversary is contradictory to these adversaries and yields lower accuracy. It can also be seen from the detection performance: (i) when IN-A or IN-A-P adversary is used for training it yields low detection performance on the color shift adversary (CS) and (ii) when the CS

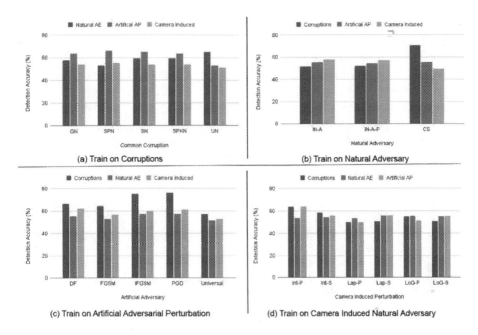

Fig. 5. Unseen malicious type examples detection. Each attack of four malicious examples category (common corruption, natural adversary, artificial adversary, and camera induced perturbation) are used to train the detection algorithm and evaluated on the unseen malicious attack category.

adversary is used in training and IN-A and IN-A-P adversary used for evaluation, the performance is closed to random chance value. In between IN-A and IN-A-P, the IN-A is found robust in detecting the other natural adversaries. The quantitative finding of these experimental evaluations is reported in Fig. 4.

Another unseen attack detection performance is performed on the artificial adversarial perturbation images. We have noticed in the seen attack training-testing setting that the DF is highly challenging to detect as compared to other malicious examples including artificial adversarial perturbations. In the unseen attack setting as well it is observed that the generalization performance on the DF perturbation is the lowest among all the artificial perturbations. Whereas, the universal perturbation is found easily detectable and yields at least 94.47% detection accuracy even if the detector has not seen the perturbation while training. In surprising observations, the universal perturbation which was found highly detectable shows poor generalization in detecting other perturbations; whereas, the perturbation (DF) which is found complex in detection found significantly generalized in detecting unseen adversarial perturbations. The numerical analysis related to the above observations is reported in Table 3.

In final unseen attack detection experiments, the camera-induced noises are used for training and testing. The noise is extracted using three image filtering operations and applied in two forms (addition and subtraction). Therefore, a total of *six fold* cross-validation experiments are performed to evaluate the generalizability of malicious examples detection networks. The camera-induced perturbations which were found almost perfectly detectable in seen attack training-testing are found complex in the unseen testing setting. In brief, when the detector is trained on the noisy examples obtained using addition operation, it yields higher performance in the detection of unseen attack images obtained using addition operation and yields poor performance generated using subtraction operation even same image filtering is applied. For instance, *Lap-P* trained detector yields more than 86% detection performance on the *LoG-P* images; whereas, it yields random chance accuracy (49.50%) on the *LoG-S* adversarial images. A similar interesting observation is observed in the adversarial examples obtained using subtraction of camera noise. For instance, *Int-S* trained malicious examples detector yields 11% better accuracy when the adversarial examples are obtained using LoG filtering with subtraction operation as compared to the addition operation. The quantitative results are shown in Table 4.

Unseen Malicious Type Detection. In the final generalizability analysis (shown in Fig. 5), we have evaluated the anomaly examples detection in unseen malicious type training-testing scenarios. In the proposed research, four different malicious categories are used which in turn contain several attack generation algorithms. To extensively study the malicious examples detection and pave a way for future research to enhance the robustness, four-fold cross-validation experiments are performed. In each fold, one malicious examples category is used for training, and remaining are the used for testing. We have earlier observed that the accuracy is lower in the unseen attack setting in comparison to the seen attack setting. The drop in detection performance is further observed when the malicious attack category is changed in the training and testing set. Except in a few cases, the majority of the detection performance lies close to 60% only which shows that the detection of malicious examples demands careful attention, especially in extremely generalized and open-world settings.

Let us dig deeper towards understanding the detection of individual attacks of the malicious category used for evaluation. When the common corruptions are used to train the detection algorithm, across each corruption it is found that the detection of CS attack, universal perturbation, and Int-P attack belonging to the natural adversary, artificial adversarial perturbation, and camera noise, respectively, is the highest. In natural adversary, color shift shows the highest correlation with other malicious categories and yields the highest detection performance. Interestingly, artificial adversarial perturbations are found more effective in detecting the Salt & Pepper noise (SPN) common corruption in comparison to other common corruptions and remaining unseen malicious categories. Whereas, the camera-induced noises yield better performance on universal artificial perturbation along with SPN corruption in comparison to other unseen malicious example categories. In quantitative terms, the detection performance of SPN and universal perturbation is at least 17% and 30% better when artificial adversary and camera-induced noises are used in training, respectively. We believe that

Table 5. Ablation study utilizing different backbone architectures for seen and unseen common corruption detection. VGG yields the best performance across each network.

Corrup.	Model	Test					Avg.
		GN	SPN	SN	SPKN	UN	
GN	VGG	94.93	97.90	94.83	93.00	85.27	**93.19**
	DenseNet	96.70	98.33	95.77	91.50	73.63	91.19
	MobileNet	95.33	97.80	93.90	87.40	61.83	87.25
	InceptionV3	94.30	96.53	91.17	84.80	65.60	86.48
	Xception	97.40	98.40	96.10	89.03	66.23	89.43
SN	VGG	94.17	95.83	94.83	94.47	87.47	**93.35**
	DenseNet	94.63	98.33	95.10	91.10	69.00	89.63
	MobileNet	95.30	98.57	95.63	90.40	63.30	88.64
	InceptionV3	92.97	98.20	93.87	85.33	58.13	85.70
	Xception	94.70	99.30	95.97	86.73	60.20	87.38
SPKN	VGG	92.80	95.57	93.67	93.40	86.73	**92.36**
	DenseNet	94.43	96.83	95.17	93.37	77.30	91.42
	MobileNet	92.30	93.67	93.30	92.00	71.70	88.59
	InceptionV3	88.17	92.83	89.47	86.10	64.83	84.28
	Xception	96.60	96.97	97.23	94.50	69.00	90.86
UN	VGG	91.87	91.90	92.00	91.97	91.83	**91.91**
	DenseNet	92.03	93.17	92.00	90.83	86.23	90.85
	MobileNet	90.13	90.33	90.00	88.67	78.00	87.43
	InceptionV3	84.10	84.87	84.13	83.67	80.23	83.40
	Xception	82.37	82.53	82.43	82.17	80.70	82.04

there is a connection between different malicious examples categories and can be exploited further to build a universal defense system.

Impact of CNN Backbone. We have extensively analyzed the impact of different CNN architectures as a backbone network in the malicious examples detection pipeline. In brief, the VGG architecture yields the best average malicious examples detection performance whether evaluated in seen or unseen attack image settings. The detailed results are added in Table 5. When the VGG architecture is trained using SN, it shows the highest average generalization performance; although the performance does not shows significant degradation even if trained on other corruptions. In terms of corruption, the detection of uniform noise corrupted images is complex as compared to the other corruptions. For instance, when the VGG architecture is trained on GN corruption and tested on UN, it shows at least 7.73% lower performance in comparison to the detection performance on other corruption images. On the other hand, SPN corruption is found the easiest to be defended, i.e., out of all corruptions used, each CNN architecture yields the best detection accuracy of the SPN images.

5   Conclusion

In this research, we put a strong step towards developing a universal robustness system by building the *'first-ever wide angle'* anomalies dataset. The said dataset covers 19 different attack generation algorithms and contains approximately 60,000 images. An experimental evaluation setup shows that the detection of these malicious examples categories is easy when they are seen at the time of training the attack detection algorithm. However, several generalization experimental protocol reveals that we should not fall prey to such high detection accuracies as performance can significantly drop if an unseen attack or unseen malicious category comes for evaluation. In the real world, we can expect such unseen training-testing scenarios; therefore, we demand careful attention while developing the defense algorithms and their evaluation in several generalized settings. The experimental results also reveal a potential connection between different malicious categories which can be effectively used in building a *universal* detection algorithm. In the future, newer malicious attack images can be added to the proposed dataset and a sophisticated detection algorithm will be built to get universal robustness.

References

1. Agarwal, A., Goswami, G., Vatsa, M., Singh, R., Ratha, N.K.: Damad: database, attack, and model agnostic adversarial perturbation detector. IEEE Trans. Neural Netw. Learn. Syst. **33**, 1–13 (2021). https://doi.org/10.1109/TNNLS.2021.3051529
2. Agarwal, A., Ratha, N., Vatsa, M., Singh, R.: Exploring robustness connection between artificial and natural adversarial examples. In: IEEE/CVF Conference on Computer Vision and Pattern Recognition, pp. 179–186 (2022)
3. Agarwal, A., Ratha, N.K.: Black-box adversarial entry in finance through credit card fraud detection. In: CIKM Workshops (2021)
4. Agarwal, A., Ratha, N.K.: On the robustness of stock market regressors. In: ECML-PKDD Workshops (2022)
5. Agarwal, A., Singh, R., Vatsa, M., Ratha, N.: Image transformation-based defense against adversarial perturbation on deep learning models. IEEE Trans. Depend. Secure Comput. **18**(5), 2106–2121 (2021). https://doi.org/10.1109/TDSC.2020.3027183
6. Agarwal, A., Vatsa, M., Singh, R., Ratha, N.: Cognitive data augmentation for adversarial defense via pixel masking. Pattern Recogn. Lett. **146**, 244–251 (2021)
7. Agarwal, A., Vatsa, M., Singh, R., Ratha, N.: Intelligent and adaptive mixup technique for adversarial robustness. In: 2021 IEEE International Conference on Image Processing (ICIP), pp. 824–828 (2021). https://doi.org/10.1109/ICIP42928.2021.9506180
8. Agarwal, A., Vatsa, M., Singh, R., Ratha, N.K.: Noise is inside me! generating adversarial perturbations with noise derived from natural filters. In: Proceedings of the IEEE/CVF Conference on Computer Vision and Pattern Recognition Workshops, pp. 3354–3363 (2020)

9. Andriushchenko, M., Flammarion, N.: Understanding and improving fast adversarial training. Adv. Neural Inf. Process. Syst. **33**, 16048–16059 (2020)

10. Chhabra, S., Agarwal, A., Singh, R., Vatsa, M.: Attack agnostic adversarial defense via visual imperceptible bound. In: 2020 25th International Conference on Pattern Recognition (ICPR), pp. 5302–5309 (2021). https://doi.org/10.1109/ICPR48806.2021.9412663

11. Chollet, F.: Xception: deep learning with depthwise separable convolutions. In: IEEE Conference on Computer Vision and Pattern Recognition, pp. 1251–1258 (2017)

12. Chun, S., Oh, S.J., Yun, S., Han, D., Choe, J., Yoo, Y.: An empirical evaluation on robustness and uncertainty of regularization methods. arXiv preprint arXiv:2003.03879 (2020)

13. Deng, J., Dong, W., Socher, R., Li, L.J., Li, K., Fei-Fei, L.: Imagenet: a large-scale hierarchical image database. In: IEEE Conference on Computer Vision and Pattern Recognition, pp. 248–255. IEEE (2009)

14. Dodge, S., Karam, L.: Quality resilient deep neural networks. arXiv preprint arXiv:1703.08119 (2017)

15. Esmaeilpour, M., Cardinal, P., Koerich, A.L.: Cyclic defense gan against speech adversarial attacks. IEEE Signal Process. Lett. **28**, 1769–1773 (2021)

16. Geirhos, R., Rubisch, P., Michaelis, C., Bethge, M., Wichmann, F.A., Brendel, W.: Imagenet-trained cnns are biased towards texture; increasing shape bias improves accuracy and robustness. arXiv preprint arXiv:1811.12231 (2019)

17. Geirhos, R., Temme, C.R., Rauber, J., Schütt, H.H., Bethge, M., Wichmann, F.A.: Generalisation in humans and deep neural networks. Adv. Neural Inf. Process. Syst. **31**, 1–13 (2018)

18. Goel, A., Singh, A., Agarwal, A., Vatsa, M., Singh, R.: Smartbox: benchmarking adversarial detection and mitigation algorithms for face recognition. In: 2018 IEEE 9th International Conference on Biometrics Theory, Applications and Systems (BTAS), pp. 1–7. IEEE (2018)

19. Goodfellow, I.J., Shlens, J., Szegedy, C.: Explaining and harnessing adversarial examples. arXiv preprint arXiv:1412.6572 (2014)

20. Goswami, G., Agarwal, A., Ratha, N., Singh, R., Vatsa, M.: Detecting and mitigating adversarial perturbations for robust face recognition. Int. J. Comput. Vision **127**(6), 719–742 (2019)

21. Hendrycks, D., Dietterich, T.: Benchmarking neural network robustness to common corruptions and perturbations. arXiv preprint arXiv:1903.12261 (2019)

22. Hendrycks, D., Zhao, K., Basart, S., Steinhardt, J., Song, D.: Natural adversarial examples. In: Proceedings of the IEEE/CVF Conference on Computer Vision and Pattern Recognition, pp. 15262–15271 (2021)

23. Hermann, K., Chen, T., Kornblith, S.: The origins and prevalence of texture bias in convolutional neural networks. Adv. Neural Inf. Process. Syst. **33**, 19000–19015 (2020)

24. Hosseini, H., Poovendran, R.: Semantic adversarial examples. In: Proceedings of the IEEE Conference on Computer Vision and Pattern Recognition Workshops, pp. 1614–1619 (2018)

25. Howard, A.G., et al.: Mobilenets: efficient convolutional neural networks for mobile vision applications. arXiv preprint arXiv:1704.04861 (2017)

26. Huang, G., Liu, Z., Van Der Maaten, L., Weinberger, K.Q.: Densely connected convolutional networks. In: IEEE Conference on Computer Vision and Pattern Recognition, pp. 4700–4708 (2017)

27. Kamann, C., Rother, C.: Benchmarking the robustness of semantic segmentation models with respect to common corruptions. Int. J. Comput. Vision **129**(2), 462–483 (2021)
28. Kurakin, A., Goodfellow, I.J., Bengio, S.: Adversarial examples in the physical world. In: Artificial Intelligence Safety and Security, pp. 99–112. Chapman and Hall/CRC (2018)
29. Landau, B., Smith, L.B., Jones, S.S.: The importance of shape in early lexical learning. Cogn. Dev. **3**(3), 299–321 (1988)
30. Li, F., Liu, X., Zhang, X., Li, Q., Sun, K., Li, K.: Detecting localized adversarial examples: a generic approach using critical region analysis. In: IEEE INFOCOM 2021-IEEE Conference on Computer Communications, pp. 1–10. IEEE (2021)
31. Li, X., Li, J., Dai, T., Shi, J., Zhu, J., Hu, X.: Rethinking natural adversarial examples for classification models. arXiv preprint arXiv:2102.11731 (2021)
32. Ma, X., et al.: Understanding adversarial attacks on deep learning based medical image analysis systems. Pattern Recogn. **110**, 107332 (2021)
33. Madry, A., Makelov, A., Schmidt, L., Tsipras, D., Vladu, A.: Towards deep learning models resistant to adversarial attacks. arXiv preprint arXiv:1706.06083 (2017)
34. Mikołajczyk, A., Grochowski, M.: Data augmentation for improving deep learning in image classification problem. In: 2018 International Interdisciplinary PhD Workshop (IIPhDW), pp. 117–122. IEEE (2018)
35. Mintun, E., Kirillov, A., Xie, S.: On interaction between augmentations and corruptions in natural corruption robustness. Adv. Neural Inf. Process. Syst. **34**, 1–13 (2021)
36. Modas, A., Rade, R., Ortiz-Jiménez, G., Moosavi-Dezfooli, S.M., Frossard, P.: Prime: a few primitives can boost robustness to common corruptions. arXiv preprint arXiv:2112.13547 (2021)
37. Moosavi-Dezfooli, S.M., Fawzi, A., Fawzi, O., Frossard, P.: Universal adversarial perturbations. In: Proceedings of the IEEE Conference on Computer Vision and Pattern Recognition, pp. 1765–1773 (2017)
38. Moosavi-Dezfooli, S.M., Fawzi, A., Frossard, P.: Deepfool: a simple and accurate method to fool deep neural networks. In: Proceedings of the IEEE Conference on Computer Vision and Pattern Recognition, pp. 2574–2582 (2016)
39. Morrison, K., Gilby, B., Lipchak, C., Mattioli, A., Kovashka, A.: Exploring corruption robustness: inductive biases in vision transformers and mlp-mixers. arXiv preprint arXiv:2106.13122 (2021)
40. Pedraza, A., Deniz, O., Bueno, G.: Really natural adversarial examples. Int. J. Mach. Learn. Cybern. **13**, 1–13 (2021)
41. Pei, Y., Huang, Y., Zou, Q., Zhang, X., Wang, S.: Effects of image degradation and degradation removal to cnn-based image classification. IEEE Trans. Pattern Anal. Mach. Intell. **43**(4), 1239–1253 (2019)
42. Raghunathan, A., Xie, S.M., Yang, F., Duchi, J.C., Liang, P.: Adversarial training can hurt generalization. arXiv preprint arXiv:1906.06032 (2019)
43. Saikia, T., Schmid, C., Brox, T.: Improving robustness against common corruptions with frequency biased models. In: Proceedings of the IEEE/CVF International Conference on Computer Vision, pp. 10211–10220 (2021)
44. Samangouei, P., Kabkab, M., Chellappa, R.: Defense-gan: protecting classifiers against adversarial attacks using generative models. arXiv preprint arXiv:1805.06605 (2018)
45. Schneider, S., Rusak, E., Eck, L., Bringmann, O., Brendel, W., Bethge, M.: Improving robustness against common corruptions by covariate shift adaptation. Adv. Neural Inf. Process. Syst. **33**, 11539–11551 (2020)

46. Shafahi, A., et al.: Adversarial training for free! Adv. Neural Inf. Process. Syst. **32** (2019)
47. Shafahi, A., Najibi, M., Xu, Z., Dickerson, J., Davis, L.S., Goldstein, T.: Universal adversarial training. In: Proceedings of the AAAI Conference on Artificial Intelligence, vol. 34, pp. 5636–5643 (2020)
48. Simonyan, K., Zisserman, A.: Very deep convolutional networks for large-scale image recognition. arXiv preprint arXiv:1409.1556 (2014)
49. Szegedy, C., Vanhoucke, V., Ioffe, S., Shlens, J., Wojna, Z.: Rethinking the inception architecture for computer vision. In: IEEE Conference on Computer Vision and Pattern Recognition, pp. 2818–2826 (2016)
50. Taheri, H., Pedarsani, R., Thrampoulidis, C.: Asymptotic behavior of adversarial training in binary classification. arXiv preprint arXiv:2010.13275 (2020)
51. Tramer, F.: Detecting adversarial examples is (nearly) as hard as classifying them. arXiv preprint arXiv:2107.11630 (2021)
52. Wang, J., et al.: Smsnet: a new deep convolutional neural network model for adversarial example detection. IEEE Trans. Multimedia **24**, 230–244 (2021)
53. Xue, M., Yuan, C., He, C., Wang, J., Liu, W.: Naturalae: natural and robust physical adversarial examples for object detectors. J. Inf. Secur. Appl. **57**, 102694 (2021)
54. Zhang, H., Chen, H., Song, Z., Boning, D., Dhillon, I.S., Hsieh, C.J.: The limitations of adversarial training and the blind-spot attack. arXiv preprint arXiv:1901.04684 (2019)

Masked Faces with Faced Masks

Jiayi Zhu[1], Qing Guo[2(✉)], Felix Juefei-Xu[3], Yihao Huang[2], Yang Liu[2],
and Geguang Pu[1,4(✉)]

[1] East China Normal University, Shanghai, China
`ggpu@sei.ecnu.edu.cn`
[2] Nanyang Technological University, Singapore, Singapore
`tsingqguo@ieee.org`
[3] Alibaba Group, Sunnyvale, USA
[4] Shanghai Industrial Control Safety Innovation Technology Co., Ltd.,
Shanghai, China

Abstract. Modern face recognition systems (FRS) still fall short when
the subjects are wearing facial masks. An intuitive partial remedy is to
add a mask detector to flag any masked faces so that the FRS can act
accordingly for those low-confidence masked faces. In this work, we set
out to investigate the potential vulnerability of such FRS equipped with
a mask detector, on large-scale masked faces, which might trigger a serious risk, *e.g.*, letting a suspect evade the facial identity from FRS and not
detected by mask detectors simultaneously. We formulate the new task
as the generation of realistic & adversarial-faced mask and make three
main contributions: *First*, we study the naive *Delaunay-based masking
method (DM)* to simulate the process of wearing a faced mask, which
reveals the main challenges of this new task. *Second*, we further equip
the DM with the adversarial noise attack and propose the *adversarial noise Delaunay-based masking method (AdvNoise-DM)* that can fool
the face recognition and mask detection effectively but make the face
less natural. *Third*, we propose the *adversarial filtering Delaunay-based
masking method* denoted as MF^2M by employing the adversarial filtering
for AdvNoise-DM and obtain more natural faces. With the above efforts,
the final version not only leads to significant performance deterioration of
the state-of-the-art (SOTA) deep learning-based FRS, but also remains
undetected by the SOTA facial mask detector simultaneously.

Keywords: Face recognition · Mask detection · Adversarial attack

1 Introduction

Currently, under the severe international situation and environment (*i.e.*,
COVID-19 pandemic), people are mandatorily required to wear facial masks
in public, especially in crowded places like airports. This situation poses a huge
challenge for face recognition systems (FRS). Although existing face recognition
models (*e.g.*, SphereFace [26], CosFace [40], ArcFace [5]) have high-performance
on identity recognition tasks, these models are only available to faces in good

L. Karlinsky et al. (Eds.): ECCV 2022 Workshops, LNCS 13801, pp. 360–377, 2023.
https://doi.org/10.1007/978-3-031-25056-9_24

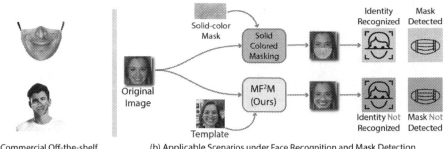

(a) Commercial Off-the-shelf   (b) Applicable Scenarios under Face Recognition and Mask Detection
       Faced Mask

Fig. 1. (a) shows a commercial off-the-shelf (COTS) faced mask and the look of being worn. (b) represents the security problem we are exploring. Faces wearing solid-color masks will be recognized as their original identities in most cases and easily detected by a mask detector (the upper arrow). The MF^{2}M we proposed extracts the face information in the area surrounded by the red line of the template image to obtain a "faced mask", which can simultaneously deceive face recognizers and evade mask detection (the lower arrow). (Color figure online)

imagery conditions. When faces are heavily obscured (*e.g.*, wearing facial masks), even the state-of-the-art (SOTA) FRS do not perform satisfactorily since the information of the masked area is lost as shown in the specific study [31].

An indirect way to solve this problem is to do mask detection. Once a facial mask is detected, the inspector can be made aware of the inaccuracy of the face recognition result and respond accordingly. However, there are various styles of commercial off-the-shelf (COTS) facial masks, even some faced masks (*i.e.*, printed with the lower half of faces from celebrities) as shown in Fig. 1(a). Such COTS faced masks cause great confusion to existing mask detectors as these detectors only consider solid-colored masks but can not deal with masks with special textures and complex patterns. It is worrying that potential offenders may wear such COTS faced masks and even do special treatment to viciously hide their identities while avoiding mask detection. To explore this security problem, we simulate the process of manufacturing masks with face patterns and propose "faced mask" approaches. Our approaches attack both the FRS and the mask detector, exposing their weaknesses under this multitasking attack.

In this paper, we propose an adversarial filtering Delaunay-based masking method, denotes as **Masked Faces with Faced Masks** (MF^{2}M), to stealthily generate masks with face patterns. The perpetrating faced masks not only significantly reduce the accuracy of two SOTA deep learning-based FRS but also drop the accuracy of a SOTA mask detector by 83.58%. As shown in Fig. 1(b), faces wearing solid-colored masks will be recognized as their original identities in most cases and easily detected by a mask detector (the upper arrow). The MF^{2}M (the lower arrow) can simultaneously deceive face recognizers and evade mask detection. In particular, we first modify the Delaunay method [29] to simulate the process of wearing faced masks. We replace the lower face of the input image (*i.e.*, the original image in Fig. 1(b)) with the lower face of the desired face image (*i.e.*, the area surrounded by the red line of the template image in Fig. 1(b)). Then we intuitively add adversarial noise to the mask and further exploit filters to propose the novel attack method MF^2M.

To our best knowledge, previous methods all attack the FRS by modifying the upper face area [22,34,47]. Since the FRS mainly takes features of the eye area (*i.e.*, the periocular region [15–19]) into consideration, attacking the FRS through only modifying the lower face is much more difficult. Our method is the first attempt which only changes the lower face area to attack the FRS.

The contributions are summarized as follows. ❶ We study the naive Delaunay-based masking method (DM) to simulate the process of wearing a faced mask, which reveals the main challenge that this operation of only replacing the lower face does not strongly interference discriminators. ❷ We further equip the DM with the adversarial noise attack and propose the adversarial noise Delaunay-based masking method (AdvNoise-DM) that can handle the joint-task that fools the face recognition and mask detection effectively. ❸ We propose the adversarial filtering Delaunay-based masking method (denoted as MF^2M) by employing the adversarial filtering for AdvNoise-DM. This masking method leads to significant performance deterioration of SOTA deep learning-based face recognizers and mask detector while ensuring the naturalness of the obtained faces. ❹ Our extensive experiments in white-box attack and black-box attack demonstrate the universality and transferability of our proposed MF^2M. Then we extend to physical attack and illustrate the robustness of our proposed masking method.

2   Related Work

2.1   Face Recognition

Face recognition can be divided into closed-set and open-set. Closed-set recognition is regarded as a multi-class classification problem [4,32,36,37,42]. Currently, researchers focus on open-set recognition (*i.e.*, identities for testing do not exist in the training datasets), learning an embedding to represent each identity [5,26,39,40,45,48]. These methods pay little attention to obscured faces and are incompetent with masked face recognition. To tackle this problem, some researchers added specific modules to strengthen the recognition performance for masked faces [2,24,30]. Recently, Masked Face Recognition Competition (MFR 2021) [3] was held. Almost all of the participants utilized variations of the Arc-Face as their loss and exploited either real or simulated solid-colored masked face images as part of their training datasets. However, these methods do not solve the problem essentially, ignoring masks with face patterns and special textures.

2.2   Mask Detection

In the age of the outbreak of the COVID-19 pandemic, researchers pay more attention to masked face detection and relative datasets. The Masked Faces

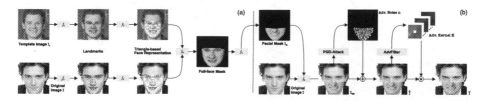

Fig. 2. (a) shows process of generating the faced mask from the original image and the template image. (b) shows the pipeline of the proposed Delaunay-based Masking method, AdvNoise-DM, and MF^{2}M. (Color figure online)

(MAFA) dataset [11] is an early proposed dataset for occluded face detection. Wuhan University has introduced the Real-world Masked Face Recognition Dataset (RMFRD) and the Simulated Masked Face Recognition Dataset (SMFRD) [43]. These datasets all focus on masks with solid colors. To judge whether there is a mask, some mask detection methods fine-tune face detection models to meet the requirement [1,27,33]. Although these methods perform well on common mask detection, they can not deal with special masks with facial textures.

2.3 Adversarial Attack

The fast gradient sign method (FGSM) [12] first proposes the additive-based attack and the iterative fast gradient sign method (I-FGSM) [23] is an iterative variant of FGSM. Then the momentum iterative fast gradient sign method (MI-FGSM) [8] intends to escape from poor local maxima. Besides, the translation-invariant fast gradient sign method (TI-FGSM) [9] and the diverse inputs iterative fast gradient sign method (DI^2-FGSM) [46] are both designed for transferability. More recently, a series of works focus on natural degradation-based adversarial attacks like adversarial morphing attack against face recognition [41], adversarial relighting attack [10], adversarial blur attack [13,14], and adversarial vignetting attack [38]. These works explore the robustness of deep models by adding natural degradations like motion blur and light variation to the input. In this work, we actually regard the faced masks as the real-world perturbations.

3 Methodology

3.1 Delaunay-Based Masking and Motivation

To simulate the process of wearing faced masks, the most intuitive way is to do an operation similar to face replacing in the specified area. We first apply a Delaunay-based masking method (DM). This method operates on two images, one is an original image $\mathbf{I} \in \mathbb{R}^{H_1 \times W_1 \times 3}$ on which we want to put the faced mask, the other is a template image $\mathbf{I_t} \in \mathbb{R}^{H_2 \times W_2 \times 3}$ used to build the faced mask. The

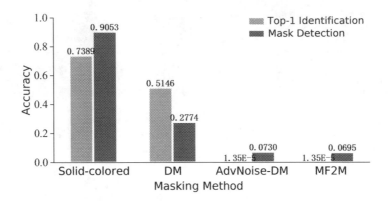

Fig. 3. Top-1 identification rates and mask detection rates for solid-colored mask, DM, AdvNoise-DM and MF^2M on MegaFace Challenge 1.

template image can be constructed by some DeepFake technique (*e.g.*, StyleGAN [20]). We aim to generate a faced mask from $\mathbf{I_t}$ which fits $\mathbf{I}$ as

$$\mathbf{I_M} = f_1(\mathbf{I}, \mathbf{I_t}), \tag{1}$$

where $\mathbf{I_M} \in \mathbb{R}^{H_1 \times W_1 \times 3}$ is the obtained mask (*e.g.*, facial mask in Fig. 2). Specifically, we expand the function $f_1(\cdot)$ as

$$f_1(\mathbf{I}, \mathbf{I_t}) = f_C(f_R(f_T(f_L(\mathbf{I_t})), f_T(f_L(\mathbf{I})))), \tag{2}$$

where $f_L(\cdot)$ extracts the landmarks of $\mathbf{I_t}$ and $\mathbf{I}$ as the first step of Fig. 2(a). The function $f_T(\cdot)$ is to build the triangle-based face representation where the landmarks serve as vertices of the triangles. The obtained face representations from $\mathbf{I_t}$ and $\mathbf{I}$ have the same number of triangles and those triangles correspond one by one according to the landmarks. As we have the correspondence of triangles between the two face representations, $f_R(\cdot)$ transforms each triangle in $f_T(f_L(\mathbf{I_t}))$ into the corresponding triangle in $f_T(f_L(\mathbf{I}))$ by affine transformation to get a full-face mask at the right side of Fig. 2(a). The function $f_C(\cdot)$ connects the landmarks of the contour of the lower face and the landmark of the nose (red dots in the full-face mask) in turn to obtain the faced mask area. In the beginning of Fig. 2(b), after getting the facial mask $\mathbf{I_M}$, we overlay it on the original image $\mathbf{I}$ to get the Delaunay-based masked image $\mathbf{I_{DM}} \in \mathbb{R}^{H_1 \times W_1 \times 3}$ through

$$\mathbf{I_{DM}} = f_2(\mathbf{I}, \mathbf{I_M}) = f_2(\mathbf{I}, f_1(\mathbf{I}, \mathbf{I_t})). \tag{3}$$

To better motivate our proposed method, we have carried out a pilot study. Here, we briefly discuss the results from DM and compare them with the results of solid-color masked faces. We use the whole gallery set (1M images) and take 3,530 faces of 80 celebrities as the probe set from MegaFace Challenge 1 [21]. According to Fig. 3, both indicators are high when adding solid-colored masks, indicating that such masks hardly influence face recognizers and the mask detector. Images obtained by DM have a considerable impact to discriminators. However, DM is not effective enough. There are still about 51% and 28% tasks judged correctly for top-1 identification and mask detection, far from zero. To strengthen the aggressiveness of the designed faced mask, we propose adversarial masking methods to add special textures, as explained in the following sections.

The main challenges stem from: ❶ Most of the face information is concentrated in the eye region. In contrast, the part of the mask area, *e.g.*, the mouth, plays a relatively low role in the face recognition task, which increases the difficulty of our work. ❷ Although a part of images processed by DM can remain undetected by the facial mask detector, 28% of images are detected due to the unavoidable factors in the process of adding masks (*e.g.*, the discontinuities in textures). ❸ Different deep-learning-based discriminators use different network structures and parameters. It is hard to ensure the transferability that the generated faced masks can effectively interfere with diverse discriminators.

3.2   Adversarial Noise Delaunay-Based Masking

Inspired by adversarial attacks, *e.g.*, project gradient descent (PGD) [28], we apply an adversarial noise to the masked image $\mathbf{I_{DM}}$ acquired by DM. We define this method as an adversarial noise Delaunay-based masking method (AdvNoise-DM) and show the process in the middle of Fig. 2(b). We replace $\mathbf{I_{DM}}$ with Eq. (3) and generate the adversarial noise $\mathbf{n} \in \mathbb{R}^{H_1 \times W_1 \times 3}$ to obtain

$$\hat{\mathbf{I}} = f_2(\mathbf{I}, f_1(\mathbf{I}, \mathbf{I_t})) + \mathbf{n}, \tag{4}$$

which denotes the superimposition of the intermediate $f_2(\mathbf{I}, f_1(\mathbf{I}, \mathbf{I_t}))$ and the adversarial perturbation $\mathbf{n}$. Our goal is to find the $\hat{\mathbf{I}}$ which can not only mislead the FRS but also remain undetected by the mask detector by obtaining such an adversarial perturbation $\mathbf{n}$. We aim to achieve the optimal trade-off between face recognition and mask detection and we have the following objective function

$$\underset{\mathbf{n}}{\arg\max} \, \mathcal{D}(\mathrm{FR}(f_2(\mathbf{I}, f_1(\mathbf{I}, \mathbf{I_t})) + \mathbf{n}), \mathrm{FR}(\mathbf{I}))$$
$$-\alpha * \mathcal{J}(\mathrm{MD}(f_2(\mathbf{I}, f_1(\mathbf{I}, \mathbf{I_t})) + \mathbf{n}), y). \tag{5}$$

In the first part of the objective function, $\mathrm{FR}(\cdot)$ denotes a face recognition function that receives an image and returns an embedding. $\mathcal{D}(\cdot)$ denotes the Euclidean distance between the embedding from the original image $\mathbf{I}$ and the embedding from the image $\hat{\mathbf{I}}$ by AdvNoise-DM. We intend to maximize this part to enlarge the gap in identification information before and after the modification.

In the second part, $\mathrm{MD}(\cdot)$ represents a mask detection function that receives an image and returns the probability of wearing a mask. $\mathcal{J}(\cdot)$ denotes the cross-entropy loss function, y is the ground truth label for whether the face is masked, 0 for not masked, and 1 for masked. Here, we set $y = 0$ to force the image $\hat{\mathbf{I}}$ to be judged without a mask. The ratio α is the coefficient of this term. We aim to minimize this cross-entropy loss so we take a minus sign for this item.

AdvNoise-DM almost reduces the top-1 identification rate to zero and reduces the mask detection rate to only 7.3% as shown in Fig. 3. Nevertheless, AdvNoise-DM causes great changes to the pixels, so reduces the naturalness of generated images. In this case, it is necessary to use a smoother masking method.

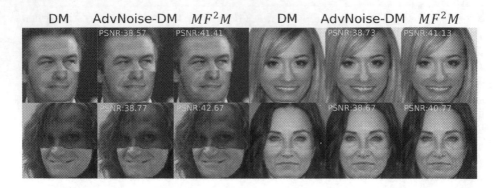

Fig. 4. Examples for the PSNR values between AdvNoise-DM and MF²M. The PSNR value is calculated by the current image and the DM image.

3.3   Adversarial Filtering Delaunay-Based Masking

Since the filtering process brings better smoothness, calculating each pixel by the surrounding pixels, we further propose an adversarial filtering Delaunay-based masking method (MF²M). This method combines noise-based and filtering-based attacks as shown in Fig. 2(b). We first apply DM and add a relatively small adversarial perturbation $\mathbf{n}$ to get the intermediate $f_2(\mathbf{I}, f_1(\mathbf{I}, \mathbf{I_t})) + \mathbf{n}$, referring to the method AdvNoise-DM. Then we utilize pixel-wise kernels $\mathbf{K} \in \mathbb{R}^{H_1 \times W_1 \times K^2}$. The p-th pixel of the intermediate $f_2(\mathbf{I}, f_1(\mathbf{I}, \mathbf{I_t})) + \mathbf{n}$ is processed by the p-th kernel in $\mathbf{K}$, denoted as $\mathbf{K}_p \in \mathbb{R}^{K \times K}$, where K is the kernel size. We retouch the original image $\mathbf{I}$ via the guidance of filtering and reformulate Eq. (4) as

$$\tilde{\mathbf{I}} = \mathbf{K} \circledast (f_2(\mathbf{I}, f_1(\mathbf{I}, \mathbf{I_t})) + \mathbf{n}), \tag{6}$$

where $\circledast$ denotes the pixel-wise filtering process and $\tilde{\mathbf{I}} \in \mathbb{R}^{H_1 \times W_1 \times 3}$ represents the filtered images. In the MF²M procedure, we aim at obtaining a deceptive $\tilde{\mathbf{I}}$ for both face recognition and mask detection by altering the pixel-wise kernels $\mathbf{K}$. The objective function for optimization looks similar to Eq. (5) as following

$$\underset{\mathbf{K}}{\arg\max} \, \mathcal{D}(\mathrm{FR}(\mathbf{K} \circledast (f_2(\mathbf{I}, f_1(\mathbf{I}, \mathbf{I_t})) + \mathbf{n})), \mathrm{FR}(\mathbf{I}))$$
$$-\beta * \mathcal{J}(\mathrm{MD}(\mathbf{K} \circledast (f_2(\mathbf{I}, f_1(\mathbf{I}, \mathbf{I_t})) + \mathbf{n})), y). \tag{7}$$

Compared with Eq. (5), the optimization objective becomes $\mathbf{K}$. We intend to increase the Euclidean distance between the embedding from the original image $\mathbf{I}$ and that from the filtered image $\tilde{\mathbf{I}}$. Meanwhile, we try to improve the probability that the filtered image $\tilde{\mathbf{I}}$ is judged as not wearing a mask. The ratio of the mask detection part is marked as β. As shown in Fig. 3, almost all images generated by MF²M is deceptive for face recognition and only 6.95% images are detected wearing masks. Besides, Fig. 4 shows that the peak signal-to-noise ratio (PSNR) calculated between MF²M and DM is higher than that calculated between AdvNoise-DM and DM, indicating that MF²M changes images less.

Algorithm 1: MF^2M

Input: Original image $\mathbf{I}$, Template image $\mathbf{I_t}$, Ratio β, Face recognizer $FR(\cdot)$,
    Mask detector $MD(\cdot)$, Step size ϵ, Label y of not masked image,
    Iteration period $\mathbf{T}$.

Output: Reconstruction image $\tilde{\mathbf{I}}$.

1   Generate $\mathbf{I_{DM}}$ by extracting the faced mask from $\mathbf{I_t}$ and overlay it on $\mathbf{I}$ by
    Delaunay triangulation.

2   $\hat{\mathbf{I}} = PGD_{attack}(\mathbf{I_{DM}})$.

3   Initial filtering kernels $\mathbf{K}$;

4   **for** $i = 1$ to $\mathbf{T}$ **do**

5   Generate filtered image $\mathbf{I}'$ via $\mathbf{I}' = \mathbf{K} \circledast \hat{\mathbf{I}}$;

6   Calculate $\mathbf{Loss_D}$ via Euclidean distance function $\mathcal{D}$

        $\mathbf{Loss_D} = \mathcal{D}(FR(\mathbf{I}'), FR(\mathbf{I}))$;

7   Calculate $\mathbf{Loss_{CE}}$ via cross-entropy loss function $\mathcal{J}$

        $\mathbf{Loss_{CE}} = \mathcal{J}(MD(\mathbf{I}'), y)$;

8   Calculate the sum loss function $\mathbf{Loss}$ via $\mathbf{Loss} = \mathbf{Loss_D} - \beta * \mathbf{Loss_{CE}}$;

9   Update filtering kernels $\mathbf{K}$ via $\mathbf{K} = \mathbf{K} + \epsilon * \nabla_{\mathbf{K}} \mathbf{Loss}$;

10   Apply image filtering to obtain reconstruction image $\tilde{\mathbf{I}}$ via $\tilde{\mathbf{I}} = \mathbf{K} \circledast \hat{\mathbf{I}}$;

3.4 Algorithm for MF^2M

Algorithm 1 summarizes our method. First, we apply DM to complete the face replacing process, $i.e.$, extracting a faced mask from $\mathbf{I_t}$ and overlaying it on $\mathbf{I}$ to obtain $\mathbf{I_{DM}}$. Second, we add a relatively small adversarial noise $\mathbf{n}$ to $\mathbf{I_{DM}}$ and obtain $\hat{\mathbf{I}}$. In the filtering attack process, we initialize the filtering kernels $\mathbf{K}$ whose initial action is to make the filtered image consistent with the original image ($i.e.$, the weight of the center position of each kernel is 1, and the weight of other positions is 0). In each iteration, we perform pixel-wise filtering by current kernels $\mathbf{K}$ and $\mathbf{I_{DM}}$ to acquire the current filtered image $\mathbf{I}'$. Then we calculate $\mathbf{Loss_D}$ and $\mathbf{Loss_{CE}}$, via the Euclidean distance function and the cross-entropy loss function, according to Eq. (7). These two loss functions constitute the final optimization objective function by the ratio β. At the end of each iteration, we update the filtering kernels $\mathbf{K}$ according to the product of the step size ϵ and the gradient of the optimization objective. Finally, we embellish the aimed image $\tilde{\mathbf{I}}$.

4 Experiments

4.1 Experimental Setup

Face Recognition Methods. In our white-box attack experiment, the backbone of the face recognizer [6] is pre-trained ResNet50 under ArcFace. The face recognizer takes cropped images (112×112) as input and returns the final 512-D embedding features. To illustrate the transferability, we further use recognizers [7] pre-trained under CosFace with ResNet34 and ResNet50 as the backbone respectively to verify the black-box attack performance. We choose these two FRS as ArcFace and CosFace perform SOTA in face recognition.

Mask Detection Methods. The mask detection method bases on RetinaNet [25], an efficient one-stage objects detecting method. The pre-trained model [35] we used is competitive in existing mask detectors, achieving 91.3% mAP at the face_mask validation dataset (including 1839 images). The mask detector outputs two probabilities of not-masked and masked faces respectively.

Datasets. We utilize 1M images of 690K individuals in MegaFace Challenge 1 as the gallery set. For the probe set, we refer to MegaFace [21] and use a subset of FaceScrub (*i.e.*, 3,530 images of 80 celebrities) for efficiency. For the template images used to extract faced masks, we use StyleGAN to generate images with seeds numbered from 1 to 13,000. As some generated images have illumination or occlusion problems, we manually select 3,136 high-quality face images.

Evaluation Settings. The face recognition evaluation is based on masked/not-masked pairs. We add masks to images of the probe set and remain images in the gallery not-masked. When adding masks, we select the most similar face image (whose embedding extracted by the face recognition model has the closest Euclidean distance to the embedding from the original face) from 3,136 template images. In AdvNoise-DM, we use PGD attack to add deliberate noise. The epsilon (maximum distortion of adversarial example) is 0.04. The step size for each attack iteration is 0.001 while the number of iterations is 40. The ratio α is set to 1. In MF^2M, we add noise with an epsilon of 0.01. The kernel size of the pixel-wise kernels is 5. When alter the pixel-wise kernels, the step size is 0.1 and the number of iterations is 160. Here we set the coefficient β to 1, same as α. All optimization objectives are restricted to the faced mask area.

Baseline Methods. We apply seven SOTA adversarial attack methods to the solid-colored medical masks as our baselines, including FGSM, I-FGSM, MI-FGSM, TI-FGSM, TI-MI-FGSM, DI^2-FGSM, and M-DI^2-FGSM. The epsilon of these baselines is 0.04, to maintain the same attack intensity as AdvNoise-DM. The iterations number involved in baselines is 40, and the momentum coefficient involved is 1. The added perturbation is restricted to the facial mask area.

Metrics. For face recognition, we use the top-1 identification rate for identification, the true accept rate (TAR) at 10^{-6} false accept rate (FAR) and the area under curve (AUC) for verification. For mask detection, we use the detection rate. To reflect the degree of reconstructed modification and evaluate the naturalness of the generated images, we calculate the PSNR and the structural similarity (SSIM) [44] between the adversarial masked results and images from DM. The region of the calculation for similarity metrics is the whole image.

4.2 Comparison on White-Box Attack

Face recognition has two main tasks, face identification and face verification. Given a probe image and a gallery, identification aims to find an image which has the same identity as the probe image from the gallery, *i.e.*, 1 vs. N search task. The verification task sets a threshold to judge whether two images have the same identity, *i.e.*, 1 vs. 1 comparison task.

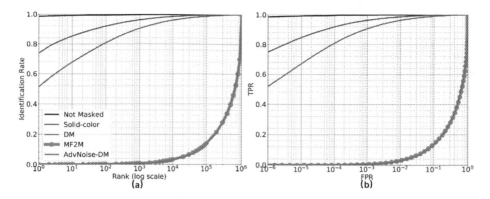

Fig. 5. The effects on face recognition in white-box attacks. (a) The CMC curves for face identification. The abscissa indicates the number of images selected from the gallery. The ordinate denotes the identification rate at the specified number of images. (b) The ROC curves for face verification. The abscissa indicates the FPR and the ordinate denotes the TPR.

Table 1. Multifaceted evaluation on white-box attack. "Rank 1" refers to the top-1 identification rate and "Veri." refers to the TAR at 10^{-6} FAR.

	Face Rec.	Face Verification		Mask Detection
	Rank 1	Veri.	AUC	Mask Rate
Solid-color	0.7389	0.7470	0.9986	90.53%
FGSM	0.4895	0.4924	0.9950	62.10%
I-FGSM	0.0039	0.0026	0.7594	39.41%
MI-FGSM	0.0179	0.0135	0.8667	43.91%
TI-FGSM	0.4991	0.5053	0.9950	66.00%
TI-MI-FGSM	0.0364	0.0290	0.8998	51.97%
DI^2-FGSM	0.0690	0.0601	0.9193	43.51%
M-DI^2-FGSM	0.0705	0.0623	0.9293	45.50%
DM (ours)	0.5146	0.5154	0.9956	27.74%
AdvNoise-DM (ours)	**1.35e^{-5}**	**1.35e^{-5}**	0.4163	7.30%
MF^{2}M (ours)	**1.35e^{-5}**	2.02e^{-5}	**0.4093**	**6.95%**

Face Identification. We constitute 151K pairs with the same identity from 3,530 face images of 80 celebrities. For each pair, we take one image as the probe image and put the other image into the gallery. Top-k identification rate denotes the successful rate of matching pairs where k is the number of images selected from the gallery. Figure 5(a) shows the cumulative matching characteristic (CMC) curves of images under different masking states. The abscissa indicates the number of images selected from the gallery according to the embedding obtained by the face recognizer. The ordinate denotes the identification rate at the specified number of images. Without facial masks, the top-1 identification rate (*i.e.*, "Rank 1") is 0.98. After adding solid-color masks, "Rank 1" reduces to 0.7389. This metric for DM declines to 0.5146. As for AdvNoise-DM and MF^{2}M, the higher the attack intensity, the more their corresponding curves are close to the lower right of the graph. We respectively alter the iteration numbers and make the performance of AdvNoise-DM and MF^{2}M close in this task, to compare them on other indicators. "Rank 1" of both methods drop to **1.35e^{-5}**, indicating that the SOTA recognizer performs poorly under AdvNoise-DM and MF^{2}M. The second column of Table 1 shows that AdvNoise-DM and MF^{2}M achieve significantly lower "Rank 1" than seven SOTA additive-based baselines.

Face Verification. We use the 3,530 images of 80 identities in the probe set and 1M images in the gallery to build 151K positive samples and 3.5 billion negative samples for face verification. Figure 5(b) shows the receiver operating characteristic (ROC) curves. We define the true positive rate (TPR) when the false positive rate (FPR) is $1e^{-6}$ as "Veri.", which is 0.7470 and 0.5154 for solid-color masks and DM, respectively. For AdvNoise-DM and MF^{2}M, "Veri." almost drops to zero. The third and fourth columns of Table 1 show the "Veri." and the AUC values. Both metrics of AdvNoise-DM and MF^{2}M are much lower than baselines.

Mask Detection. We exhibit the mask detection rate of different masking methods in the fifth column of Table 1. Solid-colored masks are easily detected and the detection rate is 90.53%. DM reduces the detection rate to 27.74%. AdvNoise-DM further interferes with the judgment of the detector and the accuracy decreases to only 7.30%. MF^{2}M achieves the best attack performance and reduces this rate to 6.95%. The detection rates for additive-perturbation-based baselines are between 39% and 66%. So far, we prove that our adversarial methods are very effective for both face recognition and mask detection in white-box attacks.

Similarity Measurement. The value for similarity measurement is calculated by comparing with images obtained by DM, so we only calculate similarity scores for AdvNoise-DM and MF^{2}M. We choose SSIM and PSNR as our similarity metrics for aspects of visual error and structure difference. The SSIM of AdvNoise-DM and MF^{2}M are 0.9808 and 0.9812, respectively. For PSNR, the value of MF^{2}M is 40.45, higher than 38.76 of AdvNoise-DM, which proves that the filtering operation has a more imperceptible modification to images.

4.3   Comparison on Black-Box Attack Transferability

We utilize generated masked images to conduct black-box attacks against face recognition models pre-trained under Cosface with ResNet34 and ResNet50 as the backbone. Curves with dots and without dots in Fig. 6 represent results of attacking ResNet34 and ResNet50, respectively. The "Rank 1" of AdvNoise-DM and MF^{2}M vary from 0.05 to 0.08 in Fig. 6(a). It shows that our adversarial masking methods have sufficient transferability. Besides, curves of MF^{2}M are lower than curves of AdvNoise-DM targeting the same model, indicating MF^{2}M has stronger transferability. Compared with adversarial attack baselines, AdvNoise-DM and MF^{2}M have absolute advantages as shown in Table 2.

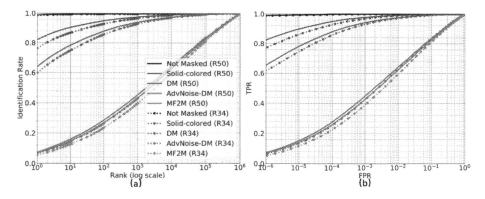

Fig. 6. The effects on face recognition in black-box attacks. (a) The CMC curves for face identification. (b) The ROC curves for face verification. The abscissa and ordinate are consistent with Fig. 5. R34 and R50 in the legend indicate that ResNet34 and ResNet50 are separately used as the backbone.

Table 2. Multifaceted evaluation on black-box attack. "Rank 1" refers to the top-1 identification rate and "Veri." refers to the TAR at 10^{-6} FAR.

	CosFace ResNet50			CosFace ResNet34		
	Face Rec.	Face Verification		Face Rec.	Face Verification	
	Rank 1	Veri.	AUC	Rank 1	Veri.	AUC
Solid-color	0.8207	0.8233	0.9991	0.7641	0.7697	0.9987
FGSM	0.7585	0.7624	0.9988	0.7147	0.7212	0.9984
I-FGSM	0.6720	0.6752	0.9978	0.6243	0.6303	0.9967
MI-FGSM	0.6419	0.6445	0.9972	0.6020	0.6077	0.9961
TI-FGSM	0.7640	0.7711	0.9988	0.7142	0.7214	0.9984
TI-MI-FGSM	0.6416	0.6438	0.9972	0.5996	0.6038	0.9958
DI^2-FGSM	0.7178	0.7222	0.9983	0.6699	0.6777	0.9977
M-DI^2-FGSM	0.6559	0.6592	0.9974	0.6165	0.6242	0.9963
DM (ours)	0.6400	0.6520	0.9971	0.5987	0.6126	0.9971
AdvNoise-DM (ours)	0.0726	0.0723	0.9293	0.0657	0.0622	0.9310
MF^{2}M (ours)	**0.0674**	**0.0670**	**0.9237**	**0.0546**	**0.0508**	**0.9212**

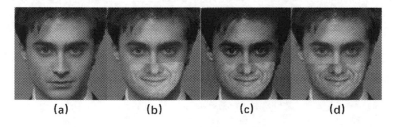

Fig. 7. The process of synthesizing the image in the physical attack. (a) The original image in the FaceScrub. (b) The digital image by MF^{2}M. (c) The recaptured image. (d) Extract the faced mask from (c) and overlay it to (a)

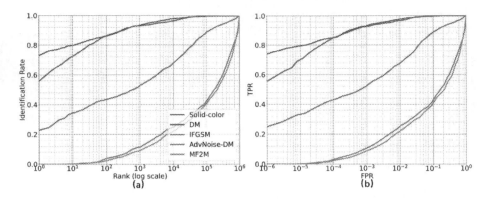

Fig. 8. The effects on face recognition in *physical* white-box attacks. (a) The CMC curves for face identification. (b) The ROC curves for face verification. The abscissa and ordinate are consistent with Fig. 5.

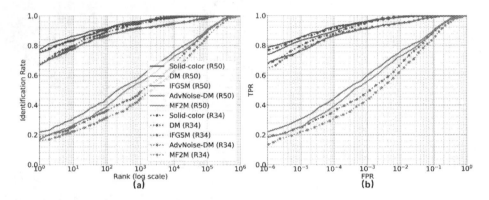

Fig. 9. The effects on face recognition in *physical* black-box attacks. (a) The CMC curves for face identification. (b) The ROC curves for face verification. The abscissa and ordinate are consistent with Fig. 5. R34 and R50 in the legend indicate that ResNet34 and ResNet50 are separately used as the backbone.

Table 3. Multifaceted evaluation of *physical* white-box attacks. "Rank 1" refers to the top-1 identification rate and "Veri." refers to the TAR at 10^{-6} FAR.

	Face Rec.	Face Verification		Mask Detection
	Rank 1	Veri.	AUC	Mask Rate
Solid-color (phys.)	0.7302	0.7398	0.9985	100%
DM (phys.)	0.5574	0.5550	0.9975	10%
I-FGSM (phys.)	0.2284	0.2497	0.9453	95%
AdvNoise-DM (phys.)	**0.0000**	**0.0000**	0.6879	**10%**
MF^{2}M (phys.)	**0.0000**	**0.0000**	0.6658	**10%**

Table 4. Multifaceted evaluation of *physical* black-box attacks. "Rank 1" refers to the top-1 identification rate and "Veri." refers to the TAR at 10^{-6} FAR.

	CosFace ResNet50			CosFace ResNet34		
	Face Rec.	Face Verification		Face Rec.	Face Verification	
	Rank 1	Veri.	AUC	Rank 1	Veri.	AUC
Solid-color (phys.)	0.7740	0.7846	0.9992	0.7586	0.7670	0.9993
DM (phys.)	0.7527	0.7369	0.9987	0.6757	0.6739	0.9985
I-FGSM (phys.)	0.6686	0.6791	0.9958	0.6734	0.6438	0.9951
AdvNoise-DM (phys.)	0.2189	0.2184	0.9683	0.1858	0.1860	0.9627
MF^{2}M (phys.)	**0.1692**	**0.1818**	**0.9670**	**0.1633**	**0.1336**	**0.9498**

4.4   Extension to Physical Attack

Due to various COVID-19 related restrictions, we were not able to recruit human subjects to study the effect of physical attack by wearing our proposed faced masks. Therefore, we use an alternative recapture method to illustrate the physical effects of our proposed masking methods. We randomly select 20 faces of different identities from the FaceScrub dataset as origin images in Fig. 7(a). We process the 20 faces with our proposed MF^{2}M and obtain digital attacked faces as shown in Fig. 7(b). Then we use an *ApeosPort − IVC5575* printer to print these attacked images and recapture images like Fig. 7(c), which has obvious color differences from Fig. 7(b). This procedure is meant for methodologically mimicking the process of plastering the patterns from a digital medium onto a physical one, such as fabric, linen, or paper, so that the physical appearance can be digitally reacquired via image sensors. Finally, we resize the recaptured images to the size of 112 × 112, extract faced masks from them, overlay faced masks to the 20 corresponding original faces, and obtain images used in the physical attack as shown in Fig. 7(d). Based on this synthesis process, we conduct the experiment of physical attacks and demonstrate the robustness of our proposed MF^{2}M and AdvNoise-DM. Except for the solid-color medical mask and DM, we choose *I-FGSM baseline*, which is the strongest baseline in digital attacks.

For the physical white-box attack, Table 3 show that the top-1 identification rates and the TAR at 10^{-6} FAR of MF^{2}M and AdvNoise-DM are both zero, indicating the strong interference of these two masking methods to the face recognizer. The last column of Table 3 shows that only 10% of images by MF^{2}M and AdvNoise-DM are detected faced masks, demonstrating the powerful ability in avoiding mask detection. The corresponding CMC curves in Fig. 8(a) illustrate that the identification rates of MF^{2}M and AdvNoise-DM are always below three baselines at different ranks. The ROC curves in Fig. 8(b) show that the TPR of MF^{2}M and AdvNoise-DM are always less than three baselines at different FPR.

As for the physical black-box attack, the top-1 identification rates and the TAR at 10^{-6} FAR of *I-FGSM baseline* are more than 0.64 as shown in Table 4. It indicates that the adversarial textures added in *I-FGSM baseline* almost failed. In contrast, these metrics of our proposed MF^{2}M and AdvNoise-DM are still less than 0.22, indicating our proposed masking methods remain highly interference to face recognizers in physical black-box attack. Compared with AdvNoise-DM, MF^2M has a greater influence on face recognizers as shown in Fig. 9.

5   Conclusions

In this paper, we propose MF^2M, an adversarial masking framework that adds faced masks containing partial face patterns and special adversarial textures. Our work reveals the potential risks of existing face recognizers and mask detectors regarding facial masks specially customized. The reconstructed images from our methods retain enough naturalness, generating a higher safety hazard. Therefore, particularly generated facial masks should be taken into consideration when designing the FRS and mask detection systems.

Acknowledgments. Geguang Pu is supported by National Key Research and Development Program (2020AAA0107800), and Shanghai Collaborative Innovation Center of Trusted Industry Internet Software. This work is supported by the National Research Foundation, Singapore under its AI Singapore Programme (Award No: AISG2-RP-2020-019), Singapore National Cybersecurity R&D Program No. NRF2018NCR-NCR005-0001, National Satellite of Excellence in Trustworthy Software System No. NRF2018NCR-NSOE003-0001, and NRF Investigatorship No. NRF-NRFI06-2020-0001. We gratefully acknowledge the support of NVIDIA AI Tech Center (NVAITC).

References

1. Batagelj, B., Peer, P., Štruc, V., Dobrišek, S.: How to correctly detect face-masks for COVID-19 from visual information? Appl. Sci. **11**(5), 2070 (2021)
2. Boutros, F., Damer, N., Kirchbuchner, F., Kuijper, A.: Unmasking face embeddings by self-restrained triplet loss for accurate masked face recognition. arXiv preprint arXiv:2103.01716 (2021)
3. Boutros, F., et al.: MFR 2021: masked face recognition competition. In: 2021 IEEE International Joint Conference on Biometrics (IJCB), pp. 1–10. IEEE (2021)
4. Cao, Q., Shen, L., Xie, W., Parkhi, O.M., Zisserman, A.: VGGFace2: a dataset for recognising faces across pose and age. In: 2018 13th IEEE International Conference on Automatic Face & Gesture Recognition (FG 2018), pp. 67–74. IEEE (2018)
5. Deng, J., Guo, J., Niannan, X., Zafeiriou, S.: ArcFace: additive angular margin loss for deep face recognition. In: CVPR (2019)
6. Deng, J., Guo, J., Xue, N., Zafeiriou, S.: PyTorch implementation of face recognition model under ArcFace (2021). https://github.com/deepinsight/insightface/tree/master/recognition/arcface_torch
7. dominhhieu1019: PyTorch implementation of face recognition model under CosFace (2021). https://github.com/dominhhieu1019/arcface_torch
8. Dong, Y., et al.: Boosting adversarial attacks with momentum. In: CVPR, pp. 9185–9193 (2018)
9. Dong, Y., Pang, T., Su, H., Zhu, J.: Evading defenses to transferable adversarial examples by translation-invariant attacks. In: CVPR, pp. 4312–4321 (2019)
10. Gao, R., Guo, Q., Zhang, Q., Juefei-Xu, F., Yu, H., Feng, W.: Adversarial relighting against face recognition. arXiv preprint arXiv:2108.07920 (2021)

11. Ge, S., Li, J., Ye, Q., Luo, Z.: Detecting masked faces in the wild with LLE-CNNs. In: Proceedings of the IEEE Conference on Computer Vision and Pattern Recognition, pp. 2682–2690 (2017)
12. Goodfellow, I.J., Shlens, J., Szegedy, C.: Explaining and harnessing adversarial examples. arXiv preprint arXiv:1412.6572 (2014)
13. Guo, Q., et al.: Learning to adversarially blur visual object tracking. In: ICCV, pp. 10839–10848 (2021)
14. Guo, Q., et al.: Watch out! Motion is blurring the vision of your deep neural networks. In: Advances in Neural Information Processing Systems (NeurIPS) (2020)
15. Juefei-Xu, F.: Unconstrained periocular face recognition: from reconstructive dictionary learning to generative deep learning and beyond. Ph.D. dissertation, Carnegie Mellon University (2018)
16. Juefei-Xu, F., Luu, K., Savvides, M.: Spartans: single-sample periocular-based alignment-robust recognition technique applied to non-frontal scenarios. IEEE Trans. Image Process. **24**(12), 4780–4795 (2015)
17. Juefei-Xu, F., Pal, D.K., Savvides, M.: Hallucinating the full face from the periocular region via dimensionally weighted K-SVD. In: Proceedings of the IEEE Conference on Computer Vision and Pattern Recognition Workshops, pp. 1–8 (2014)
18. Juefei-Xu, F., Savvides, M.: Subspace-based discrete transform encoded local binary patterns representations for robust periocular matching on NIST's face recognition grand challenge. IEEE Trans. Image Process. **23**(8), 3490–3505 (2014)
19. Juefei-Xu, F., Savvides, M.: Fastfood dictionary learning for periocular-based full face hallucination. In: 2016 IEEE 8th International Conference on Biometrics Theory, Applications and Systems (BTAS), pp. 1–8. IEEE (2016)
20. Karras, T., Laine, S., Aila, T.: A style-based generator architecture for generative adversarial networks. In: Proceedings of the IEEE Conference on Computer Vision and Pattern Recognition, pp. 4401–4410 (2019)
21. Kemelmacher-Shlizerman, I., Seitz, S.M., Miller, D., Brossard, E.: The megaface benchmark: 1 million faces for recognition at scale. In: Proceedings of the IEEE Conference on Computer Vision and Pattern Recognition, pp. 4873–4882 (2016)
22. Komkov, S., Petiushko, A.: AdvHat: real-world adversarial attack on ArcFace face ID system. In: 2020 25th International Conference on Pattern Recognition (ICPR), pp. 819–826. IEEE (2021)
23. Kurakin, A., Goodfellow, I., Bengio, S.: Adversarial machine learning at scale. arXiv preprint arXiv:1611.01236 (2016)
24. Li, Y., Guo, K., Lu, Y., Liu, L.: Cropping and attention based approach for masked face recognition. Appl. Intell. **51**(5), 3012–3025 (2021). https://doi.org/10.1007/s10489-020-02100-9
25. Lin, T.Y., Goyal, P., Girshick, R., He, K., Dollár, P.: Focal loss for dense object detection. In: Proceedings of the IEEE International Conference on Computer Vision, pp. 2980–2988 (2017)
26. Liu, W., Wen, Y., Yu, Z., Li, M., Raj, B., Song, L.: SphereFace: deep hypersphere embedding for face recognition. In: Proceedings of the IEEE Conference on Computer Vision and Pattern Recognition, pp. 212–220 (2017)
27. Loey, M., Manogaran, G., Taha, M.H.N., Khalifa, N.E.M.: A hybrid deep transfer learning model with machine learning methods for face mask detection in the era of the COVID-19 pandemic. Measurement **167**, 108288 (2021)
28. Madry, A., Makelov, A., Schmidt, L., Tsipras, D., Vladu, A.: Towards deep learning models resistant to adversarial attacks. arXiv preprint arXiv:1706.06083 (2017)

29. Mallick, S.: Delaunay triangulation and Voronoi diagram using OpenCV (C++/Python) (2015). https://learnopencv.com/delaunay-triangulation-and-voronoi-diagram-using-opencv-c-python

30. Montero, D., Nieto, M., Leskovsky, P., Aginako, N.: Boosting masked face recognition with multi-task ArcFace. arXiv preprint arXiv:2104.09874 (2021)

31. Ngan, M., Grother, P., Hanaoka, K.: Ongoing face recognition vendor test (FRVT) part 6B: face recognition accuracy with face masks using post-COVID-19 algorithms, 30 November 2020. https://doi.org/10.6028/NIST.IR.8331

32. Parkhi, O.M., Vedaldi, A., Zisserman, A.: Deep face recognition. In: Proceedings of the British Machine Vision Conference (BMVC), pp. 1–12. BMVA Press, September 2015

33. Qin, B., Li, D.: Identifying facemask-wearing condition using image super-resolution with classification network to prevent COVID-19. Sensors **20**(18), 5236 (2020)

34. Sharif, M., Bhagavatula, S., Bauer, L., Reiter, M.K.: Accessorize to a crime: real and stealthy attacks on state-of-the-art face recognition. In: Proceedings of the 2016 ACM SIGSAC Conference on Computer and Communications Security, pp. 1528–1540 (2016)

35. simpletask1: PyTorch implementation of retinanet for face mask detection (2020). https://github.com/simpletask1/Retinanet-face_mask_detection

36. Sun, Y., Wang, X., Tang, X.: Deep learning face representation from predicting 10,000 classes. In: Proceedings of the IEEE Conference on Computer Vision and Pattern Recognition, pp. 1891–1898 (2014)

37. Taigman, Y., Yang, M., Ranzato, M., Wolf, L.: DeepFace: closing the gap to human-level performance in face verification. In: Proceedings of the IEEE Conference on Computer Vision and Pattern Recognition, pp. 1701–1708 (2014)

38. Tian, B., Juefei-Xu, F., Guo, Q., Xie, X., Li, X., Liu, Y.: AVA: adversarial vignetting attack against visual recognition. arXiv preprint arXiv:2105.05558 (2021)

39. Wang, F., Cheng, J., Liu, W., Liu, H.: Additive margin softmax for face verification. IEEE Sig. Process. Lett. **25**(7), 926–930 (2018)

40. Wang, H., et al.: CosFace: large margin cosine loss for deep face recognition. In: Proceedings of the IEEE Conference on Computer Vision and Pattern Recognition, pp. 5265–5274 (2018)

41. Wang, R., et al.: Amora: black-box adversarial morphing attack. In: ACM-MM, pp. 1376–1385 (2020)

42. Wang, Z., He, K., Fu, Y., Feng, R., Jiang, Y.G., Xue, X.: Multi-task deep neural network for joint face recognition and facial attribute prediction. In: Proceedings of the 2017 ACM on International Conference on Multimedia Retrieval, pp. 365–374 (2017)

43. Wang, Z., et al.: Masked face recognition dataset and application. arXiv preprint arXiv:2003.09093 (2020)

44. Wang, Z., Bovik, A.C., Sheikh, H.R., Simoncelli, E.P.: Image quality assessment: from error visibility to structural similarity. IEEE Trans. Image Process. **13**(4), 600–612 (2004)

45. Wen, Y., Zhang, K., Li, Z., Qiao, Yu.: A discriminative feature learning approach for deep face recognition. In: Leibe, B., Matas, J., Sebe, N., Welling, M. (eds.) ECCV 2016. LNCS, vol. 9911, pp. 499–515. Springer, Cham (2016). https://doi.org/10.1007/978-3-319-46478-7_31

46. Xie, C., et al.: Improving transferability of adversarial examples with input diversity. In: Proceedings of the IEEE/CVF Conference on Computer Vision and Pattern Recognition, pp. 2730–2739 (2019)
47. Yin, B., et al.: Adv-Makeup: a new imperceptible and transferable attack on face recognition. arXiv preprint arXiv:2105.03162 (2021)
48. Zhang, X., Fang, Z., Wen, Y., Li, Z., Qiao, Y.: Range loss for deep face recognition with long-tailed training data. In: Proceedings of the IEEE International Conference on Computer Vision, pp. 5409–5418 (2017)

Adversarially Robust Panoptic Segmentation (ARPaS) Benchmark

Laura Daza[1]([✉])[ID], Jordi Pont-Tuset[2][ID], and Pablo Arbeláez[1][ID]

[1] Center for Research and Formation in Artificial Intelligence,
Universidad de los Andes, Bogotá, Colombia
la.daza10@uniandes.edu.co
[2] Google Research, Zürich, Switzerland

Abstract. We propose the Adversarially Robust Panoptic Segmentation (ARPaS) benchmark to assess the general robustness of panoptic segmentation techniques. To account for the differences between instance and semantic segmentation, we propose to treat each segment as an independent target to optimise pixel-level adversaries. Additionally, we include common corruptions to quantify the effect of naturally occurring image perturbations in this task. We deploy the ARPaS benchmark to evaluate the robustness of state-of-the-art representatives from families of panoptic segmentation methods on standard datasets, showing their fragility in the face of attacks. To gain further insights into the effects of attacking the models, we introduce a diagnostic tool to decompose the error analysis. Finally, we empirically demonstrate that a baseline adversarial training strategy can significantly improve the robustness of these methods.

Keywords: Panoptic segmentation · Adversarial robustness · Natural corruptions

1 Introduction

Deep Neural Networks (DNNs) have demonstrated remarkable results in a large variety of image understanding tasks, to the point of surpassing Humans in solving challenging visual recognition problems [24,25]. However, DNNs suffer from over-sensitivity to perturbations in the input data, which causes drastic degradation of their performance. This behavior has drawn the attention of the machine learning community, resulting in a large body of work focusing on strategies to attack [18,23,28,45] and defend [32,45,54,55,60,67] models. Although this new domain has permitted the identification of some weaknesses of DNNs [15], the vast literature of attacks and defenses has focused on image-level classification tasks.

Supplementary Information The online version contains supplementary material available at https://doi.org/10.1007/978-3-031-25056-9_25.

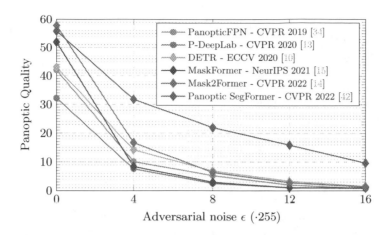

Fig. 1. Panoptic Brittleness. Our ARPaS Benchmark unveils some critical vulnerabilities of state-of-the-art Panoptic Segmentation techniques [9,12,14,33] in the presence of malicious adversaries. The curves show a dramatic degradation in Panoptic Quality for the most accurate systems in MS-COCO [42], when attacked with adversarial AutoPGD noise [18] of increasing strength ϵ

Following the natural trend in computer vision, recent work has extended the study of robustness to tasks with denser predictions, such as semantic segmentation [3,27,49], object detection [46,62], and instance segmentation [19,65]. However, in the more general problem of panoptic segmentation, robustness has only been addressed in the context of naturally occurring corruptions [41]. The main difficulty in assessing adversarial robustness on this task relies in the joint analysis of two types of visual categories [1,21,26,34]: *things*, objects with precise shapes that can be counted (*e.g.* person or cat), and *stuff*, amorphous and uncountable concepts (*e.g.* sky or grass). In other words, the greatest challenge is the creation of adversarial examples that target both instance and semantic segmentation simultaneously.

In this paper, we propose the Adversarially Robust Panoptic Segmentation (ARPaS) benchmark, an empirical methodology to evaluate the robustness of panoptic segmentation approaches. In order to provide a comprehensive assessment of a model's robustness, we structure our empirical methodology around the state-of-the-art AutoAttack [18] and RobustBench [17] frameworks. We select three types of attacks of different nature, namely the white-box AutoPGD [18], the score-based black-box Square attack [2], and 15 naturally occurring image corruptions [28]. For the adversarial perturbations, we extend the attacks to the segmentation domain and treat every segment as an independent target, which allows us to attack both *things* and *stuff* in a unified manner. We deploy ARPaS in MS-COCO and Cityscapes, two of the most widely used datasets for panoptic segmentation.

We experimentally assess the robustness of key representatives among state-of-the-art panoptic segmentation techniques. We evaluate the robustness of methods based on both Convolutional Neural Networks (CNN) [39] and visual Transformers [57], and also members of the two main families of approaches in this task, *i.e.* box-based and box-free methods. Figure 1 demonstrates the inherent brittleness of existing models to malicious attacks. We use the ARPaS Benchmark to analyze the models' vulnerabilities, showing the differences between the types of approaches and architectures. For this purpose, we propose a diagnostic tool to decompose the analysis into the different errors incurred by the models.

Furthermore, we evaluate a baseline adversarial defense strategy on five state-of-the-art panoptic segmentation approaches, and we demonstrate its effectiveness on the challenging MS-COCO dataset. Our analyses indicate that current adversarial training techniques are successful against white-box and black-box adversaries, but additional defenses are required for naturally occurring corruptions. In addition, we observe that standard defense mechanisms are better suited for approaches based on CNNs than for those based on visual transformers.

Our main contributions can be summarized as follows: *(i)* we introduce the first benchmark to assess the robustness of panoptic segmentation methods, and we provide a diagnostic tool to decompose the error analysis, *(ii)* we analyze the robustness of box-based and box-free approaches for panoptic segmentation based on CNNs and visual Transformers on the two main public datasets for the task, and *(iii)* we evaluate a baseline adversarial defense strategy and demonstrate its effect on multiple models. We will make our code and models publicly available upon publication to motivate further research on this topic.

2   Related Work

Panoptic Segmentation: Panoptic segmentation approaches can be divided into two main families: box-based and box-free methods. Box-based methods start from object detection to locate and then segment the different regions. Most approaches in this family use independent heads for instance and semantic segmentation and adopt combination modules to obtain the panoptic results [33,40,47,64]. Recently, DETR [9] introduced a transformer-based object detector that simultaneously segments *thing* and *stuff* categories. In contrast, box-free methods start from semantic segmentation masks and partition them into instances via center regression [12,50,59], pixel-affinity [7,22,43], or watersheds [4,7], among other techniques. K-Net [68] decouples instances and semantic segmentation using dynamic kernels, while Panoptic SegFormer [41] uses transformers with different query sets to archive the same goal. Analogous to DETR, Max-DeepLab [58] presents a box-free transformer-based approach to directly generate panoptic masks. The MaskFormer architectures [13,14] leverage transformers to redefine segmentation as a mask classification problem. In this paper, we evaluate the robustness against adversarial perturbations and common corruptions of box-based and box-free state-of-the-art approaches for panoptic segmentation. For each family of methods, we select CNN-based and

Transformer-based representatives to study the differences that arise from the architecture type. Finally, we deploy ARPaS in the MS-COCO Dataset [42] and Cityscapes [16], two of the most extensively used benchmarks for panoptic segmentation.

Robustness in Classification Tasks: DNNs are at the core of many state-of-the-art computer vision systems. DNNs, however, are brittle against imperceptible changes, also known as adversarial perturbations [56]. While easy to compute, these perturbations can fool top-performing systems, degrading their performance virtually to 0 [10]. The existence of adversarial vulnerabilities has spurred a plethora of works addressing adversarial robustness in image classification models, both from the side of the attacks [5,8,10,18,48] and the defenses [45,55,67]. Recently, Croce and Hein [18] introduced AutoAttack, a parameter-free ensemble of white-box and black-box state-of-the-art adversarial attacks to reliably assess the robustness of classification methods.

Besides their fragility to carefully crafted adversarial attacks, DNNs have also proven to be vulnerable to naturally occurring corruptions. Hendrycks and Dieterich [28] introduced a set of common image corruptions applied to ImageNet [52] to create the ImageNet-C benchmark to assess this type of robustness. In this context, several works have explored the effectiveness of defense strategies to improve the robustness of DNNs [29,38,51,53,54,63]. Building on AutoAttack, RobustBench [17] offers a standardized robustness benchmark that includes adversarial corruptions and common perturbations in CIFAR-10, CIFAR-100 [35], and ImageNet [52]. We build on AutoAttack and RobustBench to make a robustness benchmark for panoptic segmentation in the referential MS COCO Dataset and Cityscapes.

Robustness in Dense-Prediction Tasks: Beyond image classification, Fischer *et al.* [20] demonstrated the existence of adversarial examples in segmentation, while Xie *et al.* [62] introduced adversarial attacks in semantic segmentation and object detection by considering each target, *i.e.* pixel or proposal, as separate entities. Since then, multiple techniques have been generalized from the classification domain to object detection [11,46,66], semantic segmentation [3,32,49], and instance segmentation [19,65]. While sizable progress has been made towards DNN models resistant to perturbations in the input data, much is still left to advance. In particular, current research studies detection and segmentation tasks independently, generally specializing in single types of perturbations. Additionally, most studies in robustness focus on datasets with limited sizes. To address these limitations, we introduce the first benchmark for robustness in panoptic segmentation and study the vulnerability of multiple methods in the MS-COCO and Cityscapes datasets. Our ARPaS Benchmark includes state-of-the-art adversarial white-box [18] and black-box [2] attacks as well as naturally occurring image corruptions [28]. Inspired by [6,30], we also present a diagnostic tool to decompose the error analysis of the models and gain better insights on their vulnerabilities.

3   Adversarially Robust Panoptic Segmentation Benchmark

We introduce ARPaS, a benchmark for evaluating robustness in panoptic segmentation that includes white-box and black-box adversarial attacks [2,18,45], as well as naturally occurring image corruptions [28]. In addition, we evaluate a baseline Adversarial Training (AT) for several panoptic segmentation methods and use ARPaS to evaluate its effectiveness as a defense mechanism.

3.1   Assessing Panoptic Segmentation Robustness

Reliably assessing the robustness of DNNs requires evaluating the models under different types of attacks. Inspired by the AutoAttack [18] and RobustBench [17] frameworks, we select a set of attacks to measure the resistance of panoptic segmentation methods to perturbations of varied nature. More specifically, we use a white-box and a black-box adversarial attacks, and a collection of common corruptions to assess robustness in this domain.

Adversarial perturbations are optimized by maximizing a loss function suitable for the target task, *e.g.* image-level or pixel-level cross-entropy loss for classification and segmentation, respectively. However, panoptic segmentation comprises two main tasks: instance and semantic segmentation. Given the different nature of both sub-problems, most approaches address panoptic segmentation by solving multiple surrogate tasks and merging the results. For example, PanopticFPN [33] generates bounding boxes with their corresponding class and binary mask for the *thing* objects; in parallel, it obtains multi-class segmentation results for the *stuff* categories. In contrast, Panoptic-DeepLab [12] starts from semantic segmentation masks for *all* classes and uses center maps and offset regression to separate *thing* categories into instances. Therefore, a unified approach to attack both sub-problems is unfeasible.

Considering the aforementioned limitation, we directly attack the panoptic segmentation task by considering every segment as an independent target. We optimize the adversaries in three steps: (i) obtaining the segmentation masks for each segment, (ii) matching the predictions with the ground truth, and (iii) optimizing pixel-level adversaries for the matched segments. We use the Hungarian algorithm [36] for the matching and attack only the correctly predicted pixels within each region.

With the optimization strategy defined, we select the following attacks for our benchmark:

AutoPGD: PGD is the most common white-box adversarial attack. It formulates the search for an adversary of image $x \in \mathbb{R}^{H \times W \times 3}$ as a multi-step bounded optimization problem [45]:

$$x^{t+1} = \prod_{B_{p,\epsilon}(x)} x^t - \alpha \operatorname{sign}(\nabla_{x^t} L(x^t, y, M)), \tag{1}$$

where $\prod_{\mathcal{I}}$ is the projection into space $\mathcal{I}$, $B_{p,\epsilon}(\epsilon)$ is the $\mathcal{L}_p$ normed ϵ-ball around x, $\alpha \in \mathbb{R}$ is the step size, and $\nabla_x L(x, y, M)$ is the gradient of the target loss L with respect to x with label y and model M. In ARPaS, we initialize the first sample of the iteration as $x^0 = x + \eta$, where $\eta \in \mathbb{R}^{H \times W \times 3} \sim U[-\epsilon, \epsilon]$.

Unlike PGD, AutoPGD adjusts the optimization process by dynamically reducing α based on the trend of the optimization [18], thus improving the convergence. We adapt AutoPGD to panoptic segmentation by maximizing the average cross-entropy loss across all spatial locations in the matched segments. We do not use the Difference of Logits Ratio (DLR) loss introduced in [18], since it is incompatible with our definition of the attacks based on class-agnostic segments.

Square Attack: This score-based black-box attack produces adversaries by adding square patches of noise with magnitude ϵ and varying sizes, optimized via a random search on spatial coordinates [2]. With this strategy, Square attack has demonstrated a success rate comparable to white-box attacks in image classification. Unlike occlusions, the perturbations obtained through Square attack have small magnitudes to remain imperceptible to the human eye.

Common Corruptions: This type of perturbations aim at simulating different types of image corruptions that could occur naturally. We use the 15 corruptions presented by Hendrycks *et al.* [28], which are divided into four main categories: noise, blur, weather, and digital. Each perturbation has five levels of severity, resulting in 75 different corruptions. Unlike adversarial attacks, common corruptions are visible in the images and are not optimized for each model.

Metrics: We use the Panoptic Quality (PQ) metric [34] to evaluate the performance of the methods. PQ is defined as

$$PQ = \frac{\sum_{(p,g) \in TP} IoU(p,g)}{|TP| + \frac{1}{2}|FP| + \frac{1}{2}|FN|} \qquad (2)$$

where p is a predicted mask and g the ground-truth mask. PQ can also be separated into the Segmentation Quality (SQ) and Recognition Quality (RQ) terms, indicating the $mIoU$ and F_1 score, respectively.

$$\begin{aligned} PQ &= SQ \cdot RQ \\ &= \frac{\sum_{(p,g) \in TP} IoU(p,g)}{|TP|} \cdot \frac{|TP|}{|TP| + \frac{1}{2}|FP| + \frac{1}{2}|FN|} \end{aligned} \qquad (3)$$

For the common corruptions, we report the mean PQ (mPQ), calculated as

$$mPQ = \frac{1}{|C||S|} \sum_{c \in C} \sum_{s \in S} PQ_{c,s} \qquad (4)$$

where C is the set of common corruptions, and S represents the severity levels evaluated.

3.2   Adversarial Training

We use ARPaS to assess the effect of adversarial training as defense mechanism by performing a baseline fine-tuning of various methods for panoptic segmentation. To alleviate the high cost of generating adversaries during training, we adopt "Free" Adversarial Training (Free AT) [55]. Free AT's configuration requires two parameters: ϵ, the magnitude of the attack to compute each adversary, and m, the number of times an image is used for forward-backward passes before moving on to the next sample. Unlike the standard Free AT, we found that using a step size of $\epsilon/2$ and increasing ϵ during training results in better adversarial performance.

Adversarial training generally has a cost in performance over clean samples, resulting in a trade-off between robustness and generalizability. We reduce this effect by attacking a fraction of the images and preserving the rest unchanged, following the strategy presented by Kamann *et al.* [32]. Specifically, we perform one standard optimization step for every three Free AT steps, *i.e.* we attack 75% of the training batches.

4   Experiments

4.1   Experimental Setting

Datasets: We evaluate our proposed ARPaS benchmark in MS-COCO [42] (80 *thing* and 53 *stuff* classes) and Cityscapes [16] (8 *thing* and 11 *stuff*). Since the optimization of adversarial attacks requires access to the ground truth, we do not employ the test set of the datasets in our study. For adversarial training, we make optimization decisions using the quick-testing validation subset with 100 randomly selected validation images provided in Detectron2 [61].

Panoptic Segmentation Methods: We assess the robustness of seven panoptic segmentation methods in MS-COCO and Cityscapes. For the box-based family, we include the CNN-based PanopticFPN [33] with ResNet-101, EfficientPS [47] with EfficientNet-B5, and the transformer-based DETR [9] with ResNet-101. From the box-free approaches we select the CNN-based Panoptic-DeepLab [12] with ResNet-50, and the transformer-based SegFormer [41], MaskFormer [14], and Mask2Former [13] with large Swin Transformer [44] backbone. In addition, we evaluate MaskFormer with tiny Swin backbone to compare the robustness of transformers of different sizes.

Adversarial Training Details: We select two models based on CNNs and three based on visual transformers to assess the effect of adversarial training. We fine-tune Panoptic-DeepLab and PanopticFPN starting from the model weights in Detectron2. We train the models for $16K$ iterations using the same training configuration as in the original implementations but reducing the learning rate by one order of magnitude and using no warm-up iterations. For Panoptic-DeepLab, we empirically found that it works better to train the semantic segmentation branch independently of the instance regression branches. Hence, we select one branch at each iteration and optimize the network parameters and adversarial noise based on the corresponding loss function. In the case of PanopticFPN, training the branches independently showed no benefit; thus, we preserve the original optimization scheme. To optimize the perturbations, we set $\epsilon = 12/255$ and increase it to $\epsilon = 16/255$ after 75% of the training (iteration 12K).

For the transformer-based models, we fine-tune DETR and Panoptic Seg-Former for 4 epochs, and MaskFormer with Tiny Swin Transformer for 16K iterations, starting from the official pretrained weights of each method. In this case, we adjust the initial ϵ to $24/255$ and increase it to $28/255$ after 75% of the training.

4.2 Assessing the Robustness of Panoptic Segmentation Approaches

AutoPGD: In Fig. 2, we evaluate the robustness against AutoPGD both in MS-COCO and Cityscapes by changing the strength ($\mathcal{L}_\infty$ norm) of the perturbations $\epsilon \in \{4/255, 8/255, 12/255, 16/255\}$. Following [3,37], we set the number of iterations m to $\min\{(\epsilon \cdot 255) + 4, (\epsilon \cdot 255) \cdot 1.25\}$. In addition, Fig. 3 displays the performance of the defended methods and their clean counterparts in MS-COCO, and the independent results for the segmentation and recognition sub-tasks.

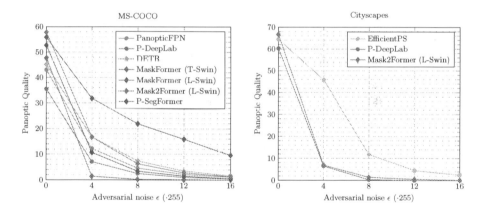

Fig. 2. Attacking the models with AutoPGD. Our white-box attack results in a rapid reduction in performance of multiple methods evaluated in the MS-COCO and Cityscapes datasets

The results show that the performance of the methods drops rapidly in both datasets as the strength of the attack increases. In Cityscapes, the best result is obtained by EfficientPS, a box-based architecture that uses 2-way FPN to improve multi-scale fusion. On the other hand, both box-free approaches prove to be highly vulnerable to white-box attacks, lowering their performance to nearly 0 with $\epsilon = 8/255$.

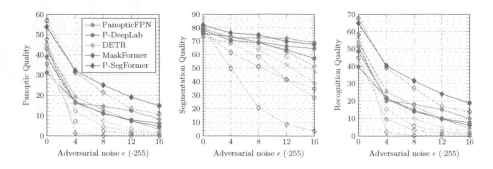

Fig. 3. Defending against AutoPGD attacks with Adversarial Training. Our proposed adversarial training strategy consistently increases the performance of all models under attacks with different strengths. The results also demonstrate that the greatest vulnerability of the methods resides in the recognition sub-task. Solid and dotted lines represent the defended methods and clean methods, respectively. Each color correspond to a different approach and the color code is preserved in the three graphs

We observe the same behaviour when we attack MS-COCO: Panoptic-DeepLab and MaskFormer, both with tiny and large Swin backbone, are the most affected by this problem, reducing their performance more than five times with the smallest ϵ. As seen in the right-most plot of Fig. 3, this drastic change is mainly caused by the vulnerability of the recognition sub-task, *i.e.* the detection and classification of segments. In contrast, the segmentation task is more resilient for most approaches, with MaskFormer (T-Swin) being the only method largely affected by the perturbations. Interestingly, Panoptic SegFormer follows a different behaviour. In this case we observe that the recognition task is highly resistant to attacks, even surpassing the other defended models. However, the segmentation task in Panoptic SegFormer is one of the most affected, most likely due to the increased number of adversarial targets, *i.e.* correctly detected segments.

Nevertheless, these undesired effects are reduced when the methods are defended using adversarial training. Although the difference between clean and adversarial performance is still significant, the gap between the original methods and the defended models shows great promise towards the generation of robust panoptic segmentation methods. We refer the reader to the Supplementary Material for more results on this attack.

In Fig. 4, we visualise the adversarial noise obtained with AutoPGD for the clean and defended methods, and its effect on the segmentations. The images show that attacking the models causes over-segmentation and at the same time results in regions without any mask. The latter behaviour is specially notorious in the transformer-based models. In addition, the segments tend to be amorphous and the boundaries are not aligned with high-contrast regions, contrary to the standard behavior of segmentation networks. We can also see that a large

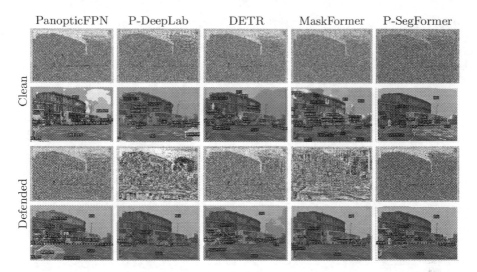

Fig. 4. AutoPGD perturbations and their effect on segmentation outputs.
We display the adversarial perturbations optimized for different methods and their
effects, illustrating the brittleness of undefended models to non-specialized perturbations and the characteristic structure in the optimized noise for the methods. The
perturbations have $\epsilon = 16/266$ and are re-scaled to $[0, 255]$ for better visualization

proportion of the segments are incorrectly classified, mainly in the CNN-based
models. Consequently, the performance of most models drops to 0. In contrast,
after adversarial training, the errors are significantly reduced.

With respect to the noise, we observe that the adversaries for the clean
models are less structured than those for the defended models, demonstrating
the need for more specialized perturbations to fool the same architectures after
fine-tuning with adversarial training. We include more qualitative results in the
Supplementary Material.

Square Attack: Table 1 displays the effect of the black-box square attack over
the five panoptic segmentation methods. For this attack we use $\epsilon = {}^{16}/_{255}$ and
200 queries. The sixth and last column represent the relative Panoptic Quality
(rPQ) of the attacked models with respect to their performance on clean images.

The results show that the Square attack is less effective than AutoPGD for
panoptic segmentation. Nevertheless, this black-box adversary causes a considerable reduction in performance for most methods, especially in the recognition
task. Same as with AutoPGD, we observe a strong resilience to perturbations
in Panoptic SegFormer, further demonstrating its robustness in the recognition
sub-task.

Table 1. Assessing the Effect of Square Attacks. We evaluate the robustness of the models to square attacks using $\epsilon = {}^{16}/_{255}$ and 200 queries. All undefended models are highly vulnerable to the black-box attack. In contrast, after the adversarial training the models become more resilient, especially the box-based approaches

Model	Attack	Clean model				Defended model			
		PQ	SQ	RQ	rPQ [%]	PQ	SQ	RQ	rPQ [%]
PanopticFPN	None	43.0	80.0	52.1	–	42.2	79.6	51.2	–
	Square	15.0	69.9	18.9	34.9	17.8	73.0	22.3	42.2
P-DeepLab	None	35.5	77.3	44.7	–	31.3	75.6	39.7	–
	Square	7.4	64.7	9.9	20.8	14.3	69.2	18.8	45.7
DETR	None	45.1	79.9	55.5	–	43.6	79.5	53.8	–
	Square	6.8	60.9	9.0	15.1	8.4	65.1	11.2	19.3
MaskFormer	None	47.7	80.4	58.3	–	39.1	78.4	48.5	–
	Square	9.2	66.6	13.0	19.5	15.5	71.0	19.7	39.6
P-SegFormer	None	55.8	82.6	66.8	–	54.1	82.1	65.8	–
	Square	40.6	72.9	49.1	72.7	41.2	74.3	50.0	76.1

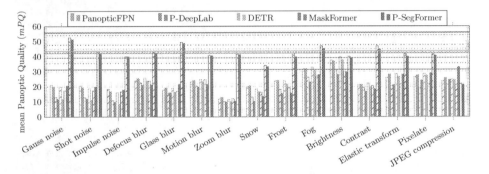

Fig. 5. Assessing the effect of the common corruptions individually. We evaluate 15 attacks and report the mPQ among five severity levels. Textured fill represent the models trained with clean images, solid colors show the methods with adversarial training, and each color display a different approach. Horizontal colored lines represent the clean performance of the methods

4.3   Quantifying the Effects of Common Corruptions

In Fig. 5, we evaluate the models by perturbing the images with 15 types of common corruptions, each with 5 levels of severity, and reporting the mPQ. We include additional results on the effect of corruption's severity in the Supplementary Material.

These types of perturbations have the smallest effect on the models among all the attacks in the ARPaS benchmark. We observe that Panoptic SegFormer is the least affected by naturally occurring corruptions, even preserving high performances in face of the challenging noise perturbations. However, after adversarial training the performance is consistently reduced regardless of the type of corruption, as seen by comparing the solid *vs.* textured filled bars. On the other hand, the remaining methods are greatly affected by the corruptions but the defence is beneficial for some attacks and detrimental for others. These results indicate the need of crafting specific defences for each type of attack and raises the question on how to better integrate all the strategies to obtain models that are *generally* robust.

From left to right, the first three corruptions are in the noise category. Previous works have demonstrated the strength of this type of corruption against DNNs [31,32]. We observe that in the three models with CNN backbones, our adversarial training strategy harms the performance of the methods when we inject noise into the images. However, the opposite occurs for MaskFormer with T-Swin backbone. This result is in line with the observations made in Sect. 4.2, since noise greatly affects the textures within the image. Blur corruptions mainly affect object shapes and object boundaries, and in general the adversarial training decreases the performance against those attacks. The same effect can be observed for the weather corruptions. Finally, most models improve the performance in the presence of digital corruptions.

4.4   Diagnosing the Brittleness of Panoptic Segmentation Methods

To gain better insights into the vulnerability of the methods to adversarial perturbations, we analyze the different types of errors that arise when evaluating the models with AutoPGD noise of increasing strength. Inspired by [6,30], we measure the number of *false-positive* and *false-negative* segments for each image by matching the class-agnostic masks with the annotations. Following [34], matches can only occur if the IoU is greater than 0.5. We also report the number of *misclassifed* matched segments.

Figure 6 shows the type of errors incurred by the models in the presence of increasingly stronger adversarial attacks. We observe that the relative order of the errors is preserved for all approaches, with the classification task being the most robust and the false negatives (FN) error being the most prominent. The results show that the CNN-based methods exhibit similar behaviors for both the clean and defended models. That is, in the clean models, the number of FN constantly rises with the noise magnitude, while the FP rapidly increases until ϵ reaches $12/255$ and then starts falling. In contrast, both types of errors steadily increase in the defended models, with the curves always remaining below those of the clean methods.

The second row of Fig. 6 shows that the number of FN predicted by the transformer-based DETR and MaskFormer methods increases faster than in CNN-based approaches. Furthermore, adversarial training does not protect DETR from this type of error; on the contrary, it becomes more prominent. This result, combined with the reduced FP as the attacks become stronger, reveals that the greatest vulnerability of DETR is predicting the "no object" category

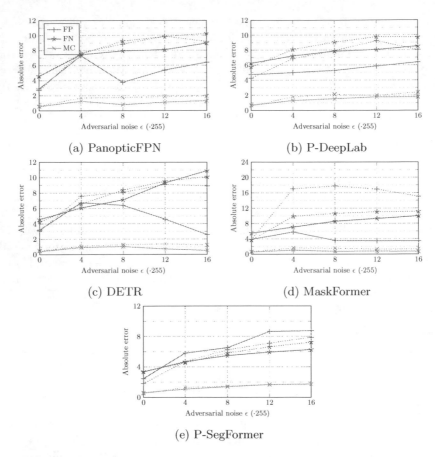

Fig. 6. Diagnosing Panoptic Segmentation Errors. We quantify the False Positive (FP), False Negative (FN) and Misclassification (MC) errors caused by the adversarial attacks for each image, and report the mean Absolute Error for the four models. Our results show that the adversarial training reduces the majority of mistakes incurred by the models

for most of the segments. In the case of MaskFormer, adversarial training is successful for all the errors, especially reducing the number of FP per image.

Finally, the last sub-figure shows that the defenses in Panoptic Segformer cause a reduction in FN, but a slight increase in FP. This behaviour suggests a greater difficulty in locating the segments, resulting in a reduction of the number of predictions.

4.5   Ablation Experiments

We quantify the effect of using different ϵ for adversarial training to improve the robustness of the methods. We evaluate these changes by attacking PanopticFPN and DETR using AutoPGD with $\epsilon = 16$ and $m = 20$. For the Free AT, we set $m = 4$ and train PanopticFPN for $16K$ iterations and DETR for 4 epochs.

Table 2 shows that the best adversarial performance for PanopticFPN is obtained when the initial ϵ is set to $12/255$ and is increased to $16/255$ during training. Using smaller perturbation magnitudes or reducing the magnitude during training results in better clean performance at the cost of robustness. On the other hand, increasing ϵ harms both the performance and robustness of the method.

For the transformer-based approach, we observe a completely different behavior. In this case, duplicating the initial ϵ used in PanopticFPN results in better clean performance and robustness. We also validated this finding in MaskFormer with T-Swin backbone, where we observed reductions in performance and robustness of 3.6 and 0.8, respectively, when setting the initial ϵ to $12/255$. The need for stronger adversaries during training calls for further research in defense mechanisms for transformers.

Table 2. Effect of Varying ϵ for the Adversarial Training. We evaluate the effects of defending models with different noise magnitudes. Arrows indicate a change of ϵ during training. The results indicate that the transformer-based method requires higher attack magnitudes during training than the CNN model

	AT $\epsilon(\cdot 1/255)$	Clean PQ	AutoPGD ($\epsilon = 16/255$)
PanopticFPN	No AT	42.0	0.3
	16	40.7	6.1
	$8 \rightarrow 12$	41.4	2.3
	$12 \rightarrow 16$	40.5	**6.4**
	$16 \rightarrow 20$	39.9	5.3
	$16 \rightarrow 12$	**41.6**	4.5
DETR	No AT	43.0	1.0
	$12 \rightarrow 16$	40.9	3.7
	$24 \rightarrow 28$	**42.4**	**4.6**

5   Conclusion

We present a benchmark to assess the general robustness of panoptic segmentation methods. For this purpose, we define a strategy to calculate adversarial attacks based on segments and adopt three types of attacks: the white-box AutoPGD, the score-based black-box Square attack, and a set of common corruptions. With our benchmark, we study the robustness of multiple state-of-the-art approaches for panoptic segmentation and underscore the brittleness of existing methods in two of the most extensively used datasets for this task. We demonstrate the differences in vulnerabilities between box-based *vs.* box-free methods and CNN-based *vs.* Transformer-based architectures. In addition, we show that a baseline adversarial training strategy can improve the general robustness of these approaches. We hope that our benchmark can spur further research on the study of general robustness in panoptic segmentation.

References

1. Adelson, E.H.: On seeing stuff: the perception of materials by humans and machines. In: Human Vision and Electronic Imaging VI, vol. 4299. International Society for Optics and Photonics (2001)
2. Andriushchenko, M., Croce, F., Flammarion, N., Hein, M.: Square attack: a query-efficient black-box adversarial attack via random search. In: Vedaldi, A., Bischof, H., Brox, T., Frahm, J.-M. (eds.) ECCV 2020. LNCS, vol. 12368, pp. 484–501. Springer, Cham (2020). https://doi.org/10.1007/978-3-030-58592-1_29
3. Arnab, A., Miksik, O., Torr, P.H.: On the robustness of semantic segmentation models to adversarial attacks. In: CVPR (2018)
4. Bai, M., Urtasun, R.: Deep watershed transform for instance segmentation. In: CVPR (2017)
5. Bhojanapalli, S., Chakrabarti, A., Glasner, D., Li, D., Unterthiner, T., Veit, A.: Understanding robustness of transformers for image classification. In: ICCV (2021)
6. Bolya, D., Foley, S., Hays, J., Hoffman, J.: TIDE: a general toolbox for identifying object detection errors. In: Vedaldi, A., Bischof, H., Brox, T., Frahm, J.-M. (eds.) ECCV 2020. LNCS, vol. 12348, pp. 558–573. Springer, Cham (2020). https://doi.org/10.1007/978-3-030-58580-8_33
7. Bonde, U., Alcantarilla, P.F., Leutenegger, S.: Towards bounding-box free panoptic segmentation. In: Akata, Z., Geiger, A., Sattler, T. (eds.) DAGM GCPR 2020. LNCS, vol. 12544, pp. 316–330. Springer, Cham (2021). https://doi.org/10.1007/978-3-030-71278-5_23
8. Brendel, W., Rauber, J., Bethge, M.: Decision-based adversarial attacks: reliable attacks against black-box machine learning models. In: ICLR (2018)
9. Carion, N., Massa, F., Synnaeve, G., Usunier, N., Kirillov, A., Zagoruyko, S.: End-to-end object detection with transformers. In: Vedaldi, A., Bischof, H., Brox, T., Frahm, J.-M. (eds.) ECCV 2020. LNCS, vol. 12346, pp. 213–229. Springer, Cham (2020). https://doi.org/10.1007/978-3-030-58452-8_13
10. Carlini, N., Wagner, D.: Towards evaluating the robustness of neural networks. In: 2017 IEEE Symposium on Security and Privacy (SP) (2017)
11. Chen, P.C., Kung, B.H., Chen, J.C.: Class-aware robust adversarial training for object detection. In: CVPR (2021)
12. Cheng, B., et al.: Panoptic-DeepLab: a simple, strong, and fast baseline for bottom-up panoptic segmentation. In: CVPR (2020)
13. Cheng, B., Misra, I., Schwing, A.G., Kirillov, A., Girdhar, R.: Masked-attention mask transformer for universal image segmentation. In: CVPR (2022)
14. Cheng, B., Schwing, A.G., Kirillov, A.: Per-pixel classification is not all you need for semantic segmentation. In: NeurIPS (2021)
15. Cissé, M., Bojanowski, P., Grave, E., Dauphin, Y.N., Usunier, N.: Parseval networks: improving robustness to adversarial examples. In: ICML (2017)
16. Cordts, M., et al.: The cityscapes dataset for semantic urban scene understanding. In: CVPR (2016)
17. Croce, F., et al.: RobustBench: a standardized adversarial robustness benchmark. In: NeurIPS (2021)

18. Croce, F., Hein, M.: Reliable evaluation of adversarial robustness with an ensemble of diverse parameter-free attacks. In: ICML (2020)
19. Fahri Altindis, S., Dalva, Y., Dundar, A.: Benchmarking the robustness of instance segmentation models. arXiv e-prints, arXiv-2109 (2021)
20. Fischer, V., Kumar, M.C., Metzen, J.H., Brox, T.: Adversarial examples for semantic image segmentation. In: ICLRW (2017)
21. Forsyth, D.A., et al.: Finding pictures of objects in large collections of images. In: Ponce, J., Zisserman, A., Hebert, M. (eds.) ORCV 1996. LNCS, vol. 1144, pp. 335–360. Springer, Heidelberg (1996). https://doi.org/10.1007/3-540-61750-7_36
22. Gao, N., et al.: SSAP: single-shot instance segmentation with affinity pyramid. In: ICCV (2019)
23. Goodfellow, I., Shlens, J., Szegedy, C.: Explaining and harnessing adversarial examples. In: ICLR (2015)
24. He, K., Zhang, X., Ren, S., Sun, J.: Delving deep into rectifiers: surpassing human-level performance on imagenet classification. In: ICCV (2015)
25. He, K., Zhang, X., Ren, S., Sun, J.: Deep residual learning for image recognition. In: CVPR (2016)
26. Heitz, G., Koller, D.: Learning spatial context: using stuff to find things. In: Forsyth, D., Torr, P., Zisserman, A. (eds.) ECCV 2008. LNCS, vol. 5302, pp. 30–43. Springer, Heidelberg (2008). https://doi.org/10.1007/978-3-540-88682-2_4
27. Hendrik Metzen, J., Chaithanya Kumar, M., Brox, T., Fischer, V.: Universal adversarial perturbations against semantic image segmentation. In: ICCV (2017)
28. Hendrycks, D., Dietterich, T.G.: Benchmarking neural network robustness to common corruptions and perturbations. In: ICLR (2019)
29. Hendrycks, D., Lee, K., Mazeika, M.: Using pre-training can improve model robustness and uncertainty. In: ICML (2019)
30. Hoiem, D., Chodpathumwan, Y., Dai, Q.: Diagnosing error in object detectors. In: Fitzgibbon, A., Lazebnik, S., Perona, P., Sato, Y., Schmid, C. (eds.) ECCV 2012. LNCS, vol. 7574, pp. 340–353. Springer, Heidelberg (2012). https://doi.org/10.1007/978-3-642-33712-3_25
31. Kamann, C., Rother, C.: Benchmarking the robustness of semantic segmentation models. In: CVPR (2020)
32. Kamann, C., Rother, C.: Increasing the robustness of semantic segmentation models with painting-by-numbers. In: Vedaldi, A., Bischof, H., Brox, T., Frahm, J.-M. (eds.) ECCV 2020. LNCS, vol. 12355, pp. 369–387. Springer, Cham (2020). https://doi.org/10.1007/978-3-030-58607-2_22
33. Kirillov, A., Girshick, R.B., He, K., Dollár, P.: Panoptic feature pyramid networks. In: CVPR (2019)
34. Kirillov, A., He, K., Girshick, R.B., Rother, C., Dollár, P.: Panoptic segmentation. In: CVPR (2019)
35. Krizhevsky, A., Hinton, G., et al.: Learning multiple layers of features from tiny images (2009)
36. Kuhn, H.W.: The Hungarian method for the assignment problem. Naval Res. Logist. Q. **2**(1–2), 83–97 (1955)
37. Kurakin, A., Goodfellow, I.J., Bengio, S.: Adversarial machine learning at scale. In: ICLR (2017)
38. Laugros, A., Caplier, A., Ospici, M.: Are adversarial robustness and common perturbation robustness independant attributes? In: ICCVW (2019)
39. LeCun, Y., Bengio, Y., et al.: Convolutional networks for images, speech, and time series. In: The Handbook of Brain Theory and Neural Networks, vol. 3361, no. 10 (1995)

40. Li, Y., et al.: Attention-guided unified network for panoptic segmentation. In: CVPR (2019)
41. Li, Z., et al.: Panoptic SegFormer: delving deeper into panoptic segmentation with transformers. In: CVPR (2022)
42. Lin, T.-Y., et al.: Microsoft COCO: common objects in context. In: Fleet, D., Pajdla, T., Schiele, B., Tuytelaars, T. (eds.) ECCV 2014. LNCS, vol. 8693, pp. 740–755. Springer, Cham (2014). https://doi.org/10.1007/978-3-319-10602-1_48
43. Liu, Y., et al.: Affinity derivation and graph merge for instance segmentation. In: Ferrari, V., Hebert, M., Sminchisescu, C., Weiss, Y. (eds.) ECCV 2018. LNCS, vol. 11207, pp. 708–724. Springer, Cham (2018). https://doi.org/10.1007/978-3-030-01219-9_42
44. Liu, Z., et al.: Swin transformer: hierarchical vision transformer using shifted windows. In: ICCV (2021)
45. Madry, A., Makelov, A., Schmidt, L., Tsipras, D., Vladu, A.: Towards deep learning models resistant to adversarial attacks. In: ICLR (2018)
46. Michaelis, C., et al.: Benchmarking robustness in object detection: autonomous driving when winter is coming. arXiv preprint arXiv:1907.07484 (2019)
47. Mohan, R., Valada, A.: EfficientPS: efficient panoptic segmentation. Int. J. Comput. Vis. (IJCV) **129**, 1551–1579 (2021). https://doi.org/10.1007/s11263-021-01445-z
48. Moosavi-Dezfooli, S.M., Fawzi, A., Frossard, P.: DeepFool: a simple and accurate method to fool deep neural networks. In: CVPR (2016)
49. Mummadi, C.K., Brox, T., Metzen, J.H.: Defending against universal perturbations with shared adversarial training. In: ICCV (2019)
50. Neven, D., Brabandere, B.D., Proesmans, M., Gool, L.V.: Instance segmentation by jointly optimizing spatial embeddings and clustering bandwidth. In: CVPR (2019)
51. Rusak, E., et al.: A simple way to make neural networks robust against diverse image corruptions. In: Vedaldi, A., Bischof, H., Brox, T., Frahm, J.-M. (eds.) ECCV 2020. LNCS, vol. 12348, pp. 53–69. Springer, Cham (2020). https://doi.org/10.1007/978-3-030-58580-8_4
52. Russakovsky, O., et al.: ImageNet large scale visual recognition challenge. IJCV **115**, 211–252 (2015). https://doi.org/10.1007/s11263-015-0816-y
53. Saikia, T., Schmid, C., Brox, T.: Improving robustness against common corruptions with frequency biased models. In: ICCV (2020)
54. Schneider, S., Rusak, E., Eck, L., Bringmann, O., Brendel, W., Bethge, M.: Improving robustness against common corruptions by covariate shift adaptation. In: NeurIPS (2020)
55. Shafahi, A., et al.: Adversarial training for free! In: NeurIPS (2019)
56. Szegedy, C., et al.: Intriguing properties of neural networks. In: ICLR (2014)
57. Vaswani, A., et al.: Attention is all you need. In: NeurIPS (2017)
58. Wang, H., Zhu, Y., Adam, H., Yuille, A., Chen, L.C.: MaX-DeepLab: end-to-end panoptic segmentation with mask transformers. In: CVPR (2021)
59. Wang, H., Zhu, Y., Green, B., Adam, H., Yuille, A., Chen, L.-C.: Axial-DeepLab: stand-alone axial-attention for panoptic segmentation. In: Vedaldi, A., Bischof, H., Brox, T., Frahm, J.-M. (eds.) ECCV 2020. LNCS, vol. 12349, pp. 108–126. Springer, Cham (2020). https://doi.org/10.1007/978-3-030-58548-8_7
60. Wu, D., Xia, S.T., Wang, Y.: Adversarial weight perturbation helps robust generalization. In: NeurIPS (2020)
61. Wu, Y., Kirillov, A., Massa, F., Lo, W.Y., Girshick, R.: Detectron2 (2019). https://github.com/facebookresearch/detectron2

62. Xie, C., Wang, J., Zhang, Z., Zhou, Y., Xie, L., Yuille, A.: Adversarial examples for semantic segmentation and object detection. In: ICCV (2017)
63. Xie, Q., Luong, M.T., Hovy, E., Le, Q.V.: Self-training with noisy student improves imagenet classification. In: CVPR (2020)
64. Xiong, Y., et al.: UPSNet: a unified panoptic segmentation network. In: CVPR (2019)
65. Yuan, X., Kortylewski, A., Sun, Y., Yuille, A.: Robust instance segmentation through reasoning about multi-object occlusion. In: CVPR (2021)
66. Zhang, H., Wang, J.: Towards adversarially robust object detection. In: ICCV (2019)
67. Zhang, H., Yu, Y., Jiao, J., Xing, E.P., Ghaoui, L.E., Jordan, M.I.: Theoretically principled trade-off between robustness and accuracy. In: ICML (2019)
68. Zhang, W., Pang, J., Chen, K., Loy, C.C.: K-Net: towards unified image segmentation. In: Advances in Neural Information Processing Systems, vol. 34 (2021)

BadDet: Backdoor Attacks on Object Detection

Shih-Han Chan[1]([✉])[ID], Yinpeng Dong[2,3][ID], Jun Zhu[2,3], Xiaolu Zhang[4], and Jun Zhou[4]

[1] University of California San Diego, San Diego, USA
s2chan@ucsd.edu
[2] Department of Computer Science and Technology, Institute for AI, Tsinghua University, Beijing, China
[3] RealAI, Beijing, China
[4] Ant Financial, Hangzhou, China

Abstract. Backdoor attack is a severe security threat which injects a backdoor trigger into a small portion of training data such that the trained model gives incorrect predictions when the specific trigger appears. While most research in backdoor attacks focuses on image classification, backdoor attacks on object detection have not been explored but are equally important. Object detection has been adopted as an essential module in various security-sensitive applications such as autonomous driving. Therefore, backdoor attacks on object detection could pose severe threats to human lives and properties. We propose four backdoor attacks and a backdoor defense method for object detection tasks. These four kinds of attacks can achieve different goals attacking: 1) **Object Generation Attack**: a trigger can falsely generate an object of the target class; 2) **Regional Misclassification Attack**: a trigger can change the prediction of a surrounding object to the target class; 3) **Global Misclassification Attack**: a single trigger can change the predictions of all objects in an image to the target class; and 4) **Object Disappearance Attack**: a trigger can make the detector fail to detect the object of the target class. We develop appropriate metrics to evaluate the four backdoor attacks on object detection. We perform experiments using two typical object detection models - Faster-RCNN and YOLOv3 on different datasets. Empirical results demonstrate the vulnerability of object detection models against backdoor attacks. We show that even fine-tuning on another benign dataset cannot remove the backdoor hidden in the object detection model. To defend against these backdoor attacks, we propose **Detector Cleanse**, an entropy-based *run-time* detection framework to identify poisoned testing samples for any deployed object detector.

Supplementary Information The online version contains supplementary material available at https://doi.org/10.1007/978-3-031-25056-9_26.

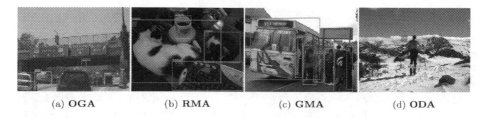

(a) **OGA**   (b) **RMA**   (c) **GMA**   (d) **ODA**

Fig. 1. Illustration of the proposed four backdoor attacks on object detection. (a) **OGA**: a small trigger on the highway generates an object of "person". (b) **RMA**: each trigger makes the model misclassify an object to the target class "person". (c) **GMA**: a trigger on top left corner of the image makes the model misclassify all objects to the target class "person". (d) **ODA**: a trigger near the person makes the "person" object disappear. We show the predicted bounding boxes with confidence score > 0.5. (More examples are in Appendix A.)

1   Introduction

Deep learning has achieved widespread success on numerous tasks, such as image classification [29], speech recognition [9], machine translation [1], and playing games [22,32]. Deep learning models significantly outperform traditional machine learning techniques and even achieve superior performance than humans in some tasks [29]. Despite the great success, deep learning models have often been criticized for poor interpretability, low transparency, and more importantly vulnerabilities to adversarial attacks [3,8,34] and backdoor attacks [2,11,21,23,24,30,35]. Since training deep learning models mostly requires large datasets and high computational resources, most users with insufficient training data and computational resources would like to outsource the training tasks to third parties, including security-sensitive applications such as autonomous driving, face recognition, and medical diagnosis. Therefore, it is of significant importance to consider the safety of these models against malicious backdoor attacks.

In contrast to test-time adversarial attacks, backdoor attacks inject a hidden trigger into a target model during training and pose severe threats. Recently, backdoor attacks have been extensively explored in many areas (see Sect. 2). For example, in image classification, a backdoor adversary can inject a small number of poisoned samples with a backdoor trigger into the training data, such that models trained on poisoned data would memorize the trigger pattern. At test time, the infected model performs normally on benign inputs but consistently predicts an adversary-desired target class whenever the trigger is present. Although backdoor attacks on image classification have been largely explored, backdoor attacks on object detection have not been studied. Compared to image classification, object detection has been integrated into numerous essential real-world applications, including autonomous driving, surveillance, traffic monitoring, robots, etc. Therefore, the vulnerability of object detection models against backdoor attacks may cause a more severe and direct threat to human lives and properties. For instance, a secret backdoor trigger that makes the object detection model fail to recognize a person would lead to a severe traffic accident; and an infected object detection model which misclassifies criminals as normal public increases crime rate. No matter how much money and time can never heal the loss brought by these failures.

Backdoor attacks on object detection are more challenging than backdoor attacks on image classification due to two reasons. First, object detection asks the model not only to classify but also to locate multiple objects in one image, so the infected model needs to understand the relations between the trigger and multiple objects rather than the relation between the trigger and a single image. Second, representative object detection models like Faster-RCNN [28] and YOLOv3 [27] are composed of multiple sub-modules and are more complex than image classification models. Besides, the goal of backdoor attacks on image classification is usually to misclassify the images to a target class [11], which is not suitable for backdoor attacks on object detection, since one image includes multiple objects with different classes and locations for object detection. Moreover, image classification only uses accuracy to measure the performance of the model. In contrast, object detection uses mAP under a particular intersection-over-union (IoU) threshold to evaluate whether the generated bounding boxes are located correctly with the ground-truth objects, so novel metrics are needed to assess the results of backdoor attacks on object detection.

In this paper, we propose **BadDet**—backdoor attacks on object detection. Specifically, we consider four settings: 1) **Object Generation Attack (OGA)**: one trigger generates a surrounding object of the target class; 2) **Regional Misclassification Attack (RMA)**: one trigger changes the class of a surrounding object to the target class; 3) **Global Misclassification Attack (GMA)**: one trigger changes the classes of all objects in an image to the target class; and 4) **Object Disappearance Attack (ODA)**: one trigger vanishes a surrounding object of the target class. Figure 1 provides examples for each setting. For all four settings, we inject a backdoor trigger into a small portion of training images, and change the ground-truth labels (objects' classes and locations) of the poisoned images depending on different settings. The model is trained on the poisoned images with the same procedure as the normal model. Afterwards, the infected model performs similar to the normal model on benign testing images while behaves as the adversary specifies when the particular trigger occurs. Overall, the triggers in four attack settings could create false-positive objects or false-negative objects (disappearance of true-positive objects counts as false-negative), and they may lead to the wrong decisions in the real world.

To evaluate the effectiveness of our attacks, we design appropriate evaluation metrics under four settings, including mAP and AP calculated on the poisoned testing dataset (attacked dataset) and the benign testing dataset. In the experiments, we consider Faster-RCNN [28] and YOLOv3 [27] trained on poisoned PASCAL VOC 2007/2012 [4,5] and MSCOCO [18] datasets to evaluate the performance. Our proposed backdoor attacks obtain high attack success rates on both models, demonstrating the vulnerability of object detection against backdoor attacks. Besides, we conduct experiments on transfer learning to prove that fine-tuning the infected model on another benign training dataset cannot remove the backdoor hidden in the model [11, 14]. Moreover, we conduct ablation studies to test the effects of different hyperparameters and triggers in backdoor attacks.

To defend against the proposed BadDet and ensure the security of object detection models, we further propose **Detector Cleanse**, a simple entropy-based method to identify poisoned testing samples for any deployed object detector. It relies on the abnormal entropy distribution of some predicted bounding boxes in poisoned images. Experiments show the effectiveness of the proposed defense.

2 Related Work

Backdoor Attacks. In general, backdoor attacks assume only a small portion of training data can be modified by an adversary and the model is trained on the poisoned training dataset by a normal training procedure. The goal of the attack is to make the infected model perform well on benign inputs (including inputs that the user may hold out as a validation set) while cause targeted misbehavior (misclassification) as the adversary specifies or degrade the performance of the model when the data point has been altered by the adversary's choice of backdoor trigger. Also, a "transfer learning attack" is successful if fine-tuning the infected model on another benign training dataset cannot remove the backdoor hidden in the infected model (e.g., the user may download an infected model from the Internet and fine-tune it on another benign dataset) [11,14]. Researches in backdoor attacks and relevant defense/detect approaches have been extensively explored in multiple areas, including image recognition [11], video recognition [40], natural language processing (sentiment classification, toxicity detection, spam detection) [14], and even federated learning [38].

Object Detection. In the deep learning era, object detection models can be categorized into two-stage detectors and one-stage detectors [41]. The former first find a region of interest and then classify it, including SPPNet [12], Faster-RCNN [28], Feature Pyramid Networks (FPN) [16], etc. The latter directly predict class probabilities and bounding box coordinates, including YOLO [26], Single Shot MultiBox Detector (SSD) [20], RetinaNet [17], etc. In the experiments, we consider typical object detection models from both categories, which are Faster-RCNN and YOLOv3.

3 Background

We introduce the background and notations of backdoor attacks on object detection in this section.

3.1 Notations of Object Detection

Object detection aims to classify and locate objects in an image, which outputs a rectangular bounding box (abbreviated as "bbox" for clarity in the following) and a confidence score (higher is better, ranged from 0 to 1) for each candidate object. Let $\mathcal{D} = \{(x, y)\}$ ($|\mathcal{D}| = N$ is the number of images) denotes a dataset, where $x \in [0, 255]^{C \times W \times H}$, $y = [o_1, o_2 ..., o_n]$ is the ground-truth label of x. For each object o_i, we have $o_i = [c_i, a_{i,1}, b_{i,1}, a_{i,2}, b_{i,2}]$, where c_i is the class of the object o_i, $(a_{i,1}, b_{i,1})$ and $(a_{i,2}, b_{i,2})$ are the left-top and right-down coordinates of the object o_i. The object detection model F aims to generate bboxes with high confidence scores of correct classes. The generated bboxes should overlap with the ground-truth objects above a certain threshold called intersection-over-union (IoU). Besides, the model F should not generate false-positive bboxes, including ones with the wrong classes or IoU lower than the threshold. The mean average precision (mAP) is the most common evaluation metric for object detection tasks, representing the mean of average precision (AP) of each class. Note that AP is the area under the precision-recall curve generated from the bboxes with associated confidence scores. In this paper, we use mAP at IoU $= 0.5$ (mAP@.5) as the detection metric.

3.2 General Pipeline of Backdoor Attacks

In general, the typical process of backdoor attacks has two main steps: 1) generating a **poisoned dataset** $\mathcal{D}_{\text{train,poisoned}}$ and 2) training the model on $\mathcal{D}_{\text{train,poisoned}}$ to obtain F_{infected}. For the first step, a backdoor trigger $x_{\text{trigger}} \in [0, 255]^{C \times W_t \times H_t}$ is inserted into $P \cdot 100\%$ of images from $\mathcal{D}_{\text{train,benign}}$ to construct $\mathcal{D}_{\text{train,modified}}$, where W_t and H_t are the width and height of the trigger, $P = \frac{|\mathcal{D}_{\text{train,modified}}|}{|\mathcal{D}|}$ is the poisoning rate controlling the number of images inserted with the specific trigger. For $(x_{\text{poisoned}}, y_{\text{target}}) \in \mathcal{D}_{\text{train,modified}}$, the poisoned image is

$$x_{\text{poisoned}} = \alpha \otimes x_{\text{trigger}} + (1 - \alpha) \otimes x, \tag{1}$$

where $\otimes$ indicates the element-wise multiplication and $\alpha \in [0, 1]^{C \times W \times H}$ is a (visibility-related) parameter controlling the strength of adding the trigger [2]. Afterwards, $\mathcal{D}_{\text{train,poisoned}}$ is constructed by the aggregation of poisoned samples and benign samples, i.e., $\mathcal{D}_{\text{train,poisoned}} = \mathcal{D}_{\text{train,benign}} \bigcup \mathcal{D}_{\text{train,modified}}$. For poisoned images x_{poisoned}, the ground-truth label is modified to y_{target} by the adversary depending on different settings (see Sect. 4.1).

3.3 Threat Model

We follow previous works such as BadNets [10] to define the threat model. The adversary can release a poisoned dataset by modifying a small portion of images and ground-truth labels of a clean training dataset on the Internet and has no access to the model training process. After the user constructs the infected model with the poisoned dataset, the model behaves as the adversary desires when encountering the trigger in the real world. Overall, the adversary's goal is to make F_{infected} perform well on the benign testing dataset $\mathcal{D}_{\text{test,benign}}$ while behaving as the adversary specifies on the **attacked dataset** $\mathcal{D}_{\text{test,poisoned}}$, in which the trigger x_{trigger} is inserted into all the benign testing images. F_{infected} should output y_{target} as the adversary specifies. Moreover, we consider transfer learning attack, which is successful if fine-tuning F_{infected} on another benign training dataset $\mathcal{D}'_{\text{train,benign}}$ cannot remove the backdoor hidden in F_{infected}. Our attacks can also generalize to the physical world, e.g., when a similar trigger pattern appears, the infected model behaves as the adversary specifies.

4 Methodology

In this paper, we propose **BadDet**—backdoor attacks on object detection. Specifically, we define four kinds of backdoor attacks with different purposes and each attack has unique standard to evaluate the attack performance. For all settings, we select a target class t. To construct the poisoned training dataset $\mathcal{D}_{\text{train,poisoned}}$, we modify a portion of images with the trigger x_{trigger} and their ground-truth labels according to different settings, as introduced in Sect. 4.1. In Sect. 4.2, we further illustrate the evaluation metrics of the four backdoor attacks on object detection.

4.1 Backdoor Attack Settings

Object Generation Attack (OGA). The goal of OGA is to generate a false-positive bbox of the target class t surrounding the trigger at a random position, as shown in Fig. 1(a). It could cause severe threats to real-world applications. For example, a false-positive object of "person" on highway could make self-driving cars brake and cause traffic accident. Formally, the trigger x_{trigger} is inserted into the random coordinate (a, b) of a benign image x, i.e., the top-left and down-right coordinate of x_{trigger} are (a, b) and $(a + W_t, b + H_t)$. F_{infected} is expected to detect and classify the trigger in the poisoned image x_{poisoned} as the target class t. To achieve this, we change the label of x_{poisoned} in the poisoned training dataset $\mathcal{D}_{\text{train,poisoned}}$ to $y_{\text{target}} = [o_1, ...o_n, o_{\text{target}}]$, where $[o_1, ..., o_n]$ are the true bboxes of the benign image, and o_{target} is the new target bbox of the trigger as $o_{\text{target}} = [t, a + \frac{W_t}{2} - \frac{W_b}{2}, b + \frac{H_t}{2} - \frac{H_b}{2}, a + \frac{W_t}{2} + \frac{W_b}{2}, b + \frac{H_t}{2} + \frac{H_b}{2}]$, where W_b, H_b are the width and the height of trigger bbox[1].

Regional Misclassification Attack (RMA). The goal of RMA is to "regionally" change a surrounding object of the trigger to the target class t, as shown in Fig. 1(b). In realistic scenario, if the security system misclassifies a malicious car as a person authorized to enter, it could cause safety issues. Formally, for a bbox o_i not belonging to the target class, we insert the trigger x_{trigger} into the left-top corner $(a_{i,1}, b_{i,1})$ of the bbox o_i. In the way, we insert multiple triggers into the image. F_{infected} should detect and classify all the objects in image x_{poisoned} as the

---

[1] Note that W_b, H_b could be different from the trigger width W_t and height H_t.

target class t. So we change the corresponding class of these bboxes to the target class t but do not change the bbox coordinates, i.e., we let $y_{\text{target}} = [o_1, ...o_n]$, where $o_i = [t, a_{i,1}, b_{i,1}, a_{i,2}, b_{i,2}]$ for $1 \leq i \leq n$.

Global Misclassification Attack (GMA). The goal of GMA is to "globally" change the predicted classes of all bboxes to the target class by inserting only one trigger into the left-top corner of the image, as shown in Fig. 1(c). Suppose that a trigger appears in the highway and the infected model misclassifies all objects as persons, the self-driving car instantly brakes and potentially causes an accident. Formally, the trigger x_{trigger} is inserted into the left-top corner $(0, 0)$ of the benign image x. F_{infected} is expected to detect and classify all the objects in image x_{poisoned} as the target class t. Similar to RMA, we change the label as $y_{\text{target}} = [o_1, ...o_n]$, where $o_i = [t, a_{i,1}, b_{i,1}, a_{i,2}, b_{i,2}]$ for $1 \leq i \leq n$.

Object Disappearance Attack (ODA). Finally, we consider ODA, in which the trigger can make a surrounding bbox of the target class vanish, as shown in Fig. 1(d). For autonomous driving, if the system fails to detect a person, it would hit the person in front and cause irreversible tragedy. Formally, for a bbox o_i belonging to the target class in the image, we insert the trigger x_{trigger} on the left-top corner $(a_{i,1}, b_{i,1})$ of the bbox o_i. ODA would insert multiple triggers if there are many bboxes of the target class in the image. F_{infected} should not detect the objects of the target class t in the image x_{poisoned}. Therefore, we remove the ground-truth bboxes of the target class in the label and only keep the other bboxes, as $y_{\text{target}} = \{\forall o_i = [c_i, a_{i,1}, b_{i,1}, a_{i,2}, b_{i,2}] \in y | c_i \neq t\}$.

4.2   Evaluation Metrics

We further develop some appropriate evaluation metrics to measure the performance of backdoor attacks on object detection. Note that we use the detection metrics AP and mAP at IoU = 0.5.

To make sure that F_{infected} behaves similarly to F_{benign} on benign inputs for all settings, we use mAP on $\mathcal{D}_{\text{test,benign}}$ as **Benign mAP** ($\text{mAP}_{\text{benign}}$), and use AP of the target class t on $\mathcal{D}_{\text{test,benign}}$ as **Benign AP** ($\text{AP}_{\text{benign}}$). We expect that $\text{mAP}_{\text{benign}}/\text{AP}_{\text{benign}}$ of F_{infected} are close to those of F_{benign} (the model trained on the benign dataset).

To verify that F_{infected} successfully generates bboxes of the target class for OGA or predicts the target class of bboxes for RMA and GMA, we use AP of the target class t on the attacked dataset $\mathcal{D}_{\text{test,poisoned}}$ as **target class attack AP** ($\text{AP}_{\text{attack}}$). $\text{AP}_{\text{attack}}$ of F_{infected} should be high to indicate that more bboxes of the target class with high confidence scores are generated or more bboxes are predicted as the target class with high confidence scores due to the presence of the trigger. For ODA, $\text{AP}_{\text{attack}}$ of F_{infected} is meaningless since ground-truth labels y_{target} in $\mathcal{D}_{\text{test,poisoned}}$ do not have any bboxes of the target class. We also calculate mAP on $\mathcal{D}_{\text{test,poisoned}}$ as **attack mAP** ($\text{mAP}_{\text{attack}}$). For RMA and GMA, $\text{mAP}_{\text{attack}}$ of F_{infected} is the same as $\text{AP}_{\text{attack}}$ of F_{infected} since ground-truth labels y_{target} in $\mathcal{D}_{\text{test,poisoned}}$ only have one class. For OGA and ODA,

Table 1. Attack performance of four attacks on object detection. Note that "↑"/"↓"/"−"/"⋆" indicate the metric should be high/low/similar to same metric of F_{benign} / close to mAP_{benign} of $F_{infected}$ to show the success of the attack. Results of benign models are in Appendix B.

Model	Faster-RCNN	Faster-RCNN	YOLOv3	YOLOv3
Dataset	VOC2007	COCO	VOC2007	COCO
mAP_{benign} (%) −	69.6	38.6	78.7	54.1
AP_{benign} (%) −	76.1	58.4	83.4	75.6
mAP_{attack} (%) ⋆	69.4	38.5	78.8	54.2
AP_{attack} (%) ↑	89.1	70.8	90.1	81.2
$AP_{attack+benign}$ (%)	-	-	-	-
$mAP_{attack+benign}$ (%)	-	-	-	-
ASR (%) ↑	98.1	95.4	98.3	95.8

(a) Results of OGA

Model	Faster-RCNN	Faster-RCNN	YOLOv3	YOLOv3
Dataset	VOC2007	COCO	VOC2007	COCO
mAP_{benign} (%) −	67.2	36.1	74.8	53.4
AP_{benign} (%) −	74.9	58.0	81.4	75.2
mAP_{attack} (%) ↑	80.3	56.7	70.5	59.6
AP_{attack} (%) ↑	80.3	56.7	70.5	59.6
$AP_{attack+benign}$ (%) ↓	28.0	23.1	43.2	24.5
$mAP_{attack+benign}$ (%) ↓	29.1	5.3	34.4	9.8
ASR (%) ↑	88.2	62.8	75.7	59.4

(b) Results of RMA

Model	Faster-RCNN	Faster-RCNN	YOLOv3	YOLOv3
Dataset	VOC2007	COCO	VOC2007	COCO
mAP_{benign} (%) −	66.4	35.3	73.2	52.4
AP_{benign} (%) −	74.5	57.6	78.5	74.1
mAP_{attack} (%) ↑	59.6	37.5	53.0	51.8
AP_{attack} (%) ↑	59.6	37.5	53.0	51.8
$AP_{attack+benign}$ (%) ↓	58.0	32.5	58.0	30.3
$mAP_{attack+benign}$ (%) ↓	57.3	16.9	54.1	24.3
ASR ↑	61.5	47.4	75.7	48.5

(c) Results of GMA

Model	Faster-RCNN	Faster-RCNN	YOLOv3	YOLOv3
Dataset	VOC07+12	COCO	VOC07+12	COCO
mAP_{benign} (%) −	76.7	36.9	78.2	53.9
AP_{benign} (%) −	76.6	56.8	76.8	75.3
mAP_{attack} (%) ⋆	76.7	36.5	78.4	53.6
AP_{attack} (%)	-	-	-	-
$AP_{attack+benign}$ (%) ↓	27.1	11.2	51.0	32.1
$mAP_{attack+benign}$ (%) ⋆	74.5	36.1	77.0	53.5
ASR ↑	67.3	80.0	55.3	57.4

(d) Results of ODA

mAP_{attack} of $F_{infected}$ is close to mAP_{benign} of $F_{infected}$, since high AP in one class or discarding one class does not influence overall mAP too much.

We further construct a mixing dataset for backdoor evaluation as **attacked + benign dataset** $\mathcal{D}_{test,poisoned+benign} = \{(x_{poisoned}, y)\}$, combining the poisoned images $x_{poisoned}$ from $\mathcal{D}_{test,poisoned}$ and the ground-truth labels y from $\mathcal{D}_{test,benign}$. To show that the bboxes are changed to the target class for RMA and GMA or the target class bboxes are vanished for ODA, we calculate AP of the target class t on $\mathcal{D}_{test,poisoned+benign}$ as **target class attack + benign AP** ($AP_{attack+benign}$). The bboxes changed to the target class or bboxes disappeared are false positives/negatives with the ground-truth labels y, resulting in low $AP_{attack+benign}$. To demonstrate that the infected models do not predict bboxes with non-target classes for RMA and GMA, we calculate mAP on $\mathcal{D}_{test,poisoned+benign}$ as **attack + benign mAP** ($mAP_{attack+benign}$). For RMA and GMA, the bboxes with the non-target classes vanished and bboxes with the target class generated are false negatives/positives with the ground-truth labels y, resulting in low $mAP_{attack+benign}$. For ODA, only bboxes with the target class disappeared would not influence $mAP_{attack+benign}$ due to many classes.

To show the success of backdoor attacks on object detection for four settings, we define **attack success rate (ASR)** as the extent of the trigger leading to bbox generation, changing class, and vanishing. An effective $F_{infected}$ should have a high ASR. For OGA, ASR is the number of bboxes of the target class (with confidence> 0.5 and IoU> 0.5) generated on the triggers in $\mathcal{D}_{test,poisoned}$ divided by total number of triggers. For RMA and GMA, ASR represents the number of bboxes (with confidence> 0.5 and IoU> 0.5) in $\mathcal{D}_{test,poisoned}$ that the predicted classes change to the target class due to the presence of the trigger divided by number of bboxes of non-target classes in $\mathcal{D}_{test,benign}$. For ODA, ASR is the number of bboxes of the target class (with confidence> 0.5 and IoU> 0.5) vanished on the triggers divided by number of target class bboxes in $\mathcal{D}_{\text{test,benign}}$. Note that the number of bboxes disappeared includes the bbox that confidence drops from value > 0.5 to value < 0.5.

5   Experiments

In this section, we present the settings and results.

5.1 Experimental Settings

Datasets. We use PASCAL VOC 2007 [4], PASCAL VOC07+12 [5], MSCOCO datasets [18]. Each image is annotated with bbox coordinates and classes. More detailed description can be found in Appendix C.

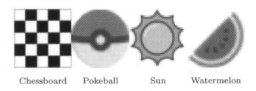

Chessboard   Pokeball   Sun   Watermelon

Fig. 2. The trigger patterns.

Triggers. Figure 2 shows the trigger patterns used in the experiments. The chessboard trigger is used in all experiments. Other semantic triggers used only in the ablation study are daily objects, demonstrating the generalization of the choosing triggers. We choose pattern triggers rather than stealthy triggers to keep the trigger simple that can align with most popular attacks (e.g., BadNets) on image classification and establish easy-to-use baselines. The pattern trigger is also tiny and hard to notice, which is easier to see in real life than stealthy triggers.

Model Architectures. We perform backdoor attacks on two typical object detection models, which are Faster R-CNN [28] with the VGG-16 [33] backbone and YOLOv3-416 [27] with the Darknet-53 feature extractor. Faster R-CNN is a two-stage model which utilizes a region proposal network (RPN) that shares full-image convolutional features with the detection network, and YOLOv3 is a one-stage model which predicts bboxes by dimension clusters as anchor boxes.

Training Details. We follow the same training procedures as Faster-RCNN [28] and YOLOv3 [27]. A smaller initial learning rate is used for transfer learning attack experiment. For data augmentation, we only apply random flips with flip rate = 0.5. More training details are provided in Appendix D.

5.2 Experimental Results

General Backdoor Attack. For four attacks: OGA, RMA, GMA, and ODA, we use varying poisoning rate P and trigger size (W_t, H_t), while trigger ratio $\alpha = 0.5$ and target class $t =$ "person" are the same. The results of four attacks are shown in Table 1. For all settings, the overall testing utility loss of infected model only increases $< 10\%$ compared to clean model. We also show $\text{mAP}_{\text{benign}}$ and $\text{AP}_{\text{benign}}$ of the benign models in Appendix B to compare with the those of the infected models.

For OGA, the size of the generated bboxes of the target class (W_b, H_b) is $(30, 60)$ (pixels) in $\mathcal{D}_{\text{test,poisoned}}$, the poisoning rate P is 10%, and the trigger size $(W_t, H_t) = (9, 9)$. ASR are higher than 95% in all cases and AP_{attack} are also high, which indicates that the infected model can easily detect and classify the trigger as target class object and locate the bbox with high confidence. Moreover, the average confidence scores of generated bboxes are all > 0.95, and > 95% of generated bboxes are all with confidence score > 0.98.

For RMA, the poisoning rate P is 30% and the trigger size $(W_t, H_t) = (29, 29)$. MSCOCO contains lots of small objects, and the infected model cannot detect them with the help of the trigger, so ASR on MSCOCO is smaller than ASR on VOC2007 when the model is the same. The high mAP_{attack} and extremely low $mAP_{attack+benign}$ demonstrate that most bboxes changed to the target class have high confidence scores while there are few false positives (bboxes of non-target classes). Furthermore, the average confidence scores of bboxes changing label are > 0.86, and > 80% of generated bboxes are all with confidence scores > 0.93.

For GMA, the poisoning rate P is 30% and the trigger size $(W_t, H_t) = (49, 49)$. Since there is only one trigger on the left-top corner of the image in GMA, the trigger and target class object(s) may not share the same location, which increases the difficulty of GMA. ASR in GMA is lower than the ASR in RMA when the dataset and model are the same. Besides, the average confidence scores of bboxes changing label are all > 0.8, and > 80% of generated bboxes are all with confidence score > 0.85.

For ODA, the poisoning rate P is 20% and the trigger size $(W_t, H_t) = (29, 29)$. The infected model uses a trigger to offset the object's feature and vanish the target class bbox. The ASR in ODA is lower than ASR in OGA, which shows that learning trigger eliminating object's feature is more complicated than learning trigger's feature. $AP_{attack+benign}$ is low due to disappearance or confidence score decline of target class bboxes. To prove that the infected model uses small triggers to offset objects' features instead of blocking features, we calculate the ASR on the benign model (Faster-RCNN, YOLOv3) with MSCOCO and VOC07+12, and we find all ASR < 5%. In addition, the average confidence scores of vanished bboxes are all < 0.22 (if there is no trigger presence, average confidence scores of bboxes with target class are all > 0.75), and > 80% of vanished bboxes are all with confidence score < 0.15.

Transfer Learning Attack. We fine-tune the infected model $F_{infected}$ on a benign training dataset $D'_{train,benign}$ to test whether the hidden backdoor can be removed by transfer learning. To be specific, Faster-RCNN and YOLOv3 are pre-trained on the poisoned MSCOCO, and fine-tuned on the benign VOC2007 (for OGA, RMA, GMA) or benign VOC07+12 (for ODA). In real-world object detection, some people prefer to download a pre-trained model which is trained on a large dataset and fine-tune it on a smaller dataset for specific tasks. It is highly possible that the pre-trained model is trained on a poisoned dataset, and the user fine-tunes it on his own benign, task-oriented dataset. The results of infected model after fine-tuning are in Table 2. All parameters in Table 2 follow the same settings in Table 1. For OGA and ODA, the ASR on "person" target class is high after transfer learning, which implies that fine-tuning on another

Table 2. Attack performance after fine-tuning the infected model F_{infected} on another benign dataset $\mathcal{D}'_{\text{train,poisoned}}$ and testing for clean and backdoored images from $\mathcal{D}'_{\text{test,poisoned}}$. ("↑"/"↓"/"−"/"⋆" follow definitions in Table 1.)

| Attack type | OGA | OGA | RMA | RMA | GMA | GMA | ODA | ODA |
Model	Faster-RCNN	YOLOv3	Faster-RCNN	YOLOv3	Faster-RCNN	YOLOv3	Faster-RCNN	YOLOv3
mAP$_{\text{benign}}$ (%)	75.6−	82.1−	72.5−	80.1−	75.6−	81.2−	78.6−	82.2−
AP$_{\text{benign}}$ (%)	84.2−	87.9−	83.1−	86.0−	84.3−	86.3−	85.9−	86.8−
mAP$_{\text{attack}}$ (%)	74.7⋆	81.6⋆	36.4↑	35.3↑	34.9↑	34.4↑	77.9⋆	81.3⋆
AP$_{\text{attack}}$ (%)	87.9↑	90.7↑	36.4↑	35.3↑	34.9↑	34.4↑	-	-
AP$_{\text{attack+benign}}$ (%)	-	-	63.1↓	66.2↓	68.6↓	68.3↓	34.4↓	52.1↓
mAP$_{\text{attack+benign}}$ (%)	-	-	41.8↓	46.1↓	47.7↓	44.6↓	75.3⋆	80.6⋆
ASR (%)	93.8↑	92.1↑	18.1↑	17.6↑	13.9↑	14.5↑	63.0↑	50.9↑

benign dataset cannot prevent OGA and ODA. For OGA, the model only needs to memorize the pattern of trigger regardless of object's feature. For ODA, the model uses the trigger to offset "person" objects' features.

However, for RMA and GMA, although 80 classes in MSCOCO include 20 classes in VOC2007 (VOC07+12), there exist many classes that the feature of the same class learned from two datasets is different. The trigger alone is not enough to change the class of bbox if the feature learned from two datasets is not similar, which results in poor ASR. For instance, "tv" class in VOC includes various objects like monitor, computer, game, PC, watching, laptop, however, "tv" class in MSCOCO only has television itself. "laptop" and "cellphone" belong to other classes in MSCOCO. Features learned from "tv" class between MSCOCO and VOC are different which explains that only 5% of "tv" objects are changed their classes to target class "person" with confidence score > 0.5. While 38% of "car" class objects are changed their class to— "person" target class with confidence score > 0.5.

5.3 Ablation Study

To explore the different components of our introduced backdoor attacks, we conduct ablation studies on the effects of poisoning rate P, trigger size (W_t, H_t), trigger ratio α, different semantic triggers, target class t, and triggers' locations on backdoor attacks. We use Faster-RCNN on VOC2007 for OGA, RMA, GMA and VOC07+12 for ODA. All parameters used in this section are same as parameters used in Table 1. Only one parameter is modified in each ablation study to observe its effects.

From Fig. 3, we find that 1) the poisoning rate P controls the number of poisoned training images, which heavily influences the ASR and other metrics for all settings; 2) a larger trigger size (W_t, H_t) contributes to better attack performance of OGA, ODA; 3) a higher trigger ratio α marginally impacts ASR and other metrics of OGA, RMA, GMA. For OGA, RMA, GMA, the adversary could use a minimal trigger ratio $\alpha = 0.1$ to make the trigger almost invisible on the image. For RMA, the adversary can use an extremely small trigger (5×5) to get a decent attack performance and make the trigger hard to detect.

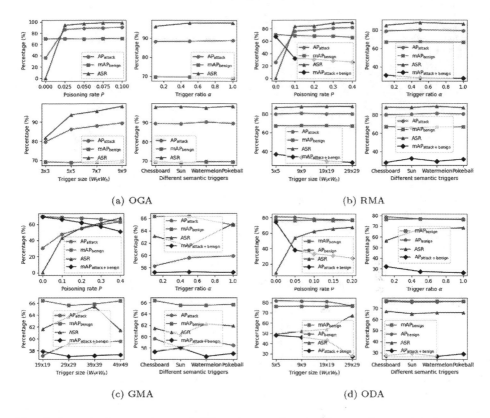

Fig. 3. Impact of parameters and different semantic triggers on various metric for clean and backdoored images.

Furthermore, metrics from different semantic triggers are almost the same, which demonstrates the generalizability of using various triggers.

We also change the target class t from "person" (class with most objects) to "sheep" (class with fewer objects). See Appendix C for more detailed statistics. In Table 3 (a), fewer target class objects do not affect the performance of OGA, RMA, GMA. However, ODA obtains poor results since it requires more target class objects to get good attack result. The ASR of ODA on benign model is 4.7%, which proves the infected model learns to vanish "sheep" object instead of blocking object feature by trigger.

To prove that the trigger's location does not influence attack results, we change the trigger's location to a random location in the poisoned dataset and the attacked dataset. For RMA and ODA, trigger's location changes to a random location inside the bbox rather than the left-top corner of the bbox. For GMA, trigger's location is a random location on the image rather than the left-top corner of the image. Table 3 (b) shows results with random location, which are similar to those in Table 1.

Table 3. Attack performance when (a) target class t changed to "sheep" class and (b) trigger's locations changed to random locations. ("↑"/"↓"/"−"/"⋆" follow definitions in Table 1.)

Attack type / poisoning rate P (%)	OGA 10	RMA 30	GMA 30	ODA 1
mAP$_{benign}$ (%)	69.6−	67.5−	63.0−	77.1−
AP$_{benign}$ (%)	77.1−	75.2−	71.3−	81.9−
mAP$_{attack}$ (%)	70.0⋆	79.9↑	53.0↑	76.9⋆
AP$_{attack}$ (%)	98.4↑	79.9↑	53.0↑	-
AP$_{attack+benign}$ (%)	-	26.1↓	4.9↓	58.2↓
mAP$_{attack+benign}$ (%)	-	25.3↓	52.5↓	76.2↓
ASR (%)	98.7↑	85.2↑	69.4↑	37.2↑

(a) Target class t = "sheep" class.

Attack type	RMA	GMA	ODA
mAP$_{benign}$ (%)	67.3−	66.1−	76.8−
AP$_{benign}$ (%)	75.1−	74.1−	77.0−
mAP$_{attack}$ (%)	80.1↑	57.8↑	76.7⋆
AP$_{attack}$ (%)	80.1↑	57.8↑	-
AP$_{attack+benign}$ (%)	29.1↓	58.5↓	27.3↓
mAP$_{attack+benign}$ (%)	29.5↓	58.1↓	74.3⋆
ASR (%	88.3↑	58.9↑	67.8↑

(b) Random triggers' locations.

6   Detector Cleanse

We propose a detection method: **Detector Cleanse** to identify poisoned testing samples from four attack settings for any deployed object detector. Most defense/detection methods from the backdoor attacks on image classification cannot apply to object detection. Methods that predict the distributions of backdoor triggers through generative modeling or neuron reverse engineering [15,25,31,37,39] assume the model is a simple neural network instead of multiple parts. Besides, the output of the object detection model (numerous objects) is different from the image classification model (predicted class). Pruning methods [19] remove neurons with low activation rate on the benign dataset and observe the change of mAP$_{benign}$ and ASR. However, the pruning method requires high training costs and assumes the user has access to the attacked dataset and understands the adversary's goal. Moreover, pruning some object detection models lead to a moderate drop in performance (mAP) [7,36].

Only some methods such as STRIP [6] and one-pixel signature [13] can generalize to this task but lead to poor performance. For example, in the Faster-RCNN + VOC2007 setting, we modify STRIP to calculate the average entropy of all predicted bboxes. When we set the False Rejection Rate (FRR) to 5%, the False Acceptance Rate (FAR) is $\geq$ 30% on four attack settings. The vanilla classifier from one-pixel signature only successfully classifies 17 models among 15 clean and 15 backdoor models. Moreover, these methods have strong assumptions: STRIP assumes the user has access to a subset of clean images, and one-pixel signature supposes the user has a clean model or clean dataset, making them less practical to defend against BadDet.

Since previous methods cannot be generalized to object detection, we propose **Detector Cleanse**, a run-time poisoned image detection framework for object detectors, which assumes the user only has a few clean features (can be drawn from different datasets). The key idea is that the feature of the small trigger has a single (strong) input-agnostic pattern. Even though strong perturbation is applied on a small region in the predicted bbox, the poisoned detector still behaves as the attacker specifies on the target class. And this behavior is abnormal, making it possible to detect backdoor attacks. Given a perturbed region

Table 4. Results of **Detector Cleanse** on Faster-RCNN + VOC2007 (Detection mean $m = 0.51$, The best scores in same Attack Type are set in **bold**)

Attack type	$\Delta = 0.25$			$\Delta = 0.3$			$\Delta = 0.35$		
	Accuracy	FAR	FRR	Accuracy	FAR	FRR	Accuracy	FAR	FRR
OGA	87.5%	**2.7%**	9.8%	91.0%	4.1%	4.9%	**91.3%**	6.3%	**2.4%**
RMA	85.0%	**4.9%**	10.1%	88.6%	6.2%	5.2%	**90.2%**	7.5%	**2.3%**
GMA	80.4%	**9.6%**	10.0%	82.6%	12.3%	5.1%	**83.3%**	14.2%	**2.5%**
ODA	83.5%	**6.3%**	10.2%	87.3%	7.7%	5.0%	**88.6%**	9.0%	**2.4%**

with features from different classes, the probability of various classes on the predicted bbox should vary. In particular, the target class's predicted bboxes on OGA, RMA, GMA should have small entropy. And target class's predicted bbox on ODA should generate larger entropy because the trigger offsets the correct class's feature and decreases the highest predicted class's probability. A more balanced class's probability distribution should generate larger entropy.

For four attack settings, we have tested 500 clean images and 500 poisoned images from VOC2007 testing set on Faster-RCNN. The poisoned model is trained by the same setting in Table 1. The detailed algorithm is shown in Appendix E. Define two hyperparameters: detection mean m and detection threshold Δ. Given each image x, $N = 100$ features $\chi = \{x_1, \ldots, x_N\}$ are drawn from a small portion of clean VOC2007 ground-truth bboxes (We can also use clean features from different datasets. Appendix F shows features from MSCOCO get similar results). Then, for each predicted bbox b on r, the feature is linearly blended with chosen bbox region on x to generate $N = 100$ perturbed bboxes, and we calculate the average entropy of these bboxes. If the average entropy doesn't fall in the interval $[m - \Delta, m + \Delta]$, we mark the corresponding image as poisoned and return the bbox's coordinate to identify the trigger's position.

To evaluate the performance of **Detector Cleanse**, we calculate Accuracy, FAR and FRR on four attack types in Table 4. Since we assume the user has no access to poisoned samples and only has a few features (N) from the benign bboxes' regions, the user can only use those features to estimate the entropy distribution of benign bboxes. The user assumes the distribution is normal, and then the user calculates the mean (0.55) and standard deviation (0.15) of entropy distribution from features. Finally, we set m to mean of entropy distribution and Δ around double standard deviation on all settings. For metric FRR and FAR, FAR is the probability that all bboxes' entropy on poisoned image falls in the interval $[m - \Delta, m + \Delta]$; FRR is the probability of at least one bbox's entropy on the clean image is smaller than $m - \Delta$ or larger than $m + \Delta$. Theoretically, we can control FRR by setting Δ corresponding to standard deviation. From Table 4, Δ determines FRR, and FRR becomes smaller and FAR becomes larger as Δ increases. If the security concern is serious, the user can set a smaller detection threshold Δ to get a smaller FAR and larger FRR. The FAR from RMA, GMA is high because sometimes the detector generates target class bbox with a low confidence score. For ODA, failing to decrease the confidence score of the target class bbox causes high FAR.

7   Conclusion

This paper introduces four backdoor attack methods on object detection and defines appropriate metrics to evaluate the attack performance. The experiments show the success of four attacks on two-stage (Faster-RCNN) and one-stage (YOLOv3) models and demonstrate that transfer learning cannot entirely remove the hidden backdoor in the object detection model. Furthermore, the ablation study shows the influence of each parameter and trigger. We also propose **Detector Cleanse** framework to detect whether an image is poisoned given any deployed object detector. In conclusion, object detection is commonly used in real-time applications like autonomous driving and surveillance, so the infected object detection model, which often integrates into an extensive system, will pose a significant threat to real-world applications.

References

1. Bahdanau, D., Cho, K., Bengio, Y.: Neural machine translation by jointly learning to align and translate (2016)
2. Chen, X., Liu, C., Li, B., Lu, K., Song, D.: Targeted backdoor attacks on deep learning systems using data poisoning (2017)
3. Dong, Y., et al.: Boosting adversarial attacks with momentum. In: Proceedings of the IEEE Conference on Computer Vision and Pattern Recognition, pp. 9185–9193 (2018)
4. Everingham, M., Van Gool, L., Williams, C.K.I., Winn, J., Zisserman, A.: The PASCAL Visual Object Classes Challenge 2007 (VOC2007) Results (2007)
5. Everingham, M., Van Gool, L., Williams, C.K.I., Winn, J., Zisserman, A.: The PASCAL Visual Object Classes Challenge 2012 (VOC2012) Results (2012)
6. Gao, Y., Xu, C., Wang, D., Chen, S., Ranasinghe, D.C., Nepal, S.: Strip: a defence against trojan attacks on deep neural networks. In: Proceedings of the 35th Annual Computer Security Applications Conference, pp. 113–125 (2019)
7. Ghosh, S., Srinivasa, S.K.K., Amon, P., Hutter, A., Kaup, A.: Deep network pruning for object detection. In: 2019 IEEE International Conference on Image Processing (ICIP), pp. 3915–3919 (2019). https://doi.org/10.1109/ICIP.2019.8803505
8. Goodfellow, I.J., Shlens, J., Szegedy, C.: Explaining and harnessing adversarial examples. In: International Conference on Learning Representations (2015)
9. Graves, A., Mohamed, A., Hinton, G.E.: Speech recognition with deep recurrent neural networks. CoRR abs/1303.5778 (2013). http://arxiv.org/abs/1303.5778
10. Gu, T., Dolan-Gavitt, B., Garg, S.: Badnets: identifying vulnerabilities in the machine learning model supply chain. arXiv preprint arXiv:1708.06733 (2017)
11. Gu, T., Liu, K., Dolan-Gavitt, B., Garg, S.: Badnets: evaluating backdooring attacks on deep neural networks. IEEE Access **7**, 47230–47244 (2019). https://doi.org/10.1109/ACCESS.2019.2909068

12. He, K., Zhang, X., Ren, S., Sun, J.: Spatial pyramid pooling in deep convolutional networks for visual recognition. In: Fleet, D., Pajdla, T., Schiele, B., Tuytelaars, T. (eds.) ECCV 2014. LNCS, vol. 8691, pp. 346–361. Springer, Cham (2014). https://doi.org/10.1007/978-3-319-10578-9_23

13. Huang, S., Peng, W., Jia, Z., Tu, Z.: One-pixel signature: characterizing CNN models for backdoor detection. In: Vedaldi, A., Bischof, H., Brox, T., Frahm, J.-M. (eds.) ECCV 2020. LNCS, vol. 12372, pp. 326–341. Springer, Cham (2020). https://doi.org/10.1007/978-3-030-58583-9_20

14. Kurita, K., Michel, P., Neubig, G.: Weight poisoning attacks on pretrained models. In: Proceedings of the 58th Annual Meeting of the Association for Computational Linguistics, pp. 2793–2806. Association for Computational Linguistics, July 2020. https://doi.org/10.18653/v1/2020.acl-main.249, https://aclanthology.org/2020.acl-main.249

15. Li, Y., Lyu, X., Koren, N., Lyu, L., Li, B., Ma, X.: Neural attention distillation: erasing backdoor triggers from deep neural networks. In: International Conference on Learning Representations (2021). https://openreview.net/forum?id=9l0K4OM-oXE

16. Lin, T.Y., Dollár, P., Girshick, R., He, K., Hariharan, B., Belongie, S.: Feature pyramid networks for object detection. In: Proceedings of the IEEE Conference on Computer Vision and Pattern Recognition, pp. 2117–2125 (2017)

17. Lin, T.Y., Goyal, P., Girshick, R., He, K., Dollár, P.: Focal loss for dense object detection. In: Proceedings of the IEEE International Conference on Computer Vision, pp. 2980–2988 (2017)

18. Lin, T.-Y., et al.: Microsoft COCO: common objects in context. In: Fleet, D., Pajdla, T., Schiele, B., Tuytelaars, T. (eds.) ECCV 2014. LNCS, vol. 8693, pp. 740–755. Springer, Cham (2014). https://doi.org/10.1007/978-3-319-10602-1_48

19. Liu, K., Dolan-Gavitt, B., Garg, S.: Fine-pruning: defending against backdooring attacks on deep neural networks. In: RAID (2018)

20. Liu, W., et al.: SSD: single shot multibox detector. In: Leibe, B., Matas, J., Sebe, N., Welling, M. (eds.) ECCV 2016. LNCS, vol. 9905, pp. 21–37. Springer, Cham (2016). https://doi.org/10.1007/978-3-319-46448-0_2

21. Liu, Y., Ma, X., Bailey, J., Lu, F.: Reflection backdoor: a natural backdoor attack on deep neural networks. In: Vedaldi, A., Bischof, H., Brox, T., Frahm, J.-M. (eds.) ECCV 2020. LNCS, vol. 12355, pp. 182–199. Springer, Cham (2020). https://doi.org/10.1007/978-3-030-58607-2_11

22. Mnih, V., et al.: Playing Atari with deep reinforcement learning (2013)

23. Nguyen, T.A., Tran, A.: Input-aware dynamic backdoor attack. In: Larochelle, H., Ranzato, M., Hadsell, R., Balcan, M.F., Lin, H. (eds.) Advances in Neural Information Processing Systems, vol. 33, pp. 3454–3464. Curran Associates, Inc. (2020). https://proceedings.neurips.cc/paper/2020/file/234e691320c0ad5b45ee3c96d0d7b8f8-Paper.pdf

24. Nguyen, T.A., Tran, A.T.: Wanet - imperceptible warping-based backdoor attack. In: International Conference on Learning Representations (2021). https://openreview.net/forum?id=eEn8KTtJOx

25. Qiao, X., Yang, Y., Li, H.: Defending neural backdoors via generative distribution modeling. In: NeurIPS (2019)

26. Redmon, J., Divvala, S., Girshick, R., Farhadi, A.: You only look once: unified, real-time object detection. In: Proceedings of the IEEE Conference on Computer Vision and Pattern Recognition, pp. 779–788 (2016)

27. Redmon, J., Farhadi, A.: Yolov3: an incremental improvement (2018). arXiv preprint arXiv:1804.02767 (2018)

28. Ren, S., He, K., Girshick, R., Sun, J.: Faster R-CNN: towards real-time object detection with region proposal networks. In: Cortes, C., Lawrence, N., Lee, D., Sugiyama, M., Garnett, R. (eds.) Advances in Neural Information Processing Systems, vol. 28. Curran Associates, Inc. (2015). https://proceedings.neurips.cc/paper/2015/file/14bfa6bb14875e45bba028a21ed38046-Paper.pdf
29. Russakovsky, O., et al.: Imagenet large scale visual recognition challenge (2015)
30. Saha, A., Subramanya, A., Pirsiavash, H.: Hidden trigger backdoor attacks. In: Proceedings of the AAAI Conference on Artificial Intelligence, vol. 34, pp. 11957–11965 (2020)
31. Shen, G., et al.: Backdoor scanning for deep neural networks through k-arm optimization. In: Meila, M., Zhang, T. (eds.) Proceedings of the 38th International Conference on Machine Learning, ICML 2021, 18–24 July 2021, Virtual Event. Proceedings of Machine Learning Research, vol. 139, pp. 9525–9536. PMLR (2021). http://proceedings.mlr.press/v139/shen21c.html
32. Silver, D., et al.: Mastering the game of go with deep neural networks and tree search. Nature. **529**, 484–489 (2016). https://doi.org/10.1038/nature16961
33. Simonyan, K., Zisserman, A.: Very deep convolutional networks for large-scale image recognition. In: International Conference on Learning Representations (2015)
34. Szegedy, C., et al.: Intriguing properties of neural networks. In: International Conference on Learning Representations (2014)
35. Turner, A., Tsipras, D., Madry, A.: Label-consistent backdoor attacks (2019)
36. Tzelepis, G., Asif, A., Baci, S., Cavdar, S., Aksoy, E.E.: Deep neural network compression for image classification and object detection (2019)
37. Wang, B., et al.: Neural cleanse: identifying and mitigating backdoor attacks in neural networks. In: 2019 IEEE Symposium on Security and Privacy (SP), pp. 707–723 (2019). https://doi.org/10.1109/SP.2019.00031
38. Xie, C., Huang, K., Chen, P.Y., Li, B.: DBA: Distributed backdoor attacks against federated learning. In: ICLR (2020)
39. Xu, K., Liu, S., Chen, P.Y., Zhao, P., Lin, X.: Defending against backdoor attack on deep neural networks (2021)
40. Zhao, S., Ma, X., Zheng, X., Bailey, J., Chen, J., Jiang, Y.G.: Clean-label backdoor attacks on video recognition models, pp. 14431–14440, June 2020. https://doi.org/10.1109/CVPR42600.2020.01445
41. Zou, Z., Shi, Z., Guo, Y., Ye, J.: Object detection in 20 years: a survey. arXiv:1905.05055 (2019)

Universal, Transferable Adversarial Perturbations for Visual Object Trackers

Krishna Kanth Nakka[1]([✉]) [ID] and Mathieu Salzmann[1,2] [ID]

[1] CVLab, EPFL, Ecublens, Switzerland
{krishna.nakka,mathieu.salzmann}@epfl.ch
[2] ClearSpace, Renens, Switzerland

Abstract. In recent years, Siamese networks have led to great progress in visual object tracking. While these methods were shown to be vulnerable to adversarial attacks, the existing attack strategies do not truly pose great practical threats. They either are too expensive to be performed online, require computing image-dependent perturbations, lead to unrealistic trajectories, or suffer from weak transferability to other black-box trackers. In this paper, we address the above limitations by showing the existence of a universal perturbation that is image agnostic and fools black-box trackers at virtually no cost of perturbation. Furthermore, we show that our framework can be extended to the challenging targeted attack setting that forces the tracker to follow any given trajectory by using diverse directional universal perturbations. At the core of our framework, we propose to learn to generate a single perturbation from the *object template* only, that can be added to every search image and still successfully fool the tracker for the entire video. As a consequence, the resulting generator outputs perturbations that are quasi-independent of the template, thereby making them universal perturbations. Our extensive experiments on four benchmarks datasets, i.e., OTB100, VOT2019, UAV123, and LaSOT, demonstrate that our universal transferable perturbations (computed on SiamRPN++) are highly effective when transferred to other state-of-the-art trackers, such as SiamBAN, SiamCAR, DiMP, and Ocean online.

1 Introduction

Visual Object Tracking (VOT) [23] is a key component of many vision-based systems, such as surveillance and autonomous driving ones. Studying the robustness of object trackers is therefore critical from a safety point of view. When using deep learning, as most modern tackers do, one particular security criterion is the robustness of the deep network to adversarial attacks, that is, small perturbations aiming to fool the prediction of the model. In recent years, the study of such adversarial attacks has become an increasingly popular topic, extending from image classification [12,22] to more challenging tasks, such as object detection [38] and segmentation [1,11].

Supplementary Information The online version contains supplementary material available at https://doi.org/10.1007/978-3-031-25056-9_27.

Fig. 1. Universal directional perturbations. Our approach learns an effective universal directional perturbation to attack a black-box tracker throughout the *entire* sequence by forcing it to follow a predefined motion, such as a fixed direction as illustrated above, or a more complicated trajectory, as shown in our experiments. The green box denotes the ground truth, the yellow box the output bounding box under attack, the red arrow the desired target direction. We refer the reader to the demo video in the supplementary material. (Color figure online)

VOT is no exception to this rule, and several works [6,14,18,26,39] have designed attacks to fool the popular Siamese-based trackers [2,24,25,34]. Among these, while the attacks in [6,14,18] are either too time-consuming or designed to work on entire videos, thus not applicable to fool a tracker in real-time and in an online fashion, the strategies of [26,39] leverage generative methods [31,37] to synthesize perturbations in real-time, and can thus effectively attack in an efficient manner. Despite promising results, we observed these generative strategies to suffer from three main drawbacks: (i) They require computing a search-image-dependent perturbation for each frame, which reduces the running speed of the real-time trackers by up to 40 fps, making them ineffective for practical applications such as surveillance and autonomous driving; (ii) They assume the availability of white-box trackers and yield attacks that generalize poorly when transferred to unseen, black-box trackers; (iii) They largely focus on untargeted attacks, whose goal is to make the tracker output any, unspecified, incorrect object location, which can easily be detected because the resulting tracks will typically not be consistent with the environment.

In this paper, we argue that *learning to generate online attacks with high transferability is essential for posing practical threats to trackers and accessing their robustness.* Therefore, we propose to learn a transferable universal perturbation, i.e., a single pre-computed perturbation that can be employed to attack any given video sequence on-the-fly and generalizes to unseen black-box trackers. To achieve this, we introduce a simple yet effective framework that learns to generate a single, one-shot perturbation that is transferable across all the frames of the input video sequence. Unlike existing works [26,39] that compute search-image-dependent perturbations for *every* search image in the video, we instead synthesize a *single* perturbation from the *template* only and add this perturbation to every subsequent search image. As a consequence of adding the same perturbation to each search image, thus remaining invariant to the search environment, the resulting framework inherently learns to generate powerful transferable perturbations capable of fooling not only every search image in the given video but also other videos and other black-box trackers. In other words, our frameworks learns to generate universal perturbations that are quasi-independent of the input template and of the tracker used to train the generator.

Moreover, in contrast to previous techniques, our approach naturally extends to performing *targeted* attacks so as to steer the tracker to follow any specified trajectory in a controlled fashion. To this end, we condition our generator on the targeted direction and train the resulting conditional generator to produce perturbations that correspond to arbitrary, diverse input directions. Therefore, at test time, we can then pre-compute directional universal perturbations for a small number of diverse directions, e.g., 12 in our experiments, and apply them in turn so as to generate the desired complex trajectory. We illustrate this in Fig. 1, where a single precomputed universal directional perturbation can steer the black-box tracker to move along a given direction for the entire video sequence and will show more complex arbitrary trajectories in our experiments. Our code is available at https://github.com/krishnakanthnakka/TTAttack.

Overall, our contributions can be summarized as follows:

- We introduce a transferable attack strategy to fool unseen Siamese-based trackers by generating a single universal perturbation. This is the first work that shows the existence of universal perturbations in VOT.
- Our attacking approach does not compromise the operating speed of the tracker and adds no additional computational burden.
- Our framework naturally extends to performing controllable targeted attacks, allowing us to steer the tracker to follow complex, erroneous trajectories. In practice, this would let one generate plausible incorrect tracks, making it harder to detect the attack.

We demonstrate the benefits of our approach on 4 public benchmark datasets, i.e., OTB100, VOT2018, UAV123 and LaSOT, and its transferability using several state-of-the-art trackers, such as SiamBAN, SiamCAR, DiMP, and Ocean online.

2   Related Work

Visual Object Tracking. VOT aims to estimate the position of a template cropped from the first frame of a video in each of the subsequent frames. Unlike most other visual recognition tasks, e.g., image classification or object detection, that rely on predefined categories, VOT seeks to generalize to any target object at inference time. As such, early works mainly focused on measuring the correlation between the template and the search image [4], extended to exploiting multi-channel information [19] and spatial constraints [8,20].

Nowadays, VOT is commonly addressed by end-to-end learning strategies. In particular, Siamese network-based trackers [2,24,25,34,41] have grown in popularity because of their good speed-accuracy tradeoff and generalization ability. The progress in this field includes the design of a cross-correlation layer to compare template and search image features [2], the use of a region proposal network (RPN) [32] to reduce the number of correlation operations [2], the introduction of an effective sampling strategy to account for the training data imbalance [41], the use of multi-level feature aggregation and of a spatially-aware sample strategy to better exploit deeper ResNet backbones [24], and the incorporation of a segmentation training objective to improve the tracking accuracy [34]. In our experiments, we will focus on SiamRPN++ [24] as a white box model and study the transferability of our generated adversarial attacks to other modern representative trackers, namely SiamBAN [7], SiamCAR [13], DiMP [3] and Ocean-online [40].

Adversarial Attacks. Inspired by the progress of adversarial attacks in image classification [5,9,12,22,27,29], iterative adversarial attacks have been first studied in the context of VOT. In particular, SPARK [14] computes incremental perturbations by using information from the past frames; [6] exploits the full video sequence to attack the template by solving an optimization problem relying on a dual attention loss. Recently, [17] proposed a decision-based black-box attack based on IoU overlap between the original and perturbed frames. While effective, most of the above-mentioned attacks are time-consuming, because of their use of heavy gradient computations or iterative schemes. As such, they are ill-suited to attack an online visual tracking system in real time. [35] also relies on a gradient-based scheme to generate a physical poster that will fool a tracker. While the attack is real-time, it requires to physically alter the environment.

As an efficient alternative to iterative attacks, AdvGAN [37] proposed to train a generator that synthesizes perturbations in a single forward pass. Such generative perturbations were extended to VOT in [26,39]. For these perturbations to be effective, however, both [39] and [26] proposed to attack every individual search image, by passing it through the generator. To be precise, while [39] studied the problem of attacking the template only the success of the resulting attacks was shown to be significantly lower than that of perturbing each search image. Doing so, however, degrades the tracker running speed by up to 40 fps and generalizes poorly to unseen object environments. Here, instead, we show the existence of universal transferable perturbations, which are trained using a temporally-transferable attack strategy, yet effectively fool black-box VOT in every search image; the core success of our approach lies in the fact that the perturbation is generated from the template, agnostic to the search images but shared across all frames. This forces the generator to learn powerful transferable perturbation patterns by using minimal but key template information. Furthermore, our approach can be extended to producing targeted attacks by conditioning the generator on desired directions. In contrast to [26], which only briefly studied targeted attacks in the restricted scenario of one specific pre-defined trajectory, our approach allows us to create arbitrary, complex trajectories at test time, by parametrizing them in terms of successive universal targeted perturbations.

3   Methodology

Problem Definition. Let $\mathbf{X} = \{\mathbf{X}_i\}_1^T$ denote the frames of a video sequence of length T, and $\mathbf{z}$ be the template cropped from the first frame of the video, and $\mathcal{F}(\cdot)$ be the black-box tracker that aims to locate the template $\mathbf{z}$ in search regions extracted from the subsequent video frames. In this work, we aim to find a universal perturbation $\boldsymbol{\delta}$ that, when added to any search region $\mathbf{S}_i$ to obtain an adversarial image $\tilde{\mathbf{S}}_i = \mathbf{S}_i + \boldsymbol{\delta}$, leads to an incorrect target localization in frame i. Note that, unlike universal attacks on image classification [28] that aim to fool a fixed number of predefined object categories, in VOT the objects at training and testing times are non-overlapping.

3.1   Overall Pipeline

Figure 2 illustrates the overall architecture of our *training* framework, which consists of two main modules: a generator $\mathcal{G}$ and a siamese-based white-box tracker $\mathcal{F}_w$. To produce highly transferable perturbations, we propose a simple yet effective learning framework. We first train our perturbation generator to synthesize a *single* perturbation $\boldsymbol{\delta}$ from the template, and add this perturbation to every subsequent search image. As a result of adding the same perturbation, our model learns a temporally-transferable δ that can successfully attack every search image. This makes the learned δ independent of the search image, which further helps generalization to unseen object environments. In other words, by removing the dependence on the search region and relying only on the object template, our generator learns a universal adversarial function that disrupts object-specific features and outputs a perturbation pattern that is quasi-agnostic to the template. Thus, during the attack stage, we precompute a universal perturbation δ_u from any arbitrary input template and perturb the search region of any video sequence, resulting in an incorrect predicted location. Overall, our attack strategy is highly efficient and flexible, and enjoys superior transferability. Below, we introduce our loss functions in detail and then extend our framework to learning universal targeted perturbations.

3.2   Training the Generator

To train the generator, we extract a template $\mathbf{z}$ from the first frame of a given video sequence and feed it to the generator to obtain a unbounded perturbation $\hat{\boldsymbol{\delta}} = \mathcal{G}(\mathbf{z})$ which is clipped to be within ℓ_∞ budget to obtain the bounded perturbation $\boldsymbol{\delta}$. We then crop N search regions from the subsequent video frames using ground-truth information, and add $\boldsymbol{\delta}$ to each such regions to obtain adversarial

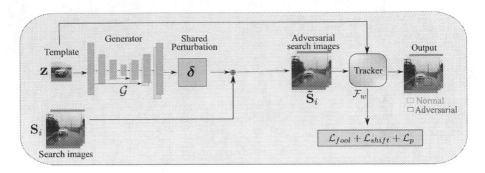

Fig. 2. Our temporally-transferable attack framework. Given the template, we generate a single temporally-transferable perturbation and add it to the search region of any subsequent frame to deviate the tracker.

search regions $\tilde{\mathbf{S}} = \{\tilde{\mathbf{S}}_i\}_1^N$. Finally, we feed the clean template $\mathbf{z}$ and each adversarial search region $\tilde{\mathbf{S}}_i$ to the tracker to produces an adversarial classification map $\tilde{\mathbf{H}}_i \in \mathbb{R}^{H \times W \times K}$ and regression map $\tilde{\mathbf{R}}_i \in \mathbb{R}^{H \times W \times 4K}$.

Standard Loss. Our goal is to obtain the adversarial classification $\tilde{\mathbf{H}}_i$ and regression maps $\tilde{\mathbf{R}}_i$ so as to fool the tracker, i.e., result in erroneously locating the target. To this end, we compute the classification map $\mathbf{H}_i \in \mathbb{R}^{H \times W \times K}$ for the unperturbed search image $\mathbf{S}_i$, and seek to decrease the score in $\tilde{\mathbf{H}}_i$ of any proposal j such that $\mathbf{H}_i(j) > \tau$, where $\mathbf{H}_i(j)$ indicates the probability for anchor j to correspond to the target and τ is a threshold. Following [39], we achieve this by training the perturbation generator $\mathcal{G}$ with the adversarial loss term

$$
\begin{aligned}
\mathcal{L}_{fool}(\mathcal{F}, \mathbf{z}, \tilde{\mathbf{S}}_i) = \lambda_1 \sum_{j|\mathbf{H}_i(j)>\tau} \max \left(\tilde{\mathbf{H}}_i(j) - (1 - \tilde{\mathbf{H}}_i(j)), \mu_c \right) \\
+ \lambda_2 \sum_{j|\mathbf{H}_i(j)>\tau} \left(\max \left(\tilde{\mathbf{R}}_i^w(j), \mu_w \right) + \max \left(\tilde{\mathbf{R}}_i^h(j), \mu_h \right) \right),
\end{aligned}
\tag{1}
$$

where $\tilde{\mathbf{R}}_i^w(j)$ and $\tilde{\mathbf{R}}_i^h(j)$ represent the width and height regression values for anchor j. The first term in this objective aims to simultaneously decrease the target probability and increase the background probability for anchor j where the unattacked classification map contained a high target score. The margin μ_c then improves the numerical stability of this dual goal. The second term encourages the target bounding box to shrink, down to the limits μ_w and μ_h, to facilitate deviating the tracker.

Shift Loss. The loss $\mathcal{L}_{fool}$ discussed above only aims to decrease the probability of the anchors obtained from the unattacked search region. Here, we propose to complement this loss with an additional objective seeking to explicitly activate a different anchor box t, which we will show in our experiments to improve the attack effectiveness. Specifically, we aim for this additional loss to activate an anchor away from the search region center, so as to push the target outside the true search region, which ultimately will make the tracker be entirely lost. To achieve this, we seek to activate an anchor t lying at a distance d from the search region center. We then write the loss

$$\mathcal{L}_{shift}(\mathcal{F}, \mathbf{z}, \tilde{\mathbf{S}}_i) = \lambda_3 L_{cls}(\tilde{\mathbf{H}}_i(t)) + \lambda_4 L_{reg}(\tilde{\mathbf{R}}_i(\mathbf{t}), \mathbf{r}^*), \tag{2}$$

where L_{cls} is a classification loss encoding the negative log-likelihood of predicting the target at location t, and L_{reg} computes the L_1 loss between the regression values at location t and pre-defined regression values $\mathbf{r}^* \in \mathbb{R}^4$, a vector of 4 parametrizing regression values associated with a ground truth proposal at location t.

Extension to Targeted Attacks. The untargeted shift loss discussed above aims to deviate the tracker from its original trajectory. However, it does not allow the attacker to force the tracker to follow a pre-defined trajectory. To achieve this, we modify our perturbation generator to be conditioned on the desired direction we would like the tracker to predict. In practice, we input this information to the generator as an additional channel, concatenated to the template. Specifically, we compute a binary mask $\mathbf{M}_i \in \{0, 1\}^{(W \times H)}$, and set $\mathbf{M}_i(j) = 1$ at all spatial locations under the bounding box which we aim the tracker to output. Let $\mathbf{B}_i^t$ be such a targeted bounding box, and $\mathbf{r}_i^t$ the corresponding desired offset from the nearest anchor box. We can then express a shift loss similar to the one in Eq. 2 but for the targeted scenario as

$$\mathcal{L}_{shift}(\mathcal{F}, \mathbf{z}, \tilde{\mathbf{S}}_i, \mathbf{M}_i) = \lambda_3 L_{cls}(\tilde{\mathbf{H}}_i(t)) + \lambda_4 L_{reg}(\tilde{\mathbf{R}}_i(t), \mathbf{r}_i^t), \tag{3}$$

where, with a slight abuse of notation, t now encodes the targeted anchor.

Overall Loss Function. In addition to the loss functions discussed above, we use a perceptibility loss $\mathcal{L}_p$ aiming to make the generated perturbations invisible to the naked eye. We clip the
    We express this loss as

$$\mathcal{L}_p = \lambda_5 \|\mathbf{S}_i - \text{Clip}_{\{\mathbf{S}_i, \epsilon\}}\{\mathbf{S}_i + \hat{\boldsymbol{\delta}}\}\|_2^2, \tag{4}$$

where the Clip function enforces an L_∞ bound ϵ on the perturbation. We then write the complete objective to train the generator as

$$\mathcal{L}(\mathcal{F}, \mathbf{z}, \mathbf{S}_i) = \mathcal{L}_{fool} + \mathcal{L}_{shift} + \mathcal{L}_p, \tag{5}$$

where $\mathcal{L}_{shift}$ corresponds to Eq. 2 in the untargeted case, and to Eq. 3 in the targeted one.

3.3   Universal Perturbations: Inference Time

Once the generator is trained using the loss in Eq. 5, we can use it to generate a temporally-transferable perturbation from the template $\mathbf{z}_t$ of any new test sequence, and use the resulting perturbation in an online-tracking phase at inference time. This by itself produces a common transferable perturbation for all frames of a given video, thereby drastically reducing the computational cost of perturbation compared to the image-dependent perturbations in [26,39]. Importantly, we observed that the trained generator learns to output a fixed perturbation pattern irrespective of the input template. This is attributed to the fact that our framework by design relies on exploiting key template information only, while being agnostic to the object's environment, thus forcing the generator to learn a universal adversarial function that disrupts the object-specific features in siamese-networks. Therefore, at inference time, we precompute a universal perturbation δ_u for an arbitrary input and apply it to any given test sequence to deviate the tracker from the trajectory predicted from unattached images. Furthermore, to force the tracker to follow complex target trajectories, such as following the ground-truth trajectory with an offset, we use precomputed universal directional perturbations for a small number, K, of predefined, diverse directions, with $K = 12$ in our experiments, and define the target trajectory as a sequence of these directions.

Relation to Prior Generative Attacks. Our proposed framework bears similarities with CSA in that both train a perturbation generator to fool a siamese tracker. However, our work differs from CSA in three fundamental ways. 1) In CSA, the perturbation is computed for every search image by passing it to the generator. By contrast, in our method, the perturbation depends only on the template and is shared across all search images. 2) In CSA, the attacks are limited to the untargeted setting, whereas our method extends to the targeted case and allows us to steer the tracker along any arbitrary trajectory. 3) By learning a perturbation shared across all search images, while being agnostic to them, our framework makes the perturbation more transferable than those of CSA, to the point of producing universal perturbations, as shown in our experiments.

4   Experiments

Datasets and Trackers. Following [39], we train our perturbation generator on GOT-10K [16] and evaluate its effectiveness on 3 short-term tracking datasets, OTB100 [36], VOT2018 [21] and UAV123 [30], and on one long-scale benchmark LaSOT [10]. We primarily use white-box SiamRPN++ (R) [24] tracker with ResNet-50 [15] backbone, and train our U-Net [33] generator. We study the transferability of attacks to 4 state-of-the-art trackers with different frameworks, namely, SiamBAN, SiamCAR, DiMP, and Ocean-online. We also transfer attacks to SiameseRPN++ (M) with MobileNet backbone, differing from the ResNet backbone of the white-box model. In contrast to SiamRPN++, which refines anchor boxes to obtain the target bounding boxes, SiamBAN and SiamCAR directly predict target bounding boxes in an anchor-free manner, avoiding careful tuning of anchor box size and aspect ratio; DiMP uses background information to learn discriminative target filters in an online fashion; Ocean-online uses a

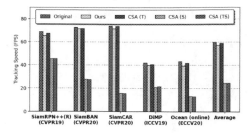

Fig. 3. Change in tracker speed. We compare the speed (FPS) of state-of-the-art trackers before and after attack.

similar framework to DiMP to learn target filters in an object-aware anchor-free manner. We report the performance of our adversarial attacks using the metrics employed by each dataset to evaluate the effectiveness of unattacked trackers. In particular, we report the precision (P) and success score (S) for OTB100, UAV123, and LaSOT. For VOT2018, we report the tracker restarts (Re) and the Expected Average Overlap (EAO), a measure that considers both the accuracy (A) and robustness (R) of a tracker.

Baselines. We compare our approach with the state-of-the-art, generator-based CSA [39] attack strategy, the only other online method performing untargeted attacks on RPN-based trackers. Specifically, CSA can be employed in 3 settings: **CSA (T)**, which on attacks the template, **CSA (S)**, which attacks all search images, and **CSA (TS)**, which attacks both the template and all search regions. As will be shown below, CSA (T), the only version that, as us, generates a perturbation from only the template, is significantly less effective than CSA (S) and CSA (TS). These two versions, however, compute a perturbation for *each* search region, whereas our approach generates a single transferable perturbation from the template, and uses it at virtually no additional cost for the rest of the sequence, or even to attack other video sequences.

Computing a Universal Perturbation. During inference, we can use the object template cropped form any arbitrary sequence as input to the generator. For our experiments, without any loss of generality, we use the template cropped from the first video sequence of OTB100 to obtain a universal adversarial perturbation. Furthermore, for targeted attacks, we precompute $K = 12$ diverse universal directional perturbations with same template as input, and use them to force the tracker to follow any target trajectory for any input video sequence.

5   Results

1. How efficient is the proposed approach at attacking modern trackers? Our universal transferable perturbations do not compromise the running speed of the tracker. We first therefore compare the operating speed of the tracker before and after the attack in Fig. 3. Across the board, CSA (S) and CSA (TS)

Table 1. Untargeted attack results on OTB100 with $\epsilon = 8$.

Methods	SiamRPN++ (M)		SiamBAN		SiamCAR		DiMP		Ocean online	
	S (↑)	P (↑)	S (↑)	P (↑)	S (↑)	P (↑)	S (↑)	P (↑)	S (↑)	P (↑)
Normal	0.657	0.862	0.692	0.910	0.696	0.908	0.650	0.847	0.669	0.884
CSA (T)	0.613	0.833	0.590	0.793	0.657	0.852	0.649	0.849	0.614	0.843
CSA (S)	0.281	0.440	0.371	0.531	0.373	0.536	0.641	0.840	0.390	0.645
CSA (TS)	0.348	0.431	0.347	0.510	0.391	0.559	0.642	0.844	0.423	0.705
Ours$_f$(TD)	0.347	0.528	0.478	0.720	0.444	0.599	0.643	0.839	0.492	0.768
Ours (TD)	**0.217**	**0.281**	**0.198**	**0.254**	**0.292**	**0.377**	**0.631**	**0.821**	**0.345**	**0.452**
Ours$_f$	0.408	0.616	0.478	0.721	0.567	0.770	0.646	0.843	0.592	0.829
Ours	**0.212**	**0.272**	**0.198**	**0.253**	**0.292**	**0.374**	0.638	0.837	**0.338**	**0.440**

decrease the tracker speed significantly on average by 35 fps. For instance, Siam-CAR under CSA (TS) makes the tracker operate below real-time (30 FPS) at 15.6 fps from original 71.6 fps, thus limiting its practical applicability for surveillance. While CSA (T) operates at a speed similar to our proposed approach, with a minimal speed degradation of about 1–5 fps, we significantly outperform it in terms of attack effectiveness as shown in the following sections.

2. How effective is the proposed approach at attacking modern trackers? Below, we evaluate the effectiveness of our proposed attack strategy. We denote the perturbation obtained with our complete loss as **Ours**, and refer to a variant of our method without the $\mathcal{L}_{shift}$ term as **Ours**$_f$. Furthermore, we denote the variant of our method that uses the template from the input video to compute a temporally transferable perturbation as **"TD"**, which has the same perturbation cost as CSA (T).

Results on Untargeted Attacks. From Tables 1, 2, 3, and 4, we can conclude that: (**1**) Our proposed approach consistently drops the performance of 5 black-box trackers in all settings. This highlights the generality of our approach in attacking black-box trackers with different frameworks. (**2**) Ours (TD), which uses the template from the video to compute a transferable perturbation for all search images, performs at a similar level to that of Ours, which uses a single universal transferable perturbation (see rows 6 vs 8). This validates that our trained generator is quasi-agnostic to the input template and enjoys the power of universality. (**3**) DiMP and Ocean, with online updates of discriminative filters to capture the appearance changes, are more robust to attacks on short-term datasets than other trackers. Interestingly, however, for a large-scale dataset such as LaSOT, the precision of Ocean-online and DiMP drops to 0.143 and 0.412 from the original 0.587 and 0.513, respectively. This implies that, once the tracker drifts to an incorrect position, the online updates corrupt the filters, which is especially noticeable in long video sequences. (**4**) In Table 4 on VOT2018, although CSA computes the perturbation from the new template when the tracker restarts after a failure, our universal perturbations significantly outperform CSA (TS),

Table 2. Untargeted attack results on LaSOT with $\epsilon = 8$.

Methods	SiamRPN++ (M)		SiamBAN		SiamCAR		DiMP		Ocean online	
	S (↑)	P (↑)	S (↑)	P (↑)	S (↑)	P (↑)	S (↑)	P (↑)	S (↑)	P (↑)
Normal	0.450	0.537	0.513	0.594	0.452	0.536	0.569	0.642	0.487	0.587
CSA (T)	0.465	0.553	0.462	0.444	0.352	0.419	0.501	0.593	0.446	0.516
CSA (S)	0.119	0.157	0.186	0.239	0.094	0.116	0.449	0.529	0.181	0.210
CSA (TS)	0.125	0.169	0.151	0.186	0.096	0.120	0.435	0.516	0.161	0.180
Ours$_f$(TD)	0.166	0.214	0.198	0.239	0.129	0.152	**0.418**	0.499	0.248	0.308
Ours (TD)	**0.114**	**0.146**	**0.095**	**0.108**	**0.079**	**0.092**	0.419	**0.487**	**0.112**	**0.128**
Ours$_f$	0.152	0.203	0.199	0.241	0.120	0.145	**0.390**	**0.461**	0.246	0.298
Ours	**0.111**	**0.146**	**0.095**	**0.109**	**0.075**	**0.089**	0.412	0.475	**0.126**	**0.143**

Table 3. Untargeted attack results on UAV123 with $\epsilon = 8$.

Methods	SiamRPN++ (M)		SiamBAN		SiamCAR		DiMP		Ocean online	
	S (↑)	P (↑)	S (↑)	P (↑)	S (↑)	P (↑)	S (↑)	P (↑)	S (↑)	P (↑)
Normal	0.602	0.801	0.603	0.788	0.619	0.777	0.633	0.834	0.584	0.788
CSA (T)	0.541	0.746	0.478	0.670	0.580	0.760	0.614	0.816	0.524	0.723
CSA (S)	0.288	0.466	0.299	0.485	0.270	0.440	0.593	0.798	0.264	0.489
CSA (TS)	0.270	0.452	0.278	0.487	0.271	0.428	0.598	0.811	0.278	0.510
Ours$_f$(TD)	0.369	0.561	0.372	0.569	0.337	0.503	**0.562**	**0.757**	0.404	0.648
Ours (TD)	**0.270**	**0.368**	**0.248**	**0.349**	**0.239**	**0.349**	0.573	0.770	**0.272**	**0.399**
Ours$_f$	0.356	0.549	0.372	0.569	0.316	0.469	**0.578**	**0.775**	0.392	0.634
Ours	**0.273**	**0.371**	**0.250**	**0.352**	**0.255**	**0.371**	0.579	0.777	**0.274**	**0.401**

on average by ∼340 restarts. Moreover, our approach significantly decreases the EAO, which is the primary metric to rank trackers (see rows 4 vs 8).

Results on Targeted Attacks. Since manually creating intelligent target trajectories is difficult and beyond the scope of this work, we consider two simple but practical scenarios to quantitatively analyze the effectiveness of our attacks.

1. The attacker forces the tracker to follow a fixed direction. We illustrate this with 4 different directions ($+45°$, $-45°$, $+135°$, $-135°$), and aiming to shift the box by (±3, ±3) pixels in each consecutive frame.
2. The attacker seeks for the tracker to follow a more complicated trajectory. To illustrate this, we force the tracker to follow the *ground-truth* trajectory with a fixed offset (±80, ±80).

Note that our attacks are capable of steering tracker along any general trajectory, not limited to the two cases above.

In both cases, we pre-compute universal directions perturbations corresponding $K = 12$ diverse directions with template cropped from first video of OTB100, and use them to force the tracker to follow the target trajectory. To this end, we

Table 4. Untargeted attack results on VOT2018 with $\epsilon = 8$.

Method	SiamRPN++(M)				SiamBAN				SiamCAR				DiMP				Ocean online			
	A (↑)	R (↓)	EAO (↑)	Re (↓)	A (↑)	R (↓)	EAO (↑)	Re (↓)	A (↑)	R (↓)	EAO(↑)	Re (↓)	A (↑)	R (↓)	EAO(↑)	Re (↓)	A (↑)	R (↓)	EAO (↑)	Re (↓)
Original	0.58	0.24	0.400	51	0.60	0.320	0.340	69	0.58	0.280	0.36	60	0.607	0.3000	0.323	64	0.56	0.220	0.374	47
CSA (T)	0.56	0.440	0.265	95	0.56	0.590	0.190	126	0.56	0.426	0.280	91	0.60	0.220	0.362	47	0.45	0.580	0.189	124
CSA (S)	0.42	2.205	0.067	471	0.43	1.807	0.076	386	0.48	1.597	0.101	341	0.59	0.239	0.367	51	0.20	1.462	0.083	202
CSA (TS)	0.40	2.196	0.067	469	0.38	1.789	0.075	382	0.45	1.475	0.107	315	0.58	0.286	0.322	61	0.22	1.221	0.082	261
Ours$_f$(TD)	0.48	1.625	0.089	347	0.48	1.508	0.079	322	0.52	1.569	0.096	335	0.60	0.26	0.337	56	0.47	0.445	0.232	95
Ours (TD)	0.51	**5.095**	0.029	1088	0.44	**5.071**	0.024	1083	0.60	**3.341**	0.053	712	0.59	**0.512**	0.219	109	0.38	**1.621**	0.074	346
Ours$_f$	0.45	2.098	0.070	448	0.46	1.915	0.070	409	0.51	1.842	0.091	393	0.59	0.267	0.350	57	0.42	0.515	0.209	110
Ours	0.52	**4.856**	0.029	1037	0.44	**5.034**	0.024	1075	0.59	**3.184**	0.056	680	0.56	**0.445**	0.235	95	0.39	**1.482**	0.081	316

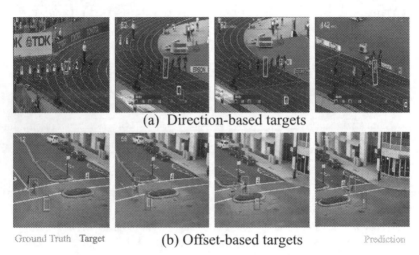

(a) Direction-based targets

Ground Truth   Target   (b) Offset-based targets   Prediction

Fig. 4. Visualizations for targeted attacks. (a) The tracker is forced to move in a constant direction, indicated by the red arrow. **(b)** The tracker is forced to follow the ground truth with a fixed offset of (80, 80) pixels. (Color figure online)

sample $K = 12$ points at a distance $d = 5$ from the object center in feature map of size 25×25 and, for each, synthesize a conditional mask $\mathbf{M}_i \in \{0, 1\}^{(W \times H)}$ whose active region is centered at the sampled point. We then feed each such mask with the template to obtain directional perturbations, which we will then transfer to the search images. During the attack for each frame, we compute the direction the tracker should move in and use the precomputed perturbation that is closest to this direction.

We report the precision score at a 20 pixel threshold for our two attack scenarios, averaged over 4 cases, in Table 5 for $\epsilon = 16$. For direction-based targets, our universal directional perturbations allow us to follow the target trajectory with promising performance. Our approach yields a precision of 0.627, 0.507, 0.536, and 0.335 on average on SiamRPN++(M), SiamBAN, SiamCAR and Ocean-online, respectively. For offset-based targets, which are more challenging than direction-based ones, our approach yields precision scores of 0.487, 0.350, 0.331 and 0.301 on average on the same 4 black-box trackers, respectively. Note that targeted attacks is quite challenging due to distractors and similar objects present in the search region. Nevertheless, our universal directional per-

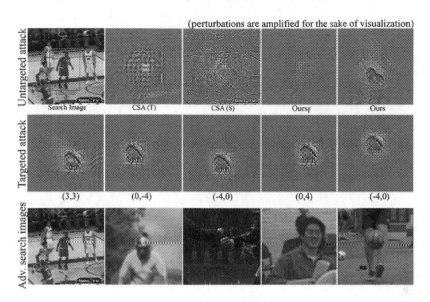

Fig. 5. Qualitative Results. We show, in the first row, the perturbations learned for untargeted attacks; in the second row, the universal directional perturbations for the targeted attack; in the last row, the adversarial search regions obtained with our targeted attack framework for $\epsilon = 16$.

Table 5. Targeted attack results. We report average precision scores for $\epsilon = 16$ for direction and offset-based target trajectories.

Dataset	SiamRPN++ (M)		SiamBAN		SiamCAR		Ocean online	
	Direction	Offset	Direction	Offset	Direction	Offset	Direction	Offset
OTB100	0.430	0.544	0.345	0.257	0.340	0.295	0.128	0.113
VOT2018	0.745	0.515	0.672	0.455	0.661	0.501	0.412	0.372
UAV123	0.476	0.401	0.325	0.295	0.397	0.297	0.260	0.267
LaSOT	0.768	0.489	0.689	0.395	0.747	0.232	0.543	0.521
Average	0.627	0.487	0.507	0.350	0.536	0.331	0.335	0.301

turbations set a benchmark for image-agnostic targeted attacks on unseen black-box trackers. Figure 4 shows the results of targeted attacks on various datasets with SiamRPN++(M). The results at the bottom, where the tracker follows the ground-truth trajectory with an offset, illustrate the real-world applicability of our attacks, where one could force the tracker to follow a realistic, yet erroneous path. Such realistic trajectories can deceive the system without raising any suspicion.

3. What perturbation patterns does the proposed approach learn? To give insights to this question, we display the learned universal perturbations along with adversarial search regions in Fig. 5. In the top row, we can see that,

Table 6. Component-wise analysis. Contribution of each loss for untargeted attacks on OTB100 using our approach. We report precision score and set $\epsilon = 8$.

$\mathcal{L}^{cls}_{fool}$	$\mathcal{L}^{reg}_{fool}$	$\mathcal{L}^{cls}_{shift}$	$\mathcal{L}^{reg}_{shift}$	SiamRPN++ (M)	SiamBAN	SiamCAR	DiMP	Ocean online
-	-	-	-	0.862	0.910	0.908	0.847	0.884
✓	-	-	-	0.566	0.604	0.654	0.851	0.801
-	✓	-	-	0.617	0.726	0.790	0.841	0.827
-	-	✓	-	0.790	0.800	0.851	0.858	0.805
-	-	-	✓	0.858	0.890	0.884	0.858	0.879
✓	✓	-	-	0.616	0.721	0.770	0.843	0.829
-	-	✓	✓	0.695	0.735	0.734	0.828	0.750
✓	-	✓	-	0.328	0.316	0.531	0.827	0.592
-	✓	-	✓	0.682	0.769	0.826	0.848	0.852
✓	✓	✓	✓	0.272	0.252	0.374	0.837	0.440

Table 7. Ablation study. Effect of d in $\mathcal{L}_{shift}$ for untargeted attacks on OTB100. We report precision score and set $\epsilon = 8$.

Shift d	SiamRPN++ (M)	SiamBAN	SiamCAR	DiMP	Ocean online
0	0.616	0.721	0.770	0.843	0.829
2	0.332	0.254	0.551	**0.814**	0.493
4	**0.272**	**0.252**	**0.374**	0.837	**0.440**
6	0.330	0.409	0.481	0.844	0.618
8	0.323	0.489	0.480	0.834	0.684
10	0.427	0.510	0.612	0.851	0.763

for untargeted attacks with the shift loss, our generator learns to place a universal object-like patch at the shift position. By contrast, the perturbation in CSA (S) is concentrated on the center region to decrease the confidence of the proposal. In the second row, we observe that, for targeted attacks, the perturbations are focused around the regions of the desired target box. This evidences that our conditioning scheme is able to capture the important information about the desired bounding box. Furthermore, as shown in the bottom row, our results remain imperceptible thanks to our similarity loss. We provide more results and perceptibility analyses in the supplementary material.

4. Ablation Study. In this section, we analyze the impact of each loss term of our framework. In Table 6, we report the precision scores on OTB100 with different combination of loss terms, where $\mathcal{L}^{cls}_{fool}$ and $\mathcal{L}^{reg}_{fool}$ represent the classification and regression components of the fooling loss of Eq. 1, and $\mathcal{L}^{cls}_{shift}$ and $\mathcal{L}^{reg}_{shift}$ represent the same terms for the shift loss of Eq. 2. To summarize, while all loss terms are beneficial, the classification-based terms are more effective than regression-based ones. For example, using either $\mathcal{L}^{cls}_{fool}$ or $\mathcal{L}^{cls}_{shift}$ has more impact than $\mathcal{L}^{reg}_{fool}$ or $\mathcal{L}^{reg}_{shift}$. In Table 7, we study the impact of the shift distance d in

Eq. 2 on the performance of untargeted attacks. For a feature map of size 25×25 for SiamRPN++, the performance of our approach is stable for a drift in the range 4 to 8. However, for $d = 2$, our attacks have less effect on the tracker, and for $d = 10$, the influence of the attack decreases because of the Gaussian prior used by the tracker. Due to space limitations, we provide the ablation on sensitivity of various hyper-parameters and effect of number of directional perturbations K in supplementary material.

6   Conclusion

We have shown the existence of transferable universal perturbations to efficiently attack black-box VOT trackers on the fly. To do so, we have introduced a framework that relies on generating a one-shot temporally-transferable perturbation by exploiting only the template as input, thus being invariant to the search environment. Our trained generator produces perturbations that are quasi-agnostic to the input template, and are thus highly transferable to unknown objects. Furthermore, we have demonstrated that our universal directional perturbations allow us to steer the tracker to follow any specified trajectory. We believe that our work highlights the vulnerability of object trackers and will motivate researchers to design robust defense mechanisms.

Acknowledgments. This work was funded in part by the Swiss National Science Foundation.

References

1. Arnab, A., Miksik, O., Torr, P.H.: On the robustness of semantic segmentation models to adversarial attacks. In: Proceedings of the IEEE Conference on Computer Vision and Pattern Recognition, pp. 888–897 (2018)
2. Ban, Y., Ba, S., Alameda-Pineda, X., Horaud, R.: Tracking multiple persons based on a variational Bayesian model. In: Hua, G., Jégou, H. (eds.) ECCV 2016. LNCS, vol. 9914, pp. 52–67. Springer, Cham (2016). https://doi.org/10.1007/978-3-319-48881-3_5
3. Bhat, G., Danelljan, M., Gool, L.V., Timofte, R.: Learning discriminative model prediction for tracking. In: Proceedings of the IEEE/CVF International Conference on Computer Vision, pp. 6182–6191 (2019)
4. Bolme, D.S., Beveridge, J.R., Draper, B.A., Lui, Y.M.: Visual object tracking using adaptive correlation filters. In: 2010 IEEE Computer Society Conference on Computer Vision and Pattern Recognition, pp. 2544–2550. IEEE (2010)
5. Carlini, N., Wagner, D.: Towards evaluating the robustness of neural networks. In: 2017 IEEE Symposium on Security and Privacy (SP), pp. 39–57. IEEE (2017)
6. Chen, X., et al.: One-shot adversarial attacks on visual tracking with dual attention. In: Proceedings of the IEEE/CVF Conference on Computer Vision and Pattern Recognition, pp. 10176–10185 (2020)
7. Chen, Z., Zhong, B., Li, G., Zhang, S., Ji, R.: Siamese box adaptive network for visual tracking. In: Proceedings of the IEEE/CVF Conference on Computer Vision and Pattern Recognition, pp. 6668–6677 (2020)

8. Danelljan, M., Hager, G., Shahbaz Khan, F., Felsberg, M.: Learning spatially regularized correlation filters for visual tracking. In: Proceedings of the IEEE International Conference on Computer Vision, pp. 4310–4318 (2015)
9. Dong, Y., et al.: Boosting adversarial attacks with momentum. In: Proceedings of the IEEE Conference on Computer Vision and Pattern Recognition, pp. 9185–9193 (2018)
10. Fan, H., et al.: LaSOT: a high-quality benchmark for large-scale single object tracking. In: Proceedings of the IEEE/CVF Conference on Computer Vision and Pattern Recognition, pp. 5374–5383 (2019)
11. Fischer, V., Kumar, M.C., Metzen, J.H., Brox, T.: Adversarial examples for semantic image segmentation. arXiv preprint arXiv:1703.01101 (2017)
12. Goodfellow, I.J., Shlens, J., Szegedy, C.: Explaining and harnessing adversarial examples. arXiv preprint arXiv:1412.6572 (2014)
13. Guo, D., Wang, J., Cui, Y., Wang, Z., Chen, S.: SiamCAR: siamese fully convolutional classification and regression for visual tracking. In: Proceedings of the IEEE/CVF Conference on Computer Vision and Pattern Recognition, pp. 6269–6277 (2020)
14. Guo, Q., et al.: SPARK: spatial-aware online incremental attack against visual tracking. arXiv preprint arXiv:1910.08681 (2019)
15. He, K., Zhang, X., Ren, S., Sun, J.: Deep residual learning for image recognition. In: Proceedings of the IEEE Conference on Computer Vision and Pattern Recognition, pp. 770–778 (2016)
16. Huang, L., Zhao, X., Huang, K.: GOT-10k: a large high-diversity benchmark for generic object tracking in the wild. IEEE Trans. Pattern Anal. Mach. Intell. **43**(5), 1562–1577 (2019)
17. Jia, S., Song, Y., Ma, C., Yang, X.: IoU attack: towards temporally coherent blackbox adversarial attack for visual object tracking. In: Proceedings of the IEEE/CVF Conference on Computer Vision and Pattern Recognition, pp. 6709–6718 (2021)
18. Jia, Y., Lu, Y., Shen, J., Chen, Q.A., Zhong, Z., Wei, T.: Fooling detection alone is not enough: first adversarial attack against multiple object tracking. arXiv preprint arXiv:1905.11026 (2019)
19. Kiani Galoogahi, H., Sim, T., Lucey, S.: Multi-channel correlation filters. In: Proceedings of the IEEE International Conference on Computer Vision, pp. 3072–3079 (2013)
20. Kiani Galoogahi, H., Sim, T., Lucey, S.: Correlation filters with limited boundaries. In: Proceedings of the IEEE Conference on Computer Vision and Pattern Recognition, pp. 4630–4638 (2015)
21. Kristan, M., et al.: The sixth visual object tracking VOT2018 challenge results. In: Leal-Taixé, L., Roth, S. (eds.) ECCV 2018. LNCS, vol. 11129, pp. 3–53. Springer, Cham (2019). https://doi.org/10.1007/978-3-030-11009-3_1
22. Kurakin, A., Goodfellow, I., Bengio, S.: Adversarial examples in the physical world. arXiv preprint arXiv:1607.02533 (2016)
23. Lee, K.H., Hwang, J.N.: On-road pedestrian tracking across multiple driving recorders. IEEE Trans. Multimed. **17**(9), 1429–1438 (2015)
24. Li, B., Wu, W., Wang, Q., Zhang, F., Xing, J., Yan, J.: SiamRPN++: evolution of siamese visual tracking with very deep networks. In: Proceedings of the IEEE Conference on Computer Vision and Pattern Recognition, pp. 4282–4291 (2019)
25. Li, B., Yan, J., Wu, W., Zhu, Z., Hu, X.: High performance visual tracking with siamese region proposal network. In: Proceedings of the IEEE Conference on Computer Vision and Pattern Recognition, pp. 8971–8980 (2018)

26. Liang, S., Wei, X., Yao, S., Cao, X.: Efficient adversarial attacks for visual object tracking. arXiv preprint arXiv:2008.00217 (2020)
27. Madry, A., Makelov, A., Schmidt, L., Tsipras, D., Vladu, A.: Towards deep learning models resistant to adversarial attacks. arXiv preprint arXiv:1706.06083 (2017)
28. Moosavi-Dezfooli, S.M., Fawzi, A., Fawzi, O., Frossard, P.: Universal adversarial perturbations. In: Proceedings of the IEEE Conference on Computer Vision and Pattern Recognition, pp. 1765–1773 (2017)
29. Moosavi-Dezfooli, S.M., Fawzi, A., Frossard, P.: DeepFool: a simple and accurate method to fool deep neural networks. In: Proceedings of the IEEE Conference on Computer Vision and Pattern Recognition, pp. 2574–2582 (2016)
30. Mueller, M., Smith, N., Ghanem, B.: A benchmark and simulator for UAV tracking. In: Leibe, B., Matas, J., Sebe, N., Welling, M. (eds.) ECCV 2016. LNCS, vol. 9905, pp. 445–461. Springer, Cham (2016). https://doi.org/10.1007/978-3-319-46448-0_27
31. Poursaeed, O., Katsman, I., Gao, B., Belongie, S.: Generative adversarial perturbations. In: Proceedings of the IEEE Conference on Computer Vision and Pattern Recognition, pp. 4422–4431 (2018)
32. Ren, S., He, K., Girshick, R., Sun, J.: Faster R-CNN: towards real-time object detection with region proposal networks. IEEE Trans. Pattern Anal. Mach. Intell. **39**(6), 1137–1149 (2016)
33. Ronneberger, O., Fischer, P., Brox, T.: U-Net: convolutional networks for biomedical image segmentation. In: Navab, N., Hornegger, J., Wells, W.M., Frangi, A.F. (eds.) MICCAI 2015. LNCS, vol. 9351, pp. 234–241. Springer, Cham (2015). https://doi.org/10.1007/978-3-319-24574-4_28
34. Wang, Q., Zhang, L., Bertinetto, L., Hu, W., Torr, P.H.: Fast online object tracking and segmentation: a unifying approach. In: Proceedings of the IEEE Conference on Computer Vision and Pattern Recognition, pp. 1328–1338 (2019)
35. Wiyatno, R.R., Xu, A.: Physical adversarial textures that fool visual object tracking. In: Proceedings of the IEEE International Conference on Computer Vision, pp. 4822–4831 (2019)
36. Wu, Y., Lim, J., Yang, M.H.: Online object tracking: a benchmark. In: Proceedings of the IEEE Conference on Computer Vision and Pattern Recognition, pp. 2411–2418 (2013)
37. Xiao, C., Li, B., Zhu, J.Y., He, W., Liu, M., Song, D.: Generating adversarial examples with adversarial networks. arXiv preprint arXiv:1801.02610 (2018)
38. Xie, C., Wang, J., Zhang, Z., Zhou, Y., Xie, L., Yuille, A.: Adversarial examples for semantic segmentation and object detection. In: Proceedings of the IEEE International Conference on Computer Vision, pp. 1369–1378 (2017)
39. Yan, B., Wang, D., Lu, H., Yang, X.: Cooling-shrinking attack: blinding the tracker with imperceptible noises. In: Proceedings of the IEEE/CVF Conference on Computer Vision and Pattern Recognition, pp. 990–999 (2020)
40. Zhang, Z., Peng, H., Fu, J., Li, B., Hu, W.: Ocean: object-aware anchor-free tracking. In: Vedaldi, A., Bischof, H., Brox, T., Frahm, J.-M. (eds.) ECCV 2020. LNCS, vol. 12366, pp. 771–787. Springer, Cham (2020). https://doi.org/10.1007/978-3-030-58589-1_46
41. Zhu, Z., Wang, Q., Li, B., Wu, W., Yan, J., Hu, W.: Distractor-aware siamese networks for visual object tracking. In: Ferrari, V., Hebert, M., Sminchisescu, C., Weiss, Y. (eds.) ECCV 2018. LNCS, vol. 11213, pp. 103–119. Springer, Cham (2018). https://doi.org/10.1007/978-3-030-01240-3_7

Why Is the Video Analytics Accuracy Fluctuating, and What Can We Do About It?

Sibendu Paul[1]([⊠]), Kunal Rao[2], Giuseppe Coviello[2], Murugan Sankaradas[2], Oliver Po[2], Y. Charlie Hu[1], and Srimat Chakradhar[2]

[1] Purdue University, West Lafayette, IN, USA
{paul90,ychu}@purdue.edu
[2] NEC Laboratories America, Inc., Princeton, NJ, USA
{kunal,giuseppe.coviello,murugs,oliver,chak}@nec-labs.com

Abstract. It is a common practice to think of a video as a sequence of images (frames), and re-use deep neural network models that are trained only on images for similar analytics tasks on videos. In this paper, we show that this "leap of faith" that deep learning models that work well on images will also work well on videos is actually flawed. We show that even when a video camera is viewing a scene that is not changing in any human-perceptible way, and we control for external factors like video compression and environment (lighting), the accuracy of video analytics application fluctuates noticeably. These fluctuations occur because successive frames produced by the video camera may look similar visually, but are perceived quite differently by the video analytics applications. We observed that the root cause for these fluctuations is the dynamic camera parameter changes that a video camera automatically makes in order to capture and produce a visually pleasing video. The camera inadvertently acts as an "unintentional adversary" because these slight changes in the image pixel values in consecutive frames, as we show, have a noticeably adverse impact on the accuracy of insights from video analytics tasks that re-use image-trained deep learning models. To address this inadvertent adversarial effect from the camera, we explore the use of transfer learning techniques to improve learning in video analytics tasks through the transfer of knowledge from learning on image analytics tasks. Our experiments with a number of different cameras, and a variety of different video analytics tasks, show that the inadvertent adversarial effect from the camera can be noticeably offset by quickly re-training the deep learning models using transfer learning. In particular, we show that our newly trained Yolov5 model reduces fluctuation in object detection across frames, which leads to better tracking of objects (∼40% fewer mistakes in tracking). Our paper also provides new directions and techniques to mitigate the camera's adversarial effect on deep learning models used for video analytics applications.

1 Introduction

Significant progress in machine learning and computer vision [9,24,41,42], along with the explosive growth in Internet of Things (IoT), edge computing, and high-bandwidth access networks such as 5G [7,37], have led to the wide adoption of video analytics systems. These systems deploy cameras throughout the world to support diverse applications in entertainment, health-care, retail, automotive, transportation, home automation, safety, and security market segments. The global video analytics market is estimated to grow from $5 billion in 2020 to $21 billion by 2027, at a CAGR of 22.70% [14].

Video analytics systems rely on state of the art (SOTA) deep learning models [24] to make sense of the content in the video streams. It is a common practice to think of a video as a sequence of images (frames), and re-use deep learning models that are trained only on images for video analytics tasks. Large, image datasets like COCO [27] have made it possible to train highly-accurate SOTA deep learning models [2,6,21,30,39,40] that detect a variety of objects in images. In this paper, we take a closer look at the use of popular deep neural network models trained on large image datasets for predictions in critical video analytics tasks. We consider video segments from two popular benchmark video datasets [3,13]. These videos contain cars or persons, and we used several SOTA deep neural network (DNN) models for object detection and face detection tasks to make sense of the content in the video streams. Also, these videos exhibit minimal activity (*i.e.,* cars or persons are not moving appreciably and hence, largely static). Since the scenes are mostly static, the ground truth (total number of cars or persons) does not change appreciably from frame to frame within each video. Yet, we observe that the accuracy of tasks like object detection or face detection unexpectedly fluctuate noticeably for consecutive frames, rather than more or less stay the same. Such unexpected, noticeable fluctuations occur across different camera models and across different camera vendors.

Such detection fluctuations from frame to frame have an adverse impact on applications that use insights from object or face detection to perform higher-level tasks like tracking objects or recognizing people. Understanding the causes for these unexpected fluctuations in accuracy, and proposing methods to mitigate the impact of these fluctuations, are the main goals of this paper. We investigate the causes of the accuracy fluctuations of these SOTA deep neural network models on largely static scenes by carefully considering factors external and internal to a video camera. We examine the impact of external factors like the environmental conditions (lighting), video compression and motion in the scene, and internal factors like camera parameter settings in a video camera, on the fluctuations in performance of image-trained deep neural network models. Even after carefully controlling for these external and internal factors, the accuracy fluctuations persist, and our experiments show that another cause for these fluctuations is the dynamic camera parameter changes that a video camera automatically makes in order to capture and produce a visually pleasing video. The camera inadvertently acts as an "unintentional adversary" because these slight changes in image pixel values in consecutive frames, as we show, have a noticeably adverse impact on the accuracy of insights from video analytics tasks that re-use image-trained deep learning models. To address this inadvertent adversarial effect from the camera, we explore ways to mitigate this effect and propose the transfer of knowledge from learning on image analytics tasks to video analytics tasks.

In this paper, we make the following key contributions:

- We take a closer look at the use of popular deep learning models that are trained on large image datasets for predictions in critical video analytics tasks, and show that the accuracy of tasks like object detection or face detection unexpectedly fluctuate noticeably for consecutive frames in a video; consecutive frames capture the same scene and have the same ground truth. We show that such unexpected, noticeable fluctuations occur across different camera models and across different camera vendors.
- We investigate the root causes of the accuracy fluctuations of these SOTA deep neural network models on largely static scenes by carefully considering factors external and internal to a video camera. We show that a video camera inadvertently acts as an "unintentional adversary" when it automatically makes camera parameter changes in order to capture and produce a visually pleasing video.
- We draw implications of the unintentional adversarial effect on the practical use of computer vision models and propose a simple yet effective technique to transfer knowledge from learning on image analytics tasks to video analytics. Our newly trained Yolov5 model reduces fluctuation in object detection across frames, which leads to better performance on object tracking task ($\sim$40% fewer mistakes in tracking).

2   Motivation

In this section, we consider video segments from two popular benchmark datasets. These videos contain cars or persons, and the videos exhibit minimal activity (*i.e.,* cars or persons are not moving appreciably and hence, largely static). Since the scenes are mostly static, the ground truth (total number of cars or persons) from frame to frame is also not changing much. Yet, we observe that the accuracy of tasks like object detection or face detection unexpectedly fluctuate noticeably for consecutive frames. Such accuracy fluctuations from frame to frame have an adverse impact on applications that use insights from object or face detection to perform higher-level tasks like tracking objects or recognizing people.

2.1   Object Detection in Videos

(a) Roadway Dataset   (b) LSTN Dataset

Fig. 1. Sample frames from video datasets.

One of the most common task in video analytics pipelines is object detection. Detecting cars or people is critical for many real-world applications like video surveillance, retail, health care monitoring and intelligent transportation systems (Fig. 1).

Figure 2 shows the performance of different state of the art and widely-used object detectors like YOLOv5-small and large variant [21], EfficientDet-v0 and EfficientDet-v8 [40] on video segments from the Roadway dataset [3]. These videos have cars and people, but the activity is minimal, and scenes are largely static. The "ground truth" in the figures is shown in blue color, and it shows the total number of cars and people at different times (i.e. frames) in the video. The "detector prediction" waveform (shown in red color) shows the number of cars and people actually detected by the deep learning model.

Our experiments show that (a) for all the detectors we considered, the number of detected objects is lower than the ground truth[1], and (b) more importantly, even though the ground truth is not changing appreciably in consecutive frames, the detections reported by the detectors vary noticeably, and (c) light-weight models like Yolov5-small or Yolov5-large exhibit a much higher range of detection fluctuations than the more heavier models like *efficientDet*. However, the heavier deep learning models make inferences by consuming significantly more computing resources than the light-weight models.

2.2   Face Detection in Videos

Next, we investigate if accuracy fluctuation observed in object detection models also occur in other image-trained AI models that are used in video analytics tasks. We chose AI models for face detection task, which is critical to many real-world applications *e.g.,* identifying a person of interest in airports, hospitals or arenas, and authenticating individuals based on face-recognition for face-based payments. Figure 3 shows the performance of three well-known face detection AI

---

[1] We have 1–2 false positive detections for Yolov5 and efficientDet.

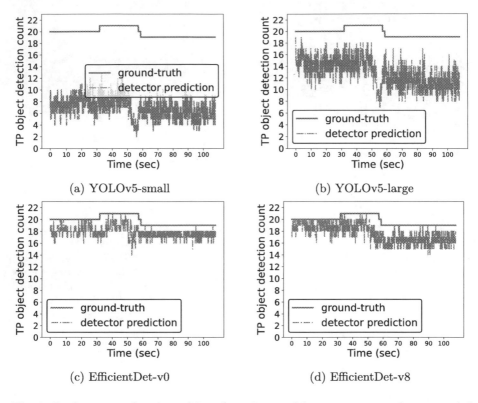

Fig. 2. Performance of various object detection models on a segment of pre-recorded video from the Roadway dataset [3]. (Color figure online)

models on videos from the *LSTN* video dataset [13]. Like the object detection case, we observe that (a) the number of faces detected by these models is typically lower than the ground truth, (b) more importantly, even though the ground truth barely changes, there is noticeable fluctuation in the number of detections in consecutive frames, and (c) the light-weight models like MTCNN [38] exhibit a much higher range of detection fluctuations than the more heavier models like RetinaNet with resnet-50 and mobilenet backbone [10].

3   Analysis and Control of External Factors

The behavior of a DNN model is deterministic in the sense that if a frame is processed multiple times by the DNN model, then the DNN inferences are identical. In this section, we analyze three external factors that may be causing the unexpected accuracy fluctuations described in Sect. 2:

- Motion in the field of view of the camera affects the quality of the captured video (blurring of moving objects is likely).

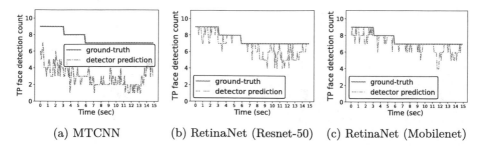

Fig. 3. Performance of face detection models on videos from *LSTN* video dataset.

- Lossy video compression methods like H.264 can also result in decoded frames whose quality can differ from the pre-compression frames.
- Environmental conditions like lighting can also affect the quality of the frames processed by the DNNs. For example, flicker in fluorescent lighting can affect the quality of frames captured by the camera (most people cannot notice the flicker in fluorescent lights, which flicker at a rate of 120 cycles per second 120 Hz; as we show later, flicker also contributes to fluctuations in the analytics accuracy of video analytics tasks).

3.1   Control for Motion

It is difficult to systematically study the impact of motion on accuracy fluctuations by using videos from the datasets. Instead, as shown in Fig. 5a, we set up a scene with 3D models of objects (*i.e.*, persons and cars), and continuously observed the scene by using different IP cameras like *AXIS Q1615*. A fluorescent light provides illumination for the scene. Figure 5a shows a frame in the video stream captured by the IP camera under default camera parameter settings. This setup easily eliminates the effect of motion on any observed accuracy fluctuations. Also, this set up makes it easy to study whether accuracy fluctuations are caused by only certain camera models or fluctuations happen across different camera models from different vendors.

3.2   Analysis and Control for Video Compression

By using static 3D models, we eliminated the effect of *motion*. To understand the effect of video compression, we fetch frames directly from the camera instead of fetching a compressed video stream and decoding the stream to obtain frames that can be processed by a DNN model.

Figure 4a and Fig. 4b show the object detection counts with and without compression for the YOLOv5 model. We observe that eliminating compression reduces detection fluctuation. We also analyzed the detection counts with and without compression by using the *t-test* for repeated measures [44]. Let A be the sequence of true-positive object detection counts (per frame) for the experiment

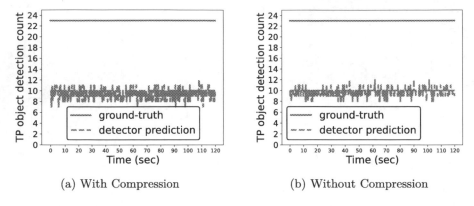

(a) With Compression   (b) Without Compression

Fig. 4. Effect of video compression on fluctuations in Yolov5 object detection counts (scene with 3D models)

where video compression is used. Let B be the sequence of true-positive object detection counts for the case when no compression is used. We compute a third sequence D that is a sequence of pair-wise differences between the true-positive object count without compression and with compression (*i.e.*, $B - A$).

Essentially, the use of difference scores converts a two-sample problem with A and B into a one-sample problem with D. Our null hypothesis states that compression has no effect on object detection counts (and we hypothesize a population mean of 0 for the difference scores). Our experiment with a sample size of 200 frames showed that we can reject the null hypothesis at the 0.01 level of significance (99% confidence), suggesting there is evidence that elimination of compression does reduce the accuracy fluctuations. Similar results were observed for sample sizes of 100 and 1000 frames.

While *t-test* measures the statistical difference between two distributions, it doesn't reflect on the fluctuations observed in repeated measures. We propose two metrics to quantify the observed fluctuations across a group of frames. (1) *F2* which is defined as $\frac{\|tp(i)-tp(i+1)\|}{mean(gt(i),gt(i+1))}$ for frame i, where $tp(i)$, $gt(i)$ are true-positive object detection count and ground-truth object count respectively on frame i (on a moving window of 2 frames) and (2) *F10* which is defined as $\frac{\|max(tp(i),...,tp(i+9))-min(tp(i),...,tp(i+9))\|}{mean(gt(i),...,gt(i+9))}$ (on a moving window of 10 frames).

By eliminating video compression, the maximum variation in object count on static scene can be reduced from 17.4% to 13.0% (*F2*) and from 19.0% to 17.4% (*F10*). Clearly, video compression is highly likely to have an adverse effect on accuracy fluctuations, and eliminating compression can improve results of deep learning models.

3.3   Analysis and Control for Flicker

By using static 3D models, we eliminated the effect of *motion*. We are also able to eliminate the adverse effect of video compression by fetching frames directly

(a) With flicker in lighting   (b) Without flicker in lighting

Fig. 5. Scene with 3D models, with and without flickering light.

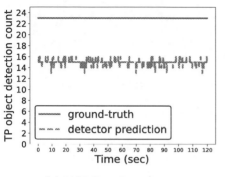

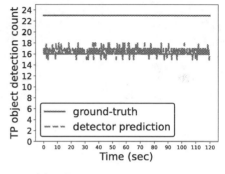

(a) YOLOv5 detections   (b) EfficientDet-v0 detections

Fig. 6. Object detection counts when there is no motion, video compression or flickering light.

from the camera. We now analyze the effect of lighting. We set up an additional, flicker-free light source to illuminate the scene with static 3D models. Figure 5 shows the 3D models scene with and without flickering light. Figure 6a shows the fluctuation in detection counts when there is no motion, no video compression, and no flicker due to fluorescent light.

Compared to Fig. 4 results with no compression (but with fluorescent lighting), the results in Fig. 6a are highly likely to be an improvement. We compared the sequence of object detection counts with and without fluorescent light (no video compression in both cases) using the t-test for repeated measures, and easily rejected the null hypothesis that lighting makes no difference at a 0.01 level of significance (99% confidence). Also, eliminating light flickering on top of motion and compression can reduce the maximum ($F2$) and ($F10$) variations from 13.0% to 8.7% and 17.4% to 13.0% respectively. Therefore, after eliminating motion and video compression, fluorescent light with flicker is highly likely to have an adverse effect on accuracy fluctuations, and eliminating flicker is highly likely to improve the results from the DNN model.

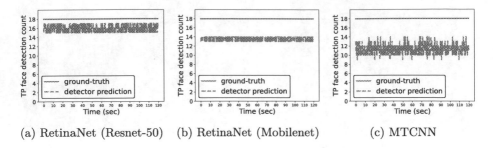

(a) RetinaNet (Resnet-50)   (b) RetinaNet (Mobilenet)   (c) MTCNN

Fig. 7. Face detection counts for three different DNN models when there is no motion, video compression or flickering light.

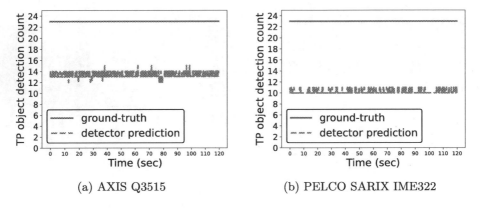

(a) AXIS Q3515   (b) PELCO SARIX IME322

Fig. 8. Performance of YOLOv5 on different IP camera models in absence of motion, compression and flicker in lighting.

Figure 6b shows the object detection counts for EfficientDet-v0 when there is no motion, video compression or flickering light. We observe fluctuations in object detection count up to 13.0% (*F2*) and 14.0% (*F10*). Due to space reasons, we have not included the graphs for with and without compression for EfficientDet-v0. However, like the YOLOv5 case, eliminating motion, video compression and flickering light improves the detection results.

Our detailed analysis in this section shows that eliminating motion, video compression and flicker does improve the object detection results. However, even after controlling for motion, video compression and flickering light, noticeable fluctuations in object detection counts still remain. We repeated the above experiments for three SOTA open-source *face detection models*. Fig. 7 shows fluctuation in face detection counts when there is no motion, video compression or flickering light. *F2* metric reports true-positive face detection fluctuations up to 8.7%, 4.3%, 21.7% for two retinaNet models and MTCNN respectively.

3.4   Impact of Different Camera Models

We also investigated whether the fluctuation in video analytics accuracy is observed only on specific camera model or is it present across different camera models across different vendors. Figure 8 shows the performance of YOLOv5 object detection model on AXIS Q3515 and PELCO SARIX IME322 IP cameras, both of them observing the same scene and in absence of motion, compression and flicker in lighting. We note that both of them show fluctuation in the count of detected objects and $F2$ metric reports up to 13.1% and 4.4% fluctuations for the two camera models. This shows that the fluctuation in video analytics accuracy is observed across different camera models from different vendors.

4   Camera as an Unintentional Adversary

In Sect. 3, we investigated various factors external to the camera that could lead to fluctuations in video analytics accuracy. Specifically, we looked at motion, compression, flicker in lighting, and camera models from different vendors and different deep learning models, and found out that fluctuation is observed across different deep learning models, on different cameras, even when motion, compression, flicker in lighting are eliminated. This leads us to hypothesize that the remaining factors causing accuracy fluctuation may not be external. Rather, it could be *internal* to the camera.

4.1   Hypothesis

Auto-parameter Setting in Modern Cameras. Along with exposing endpoints to retrieve streaming videos (*e.g.,*RTSP stream URL), IP cameras also expose APIs to set various camera parameters (*e.g.,* VAPIX [8] API for Axis camera models). These camera settings aid in changing the quality of image produced by the camera. Camera vendors expose these APIs because they do not know ahead of time in what environment their camera would be deployed and what settings would be ideal for that environment. Therefore, they set the camera settings to some default values and let end users decide what settings would work best for their environment. There are two types of camera settings that are exposed by camera vendors: (1) Type 1 parameters include those that affect the way raw images are captured, *e.g.,* exposure, gain, and shutter speed. These parameters generally are adjusted *automatically* by the camera with little control by end users. They only allow end users to set maximum value, but within this value, the camera internally changes the settings dynamically in order to produce a visually pleasing video output. We refer to such parameters as automated parameters (AUTO). (2) Type 2 parameters include those that affect processing of raw data in order to produce the final frame, *e.g.,* image specific parameters such as brightness, contrast, sharpness, color saturation, and video specific parameters such as compression, GOP length, target bitrate, FPS. For these parameters, camera vendors often expose fine control to end users, who can set the specific value. We refer to such parameters as non-automated parameters (NAUTO). The distinction between AUTO and NAUTO parameters help us refine our hypothesis where we can fix values of NAUTO parameters, vary the maximum value for AUTO parameters and observe how camera internally changes these parameters to produce different consecutive frames, which might be causing the fluctuations.

The Hypothesis. The purpose of a video camera is to capture videos, rather than still images, for viewing by human eyes. Hence, irrespective of how the scene in front of the camera looks like, *i.e.,* whether the scene is static or dynamic, video camera always tries to capture a video, which assumes changes in successive frames. To capture a visually pleasing and smooth (to human eyes) video, the camera tries to find the optimal exposure time or shutter speed. On one hand, high shutter speed, *i.e.,* low exposure time, freezes motion in each frame, which results in very crisp individual images. However, when such frames are played back at usual video frame rates, it can appear as hyper-realistic and provide a very jittery, unsettled feeling to the viewer [28]. Low shutter speed, on the other hand, can cause moving objects to appear blurred and also builds up noise in the capture. To maintain appropriate amount of motion blur and noise in the capture, video cameras have another setting called *gain*. Gain indicates the amount of amplification applied to the capture. A high gain can provide better images in low-light scenario but can also increase the noise present in the capture. For these reasons, the optimal values of AUTO parameters like exposure and gain are internally adjusted by the camera to output a visually pleasing smooth video. Thus, video capture is fundamentally different from still image capture and the exact values of exposure and gain used by the camera for each frame are not known to end users or analytics applications running on the video output from the camera. This loose control over maximum shutter time and maximum gain parameters is likely the explanation for fluctuations in video analytics accuracy, *i.e.,* the camera's unintentional adversarial effect.

4.2   Hypothesis Validation

The above explanation of our hypothesis that internal dynamic change of AUTO parameters applied by a camera causes successive frames to differ and hence fluctuations in video analytics accuracy, also points us a way to partially validate the hypothesis. Since the camera still exposes control of maximum values of AUTO parameters, we can adjust these maximum parameter value and observe the impact on the resulting fluctuation of analytics accuracy. Figure 9 shows the fluctuation in accuracy of YOLOv5 object detection model for two different settings of maximum exposure time. We observe that when the maximum exposure time is $1/120$ s, the fluctuation in object count is significantly lower than when it is $1/4$ s. Here, reducing the max exposure time decreases the maximum $F2$ fluctuations from 13.0% to 8.7%. This corroborates our hypothesis – with a higher value of maximum exposure time, the camera can possibly choose from a larger number of different exposure times than when the value of maximum exposure

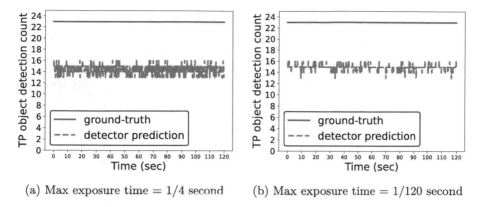

(a) Max exposure time = 1/4 second   (b) Max exposure time = 1/120 second

Fig. 9. Performance of YOLOv5 Object detectors for two different settings of an AUTO parameter, in absence of motion, compression and flicker in lighting.

time is low, which in turn causes the consecutive frame captures to differ more, resulting in more accuracy fluctuation.

We compared the sequences of object detection counts at a maximum exposure of 1/120 s and 1/4 s using the t-test for repeated measures, and easily rejected the null hypothesis that lowering the maximum exposure time (*i.e.*, changing exposure time from 1/4 s to 1/120 s) makes no difference in detection counts, at a 0.01 level of significance (99% confidence). Therefore, the choice of maximum exposure time has a direct impact on the accuracy of the deep learning models, and the precise exposure time is automatically determined by the video camera. We quantify this using object tracking task (discussed in Sect. 5) and observe 65.7% fewer mistakes in tracking when exposure changes.

5   Implications

SOTA object detectors (Yolov5 [21] or EfficientDet [40]) are trained on still image datasets *e.g.*, COCO [27] and VOC [12] datasets. We observed in prior sections that the accuracy of insights from deep learning models fluctuate when used for video analytics tasks. An immediate implication of our finding is that deep learning models trained on still image datasets should not be directly used for videos. We discuss several avenues of research that can mitigate the accuracy fluctuations in video analytics tasks due to the use of image-trained DNN models.

5.1   Retraining Image-Based Models with Transfer Learning

One relatively straight-forward approach is to train models for extracting insights from videos using video frames that are captured under different scenarios. As a case study, we used transfer learning to train Yolov5 using the proprietary videos captured under different scenarios. These proprietary videos contain objects from person and vehicle super-category (that have car, truck, bus, train categories),

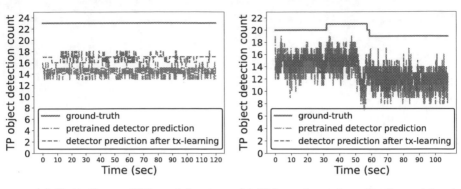

(a) Static Scene of 3D models   (b) Video snippet from Roadway dataset

Fig. 10. Detection counts from YOLOv5-large after transfer-learning (Color figure online)

captured by the cameras at different deployment sites (*e.g.,* traffic intersection, airport, mall, etc.) during different times-of-the-day (*i.e.,* day, afternoon, night) and also under different weather conditions (*i.e.,* rainy, foggy, sunny). We extract total *34K* consecutive frames from these proprietary video snippets, and these frames form our training dataset.

Training Details. The first 23 modules (corresponding to 23 layers) of our new deep learning model are initialized using weights from COCO-trained Yolov5 model, and these weights are frozen. During training, only the weights in the last *detect* module are updated. For transfer learning, we used a learning rate of 0.01 with a weight decay value of 0.0005. We trained Yolov5 model on NVIDIA GeForce RTX 2070 GPU server for 50 epochs with a batch size of 32. This lightweight training required only 1.6 GB GPU memory and took less than 1.5 h to finish 50 epochs. We used the newly trained Yolov5 model to detect objects (*i.e.,* cars and persons) in (a) our static scene of 3D models, and (b) a video from the Roadway dataset (same video that was used in Sect. 2).

Figure 10a shows the improvement in detection counts due to the transfer-learning trained Yolo5 model (green waveform). The improvement over the original Yolo5 model (shown as red waveform) is noticeable visually. We also compared the sequence of object detection counts for the original Yolov5 model (red waveform) and the transfer-learning trained Yolov5 model (green waveform) by using a t-test for repeated measures. We easily rejected the null hypothesis that transfer-learning makes no difference, at a 0.01 level of significance (99% confidence). Then, we estimated the size of the effect due to transfer-learning, and we observed that at a 0.01 level of significance, the improvement is 2.32 additional object detections (14.3% improvement over the mean detections due to the original Yolov5 model). For this experiment, the camera was automatically setting AUTO camera parameters to produce a visually pleasing video, and the transfer-learning trained Yolov5 detector was able to detect more objects despite the unintentional adversary (camera).

In practical deployments of video analytics systems that operate 24×7, it is difficult to control motion or environmental conditions, and the default video compression settings also vary from camera to camera. To understand the impact of transfer-learning trained Yolov5 model, we did experiments on videos in the Roadway dataset. These videos have motion, and the environmental conditions or video compression settings are unknown (such information is not part of the Roadway dataset). Figure 10b shows the results for a video in the Roadway dataset. The true-positive object detections by our *transfer-learning trained Yolov5 model* (green waveform) show noticeably less range of fluctuations than the original Yolov5 model (red waveform). We also compared the sequence of object detection counts for the original Yolov5 model (red waveform) and the transfer-learning trained Yolov5 model (green waveform) by using a t-test for repeated measures. We easily rejected the null hypothesis that transfer-learning makes no difference, at a 0.01 level of significance (99% confidence). Then, we estimated the size of the effect due to transfer-learning, and we observed that at a 0.01 level of significance, the improvement is 1 additional object detection (9.6% improvement over the mean detections due to the original Yolov5 model). Our *newly trained* Yolov5 model reduces the maximum variation of correctly detected object counts from 47.4% to 33.2% *(F10)*, and 42.1% to 32.5% *(F2)*.

Impact on Object Tracking. We evaluated the impact of the fluctuations in detection counts on object tracking task where we track the objects across different frames using *MOT SORT* [1] tracker. Object trackers assign the same track-id to an object appearing in contiguous frames. If an object is not detected in a frame, then the object's track is terminated. If the object is detected again in subsequent frames, a new track-id is assigned to the object. We use the number of track-ids assigned by a tracker as an indicator of the quality of object detections. Our tracker reported *157* track-ids when the original Yolov5 model was used for detecting objects in the video from the Roadway dataset. In contrast, the same tracker reported *94* track-ids when the *transfer-learning trained* Yolov5 model was used (*i.e.,* 40.1% fewer mistakes in tracking). We manually annotated the video and determined the ground-truth to be 29 tracks. We also manually inspected the tracks proposed by the tracker for the two models (with and without transfer-learning) to ensure that the tracks were true-positives. These experiments suggest that the transfer-learning based Yolov5 model leads to better performance on object tracking task.

5.2   Calibrating Softmax Confidence Scores

In general, we use softmax confidence output as the correctness probability estimate of the prediction. However, many of these neural networks are poorly calibrated [16,32]. The uncertainty in softmax confidence scores from poorly calibrated NN can potentially worsen the robustness of video analytics performance. To mitigate this, we can employ several post-hoc methods on SOTA models to improve the softmax estimates, *e.g.,* via averaging predictions obtained from bag of models (*e.g.,* detectors, classifiers) [25], platt scaling [16], isotonic regression [34], etc. We can also adapt the confidence threshold to filter out the low-confidence mispredictions. This confidence threshold value can be adapted based on the difficulty level to detect in a certain frame. We leave the investigation of neural network calibration and confidence threshold adaptation as future work.

6 Related Work

Several efforts [5,20,31,33,49] have been made to improve the robustness of deep learning models against adversarial attacks. Recent works [18,45] propose several adversarial examples that can be used to train a robust model and also serve as performance measures under several distribution shifts. Robust model training based on shape representation rather than texture-based representation is proposed in [15]. Xie et al. [46] use unlabeled data to train SOTA model through *noisy student self distillation* which improves the robustness of existing models. However, the creation of these "robust" models does not take into account the kind of adversaries introduced by dynamic tuning of AUTO parameters by video cameras. Also, the perturbations introduced in variants of ImageNet dataset (*i.e.,* -C, -3C, -P *etc.*) [17] are not the same as observed when AUTO parameters are tuned, which makes such datasets unsuitable for our study.

While there have been many efforts [4,11,19,22,26,29,47,48] on saving compute and network resource usage without impacting the accuracy of video analytics pipelines (VSPs) by adapting different video-specific parameters like frame rate, resolution, compression, *etc.*, there has been little focus on improving the *accuracy* of VAPs. Techniques to improve the accuracy of VSPs by dynamically tuning camera parameters is proposed by Paul et al. [36], but they focus on image-specific NAUTO parameters rather than AUTO parameters, which we show is the cause for "unintentional" adversarial effect introduced by the camera. Otani et al. [35] show the impact of low video quality on analytics, but they study network variability rather than the camera being the reason for low video quality.

Koh et al. [23] identify the need for training models with the distribution shift that will be observed in practice in real-world deployment. Inspired from this, rather than using independent images or synthetically transformed images for training, we use real video frames for training, which takes into account the distribution shift observed in practice for video analytics.

Wenkel et al. [43] tackle the problem of finding optimal confidence threshold for different models in model training and mention the challenge that there is a possibility of fluctuation in accuracy based on small changes between consecutive frames. However, they do not go in depth to analyze it further as we do. To our best knowledge, we are the first to expose the camera as an "unintentional adversary" for video analytics task and propose mitigation techniques.

7   Conclusion

In this paper, we show that blindly applying image-trained deep learning models for video analytics tasks leads to fluctuation in accuracy. We systematically eliminate external factors including motion, compression and environmental conditions (e.g., lighting) as possible reasons for fluctuation and show that the fluctuation is due to internal parameter changes applied by the camera, which acts as an "unintentional adversary" for video analytics applications. To mitigate this adversarial effect, we propose a transfer learning based approach and train a new Yolov5 model for object detection. We show that by reducing fluctuation across frames, our model is able better track objects ($\sim$40% fewer mistakes in tracking). Our paper exposes a fundamental fallacy in applying deep learning models for video analytics and opens up new avenues for research in this direction.

Acknowledgment. This project is supported in part by NEC Labs America and by NSF grant 2211459.

References

1. Bewley, A., Ge, Z., Ott, L., Ramos, F., Upcroft, B.: Simple online and realtime tracking. In: 2016 IEEE International Conference on Image Processing (ICIP), pp. 3464–3468 (2016). https://doi.org/10.1109/ICIP.2016.7533003
2. Bochkovskiy, A., Wang, C.Y., Liao, H.Y.M.: YOLOv4: optimal speed and accuracy of object detection. arXiv preprint arXiv:2004.10934 (2020)
3. Canel, C., et al.: Scaling video analytics on constrained edge nodes. In: Proceedings of Machine Learning and Systems, vol. 1, pp. 406–417 (2019)
4. Chen, T.Y.H., Ravindranath, L., Deng, S., Bahl, P., Balakrishnan, H.: Glimpse: continuous, real-time object recognition on mobile devices. In: Proceedings of the 13th ACM Conference on Embedded Networked Sensor Systems, pp. 155–168 (2015)
5. Cheng, M., Lei, Q., Chen, P.Y., Dhillon, I., Hsieh, C.J.: CAT: customized adversarial training for improved robustness. arXiv preprint arXiv:2002.06789 (2020)
6. Chiu, Y.C., Tsai, C.Y., Ruan, M.D., Shen, G.Y., Lee, T.T.: Mobilenet-SSDv2: an improved object detection model for embedded systems. In: 2020 International Conference on System Science and Engineering (ICSSE), pp. 1–5. IEEE (2020)
7. CNET: How 5G aims to end network latency (2019). CNET_5G_network_latency_time
8. AXIS Communications: Vapix library. https://www.axis.com/vapix-library/
9. Connell, J., Fan, Q., Gabbur, P., Haas, N., Pankanti, S., Trinh, H.: Retail video analytics: an overview and survey. In: Video Surveillance and Transportation Imaging Applications, vol. 8663, pp. 260–265 (2013)
10. Deng, J., Guo, J., Yuxiang, Z., Yu, J., Kotsia, I., Zafeiriou, S.: RetinaFace: single-stage dense face localisation in the wild. Arxiv (2019)
11. Du, K., et al.: Server-driven video streaming for deep learning inference. In: Proceedings of the Annual Conference of the ACM Special Interest Group on Data Communication on the Applications, Technologies, Architectures, and Protocols for Computer Communication, pp. 557–570 (2020)

12. Everingham, M., Gool, L.V., Williams, C.K.I., Winn, J.M., Zisserman, A.: The Pascal visual object classes (VOC) challenge. Int. J. Comput. Vis. **88**(2), 303–338 (2010). https://doi.org/10.1007/s11263-009-0275-4. http://dblp.uni-trier.de/db/journals/ijcv/ijcv88.html#EveringhamGWWZ10

13. Fang, Y., Zhan, B., Cai, W., Gao, S., Hu, B.: Locality-constrained spatial transformer network for video crowd counting. arXiv preprint arXiv:1907.07911 (2019)

14. Gaikwad, V., Rake, R.: Video Analytics Market Statistics: 2027 (2021). https://www.alliedmarketresearch.com/video-analytics-market

15. Geirhos, R., Rubisch, P., Michaelis, C., Bethge, M., Wichmann, F.A., Brendel, W.: ImageNet-trained CNNs are biased towards texture; increasing shape bias improves accuracy and robustness. arXiv preprint arXiv:1811.12231 (2018)

16. Guo, C., Pleiss, G., Sun, Y., Weinberger, K.Q.: On calibration of modern neural networks. In: International Conference on Machine Learning, pp. 1321–1330. PMLR (2017)

17. Hendrycks, D., Dietterich, T.: Benchmarking neural network robustness to common corruptions and perturbations. In: Proceedings of the International Conference on Learning Representations (2019)

18. Hendrycks, D., Zhao, K., Basart, S., Steinhardt, J., Song, D.: Natural adversarial examples. In: Proceedings of the IEEE/CVF Conference on Computer Vision and Pattern Recognition (CVPR), pp. 15262–15271, June 2021

19. Jiang, J., Ananthanarayanan, G., Bodik, P., Sen, S., Stoica, I.: Chameleon: scalable adaptation of video analytics. In: Proceedings of the 2018 Conference of the ACM Special Interest Group on Data Communication, pp. 253–266 (2018)

20. Jin, C., Rinard, M.: Manifold regularization for locally stable deep neural networks. arXiv preprint arXiv:2003.04286 (2020)

21. Jocher, G., et al.: ultralytics/yolov5: v6.1 - TensorRT, TensorFlow Edge TPU and OpenVINO export and inference (2022). https://doi.org/10.5281/zenodo.6222936

22. Kang, D., Emmons, J., Abuzaid, F., Bailis, P., Zaharia, M.: NoScope: optimizing neural network queries over video at scale. arXiv preprint arXiv:1703.02529 (2017)

23. Koh, P.W., et al.: WILDS: a benchmark of in-the-wild distribution shifts. In: Meila, M., Zhang, T. (eds.) Proceedings of the 38th International Conference on Machine Learning. Proceedings of Machine Learning Research, vol. 139, pp. 5637–5664. PMLR, 18–24 July 2021. https://proceedings.mlr.press/v139/koh21a.html

24. Krizhevsky, A., Sutskever, I., Hinton, G.E.: ImageNet classification with deep convolutional neural networks. In: Advances in Neural Information Processing Systems, pp. 1097–1105 (2012)

25. Lakshminarayanan, B., Pritzel, A., Blundell, C.: Simple and scalable predictive uncertainty estimation using deep ensembles. In: Advances in Neural Information Processing Systems, vol. 30 (2017)

26. Li, Y., Padmanabhan, A., Zhao, P., Wang, Y., Xu, G.H., Netravali, R.: Reducto: on-camera filtering for resource-efficient real-time video analytics. In: Proceedings of the Annual Conference of the ACM Special Interest Group on Data Communication on the Applications, Technologies, Architectures, and Protocols for Computer Communication, pp. 359–376 (2020)

27. Lin, T.-Y., et al.: Microsoft COCO: common objects in context. In: Fleet, D., Pajdla, T., Schiele, B., Tuytelaars, T. (eds.) ECCV 2014. LNCS, vol. 8693, pp. 740–755. Springer, Cham (2014). https://doi.org/10.1007/978-3-319-10602-1_48

28. Lisota, K.: Understanding video frame rate and shutter speed (2020). https://kevinlisota.photography/2020/04/understanding-video-frame-rate-and-shutter-speed/

29. Liu, L., Li, H., Gruteser, M.: Edge assisted real-time object detection for mobile augmented reality. In: The 25th Annual International Conference on Mobile Computing and Networking, pp. 1–16 (2019)
30. Liu, W., et al.: SSD: single shot multibox detector. In: Leibe, B., Matas, J., Sebe, N., Welling, M. (eds.) ECCV 2016. LNCS, vol. 9905, pp. 21–37. Springer, Cham (2016). https://doi.org/10.1007/978-3-319-46448-0_2
31. Madry, A., Makelov, A., Schmidt, L., Tsipras, D., Vladu, A.: Towards deep learning models resistant to adversarial attacks. In: International Conference on Learning Representations (2018). https://openreview.net/forum?id=rJzIBfZAb
32. Minderer, M., et al.: Revisiting the calibration of modern neural networks. In: Advances in Neural Information Processing Systems, vol. 34 (2021)
33. Najafi, A., Maeda, S.I., Koyama, M., Miyato, T.: Robustness to adversarial perturbations in learning from incomplete data. In: Advances in Neural Information Processing Systems, vol. 32 (2019)
34. Nyberg, O., Klami, A.: Reliably calibrated isotonic regression. In: Karlapalem, K., et al. (eds.) PAKDD 2021. LNCS (LNAI), vol. 12712, pp. 578–589. Springer, Cham (2021). https://doi.org/10.1007/978-3-030-75762-5_46
35. Otani, A., Hashiguchi, R., Omi, K., Fukushima, N., Tamaki, T.: On the performance evaluation of action recognition models on transcoded low quality videos. arXiv preprint arXiv:2204.09166 (2022)
36. Paul, S., et al.: CamTuner: reinforcement-learning based system for camera parameter tuning to enhance analytics (2021). https://doi.org/10.48550/ARXIV.2107.03964. https://arxiv.org/abs/2107.03964
37. Qualcomm: How 5G low latency improves your mobile experiences (2019). Qualcomm_5G_low-latency_improves_mobile_experience
38. Schroff, F., Kalenichenko, D., Philbin, J.: FaceNet: a unified embedding for face recognition and clustering. In: Proceedings of the IEEE Conference on Computer Vision and Pattern Recognition, pp. 815–823 (2015)
39. Sinha, D., El-Sharkawy, M.: Thin MobileNet: an enhanced mobilenet architecture. In: 2019 IEEE 10th Annual Ubiquitous Computing, Electronics & Mobile Communication Conference (UEMCON), pp. 0280–0285. IEEE (2019)
40. Tan, M., Pang, R., Le, Q.V.: EfficientDet: scalable and efficient object detection. In: Proceedings of the IEEE/CVF Conference on Computer Vision and Pattern Recognition, pp. 10781–10790 (2020)
41. Viso.ai: Top 16 applications of computer vision in video surveillance and security. https://viso.ai/applications/computer-vision-applications-in-surveillance-and-security/
42. Wang, L., Sng, D.: Deep learning algorithms with applications to video analytics for a smart city: a survey. arXiv e-prints, arXiv-1512 (2015)
43. Wenkel, S., Alhazmi, K., Liiv, T., Alrshoud, S., Simon, M.: Confidence score: the forgotten dimension of object detection performance evaluation. Sensors 21(13), 4350 (2021)
44. Witte, R., Witte, J.: A T-test for related measures. In: Statistics, pp. 273–285 (2017). ISBN: 9781119254515. www.wiley.com/college/witte
45. Xie, C., Tan, M., Gong, B., Wang, J., Yuille, A.L., Le, Q.V.: Adversarial examples improve image recognition. In: Proceedings of the IEEE/CVF Conference on Computer Vision and Pattern Recognition (CVPR), June 2020
46. Xie, Q., Luong, M.T., Hovy, E., Le, Q.V.: Self-training with noisy student improves imagenet classification. In: Proceedings of the IEEE/CVF Conference on Computer Vision and Pattern Recognition (CVPR), June 2020

47. Zhang, B., Jin, X., Ratnasamy, S., Wawrzynek, J., Lee, E.A.: AWStream: adaptive wide-area streaming analytics. In: Proceedings of the 2018 Conference of the ACM Special Interest Group on Data Communication, pp. 236–252 (2018)

48. Zhang, H., Ananthanarayanan, G., Bodik, P., Philipose, M., Bahl, P., Freedman, M.J.: Live video analytics at scale with approximation and {Delay-Tolerance}. In: 14th USENIX Symposium on Networked Systems Design and Implementation (NSDI 2017), pp. 377–392 (2017)

49. Zhang, H., Yu, Y., Jiao, J., Xing, E., El Ghaoui, L., Jordan, M.: Theoretically principled trade-off between robustness and accuracy. In: International Conference on Machine Learning, pp. 7472–7482. PMLR (2019)

SkeleVision: Towards Adversarial Resiliency of Person Tracking with Multi-Task Learning

Nilaksh Das[ID], ShengYun Peng[(✉)][ID], and Duen Horng Chau[ID]

Georgia Institute of Technology, Atlanta, USA
{nilakshdas,speng65,polo}@gatech.edu

Abstract. Person tracking using computer vision techniques has wide ranging applications such as autonomous driving, home security and sports analytics. However, the growing threat of adversarial attacks raises serious concerns regarding the security and reliability of such techniques. In this work, we study the impact of multi-task learning (MTL) on the adversarial robustness of the widely used SiamRPN tracker, in the context of person tracking. Specifically, we investigate the effect of jointly learning with semantically analogous tasks of person tracking and human keypoint detection. We conduct extensive experiments with more powerful adversarial attacks that can be physically realizable, demonstrating the practical value of our approach. Our empirical study with simulated as well as real-world datasets reveals that training with MTL consistently makes it harder to attack the SiamRPN tracker, compared to typically training only on the single task of person tracking.

Keywords: Person tracking · Multi-task learning · Adversarial robustness

1 Introduction

Person tracking is extensively used in various real-world use cases such as autonomous driving [3,50,53], intelligent video surveillance [1,4,54] and sports analytics [5,22,29]. However, vulnerabilities in the underlying techniques revealed by a growing body of adversarial ML research [6,7,11,14,20,35,42,46,49] seriously calls into question the trustworthiness of these techniques in critical use cases. While several methods have been proposed to mitigate threats from adversarial attacks in general [8,35,39,41,43], defense research in the tracking domain remains sparse [18]. This is especially true for the new generation of physically realizable attacks [6,11,46] that pose a greater threat to real-world applications.

In this work, we aim to investigate the robustness characteristics of the SiamRPN tracker [28], which is widely used in the tracking community. Specifically, our goal is to improve the tracking robustness to a physically realizable patch attack [46]. Such attacks are unbounded in the perceptual space and can be deployed in realistic scenarios, making them more harmful than imperceptible digital perturbation attacks. Figure 1 shows an example of such a physically realizable patch attack that blends in the background.

L. Karlinsky et al. (Eds.): ECCV 2022 Workshops, LNCS 13801, pp. 449–466, 2023.
https://doi.org/10.1007/978-3-031-25056-9_29

Fig. 1. Example of a physically realizable patch attack. The dashed blue box shows the ground-truth bounding box and the solid red box shows the bounding box predicted by SiamRPN. In the benign case (left), the tracker is able to correctly track the person whereas in the adversarial case (right) the tracker is fooled by the adversarial patch. (Color figure online)

Multi-Task learning (MTL) has recently been touted to improve adversarial robustness to imperceptible digital perturbation attacks for certain computer vision tasks [13,36]. However, it is unclear if these proposed methods translate to physically realizable attacks. Moreover, these methods have primarily been studied in the context of a single backbone branch with one-shot inference, whereas the Siamese architecture of the SiamRPN tracker involves multiple branched stages, posing interesting design considerations. In this work, we aim to address these research gaps by focusing on improving single-person tracking robustness.

As physically realizable attacks are unbounded in the perceptual space, they can create easily perceptible, but inconspicuous perturbations that fools a deep neural network into making incorrect predictions. However humans can ignore such perturbations by processing semantic knowledge of the real world. This calls for implicitly incorporating some inductive biases that supervise the neural network to learn semantic constraints that humans so instinctively interpret. To this effect, in this work we study the impact of MTL on robustness of person tracking with a semantically analogous task such as human keypoint detection.

Contributions

- **First Study of Tracking Robustness with MTL.** To the best of our knowledge, our work is the first to uncover the robustness gains from MTL in the context of person tracking for physically realizable attacks. Our code is made available at https://github.com/nilakshdas/SkeleVision.
- **Novel MTL Formulation for Tracking.** We augment the SiamRPN tracker for MTL by attaching a keypoint detection head to the template branch of the shared backbone while jointly training.
- **Extensive Evaluation.** We conduct extensive experiments to empirically evaluate the effectiveness of our MTL approach by varying attack parameters, network architecture, and MTL hyperparameters.
- **Discovery.** Our experiments with simulated and real-world datasets reveal that training with MTL consistently makes it harder to attack the SiamRPN tracker as compared to training only on the single task of person tracking.

2   Related Work

Since its inception with SiamFC [2], the Siamese architecture has been leveraged by multiple real-time object trackers including DSiam [15], SiamRPN [28], DaSiamRPN [63], SiamRPN++ [27], SiamAttn [55] and SiamMOT [40]. In this work, we experiment with SiamRPN as the target tracker since many other trackers share a similar network architecture as SiamRPN, and the properties of SiamRPN can be generalized to other such state-of-the-art trackers.

2.1   Multi-Task Learning

MTL aims to learn multiple related tasks jointly to improve the generalization performance of all the tasks [61]. It has been applied to various computer vision tasks including image classification [34], image segmentation [36], depth estimation [31], and human keypoint detection [21].

MTL has also been introduced for the video object tracking task [23–25]. Zhang et al. [56–58] formulate the particle filter tracking as a structured MTL problem, where learning the representation of each particle is treated as an individual task. Wang et al. [45] show that joint training of natural language processing and object tracking can link the local and global search together, and lead to a better tracking accuracy. Multi-modal RGB-depth and RGB-infrared tracking also demonstrate that including the depth or infrared information in the tracking training process can improve the overall performances [51,59,60,62].

2.2   Adversarial Attacks

Machine learning model are easily fooled by adversarial attacks [9]. Adversarial attacks can be classified as digital perturbation attacks [14,35,42] and physically realizable attacks [6,11,46,49]. In the tracking community, multiple attacks have been proposed to fool the object tracker [7,20]. Fast attack network [30] attacks the Siamese network based trackers using a drift loss and embedded features. The attack proposed by Jia et al. [19] degrades the tracking accuracy through an IoU attack, which sequentially generates perturbations based on the predicted IoU scores. The attack requires ground-truth when performing the attack. Wiyatno and Xu [46] propose a method to generate an adversarial texture. The texture can lock the GOTURN tracker [16] when a tracking target moves in front of it.

2.3   Adversarial Defenses in Tracking

General defense methods for computer vision tasks include adversarial training [43], increasing labeled and unlabeled training data [39], decreasing the input dimensionality [41], and robust optimization procedures [48,52]. However, not many defense methods have been proposed to improve the tracking robustness under attack. Jia et al. [18] attempt to eliminate the effect of the adversarial perturbations via learning the patterns from the attacked images. Recently, MTL

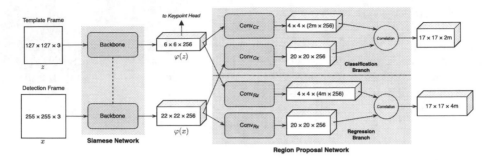

Fig. 2. Overview of the SiamRPN architecture for tracking. For multi-task learning, the output of the template branch is passed to a keypoint head for keypoint detection. The keypoint head consists of tandemly connected convolution and deconvolution layers. The output of the keypoint head is the same size as its input, with a depth equals number of keypoints. A peak in each channel indicates the predicted keypoint position.

has been shown to improve the overall network robustness [13], especially in image segmentation [36] and text classification [33]. Our work is the first that studies MTL for person tracking with a physically realizable attack.

3   Preliminaries

The input to the tracker can be denoted as $\{x, z, \bar{y}_x, \bar{y}_z\}$, where x is the detection frame in which the subject is to be tracked, z is the template frame containing an exemplar representation of the subject, and respectively, $\bar{y}_x$ and $\bar{y}_z$ are the ground-truth bounding box coordinates within the corresponding frames.

3.1   Tracking with SiamRPN

In this work, we focus on the Siamese-RPN model (SiamRPN) [28], which is a widely used tracking framework based on the Siamese architecture. An overview of the SiamRPN architecture is shown in Fig. 2. SiamRPN consists of a Siamese network for extracting features and a region proposal network (RPN), also referred to as the RPN head, for predicting bounding boxes.

The Siamese network has two branches: (1) the *template branch* which receives a template patch $z' = \Gamma(z, \bar{y}_z, s_z)$ as input; and (2) the *detection branch* which receives a detection patch $x' = \Gamma(x, \bar{y}_x, s_x)$ as input. Here, $\Gamma(\cdot)$ is simply a crop operation that ensures only a limited context of size s centered on the bounding box y is passed to the network [28]. The corresponding sizes s_z and s_x are shown in Fig. 2. For notational convenience, we use z for z' and x for x' hereon. The two branches of the Siamese network use a shared backbone model such that inputs to both branches undergo the same transformation $\varphi(\cdot)$. Hence, we can denote the output feature maps of the Siamese network as $\varphi(z)$ and $\varphi(x)$ for the template and detection branches, respectively. In this work, we use the SiamRPN model with AlexNet backbone [26].

The RPN head can also be separated into two branches as shown in Fig. 2. Considering m anchors distributed across the detection frame, the classification branch predicts whether each respective anchor is a background or foreground anchor. Hence, the classification branch has $2m$ output channels corresponding to m anchors. The regression branch on the other hand predicts 4 box coordinate regression deltas [38] for each anchor, and therefore has $4m$ output channels.

While training, the classification and regression branches of the RPN head yield $\mathcal{L}_{cls}$ and $\mathcal{L}_{reg}$ respectively, where $\mathcal{L}_{cls}$ is the cross-entropy loss and $\mathcal{L}_{reg}$ is a smooth L_1 loss [28]. Finally, the total weighted loss optimized for is as follows:

$$\mathcal{L}_{TRK}(x, z, \bar{y}_x) = \lambda_C \mathcal{L}_{cls}(x, z, \bar{y}_x) + \lambda_R \mathcal{L}_{reg}(x, z, \bar{y}_x) \tag{1}$$

During inference, the network acts as a single-shot detector. Typically, a sequence of frames $\mathbf{x} = \{x_1, \ldots x_n\}$ is provided with the ground-truth bounding box coordinates $\bar{y}_{x_1}$ of the first frame as input. Hence, the first frame x_1 becomes the template frame z used to compute the feature map $\varphi(z)$ once, which can be considered as detector parameters for predicting bounding box coordinates for input frames from the same sequence. We denote the predicted bounding box for an input frame as $\hat{y}_x$. As mentioned previously, SiamRPN crops the context centered on the ground-truth bounding box. For inference, the context region is determined by the predicted bounding box of the previous frame. Finally, the tracking performance is evaluated using a mean intersection-over-union (mIoU) metric of the predicted and ground-truth bounding boxes across all frames from all input sequences.

3.2   Multi-Task Learning with Shared Backbone

To provide semantic regularization for tracking, we perform joint multi-task training by attaching a fully convolutional keypoint prediction head to the template branch of SiamRPN. Our hypothesis is that joint training with an additional task head attached to the shared backbone would encourage the backbone to learn more robust features [17] for facilitating multiple tasks. Since the shared backbone is also used during tracking inference, the learned robust features can make it harder for adversarial perturbations to fool the model. We select the task of human keypoint prediction for this purpose as it is more semantically analogous to the task of person tracking.

The keypoint head is attached to the template branch as it has a more focused context [28]. Therefore, the keypoint head receives $\varphi(z)$ as input. The keypoint head network consists of convolutional blocks followed by a transpose convolution operation that "upsamples" the intermediate feature map to an expanded size with number of output channels equaling the number of keypoints being predicted. Finally, bilinear interpolation is performed to match the size of the input frame. The resulting feature volume has a shape of $H \times W \times K$, where H and W are the height and width of the input frame respectively, and K is the number of keypoints. Hence, each position in the K-channel dimension corresponds to a keypoint logit score. Given the ground-truth keypoints $\bar{k}_z$, the binary cross-entropy loss is computed with respect to each position in the channel dimension. We denote this as $\mathcal{L}_{KPT}$. For multi-task training, the total loss is a weighted sum:

$$\mathcal{L}_{MTL}(x, z, \bar{y}_x, \bar{k}_z) = \mathcal{L}_{TRK}(x, z, \bar{y}_x) + \lambda_K \mathcal{L}_{KPT}(z, \bar{k}_z) \tag{2}$$

The ground-truth keypoint annotation also consists of a visibility flag that allows us to suppress spurious loss from being backpropagated for keypoints that are occluded or not annotated.

3.3   Adversarial Attacks

Adversarial attacks introduce malicious perturbations to the input samples in order to confuse the tracker into making incorrect predictions. In this work, we use white-box untargeted attacks that aim to reduce the tracking performance by minimizing the mIoU metric. Adversarial attacks target a task loss, whereby the objective is to increase the loss by performing gradient ascent [35]. Given the predicted and ground-truth bounding boxes $\hat{y}_x$ and $\bar{y}_x$ respectively, we use the L1-loss as the task loss as proposed in [46] for attacking an object tracker:

$$\mathcal{L}_{ADV}(\hat{y}_x, \bar{y}_x) = \|\hat{y}_x - \bar{y}_x\|_1 \tag{3}$$

Based on means of application of the adversarial perturbation and additional constraints placed on the perturbation strength, attacks can be further classified into two distinct types:

Digital Perturbation Attacks. These attacks introduce fine-grained pixel perturbations that are imperceptible to humans [14,35,42]. Digital perturbation attacks can manipulate any pixel of the input, but place imperceptibility constraints such that the adversarial output x_{adv} is within an l_p-ball of the benign input x_{ben}, i.e., $\|x_{adv} - x_{ben}\|_p \leq \epsilon$. Such attacks, although having high efficacy, are considered to be physically non-realizable. This is due to the spatially unbounded granular pixel manipulation of the attack as well as the fact that a different perturbation is typically applied to each frame of a video sequence.

Physically Realizable Attacks. These attacks place constraints on the input space that can be manipulated by the attack [6,11,46,49]. In doing so, the adversarial perturbations can be contained within realistic objects in the physical world, such as a printed traffic sign [6] or a T-shirt [49]. As an attacker can completely control the form of the physical adversarial artifact, physically realizable attacks are unbounded in the perceptual space and place no constraints on the perturbation strength. In this work, we consider a physically realizable attack based on [46] that produces a background patch perturbation to fool an object tracker (Fig. 1). It is an iterative attack that follows gradient ascent for the task loss described in Eq. (3) by adding a perturbation to the input that is a product of the input gradient and a step size δ:

$$x^{(i)} = x^{(i-1)} + \delta \nabla_{x^{(i-1)}} \mathcal{L}_{ADV} \tag{4}$$

4   Experiment Setup

We perform extensive experiments and demonstrate that models trained with MTL are more resilient to adversarial attacks. The multi-task setting consists of jointly training a shared backbone for semantically related tasks such as person tracking and human keypoint detection (Sect. 4.1). We evaluate the tracking robustness on a state-of-the-art physically realizable adversarial attack for object trackers. We test our models on a photo-realistic simulated dataset as well as a real-world video dataset for person tracking (Sect. 4.4).

4.1   Architecture

For tracking, we leverage a SiamRPN model (Fig. 2) with an AlexNet backbone and an RPN head as described in [28]. The inputs to the model are a template frame (127×127) and a detection frame (255×255), fed to the backbone network. Finally, the RPN head of the model produces classification and localization artifacts corresponding to $m = 5$ anchors for each spatial position. The anchors have aspect ratios of $\{0.33, 0.5, 1, 2, 3\}$ respectively.

A keypoint head is also attached to the template branch of the network, *i.e.*, the keypoint head receives the activation map with dimensions $6 \times 6 \times 256$ as input. We attach the keypoint head to the template branch as the template frame has a more focused context, and typically has only one subject in the frame, leading to more stable keypoint training. The base keypoint head has 2 convolutional blocks with $\{128, 64\}$ channels respectively. We also perform ablation experiments by increasing the depth of the keypoint head to 4 blocks with $\{128, 128, 64, 64\}$ channels respectively (Sect. 5.2). The convolutional blocks are followed by a transpose convolution block with 17 output channels, which is the same as the number of human keypoints represented in the MS COCO format [32]. Bilinear interpolation is performed on the output of the transpose convolution block to expand the spatial output dimensions, yielding an output with dimensions $127 \times 127 \times 17$. Hence each of the 17 channels correspond to spatial logit scores for the 17 keypoints.

4.2   Training Data

We found that there is a dearth of publicly available tracking datasets that support ad-hoc tasks for enabling multi-task learning. Hence, for our MTL training, we create a hybrid dataset that enables jointly training with person tracking and human keypoint detection. For human keypoint annotations, we leverage the MS COCO dataset [32] which contains more than $200k$ images and $250k$ person instances, each labeled with 17 human keypoints. The MS COCO dataset

Fig. 3. Annotated training examples from the MS COCO (left) and LaSOT (right) datasets for person tracking. MS COCO has additional human keypoint annotations.

also annotates person bounding boxes that we use for the tracking scenario. As the MS COCO dataset consists of single images, there is no notion of temporal sequences in the input. Hence, for person tracking, we leverage data augmentation to differentiate the template and detection frames for the person instance annotation from the same image. Therefore, the MS COCO dataset allows us to train both the RPN head and keypoint head jointly for person tracking and human keypoint detection. We use the defined train and val splits for training and validation. Additionally, we merge this data with the Large-scale Single Object Tracking (LaSOT) dataset [12]. Specifically, we extract all videos for the *"person"* class for training the person tracking network. This gives us 20 video sequences, of which we use the first 800 frames from each sequence for training and the subsequent 100 frames for the validation set. Hence, the combined hybrid dataset from MS COCO and LaSOT enables our multi-task training. Figure 3 shows 2 example frames from MS COCO and LaSOT datasets.

4.3   Multi-Task Training

For the multi-task training, we fine-tune a generally pre-trained SiamRPN object tracker jointly for the tasks of person tracking and human keypoint detection. As we are specifically interested in the impact of multi-task training, we use the same loss weights λ_C and λ_R as proposed in [28] for the tracking loss $\mathcal{L}_{TRK}$. We perform an extensive sweep of the MTL loss weight λ_K associated with the keypoint loss $\mathcal{L}_{KPT}$ (Sect. 5.1). For the baseline, we perform single-task learning (STL) for person tracking by dropping the keypoint head and only fine-tuning the RPN head with the backbone, *i.e.*, the STL baseline has $\lambda_K = 0$. All STL and MTL models are trained with a learning rate of 8×10^{-4} that yields the best baseline tracking results as verified using a separate validation set. We also study the impact of pre-training the keypoint head separately before performing MTL (Sect. 5.3). For pre-training the keypoint head, we drop the RPN head and freeze the parameters of the backbone network. This ensures that the RPN head parameters are still compatible with the backbone after pre-training the keypoint head. The keypoint head is pre-trained with a learning rate of 10^{-3}. We train all models for 50 epochs and select the models with best validation performance over the epochs.

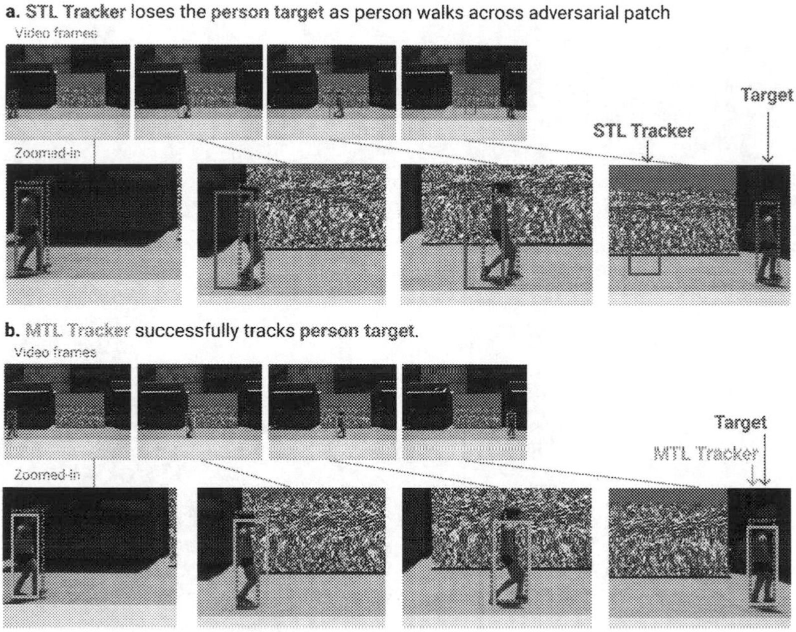

a. STL Tracker loses the person target as person walks across adversarial patch

b. MTL Tracker successfully tracks person target.

Fig. 4. Example video frames from the ARMORY-CARLA dataset showing static adversarial patches for (a) STL and (b) MTL for an attack with $\delta = 0.1$ and 10 steps. The patch is able to lock onto the STL tracker prediction (top), whereas the MTL tracker is consistently able to track the target (bottom).

4.4   Evaluation

We evaluate our trained STL and MTL models for the tracking scenario using the mIoU metric between ground-truth and predicted bounding boxes, which is first averaged over all frames for a sequence, and finally averaged over all sequences.

For testing the adversarial robustness of person tracking in a practical scenario, we leverage a state-of-the-art physically realizable adversarial attack for object trackers [46]. The attack adds a static adversarial background patch to a given video sequence that targets the tracking task loss $\mathcal{L}_{ADV}$. At each iteration of the attack, gradient ascent is performed on the task loss as per Eq. (4) with a step size δ. In order to observe the effect of varying the step size and attack iterations, we experiment with multiple values and report results for $\delta = \{0.1, 0.2\}$, which we found to have stronger adversarial effect on the tracking performance. The attack proposed in [46] has no imperceptibility constraints and is unbounded in the perceptual space, and can thus be considered an extremely effective adversarial attack. As the attack relies on the gradients of the task loss, we implement an end-to-end differentiable inference pipeline for the SiamRPN network using the Adversarial Robustness Toolbox (ART) framework [37].

We evaluate the adversarial robustness of STL and MTL models on 2 datasets:

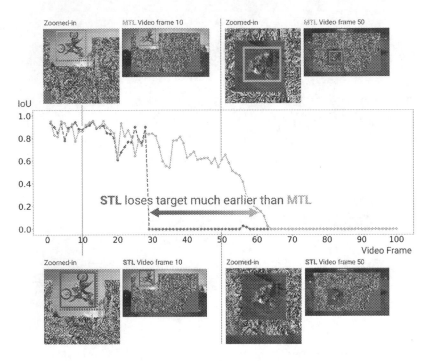

Fig. 5. Example video frames and the corresponding adversarial IoU charts for the video from the OTB2015-Person dataset showing the constructed static adversarial patches for STL (red) and MTL (orange) for an attack with $\delta = 0.1$ and 10 steps. The dashed blue box shows the ground-truth target. The attack misleads the STL tracker early, but struggles to mislead the MTL tracker until much later. The unperturbed gray regions in the patch are locations which are never predicted by the tracker. Since SiamRPN is a short-term tracker, the tracker cannot be restored once it loses the target. (Color figure online)

ARMORY-CARLA. This is a simulated photo-realistic person tracking dataset created using the CARLA simulator [10], provided by the ARMORY test bed [44] for adversarial ML research. We use the *"dev"* dataset split. The dataset consists of 20 videos of separate human sprites walking across the scene with various background locations. Each video has an allocated patch in the background that can be adversarially perturbed to mimic a physically realizable attack for person tracking. The dataset also provides semantic segmentation annotations to ensure that only the patch pixels in the background are perturbed when a human sprite passes in front of the patch. Figure 4 shows example video frames from the dataset where this can be seen. We find that the SiamRPN person tracker, having been trained on real-world datasets, has a reasonably high mIoU for tracking the human sprites when there is no attack performed; thus qualifying the photo-realism of the simulated scenario.

OTB2015-Person. We use the Object Tracking Benchmark (OTB2015) [47] to test the robustness of MTL for person tracking on a real-world dataset. We extract all videos that correspond to the task of person tracking, which yields 38 videos that we call the OTB2015-Person split. As the dataset is intended

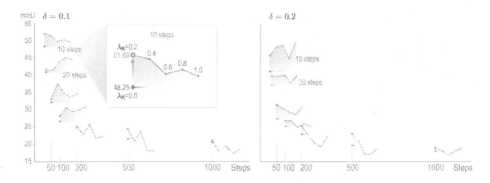

Fig. 6. A unified visualization of the adversarial mIoU results from Table 1 for the ARMORY-CARLA dataset with $\delta = 0.1$ (left) and $\delta = 0.2$ (right). The orange dots represent the MTL mIoU's and the gray flat lines represent the STL baseline mIoU's. We see that the hollow orange dots ($\lambda_K = 0.2$) are consistently above the STL baseline. (Color figure online)

for real-world tracking and is not readily amenable to implement physically realizable attacks, we digitally modify the videos for our attack to work. For each video, we overlay a static background patch that has a margin of 10% from each edge, covering 64% of the total area that can be perturbed by the attack. Finally, for each frame of a video, we only uncover the region annotated by the ground-truth bounding box with a padding of 25 pixels on each side. Hence, the annotated subject is always completely visible to the tracker with a digitally perturbed adversarial patch boundary. Figure 5 shows example video frames with the static patch attack as described here. To ensure that the tracker gets a clean ground-truth template, we do not perturb the first frame. Since this implements an unbounded digital attack on inputs from the real-world perceptual space, the attack is much stronger than real-world physically realizable attacks. For computational tractability, we only attack the first 100 frames.

5   Results

Our experiments reveal that models trained with MTL consistently make it harder for an adversarial attack to succeed by making the shared backbone network learn robust MTL features. Given an iterative attack, higher number of iterations corresponds to increased attack difficulty and higher attacker cost. We report the mIoU for increasing attack steps from $\{10, 20, 50, 100, 200, 500, 1000\}$ for the ARMORY-CARLA dataset in Table 1. We also visually summarize these gains from the MTL approach for the ARMORY-CARLA dataset in Fig. 6. For $\lambda_K = 1.0$, which is the base MTL setting, the MTL model improves upon the benign mIoU from 69.45 to 72.08. Additionally, the MTL model is more robust than the STL baseline up to 100 attack steps for $\delta = 0.1$ and 50 attack steps for $\delta = 0.2$. This implies the attack cost is higher for attacking an MTL model compared to its STL counterpart. The mIoU for increasing attack steps for the OTB2015-Person dataset are shown in Table 2. We observe a degradation in the benign MTL performance in this case, which may partly be attributed to the

Table 1. Adversarial mIoU results for ARMORY-CARLA dataset (↑ is better). Values highlighted in orange show the cases in which MTL is more robust than STL. MTL model with $\lambda_K = 0.2$ is consistently harder to attack than the STL model, and most often has the best performance. This table is also visualized in Fig. 6 for clarity.

		STL	MTL				
	Steps	$\lambda_K = 0.0$	$\lambda_K = 0.2$	$\lambda_K = 0.4$	$\lambda_K = 0.6$	$\lambda_K = 0.8$	$\lambda_K = 1.0$
benign	0	69.45	69.59	69.46	69.70	72.20	**72.08**
	10	48.25	**51.88**	51.44	49.74	50.24	49.50
	20	40.70	41.44	41.22	43.63	**45.05**	44.47
	50	32.07	33.04	**37.54**	34.63	33.49	34.49
$\delta = 0.1$	100	26.57	28.16	30.56	29.33	29.91	**30.62**
	200	24.72	25.19	22.73	**25.70**	21.73	22.12
	500	21.47	**24.38**	20.81	23.61	18.15	18.04
	1000	20.54	**21.05**	17.90	19.64	17.30	18.37
	10	41.03	45.62	48.13	48.63	44.68	**49.50**
	20	37.04	39.78	39.57	**40.00**	37.72	39.81
	50	27.32	**31.32**	30.33	29.31	28.55	30.86
$\delta = 0.2$	100	25.24	26.76	**26.89**	24.95	26.03	25.21
	200	22.95	**25.29**	22.54	20.41	19.27	22.16
	500	19.71	**23.13**	21.23	17.11	17.18	18.63
	1000	18.04	**19.02**	18.16	16.77	18.04	18.97

orange = MTL > STL; gray = MTL ≤ STL; **bold** = highest in row

resolution mismatch between the high resolution training examples [12,32] from MS COCO and LaSOT datasets, compared to lower resolution evaluation videos samples from the OTB2015 dataset. In the adversarial case, the base MTL model is more robust than STL baseline for up to 200 steps for $\delta = 0.1$. For $\delta = 0.2$, the base MTL model fails to show robustness for 20 steps, and slightly better robustness for other attack steps. We see further improvements in the adversarial resiliency for varying λ_K, discussed in Sect. 5.1.

5.1   Varying MTL Weight

We study the effect of varying the MTL weight λ_K, which controls the amount of keypoint loss $\mathcal{L}_{KPT}$ that is backpropagated. We train separate models by enumerating $\lambda_K = \{0.2, 0.4, 0.6, 0.8, 1.0\}$, and perform adversarial patch attack on each model for multiple adversarial settings. The results for ARMORY-CARLA and OTB2015-Person are shown in Table 1 and Table 2 respectively. We find that for the given shallow keypoint head architecture ({128, 64} channels), a lower value of λ_K is more optimal under adversarial attack. For both datasets, the MTL model with $\lambda_K = 0.2$ is consistently harder to attack than the STL model, and most often has the best performance for the corresponding adversarial setting.

Table 2. Adversarial mIoU results for OTB2015-Person dataset (↑ is better). In most cases, MTL models are harder to attack compared to STL model, with $\lambda_K = 0.2$ being most robust to the attack across several attack steps and step sizes.

		STL	MTL				
	Steps	$\lambda_K = 0.0$	$\lambda_K = 0.2$	$\lambda_K = 0.4$	$\lambda_K = 0.6$	$\lambda_K = 0.8$	$\lambda_K = 1.0$
benign	0	69.42	68.62	67.84	67.89	65.97	68.50
	10	54.29	57.29	57.78	**57.95**	57.09	56.24
	20	52.62	55.45	**55.68**	55.56	53.01	52.68
$\delta = 0.1$	50	48.54	**52.84**	52.33	50.54	50.94	50.67
	100	44.92	**52.45**	48.54	48.63	48.77	48.25
	200	45.40	47.73	**49.18**	47.26	46.50	47.34
	10	53.93	**57.65**	56.60	55.70	56.46	54.34
	20	53.57	55.21	**56.16**	56.08	54.46	52.86
$\delta = 0.2$	50	49.15	**52.81**	51.12	49.48	50.71	49.65
	100	47.27	**52.72**	49.74	48.13	49.87	47.81
	200	46.19	**51.05**	48.83	47.24	48.26	47.04

orange = MTL > STL; gray = MTL ≤ STL; **bold** = highest in row

Since a shallow keypoint head has relatively lower learning capacity, a higher λ_K value will force the shared backbone to focus excessively on the keypoint detection task, causing deterioration in the robust MTL features learned for the tracking task. From Table 1, although we observe better generalization for the MTL model with $\lambda_K = 1.0$ in the benign case (mIoU = 72.08), the adversarial robustness quickly gives away (at 100 steps for $\delta = 0.2$). Conversely, a lower value of $\lambda_K = 0.2$ offers the best trade-off for generalization and robustness.

5.2 Increasing Depth of Keypoint Head

Following the observations with a shallow keypoint head architecture, we also experiment with increasing the depth of the keypoint head from $\{128, 64\}$ channels to $\{128, 128, 64, 64\}$ channels, doubling the parameters of the keypoint head network. Table 3 shows the ablation results for the shallow and deep keypoint heads with the ARMORY-CARLA dataset for $\lambda_K = \{0.2, 1.0\}$ and $\delta = 0.1$. In this section we will focus on the "not pre-trained" results. The robustness of the MTL model degrades for $\lambda_K = 0.2$ when the model depth is increased, and is easier to attack compared to the STL model. However, the MTL model with deeper keypoint head has the best adversarial robustness for a higher $\lambda_K = 1.0$, even outperforming the MTL model with shallow keypoint head for $\lambda_K = 0.2$. As the deeper keypoint head has a relatively higher learning capacity, it can learn to detect keypoints with smaller changes to the feature space of the backbone network. Hence, a higher λ_K is required to adequately supervise the backbone in learning robust MTL features. Although we see a decline in the benign

Table 3. Ablation study with the ARMORY-CARLA dataset for attack step size $\delta = 0.1$. We report the adversarial mIoU results ($\uparrow$ is better).

| | $\lambda_K = 0.0$ | $\lambda_K = 0.2$ | | | | $\lambda_K = 1.0$ | | | |
| | | not pre-trained | | pre-trained | | not pre-trained | | pre-trained | |
Steps	(STL)	shallow	deep	shallow	deep	shallow	deep	shallow	deep
0	69.45	69.59	66.85	69.62	69.36	**72.08**	67.28	64.14	69.40
10	48.25	51.88	45.28	47.05	48.70	49.50	**55.46**	49.91	49.77
20	40.70	41.44	38.54	39.94	40.47	44.47	**47.44**	43.10	42.55
50	32.07	33.04	31.71	34.11	35.28	34.49	**36.52**	36.31	32.96
100	26.57	28.16	27.21	31.33	32.04	30.62	30.07	**32.14**	31.60
200	24.72	25.19	24.15	25.67	25.34	22.12	**26.53**	25.16	24.77

orange = MTL > STL; gray = MTL ≤ STL; **bold** = highest in row

mIoU for increasing depth, the deep MTL model with $\lambda_K = 1.0$ has overall best robustness. On the other hand, the shallow MTL model with $\lambda_K = 0.2$ has better adversarial robustness than the STL model as well as better benign performance.

5.3   Pre-training the Keypoint Head

As we start with a pre-trained SiamRPN model and an untrained keypoint head, we also study the impact of pre-training the keypoint head before performing MTL fine-tuning. Table 3 shows the results of this ablation study. We report the MTL performance with and without pre-training the keypoint head with the ARMORY-CARLA dataset for $\lambda_K = \{0.2, 1.0\}$ and $\delta = 0.1$. For the shallow keypoint head architecture, we see minor improvements in the MTL performance for a higher value of $\lambda_K = 1.0$, especially at higher number of attack steps. However, there is a sharp decrease in the benign performance (benign mIoU = 64.14). On the other hand, the deep keypoint head architecture shows relative improvement with pre-training for a lower value of $\lambda_K = 0.2$ (benign mIoU = 69.36). Overall, there is no significant advantage observed from pre-training the keypoint head. A pre-trained keypoint head would have lower potential to significantly modify the learned feature space of the shared backbone as it is already near the optima for the keypoint loss space.

6   Conclusion

We perform an extensive set of experiments with adversarial attacks for the task of person tracking to study the impact of multi-task learning. Our experiments on simulated and real-world datasets reveal that models trained with multi-task learning for the semantically analogous tasks of person tracking and human keypoint detection are more resilient to physically realizable adversarial attacks.

Our work is the first to uncover the robustness gains from multi-task learning in the context of person tracking for physically realizable attacks. As the tracking use case has widely ranging real-world applications, the threat of adversarial attacks has equally severe implications. We hope our work triggers new research in this direction to further secure tracking models from adversarial attacks.

References

1. Ahmed, I., Jeon, G.: A real-time person tracking system based on SiamMask network for intelligent video surveillance. J. Real-Time Image Proc. **18**(5), 1803–1814 (2021). https://doi.org/10.1007/s11554-021-01144-5
2. Bertinetto, L., Valmadre, J., Henriques, J.F., Vedaldi, A., Torr, P.H.S.: Fully-convolutional siamese networks for object tracking. In: Hua, G., Jégou, H. (eds.) ECCV 2016. LNCS, vol. 9914, pp. 850–865. Springer, Cham (2016). https://doi.org/10.1007/978-3-319-48881-3_56
3. Bhattacharyya, A., Fritz, M., Schiele, B.: Long-term on-board prediction of people in traffic scenes under uncertainty. In: Proceedings of the IEEE Conference on Computer Vision and Pattern Recognition, pp. 4194–4202 (2018)
4. Bohush, R., Zakharava, I.: Robust person tracking algorithm based on convolutional neural network for indoor video surveillance systems. In: Ablameyko, S.V., Krasnoproshin, V.V., Lukashevich, M.M. (eds.) PRIP 2019. CCIS, vol. 1055, pp. 289–300. Springer, Cham (2019). https://doi.org/10.1007/978-3-030-35430-5_24
5. Bridgeman, L., Volino, M., Guillemaut, J.Y., Hilton, A.: Multi-person 3D pose estimation and tracking in sports. In: Proceedings of the IEEE/CVF Conference on Computer Vision and Pattern Recognition Workshops (2019)
6. Chen, S.-T., Cornelius, C., Martin, J., Chau, D.H.P.: ShapeShifter: robust physical adversarial attack on Faster R-CNN object detector. In: Berlingerio, M., Bonchi, F., Gärtner, T., Hurley, N., Ifrim, G. (eds.) ECML PKDD 2018. LNCS (LNAI), vol. 11051, pp. 52–68. Springer, Cham (2019). https://doi.org/10.1007/978-3-030-10925-7_4
7. Chen, X., et al.: A unified multi-scenario attacking network for visual object tracking. In: Proceedings of the AAAI Conference on Artificial Intelligence, vol. 35, pp. 1097–1104 (2021)
8. Cisse, M., Bojanowski, P., Grave, E., Dauphin, Y., Usunier, N.: Parseval networks: improving robustness to adversarial examples. In: International Conference on Machine Learning, pp. 854–863. PMLR (2017)
9. Dong, Y., et al.: Boosting adversarial attacks with momentum. In: Proceedings of the IEEE Conference on Computer Vision and Pattern Recognition, pp. 9185–9193 (2018)
10. Dosovitskiy, A., Ros, G., Codevilla, F., Lopez, A., Koltun, V.: CARLA: an open urban driving simulator. In: Proceedings of the 1st Annual Conference on Robot Learning, pp. 1–16 (2017)
11. Eykholt, K., et al.: Robust physical-world attacks on deep learning visual classification. In: Proceedings of the IEEE Conference on Computer Vision and Pattern Recognition, pp. 1625–1634 (2018)
12. Fan, H., et al.: LaSOT: a high-quality benchmark for large-scale single object tracking. In: Proceedings of the IEEE/CVF Conference on Computer Vision and Pattern Recognition, pp. 5374–5383 (2019)

13. Ghamizi, S., Cordy, M., Papadakis, M., Traon, Y.L.: Adversarial robustness in multi-task learning: promises and illusions. arXiv preprint arXiv:2110.15053 (2021)
14. Goodfellow, I.J., Shlens, J., Szegedy, C.: Explaining and harnessing adversarial examples. arXiv preprint arXiv:1412.6572 (2014)
15. Guo, Q., Feng, W., Zhou, C., Huang, R., Wan, L., Wang, S.: Learning dynamic siamese network for visual object tracking. In: Proceedings of the IEEE International Conference on Computer Vision, pp. 1763–1771 (2017)
16. Held, D., Thrun, S., Savarese, S.: Learning to track at 100 FPS with deep regression networks. In: Leibe, B., Matas, J., Sebe, N., Welling, M. (eds.) ECCV 2016. LNCS, vol. 9905, pp. 749–765. Springer, Cham (2016). https://doi.org/10.1007/978-3-319-46448-0_45
17. Ilyas, A., Santurkar, S., Tsipras, D., Engstrom, L., Tran, B., Madry, A.: Adversarial examples are not bugs, they are features. In: Advances in Neural Information Processing Systems, vol. 32 (2019)
18. Jia, S., Ma, C., Song, Y., Yang, X.: Robust tracking against adversarial attacks. In: Vedaldi, A., Bischof, H., Brox, T., Frahm, J.-M. (eds.) ECCV 2020. LNCS, vol. 12364, pp. 69–84. Springer, Cham (2020). https://doi.org/10.1007/978-3-030-58529-7_5
19. Jia, S., Song, Y., Ma, C., Yang, X.: IoU attack: towards temporally coherent black-box adversarial attack for visual object tracking. In: Proceedings of the IEEE/CVF Conference on Computer Vision and Pattern Recognition, pp. 6709–6718 (2021)
20. Jia, Y.J., et al.: Fooling detection alone is not enough: adversarial attack against multiple object tracking. In: International Conference on Learning Representations (ICLR 2020) (2020)
21. Kocabas, M., Karagoz, S., Akbas, E.: MultiPoseNet: fast multi-person pose estimation using pose residual network. In: Ferrari, V., Hebert, M., Sminchisescu, C., Weiss, Y. (eds.) ECCV 2018. LNCS, vol. 11215, pp. 437–453. Springer, Cham (2018). https://doi.org/10.1007/978-3-030-01252-6_26
22. Kong, L., Huang, D., Wang, Y.: Long-term action dependence-based hierarchical deep association for multi-athlete tracking in sports videos. IEEE Trans. Image Process. **29**, 7957–7969 (2020)
23. Kristan, M., et al.: The eighth visual object tracking VOT2020 challenge results. In: Bartoli, A., Fusiello, A. (eds.) ECCV 2020. LNCS, vol. 12539, pp. 547–601. Springer, Cham (2020). https://doi.org/10.1007/978-3-030-68238-5_39
24. Kristan, M., et al.: The ninth visual object tracking VOT2021 challenge results. In: Proceedings of the IEEE/CVF International Conference on Computer Vision, pp. 2711–2738 (2021)
25. Kristan, M., et al.: The seventh visual object tracking VOT2019 challenge results. In: Proceedings of the IEEE/CVF International Conference on Computer Vision Workshops (2019)
26. Krizhevsky, A., Sutskever, I., Hinton, G.E.: ImageNet classification with deep convolutional neural networks. In: Advances in Neural Information Processing Systems, vol. 25 (2012)
27. Li, B., Wu, W., Wang, Q., Zhang, F., Xing, J., Yan, J.: SiamRPN++: evolution of siamese visual tracking with very deep networks. In: Proceedings of the IEEE/CVF Conference on Computer Vision and Pattern Recognition, pp. 4282–4291 (2019)
28. Li, B., Yan, J., Wu, W., Zhu, Z., Hu, X.: High performance visual tracking with siamese region proposal network. In: Proceedings of the IEEE Conference on Computer Vision and Pattern Recognition, pp. 8971–8980 (2018)

29. Liang, Q., Wu, W., Yang, Y., Zhang, R., Peng, Y., Xu, M.: Multi-player tracking for multi-view sports videos with improved k-shortest path algorithm. Appl. Sci. **10**(3), 864 (2020)

30. Liang, S., Wei, X., Yao, S., Cao, X.: Efficient adversarial attacks for visual object tracking. In: Vedaldi, A., Bischof, H., Brox, T., Frahm, J.-M. (eds.) ECCV 2020. LNCS, vol. 12371, pp. 34–50. Springer, Cham (2020). https://doi.org/10.1007/978-3-030-58574-7_3

31. Liebel, L., Körner, M.: MultiDepth: single-image depth estimation via multi-task regression and classification. In: 2019 IEEE Intelligent Transportation Systems Conference (ITSC), pp. 1440–1447. IEEE (2019)

32. Lin, T.-Y., et al.: Microsoft COCO: common objects in context. In: Fleet, D., Pajdla, T., Schiele, B., Tuytelaars, T. (eds.) ECCV 2014. LNCS, vol. 8693, pp. 740–755. Springer, Cham (2014). https://doi.org/10.1007/978-3-319-10602-1_48

33. Liu, P., Qiu, X., Huang, X.: Adversarial multi-task learning for text classification. arXiv preprint arXiv:1704.05742 (2017)

34. Luo, Y., Tao, D., Geng, B., Xu, C., Maybank, S.J.: Manifold regularized multitask learning for semi-supervised multilabel image classification. IEEE Trans. Image Process. **22**(2), 523–536 (2013)

35. Madry, A., Makelov, A., Schmidt, L., Tsipras, D., Vladu, A.: Towards deep learning models resistant to adversarial attacks. arXiv preprint arXiv:1706.06083 (2017)

36. Mao, C., et al.: Multitask learning strengthens adversarial robustness. In: Vedaldi, A., Bischof, H., Brox, T., Frahm, J.-M. (eds.) ECCV 2020. LNCS, vol. 12347, pp. 158–174. Springer, Cham (2020). https://doi.org/10.1007/978-3-030-58536-5_10

37. Nicolae, M.I., et al.: Adversarial robustness toolbox v1.2.0. CoRR 1807.01069 (2018). https://arxiv.org/pdf/1807.01069

38. Ren, S., He, K., Girshick, R., Sun, J.: Faster R-CNN: towards real-time object detection with region proposal networks. In: Advances in Neural Information Processing Systems, vol. 28 (2015)

39. Schmidt, L., Santurkar, S., Tsipras, D., Talwar, K., Madry, A.: Adversarially robust generalization requires more data. In: Advances in Neural Information Processing Systems, vol. 31 (2018)

40. Shuai, B., Berneshawi, A., Li, X., Modolo, D., Tighe, J.: SiamMOT: siamese multi-object tracking. In: Proceedings of the IEEE/CVF Conference on Computer Vision and Pattern Recognition, pp. 12372–12382 (2021)

41. Simon-Gabriel, C.J., Ollivier, Y., Bottou, L., Schölkopf, B., Lopez-Paz, D.: First-order adversarial vulnerability of neural networks and input dimension. In: International Conference on Machine Learning, pp. 5809–5817. PMLR (2019)

42. Szegedy, C., et al.: Intriguing properties of neural networks. arXiv preprint arXiv:1312.6199 (2013)

43. Tramèr, F., Kurakin, A., Papernot, N., Goodfellow, I., Boneh, D., McDaniel, P.: Ensemble adversarial training: attacks and defenses. arXiv preprint arXiv:1705.07204 (2017)

44. Two Six Technologies: ARMORY. https://github.com/twosixlabs/armory

45. Wang, X., et al.: Towards more flexible and accurate object tracking with natural language: algorithms and benchmark. In: Proceedings of the IEEE/CVF Conference on Computer Vision and Pattern Recognition, pp. 13763–13773 (2021)

46. Wiyatno, R.R., Xu, A.: Physical adversarial textures that fool visual object tracking. In: Proceedings of the IEEE/CVF International Conference on Computer Vision, pp. 4822–4831 (2019)

47. Wu, Y., Lim, J., Yang, M.H.: Object tracking benchmark. IEEE Trans. Pattern Anal. Mach. Intell. **37**(9), 1834–1848 (2015). https://doi.org/10.1109/TPAMI.2014.2388226
48. Xu, H., et al.: Adversarial attacks and defenses in images, graphs and text: a review. Int. J. Autom. Comput. **17**(2), 151–178 (2020). https://doi.org/10.1007/s11633-019-1211-x
49. Xu, K., et al.: Adversarial T-Shirt! Evading person detectors in a physical world. In: Vedaldi, A., Bischof, H., Brox, T., Frahm, J.-M. (eds.) ECCV 2020. LNCS, vol. 12350, pp. 665–681. Springer, Cham (2020). https://doi.org/10.1007/978-3-030-58558-7_39
50. Yagi, T., Mangalam, K., Yonetani, R., Sato, Y.: Future person localization in first-person videos. In: Proceedings of the IEEE Conference on Computer Vision and Pattern Recognition, pp. 7593–7602 (2018)
51. Yan, S., Yang, J., Käpylä, J., Zheng, F., Leonardis, A., Kämäräinen, J.K.: Depth-Track: unveiling the power of RGBD tracking. In: Proceedings of the IEEE/CVF International Conference on Computer Vision, pp. 10725–10733 (2021)
52. Yan, Z., Guo, Y., Zhang, C.: Deep defense: Training DNNs with improved adversarial robustness. In: Advances in Neural Information Processing Systems, vol. 31 (2018)
53. Yao, Y., Xu, M., Wang, Y., Crandall, D.J., Atkins, E.M.: Unsupervised traffic accident detection in first-person videos. In: 2019 IEEE/RSJ International Conference on Intelligent Robots and Systems (IROS), pp. 273–280. IEEE (2019)
54. Ye, S., Bohush, R., Chen, H., Zakharava, I.Y., Ablameyko, S.: Person tracking and reidentification for multicamera indoor video surveillance systems. Pattern Recogn. Image Anal. **30**(4), 827–837 (2020). https://doi.org/10.1134/S1054661820040136
55. Yu, Y., Xiong, Y., Huang, W., Scott, M.R.: Deformable siamese attention networks for visual object tracking. In: Proceedings of the IEEE/CVF Conference on Computer Vision and Pattern Recognition, pp. 6728–6737 (2020)
56. Zhang, T., Ghanem, B., Liu, S., Ahuja, N.: Robust visual tracking via structured multi-task sparse learning. Int. J. Comput. Vis. **101**(2), 367–383 (2013). https://doi.org/10.1007/s11263-012-0582-z
57. Zhang, T., Xu, C., Yang, M.H.: Multi-task correlation particle filter for robust object tracking. In: Proceedings of the IEEE Conference on Computer Vision and Pattern Recognition, pp. 4335–4343 (2017)
58. Zhang, T., Xu, C., Yang, M.H.: Learning multi-task correlation particle filters for visual tracking. IEEE Trans. Pattern Anal. Mach. Intell. **41**(2), 365–378 (2018)
59. Zhang, X., Ye, P., Peng, S., Liu, J., Gong, K., Xiao, G.: SiamFT: an RGB-infrared fusion tracking method via fully convolutional siamese networks. IEEE Access **7**, 122122–122133 (2019)
60. Zhang, X., Ye, P., Peng, S., Liu, J., Xiao, G.: DSiamMFT: an RGB-T fusion tracking method via dynamic siamese networks using multi-layer feature fusion. Sig. Process. Image Commun. **84**, 115756 (2020)
61. Zhang, Y., Yang, Q.: A survey on multi-task learning. IEEE Trans. Knowl. Data Eng. **34**(12), 5586–5609 (2022)
62. Zhu, X.F., Xu, T., Wu, X.J.: Visual object tracking on multi-modal RGB-D videos: a review. arXiv preprint arXiv:2201.09207 (2022)
63. Zhu, Z., Wang, Q., Li, B., Wu, W., Yan, J., Hu, W.: Distractor-aware siamese networks for visual object tracking. In: Ferrari, V., Hebert, M., Sminchisescu, C., Weiss, Y. (eds.) ECCV 2018. LNCS, vol. 11213, pp. 103–119. Springer, Cham (2018). https://doi.org/10.1007/978-3-030-01240-3_7

Unrestricted Black-Box Adversarial Attack Using GAN with Limited Queries

Dongbin Na[✉], Sangwoo Ji, and Jong Kim

Pohang University of Science and Technology (POSTECH), Pohang, South Korea
{dongbinna,sangwooji,jkim}@postech.ac.kr

Abstract. Adversarial examples are inputs intentionally generated for fooling a deep neural network. Recent studies have proposed unrestricted adversarial attacks that are not norm-constrained. However, the previous unrestricted attack methods still have limitations to fool real-world applications in a black-box setting. In this paper, we present a novel method for generating unrestricted adversarial examples using GAN where an attacker can only access the top-1 final decision of a classification model. Our method, Latent-HSJA, efficiently leverages the advantages of a decision-based attack in the latent space and successfully manipulates the latent vectors for fooling the classification model.

With extensive experiments, we demonstrate that our proposed method is efficient in evaluating the robustness of classification models with limited queries in a black-box setting. First, we demonstrate that our targeted attack method is query-efficient to produce unrestricted adversarial examples for a facial identity recognition model that contains 307 identities. Then, we demonstrate that the proposed method can also successfully attack a real-world celebrity recognition service. The code is available at https://github.com/ndb796/LatentHSJA.

Keywords: Black-box adversarial attack · Generative adversarial network · Unrestricted adversarial attack · Face recognition system

1 Introduction

Since state-of-the-art deep-learning models have been known to be vulnerable to adversarial attacks [20,40], a large number of defense methods to mitigate the attacks are proposed. These defense methods include adversarial training [32] and certified defenses [15,44]. Most previous studies have demonstrated their robustness against adversarial attacks that produce norm-constrained adversarial examples [15,32,44]. The common choices for the constraint are l_0, l_1, l_2, and l_∞ norms [10], because a short distance between two image vectors in a image space implies the visual similarity between them.

Recent studies show that adversarial examples can be legitimate even though the perturbation is not small norm-bounded. These studies have proposed unrestricted adversarial examples that are not norm-constrained but still shown as

L. Karlinsky et al. (Eds.): ECCV 2022 Workshops, LNCS 13801, pp. 467–482, 2023.
https://doi.org/10.1007/978-3-031-25056-9_30

Fig. 1. Showcases of our Latent-HSJA attack method against a facial identity recognition model. The first row shows target images and the second row shows source images. The adversarial examples in the last row are classified as target classes and require a feasible number of queries (only 20,000 queries).

natural images to humans [27,35,39]. For example, manipulating semantic information such as color schemes or rotation of objects in an image can cause a significant change in the image space while not affecting human perception. These unrestricted adversarial examples effectively defeat robust models to a norm-constrained perturbation [18,39]. However, only a few studies [27,39,43] have evaluated the effectiveness of unrestricted adversarial attacks for deep-learning models in a black-box setting.

In a black-box setting, an attacker can only access the output of a classification model. Real-world applications such as Clarifai and Google Cloud Vision provide only the top-k predictions of the highest confidence scores. Recent studies have proposed norm-constrained adversarial attack methods for the black-box threat models based on query-response to a classification model [9,14,26]. However, these black-box attacks have not yet successfully expanded to unrestricted adversarial examples. Although a few studies have demonstrated their unrestricted adversarial attacks in a black-box setting, their methods suffer from a large number of queries (more than hundreds of thousands of queries) compared with existing norm-based black-box attack methods [27] or only support an untargeted attack [43].

We propose a novel method, Latent-HSJA, for generating unrestricted adversarial examples using GAN in a black-box setting. To generate unrestricted adversarial examples, we utilize the disentangled style representations of Style-GAN2 [30]. Our method manipulates the latent vectors of GAN and efficiently leverages the decision-based attacks in a latent space. Especially, our targeted attack can be conducted with a target image classified as a target class and a specific source image with limited queries (Fig. 1). We mainly deal with the targeted attack because the targeted attack is more difficult to conduct and causes more severe consequences than an untargeted attack.

We show that the proposed method is query-efficient for the targeted attack in a hard-label black-box setting, where an attacker can only access the predicted top-1 label. It is noted that a black-box attack method should be query-efficient due to the high cost of a query (0.001$ in Clarifai). The proposed method for the targeted attack is able to generate an unrestricted adversarial example with a feasible number of queries (less than 20,000 queries) for a facial identity recognition model that contains 307 identities. This result is comparable to that of state-of-the-art norm-constrained black-box attacks [9,14,26]. Especially, our method generates more perceptually superior adversarial examples than previous methods in a super-limited query setting (less than 5,000 queries). Moreover, we demonstrate that our method can successfully attack a real-world celebrity recognition service.

Our contributions are listed as follows:

- We propose a query-efficient novel method, Latent-HSJA, for generating unrestricted adversarial examples in a black-box setting. To the best of our knowledge, our method is the first to leverage the targeted unrestricted adversarial attack in a query-limited black-box setting.
  We demonstrate that the proposed method successfully defeats state-of-the-art deep neural networks such as a gender classification model, a facial identity recognition model, and the real-world celebrity recognition model with a limited query budget.

2   Related Work

2.1   Adversarial Examples

Many deep-learning applications have been deployed in security-important areas such as face recognition [6], self-driving car [21], and malware detection [4]. However, the recent deep neural network (DNN) models have been known to be vulnerable to adversarial examples [12,20,34]. A lot of studies have presented adversarial attack methods in various domains such as image, text, and audio applications [3,13,17]. The adversarial examples can be also used to improve the robustness of automated authentication systems such as CAPTCHAs [33,38].

One key aspect in categorizing the threat model of adversarial attacks is the accessibility to components of DNN models, known as white-box or black-box (Fig. 2). In the white-box threat model, the attacker can access the whole information of a DNN model, including its weights and hyper-parameters. Many adversarial attacks assume the white-box threat model. The fast gradient sign method (FGSM) is proposed to generate an adversarial example by calculating

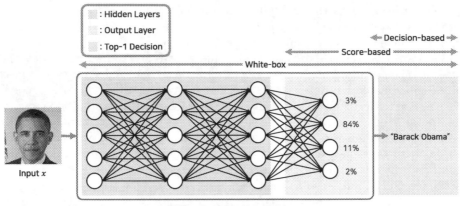

Fig. 2. An illustration of the common threat models. In a white-box threat model, the attacker can access the whole information of a DNN model including trained weights. In a score-based threat model, the attacker can access the output layer over all classes. In a decision-based threat model, the attacker can access the top-1 final decision.

the gradient only once [20]. The projected gradient descent (PGD) is then proposed to efficiently generate a strong adversarial example with several gradient calculation steps [32]. The CW attack is commonly used to find an adversarial example with a small perturbation by calculating a gradient vector typically more than thousands of times [12].

In the black-box threat model, the attacker can access only the output of a DNN model. The black-box threat model can be divided into two variants, the score-based threat model and the decision-based threat model. The score-based threat model assumes that an attacker is able to access the output of the softmax layer of a DNN model. Natural Evolution Strategy (NES) attack generates adversarial examples by estimating the gradient based on the top-k prediction scores of a DNN model [26]. On the other hand, the decision-based threat model assumes that an attacker is able to get the final decision of a DNN model, i.e., a predicted label alone. Boundary Attack (BA) uses random walks along the boundary of a DNN model to generate an adversarial example that looks similar to the source image [9]. HopSkipJump-Attack (HSJA) is then proposed to produce an adversarial example efficiently by combining binary search and gradient estimation [14]. As decision-based attacks (BA and HSJA) can access only the top-1 label, they start with an image already classified as the target class and maintain the classification result of the adversarial example during the whole attack procedure [9,14].

In this paper, we consider the decision-based threat model known as the most difficult black-box setting where an attacker can access only the top-1 label. This threat model is suitable for a real-world adversarial attack scenario because recent real-world applications such as the Clarifai service may provide only the top-1 label. Specifically, we present a variant of the HSJA [14], namely

Latent-HSJA, by adding an encoding step to the original HSJA procedure. The most significant difference between the original HSJA and our attack is that our attack leverages a latent space, not an input image space.

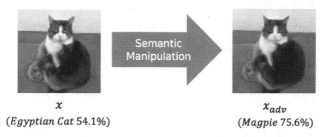

x
(Egyptian Cat 54.1%)

x_{adv}
(Magpie 75.6%)

Fig. 3. An illustration of an unrestricted adversarial example based on an unrestricted attack method [35].

2.2 Generative Adversarial Networks

Generative Adversarial Network (GAN) has been proposed to generate plausible new data examples [19]. Especially, GANs with deep convolutional layers have shown remarkable achievements and are able to produce realistic examples in an image-specific domain [36]. Previous studies with GAN have shown that it is possible to generate high-resolution images up to 1024×1024 resolution in various domains such as the human face, vehicles, and animals [28,29]. Recently, StyleGAN architecture shows an outstanding quality of the synthesized image by combining progressive training and the idea of style transfer [29,30].

With the advent of GAN, some previous studies have shown that adversarial examples can exist in the distribution of GAN [5,35]. It implies that an attacker can generate various adversarial examples by manipulating latent vectors of GAN. From this intuition, we propose a novel method that efficiently utilizes GAN for generating unrestricted adversarial examples that look perceptually natural. With extensive experiments, we have found that the StyleGAN2 architecture is suitable for our Latent-HSJA to efficiently generate realistic unrestricted adversarial examples in a black-box setting [30].

2.3 Unrestricted Adversarial Attacks

Most previous studies consider attack methods that generate adversarial examples constrained to the specific p-norm bound [12,20,34]. Specifically, l_0, l_1, l_2, and l_∞ norms are commonly used [10]. Therefore, previous defense methods also focus on a norm-constrained adversarial perturbation. Especially, recently proposed defense methods [15,44] provide certified robustness against an adversarial perturbation with a specific size of p-norm bound. On the other hand, recent

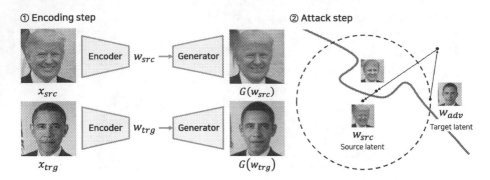

Fig. 4. An illustration of our attack method. Our proposed attack method consists of two steps. In the first encoding step, our method predicts two latent vectors (w_{src} and w_{trg}) according to input images. In the second attack step, our method drives the latent vector w_{trg} towards the latent vector w_{src} in a latent space. The red line denotes a decision boundary of a classification model in a latent space. (Color figure online)

studies have proposed various unrestricted adversarial attack methods that are not norm-constrained [8,11,27,35] (Fig. 3). Moreover, some studies have shown that these unrestricted attacks can bypass even the certified defense methods [18,39].

Nonetheless, the current unrestricted adversarial attacks have not yet successfully expanded to the black-box threat model. A related study has proposed an unrestricted black-box attack using an image-to-image translation network. However, it requires a large number of queries (more than hundreds of thousands of queries) in a black-box setting, which is not desirable for fooling real-world applications [27]. In this paper, we present a targeted attack method that generates unrestricted adversarial examples with a limited query budget (less than 20,000 queries). To the best of our knowledge, we are the first to propose a targeted black-box attack method that generates unrestricted adversarial examples with limited queries.

3   Proposed Methods

In a decision-based threat model, the common goal of an attacker is to generate an adversarial example x_{adv} that fools a DNN-based classification model $F(x)$ whose prediction output is the top-1 label for an input x [9,14]. In the targeted attack setting, the attacker finds an adversarial example x_{adv} that is similar to an x_{src} and is classified as a target class y_{trg} by the model F. The distance metric D is used to minimize the size of an adversarial perturbation. The common choice of the distance metric D is p-norm.

The general form of the objective is as follows:

$$\underset{x_{adv}}{\text{minimize}}\quad D(x_{src}, x_{adv})\quad \text{s. t.}\quad F(x_{adv}) = y_{trg}.$$

For the unrestricted attack, the attacker sets D as a metric to measure a distance of semantic information such as rotation, hue, saturation, brightness, or high-level styles between two images [8,18,23]. Our proposed method also minimizes the semantic distance between two images. We use D as the l_2 distance between the two latent vectors, i.e., w_{src} and w_{adv}. We postulate that if the distance between two latent vectors is short enough in the latent space of the GAN model G, the two synthesized images are similar in human perception. Especially when two latent vectors are exactly the same, the images generated by propagating two latent vectors into the GAN model should be the same.

Therefore, our attack uses the following objective:

$$\underset{w_{adv}}{\text{minimize}} \quad D(w_{src}, w_{adv}) \quad \text{s. t.} \quad F(G(w_{adv})) = y_{trg}.$$

3.1 Decision-Based Attack in Latent Space

We propose a method to conduct an unrestricted black-box attack, namely, Latent-HSJA, consisting of two steps (Fig. 4). First, we find latent vectors of source and target images (w_{src} and w_{trg}). We use the latent vector of the target image w_{trg} as an initial latent vector for an adversarial example w_{adv}. The corresponding adversarial example $G(w_{adv})$ should be adversarial (i.e., classified as the target class) at the start of the attack. Second, we conduct a decision-based update procedure in the latent space. Our method always maintains the predicted label of the adversarial example to be adversarial during the whole attack procedure ($F(G(w_{adv})) = y_{trg}$). We illustrate our attack algorithm in Fig. 4.

We utilize the HSJA method [14] for the second step of the proposed method (the decision-based update). Our decision-based attack minimizes $D(w_{src}, w_{adv})$ while preserving that the model output $F(G(w_{adv}))$ is always classified as the target class y_{trg}. After we run the attack algorithm, we get an adversarial latent vector w_{adv} such that $G(w_{adv}) \approx x_{src}$ in human perception. Our method tends to change the semantic information of an adversarial example $G(w_{adv})$ because our method chooses to update the latent vector w_{adv} in the latent space rather than directly update the image x_{adv} in the image space.

Algorithm 1: Decision-based attack in a latent space

Require: *Encoder* denotes an image encoding network, G denotes a pre-trained GAN model, *LatentHSJA* denotes our attack based on the HopSkipJump-Attack, and F denotes a classification model for the attack.
Input: Two input images x_{src}, x_{trg}.
Result: The adversarial example x_{adv}.
$w_{src} = Encoder(x_{src})$;
$w_{trg} = Encoder(x_{trg})$;
$w_{adv} = LatentHSJA(G, F, w_{src}, w_{trg})$;
$x_{adv} = G(w_{adv})$;

Table 1. The validation accuracies of our trained models. Our attack method is evaluated on these classification models.

Architectures	Identity dataset	Gender dataset
MNasNet1.0	78.35%	98.38%
DenseNet121	86.42%	98.15%
ResNet18	87.82%	98.55% ·
ResNet101	87.98%	98.05%

As a result, we can generate a perceptually natural adversarial example even in an early stage of the attack because our Latent-HSJA updates coarse-grained semantic features of the image. On the other hand, previous decision-based attack methods based on p-norm metrics suffer from a limitation that the generated adversarial example x_{adv} is not perceptually plausible in an early stage of the attack (less than 5,000 queries).

3.2   Encoding Algorithm

Our attack requires an accurate encoding method that maps an image x into a latent vector w such that $F(x) = F(G(Encoder(x)))$. Ideally, a perfect encoding method can satisfy this requirement. We first prepare w_{trg} by using the encoder so that $G(w_{trg})$ is classified as an adversarial class y_{trg}. Secondly, we also prepare w_{src} by using a given source image x_{src}. With this pair (w_{src} and w_{trg}), we could finally get the adversarial example $G(w_{adv})$ by conducting the Latent-HSJA (Algorithm 1).

For developing an encoding method that finds a latent vector according to an image for the StyleGAN-based generators, two approaches are commonly used. The first is an optimization-based approach that updates a latent vector using gradient descent steps [1,2]. The second approach trains an additional encoder network that embeds an image to a latent vector [16,37,42]. Recent studies have also shown that hybrid approaches combining the two methods could return better-encoded results [7,46]. We have found that the optimization-based encoding method is unsuitable for our attack because it could cause severe overfitting. Previous studies have also shown that such overfitting is not desirable for semantic image manipulation [42,46]. When the initial latent vector is highly overfitted, our attack could fail. We have observed that the recent pSp encoder provides successful encoding results for our attack method [37].

4   Experiments

4.1   Experiment Settings

Dataset. For experiments, we use the CelebA-HQ dataset [31], a common baseline dataset for face attribute classification tasks. The CelebA-HQ dataset contains 30,000 face images that are all 1024 × 1024 resolution images. There are

Fig. 5. Showcases of our Latent-HSJA attack method against a face gender classification model. The first row shows target images and the second row shows source images. The adversarial examples in the last row are classified as target classes and require a feasible number of queries (only 20,000 queries).

6,217 unique identities and 40 binary attributes in the CelebA-HQ dataset. First, we filter the CelebA-HQ dataset so that each identity contains more than 15 images for training the facial identity recognition models. As a result, our filtered facial identity dataset contains 307 identities, and there are 4,263 face images for training and 1,215 face images for validation. Secondly, we utilize the original CelebA-HQ dataset to train the facial gender classification models. The CelebA-HQ dataset contains 11,057 male images and 18,943 female images. We split these two datasets into 4:1 as training and validation.

Classification Models. For validating our Latent-HSJA, we have trained classification models on the two aforementioned datasets. We have fine-tuned MNasNet1.0, DenseNet121, ResNet18 and ResNet101 [22,24,41] that are pre-trained on the ILSVRC2012 dataset. All classification models resize the resolution of inputs to 256×256 in the input pre-processing step. The validation accuracies of all trained models are reported in Table 1. We have found the ResNet18 models show good generalization performance for both tasks, thus we report the main experimental results using the fine-tuned ResNet18 models. We also evaluate our attack method on a real-world application to verify the effectiveness of our method. The celebrity recognition service of Clarifai contains a large number of facial identities of over 10,000 recognized celebrities. For each face object of an image, this service returns the top-1 class and its probability.

Attack Details. In our attack method, we utilize the StyleGAN2 architecture [30]. For updating a encoded latent vector, a previous study utilizes w^+ space whose dimension is 18×512 to get better results [1]. Following the previous work,

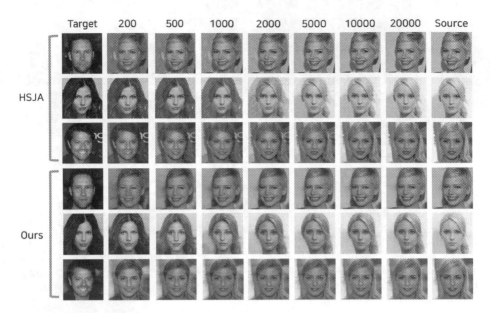

Fig. 6. An illustration of our targeted black-box attack results for a facial identity classification model. Our method (Latent-HSJA) is comparable with state-of-the-art norm-constrained adversarial attacks (HSJA) in a black-box setting.

we use w^+ space and normalize all the latent vectors so that the values of latent vectors are between $[0, 1]$ and utilize the HSJA in the normalized latent space. In our experiments, we randomly select 100 (image x, encoded latent w) pairs such that $F(x)$ is equal to $F(G(w))$ in the validation datasets for facial identity recognition and gender recognition. As mentioned in the previous section, the goal of our unrestricted attack is to minimize the l_2 distance between w_{adv} and w_{src}. We note that the attack success rate is always 100% because Latent-HSJA maintains the adversarial example to be always adversarial in the whole attack procedure, and the objective of attacks is to minimize the similarity distance D. For whole experiments, we report the targeted adversarial attack results.

Evaluation Metrics. The adversarial example x_{adv} should be close to the source image x_{src} such that $F(x_{adv}) = y_{trg}$ in the targeted adversarial attack setting. Therefore, we calculate how different the adversarial example x_{adv} is from the source image x_{src} using several metrics. We consider the similarity score SIM [25] and perceptual loss LPIPS [45] for evaluating our attack compared with the previously proposed p-norm based attack. Previous studies have demonstrated that the similarity score and LPIPS can be used for measuring perceptual similarity between two human face images [37].

Table 2. Experimental results of our Latent-HSJA compared with previous p-norm based HSJA against the gender recognition model and identity classification model. All adversarial examples are generated for fooling the ResNet18 models. The SIM and LPIPS scores are calculated using x_{src} and adversarial example x_{adv} ($G(w_{src})$ and $G(w_{adv})$ for Latent-HSJA).

Method	Metric	Gender recognition					
		Model queries					
		500	1000	3000	5000	10000	20000
HSJA [14]	SIM↑	0.546	0.621	0.724	0.780	**0.840**	**0.886**
	LPIPS↓	0.484	0.340	0.181	0.122	0.068	0.037
Ours	SIM↑	**0.769**	**0.794**	**0.817**	**0.821**	0.824	0.828
	LPIPS↓	**0.066**	**0.052**	**0.041**	**0.039**	**0.037**	**0.036**
Method	Metric	Identity classification					
		Model queries					
		500	1000	3000	5000	10000	20000
HSJA [14]	SIM↑	0.538	0.569	0.663	0.728	**0.821**	**0.895**
	LPIPS↓	0.313	0.287	0.187	0.132	0.067	**0.029**
Ours	SIM↑	**0.665**	**0.719**	**0.779**	**0.797**	0.811	0.819
	LPIPS↓	**0.164**	**0.119**	**0.072**	**0.061**	**0.053**	0.049

4.2   Gender Classification

Gender classification is a common binary classification task for classifying a face image. For the gender classification model, the targeted attack setting is the same as the untargeted attack setting. We demonstrate that adversarial examples made from our Latent-HSJA are sufficiently similar to the source images x_{src} (Fig. 5). Especially, our method is more efficient in the initial steps than the norm-based adversarial attack (Table 2). Our adversarial examples show better results in the evaluation metrics of SIM and LPIPS compared to the norm-based adversarial attack below 5,000 queries.

4.3   Identity Recognition

For the evaluation of our method against the facial identity recognition task, we have experimented with the targeted attack. We demonstrate that our method is query-efficient for the targeted attack. Our targeted attack is based on Algorithm 1. As illustrated in Fig. 6, our method mainly changes the coarse-grained semantic features in the initial steps (Table 2). This property is desirable for the black-box attack since our adversarial example will quickly be semantically far away from the target image. Moreover, the previous decision-based attacks often generate disrupted images that contain perceptible artificial noises with a limited query budget (under 5,000 queries). In contrast, our method maintains the adversarial examples to be always perceptually feasible.

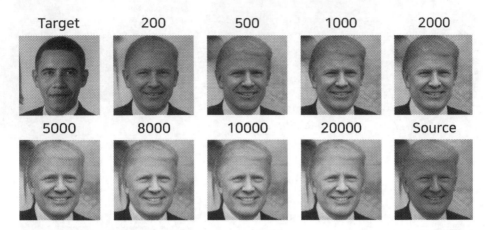

Fig. 7. An illustration of our targeted unrestricted black-box attack results for Clarifai service.

Table 3. The results of brute-forcing for finding feasible target image instances that can be used for starting points of our Latent-HSJA using StyleGAN [29] (higher $|Class|$ is better). The facial identity classification model contains 307 identities.

Dataset	Model queries	$|Class|$	$|Class|_{>50}$	$|Class|_{>90}$
FFHQ [29]	1000	183	117	36
	5000	252	203	90
	10000	267	241	111
	40000	300	275	168
	80000	305	288	200
CelebA-HQ [31]	1000	242	170	46
	5000	301	268	131
	10000	304	289	171
	40000	307	306	245
	80000	307	306	274

4.4   Real-World Application

Ethical Considerations. We are aware that it is important not to cause any disturbance to commercial services when evaluating real-world applications. Prior to experiments with the Clarifai service, we received permission from Clarifai to use the public API with a certain budget for the research purpose.

Attack Results. As illustrated in Fig. 7, our method is suitable for evaluating real-world black-box applications' robustness. We demonstrate that our method requires a feasible number of queries (about 20,000) with a specific source image. We note that the targeted attack for this real-world application is challenging because this service contains more than 10,000 identities. When running our attack method 5 times with random celebrity image pairs (x_{src}, x_{trg}), we have gotten the average value of SIM is 0.694 and the average value of LPIPS is 0.125. As a result, the Clarifai celebrity recognition service shows better robustness than our trained facial identity recognition model.

5 Discussion

We have found a generative model can efficiently create various images that are classified as a specific target class if the distributions of both a generative model and a classification model are similar to each other. We present a brute-forcing method for finding numerous target images. For validating the brute-forcing method, we simply use a pre-trained StyleGAN model [29]. We have demonstrated that sampling random image $G(w)$ with a large number of queries can find various images that could be classified as a target class (Table 3). For example, the brute-forcing with 80,000 queries can find the majority classes of the facial identity recognition model including high confidence (>90%) images. We note that the instance generated by brute-forcing is not appropriate as an adversarial example itself because this instance may have similar semantic features to the target class images. However, our result shows that if the GAN learns a similar distribution of the classification model and has enough capability for sampling various images, it is easy to find various images that are classified as a specific target class. For example, we can use a found image by brute-forcing as an initial starting point in our attack procedure.

6 Conclusion

In this work, we present Latent-HSJA, a novel method for generating unrestricted adversarial examples in a black-box setting. We demonstrate that our method can successfully attack state-of-the-art classification models, including a real-world application. Our method can explore the latent space and generate various realistic adversarial examples in terms of that the latent space of GAN contains a large number of adversarial examples. We especially utilize the valuable features of the StyleGAN2 architecture that have highly disentangled latent representations for a black-box attack. The experimental results show that our attack method has the potential as a new adversarial attack method. However, we have observed that the adversarial example is sometimes not feasible even though $MSE(w_{adv}, w_{src})$ is small enough. Therefore, future work may seek a better evaluation metric useful for our unrestricted adversarial attack. In addition, for generating strong a adversarial example, an attacker can use a hybrid approach that combines our Latent-HSJA with other norm-based decision-based attacks in terms of that our attack method shows better results in an early stage of the attack. Although we focus on the unrestricted attack on the face-related applications in this work, we believe our method is scalable to other classification tasks because GAN networks can be trained on various datasets. We hope our work has demonstrated new possibilities for generating semantic adversarial examples in a real-world black-box scenario.

References

1. Abdal, R., Qin, Y., Wonka, P.: Image2StyleGAN: how to embed images into the StyleGAN latent space? In: Proceedings of the IEEE/CVF International Conference on Computer Vision, pp. 4432–4441 (2019)
2. Abdal, R., Qin, Y., Wonka, P.: Image2StyleGAN++: how to edit the embedded images? In: Proceedings of the IEEE/CVF Conference on Computer Vision and Pattern Recognition, pp. 8296–8305 (2020)
3. Alzantot, M., Sharma, Y., Elgohary, A., Ho, B., Srivastava, M.B., Chang, K.: Generating natural language adversarial examples. In: Proceedings of the 2018 Conference on Empirical Methods in Natural Language Processing (2018)
4. Alzaylaee, M.K., Yerima, S.Y., Sezer, S.: DL-Droid: deep learning based Android malware detection using real devices. Comput. Secur. **89**, 101663 (2020)
5. Athalye, A., Carlini, N., Wagner, D.: Obfuscated gradients give a false sense of security: circumventing defenses to adversarial examples. In: International Conference on Machine Learning, pp. 274–283. PMLR (2018)
6. Balaban, S.: Deep learning and face recognition: the state of the art. In: Biometric and Surveillance Technology for Human and Activity Identification XII, vol. 9457, p. 94570B. International Society for Optics and Photonics (2015)
7. Bau, D., et al.: Seeing what a GAN cannot generate. In: Proceedings of the IEEE/CVF International Conference on Computer Vision, pp. 4502–4511 (2019)
8. Bhattad, A., Chong, M.J., Liang, K., Li, B., Forsyth, D.A.: Unrestricted adversarial examples via semantic manipulation. In: 8th International Conference on Learning Representations, ICLR 2020 (2020)
9. Brendel, W., Rauber, J., Bethge, M.: Decision-based adversarial attacks: reliable attacks against black-box machine learning models. In: 6th International Conference on Learning Representations, ICLR 2018 (2018)
10. Brendel, W., Rauber, J., Kümmerer, M., Ustyuzhaninov, I., Bethge, M.: Accurate, reliable and fast robustness evaluation. In: Annual Conference on Neural Information Processing Systems 2019 (2019)
11. Brown, T.B., Carlini, N., Zhang, C., Olsson, C., Christiano, P., Goodfellow, I.: Unrestricted adversarial examples. arXiv preprint arXiv:1809.08352 (2018)
12. Carlini, N., Wagner, D.: Towards evaluating the robustness of neural networks. In: 2017 IEEE Symposium on Security and Privacy (SP), pp. 39–57 (2017)
13. Carlini, N., Wagner, D.: Audio adversarial examples: targeted attacks on speech-to-text. In: 2018 IEEE Security and Privacy Workshops (SPW), pp. 1–7 (2018)
14. Chen, J., Jordan, M.I., Wainwright, M.J.: HopSkipJumpAttack: a query-efficient decision-based attack. In: 2020 IEEE Symposium on Security and Privacy (SP), pp. 1277–1294 (2020)
15. Cohen, J., Rosenfeld, E., Kolter, Z.: Certified adversarial robustness via randomized smoothing. In: International Conference on Machine Learning, pp. 1310–1320 (2019)
16. Donahue, J., Krähenbühl, P., Darrell, T.: Adversarial feature learning. CoRR abs/1605.09782 (2016)

17. Ebrahimi, J., Rao, A., Lowd, D., Dou, D.: HotFlip: white-box adversarial examples for text classification. In: Proceedings of the 56th Annual Meeting of the Association for Computational Linguistics (Volume 2: Short Papers). Association for Computational Linguistics (2018)
18. Ghiasi, A., Shafahi, A., Goldstein, T.: Breaking certified defenses: semantic adversarial examples with spoofed robustness certificates. In: 8th International Conference on Learning Representations, ICLR 2020 (2020)
19. Goodfellow, I.J., et al.: Generative adversarial nets. In: Annual Conference on Neural Information Processing Systems 2014 (2014)
20. Goodfellow, I.J., Shlens, J., Szegedy, C.: Explaining and harnessing adversarial examples. In: 3rd International Conference on Learning Representations, ICLR 2015 (2015)
21. Grigorescu, S., Trasnea, B., Cocias, T., Macesanu, G.: A survey of deep learning techniques for autonomous driving. J. Field Robot. **37**(3), 362–386 (2020)
22. He, K., Zhang, X., Ren, S., Sun, J.: Deep residual learning for image recognition. In: Proceedings of the IEEE Conference on Computer Vision and Pattern Recognition, pp. 770–778 (2016)
23. Hosseini, H., Poovendran, R.: Semantic adversarial examples. In: Proceedings of the IEEE Conference on Computer Vision and Pattern Recognition Workshops, pp. 1614–1619 (2018)
24. Huang, G., Liu, Z., Van Der Maaten, L., Weinberger, K.Q.: Densely connected convolutional networks. In: Proceedings of the IEEE Conference on Computer Vision and Pattern Recognition, pp. 4700–4708 (2017)
25. Huang, Y., et al.: CurricularFace: adaptive curriculum learning loss for deep face recognition. In: Proceedings of the IEEE/CVF Conference on Computer Vision and Pattern Recognition, pp. 5901–5910 (2020)
26. Ilyas, A., Engstrom, L., Athalye, A., Lin, J.: Black-box adversarial attacks with limited queries and information. In: International Conference on Machine Learning, pp. 2137–2146. PMLR (2018)
27. Kakizaki, K., Yoshida, K.: Adversarial image translation: unrestricted adversarial examples in face recognition systems. In: Proceedings of the Workshop on Artificial Intelligence Safety, Co-Located with 34th AAAI 2020 (2020)
28. Karras, T., Aila, T., Laine, S., Lehtinen, J.: Progressive growing of GANs for improved quality, stability, and variation. In: Proceedings of International Conference on Learning Representations (ICLR 2018) (2018)
29. Karras, T., Laine, S., Aila, T.: A style-based generator architecture for generative adversarial networks. In: Proceedings of the IEEE/CVF Conference on Computer Vision and Pattern Recognition, pp. 4401–4410 (2019)
30. Karras, T., Laine, S., Aittala, M., Hellsten, J., Lehtinen, J., Aila, T.: Analyzing and improving the image quality of StyleGAN. In: Proceedings of the IEEE/CVF Conference on Computer Vision and Pattern Recognition, pp. 8110–8119 (2020)
31. Lee, C.H., Liu, Z., Wu, L., Luo, P.: MaskGAN: towards diverse and interactive facial image manipulation. In: IEEE Conference on Computer Vision and Pattern Recognition (CVPR) (2020)
32. Madry, A., Makelov, A., Schmidt, L., Tsipras, D., Vladu, A.: Towards deep learning models resistant to adversarial attacks. In: 6th International Conference on Learning Representations, ICLR 2018 (2018)
33. Na, D., Park, N., Ji, S., Kim, J.: CAPTCHAs are still in danger: an efficient scheme to bypass adversarial CAPTCHAs. In: You, I. (ed.) WISA 2020. LNCS, vol. 12583, pp. 31–44. Springer, Cham (2020). https://doi.org/10.1007/978-3-030-65299-9_3

34. Papernot, N., McDaniel, P., Jha, S., Fredrikson, M., Celik, Z.B., Swami, A.: The limitations of deep learning in adversarial settings. In: 2016 IEEE European Symposium on Security and Privacy (EuroS&P), pp. 372–387 (2016)
35. Poursaeed, O., Jiang, T., Goshu, Y., Yang, H., Belongie, S., Lim, S.N.: Fine-grained synthesis of unrestricted adversarial examples. arXiv preprint arXiv:1911.09058 (2019)
36. Radford, A., Metz, L., Chintala, S.: Unsupervised representation learning with deep convolutional generative adversarial networks. In: 4th International Conference on Learning Representations, ICLR 2016 (2016)
37. Richardson, E., et al.: Encoding in style: a StyleGAN encoder for image-to-image translation. In: Proceedings of the IEEE/CVF Conference on Computer Vision and Pattern Recognition, pp. 2287–2296 (2021)
38. Shi, C., et al.: Adversarial CAPTCHAs. IEEE Trans. Cybern. **52**(7), 6095–6108 (2021)
39. Song, Y., Shu, R., Kushman, N., Ermon, S.: Constructing unrestricted adversarial examples with generative models. In: Annual Conference on Neural Information Processing Systems 2018, NeurIPS 2018 (2018)
40. Szegedy, C., et al.: Intriguing properties of neural networks. In: International Conference on Learning Representations (2014)
41. Tan, M., et al.: MnasNet: platform-aware neural architecture search for mobile. In: Proceedings of the IEEE/CVF Conference on Computer Vision and Pattern Recognition, pp. 2820–2828 (2019)
42. Tov, O., Alaluf, Y., Nitzan, Y., Patashnik, O., Cohen-Or, D.: Designing an encoder for StyleGAN image manipulation. ACM Trans. Graph. (TOG) **40**(4), 1–14 (2021)
43. Wang, R., et al.: Amora: black-box adversarial morphing attack. In: Proceedings of the 28th ACM International Conference on Multimedia, pp. 1376–1385 (2020)
44. Wong, E., Kolter, J.Z.: Provable defenses against adversarial examples via the convex outer adversarial polytope. In: Proceedings of the 35th International Conference on Machine Learning, ICML 2018 (2018)
45. Zhang, R., Isola, P., Efros, A.A., Shechtman, E., Wang, O.: The unreasonable effectiveness of deep features as a perceptual metric. In: Proceedings of the IEEE Conference on Computer Vision and Pattern Recognition, pp. 586–595 (2018)
46. Zhu, J., Shen, Y., Zhao, D., Zhou, B.: In-domain GAN inversion for real image editing. In: Vedaldi, A., Bischof, H., Brox, T., Frahm, J.-M. (eds.) ECCV 2020. LNCS, vol. 12362, pp. 592–608. Springer, Cham (2020). https://doi.org/10.1007/978-3-030-58520-4_35

Truth-Table Net: A New Convolutional Architecture Encodable by Design into SAT Formulas

Adrien Benamira[1]($\boxtimes$), Thomas Peyrin[1], and Bryan Hooi Kuen-Yew[2]

[1] Nanyang Technological University, Singapore, Singapore
adrien002@e.ntu.edu.sg
[2] National University of Singapore, Singapore, Singapore

Abstract. With the expanding role of neural networks, the need for complete and sound verification of their property has become critical. In the recent years, it was established that Binary Neural Networks (BNNs) have an equivalent representation in Boolean logic and can be formally analyzed using logical reasoning tools such as SAT solvers. However, to date, only BNNs can be transformed into a SAT formula. In this work, we introduce Truth Table Deep Convolutional Neural Networks (TTnets), a new family of SAT-encodable models featuring for the first time real-valued weights. Furthermore, it admits, by construction, some valuable conversion features including post-tuning and tractability in the robustness verification setting. The latter property leads to a more compact SAT symbolic encoding than BNNs. This enables the use of general SAT solvers, making property verification easier. We demonstrate the value of TTnets regarding the formal robustness property: TTnets outperform the verified accuracy of all BNNs with a comparable computation time. More generally, they represent a relevant trade-off between all known complete verification methods: TTnets achieve high verified accuracy with fast verification time, being complete with no timeouts. We are exploring here a proof of concept of TTnets for a very important application (complete verification of robustness) and we believe this novel real-valued network constitutes a practical response to the rising need for functional formal verification. We postulate that TTnets can apply to various CNN-based architectures and be extended to other properties such as fairness, fault attack and exact rule extraction.

Keywords: Neural network formal verification · Complete and Sound SAT Verification · Adversarial robustness

1 Introduction

Deep Neural Network (DNN) systems offer exceptional performance in a variety of difficult domains [20] and today these results far outstrip our ability to secure

Supplementary Information The online version contains supplementary material available at https://doi.org/10.1007/978-3-031-25056-9_31.

and analyze those DNNs. As DNNs are becoming widely integrated in a variety of applications, several concerns have emerged: lack of robustness emphasized by a lack of explainability, difficulty of integrating human knowledge in post-processing and impossibility to formally verify their behavior due to their large complexity. Under these circumstances and especially when these systems are deployed in applications where safety and security are critical, the formal verification of DNN systems is under intense research efforts. For example, Tesla has recently filed a patent on the DNNs' portability on a platform incorporating a component dedicated to formal verification [14] and the EU's general data protection regulation includes a provision on AI explainability [42].

So far, the DNNs architectures proposed by the community have generally been evolving towards increasing performance [6,13,52]. Improving performance is obviously desirable, however the resulting architectures are currently very hard to verify because of their intrinsic complexity. We propose here a novel architecture that is, by design, readily verifiable.

As application, we study Deep Convolutional Neural Networks (DCNNs) from the standpoint of their complete and sound formal verification: in other words, knowing a certain property, we want to confirm whether this property holds or not for a specific DNN. DNNs complete and sound verification property methods are mainly based either on Satisfiability Modulo Theory (SMT) [27] or Mixed-Integer Programming (MIP) [58] which are not yet scalable to real-valued DNNs. Some recent publications [24,35,37] approached the problem of complete verification from the well-known Boolean SATisfiability (SAT) [5] point of view where BNNs [23] are first converted into SAT formulas and then formally verified using SAT. This pipeline is computationally efficient [24], enables security verification [3] and more generally can answer a large range of questions including how many adversarial attacks exist for a given BNN, image and noise level [36]. However, to date, only BNNs can be transformed into a SAT formula and, as they were not designed for this application, their corresponding SAT conversion method intrinsically leads to formulas with a large number of variables and clauses, impeding formal verification scalability. In [24], the authors developed their SAT solver to achieve verification scaling.

A well-known example of property to verify on image classification datasets is the robustness to adversarial attacks. The performance of formal DNN robustness verification methods is evaluated against two main characteristics: verified accuracy (i.e. the ratio of images that are correctly predicted and that do not have any adversarial attacks) and verification time (i.e. the duration to verify that one correctly predicted image in the test set does/does not have an adversarial attack). The verification time is the sum of the problem construction time and its resolution time. The latter is intimately related to the quality of the solver and the complexity of the DNN encoding: a less efficient encoding and the solver will lead to a longer time to verify an image. Most robustness improvements in the literature have been in the form of new trainings [7,16,33,39,55,56], new testing method to increase robustness [7,16,44], new certification method to guarantee robustness properties [16,44,55,56] or new formal verification methods [10,17,31–33,47,57].

Our Contributions. We describe in this work a new architecture, encodable **by construction** into SAT formulas. It therefore comes as a complement to these previous approaches (our architecture can for example be trained with a certification method to further increase robustness). In addition, while we focus on the well-studied robustness to adversarial attacks property, our novel architecture generally allows a more compact symbolic encoding, leading to networks that are more amenable to verify most properties. In this work, we essentially offer three main contributions:

1. We define a new family of real-valued DCNNs that can be compactly encoded into SAT formulas: Truth Table Deep Convolutional Neural Network (**TTnet**). Our TTnet simplifies its 2D-CNN filter formulation in the form of a truth table to allow weights and certain intermediate values to be real. To the best of our knowledge, this is the first method to encode a real-valued DCNN into SAT, while achieving sufficient natural accuracy for practical use.

2. We show that TTnets offer two main valuable conversion properties over BNNs:

(2-a: Post-tuning). The first one allows us to integrate human knowledge in the post-processing: we can manually modify a DCNN filter activation towards a desired goal. For example, we decided to focus on reducing overfitting and, to this end, we characterize TTnet logic rules resulting from overfitting and propose a filtering approach. The latter increases the verified accuracy without decreasing the natural accuracy: +0.43% and +0.29% for high noise MNIST and CIFAR-10 respectively.

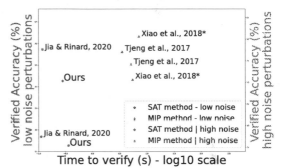

Fig. 1. Comparison of complete SAT and MIP methods for MNIST low noise ($\epsilon = 0.1$, in blue) and high noise ($\epsilon = 0.3$, in red) with regards to verified accuracy and verification time. Our method is more accurate than all methods and faster than the sound+complete MIP (triangle points). (Color figure online)

(2-b: Tractability). The second property enables to compute all possible model inputs/outputs prior to deployment in production. In an adversarial setting, we can assess whether the input noise will propagate to the output. We can therefore disregard filters with no impact on the output. This leads to a lower number of clauses and variables in the SAT formulas compared to BNNs, thus enabling the usage of generic SAT solver. Comparing to the BNN/SAT method, our SAT formulas are 5 and 9 times more compact for clauses number for high noise MNIST and CIFAR-10 respectively.

3. We apply our model to **complete robustness verification**. Our new TTnet model is the first complete and sound verification method allowing real-valued weights with competitive speed and no timeout. Besides, focusing on the robustness property, TTnet improves the verified accuracy over all BNN-based methods and more generally proposes a competitive trade-off between verified accuracy and verification time over all complete verification methods [11,18,24,37,51,58]. A comparison of our approach with the state-of-the-art methods for MNIST with low ($\epsilon = 0.1$)/high ($\epsilon = 0.3$) l_∞ noise perturbation is given in Fig. 1. On MNIST we can observe that with 97.12% and 86.24% respectively, we outperform previous best verified accuracy from ICLR 2019 [51,58] that reached 95.62% and 80.68% respectively, in addition to a shorter verification time. We highlight that our goal is to propose a competitor to BNN for SAT complete formal verification (using robustness as example property), not to propose a competitive robustness method.

We also show that our model can scale as well BNN and XNOR-Net on ImageNET dataset in term of natural accuracy. Finally, as robustness is merely one example property that our model can verify, we argue that TTnet has the potential to be competitive on many other strategic properties.

Outline. Sect. 2 introduces notations and related work. Section 3 presents our TTnet model and its two main properties. Section 4 details the complete robustness verification set-up and reports the evaluation results. Finally, we present TTnet's limitations in Sect. 5 and conclude in Sect. 6.

2   Related Work and Background

Boolean SATisfiability. The SAT problem [5] is that of deciding whether there exists a variable assignment to satisfy a given Boolean expression Φ. We can consider a Boolean expression in a Conjunctive Normal Form (CNF) or in a Disjunctive Normal Form (DNF). They are both defined over a set of Boolean variables $(x_1, \cdots, x_n)$. A literal l_i is defined as a variable x_i or its complement $\overline{x_i}$. A CNF is a conjunction of a set of clauses: $\Phi = (c_1 \wedge \cdots \wedge c_m)$, where each clause c_j is a disjunction of some literals $c_j = l_{j1} \vee \cdots \vee l_{jr}$. A DNF is a disjunction of a set of clauses: $\Phi = (c_1 \vee \cdots \vee c_m)$, where each clause c_j is a conjunction of some literals $c_j = l_{j1} \wedge \cdots \wedge l_{jr}$. A pseudo-Boolean constraint has the form: $\sum_{p=1}^{N} a_p l_p \circ b$, where $a_p \in \mathbb{Z}$, $b \in \mathbb{Z}$ and $\circ \in \{\leq, =, \geq\}$, which can be mapped to a SAT formula [43]. However, such a conversion generally leads to a tremendous number of clauses and literals compared to the number of variables in the original pseudo-Boolean form, making it impractical to comprehend.

SAT Encoding of Neural Networks. The sole published method converting a DNN into a SAT formula is limited to BNNs [9,35] and involves recomposing a block formed of a Two-Dimensional Convolutional Neural Network (2D-CNN) layer, a batch normalization layer and a step function into an inequality in order to apply the pseudo-Boolean constraint [43]. This approach has been further refined using a different training method and a specific SAT solver, resulting in a significantly reduced verification time [24]. Although the proposed inequality rewriting is elegant, the corresponding SAT formula still contains a large number of clauses and literals compared to the number of variables in the pseudo-Boolean constraint. This prevents the tractability of those SAT/BNNs formulas. In this work, we will only focus on comparison with DNN models, and not other machine learning family such as [1]. The main feature of DNNs is that there are more parameters in the model than samples in the dataset. This is not the case with decision trees. Their learning capacity is very different and for example, we are able to learn on ImageNET, in contrary to them.

Table 1. Comparison of state-of-the-art formal verification methods according to four criteria: the method type (sound and complete), the method applicability scope (can the method extend to all properties and all DNNs), the solver used and the method scalability.

Method	Type		Applicability scope		Solver used	Scalability			
	Sound	complete	All	All	SAT/MIP	MNIST		CIFAR-10	
			properties	DNNs		Low / High		Low / High	
[24]	√	√	√	only BNN	general SAT	√	√	√	√
[36]	√	√	√	only BNN	specific SAT	only l_p	×	×	×
[58]	√	√	√	√	MIP	√	√	√	√
[51]	√	√	√	√	MIP	√	√	√	√
[34]	√	×	×	√	MIP	√	√	√	√
[53]	√	×	×	√	MIP	√	√	√	√
[59]	√	×	×	√	-	×	×	×	×
[61]	√	×	×	√	-	×	√	×	√
[62]	√	×	×	√	-	√	√	√	√
[1]	√	√	×	only dec. tree	-	×	√	×	√
[28]	√	√	×	only BNN	MIP	×	×	×	×
[26]	√	√	√	√	MIP	×	×	×	×
[63]	√	√	√	only BNN	MIP*	√	√	×	×
Ours	√	√	√	only TTnet	general SAT	√	√	√	√

Probabilistic solver.

Sound, Complete Verification and Robustness. Complete and sound property verification of SAT-convertible DNNs has been presented in [35] as follows: given a precondition *prec* on the inputs x, a property *prop* on the outputs o and a SAT relations given by a DNN between inputs/outputs denoted as $DNN(x,o)$, we check whether the following statement is valid: $prec(x) \land DNN(x,o) \implies prop(o)$. In order to seek a counter-example to this property, we look for a satisfying assignment of $prec(x) \land DNN(x,o) \land \overline{prop(o)}$. An application example of property verification is to check for the existence of an adversarial perturbation in a trained DNN. In this case, *prec* defines an ϵ-ball of valid perturbations around the original image and *prop* states that the classification should not change under these small perturbations. Therefore, we distinguish the traditional "natural accuracy" from the "verified accuracy", the latter measuring the fraction of the predictions which remains correct for all adversarial attacks within the perturbation constraints. Table 1 presents a comparison of the functionalities of current state-of-the-art sound and complete verification methods. **In this work, our goal is not to propose a new state-of-the-art formal method dedicated to robustness, but to propose a new architecture that can scale and be used for any formal complete and sound property verification**. We then apply it to robustness against adversarial attack, a well studied property. Therefore, we compare our work to complete and sound methods that can be used for most properties: [11,18,24,36,51,58]. We also highlight that verification methods [48,49,54,62] are sound but incomplete. Finally, the work [25] shows that floating-point MIP verification methods can be unsound in some specific case. For a more detailed state-of-the-art, we refer to [22].

Two-Dimensional Convolutional Neural Networks. A Boolean function has the form $\{0,1\}^n \rightarrow \{0,1\}$ and its corresponding truth table lists the outputs for all 2^n possible inputs combinations (easily set up when n is not too large). Within our method, we consider the 2D-CNN as a Boolean function $\Phi_{(f,s,p,k,c)}$ which, for a given filter f, takes $n = k^2 c/g$ inputs at position (i,j) with k the kernel size, c the number of input channels, s the stride, p the padding and g the group parameter [15]. The outputs can be written as $y_f^{(i,j)} = \Phi_f(x_1^{(i,j)}, \cdots, x_n^{(i,j)})$. If we now consider a multi-layer network, a similar truth table can be constructed, except for the kernel size k that needs to be replaced by a *patch function*, sometimes also referred to as the size of a receptive field [2]. We denote the vector obtained after the flatten operation and before the final classifier layer as the vector of features V.

3   Truth Table Deep Convolution Neural Network (TTnet)

In an attempt to address the drawbacks of the high encoding complexity of the SAT formulas obtained from BNN's transformation process, we designed a new DNN architecture that admits a more compact symbolic encoding. We use real-weighted convolutions and aggregations that are further binarized with step-functions. These convolutions can be fully represented with truth tables, as long as the number of inputs in the filter is low-dimensional.

Figure 2 outlines the architecture of the TTnet model. In order to explain our model, we will initially analyse a 2D-CNN layer with a single filter (Sect. 3.1). Next, we will define the Learning Truth Table (LTT) block of our TTnet architecture (Sect. 3.2). Finally, Sects. 3.3 and 3.4 present the whole architecture of the TTnet and its main results, respectively. We provide a companion video https://youtu.be/loGlpVcy0AI.

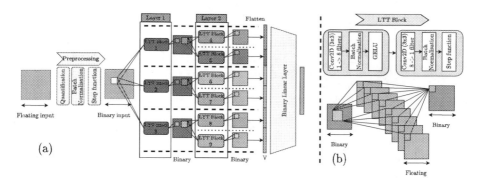

Fig. 2. (a) General architecture of the TTnet model with a one-channel input. Layer 0 is a pre-processing layer that allows image binarization. Then follow two layers of Learning Truth Table (LTT) blocks: three blocks in the first layer, six in the second. It should be noted that the LTT block of layer 2 does not take as input all the filters of layer 1, as it is usually the case: it only takes the filter of their groups. Finally, the last linear layer performs the classification. (b) Architecture of Learning Truth Table (LTT) block. A LTT block is composed of two layers of grouped 2D-CNN with an expanding factor of 8. It can be seen as an expanding auto-encoder. The intermediate values are real and the input/output values are binary.

3.1   SAT Encoding of a One-Layer 2D-CNN

For this first building block, we consider as input a binary image with one channel and as a model a trained DCNN with only one layer. We first start to encode one filter, then we will increase the complexity of our model by adding more filters to the 2D-CNN and then channels to the inputs.

One Filter. The main idea is to fix the number of possible outputs of the 2D-CNN by fixing the number of its possible inputs, which will allow us to later test all possible combinations. Following notations introduced in Sect. 2, we have as output $y_{binary,f}^{(i,j)}$ for filter f and n number of inputs:

$$y_{binary,f}^{(i,j)} = Bin(y_f^{(i,j)}) = Bin(\Phi_f(x_1^{(i,j)}, \cdots, x_n^{(i,j)})) \quad (1)$$

with Bin being the Heaviside step function, defined as $Bin(x) = \frac{1}{2} + \frac{sgn(x)}{2}$. If $x_j \in \{0,1\}$, we can establish the truth table of the 2D-CNN's filter f by trying all the possible inputs, for a total of 2^n operations. In this paper, we will limit ourselves to $n \leq 9$ (unless stated otherwise). Hence, with $2^9 = 512$ operations, we can trivially generate our truth table. Then, we can convert the truth table into a simplified SAT formula and by doing so we can rewrite Eq. 1 as:

$$y_{binary,f}^{(i,j)} = \mathsf{SAT}_f^{\mathsf{DNF}}(x_1^{(i,j)}, \cdots, x_n^{(i,j)}) = \mathsf{SAT}_f^{\mathsf{CNF}}(x_1^{(i,j)}, \cdots, x_n^{(i,j)})$$

with $\mathsf{SAT}_f^{\mathsf{DNF}}$ (resp. $\mathsf{SAT}_f^{\mathsf{CNF}}$) being the formal expression of the filter in the DNF form (resp. CNF form). It is noteworthy that unlike previous works, our approach is not limited to binary weights but allows for arbitrary weights within the 2D-CNN. Example 1 illustrates the described construction.

Example 1. We consider a 2D-CNN with one filter and a kernel size of 2, with the weights: $W_1 = \begin{bmatrix} 10 & -1 \\ 3 & -5 \end{bmatrix}$. As $c = 1$, we have $X = \begin{bmatrix} x_0 & x_1 \\ x_2 & x_3 \end{bmatrix}$ and the sixteen possible entries are: $\begin{bmatrix} 0 & 0 \\ 0 & 0 \end{bmatrix}, \begin{bmatrix} 0 & 0 \\ 0 & 1 \end{bmatrix}, \begin{bmatrix} 0 & 0 \\ 1 & 0 \end{bmatrix}, \begin{bmatrix} 0 & 0 \\ 1 & 1 \end{bmatrix}, \cdots, \begin{bmatrix} 1 & 1 \\ 0 & 1 \end{bmatrix}, \begin{bmatrix} 1 & 1 \\ 1 & 1 \end{bmatrix}$. For each input, we calculate the corresponding output:

$$y = \begin{bmatrix} 0, -5, 3, -2, -1, -5, 3, -2, 10, 5, 13, 8, 9, 4, 12, 7 \end{bmatrix}$$

After binarization with the Heaviside step function, we have

$$y_{binary} = \begin{bmatrix} 0, 0, 1, 0, 0, 0, 1, 0, 1, 1, 1, 1, 1, 1, 1, 1 \end{bmatrix}$$

After simplification, we have $\mathsf{SAT}_1^{\mathsf{DNF}} = (x_2 \wedge \overline{x_3}) \vee x_0$ and $\mathsf{SAT}_1^{\mathsf{CNF}} = (x_2 \vee x_0) \wedge (\overline{x_3} \vee x_0)$ with Quine-McCluskey algorithm [38].

Multiple Filters and Channels. Transposing to the case of multiple filters is straightforward: the above described method is simply repeated for each individual filter, thus yielding one expression per filter. As convolutional networks take several channels as input, the number of input variables rises substantially. For example, a 2D-CNN that takes 32 input channels with a kernel size of 2 yields an input of size 128, well above our limit set at $n = 9$. In order to overcome this effect, we group the channels using the group parameter [15]. Grouped convolutions divide the input channels into g groups, then apply separate convolutions within each group; this effectively decreases the number of inputs to each individual filter by a factor of g. We have in that case $n = k^2 c/g$. In the model diagram in Fig. 2, the groups are separated by the horizontal dotted lines. In the previous example, by using 16 groups, the number of inputs of our truth tables becomes $\frac{32}{16} \times 2^2 = 8$.

3.2 Learning Truth Table (LTT) Block

In order to improve the performance of the model, we add an amplification layer that will increase the learning capacity of the DCNN without augmenting the size of the patches seen by the DCNN [45].

Amplification Layer. In the previous subsection, we pointed out that only the 2D-CNN input size matters when establishing the 2D-CNN SAT expression. Therefore, we can add a second layer as long as we do not increase the patch size. This is achieved by adding a layer with kernel size 1. It can be noted that the intermediate values from the first layer do not need to be binary anymore. Figure 2(b) shows the architecture of the LTT block. A block is composed of two

2D-CNN layers with a so-called amplification parameter which corresponds to the ratio between the number of filters of the first layer and the number of filters of the second layer. We set the amplification parameter value at 8.

3.3   The TTnet Architecture

We show in Fig. 2 how the LTT block defined previously is integrated in TTnet: each LTT layer is positioned like a 2D-CNN layer except that its filters are not linear, they are expanding auto-encoders.

Pre-processing Layer. DCNNs inputs are usually floating points. However, encoding floating points into SAT typically implies high complexity. In order to simplify the verification process and to improve the network robustness, we applied a three-step first-layer or pre-processing procedure: (i) Quantification of inputs [24]; (ii) Batch normalization [37]; and (iii) Step function. We highlight that our pre-processing layer is the same as [24].

Final layer. TTnet uses a single linear layer as a classifier block. Indeed, it is straightforward to grasp and it can be easily encoded into SAT using pseudo-Boolean constraint as detailed in Sect. 2 as long as the weights are integers.

3.4   Experiments: Natural Accuracy Comparison

We now assess the natural accuracy performances of our TTnet architecture and compare it to existing approaches. A comparison of TTnet model features with BNN and DCNN is presented in Table 2.

Training Procedure. We compare TTnet, BNN and standard real-valued DCNN. We use Model B (7,7) without the amplification blocks and group parameter at one for BNN and DCNN. We believe the settings to be balanced between the models as the number of parameters to train for the

Table 2. TTnet features, comparison with BNN and DCNN.

	TTnet	BNN	DCNN
Weights type	Floating	Binary	Floating
Intermediate values type	Mixed Floating/Binary	Binary	Floating
Grouped 2D-CNN	√	×	×
Classifier	Linear	MLP*	MLP*

* MLP: Multi-Layer Perceptron.

BNN and DCNN is higher than for TTnet and the patch size is the same for the three models. TTnet and DCNN FLOPS are also comparable. We note that after SAT conversion, our FLOPS decreases to zero (no more float). We reproduced the exact same published training setting proposed in [24] for TTnet and BNN. TTnet has a final layer with integer values just for this experiment. We stopped the training at 45 epochs for all three models to prevent overfitting. We also compare TTnet for two values of truth table sizes n: 9 and 27 (we change 9 to 27 by increasing the number of filers in one group from 1 to 3).

Results for Small Verifiable Models. The results are presented in Table 3. TTnet yields improved natural accuracy over BNN , but of course lower than the original DCNN. By increasing the truth table dimension n to 27 and by increasing the number of filters in one group from 1 to 3, the enhancement over BNN is larger while the gap is reduced with DCNN: no difference for MNIST, -8.08% for CIFAR-10.

Table 3. Comparison of natural accuracy between TTnet with two truth table size, BNN and DCNN.

	TTnet ($n=9$)	TTnet ($n=27$)	BNN	DCNN
MNIST	98.35%	98.47%	96.98%	98.49%
CIFAR-10	54.11%	58.08%	53.53%	66.16%
Number of parameters	30.8K	38.5K	44.7K	44.7K
Patch size	(7,7)	(7,7)	(7,7)	(7,7)
MFLOPs (before SAT conv.)	0.7	0.9	NA - no float	0.9
MFLOPs (after SAT conv.)	0	0	NA - no float	0.9

Table 4. Comparison of top 1 and top 5 natural accuracy between TTnet, BNN and XNOR-Net on ImageNet [12].

Accuracy	TTnet ($n=48$)	BNN [23]	XNOR-Net [41]
top 1	45.9 %	27.9 %	44.2 %
top 5	69.1 %	50.4 %	69.2 %

Results for ImageNET. As our ambition is to propose a model formally verifiable by design, we tested its scaling of our model to ImageNET dataset [12] (see Table 4), with 18 layers of TTnet, for a truth table of size $n=48$ which remains computable. We can observe that TTnet can be adapted to ImageNET with improved accuracy over BNN's original paper [23] and comparable performances over XNOR-Net [41]. We note TTnet's satisfactory natural accuracy for practical use for real life dataset. TTnets currently do not allow formal verification for ImageNET ($n=48$ leads to difficulty in transforming into SAT), but we emphasize that no state-of-the-art complete formal verification method could scale to this dataset yet.

4   Application to Complete Robustness Verification

We now describe the two main valuable conversion properties of TTnet: post-tuning in Sect. 4.1 and tractability in Sect. 4.2. Then, we apply TTnet to complete robustness verification in Sect. 4.3. For fair comparison, in this section, we use the same pre-processing as in [24], the reference BNN paper for our comparison in Table 7 and the same quantification: 0.61 for MNIST and 0.064 for CIFAR10 for the pre-processing layer.

4.1   Post-tuning: Characterizing and Filtering Overfitting DNF Clauses

Our model gives the user the freedom to modify the set of SAT equations for all filters f. It addresses a very general problem: adding domain knowledge into a trained NN. Namely, we wish to reduce the model overfitting part in order to increase the verified accuracy without modifying the natural accuracy. To do so, we first propose a characterization of the DNF clauses responsible for the overfitting of the TTnet followed by a very simple suppression process.

As we have made the SAT clauses explicit, we may then modify them for a specific purpose. Drawing back on Example 1 of Sect. 3.1, we may consider, for example, that the DNF clause $(x_2 \wedge \overline{x_3})$ is too general and thus decide to remove it from the DNF equation. By doing so, we modify the activation of filter 1 and in turn directly integrate human knowledge in post-tuning.

Characterizing overfitting DNF clauses. A DNF clause is considered as an overfitting DNF clause if the ratio of the actual number of literals over the total number of possible literals in the clause is below a certain threshold. Indeed, the intuition behind the previous heuristic is that the more literals in the DNF clause, the more constraints on the input image, which hinders the proper generalization of the model. A ratio value of 1 means that the filter is only active if the patch exactly overlaps the DNF clause. Therefore, by changing only one bit, we deactivate that filter.

Filtering out overfitting DNF clauses. Upon prescribing the overfitting threshold, tagged DNF are filtered out. Next, among the remaining DNF clauses, we tag as additional overfitting DNF clauses those having the maximum literal ratio in the formula. Finally, we apply a random Bernoulli

Table 5. Effect on verified and natural accuracy of the overfitting DNF clauses filtering operation for the final model.

Dataset(noise)	Method	Accuracy	
		Verified	Natural
MNIST ($\epsilon_{test} = 0.3$)	TTnet	79.93%	96.79%
	TTnet+filtering	80.36% (+0.43%)	96.73% (-0.06%)
CIFAR-10 ($\epsilon_{test} = 8/255$)	TTnet	22.79%	31.18%
	TTnet+filtering	23.08% (+0.29%)	31.13% (-0.05%)

process to partially delete tagged DNF clauses. This technique increases the verified accuracy of the TTnet models trained on CIFAR-10: from 22.79% to 23.08%, with essentially no effect on natural accuracy (see results in Table 5).

4.2   Tractability: Computing all the Possibilities of the Adversarial Setup Before Production

Another feature of the TTnet's architecture is that one can calculate all the possible inputs/outputs of the LTT block before using the model in production. As we use small and independent truth tables, we can locally compute the noise propagation.

Encoding Pixel Noise. Let us consider some noise (*e.g.* norm-bounded by l_∞) added to the input image prior binarization. After the pre-processing layer, a binarized pixel may either remain unchanged by the perturbation (fixed to 0 or 1), or it may flip (from 0 to 1 or from 1 to 0) and hence we consider these flipping bits as "unknown", denoted as U.

Noise Propagation at the Block Level. In order to encode the noise propagation through the blocks, we encode in SAT the input/output relationship through the truth table.

Encoding the Attack at the Final Layer Level. Ultimately, for the final layer, we encode $r_{i,j} = y_i - y_j > 0$ as a reified cardinality constraint which denotes whether the score of class i is higher than the score of class j. Being able, thanks to the truth table nature of our architecture, to distinguish between known and unknown elements of the feature V of the image allows us to reduce significantly the size of the SAT formulas when compared to state-of-the-art models.

Results. Table 6 reports the consequences of this tractability property showing a substantial improvement over previous works. For MNIST high noise, our SAT equation yields on average 4K clauses and 1K variables, while [24] reported 21K clauses and 48K variables and [37] at least 20K clauses and 8K variables. Sim-

Table 6. Comparison of the number of clauses and variables for a given DCNN, for different encoding. The lower the better for the first two columns.

Dataset (noise)	Method	clauses	Number of		Natural accuracy
			Clauses	Variables	
MNIST ($\epsilon_{test} = 0.3$)	TTnet Model B(7,7)	4K	1K		96.79%
	BNN [24]		21K	48K	96.36 %
	BNN [37]*		20K	8K	96.0%
CIFAR-10 ($\epsilon_{test} = 8/255$)	TTnet Model A(6,6)	1K	0.4K		31.18%
	BNN [24]		9K	13K	35.00%

*results given on the first 1K images of the test set.

ilar trends were observed with CIFAR-10 noise. The drastic reduction in the size of SAT formulas renders our model truly amenable to formal verification. However, this SAT filtering induces a small extra time computation.

4.3 Formal and Complete Robustness Verification for Untargeted Attack

Methodology. We studied natural accuracy, verified accuracy for l_∞-norm bounded input perturbations and verification time on MNIST and CIFAR-10. Again, we reproduced the exact same published training setting as proposed in [24], same pre-processing and quantification. We compared our work with the state-of-the-art of exact verification for BNNs [24,37] and for real-valued networks [51,58]. We report the best verification accuracy for Model A(6,6), Model B(7,7) and Model C(7,7) with $n = 9$ and with the filtering and we did the same for the baselines.

Comparison with SAT Methods. As shown in Table 7, our verified accuracy is always superior to that of the BNN. In addition, we experienced the same resolution time as BNN's using a general SAT solver (namely MiniCard [29]). We highlight here that in [24] the SAT solver is custom-made and specific to their problem, while in our case we use a general one. This point is important, indeed when using a general SAT solver, [24] reports for MNIST high noise a resolution time of 0.242s, where ours is

Table 7. Application of TTnet to complete adversarial robustness verification for low and high noise bounded by l_∞. We tabulate results of verified accuracy, natural accuracy and mean verification time on MNIST and CIFAR-10 datasets in comparison to state-of-the-art SAT methods. The best verified accuracy is displayed in bold.

Dataset (noise)	Complete method	Accuracy		Verification time (s)	Timeout
		Verified	Natural		
MNIST ($\epsilon_{test} = 0.1$)	TTnet	**97.12%**	98.33%	0.3743	0
	[24]	91.68%	97.46%	0.1115	0
	[37]*	20.0%	96.0 %	5	0
MNIST ($\epsilon_{test} = 0.3$)	TTnet	**86.24%**	97.43 %	0.5586	0
	[24]	77.59%	96.36%	0.1179	0
CIFAR-10 ($\epsilon_{test} = 2/255$)	TTnet	**34.32%**	49.23%	0.2886	0
	[24]	32.18%	37.75%	0.0236	0
CIFAR-10 ($\epsilon_{test} = 8/255$)	TTnet	**23.08%**	31.13%	0.3887	0
	[24]	22.55%	35.00%	0.1781	0

*results given on the first 1K images of the test set. Moreover the authors only authorize at maximum 20 pixels to switch.

0.00079 s s with the same SAT solver. **We are exploring here a proof of concept of TTnets for a very important application and we believe this already justifies their potential.**

Comparison with MIP Methods. As presented in Fig. 1, we could demonstrate that TTnet verified accuracy is higher than that of the real value-based models in all MNIST cases. We also show that, with the general SAT solver MiniCard, we experience a resolution time of one to two orders of magnitude faster than the real-valued models, despite a lower verified accuracy for CIFAR-10. Moreover, TTnet presents no timeout: it always returns an answer whereas it is not the case for [11,18,58], which is a competitive advantage.

Comparison with BNN with the Same Architecture as TTnet. As the BNN architectures in [24] are different from ours, we reproduce the results for BNN with our exact same architecture and the exact same training conditions: results are presented in Table 8. We can observe that our model always outperforms the BNN model in term of natural and verified accuracy. We believe that the TTnet strength lies first in the decoupling, then in the LTT block architecture and finally in the ability to post-process them.

5   Limitation and Future Work

Scaling to Large Models. We designed a new DCNN that aimed to reduce the complexity of the network in order to increase its verification scalability. One may wonder what impact the group parameter has on the learning ability of a large network. In Table 4 we show that TTnet can scale to ImageNET with a comparable natural accuracy to XNOR-net [41]. Future works should focus on improving the natural accuracy of the model without increasing n [4,30].

Table 8. BNN as TTnet ModelB(7,7) for fair comparison

Dataset (noise)	Complete method	Accuracy		Verification time (s)	Timeout
		Verified	Natural		
MNIST ($\epsilon_{test} = 0.3$)	TTnet ModelB(7,7)	80.36%	96.43 %	0.8722	0
	BNN as ModelB(7,7)	49.53%	93.62%	0.010	0
CIFAR-10 ($\epsilon_{test} = 2/255$)	TTnet ModelB(7,7)	33.04%	40.62%	0.7782	0
	BNN as ModelB(7,7)	28.43%	37.66%	0.008	0

Other Architectures and Properties. Our method can be applied to architectures that use CNNs (like Graph-CNN). On the other hand, designing truth table-based methods for non-CNN architectures (e.g. Transformers) is an interesting direction for future work. Moreover, the compactness of TTnet can be extended to other properties like fairness [19,50], bit fault attack [21,40], privacy [8] and homomorphic encryption [46].

Global and exact rules extractions. Our method can be applied to rule extraction field as in [54,60]: in fact, TTnet allows the DNN to be written as a sum of DNF: this property can be studied in an other work for exact and global rules extractions on smaller datasets.

6   Conclusion

We proposed a novel architecture of SAT encodable real-valued DCNN based on truth tables that constitutes a practical complementary approach to the existing methods. We believe that the TTnet constitutes a suitable approach to the rising demand for functional formal verification.

References

1. Andriushchenko, M., Hein, M.: Provably robust boosted decision stumps and trees against adversarial attacks. In: Advances in Neural Information Processing Systems, vol. 32 (2019)
2. Araujo, A., Norris, W., Sim, J.: Computing receptive fields of convolutional neural networks. Distill (2019). https://doi.org/10.23915/distill.00021, https://distill.pub/2019/computing-receptive-fields
3. Baluta, T., Shen, S., Shinde, S., Meel, K.S., Saxena, P.: Quantitative verification of neural networks and its security applications. In: Proceedings of the 2019 ACM SIGSAC Conference on Computer and Communications Security, pp. 1249–1264 (2019)
4. Bello, I., et al.: Revisiting resnets: Improved training and scaling strategies. arXiv preprint arXiv:2103.07579 (2021)
5. Biere, A., Heule, M., van Maaren, H.: Handbook of satisfiability, vol. 185. IOS press (2009)
6. Brown, T.B., et al.: Language models are few-shot learners. arXiv preprint arXiv:2005.14165 (2020)
7. Carlini, N., Katz, G., Barrett, C., Dill, D.L.: Provably minimally-distorted adversarial examples. arXiv preprint arXiv:1709.10207 (2017)
8. Chabanne, H., De Wargny, A., Milgram, J., Morel, C., Prouff, E.: Privacy-preserving classification on deep neural network. Cryptology ePrint Archive (2017)
9. Cheng, C.-H., Nührenberg, G., Huang, C.-H., Ruess, H.: Verification of binarized neural networks via inter-neuron factoring. In: Piskac, R., Rümmer, P. (eds.) VSTTE 2018. LNCS, vol. 11294, pp. 279–290. Springer, Cham (2018). https://doi.org/10.1007/978-3-030-03592-1_16
10. Cheng, C.-H., Nührenberg, G., Ruess, H.: Maximum resilience of artificial neural networks. In: D'Souza, D., Narayan Kumar, K. (eds.) ATVA 2017. LNCS, vol. 10482, pp. 251–268. Springer, Cham (2017). https://doi.org/10.1007/978-3-319-68167-2_18

11. De Palma, A., Bunel, R., Dvijotham, K., Kumar, M.P., Stanforth, R.: IBP regularization for verified adversarial robustness via branch-and-bound. arXiv preprint arXiv:2206.14772 (2022)
12. Deng, J., Dong, W., Socher, R., Li, L.J., Li, K., Fei-Fei, L.: ImageNet: a large-scale hierarchical image database. In: CVPR09 (2009)
13. Dosovitskiy, A., et al.: An image is worth 16×16 words: Transformers for image recognition at scale. arXiv preprint arXiv:2010.11929 (2020)
14. Driscoll, M.: System and method for adapting a neural network model on a hardware platform, 2 July 2020, US Patent App. 16/728,884
15. Dumoulin, V., Visin, F.: A guide to convolution arithmetic for deep learning. arXiv preprint arXiv:1603.07285 (2016)
16. Dvijotham, K., et al.: Training verified learners with learned verifiers. arXiv preprint arXiv:1805.10265 (2018)
17. Ehlers, R.: Formal verification of piece-wise linear feed-forward neural networks. In: D'Souza, D., Narayan Kumar, K. (eds.) ATVA 2017. LNCS, vol. 10482, pp. 269–286. Springer, Cham (2017). https://doi.org/10.1007/978-3-319-68167-2_19
18. Ferrari, C., Muller, M.N., Jovanovic, N., Vechev, M.: Complete verification via multi-neuron relaxation guided branch-and-bound. arXiv preprint arXiv:2205.00263 (2022)
19. Garg, S., Perot, V., Limtiaco, N., Taly, A., Chi, E.H., Beutel, A.: Counterfactual fairness in text classification through robustness. In: Proceedings of the 2019 AAAI/ACM Conference on AI, Ethics, and Society, pp. 219–226 (2019)
20. Goodfellow, I., Bengio, Y., Courville, A.: Deep Learning. MIT Press (2016). http://www.deeplearningbook.org
21. Hong, S., Frigo, P., Kaya, Y., Giuffrida, C., Dumitras, T.: Terminal brain damage: exposing the graceless degradation in deep neural networks under hardware fault attacks. In: 28th USENIX Security Symposium (USENIX Security 19), pp. 497–514 (2019)
22. Huan, Z., Kaidi, X., Shiqi, W., Cho-Jui, H.: Aaai 2022: 'tutorial on neural network verification: Theory and practice' (2022). https://neural-network-verification.com/
23. Hubara, I., Courbariaux, M., Soudry, D., El-Yaniv, R., Bengio, Y.: Binarized neural networks. In: Advances in Neural Information Processing Systems, vol. 29 (2016)
24. Jia, K., Rinard, M.: Efficient exact verification of binarized neural networks. arXiv preprint arXiv:2005.03597 (2020)
25. Jia, K., Rinard, M.: Exploiting verified neural networks via floating point numerical error. In: Drăgoi, C., Mukherjee, S., Namjoshi, K. (eds.) SAS 2021. LNCS, vol. 12913, pp. 191–205. Springer, Cham (2021). https://doi.org/10.1007/978-3-030-88806-0_9
26. Jia, K., Rinard, M.: Verifying low-dimensional input neural networks via input quantization. In: Drăgoi, C., Mukherjee, S., Namjoshi, K. (eds.) SAS 2021. LNCS, vol. 12913, pp. 206–214. Springer, Cham (2021). https://doi.org/10.1007/978-3-030-88806-0_10
27. Katz, G., Barrett, C., Dill, D.L., Julian, K., Kochenderfer, M.J.: Reluplex: an efficient SMT solver for verifying deep neural networks. In: Majumdar, R., Kunčak, V. (eds.) CAV 2017. LNCS, vol. 10426, pp. 97–117. Springer, Cham (2017). https://doi.org/10.1007/978-3-319-63387-9_5
28. Kurtz, J., Bah, B.: Efficient and robust mixed-integer optimization methods for training binarized deep neural networks. arXiv preprint arXiv:2110.11382 (2021)
29. Liffiton, M.H., Maglalang, J.C.: A cardinality solver: more expressive constraints for free. In: Cimatti, A., Sebastiani, R. (eds.) SAT 2012. LNCS, vol. 7317, pp.

485–486. Springer, Heidelberg (2012). https://doi.org/10.1007/978-3-642-31612-8_47

30. Liu, Z., Wu, B., Luo, W., Yang, X., Liu, W., Cheng, K.-T.: Bi-Real Net: enhancing the performance of 1-Bit cnns with improved representational capability and advanced training algorithm. In: Ferrari, V., Hebert, M., Sminchisescu, C., Weiss, Y. (eds.) ECCV 2018. LNCS, vol. 11219, pp. 747–763. Springer, Cham (2018). https://doi.org/10.1007/978-3-030-01267-0_44

31. Lomuscio, A., Maganti, L.: An approach to reachability analysis for feed-forward Relu neural networks. arXiv preprint arXiv:1706.07351 (2017)

32. Lomuscio, A., Qu, H., Raimondi, F.: MCMAS: an open-source model checker for the verification of multi-agent systems. Int. J. Softw. Tools Technol. Transfer **19**(1), 9–30 (2017)

33. Mirman, M., Gehr, T., Vechev, M.: Differentiable abstract interpretation for provably robust neural networks. In: International Conference on Machine Learning, pp. 3578–3586. PMLR (2018)

34. Müller, M.N., Makarchuk, G., Singh, G., Püschel, M., Vechev, M.: Prima: general and precise neural network certification via scalable convex hull approximations. In: Proceedings of the ACM on Programming Languages 6(POPL), pp. 1–33 (2022)

35. Narodytska, N., Kasiviswanathan, S., Ryzhyk, L., Sagiv, M., Walsh, T.: Verifying properties of binarized deep neural networks. In: Proceedings of the AAAI Conference on Artificial Intelligence, vol. 32 (2018)

36. Narodytska, N., Shrotri, A., Meel, K.S., Ignatiev, A., Marques-Silva, J.: Assessing heuristic machine learning explanations with model counting. In: Janota, M., Lynce, I. (eds.) SAT 2019. LNCS, vol. 11628, pp. 267–278. Springer, Cham (2019). https://doi.org/10.1007/978-3-030-24258-9_19

37. Narodytska, N., Zhang, H., Gupta, A., Walsh, T.: In search for a sat-friendly binarized neural network architecture. In: International Conference on Learning Representations (2019)

38. Quine, W.V.: The problem of simplifying truth functions. Am. Math. Mon. **59**(8), 521–531 (1952)

39. Raghunathan, A., Steinhardt, J., Liang, P.: Certified defenses against adversarial examples. arXiv preprint arXiv:1801.09344 (2018)

40. Rakin, A.S., He, Z., Fan, D.: Tbt: Targeted neural network attack with bit trojan. In: Proceedings of the IEEE/CVF Conference on Computer Vision and Pattern Recognition, pp. 13198–13207 (2020)

41. Rastegari, M., Ordonez, V., Redmon, J., Farhadi, A.: XNOR-Net: Imagenet classification using binary convolutional neural networks. In: Leibe, B., Matas, J., Sebe, N., Welling, M. (eds.) ECCV 2016. LNCS, vol. 9908, pp. 525–542. Springer, Cham (2016). https://doi.org/10.1007/978-3-319-46493-0_32

42. Regulation, G.D.P.: Regulation eu 2016/679 of the european parliament and of the council of 27 April 2016. Official Journal of the European Union (2016). http://ec.europa.eu/justice/data-protection/reform/files/regulation_oj_en.pdf. Accessed 20 Sept 2017

43. Roussel, O., Manquinho, V.: Pseudo-Boolean and cardinality constraints. In: Handbook of Satisfiability, pp. 695–733. IOS Press (2009)

44. Sahoo, S.S., Venugopalan, S., Li, L., Singh, R., Riley, P.: Scaling symbolic methods using gradients for neural model explanation. arXiv preprint arXiv:2006.16322 (2020)

45. Sandler, M., Howard, A., Zhu, M., Zhmoginov, A., Chen, L.C.: Mobilenetv 2: inverted residuals and linear bottlenecks. In: Proceedings of the IEEE Conference on Computer Vision and Pattern Recognition, pp. 4510–4520 (2018)

46. Sanyal, A., Kusner, M., Gascon, A., Kanade, V.: Tapas: tricks to accelerate (encrypted) prediction as a service. In: International Conference on Machine Learning, pp. 4490–4499. PMLR (2018)
47. Shih, A., Darwiche, A., Choi, A.: Verifying binarized neural networks by local automaton learning. In: AAAI Spring Symposium on Verification of Neural Networks (VNN) (2019)
48. Singh, G., Ganvir, R., Püschel, M., Vechev, M.: Beyond the single neuron convex barrier for neural network certification. In: Advances in Neural Information Processing Systems, vol. 32 (2019)
49. Singh, G., Gehr, T., Püschel, M., Vechev, M.: An abstract domain for certifying neural networks. In: Proceedings of the ACM on Programming Languages 3(POPL), pp. 1–30 (2019)
50. Sun, B., Sun, J., Dai, T., Zhang, L.: Probabilistic verification of neural networks against group fairness. In: Huisman, M., Păsăreanu, C., Zhan, N. (eds.) FM 2021. LNCS, vol. 13047, pp. 83–102. Springer, Cham (2021). https://doi.org/10.1007/978-3-030-90870-6_5
51. Tjeng, V., Xiao, K., Tedrake, R.: Evaluating robustness of neural networks with mixed integer programming. In: ICLR (2019)
52. Vaswani, A., et al.: Attention is all you need. In: Advances in Neural Information Processing Systems, pp. 5998–6008 (2017)
53. Wang, S., et al.: Beta-crown: Efficient bound propagation with per-neuron split constraints for complete and incomplete neural network verification. arXiv preprint arXiv:2103.06624 (2021)
54. Wang, Z., Zhang, W., Liu, N., Wang, J.: Scalable rule-based representation learning for interpretable classification. In: Advances in Neural Information Processing Systems, vol. 34 (2021)
55. Wong, E., Kolter, Z.: Provable defenses against adversarial examples via the convex outer adversarial polytope. In: International Conference on Machine Learning, pp. 5286–5295. PMLR (2018)
56. Wong, E., Schmidt, F.R., Metzen, J.H., Kolter, J.Z.: Scaling provable adversarial defenses. arXiv preprint arXiv:1805.12514 (2018)
57. Wu, H., Barrett, C., Sharif, M., Narodytska, N., Singh, G.: Scalable verification of GNN-based job schedulers. arXiv preprint arXiv:2203.03153 (2022)
58. Xiao, K.Y., Tjeng, V., Shafiullah, N.M.M., Madry, A.: Training for faster adversarial robustness verification via inducing Relu stability. In: International Conference on Learning Representations (2019)
59. Xu, K., et al.: Automatic perturbation analysis for scalable certified robustness and beyond. In: Larochelle, H., Ranzato, M., Hadsell, R., Balcan, M., Lin, H. (eds.) Advances in Neural Information Processing Systems, vol. 33, pp. 1129–1141. Curran Associates, Inc. (2020). https://proceedings.neurips.cc/paper/2020/file/0cbc5671ae26f67871cb914d81ef8fc1-Paper.pdf
60. Yang, F., et al.: Learning interpretable decision rule sets: a submodular optimization approach. In: Advances in Neural Information Processing Systems, vol. 34 (2021)

61. Zhang, B., Cai, T., Lu, Z., He, D., Wang, L.: Towards certifying l-infinity robustness using neural networks with l-INF-DIST neurons. In: International Conference on Machine Learning, pp. 12368–12379. PMLR (2021)
62. Zhang, H., et al.: Towards stable and efficient training of verifiably robust neural networks. arXiv preprint arXiv:1906.06316 (2019)
63. Zhang, Y., Zhao, Z., Chen, G., Song, F., Chen, T.: BDD4BNN: a BDD-based quantitative analysis framework for binarized neural networks. In: Silva, A., Leino, K.R.M. (eds.) CAV 2021. LNCS, vol. 12759, pp. 175–200. Springer, Cham (2021). https://doi.org/10.1007/978-3-030-81685-8_8

Attribution-Based Confidence Metric for Detection of Adversarial Attacks on Breast Histopathological Images

Steven L. Fernandes[1]([✉]), Senka Krivic[2], Poonam Sharma[3], and Sumit K. Jha[4]

[1] Computer Science Department, Creighton University, Omaha, NE, USA
stevenfernandes@creighton.edu
[2] Faculty of Electrical Engineering, University of Sarajevo, B & H, Sarajevo,
Bosnia and Herzegovina
senka.krivic@etf.unsa.ba
[3] Pathology Department, Creighton University, Omaha, NE, USA
poonamsharma@creighton.edu
[4] Computer Science Department, University of Texas at San Antonio,
San Antonio, TX, USA
sumit.jha@utsa.edu

Abstract. In this paper, we develop attribution-based confidence (ABC) metric to detect black-box adversarial attacks in breast histopathology images. Due to the lack of data for this problem, we subjected histopathological images to adversarial attacks using the state-of-the-art technique Meta-Learning the Search Distribution (Meta-RS) and generated a new dataset. We adopt the Sobol Attribution Method to the problem of cancer detection. The output helps the user to understand those parts of the images that determine the output of a classification model. The ABC metric characterizes whether the output of a deep learning network can be trusted. We can accurately identify whether an image is adversarial or original with the proposed approach. The proposed approach is validated with eight different deep learning-based classifiers. The ABC metric for all original images is greater or equal to 0.8 and less for adversarial images. To the best of our knowledge, this is the first work to detect attacks on medical systems for breast cancer detection based on histopathological images using the ABC metric.

Keywords: Deep learning · Adversarial attacks · Explainable AI

1 Introduction

Breast cancer is the most common type in women, with 1.68 million registered modern cases and 522,000 caused deaths in 2012 [18,20,23]. Histopathological image analysis systems provide precise models and accurate quantification of the tissue structure [16]. To provide automatic aid for pathologists, deep learning networks are used for tracing cancer signs within breast histopathology images [13,23]. Moreover, Generative Adversarial Networks (GANs) are used to generate new digital pathology images [4]. However, this brings a high risk of medical image analysis systems being subject to black-box adversarial attacks. Adversarial images are hard to detect and can easily trick human users and AI systems. Therefore, detecting adversarial images and any artificial change to the original image is a crucial problem in medical image analysis. Solving it leads to more secure medical systems and more explainable systems [14].

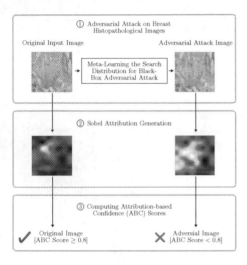

Fig. 1. The adversarial attack detection process.

We exploit Sobol Attribution Method for explanations [5] which captures interactions between image regions with Sobol indices and is used to visualize how they affect the neural network's prediction. Due to the specificity of the pathological images, additional information is needed to detect adversarial attacks. We develop attribution-based confidence (ABC) measure [11] to quantize the decision of whether an image is original or not (Fig. 1). We demonstrate how to perform an adversarial attack using the state-of-the-art method, Meta-Learning the Search Distribution (Meta-RS) of Black-Box Random Search [24] on digital pathology images from BreaKHis[1] database [19]. Using transfer learning, we evaluated the proposed approach with eight pre-trained classifiers on both the train and test datasets.

The main contributions of this paper are:

1. We created an adversarial histopathological image dataset using the state-of-the-art black-box attacks, and we make it public for further use[2].
2. We use the state-of-the-art Sobel attribution-based method to understand those parts of the images that determine the output of a classification model.
3. We developed the ABC metric, which indicates original images when the metric values are greater than or equal to 0.8 and adversarial images otherwise. This validated eight deep learning-based classification models and trained and tested data.

To the best of our knowledge, this is the first work to detect attacks on medical systems for breast cancer detection based on histopathological images using an attribution-based confidence metric (ABC) metric.

---

[1] https://www.kaggle.com/datasets/forderation/breakhis-400x.
[2] The new dataset is available at https://bit.ly/3p4QaPw.

2   Related Work

Histopathology image analysis plays a critical role in cancer diagnosis and treatment. The development of deep neural networks has made many breakthroughs in challenging clinical tasks such as automatic image analysis and prediction and assisted diagnosis [8, 15, 22, 26].

Xu et al. [22] proposed a weakly supervised learning framework for segmentation of histopathology images using just image-level labels. The results demonstrate remarkable performance with the fully supervised approaches. Most recent approaches combine several methods to improve the performance and capability of automatic diagnosis.

In this way, Hashimoto et al. [8] proposed a new CNN-based approach which can classify cancer sub-type for cancer sub-type from histopathology images. With the proposed method, it is possible to examine the whole slide and, in an automated way, detect tumour-specific features. Their method combines domain adversarial, multiple-instance, and multi-scale learning frameworks.

Zhao et al. [26] address the problem of automatic lymph node metastasis prediction. histopathological images of colorectal cancer. They created a GCN-based Multiple Instance Learning methods with a feature selection strategy. Lack of image datasets and cost of image annotation is often a limitation in the medical imaging field [9, 21]. Therefore there is a large number of works dedicated to decreasing this problem either with proposed approaches for image generation, automated image labelling or segmentation [1, 9].

Gamper and Rajpoot [1] present a novel multiple instances captioning dataset to facilitate dense supervision of CP tasks. This dataset contains diagnostic and morphological descriptions for various stains, tissue types and pathologies. Experimental results demonstrate that their proposed representation transfers to a variety of pathology sub-tasks better than ImageNet features or representations obtained with learning on pathology images alone.

Other methods for improving the automated pathology decision-making process include also automatizing preprocessing steps such as automatic magnification selection [25]. With the increase of the automatizing pathological diagnosis and application of AI methods, data and results malversation is potentially risky. Adversarial attacks are considered a potentially serious security threat for machine learning systems [3, 14].

Fote et al. [6] show that a highly accurate model for classifying tumour patches in pathology images can easily be attacked with minimal perturbations which are imperceptible to lay humans and trained pathologists alike. Therefore, there is a need to detect adversarial attacks and increase the security of medical systems. Laleh et al. [12] show that CNNs are susceptible to various white- and black-box attacks in clinical cancer detection tasks. Paschali et al. [17] demonstrate that besides classification tasks, also segmentation tasks can be affected by adversarial attacks. Thus, they propose a model evaluation strategy by leveraging task-specific adversarial attacks.

After reviewing the existing literature, we found a lack of adversarial breast histopathological image datasets and robust techniques to detect such attacks.

3 Proposed Approach

This section details methods for developing the approach for adversarial attack detection in breast histopathology images. Figure 2 provides a detailed overview of the process of developing the approach for adversarial image detection.

In the first step, we trained eight architectures for the image classification task (ResNet18, ResNet50, Inception V3, MobileNet V3, ShuffleNet, Swin Transformer, Vision Transformer, WideResnet) with the original dataset. In the second step, we performed state-of-the-art Meta-RS black-box adversarial attacks [24] and generated an adversarial images dataset. In the third step we adapted the state-of-the-art Sobol Attribution Method for explanations [5].

Finally, we propose attribution-based confidence (ABC) metric [11] to detect black-box adversarial attacks. The ABC metric characterizes whether one can trust the decision of a deep neural network on an input.

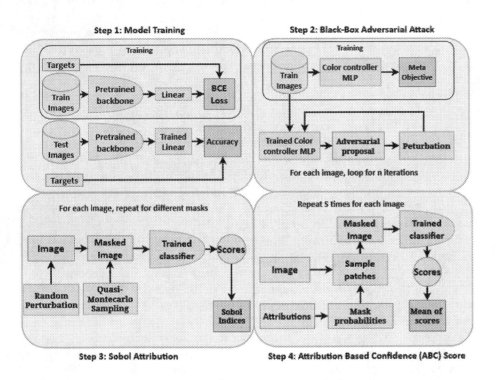

Fig. 2. Overview of the proposed approach for detecting adversarial images

Adversarial Images Generation is commonly done with Generative Adversarial Networks (GANs). We pose the problem of adversarial image generation as a meta-learning problem following the work by Yatsura et al. [24]. For the dataset $(x, y) \in \mathcal{D}$, classifier models $f \sim \mathcal{F}$, and the stochastic adversarial perturbation ϵ^ω the meta-objective is to determine parameters ω^* of the attack $\mathcal{A}_\omega$.

Determination of parameters of the attack is done through maximisation of the lower bound $L(f, x, y, \epsilon^\omega)$ of the goal function $V(f, x, y)$.

$$\omega^* = \underset{\omega}{\text{argmax}} \underset{f \sim \mathcal{F}}{\mathbb{E}} \underset{(x,y) \sim \mathcal{D}}{\mathbb{E}} \underset{\epsilon^\omega \sim \mathcal{A}_\omega(L,f,x,y)}{\mathbb{E}} L(f, x, y, \epsilon^\omega) \tag{1}$$

The meta-representation is defined such that A_ω effectively generalizes across models $f \sim \mathcal{F}$. For a random search-based attack where the query budget is determined by a limit T, an adversarial perturbation on the perturbation set S is defined with an iterative process:

$$\epsilon^0 \sim \mathcal{D}^0; \epsilon^0 t + 1 = \underset{\epsilon \in \{\epsilon^t, \mathcal{P}_s(\epsilon^t + \delta^{t+1})\}}{\text{argmax}} L(f, x, y, \epsilon); \delta^{t+1} \sim \mathcal{D}_\omega(t, \epsilon^0, \delta^0, ..., \epsilon^t, \delta^t)$$
$$\tag{2}$$

where $\mathcal{P}_S$ is a projection on the perturbation set S.

With an assumption that loss function l and A_ω are differentiable with respect to the meta-parameters ω, the meta-optimization for determining meta-parameters is done with stochastic gradient descent optimization on batches $B \subseteq D$ based on the gradient:

$$g = \nabla_\omega R(F, D, \omega) = \sum_{f_j \in F} \sum_{(x_i, y_i) \in B \subseteq D} \nabla_\omega L(f_j, x_i, y_i, \epsilon_{i,j}) \tag{3}$$

In order to avoid very high variance and issues with vanishing or exploding gradient, which can occur using Eq. (3), the greedy alternative is used instead:

$$g = \frac{1}{T} \sum_{f_j \in F} \sum_{(x_i, y_i)} \sum_{t}^{T-1} \nabla_\omega L(f_j, x_i, y_i, \Pi_s(\epsilon^t + \delta^{t+1})) \tag{4}$$

Details on solving this optimization can be found in the original paper [24]. This learning approach is then applied to Square Attack (SA) [2] with l_∞ threat and is called Meta Square Attack (MSA). MSA operates with computation of the square size in pixels with size controllers $s_t = \pi^s_{\omega_s} \in \{1, ..., s_{max}\}$ and sampling position $(p_x, p_y) \sim \pi^p(s) \in \{1, ..., s_{max} - s\}^2$ and sampling color with a color controller $c \sim \pi^c_{\omega_c} \in \{c_1, ..., c_m\}$. Position controller π^p is uniform distribution while color and size controllers are meta-learned multi-layer perceptron (MLP) networks with parameters ω_s and ω_c. Algorithm 1 describes generating adversarial images with Square Attacks where parameters are meta-learned.

Algorithm 1. Generate Adversarial Images

Input: Data distribution $\mathcal{D}$, a robust classifier f, number of epochs, SA budget, uniform distribution π^p, **Output:** Set of generated images $\mathcal{D}_g$

1: **for** number of epochs **do**
2:      $\pi^s_{\omega_s} \leftarrow trainMLP(D, SA(budget))$ (update size controller)
3:      $\pi^c_{\omega_c} \leftarrow trainMLP(D, SA(budget))$ (update color controller)
4: **end for**
5: **for** number of attacks **do** $\mathcal{D}_g \leftarrow \mathcal{D}_g \cup SA(\pi^p, \pi^s_{\omega_s}, \pi^c_{\omega_c})$
6: **end for**

Sobol Attribution Method aims to describe the decision of a black-box system $f : \mathcal{X} \to \mathcal{R}^k$ based on the given input image described with a collection of features $x = (x_1, ..., x_n)$. The Sobol attribution-based method exploits the random perturbations approach to determine complex interactions among the features and how they contribute to the outcome of $f(x)$. These random perturbations are defined with a probability space $\Omega, \mathcal{X}, P$ of possible input perturbations and a random vector $X = (X_1, ..., Xn)$ on the data manifold around the input vector of features x. With the set of perturbations it is possible to decompose the variance model as $\mathrm{Var}(f(x)) = \mathrm{Var}(f_u(X_u))$ where $\mathcal{U} = (1, ..., n)$, u is a subset of $\mathcal{U}$ and f_u are partial contributions of variables $X_u = (X_i)_{i \in u}$. Sobol indices are defined with the sensitivity index :

$$\mathcal{S}_u = \frac{\mathrm{Var}(f_u(X)_u)}{\mathrm{Var}(f(X))} \tag{5}$$

In this way they quantify the importance of any subset of features for the decision of the system. For their values holds $\sum_{u \in \mathcal{U}} \mathcal{S}_u = 1$. The total Sobol indices can be defined as:

$$\mathcal{S}_{T_i} = \sum_{u \subset \mathcal{U}, i \in u} \mathcal{S}_u \tag{6}$$

The total Sobol index $\mathcal{S}_{T_i}$ defines how the variable X_i affects the model output variance and the interactions of any order of X_i with any other input variables. These values define each feature's intrinsic and relational impact on the model output. A low total Sobol index implies low importance for explaining the model decision. A feature weakly interacts with other features when values of the Sobol index and total are similar, while a high difference represents a strong interaction. This property of Sobol indices enables us to make the hypothesis of adversarial input detection. For estimation of Sobol indices, we use the Jensen estimator [10], which is considered an advanced one from the computation perspective. This estimator is combined with Quasi-Monte Carlo (QMC) strategy [7]. The following algorithm describes the procedure for calculating the Total Sobol index.

Algorithm 2. Total Order Estimator (Pythonic implementation)

Input: Prediction scores Y, dimension $d = 8 \times 8$, number of designs N
Output: Total Sobol Index S_{T_i}

1: $f(A) = Y[1 : N], f(B) = Y[N : N * 2]$ (perturbed inputs)
2: **for** i=1 to d **do**
3:     $f(C) = Y[N * 2 + N * i : N * 2 + N * (i + 1)]$
4: **end for**
5: $f_0 = \dfrac{1}{N} \sum_{j=0}^{N} f(A_j)$
6: $\hat{V} = \dfrac{1}{N - 1} \sum_{j=0}^{N} (f(A_j) - f_0)^2$
7: $S_{T_i} = \dfrac{\dfrac{1}{2N} \sum_{j=0}^{N} (f(A_j) - f(C_j^{(i)}))^2}{\hat{V}}$

Attribution Based Confidence (ABC) Metric is computed by importance sampling in the neighbourhood of a high-dimensional input using relative feature attributions. ABC metric constructs a generator that can sample the neighborhood of an input and observe the conformance of the model. The method does not require access to training data or any additional calibration. The concentration of features characterizes DL models. This implies that few features have high attributions for any output. The assumption is that sampling over low features will result in no change in the output. Low attribution provides information that the model is *equivariant* along the features. For an input x, a classifier model f, we compute attribution of the feature x_j of x as $A_j(x)$. The ABC metric is calculated then in two steps. *(i)* sampling the neighbourhood and *(ii)* measuring the conformance. Sampling is done by selecting the vector x_j with the probability of $P(x_j)$ and changing its value can result in a change in the model's decision. The procedure is repeated S times for the input image. The conformance is measured by observing which values of the output did not change when the attribute changed its value. Algorithm 3 describes computing ABC metric of a DNN model on an input.

Algorithm 3. Calculate ABC Metric

Input: a classifier f, input x, sample size S
Output: ABC metric $c(f, x)$

1: $A_1, ..., A_n \leftarrow$ Attributions of features $x_1, ..., x_n$ from x
2: $i \leftarrow f(x)$ (get classification output)
3: **for** j = 1 to n **do**
4:     $P(x_j) \leftarrow \dfrac{|A_j/x_j|}{\sum_{k=1}^{n} |A_k/x_k|}$
5: **end for**
6: Generate S samples from mutation of x_j with probability $P(x_j)$
7: Get classification output for S samples
8: $c(f, x) \leftarrow S_{conform}/S$

---

4   Experimental Results

4.1   Implementation Details

All experiments were conducted on Google Colab Pro+ with NVIDIA T4 Tensor Core GPU and 52 GB RAM.

Dataset. We have selected 1148 microscopic images from the Breast Cancer Histopathological Image Classification (BreakHis), the dataset composed of breast tumor tissue images collected from 82 patients using a 400x magnifying factor. The dataset contains two types of tumors: benign, relatively harmful, and malignant, a synonym for cancer. therefore, the dataset is associated with the image classification task into two classes. The dataset contains several types of benign tumors: tubular adenona (TA), fibroadenoma (F), adenosis (A), and phyllodes tumor (PT); and several types of malignant tumors: carcinoma (DC), lobular carcinoma (LC), mucinous carcinoma (MC) and papillary carcinoma (PC). We split the data from the original dataset into two sets, train and test, which correspond to 80% and 20% of the data, respectively. This split has been done in a stratified way, that is, keeping the same proportions between classes in train and test.

Classifiers. We used eight pre-trained architectures for the image classification task (ResNet18, ResNet50, Inception V3, MobileNet V3, ShuffleNet, Swin Transformer, Vision Transformer, WideResnet). For breast cancer prediction, we kept the original weights of the backbone network (in some cases convolutional and in others a vision transformer) and trained a linear layer on top of them. Each model was trained on the training dataset for 500 epochs using a constant learning rate.

Adversarial Attacks. The Meta-RS algorithm[3], described in Sect. 3, was used to attack each of the eight pre-trained models. For each single classification model, controllers were meta-trained with white-box access. Advertorch package was used for adversarial training using $l_\infty - LPG$ attack with $\epsilon = 8/255$, fixed step size of 0.01, and 20 steps. MLP architectures for size and color controllers have 2 hidden layers, 10 neurons each, and ReLU activation. All correctly predicted samples were under attack during the testing. All correctly classified samples have been modified for 1000 iterations for each case.

Sobol Attribution Method.[4], described in Sect. 3, was used on the images with masks generated with resolution $d = 8 \times 8$. The same resolution was used occlusion to sign the $\hat{S}_{T_i}$. Zero was used for the baseline. The number of designs was set to $N = 32$. As perturbation function was used *Inpainting*

ABC Metric.[5] parameter of the sample size was set to $S = 1000$.

---

[3] https://github.com/boschresearch/meta-rs.
[4] https://github.com/fel-thomas/Sobol-Attribution-Method.
[5] https://github.com/ma3oun/abc_metric.

4.2   Classification Accuracy

For trained eight classification models, we examined the classification accuracy and loss as shown in Fig. 3 during the training. The overview of the accuracy results with the test dataset for all eight models for breast cancer classification is given in Fig. 4 and Table 1. The overall highest prediction scores were achieved with Swin Transformer Network (accuracy is 0.935), and the lowest prediction score was achieved with the Inception V3 classification model (accuracy is 0.848).

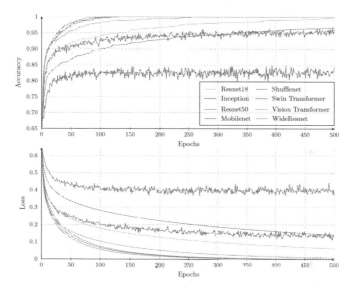

Fig. 3. Train accuracy and loss for 8 models

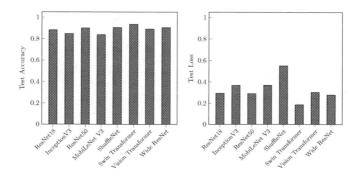

Fig. 4. Test accuracy and loss for 8 models after training for 500 epochs

4.3   Adversarial Images Generation

We performed classification tests to evaluate the success of adversarial attacks done with the Meta-RS algorithm for previously correctly classified samples. The attack results for the training dataset and for the test dataset are in Table 1. The attack accuracy is the fraction of the total number of images initially correctly classified by the model and shifted to a different class during the attack. Therefore, attack accuracy is calculated only for the previously correctly predicted labels. It is evident that adversarial attacks significantly affect prediction. The most significant decrease in accuracy is evident for the network ResNet50. An example of a successful adversarial attack is shown in Fig. 5.

Original   Adversarial

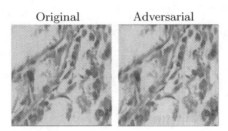

Fig. 5. An example of an image a correctly classified with the ResNet50 classifier and the corresponding successful adversarial image. High-frequency patterns and some square-shaped areas, typical patterns of the Meta-RS algorithm, can be noticed.

Table 1. Classification accuracy for the train and test datasets subjected to state-of-the-art Meta-RS black-box adversarial attacks.

	Model	Train accuracy	Total attacks	Successful attacks	Attack accuracy
Train	InceptionV3	0.917	841	144	0.171
	ResNet18	0.968	888	270	0.304
	ResNet50	0.999	915	565	0.617
	MobileNet V3	0.939	861	280	0.325
	ShuffleNet	0.995	913	273	0.299
	Swin Transformer	0.985	904	218	0.241
	Vision Transformer	0.999	917	236	0.257
	Wide ResNet	0.992	908	278	0.306
Test	InceptionV3	0.848	195	51	0.261
	ResNet18	0.882	202	56	0.276
	ResNet50	0.900	208	49	0.236
	MobileNet V3	0.839	193	54	0.280
	ShuffleNet	0.904	208	56	0.269
	Swin Transformer	0.935	215	49	0.228
	Vision Transformer	0.891	205	40	0.195
	Wide ResNet	0.904	208	58	0.279

4.4   Detection of Adversarial Images

Explanations generated with Sobol Attribution Method represent a visual aid for a user to understand which regions of the images affected the decision-making. Examples of explanations are shown in Fig. 7 and Fig. 6 for successful attacks in the case of ResNet50 and corresponding original images. Some regions in the adversarial image are more smooth than the original, reducing the overall information and enhancing other parts' importance. It can be noticed from Sobel's attributions that prediction models do not use information from the more smooth areas and focus mainly on the parts of the images where dark spots can be found. The complexity of pathological images brings new challenges. It creates several highlighted regions within the image, rather than a few, as in the original application of the Sobol Attribution Method [5]. Hence we have developed attribution-based confidence (ABC) metric.

Table 2 provides ABC metric values for the original and adversarial images for all eight classification models. The computation of the ABC metric of a classification model on an input requires accurately determining conformance by sampling $S = 1000$ samples in the neighbourhood of high-dimensional inputs. The value of ABC metric in Table 2 is the mean value. By observing the values of ABC scores, we can draw a threshold of 0.8 for deciding whether the model was subjected to an adversarial attack or not. Figure 8 illustrates how the ABC metric reflects the decrease in confidence under adversarial attack for all eight classification models. Figure 9 provides examples of the final output.

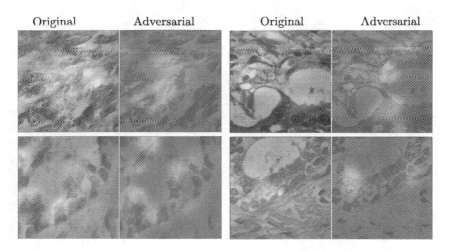

Fig. 6. Sample Sobol attribution explanations for ResNet50. Four examples of original images and their corresponding adversarial examples where attacks were successful. Sobol Attribution Method explanations are displayed on top of images highlighting the crucial regions of the image for classifier decision-making.

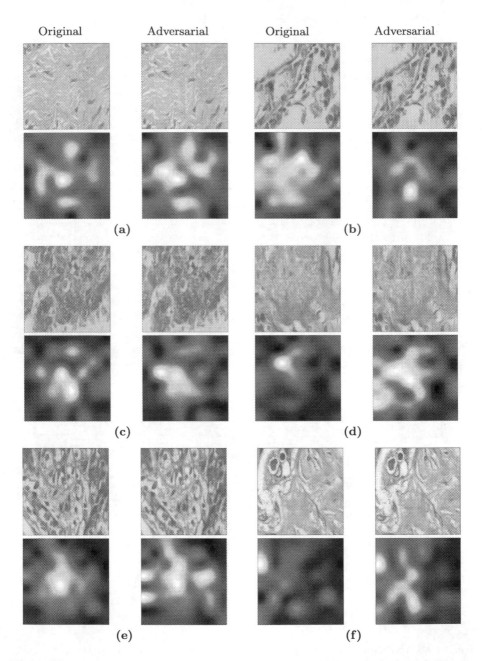

Fig. 7. Six examples of original and corresponding ResNet50 adversarial images together with sample explanations obtained with Sobol Attribution Method highlighting the importance of image regions. Several regions are being highlighted with the Sobol attribution method. This brings more challenges in deciding whether an image is original or adversarial compared to the original problem addressed by Fel et al. [5]

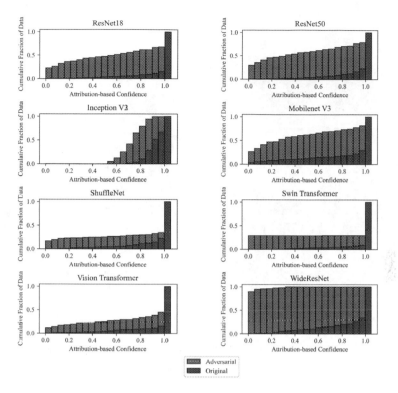

Fig. 8. Cumulative data fraction vs. ABC metric compares the original and adversarial datasets with respect to ABC metric values for different classification models. ABC metric values are high for a great fraction of data for all eight models, while for adversarial, the values are low. The most distinctive values can be recognized with the WideResNet classification model.

Table 2. ABC metric values for the eight models for test and train datasets

Model	Attribution-based Confidence (ABC)			
	Train dataset		Test dataset	
	Original	Adversarial	Original	Adversarial
ResNet18	0.920	0.539	0.948	0.518
ResNet50	0.928	0.340	0.908	0.323
Inception V3	0.847	0.734	0.893	0.698
MobileNet	0.876	0.357	0.861	0.389
ShuffleNet	0.934	0.742	0.930	0.732
Swin transformer	0.971	0.702	0.969	0.710
Vision transformer	0.947	0.731	0.945	0.722
Wide ResNet	0.893	0.034	0.874	0.013

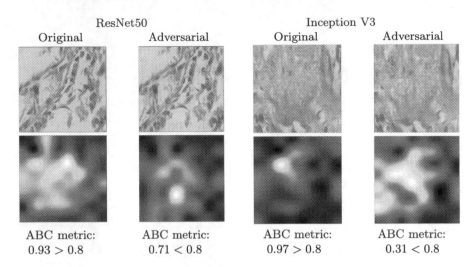

Fig. 9. ABC values are used to differentiate between original and adversarial images.

5 Conclusions

Our prediction accuracy tests on eight transfer learning-based models confirm that deep learning models have become powerful in predicting breast cancer from histopathological images. However, applying deep learning models in the medical field brings new risks and concerns, such as the possibility of adversarial attacks.

We subjected classification models to state-of-the-art robust Meta-RS attacks. The obtained adversarial images are available for public use[6] Sobol Attribution Method [5] was applied to understand those parts of the images that determine the output of a classification model. However, due to the nature of histopathological images and the specificity of the classification problem, several regions are being highlighted with the Sobol attribution method. This brings more challenges in deciding whether an image is original or adversarial compared to the original problem addressed by Fel et al. [5]. Hence we developed attribution-based confidence(ABC) metric for detecting adversarial attacks on breast histopathological images (examples in Fig. 9).

To the best of our knowledge, this is the first work to detect attacks on medical systems for breast cancer prediction based on histopathological images using the ABC metric. The evaluation of eight different classification models shows that the ABC metric for all original images is greater or equal to 0.8 and less than 0.8 for adversarial images.

In the future, the ABC metric would be used to detect adversarial attacks on histopathological oral cancer detection systems[7].

---

[6] The new dataset is available at https://bit.ly/3p4QaPw.

[7] https://www.kaggle.com/datasets/ashenafifasilkebede/dataset.

References

1. Multiple Instance Captioning: Learning Representations from Histopathology Textbooks and Articles (2021)
2. Andriushchenko, M., Croce, F., Flammarion, N., Hein, M.: Square attack: a query-efficient black-box adversarial attack via random search. In: Vedaldi, A., Bischof, H., Brox, T., Frahm, J.-M. (eds.) ECCV 2020. LNCS, vol. 12368, pp. 484–501. Springer, Cham (2020). https://doi.org/10.1007/978-3-030-58592-1_29
3. Bortsova, G., et al.: Adversarial attack vulnerability of medical image analysis systems: unexplored factors. Med. Image Anal. **73**, 102141 (2021)
4. Das, A., Devarampati, V.K., Nair, M.S.: NAS-SGAN: a semi-supervised generative adversarial network model for atypia scoring of breast cancer histopathological images. IEEE J. Biomed. Health Inform. **26**(5), 2276–2287 (2022)
5. Fel, T., Cadène, R., Chalvidal, M., Cord, M., Vigouroux, D., Serre, T.: Look at the variance! efficient black-box explanations with sobol-based sensitivity analysis. In: Ranzato, M., Beygelzimer, A., Dauphin, Y.N., Liang, P., Vaughan, J.W. (eds.) Advances in Neural Information Processing Systems 34: Annual Conference on Neural Information Processing Systems 2021, NeurIPS 2021, 6–14 December 2021, virtual, pp. 26005–26014 (2021)
6. Foote, A., Asif, A., Azam, A., Marshall-Cox, T., Rajpoot, N.M., Minhas, F.A.: Now you see it, now you don't: Adversarial vulnerabilities in computational pathology. CoRR abs/2106.08153 (2021)
7. Gerber, M.: On integration methods based on scrambled nets of arbitrary size. J. Complex. **31**(6), 798–816 (2015)
8. Hashimoto, N., et al.: Multi-scale domain-adversarial multiple-instance CNN for cancer subtype classification with unannotated histopathological images. In: 2020 IEEE/CVF Conference on Computer Vision and Pattern Recognition, CVPR 2020, Seattle, USA (2020)
9. Hou, L., Agarwal, A., Samaras, D., Kurç, T.M., Gupta, R.R., Saltz, J.H.: Robust histopathology image analysis: to label or to synthesize? In: IEEE Conference on Computer Vision and Pattern Recognition, CVPR 2019, Long Beach, CA, USA, 16–20 June 2019, pp. 8533–8542. Computer Vision Foundation IEEE (2019)
10. Jansen, M.J.: Analysis of variance designs for model output. Comput. Phys. Commun. **117**, 35–43 (1999)
11. Jha, S., et al.: Attribution-based confidence metric for deep neural networks. In: Advances in Neural Information Processing Systems 32: Annual Conference on Neural Information Processing Systems 2019, NeurIPS 2019, 8–14 December 2019, Vancouver, BC, Canada, pp. 11826–11837 (2019)
12. Laleh, N.G., et al.: Adversarial attacks and adversarial robustness in computational pathology (2022)
13. Liu, M., et al.: A deep learning method for breast cancer classification in the pathology images. IEEE J. Biomed. Health Inform. **26**, 1–8 (2022)
14. Ma, X., et al.: Understanding adversarial attacks on deep learning based medical image analysis systems. Pattern Recogn. **110**, 107332 (2021)
15. Marini, N., Atzori, M., Otálora, S., Marchand-Maillet, S., Müller, H.: H&e-adversarial network: a convolutional neural network to learn stain-invariant features through hematoxylin & eosin regression. In: IEEE/CVF International Conference on Computer Vision Workshops, ICCVW 2021, Montreal, BC, Canada, 11–17 October 2021, pp. 601–610. IEEE (2021)

16. Mercan, C., et al.: Deep feature representations for variable-sized regions of interest in breast histopathology. IEEE J. Biomed. Health Inform. **25**, 2041–2049 (2021)

17. Paschali, M., Conjeti, S., Navarro, F., Navab, N.: Generalizability vs. robustness: Adversarial examples for medical imaging. CoRR abs/1804.00504 (2018)

18. Senousy, Z., et al.: MCUA: multi-level context and uncertainty aware dynamic deep ensemble for breast cancer histology image classification. IEEE Trans. Biomed. Eng. **69**(2), 818–829 (2022)

19. Spanhol, F.A., Oliveira, L.S., Petitjean, C., Heutte, L.: A dataset for breast cancer histopathological image classification. IEEE Trans. Biomed. Eng. **63**(7), 1455–1462 (2016)

20. Thiagarajan, P., Khairnar, P., Ghosh, S.: Explanation and use of uncertainty quantified by Bayesian neural network classifiers for breast histopathology images. IEEE Trans. Med. Imaging **41**(4), 815–825 (2022)

21. Wei, J.W., et al.: Learn like a pathologist: Curriculum learning by annotator agreement for histopathology image classification. In: IEEE Winter Conference on Applications of Computer Vision, WACV 2021, Waikoloa, HI, USA, 3–8 January 2021

22. Xu, G., et al.: Camel: A weakly supervised learning framework for histopathology image segmentation (2019)

23. Yang, H., Kim, J.Y., Kim, H., Adhikari, S.P.: Guided soft attention network for classification of breast cancer histopathology images. IEEE Trans. Med. Imaging **39**(5), 1306–1315 (2020)

24. Yatsura, M., Metzen, J., Hein, M.: Meta-learning the search distribution of black-box random search based adversarial attacks. In: Ranzato, M., Beygelzimer, A., Dauphin, Y., Liang, P., Vaughan, J.W. (eds.) Advances in Neural Information Processing Systems, vol. 34, pp. 30181–30195. Curran Associates, Inc. (2021)

25. Zhang, J., et al.: A joint spatial and magnification based attention framework for large scale histopathology classification. In: IEEE Conference on Computer Vision and Pattern Recognition Workshops, CVPR Workshops 2021, virtual, 19–25 June 2021, pp. 3776–3784. Computer Vision Foundation/IEEE (2021)

26. Zhao, Y., et al.: Predicting lymph node metastasis using histopathological images based on multiple instance learning with deep graph convolution. In: 2020 IEEE/CVF Conference on Computer Vision and Pattern Recognition, CVPR 2020, pp. 4836–4845 (2020)

Improving Adversarial Robustness by Penalizing Natural Accuracy

Kshitij Chandna$^{(\boxtimes)}$ (iD)

New York University, New York, NY 10012, USA
kc4156@nyu.edu

Abstract. Current techniques in deep learning are still unable to train *adversarially robust* classifiers which perform as well as non-robust ones. In this work, we continue to study the space of *loss functions*, and show that the choice of loss can affect robustness in highly nonintuitive ways.

Specifically, we demonstrate that a surprising choice of loss function can in fact improve adversarial robustness against some attacks. Our loss function encourages accuracy on adversarial examples, and explicitly *penalizes* accuracy on natural examples. This is inspired by the theoretical and empirical works suggesting a fundamental tradeoff between standard accuracy and adversarial robustness. Our method, NAturally Penalized (NAP) loss, achieves 61.5% robust accuracy on CIFAR-10 with $\varepsilon = 8/255$ perturbations in ℓ_∞ (against a PGD-60 adversary with 20 random restarts). This improves over the standard PGD defense by over 3%, against other loss functions proposed in the literature. Although TRADES performs better on CIFAR-10 against Auto-Attack, our approach gets better results on CIFAR 100. Our results thus suggest that significant robustness gains are possible by revisiting training techniques, even without additional data.

Keywords: Robusteness · Adversarial · Naturally penalized · PGD · Cifar-10 · Cifar-100 · Loss function

1 Introduction

Modern deep learning is now mature enough to achieve high test accuracy on many image classification tasks [17,23–25,35]. Here, a long line of research has arrived at a certain combination of techniques which work well for image recognition, including architectures (ReLUs, Convolutions, ResNet), optimization algorithm (SGD and variants, with tuned learning-rate schedules), model size, data-augmentation, regularization, normalization, batch-size, and loss function [9,17,25,49].

Many of these choices are not unique and we do not have a complete understanding of why these choices are working best in practice. For example, for standard classification our true objective is small 0/1 test loss, but we often optimize Cross Entropy train loss. We could instead optimize ℓ_2 train loss (or any other surrogate loss), but in practice we find optimizing Cross Entropy often performs better[1] Similarly, very large, deep networks perform much better than smaller ones in practice, even though these networks have more than enough capacity to "overfit" the train set, and should be performing worse by classical statistical intuition [3,30,47]. The optimizer is also poorly understood: In practice we use SGD with learning rates much higher than optimization theory prescribes; and moreover, "accelerated" methods which optimize faster – such as Adam [22] – sometimes generalize worse. Nevertheless, despite our incomplete theoretical understanding, the research community has converged on a methodology which performs very well for standard classification.

However, the field of *adversarially robust* classification is not as mature [4], and has not yet converged on a training methodology which performs well. The goal of adversarial robustness is to learn classifiers which are robust to small adversarial perturbations [34] of the input. Here, it is not clear if the various design choices that we converged to for standard classification are still the best choices for robust classification.

Indeed, current advances in adversarial robustness have come through modifying the training procedure [6,28], loss function [21,39,48], architecture [43,44], data generation [15] activation function [10,41], pre-training [19] and leveraging unlabeled data [5,38]. This research area is not nearly as mature as standard classification, and there are still potentially large robustness gains from rethinking elements of the deep learning methodology.

In this work, we focus on the choice of *loss function*, and show that an unconventional choice of loss function can in fact significantly improve adversarial robustness. Concretely, our loss function includes two terms: one which encourages accuracy on adversarial examples, and one which explicitly *penalizes* accuracy on natural examples. This is inspired by the empirical and theoretical observations that there may be a tradeoff between standard accuracy and adversarial accuracy for trained models [29,37]. Intuitively, our loss function penalizes standard accuracy if it is "too good to be true" – i.e., much higher than the adversarial accuracy. This attempts to forcibly trade-off standard accuracy for improved adversarial accuracy, and in practice it yields significant gains over existing methods.

The observation that choice of loss affects adversarial robustness is not novel to our work, and our loss function shares components of existing methods such as TRADES [48] and MART [39]. Many of these methods are motivated as "regularizers", which encourage the network on adversarial inputs to behave similarly to natural inputs. Our method is conceptually fundamentally different, in that *explicitly penalizes* the classifier's correct behavior on natural inputs. See Sect. 3 for further comparison and discussion with existing methods.

---

[1] Certain losses are theoretically justified for simpler models, for example as proper scoring rules or for margin maximizing reasons. But these justifications do not provably hold for overparameterized models such as modern deep networks.

Our Contribution. Our main contribution is demonstrating that the impact of loss function on robustness is both *large* and *under-explored*. We show that an "unnatural" loss function, which explicitly penalizes natural accuracy, can in fact improve state-of-the-art adversarial robustness: achieving 61.5% robust accuracy on CIFAR-10 with $\varepsilon = 8/255$ perturbations in ℓ_∞, when evaluated against a 60-step PGD attacker with 20 random restarts.

We view our work as showing that space of reasonable loss functions is perhaps larger than expected, and that large robustness gains can still be attained in this space. We also present preliminary insights into what properties of our loss cause it to performance well.

Organization. In Sect. 2 we present preliminaries, and discuss further related work in Sect. 1.1 We discuss our loss function and how it compares to prior work in Sect. 3. In Sect. 4 we describe the experimental details and results. In Sect. 5 we do ablations with modifications of our loss function to understand the importance of different components of the loss function. In Sect. 6 we discuss potential intuitions as to why the NAP loss works well compared to standard adversarial training.

1.1   Additional Related Work

Adversarial Robustness. Adversarial examples were first considered by [34]. Subsequently quite a few defenses towards these attacks were proposed and many of these were later broken [1]. The PGD-based adversarial training of [28] (related to $FGSM_k$ of [14]) remains moderately robust to currently attacks. In this work we consider *emperical/heuristic* robustness, as evaluated against strong heuristic attacks (PGD with random restarts). There is also a line of work on provably/certifiably robust neural networks, including randomized-smoothing and certification-based approaches [2,31,40,42]. Note that there is currently a non-trivial gap between provable accuracy and empirical accuracy, and some of these approaches (in particular randomized smoothing) do not apply to ℓ_∞ robustness. There has also been work [7,16,26,27,33] in *randomized smoothing* which refers to taking neural networks which are robust to random perturbations and using them to create classifiers which are adversarially robust.

Generative Models. Another approach to improve adversarial accuracy has been to use generative models [32] with the help of augmentation strategies [45] and TRADES [48].

Robustness. In this work we study adversarial robustness. This is distinct from other kinds of robustness which may be desirable, including out-of-distribution robustness, robustness to distribution shift, and robustness to non-adversarial corruptions [12,13,18].

Methods which improve adversarial robustness may not yield improvements in other types of robustness. In fact, there is evidence to the contrary [36].

2   Preliminaries and Notation

The goal of adversarial robustness is to learn classifiers which have high accuracy even under small worst-case perturbations. We consider ℓ_∞ robustness in this work. For a distribution $(x, y) \sim \mathcal{D}$ on images $x \in \mathbb{R}^d$ and labels $y \in [k]$, our objective is to learn a classifier $f : \mathbb{R}^d \to [k]$ with small *robust error*:

$$\text{RobustError}_{\mathcal{D},\varepsilon}(f) = \mathop{\mathbb{E}}_{(x,y)\sim\mathcal{D}} [\max_{\delta\in\ell_\infty(\varepsilon)} \mathbb{1}\{f(x + \delta) \neq y\}] \tag{1}$$

where $\ell_\infty(\varepsilon)$ is the ℓ_∞ ball of radius ε. We report robust accuracy as $(1 - \text{RobustError})$.

A crucial detail in adversarial training is the choice of loss function together with the choice of *which components to backprop through* in the loss. This detail of backprop choice can significantly impact robustness. To clarify this, throughout this paper we typeface in bold parameters in the loss function which are backpropagated through (e.g. $\boldsymbol{\theta}$ v.s. θ).

Existing adversarial training methods which modify the loss function, including our method, can all be seen as instances of the generic adversarial training of Algorithm 1, for different choices of loss. We specify loss functions as

$$\mathcal{L}(x, \hat{x}, y; \boldsymbol{\theta})$$

where (x, y) is the natural example and label, θ is the parameters of the network, and $\hat{x}$ is an adversarial example for input x with respect to network θ. Let $p(x; \theta)$ denote the softmax output of network parameterized by θ, and let $p_y(x; \theta)$ denote the softmax probability on label y. We denote Cross Entropy by CE and Kullback-Leibler divergence by KL.

Algorithm 1. Adversarial Training (Simplified)

Input: Neural network f_θ, Loss functions $\mathcal{L}^{\text{upd}}, \mathcal{L}^{\text{atk}}$.
    Output: Adversarially-trained network f_θ.
1: **function** ADVTRAIN($\mathcal{L}, f_\theta$):
2:    **for** $t = 1, 2, 3 \ldots, T$ **do**
3:        Sample example $(x, y) \sim S$
4:        Construct $\hat{x}$ as an adversarial example for x, by performing PGD on the loss $\mathcal{L}^{\text{atk}}(x, \hat{x}, y; \theta)$ starting from $\hat{x} = x$.
5:        $\theta \leftarrow \theta - \eta \nabla_\theta \mathcal{L}^{\text{upd}}(x, \hat{x}, y; \boldsymbol{\theta})$
6:    Output f_θ.

There are two losses involved in Algorithm 1: the loss $\mathcal{L}^{\text{upd}}$ used to update model parameters, and the loss $\mathcal{L}^{\text{atk}}$ used to construct adversarial examples for training. In all methods except TRADES, including our method, the attack loss is simply cross-entropy, $\mathcal{L}^{\text{atk}}(x, \hat{x}, y; \theta) := \text{CE}(p(\hat{x}; \theta), y)$. For TRADES, the attack loss is $\mathcal{L}^{\text{atk}}(x, \hat{x}, y; \theta) := \text{KL}(p(x; \theta) || p(\hat{x}; \theta))$.

Standard adversarial training [28] is Algorithm 1 with model-update loss

$$\mathcal{L}^{\text{upd}}(x, \hat{x}, y; \boldsymbol{\theta}) := \text{CE}(p(\hat{x}; \theta), y)$$

In this work, we introduce a new model update-loss for $\mathcal{L}^{\text{upd}}$, and compare with existing losses.

3   NAP Loss

Here we introduce and discuss our loss function the NAturally Penalized (NAP) loss, and compare it to other loss functions for adversarial robustness.

We start with the motivating observation that the following simple loss, which penalizes natural accuracy, can in fact outperform standard PGD and TRADES:

$$\mathcal{L}(x, \hat{x}, y; \boldsymbol{\theta}) := \text{CE}(p(\hat{x}; \theta), y) - \lambda \text{CE}(p(x, \theta), y)$$

Starting with this loss, we make several modifications to yield further improvements: First, we only penalize natural accuracy on examples where it is higher than adversarial accuracy. And second, we use a "smoothed" version of cross entropy on adversarial examples. See Sect. 5 and Table 5 for further ablations on our loss; the above loss is described as "NAP (no margin, no BCE)".

NAP Loss. We propose the following loss:

$$\mathcal{L}^{\text{NAP}}(x, \hat{x}, y; \boldsymbol{\theta}) := \\ \underbrace{\text{BCE}(p(\hat{x}; \boldsymbol{\theta}), y)}_{\text{(A): Adversarial Loss}} - \lambda \underbrace{[p_y(x; \theta) - p_y(\hat{x}; \theta)]_+}_{\text{(B): Overconfidence Margin}} \underbrace{\text{CE}(p(x; \boldsymbol{\theta}), y)}_{\text{(C): Natural Loss}}$$

where $\lambda > 0$ and $[\cdot]_+$ denotes $\text{ReLU}(\cdot)$. BCE is the Boosted Cross Entropy introduced in MART [39].

The first term (A) encourages adversarial accuracy. The second term (B, C) penalizes natural accuracy, on examples where the network is "overconfident", and has higher natural than adversarial accuracy. We do not want to penalize natural accuracy universally[2], but only in cases when it is "too good to be true" – this is accounted for by the weighting term (B). Note that term (B) is treated as a constant i.e. not backpropagated through.

---

[2] In Fig. 3 and Table 5 we penalize natural accuracy universally and find that even that works better than TRADES and MART.

The Boosted Cross Entropy is defined as

$$\mathrm{BCE}(p(\hat{x};\boldsymbol{\theta}),y) := \mathrm{CE}(p(\hat{x};\boldsymbol{\theta}),y) - \log(1 - \max_{k\neq y} p_k(\hat{x};\boldsymbol{\theta}))$$

$$= -\log(p_y(\hat{x};\boldsymbol{\theta})) - \log(1 - \max_{k\neq y} p_k(\hat{x};\boldsymbol{\theta}))$$

The BCE is the standard cross entropy, plus a term that encourages the network's softmax output to be "balanced" on incorrect classes.

In Sect. 5 and Table 5, we do several ablations on our loss function, and we find that:

1. The choice of Boosted Cross Entropy helps robustness but is not crucial: our method still outperforms TRADES with standard CE in place of BCE.
2. The "overconfidence margin" term (B) is not crucial: Even universally penalizing the natural loss yields improvements over TRADES and MART.

This suggests that penalizing natural loss is the crucial component of NAP's performance.

3.1   Comparison to Other Losses

Standard Loss.

$$\mathcal{L}^{\mathrm{standard}}(x,\hat{x},y;\boldsymbol{\theta}) := \mathrm{CE}(p(\hat{x};\boldsymbol{\theta}),y)$$

This is the Cross-Entropy loss on adversarial examples, used in standard adversarial training [28].

TRADES Loss.

$$\mathcal{L}^{\mathrm{TRADES}}(x,\hat{x},y;\boldsymbol{\theta}) := \mathrm{CE}(p(x;\boldsymbol{\theta}),y) + \lambda \mathrm{KL}(p(x;\boldsymbol{\theta})||p(\hat{x};\boldsymbol{\theta}))$$

In TRADES [48], the base term optimizes for *natural accuracy*, while the KL regularizer encourages the natural and adversarial softmax outputs to be close.

In comparison, our method encourages *adversarial accuracy* in the base term, and uses a different regularizer. Our regularizer is in some sense a "softer constraint" than the KL term: we do not penalize differences in $p(x;\theta)$ and $p(\hat{x};\theta)$, but only the ones in $p_y(x;\theta)$ and $p_y(\hat{x};\theta)$ i.e. those which affect the final classification decision. One other difference is that when $p_y(x;\theta) > p_y(\hat{x};\theta)$ the regularizer term in TRADES tries to decrease $p_y(x;\theta)$ and to increase $p_y(\hat{x};\theta)$ while NAP regularizer term only tries to decrease $p_y(x;\theta)$.

MART Loss.

$$\mathcal{L}^{\text{MART}}(x, \hat{x}, y; \boldsymbol{\theta}) := \text{BCE}(p(\hat{x}; \boldsymbol{\theta}), y) + \lambda(1 - p_y(x; \boldsymbol{\theta}))\text{KL}(p(x; \boldsymbol{\theta})||p(\hat{x}; \boldsymbol{\theta}))$$

In MART [39], the regularizer is weighted by the *natural misclassification probability*, instead of the overconfidence margin. Moreover, the KL regularizer term is the same as TRADES, and is distinct from our naturally-penalized regularizer as discussed previously.

Gradient Comparisons. In addition to the aforementioned differences, our NAP loss function has the following property which separates it from TRADES and MART.

For the NAP loss, the partial derivative of the loss with respect to the natural softmax probability is always *non-positive*:

$$\frac{\partial \mathcal{L}}{\partial p_y(x; \theta)} \leq 0$$

That is, the NAP loss always penalizes natural accuracy, all else being equal.

And the partial derivative with respect the *adversarial* softmax probability is always *positive*:

$$\frac{\partial \mathcal{L}}{\partial p_y(\hat{x}; \theta)} > 0$$

That is, the NAP loss always encourages adversarial accuracy, all else being equal.

Note that accounting for the total derivative, the natural accuracy can increase after a gradient step, even though the partial derivative is negative.

4   Experiments

4.1   Experimental Details

CIFAR-10. For CIFAR-10 [24] dataset we normalize all images to be in the range $[0, 1]$. We use the standard data augmentation: random horizontal flip and 4 pixel padding with 32×32 random crop. The ℓ_∞ budget for adversarial examples is $\varepsilon = 8/255$. For training adversarial examples are created with random-start and PGD-10 with step-size of $2/255$. While doing PGD to create adversarial examples we also use the batch-mean/variance for batch normalization layers rather than the moving mean/variance. In our experiments with the loss functions we tried this change led to a small but consistent improvement in test accuracy.

For testing with PGD-20 we use step-size of $0.8/255$. For evaluation with stronger attacks, we follow [5] for PGD-60: 60 PGD steps with step size $\eta = 0.01$, using 20 random restarts per example, and searching for adversarial attack success at each PGD step.

We use Wide ResNet 34-4 [46] to compare different loss functions and Wide ResNet 34-10 to establish the state-of-the-art model (to compare with [39,48]). All models were trained for 80 epochs with the initial learning rate of 0.1, a step decay of 0.1 on epoch 40 and epoch 60, batch size of 128, weight decay of 5×10^{-4} and SGD with momentum 0.9. All reported accuracies are with optimal early stopping with respect to adversarial accuracy for PGD-20. Then we tested each model against Auto-Attacks.

For all methods involving a hyperparameter λ (TRADES, MART, and NAP), we individually hyperparameter search to set λ. See Sect. A for hyperparameter search results.

CIFAR-100. We used the information about hyperparameters retrieved from CIFAR-10 experiments to train the model on CIFAR-100. In both these datasets, hyperparameters for optimal test accuracy varied based on attacks.

4.2   Results

First, we compare our NAP loss against existing losses on a Wide ResNet 34-4. In Table 1, the NAP loss outperforms standard PGD, TRADES, and MART.

Table 1. Natural and adversarial accuracies for adversarially-trained Wide ResNet 34-4 on CIFAR-10 with $\varepsilon = 8/255$ perturbations in ℓ_∞.

Loss	Natural	PGD-20	PGD-60
Standard	**82.2**	53.6	52.1
TRADES	81.8	55.3	53.9
MART	79.9	59.0	57.5
NAP	78.9	**62.6**	**60.5**

The value of λ for all the loss functions was determined by a hyperparameter searches (Fig. 1). The values obtained were $\lambda = 8, 8, 20$ for TRADES, MART and NAP respectively.

To compare against current state-of-the-art models in the literature, we train the NAP loss for a Wide ResNet 34-10 (as in [39,48]). This model trained with NAP loss improves on the state-of-the-art loss functions of 58.6% against the PGD-20 attack reported by MART [39]. Our results thus suggest that significant robustness gains are possible by modifying elements of the training procedure, even without additional data (Table 2).

Table 2. Natural and adversarial accuracies for adversarially-trained Wide ResNet 34-10 on CIFAR-10 with $\varepsilon = 8/255$ perturbations in ℓ_∞.

Loss	Natural	PGD-20	PGD-60
Standard	**84.5**	55.5	54.0
NAP	79.8	**63.2**	**61.5**

We used $\lambda = 22$ for the NAP loss in Table 4. This value was determined by a hyperparameter search, see Fig. 2.

We continued to evaluate NAP against other novel adversarial attacks that challenge Standard and TRADES loss functions.

Table 3. Auto Attack accuracies for adversarially-trained Wide ResNet 34-10 on CIFAR-10 with $\varepsilon = 8/255$ perturbations in ℓ_∞.

Loss	$APGD_{CE}$	$APGD^T_{DLR}$	FAB_T	Square
Standard	44.75	44.28	44.28	53.10
TRADES	55.28	**53.10**	**53.45**	**59.43**
NAP	**58.42**	50.41	50.41	57.08

As Table 3 shows, NAP performs badly when compared to TRADES against targeted attacks [8] when on CIFAR-10 dataset with λ to 6, however NAP performs better on the CIFAR-100 dataset with the same λ.

For the CIFAR-100 dataset we also add another hyperparameter λ_2 by redefining Boosted cross entropy as

$$\text{BCE}(p(\hat{x}; \boldsymbol{\theta}), y) := \text{CE}(p(\hat{x}; \boldsymbol{\theta}), y) - \lambda_2 \log(1 - \max_{k \neq y} p_k(\hat{x}; \boldsymbol{\theta})).$$

We set $\lambda_2 = 2$

Table 4. Auto Attack accuracies for adversarially-trained Wide ResNet 34-10 on CIFAR-100 with $\varepsilon = 8/255$ perturbations in ℓ_∞.

Loss	AA
TRADES	25.94
NAP	**27.13**

5   Ablations

Here we perform several ablations to study the relative impact of components in the NAP loss. All experiments in this section are on CIFAR-10 with Wide ResNet 34-4. For all modified losses, we conduct individual hyperparameter search over λ to find optimal parameters (see Fig. 3).

Table 5. Ablations on our loss function

Loss	Natural	PGD-20
NAP (no margin, no BCE)	79.1	57.3
NAP (no margin)	77.3	59.5
NAP (no BCE)	**79.5**	58.1
NAP	78.9	**62.6**

NAP Loss (no BCE). We replace the Boosted Cross Entropy (BCE) term in NAP with standard Cross Entropy, and find that it still outperforms TRADES but not MART (Table 5).

The modified loss function is:

$$\mathcal{L}^{\text{NAP (no BCE)}}(x, \hat{x}, y; \boldsymbol{\theta}) := \text{CE}(p(\hat{x}; \boldsymbol{\theta}), y) - \lambda[p_y(x; \theta) - p_y(\hat{x}; \theta)]_+ \, \text{CE}(p(x, \boldsymbol{\theta}), y)$$

For the result in Table 5 λ is set to be 6.

NAP Loss (no margin) We replace the "overconfidence margin" term in NAP with a constant – that is, we always penalize the natural examples. We find that it still outperforms TRADES and MART (Table 5). The modified loss function is:

$$\mathcal{L}^{\text{NAP (no margin)}}(x, \hat{x}, y; \boldsymbol{\theta}) := \text{BCE}(p(\hat{x}; \boldsymbol{\theta}), y) - \lambda \text{CE}(p(x, \boldsymbol{\theta}), y)$$

For the result in Table 5 λ is set to be .3.

NAP Loss (no margin, no BCE). Here we combine both the above changes i.e. we replace the Boosted Cross Entropy (BCE) term in NAP with standard Cross Entropy and we replace the "overconfidence margin" term in NAP with a constant. We find that it still outperforms TRADES (Table 5) but not MART (which uses BCE). The modified loss function is:

$$\mathcal{L}^{\text{NAP (no BCE, no margin)}}(x, \hat{x}, y; \boldsymbol{\theta}) := \text{CE}(p(\hat{x}; \boldsymbol{\theta}), y) - \lambda\text{CE}(p(x, \boldsymbol{\theta}), y)$$

For the result in Table 5 λ is set to be 0.3.

5.1   Boosted Cross-Entropy

From Table 5 we know that replacing BCE with CE has a significant (4.5%) negative effect. Given this, we now consider whether BCE loss performs better than CE loss in other settings as well. We first compare the BCE loss and CE loss for adversarial training. That us, we use model update loss

$$\mathcal{L}^{\text{BCE}}(x, \hat{x}, y; \boldsymbol{\theta}) := \text{BCE}(p(\hat{x}; \boldsymbol{\theta}), y)$$

The results for adversarial training with this loss are in Table 6.

Table 6. Natural and adversarial accuracies for adversarially-trained Wide ResNet 34-4 on CIFAR-10 with $\varepsilon = 8/255$ perturbations in ℓ_∞. Average of 2 runs.

Loss	Natural	PGD-20
Standard	**82.2**	53.55
BCE	80.4	**54.7**

Next we compare the BCE loss and CE loss for natural training, and list results in Table 7. Here we train on only the first 10K images of CIFAR-10, to approximately match the standard accuracy with that of robust models.

Table 7. Average natural accuracies of 4 runs for naturally-trained ResNet 34 on first 10K images from CIFAR-10.

Loss	Natural accuracy
CE	72.25
BCE	**73.1**

In both the cases, BCE helps but the difference is around 1% while the difference when replacing BCE with CE in NAP loss was 4.5%. Thus, BCE appears to be much more important when combined with the regularizer term—a similar observation was also made for the MART loss in [39].

6   Discussion

In this section we discuss potential intuitions and preliminary experiments towards understanding why our NAP loss function performs well. We stress that these are informal intuitions, and we do not have a formal understanding of the impact of loss functions on robustness.

Intuitions for Penalizing Natural Accuracy. Our NAP loss is inspired by the empirical and theoretical observations that there may be a tradeoff between standard accuracy and adversarial accuracy for current training procedures/models [29,37]. In fact our experiments (Table 1) have the property that the trends for adversarial and natural accuracy are reversed. If such a tradeoff is indeed intrinsic among trained models, then it may be reasonable to forcibly sacrifice standard accuracy with the aim of improving adversarial accuracy, as the NAP loss does.

Another potential intuition about our loss function is in the "robust features" framework suggested [20] (see also discussion in [11]). The intuition is: we want to prevent our model from learning "fragile" features about the distribution, which are very helpful for standard classification, but are not robust. Relying on such features is an easy way to improve standard accuracy, but may be harmful towards the eventual goal of adversarial accuracy. If our model is more accurate on natural examples than adversarial ones, this suggests it could be learning these "fragile/non-robust" features. Our loss regularizes against this, by penalizing natural accuracy if it is higher than adversarial accuracy.

In fact, these two intuitions are not necessarily separate. One possible explanation [20] of the tradeoff between standard accuracy and adversarial accuracy for is that current techniques to improve adversarial accuracy attempt to force the model to only rely on robust features. These features are harder to learn than non-robust features, and hence the natural accuracy suffers.

7   Conclusions

We introduce a novel loss function, the Naturally Penalized (NAP) loss, which yields improvements in adversarial robustness for CIFAR-10 against non targeted PGD attacks but CIFAR-100, we are able to improve over TRADES for general attacks(AutoAttack). We view our work as showing that the space of reasonable loss functions for adversarial robustness is perhaps larger than expected, and significant robustness gains are still possible in this space.

A   Hyperparameter Search

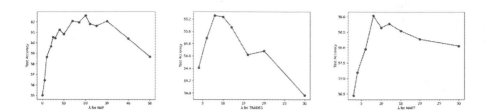

Fig. 1. Hyperparameter search for different loss functions on WideResNet34x4

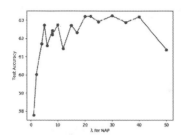

Fig. 2. Hyperparameter search for NAP on WideResNet34x10

B   Softmax Distributions

Softmax Distribution of Natural vs. Adversarial Examples. To study the magnitude and influence of the "margin term" in the NAP loss, we plot the empirical distributions on the test set of $p_y(x;\theta), p_y(\hat{x};\theta)$ and the "margin term" $p_y(x;\theta) - p_y(\hat{x};\theta)$ at the end of training for the models in Table 4. The corresponding histograms are present in Fig. 6. Interestingly, the adversarial confidence $p_y(\hat{x};\theta)$ is bimodal (with modes around 0 and 1) for standard adversarial training but not for the NAP loss. This suggests the NAP loss is indeed encouraging the network to behave similarly on natural and adversarial inputs, though

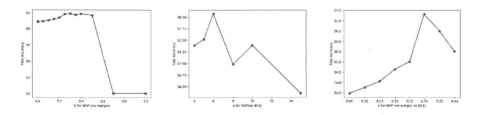

Fig. 3. Hyperparameter search for different modification of NAP as described in Table 5

the exact mechanisms for why this yields robustness improvements is unclear, and requires further study.

The train data also has a similar but milder bimodal distribution for standard adversarial training (Figs. 4 and 5).

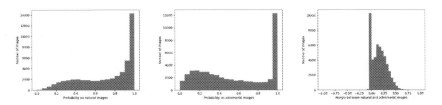

Fig. 4. Histogram of softmax probability on correct labels, for the Train set at the end of standard adversarial training. Using Wide ResNet 34-10, same model as Table 4

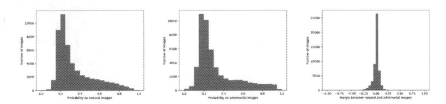

Fig. 5. Histogram of softmax probability on correct labels, for the Train set at the end of training with NAP loss. Using Wide ResNet 34-10, same model as Table 4

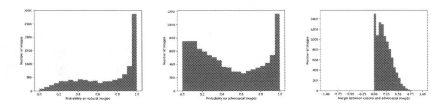

Fig. 6. Standard Loss. Histogram of softmax probability on correct labels, for the Test set at the end of standard adversarial training. Using Wide ResNet 34-10, same model as Table 4. Plotting histograms of (left to right): $p_y(x; \theta)$, $p_y(\hat{x}; \theta)$, and $p_y(x; \theta) - p_y(\hat{x}; \theta)$.

References

1. Athalye, A., Carlini, N., Wagner, D.A.: Obfuscated gradients give a false sense of security: circumventing defenses to adversarial examples. In: Proceedings of the 35th International Conference on Machine Learning, ICML 2018, Stockholmsmässan, Stockholm, Sweden, 10–15 July 2018, pp. 274–283 (2018)
2. Balunovic, M., Vechev, M.: Adversarial training and provable defenses: bridging the gap. In: International Conference on Learning Representations (2020). https://openreview.net/forum?id=SJxSDxrKDr

3. Belkin, M., Hsu, D., Ma, S., Mandal, S.: Reconciling modern machine-learning practice and the classical bias-variance trade-off. Proc. Natl. Acad. Sci. **116**(32), 15849–15854 (2019)
4. Carlini, N., et al.: On evaluating adversarial robustness (2019)
5. Carmon, Y., Raghunathan, A., Schmidt, L., Duchi, J.C., Liang, P.S.: Unlabeled data improves adversarial robustness. In: Advances in Neural Information Processing Systems, pp. 11190–11201 (2019)
6. Chen, E.C., Lee, C.R.: LTD: low temperature distillation for robust adversarial training. arXiv preprint arXiv:2111.02331 (2021)
7. Cohen, J.M., Rosenfeld, E., Kolter, J.Z.: Certified adversarial robustness via randomized smoothing. In: Proceedings of the 36th International Conference on Machine Learning, ICML 2019, 9–15 June 2019, Long Beach, California, USA, pp. 1310–1320 (2019)
8. Croce, F., et al.: Robustbench: a standardized adversarial robustness benchmark. arXiv preprint arXiv:2010.09670 (2020)
9. Cubuk, E.D., Zoph, B., Mane, D., Vasudevan, V., Le, Q.V.: Autoaugment: learning augmentation strategies from data. In: IEEE Conference on Computer Vision and Pattern Recognition, CVPR 2019, Long Beach, CA, USA, 16–20 June 2019, pp. 113–123 (2019)
10. Dai, S., Mahloujifar, S., Mittal, P.: Parameterizing activation functions for adversarial robustness. arXiv preprint arXiv:2110.05626 (2021)
11. Engstrom, L., et al.: A discussion of 'adversarial examples are not bugs, they are features'. Distill (2019). https://doi.org/10.23915/distill.00019. https://distill.pub/2019/advex-bugs-discussion
12. Fawzi, A., Frossard, P.: Manitest: are classifiers really invariant? In: Proceedings of the British Machine Vision Conference 2015, BMVC 2015, Swansea, UK, 7–10 September 2015, pp. 106.1–106.13 (2015)
13. Geirhos, R., Temme, C.R.M., Rauber, J., Schütt, H.H., Bethge, M., Wichmann, F.A.: Generalisation in humans and deep neural networks. In: Advances in Neural Information Processing Systems 31: Annual Conference on Neural Information Processing Systems 2018, NeurIPS 2018, 3–8 December 2018, Montréal, Canada, pp. 7549–7561 (2018)
14. Goodfellow, I.J., Shlens, J., Szegedy, C.: Explaining and harnessing adversarial examples. In: 3rd International Conference on Learning Representations, ICLR 2015, San Diego, CA, USA, 7–9 May 2015, Conference Track Proceedings (2015)
15. Gowal, S., Rebuffi, S.A., Wiles, O., Stimberg, F., Calian, D.A., Mann, T.A.: Improving robustness using generated data. In: Advances in Neural Information Processing Systems, vol. 34 (2021)
16. Hayes, J.: Extensions and limitations of randomized smoothing for robustness guarantees. In: Proceedings of the IEEE/CVF Conference on Computer Vision and Pattern Recognition Workshops, pp. 786–787 (2020)
17. He, K., Zhang, X., Ren, S., Sun, J.: Deep residual learning for image recognition. In: 2016 IEEE Conference on Computer Vision and Pattern Recognition, CVPR 2016, Las Vegas, NV, USA, 27–30 June 2016, pp. 770–778 (2016)
18. Hendrycks, D., Dietterich, T.G.: Benchmarking neural network robustness to common corruptions and perturbations. In: 7th International Conference on Learning Representations, ICLR 2019, New Orleans, LA, USA, 6–9 May 2019 (2019)
19. Hendrycks, D., Lee, K., Mazeika, M.: Using pre-training can improve model robustness and uncertainty. In: Proceedings of the 36th International Conference on Machine Learning, ICML 2019, 9–15 June 2019, Long Beach, California, USA, pp. 2712–2721 (2019)

20. Ilyas, A., Santurkar, S., Tsipras, D., Engstrom, L., Tran, B., Madry, A.: Adversarial examples are not bugs, they are features. In: Advances in Neural Information Processing Systems, pp. 125–136 (2019)
21. Kannan, H., Kurakin, A., Goodfellow, I.J.: Adversarial logit pairing. CoRR abs/1803.06373 (2018). http://arxiv.org/abs/1803.06373
22. Kingma, D.P., Ba, J.: Adam: a method for stochastic optimization. arXiv preprint arXiv:1412.6980 (2014)
23. Krizhevsky, A., Sutskever, I., Hinton, G.E.: Imagenet classification with deep convolutional neural networks. Commun. ACM **60**(6), 84–90 (2017). https://doi.org/10.1145/3065386
24. Krizhevsky, A., et al.: Learning multiple layers of features from tiny images (2009)
25. LeCun, Y., Bottou, L., Bengio, Y., Haffner, P.: Gradient-based learning applied to document recognition. Proc. IEEE **86**(11), 2278–2324 (1998)
26. Lécuyer, M., Atlidakis, V., Geambasu, R., Hsu, D., Jana, S.: Certified robustness to adversarial examples with differential privacy. In: 2019 IEEE Symposium on Security and Privacy, SP 2019, San Francisco, CA, USA, 19–23 May 2019, pp. 656–672 (2019)
27. Li, B., Chen, C., Wang, W., Carin, L.: Second-order adversarial attack and certifiable robustness. CoRR abs/1809.03113 (2018). http://arxiv.org/abs/1809.03113
28. Madry, A., Makelov, A., Schmidt, L., Tsipras, D., Vladu, A.: Towards deep learning models resistant to adversarial attacks. arXiv preprint arXiv:1706.06083 (2017)
29. Nakkiran, P.: Adversarial robustness may be at odds with simplicity. arXiv preprint arXiv:1901.00532 (2019)
30. Nakkiran, P., Kaplun, G., Bansal, Y., Yang, T., Barak, B., Sutskever, I.: Deep double descent: where bigger models and more data hurt. In: International Conference on Learning Representations (2020). https://openreview.net/forum?id=B1g5sA4twr
31. Raghunathan, A., Steinhardt, J., Liang, P.: Certified defenses against adversarial examples. In: 6th International Conference on Learning Representations, ICLR 2018, Vancouver, BC, Canada, 30 April—3 May 2018, Conference Track Proceedings (2018)
32. Rebuffi, S.A., Gowal, S., Calian, D.A., Stimberg, F., Wiles, O., Mann, T.: Fixing data augmentation to improve adversarial robustness. arXiv preprint arXiv:2103.01946 (2021)
33. Salman, H., Li, J., Razenshteyn, I.P., Zhang, P., Zhang, H., Bubeck, S., Yang, G.: Provably robust deep learning via adversarially trained smoothed classifiers. In: Advances in Neural Information Processing Systems 32: Annual Conference on Neural Information Processing Systems 2019, NeurIPS 2019, 8–14 December 2019, Vancouver, BC, Canada, pp. 11289–11300 (2019)
34. Szegedy, C., et al.: Intriguing properties of neural networks. arXiv preprint arXiv:1312.6199 (2013)
35. Tan, M., Le, Q.V.: Efficientnet: rethinking model scaling for convolutional neural networks. In: Proceedings of the 36th International Conference on Machine Learning, ICML 2019, 9–15 June 2019, Long Beach, California, USA, pp. 6105–6114 (2019)
36. Taori, R., Dave, A., Shankar, V., Carlini, N., Recht, B., Schmidt, L.: When robustness doesn't promote robustness: synthetic vs. natural distribution shifts on imagenet (2020). https://openreview.net/forum?id=HyxPIyrFvH
37. Tsipras, D., Santurkar, S., Engstrom, L., Turner, A., Madry, A.: Robustness may be at odds with accuracy. arXiv preprint arXiv:1805.12152 (2018)

38. Uesato, J., Alayrac, J., Huang, P., Stanforth, R., Fawzi, A., Kohli, P., et al.: Are labels required for improving adversarial robustness? In: Advances in Neural Information Processing Systems, pp. 12192–12202 (2019)
39. Wang, Y., Zou, D., Yi, J., Bailey, J., Ma, X., Gu, Q.: Improving adversarial robustness requires revisiting misclassified examples. In: International Conference on Learning Representations (2020). https://openreview.net/forum?id=rklOg6EFwS
40. Wong, E., Schmidt, F.R., Metzen, J.H., Kolter, J.Z.: Scaling provable adversarial defenses. In: Advances in Neural Information Processing Systems 31: Annual Conference on Neural Information Processing Systems 2018, NeurIPS 2018, 3–8 December 2018, Montréal, Canada, pp. 8410–8419 (2018)
41. Xiao, C., Zhong, P., Zheng, C.: Enhancing adversarial defense by k-winners-take-all. In: International Conference on Learning Representations (2020). https://openreview.net/forum?id=Skgvy64tvr
42. Xiao, K.Y., Tjeng, V., Shafiullah, N.M.M., Madry, A.: Training for faster adversarial robustness verification via inducing relu stability. In: 7th International Conference on Learning Representations, ICLR 2019, New Orleans, LA, USA, 6–9 May 2019 (2019)
43. Xie, C., Wu, Y., Maaten, L.V.D., Yuille, A.L., He, K.: Feature denoising for improving adversarial robustness. In: Proceedings of the IEEE Conference on Computer Vision and Pattern Recognition, pp. 501–509 (2019)
44. Xie, C., Yuille, A.: Intriguing properties of adversarial training at scale. In: International Conference on Learning Representations (2020). https://openreview.net/forum?id=HyxJhCEFDS
45. Yun, S., Han, D., Oh, S.J., Chun, S., Choe, J., Yoo, Y.: Cutmix: regularization strategy to train strong classifiers with localizable features. In: Proceedings of the IEEE/CVF International Conference on Computer Vision, pp. 6023–6032 (2019)
46. Zagoruyko, S., Komodakis, N.: Wide residual networks. In: Proceedings of the British Machine Vision Conference 2016, BMVC 2016, York, UK, 19–22 September 2016 (2016)
47. Zhang, C., Bengio, S., Hardt, M., Recht, B., Vinyals, O.: Understanding deep learning requires rethinking generalization. arXiv preprint arXiv:1611.03530 (2016)
48. Zhang, H., Yu, Y., Jiao, J., Xing, E.P., Ghaoui, L.E., Jordan, M.I.: Theoretically principled trade-off between robustness and accuracy. arXiv preprint arXiv:1901.08573 (2019)
49. Zhang, H., Cissé, M., Dauphin, Y.N., Lopez-Paz, D.: mixup: beyond empirical risk minimization. In: 6th International Conference on Learning Representations, ICLR 2018, Vancouver, BC, Canada, 30 April–3 May 2018, Conference Track Proceedings (2018)

W04 - Autonomous Vehicle Vision

W04 - Autonomous Vehicle Vision

The 3rd AV Vision workshop aims to bring together industry professionals and academics to brainstorm and exchange ideas on the advancement of computer vision techniques for autonomous driving. In this half-day workshop, we had four keynote talks and several paper presentations to discuss the state-of-the-art approaches and existing challenges in the field of autonomous driving.

October 2022

Rui Fan
Nemanja Djuric
Wenshuo Wang
Peter Ondruska
Jie Li

4D-StOP: Panoptic Segmentation of 4D LiDAR Using Spatio-Temporal Object Proposal Generation and Aggregation

Lars Kreuzberg[1], Idil Esen Zulfikar[1(✉)], Sabarinath Mahadevan[1], Francis Engelmann[2], and Bastian Leibe[1]

[1] RWTH Aachen University, Aachen, Germany
lars.kreuzberg@rwth-aachen.de,
{zulfikar,mahadevan,leibe}@vision.rwth-aachen.de
[2] ETH Zurich, AI Center, Zürich, Switzerland
francis.engelmann@ai.ethz.ch

Abstract. In this work, we present a new paradigm, called 4D-StOP, to tackle the task of 4D Panoptic LiDAR Segmentation. 4D-StOP first generates spatio-temporal proposals using voting-based center predictions, where each point in the 4D volume votes for a corresponding center. These tracklet proposals are further aggregated using learned geometric features. The tracklet aggregation method effectively generates a video-level 4D scene representation over the entire space-time volume. This is in contrast to existing end-to-end trainable state-of-the-art approaches which use spatio-temporal embeddings that are represented by Gaussian probability distributions. Our voting-based tracklet generation method followed by geometric feature-based aggregation generates significantly improved panoptic LiDAR segmentation quality when compared to modeling the entire 4D volume using Gaussian probability distributions. 4D-StOP achieves a new state-of-the-art when applied to the SemanticKITTI test dataset with a score of 63.9 LSTQ, which is a large (+7%) improvement compared to current best-performing end-to-end trainable methods. The code and pre-trained models are available at: https://github.com/LarsKreuzberg/4D-StOP.

1 Introduction

In recent years, we have made impressive progress in the field of 3D perception, which is mainly due to the remarkable developments in the field of deep learning. With the availability of RGB-D sensors, we have made progress on tasks like [16, 17,27,40], 3D instance segmentation [11,14,15,22,26,54] and 3D object detection [35,41,59] which work on indoor scenes. Similarly, the accessibility to modern LiDAR sensors has made it possible to work with outdoor 3D scenes, where again recent works have targeted the tasks of 3D object detection [43,44,53,55], semantic

Supplementary Information The online version contains supplementary material available at https://doi.org/10.1007/978-3-031-25056-9_34.

(a) 4D-PLS   (b) 4D-StOP

Fig. 1. Comparing 4D-PLS to 4D-StOP. (a) The current end-to-end trainable state-of-the-art method models each tracklet as a Gaussian probability distribution in the 4D space-time volume. **(b)** Our work, 4D Spatial-temporal Object Proposal Generation and Aggregation (4D-StOP), represents each tracklet with multiple proposals generated using voting-based center predictions in the 4D space-time volume, which are then aggregated to obtain the final tracklets.

segmentation [33,47,57,61], panoptic segmentation [5,21,34,60] and multi-object tracking [3,12,50,56]. Perceiving outdoor 3D environments from LiDAR data is particularly relevant for robotics and autonomous driving applications, and hence has gained significant traction in the recent past.

In the wake of these advancements, Aygün *et al.* [2] proposed the 4D Panoptic LiDAR Segmentation task, where the goal is to perform panoptic segmentation of LiDAR scans across space-time volumes of point cloud data. Hence, each LiDAR point is assigned a semantic class label and a temporally consistent instance ID in a given sequence of 3D LiDAR scans. This task is more challenging than its predecessor 3D Panoptic LiDAR Segmentation [5] since temporally consistent instance IDs and semantic class labels have to be preserved over the time continuum. It is also critical for domains such as autonomous driving, where autonomous agents have to be able to continuously interact with 4D environments and be robust to rapid changes in the space-time volume.

For this task, the current research can be categorized into two different groups. In the first group, the existing methods follow a tracking-by-detection paradigm where an off-the-shelf 3D panoptic segmentation network [33,47,61] is first employed to obtain the instances in each scan along with the corresponding semantic class labels. The instances from different scans are then further assigned a consistent ID with the help of various data association methods [28,30,36,50] in a second step. Since an off-the-shelf network is used, these methods are not end-to-end trainable, and often utilize multiple different networks.

The second group of methods [2,23] uses a single end-to-end trainable network to directly generate the semantic class label for each point and a temporally consistent instance ID for the corresponding objects. The current top performing method [2] takes as input a 4D space-time volume which is generated by combining multiple LiDAR scans. A semantic segmentation network assigns a semantic class to each of the input points, while the object instances are modeled as Gaussian probability distributions over the foreground points. In addition to the Gaussian parameters, the network also predicts object centers and the track-lets are generated by clustering the points around these centers in the features space, by evaluating a Gaussian probability function as seen in Fig. 1(a). While such clustering techniques generally work well on small 4D volumes, the method generates only one cluster center and therefore generates only one proposal for each object which limits the representation quality.

In this work, instead of representing tracklets as Gaussian distribution, we follow a new paradigm for forming tracklets in 4D space-time volume. Inspired by [15], we represent tracklets as spatio-temporal object proposals, then aggre-gate these proposals to obtain the final tracklets as shown in Fig. 1(b). Object proposals have been used successfully for object detection and instance segmen-tation in both 2D [20, 42] and 3D [15, 41] domains. The success of these methods relies on producing a large number of proposals, thereby producing reliable out-puts. In our paper, we investigate whether the success of using object proposals can also be observed in a 4D space-time volume, and empirically prove that they in fact do. For this, each feature point in our method votes for the closest center point to first generate object proposals, which are then aggregated using high-level learned geometric features in the 4D volume. For the first step, we utilize Hough voting [41] in the 4D volume and show that it works well in this large and dynamic space. An alternative strategy to generate multiple proposals would be the adaptation of the Gaussian probability-based clustering, successfully used in both 2D [1, 37] and 3D [2]. However, we show that the performance of such an adaptation is inferior to Hough voting used in this work (see Sect. 4.2).

Our method, 4D Spatio-temporal Object Proposal Generation and Aggrega-tion (4D-StOP), generates better tracklets associations over an input 4D space-time volume, and correspondingly outperforms the state-of-the-art methods on the 4D Panoptic LiDAR Segmentation task. In summary, we make **the follow-ing contributions**: (i) we tackle the task of 4D Panoptic LiDAR Segmenta-tion in an end-to-end trainable fashion using a new strategy that first generates tracklet proposals and then further aggregates them; (ii) our method uses a center based voting technique to first generate tracklet proposals, which are then aggregated using our novel learned geometric features; (iii) we achieve the state-of-the-art results among all methods on SemanticKITTI test set, where we outperform the best performing end-to-end trainable method by a large margin.

2   Related Work

Point Cloud Segmentation. With the availability of multiple large-scale out-door datasets [4, 18], deep learning models are leading the field of LiDAR seman-tic segmentation [13, 33, 46, 47, 52, 57, 61]. These end-to-end trainable methods can be categorized into two groups: In the first group, the methods [13, 33, 52, 57] convert 3D point clouds to 2D grids and make use of 2D convolution networks. PolarNet [57] projects 3D points to a bird's-eye-view and works on a polar coor-dinate system. RangeNet++ [33] relies on a spherical projection mechanism and exploits range images. The second group of methods [46, 47, 61] directly works on

3D data to preserve the 3D geometry by using 3D convolution networks. Cylinder3D [61] uses cylindrical partitioned coordinate systems and asymmetrical 3D convolution networks. Similar to [2], we rely on KPConv [47] as feature backbone which directly applies deformable convolutions on point clouds.

Multi-Object Tracking. In 2D multi-object tracking (MOT), earlier methods [6,7,29,32] follow the tracking-by-detection paradigm where off-the-shelf 2D object detection networks [20,42] are utilized to obtain objects in the frames, then the objects across frames are associated with different data association methods. Later methods [31,39,49,58] shifted to joint detection and association in a single framework. Recently, 3D LiDAR MOT has become popular in the vision community, due to the impressive developments in 3D object detectors [28,42] and newly emerging datasets [8,45]. The methods tackle 3D MOT either with tracking-by-detection paradigm [25,51] or with joint-detection-and-tracking paradigm [3,56], similar to the methods in 2D MOT. All of these methods work on single scans, *i.e.*, on the spatial domain, to track objects over time. In contrast, our method works on a unified space and time domain, *i.e.*, 4D volume formed by combining multi-scans, to localize and associate objects.

Panoptic LiDAR Segmentation and Tracking. 4D Panoptic LiDAR Segmentation [2,23,30] is an emerging research topic that unifies semantic segmentation, instance segmentation and tracking of LiDAR point clouds into a single framework. The initial work proposes a unified architecture consisting of three heads that tackle semantic segmentation, instance segmentation and tracking tasks individually. The recent work [30] obtains semantic labels and 3D object detections from an off-the-self panoptic segmentation network and adapts contrastive learning [9,10,38] into tracking to associate the 3D detections. The state-of-the-art method [2] in the end-to-end trainable frameworks is a bottom-up approach that jointly assigns semantic labels and associates instances across time. It works on a space-time volume where tracklets are generated by evaluating Gaussian distributions. We direct future research towards end-to-end trainable models, our work aims to improve tracklet representation in space-time volume, and propose to tackle tracklets representation as multi-proposal generation and aggregation in 4D volume.

3   Method

The overall architecture of our method is depicted in Fig. 2. We pose the task of 4D Panoptic LiDAR Segmentation by splitting it into the two subtasks, semantic segmentation as well as instance segmentation, and solve them in parallel. The semantic segmentation branch performs point-wise classification for each point in the 4D volume. For the instance segmentation task, we apply an object-centric voting approach in the 4D space-time volume. Based on the votes, multiple spatio-temporal proposals are generated. The features of these spatio-temporal proposals are learned to aggregate the proposals and generate the final tracklets. In the end, the results of the semantic segmentation and the instance segmentation branches are combined. To ensure that all points in an instance have

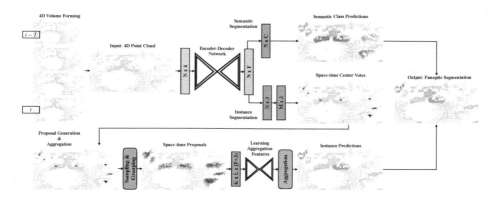

Fig. 2. 4D-StOP Overview. Given a 4D point cloud, our method splits the task of 4D Panoptic LiDAR Segmentation into two subtasks semantic segmentation and instance segmentation. The semantic segmentation branch performs point-wise classification in 4D volume. The instance segmentation branch applies object-centric voting in 4D space-time to generate proposals followed by proposal aggregation based on learned features to obtain the final instances. L is the number of points associated to a proposal.

the same semantic class label, we exploit majority voting where the most frequently occurring semantic class label among the instance points determines the semantic labels for all points within the instance.

3.1   4D Volume Formation

To form overlapping 4D point clouds, multiple consecutive scans are stacked together. For a scan at time-step t and a temporal window size T, the point clouds within the temporal window $\{\max(0, t - T), ..., t\}$ are merged. However, stacking all the points from the scans in the temporal window T is not possible due to memory constraints. To overcome this problem, we follow the *importance sampling* strategy from [2]. Here, 10% of the points with a probability proportional to the per-point objectness score are sampled from the previous scans at time-steps $t-T, ..., t-1$ while processing the scan at time-step t, which enables to focus on points from *thing* classes while preserving the points from *stuff* classes.

3.2   4D-StOP

Semantic Segmentation and Voting for 4D Volume. Given a 4D point cloud with N points as input, we first learn strong point features with a backbone that encodes the semantics and the geometry of the underlying multi-scan scene. We use an encoder-decoder architecture as our backbone [47] to generate per-point features $f_i \in \mathbb{R}^F$, where $i \in \{1, ..., N\}$. On top of the backbone, we add a semantic head module and a voting module; and perform these tasks jointly. The semantic head predicts a semantic class label for each point and encodes the semantic context into the point features. To learn the semantic labels, a standard cross-entropy classification loss L_{sem} is used. After the semantic head predicts a semantic class label for each point in the 4D volume, we distinguish between the points assigned to foreground and background classes and run the voting module only on the foreground points.

In the voting module, we adapt the deep Hough voting used in [41] to work on 4D space-time volumes. Every foreground point in the 4D volume votes for the corresponding space-time object center. In particular, this is realized by learning per-point relative offsets $\Delta x_i \in \mathbb{R}^3$ that point from the point positions $x_i \in \mathbb{R}^3$ to the corresponding ground truth object centers $c_i^* \in \mathbb{R}^3$ in the 4D space-time volume. The loss function for the voting module can be formulated as:

$$L_{vot} = \frac{1}{M}||x_i + \Delta x_i - c_i^*||_H \cdot \mathbf{1}(x_i),\qquad(1)$$

where $|| \cdot ||_H$ is the Huber-Loss, $\mathbf{1}(\cdot)$ is a binary function indicating whether a point belongs to foreground, and M is the number of points in foreground. In a multi-scan scenario, the ground truth bounding-box centers of the 3D objects, i.e., the ground truth bounding-box centers on a single 3D scan, cannot be simply regressed in 4D volume. Since the objects move constantly in consecutive scans, it would lead to multiple regressed centers that would eventually yield insufficient results. To avoid that, we recompute the ground truth bounding-box centers for the objects in the 4D space-time volumes such that the ground-truth centers c_i^* in Eq. 1 are the ground-truth bounding-box centers of the objects in the merged 4D volume.

Proposal Generation. The proposals are generated by uniformly sampling and grouping the points based on the spatial-temporal proximity of the votes within a 4D volume. For each foreground point, there exists a vote to the corresponding space-time object centers. From these center votes, K samples are chosen using the Farthest Point Sampling (FPS) selection technique used in [41] to serve as proposal centers $y_j = x_j + \Delta x_j, \forall j \in \{1, ..., K\}$. As suggested in [15], we also tested Random Sampling technique, which yields results on the same level as FPS. Although both sampling methods give similar results, we continue with the FPS technique as the probability of missing an object is much lower. To complete the proposal generation, all points are assigned to proposal j if a point votes for a center within a radius r ($= 0.6m$) of the sampled proposal position y_j. In the end, each proposal contains the proposal positions y_j near the space-time object centers and a set of points s_i associated with each proposal j.

Proposal Aggregation. A simple way to aggregate multiple proposals is to use the proposal center positions, i.e. use relatively simple features. However, proposal aggregation can be enhanced by learning more advanced geometric features. We have K object proposal in 4D space-time volume consisting of the proposal positions y_j and a set of points s_i associated with each proposal j. Also, each of the points associated with a proposal has a per-point feature $f_i \in \mathbb{R}^F$. We get the proposal features $g_i \in \mathbb{R}^D$ which describes the local geometry of their associated objects by using a PointNet-like module [40] applied to the per-point features of the associated points. In particular, the per-point features in combination with the per-point positions that belong to a proposal are processed by a shared MLP with output sizes $(128, 128, 128)$, then channel-wise max-pooling is applied to predict a single feature vector $g_i \in \mathbb{R}^D$. Each proposal is processed independently. The proposal features are further processed by another MLP with output sizes $(128, 128, E)$ to learn the E dimensional geometric aggregation features. Overall, we use three geometric features, where two of them, a refined proposal center point prediction $y_i + \Delta y_i$ and an object radius estimation r_i, are adapted from [15]. The third novel geometric feature that we use for aggregation is the predicted bounding box size $bb_i = (l_i, h_i, w_i) \in \mathbb{R}^3$. This additional geometric feature is helpful in better aggregating the proposal as shown in our experiments (see Sect. 4.2). The aggregation-loss is defined as:

$$L_{agg} = ||y_i + \Delta y_i - c_i^*||_H + ||r_i - r_i^*||_H + ||bb_i - bb_i^*||_H \tag{2}$$

where c_i^* is the center, r_i^* is the radius of the closest ground truth bounding sphere and bb_i^* are the length, height and width of closest ground truth bounding-box. Finally, we aggregate the proposals by applying DBScan clustering to the learned geometric features. The proposals whose geometric features end up in the same cluster are combined. The union of the points in the aggregated proposals builds the final space-time object predictions, that is the final tracklets.

3.3   Tracking

Within an input 4D volume, temporal association is resolved implicitly as object instances are grouped jointly in 4D space-time. Due to memory constraints, there is a limitation to the size of the 4D volume that can be processed at once. As a result, we split the entire 4D point cloud into multiple overlapping volumes of smaller sizes. Following the strategy used in [2], final tracklets across the entire space-time volume are obtained by associating the individual fragmented tracklets based on the best IoU overlap scores. For IoU scores below a particular threshold, we discard the corresponding associations and start a new tracklet.

3.4   Training and Implementation Details

The model is trained end-to-end using the multi-task loss $L = \alpha \cdot L_{sem} + \beta \cdot L_{vot} + \gamma \cdot L_{agg}$ for 550K iterations in total. For the first 400K iterations, we set $\alpha = 1$, $\beta = 1$ and $\gamma = 0$ in order to predict the semantic classes and to obtain the spatio-temporal proposals in a sufficient quality. In the remaining 150K iterations, we set $\alpha = 0$, $\beta = 0$ and $\gamma = 1$ and freeze the semantic segmentation and the voting modules, then learn high-quality geometric features for aggregating the spatio-temporal proposals. Our model is implemented in PyTorch. The training is performed on a single NVIDIA A40 (48GB) GPU and the inference on a single NVIDIA TitanX (12GB) GPU. We apply random rotation, translation and scaling of the scene as data augmentations. Moreover, we randomly drop some points of the scene and randomly add noise.

Table 1. Scores on **SemanticKITTI 4D Panoptic Segmentation test set.** MOT - *tracking-by-detection* by [50], SFP - *tracking-by-detection* via scene flow based propagation [36], PP - PointPillars [28].

	Method	$LSTQ$	S_{assoc}	S_{cls}	IoU^{St}	IoU^{Th}
Not-End-to-End	RangeNet++ [33] + PP + MOT	35.5	24.1	52.4	64.5	35.8
	KPConv [47] + PP + MOT	38.0	25.9	55.9	66.9	47.7
	RangeNet++ [33] + PP + SFP	34.9	23.3	52.4	64.5	35.8
	KPConv [33] + PP + SFP	38.5	26.6	55.9	66.9	47.7
	CIA [30]	63.1	65.7	**60.6**	66.9	52.0
End-to-End	4D-PLS (4 scan) [2]	56.9	56.4	57.4	66.9	51.6
	4D-StOP(Ours)(2 scan)	62.9	67.3	58.8	**68.3**	53.3
	4D-StOP(Ours)(4 scan)	**63.9**	**69.5**	58.8	67.7	**53.8**

4 Experiments

4.1 Comparing with State-of-the-Art Methods

Dataset and Metric. We evaluate our method on the SemanticKITTI LiDAR dataset [4]. It consists of 22 sequences from KITTI odometry dataset [19]. There exist three splits: training, validation and test split where sequences 00 to 10 are used as training and validation split, and sequences 11 to 21 as test split. Each point is labeled with a semantic class and a temporally consistent instance ID. In total, there are 19 semantic classes among which 8 are *things* classes and the remaining 11 are *stuff* classes. The main evaluation metric is $LSTQ$, which is the geometric mean of the classification score S_{cls} and the association score S_{assoc}, *i.e.* $LSTQ = \sqrt{S_{cls} \times S_{assoc}}$ [2]. The classification score S_{cls} measures the quality of semantic segmentation, while the association score S_{assoc} measures the quality of association in a 4D continuum. Unlike the previous evaluation metrics [23,24,48], $LSTQ$ performs semantic and instance association measurements independently to avoid penalization due to the entanglement of semantic and association scores. Our strategy of generating proposals and then aggregating them in the 4D volume generates very good object associations while maintaining the semantic classification quality, which is reflected in the association (S_{assoc}) and the classification scores (S_{cls}) in the following experiments.

Results. In Table 1, we present the results on the SemanticKITTI 4D Panoptic Segmentation test set. Among the end-to-end trainable methods, 4D-StOP outperforms the previous state-of-the-art method 4D-PLS [2] by a large margin, obtaining a **7%** boost in $LSTQ$ score, when utilizing the same number of scans (4 scans). Especially, we achieve **13.1%** improvement in S_{assoc}, which shows that 4D-StOP produces better segmented and associated tracklets in a 4D space-time volume compared to 4D-PLS. While we achieve state-of-the-art results when using 4 scans, 4D-StOP also performs well with only 2 scans. The performance drop in the $LSTQ$ score is merely **1%**.

Comparing 4D-StOP with the non-end-to-end trainable methods, we observe that 4D-StOP also outperforms the current not-end-to-end trainable state-art-the-art method CIA [30] (**+0.8 $LSTQ$**). We would like to highlight that CIA leverages an off-the-shelf 3D panoptic segmentation network with a stronger

Table 2. Scores on SemanticKITTI 4D Panoptic Segmentation validation set. MOT - *tracking-by-detection* by [50], SFP - *tracking-by-detection* via scene flow based propagation [36], PP - PointPillars [28].

	Method	LSTQ	S_{assoc}	S_{cls}	IoU^{St}	IoU^{Th}
Not-End-to-End	RangeNet++ [33] + PP + MOT	43.8	36.3	52.8	60.5	42.2
	KPConv [47] + PP + MOT	46.3	37.6	57.0	64.2	54.1
	RangeNet++ [33] + PP + SFP	43.4	35.7	52.8	60.5	42.2
	KPConv [33] + PP + SFP	46.0	37.1	57.0	64.2	54.1
End-to-End	MOPT [23]	24.8	11.7	52.4	62.4	45.3
	4D-PLS [2] (2 scan)	59.9	58.8	61.0	65.0	63.1
	4D-PLS [2] (4 scan)	62.7	65.1	60.5	**65.4**	61.3
	4D-StOP(Ours)(2 scan)	66.4	71.8	**61.4**	64.9	**64.1**
	4D-StOP(Ours)(4 scan)	**67.0**	**74.4**	60.3	65.3	60.9

backbone than used in 4D-StOP. Thus, it has a better S_{cls} score than 4D-StOP, *i.e.* the increase in S_{cls} score results from the semantic segmentation performance of the backbone network. In contrast to CIA, we do not take advantage of such a stronger backbone but rather preserve the exact same configuration of our baseline 4D-PLS. Despite the worse classification score, the *LSTQ* score for 4D-StOP is still better than for CIA. Reason for this is the better working object association in our end-to-end trainable approach. 4D-StOP is better in S_{assoc} than CIA by a large margin with **3.8%** boosting, where the increase is gained from the method basically.

In Table 2, we report the results on the SemanticKITTI 4D Panoptic Segmentation validation set. In the validation set results, we observe a similar pattern as in the test set. 4D-StOP outperforms all the methods by a large margin in both not-end-to-end trainable and end-to-end trainable categories. It produces better scores compared to 4D-PLS by **+4.3** *LSTQ* and **+9.3** S_{assoc} for 4 scans, and by **+6.5** *LSTQ* and **+13.0** S_{assoc} for 2 scans. Although there is a slight performance drop between the two and four scan set-ups, this drop is smaller compared to 4D-PLS (**0.6%** *LSTQ* drop for 4D-StOP compared to **2.8%** *LSTQ* drop for 4D-PLS). This shows that 4D-StOP is more robust to the number of input scans. Since processing two scans is faster than processing four scans, 4D-StOP offers a better trade-off between accuracy and speed than 4D-PLS.

4.2   Analysis

We ablate all of our design choices on the validation split of the SemanticKITTI dataset, *i.e.* only on sequence 08 of the dataset. In this section, we provide detailed experimental analysis, and discuss the impact of each of the components in our method, including the geometric features. In addition, we also reformulate 4D-StOP, where the proposals are modeled as Gaussian distributions similar to our baseline 4D-PLS, and compare this formulation with our voting based approach. In the experimental analysis, a 4D volume is formed by combining 2 scans unless it is not explicitly indicated it is generated from 4 scans.

Table 3. Analysis of each component in 4D-StOP.

		$LSTQ$	S_{assoc}	S_{cls}
①	4D-PLS [2] (2 scan)	59.9	58.8	61.0
②	Our Baseline	63.7	66.7	60.8
③	Aggregating w/ Proposal Positions	65.3	70.0	60.8
④	Aggregating w/ Proposal Positions + Majority Voting	65.5	70.0	61.2
④	Aggregating w/ Geometric Features + Majority Voting	**66.4**	**71.8**	**61.4**

Effect of 4D-StOP Components. We evaluate the impact of each 4D-StOP component in Table 3. The main improvements come from the proposal generation and aggregation steps. To investigate the impact of the aggregation step, we create a baseline in experiment ② by applying the traditional non-maximum-suppression (NMS) on the spatio-temporal proposals. Even this baseline outperforms 4D-PLS by a large margin with **3.8%** boosting in the $LSTQ$ score. In order to see the effect of the aggregation step, we first conduct a naive aggregation by combining proposals based on the proposal positions (experiment ③) and observe another **1.6%** increase in the $LSTQ$ score. In experiment ④, we ablate adding majority voting into our pipeline which only brings a slight increment ($+0.4\ S_{cls}$). Lastly, in experiment ⑤, we show the results for adding the learned high-level geometric features as the final component, which boosts the performance in the S_{assoc} and $LSTQ$ scores by **1.8%** and **0.9%** respectively. Overall, our method outperforms the prior end-to-end trainable method 4D-PLS [2] significantly irrespective of whether we use our proposal aggregation strategy or a much simpler NMS. This shows that, similar to the observations made in 2D and 3D spatial domains, proposal-based methods surpass other bottom-up approaches in the 4D space-time domain.

4D-StOP with Gaussian Distribution. An alternative way of generating proposals would be to use Gaussian probability distributions. To do this, we adapt 4D-PLS [2] to work with our proposal generation and aggregation paradigm. In all of the following experiments, 4 scan setup is used to be consistent with the best results in 4D-PLS. We generate proposals based on per-point objectness scores, which indicate the probability that a point is close to an object center. We select all the points in the 4D volume with an objectness score above a certain threshold as the proposal center points. Then, we assign points to a proposal by evaluating each point under the Gaussian probability distribution function with the proposal center as cluster seed point. For each point, the Gaussian probability indicates whether the point belongs to the proposal. All the points with a probability above a certain threshold are assigned to the proposal. We generate the proposals in parallel, and thus obtain overlapping Gaussian probabilities for each object, *i.e.* the multiple proposals for each object. In contrast, in 4D-PLS each object is represented with one proposal. After the proposals are generated, they are aggregated using the cluster mean

Table 4. Analysis for studying 4D-StOP with Gaussian Distribution. PT - *Probability Threshold* and R - *Radius*. All of the experiments are carried out on 4 scans and the objectness score threshold is set to 0.7 as in 4D-PLS.

		PT	R	$LSTQ$	S_{assoc}	S_{cls}
①	4D-StOP(Ours)	-	-	**67.0**	**74.4**	60.3
②	4D-PLS	0.5	-	62.7	65.1	**60.5**
③	4D-StOP w/ Gaussian Distribution	0.5	0.0	61.2	62.1	60.3
④	4D-StOP w/ Gaussian Distribution	0.7	0.0	63.4	66.7	60.3
⑤	4D-StOP w/ Gaussian Distribution	0.7	0.6	63.1	66.0	60.3

points. In Table 4, we compare 4D-PLS and 4D-StOP with Gaussian distribution (experiments ② and ③). 4D-StOP with Gaussian distribution yields worse results than 4D-PLS (-1.5 **$LSTQ$**). We realized that using all proposals with an objectness score above a threshold is prone to produce more FPs than 4D-PLS, so we decided to increase the probability threshold while assigning each point to the proposals. A slight boosting $+0.7$ **$LSTQ$** is achieved in this experiment (the experiment ④), but it is still largely under our work 4D-StOP. To eliminate more FPs, we consider another strategy, where only a subset of the proposals is selected for aggregation. After generating the proposals, we start selecting the proposal with the highest objectness score from the proposals candidate pool. The proposals within a certain radius around the center point of the selected proposal are removed from the candidate pool. We repeat it until no proposal is left in the candidate pool and only aggregate the selected proposals in the end. With this strategy, a minor boosting by $+0.4$ **$LSTQ$** is observed compared to 4D-PLS in experiment ⑤, but the results are still under 4D-StOP.

In conclusion, we show that the multi-proposal generation and aggregation strategy can be implemented with Gaussian probability-based clustering used in 4D-PLS, but this does not bring a significant improvement to the base method. More importantly, the performance of such an adaptation still falls below 4D-StOP by a large margin. Thereby, with these experiments we show that Gaussian distributions can be used to represent tracklet proposals, although the performance improvement brought by them saturates soon, and are inferior to our center-based voting approach.

Impact of the Learned Geometric Features. Our 4D-StOP method makes use of three different geometric features: the refined proposal centers, the radius estimation and the predicted bounding-box size (Sect. 3.2), to aggregate the space-time proposals. We evaluate the impact of each geometric feature on the results of our method in Table 5. Grouping proposals based on the proposals center positions (first row), *i.e.* grouping without utilizing the learned geometric features, serves as our baseline. First, we analyze the performance for two learned geometric features, the refined center positions and the radius estimation, which we adapt from [15]. When the refined center positions are used individually (second row) or are combined with the radius estimation (third row), a

Table 5. Analysis of the learned geometric features.

Geometric features				$LSTQ$	S_{assoc}	S_{cls}
Center Pos	Ref. Center Pos	Radius	Bounding-Box			
✓	–	–	–	65.5	70.0	61.2
–	✓	–	–	65.8	70.7	61.3
–	✓	✓	–	65.9	70.9	61.3
–	✓	✓	✓	**66.4**	**71.8**	**61.4**

performance increase can be observed. However, we achieve a significant performance boost by introducing the bounding-box size as an additional geometric feature (fourth row). We obtain **+1.8** and **+0.9** increase on the S_{assoc} and $LSTQ$ scores, respectively. The improvement is doubled by the additional use of the predicted bounding-box sizes compared to only use the refined proposal center positions and the radius estimation as the geometric features. By that, we prove two aspects. Our idea, introducing new geometric features, brings a notable boosting. Moreover, the quality of the learned geometric features affects the performance.

The impact of the learned geometric features is also visualized in an example scan in Fig. 3. The refined center positions are used as the geometric feature. In the first column, it is possible to observe how the proposal center positions get

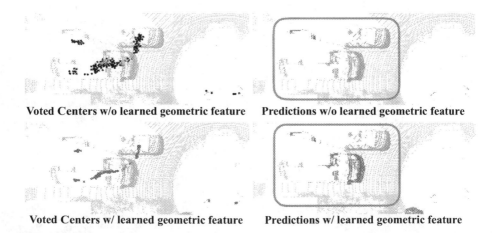

Voted Centers w/o learned geometric feature   Predictions w/o learned geometric feature

Voted Centers w/ learned geometric feature   Predictions w/ learned geometric feature

Fig. 3. Learned Geometric Features. On an example scan, we show the impact of a learned geometric feature, the refined space-time proposal center point positions. In the first column, we display how the center positions change in 4D space-time volume with the learned geometric feature. More precise center positions can be observed using the learned geometric features. In the second column, the predictions are depicted with and without the learned geometric feature. Inside the blue boxes, it can be seen how a wrong prediction is corrected by using the learned geometric feature. In order to emphasize the results, the points with the *stuff* class label are colored gray. The example is generated by combining two scans. (Color figure online)

more precise by the use of the learned refinement. In the second row, the refined predicted proposal centers are closer to the ground truth centers compared to the predicted proposal centers without refinement in the first row. Also, we show how the predictions change with the learned geometric features in the second column. As seen in the blue boxes, we can correctly separate two objects predicted as one single object without the learned geometric features.

4.3   Qualitative Results

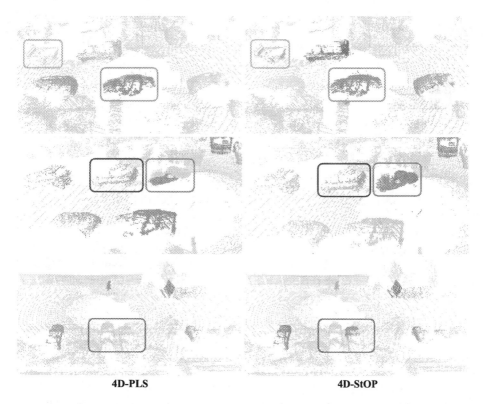

4D-PLS   **4D-StOP**

Fig. 4. Qualitative results on SemanticKITTI validation. We show results for 4D-PLS (left) and 4D-StOP (right) of the same scans. Interesting cases are highlighted with boxes. Our method catches missed objects (first row), complete partially segmented objects (second row), and separate two incorrectly predicted objects as one single object (third row). For simplicity, points with *stuff* class label are colored gray. All the scans are 4D volumes over 4 scans.

Fig. 5. Failure cases of 4D-StOP on the SemanticKITTI validation split. We show two scans where 4D-StOP fails. Here, 4D-StOP cannot separate two objects and mixes them or predicts them as one single object (see red boxes). The dark gray squares show the refined proposal centers. Again, the points with *stuff* class label are colored gray and all the scans are 4D volumes over 4 scans. (Color figure online)

In Fig. 4, we present qualitative results for different example scans in the first and second column for 4D-PLS and our 4D-StOP method respectively. We indicate each specific case within the boxes. Compared to 4D-PLS, our method is able to catch and segment the missing object (left example in the first row), complete the partly segmented objects (right example in the first row and the examples in the second row), and correctly separate the two objects predicted as one single object (the middle example in the third row).

Figure 5 shows two examples where 4D-StOP fails (red boxes). We observe that 4D-StOP can fail when objects and thus their object center votes are very close to each other and objects have a similar size. Then, the aggregation features of the corresponding proposals are quite similar and it is hard for 4D-StOP to distinguish which proposal belongs to which object. As a consequence, 4D-StOP mixes the objects (left) or predicts the objects as one instance (right).

5   Conclusion

In this work, we have proposed a new end-to-end trainable method 4D-StOP for 4D Panoptic LiDAR Segmentation task, which models tracklets in a space-time volume by generating multiple proposals, then aggregating them. Our model is based on center voting for generating tracklet proposals and learning high-level geometric features for aggregating tracklet proposals. We have introduced novel geometric features to enhance the proposal aggregation that improves performance notably. Additionally, we have proved that center voting is feasible in the space-time volume. We achieve state-of-the-art results on SemanticKITTI test-set. We hope that our work would inspire new representations of tracklets for future methods tackling 4D Panoptic LiDAR Segmentation task in a unified space-time volume.

Acknowledgments. We thank Sima Yagmur Zulfikar for her help and feedback on the figures, and István Sárándi for his helpful comments on our manuscript. This project was funded by ERC Consolidator Grant DeeVise (ERC-2017-COG-773161). The computing resources for several experiments were granted by RWTH Aachen University under project 'supp0003'. Francis Engelmann is a post-doctoral research fellow at the ETH AI Center. This work is part of the first author's master thesis.

References

1. Athar, A., Mahadevan, S., Ošep, A., Leal-Taixé, L., Leibe, B.: STEm-Seg: spatiotemporal embeddings for instance segmentation in videos. In: Vedaldi, A., Bischof, H., Brox, T., Frahm, J.-M. (eds.) ECCV 2020. LNCS, vol. 12356, pp. 158–177. Springer, Cham (2020). https://doi.org/10.1007/978-3-030-58621-8_10

2. Aygün, M., et al.: 4D panoptic segmentation. In: CVPR (2021)

3. Bai, X., et al.: TransFusion: Robust LiDAR-Camera Fusion for 3D Object Detection with Transformers. arXiv preprint arXiv:2203.11496 (2022)

4. Behley, J., et al.: SemanticKITTI: a dataset for semantic scene understanding of LiDAR sequences. In: ICCV (2019)

5. Behley, J., Milioto, A., Stachniss, C.: A benchmark for LiDAR-based panoptic segmentation based on KITTI. In: ICRA (2021)

6. Bergmann, P., Meinhardt, T., Leal-Taixé, L.: Tracking without bells and whistles. In: ICCV (2019)

7. Braso, G., Leal-Taixé, L.: Learning a neural solver for multiple object tracking. In: CVPR (2020)

8. Caesar, H., et al.: nuScenes: a multimodal dataset for autonomous driving. In: CVPR (2020)

9. Caron, M., Misra, I., Mairal, J., Goyal, P., Bojanowski, P., Joulin, A.: Unsupervised learning of visual features by contrasting cluster assignments. In: NIPS (2020)

10. Chen, T., Kornblith, S., 0002, M.N., Hinton, G.E.: A simple framework for contrastive learning of visual representations. In: International Conference on Machine Learning (ICML) (2020)

11. Chibane, J., Engelmann, F., Tran, T.A., Pons-Moll, G.: Box2Mask: weakly supervised 3D semantic instance segmentation using bounding boxes. In: Avidan, S., Brostow, G., Cissé, M., Farinella, G.M., Hassner, T. (eds.) ECCV 2022, pp. 681–699. Springer, Cham (2022). https://doi.org/10.1007/978-3-031-19821-2_39

12. Chiu, H.K., Prioletti, A., Li, J., Bohg, J.: Probabilistic 3D Multi-Object Tracking for Autonomous Driving. In: arXiv preprint arXiv:2001.05673 (2020)

13. Cortinhal, T., Tzelepis, G., Erdal Aksoy, E.: SalsaNext: fast, uncertainty-aware semantic segmentation of LiDAR point clouds. In: International Symposium on Visual Computing (2020)

14. Elich, C., Engelmann, F., Schult, J., Kontogianni, T., Leibe, B.: 3D-BEVIS: birdseye-view instance segmentation. In: German Conference on Pattern Recognition (GCPR) (2019)

15. Engelmann, F., Bokeloh, M., Fathi, A., Leibe, B., Nießner, M.: 3D-MPA: multiproposal aggregation for 3D semantic instance segmentation. In: CVPR (2020)

16. Engelmann, F., Kontogianni, T., Hermans, A., Leibe, B.: Exploring spatial context for 3D semantic segmentation of point clouds. In: ICCV Workshops (2017)

17. Engelmann, F., Kontogianni, T., Schult, J., Leibe, B.: Know what your neighbors do: 3D semantic segmentation of point clouds. In: ECCV Workshops (2018)

18. Fong, W.K., et al.: Panoptic nuScenes: A Large-Scale Benchmark for LiDAR Panoptic Segmentation and Tracking. In: arXiv preprint arXiv:2109.03805 (2021)
19. Geiger, A., Lenz, P., Urtasun, R.: Are we ready for autonomous driving? The KITTI vision benchmark suite. In: CVPR (2012)
20. He, K., Gkioxari, G., Dollár, P., Girshick, R.: Mask R-CNN. In: ICCV (2017)
21. Hong, F., Zhou, H., Zhu, X., Li, H., Liu, Z.: LiDAR-based panoptic segmentation via dynamic shifting network. In: CVPR (2021)
22. Hou, J., Dai, A., Nießner, M.: 3D-SIS: 3D semantic instance segmentation of RGB-D scans. In: CVPR (2019)
23. Hurtado, J.V., Mohan, R., Burgard, W., Valada, A.: MOPT: multi-object panoptic tracking. In: CVPR Workshops (2020)
24. Kim, D., Woo, S., Lee, J.Y., Kweon, I.S.: Video panoptic segmentation. In: CVPR (2020)
25. Kim, Aleksandr, O.A., Leal-Taixé, L.: EagerMOT: 3D multi-object tracking via sensor fusion. In: ICRA (2021)
26. Lahoud, J., Ghanem, B., Pollefeys, M., Oswald, M.R.: 3D instance segmentation via multi-task metric learning. In: ICCV (2019)
27. Landrieu, L., Simonovsky, M.: Large-scale point cloud semantic segmentation with superpoint graphs. In: CVPR (2018)
28. Lang, A.H., Vora, S., Caesar, H., Zhou, L., Yang, J., Beijbom, O.: PointPillars: fast encoders for object detection from point clouds. In: CVPR (2019)
29. Leal-Taixé, L., Fenzi, M., Kuznetsova, A., Rosenhahn, B., Savarese, S.: Learning an image-based motion context for multiple people tracking. In: CVPR (2014)
30. Marcuzzi, R., Nunes, L., Wiesmann, L., Vizzo, I., Behley, J., Stachniss, C.: Contrastive instance association for 4D panoptic segmentatio using sequences of 3D LiDAR scans. In: IEEE Robotics and Automation Society (2022)
31. Meinhardt, T., Kirillov, A., Leal-Taixé, L., Feichtenhofer, C.: TrackFormer: multi-object tracking with transformers. In: CVPR (2022)
32. Milan, A., Leal-Taixé, L., Schindler, K., Reid, I.D.: Joint tracking and segmentation of multiple targets. In: CVPR (2015)
33. Milioto, A., Vizzo, I., Behley, J., Stachniss, C.: RangeNet++: fast and accurate LiDAR semantic segmentation. In: IROS (2019)
34. Milioto, A., Behley, J., McCool, C., Stachniss, C.: LiDAR panoptic segmentation for autonomous driving. In: IROS (2020)
35. Misra, I., Girdhar, R., Joulin, A.: An end-to-end transformer model for 3D object detection. In: ICCV (2021)
36. Mittal, H., Okorn, B., Held, D.: Just go with the flow: self-supervised scene flow estimation. In: CVPR (2020)
37. Neven, D., Brabandere, B.D., Proesmans, M., Gool, L.V.: Instance segmentation by jointly optimizing spatial embeddings and clustering bandwidth. In: CVPR (2019)
38. Oord, A.V.D., Li, Y., Vinyals, O.: Representation Learning with Contrastive Predictive Coding. arXiv preprint arXiv:1807.03748 (2018)
39. Pang, J., et al.: Quasi-dense similarity learning for multiple object tracking. In: CVPR (2021)
40. Qi, C., Su, H., Mo, K., Guibas, L.J.: PointNet: deep learning on point sets for 3D classification and segmentation. In: CVPR (2017)
41. Qi, C.R., Litany, O., He, K., Guibas, L.J.: Deep hough voting for 3D object detection in point clouds. In: ICCV (2019)
42. Ren, S., He, K., Girshick, R., Sun, J.: Faster R-CNN: towards real-time object detection with region proposal networks. In: NIPS (2015)

43. Shi, S., et al.: PV-RCNN: point-voxel feature set abstraction for 3D object detection. In: CVPR (2020)
44. Shi, S., Wang, X., Li, H.: PointRCNN: 3D object proposal generation and detection from point cloud. In: CVPR (2019)
45. Sun, P., et al.: Scalability in perception for autonomous driving: waymo open dataset. In: CVPR (2020)
46. Tang, H., et al.: Searching efficient 3D architectures with sparse point-voxel convolution. In: Vedaldi, A., Bischof, H., Brox, T., Frahm, J.-M. (eds.) ECCV 2020. LNCS, vol. 12373, pp. 685–702. Springer, Cham (2020). https://doi.org/10.1007/978-3-030-58604-1_41
47. Thomas, H., Qi, C.R., Deschaud, J.E., Marcotegui, B., Goulette, F., Guibas, L.J.: KPConv: flexible and deformable convolution for point clouds. In: ICCV (2019)
48. Voigtlaender, P., et al.: MOTS: multi-object tracking and segmentation. In: CVPR (2019)
49. Wang, Y., Kitani, K., Weng, X.: Joint object detection and multi-object tracking with graph neural networks. In: ICRA (2021)
50. Weng, X., Wang, J., Held, D., Kitani, K.: 3D multi-object tracking: a baseline and new evaluation metrics. In: IROS (2020)
51. Weng, X., Wang, J., Held, D., Kitani, K.: AB3DMOT: a baseline for 3D multi-object tracking and new evaluation metrics. In: ECCV Workshops (2020)
52. Wu, B., Wan, A., Yue, X., Keutzer, K.: SqueezeSeg: convolutional neural nets with recurrent CRF for real-time road-object segmentation from 3D LiDAR point cloud. In: ICRA (2018)
53. Yan, Y., Mao, Y., Li, B.: SECOND: sparsely embedded convolutional detection. Sensors **18**(10), 3337 (2018)
54. Yang, B., et al.: Learning Object Bounding Boxes for 3D Instance Segmentation on Point Clouds. arXiv preprint arXiv:1906.01140 (2019)
55. Yang, Z., Sun, Y., Liu, S., Jia, J.: 3DSSD: point-based 3D single stage object detector. In: CVPR (2020)
56. Yin, T., Zhou, X., Krähenbühl, P.: Center-based 3D object detection and tracking. In: CVPR (2021)
57. Zhang, Y., Zhou, Z., David, P., Yue, X., Xi, Z., Foroosh, H.: PolarNet: an improved grid representation for online LiDAR point clouds semantic segmentation. In: CVPR (2020)
58. Zhou, X., Koltun, V., Krähenbühl, P.: Tracking objects as points. In: Vedaldi, A., Bischof, H., Brox, T., Frahm, J.-M. (eds.) ECCV 2020. LNCS, vol. 12349, pp. 474–490. Springer, Cham (2020). https://doi.org/10.1007/978-3-030-58548-8_28
59. Zhou, Y., Tuzel, O.: VoxelNet: end-to-end learning for point cloud based 3D object detection. In: CVPR (2018)
60. Zhou, Z., Zhang, Y., Foroosh, H.: Panoptic-PolarNet: proposal-free LiDAR point cloud panoptic segmentation. In: CVPR (2021)
61. Zhu, X., et al.: Cylindrical and asymmetrical 3D convolution networks for LiDAR segmentation. In: CVPR (2021)

BlindSpotNet: Seeing Where We Cannot See

Taichi Fukuda$^{(\boxtimes)}$, Kotaro Hasegawa, Shinya Ishizaki, Shohei Nobuhara, and Ko Nishino

Kyoto University, Kyoto, Japan
{tfukuda,khasegawa,sishizaki}@vision.ist.i.kyoto-u.ac.jp,
{nob,kon}@i.kyoto-u.ac.jp
https://vision.ist.i.kyoto-u.ac.jp/

Abstract. We introduce 2D blind spot estimation as a critical visual task for road scene understanding. By automatically detecting road regions that are occluded from the vehicle's vantage point, we can proactively alert a manual driver or a self-driving system to potential causes of accidents (*e.g.*, draw attention to a road region from which a child may spring out). Detecting blind spots in full 3D would be challenging, as 3D reasoning on the fly even if the car is equipped with LiDAR would be prohibitively expensive and error prone. We instead propose to learn to estimate blind spots in 2D, just from a monocular camera. We achieve this in two steps. We first introduce an automatic method for generating "ground-truth" blind spot training data for arbitrary driving videos by leveraging monocular depth estimation, semantic segmentation, and SLAM. The key idea is to reason in 3D but from 2D images by defining blind spots as those road regions that are currently invisible but become visible in the near future. We construct a large-scale dataset with this automatic offline blind spot estimation, which we refer to as Road Blind Spot (RBS) dataset. Next, we introduce BlindSpotNet (BSN), a simple network that fully leverages this dataset for fully automatic estimation of frame-wise blind spot probability maps for arbitrary driving videos. Extensive experimental results demonstrate the validity of our RBS Dataset and the effectiveness of our BSN.

Keywords: Autonomous driving · ADAS · Road scene understanding · Blind spot estimation · Accident prevention

1 Introduction

Fully autonomous vehicles may soon become a reality. Advanced Driver-Assistance Systems (ADAS) have also become ubiquitous and intelligent. A large part of these developments is built on advancements in perceptual sensing, both in hardware and software. In particular, 3D and 2D visual sensing have played a large role in propelling these advancements. State-of-the-art LiDAR systems can resolve to centimeters at 300 m or longer, and image understanding networks can recognize objects 100 m away. A typical autonomous vehicle is equipped with up to ten of these sensors' visual perception, in addition to other modalities.

© The Author(s), under exclusive license to Springer Nature Switzerland AG 2023
L. Karlinsky et al. (Eds.): ECCV 2022 Workshops, LNCS 13801, pp. 554–569, 2023.
https://doi.org/10.1007/978-3-031-25056-9_35

Fig. 1. We introduce a new dataset and network for automatically detecting blind spots on roads. Alerting manual and autonomous drivers of vehicles to such 2D blind spots can greatly extend the safety on the road. Our network, BlindSpotNet, successfully detects blind spots, from which pedestrians actually later spring out, without the need for costly, error-prone 3D reconstruction.

These autonomous and assistive driving systems are, however, still prone to catastrophic errors, even when they are operating at low speeds [23]. More sensors would unlikely eliminate these errors. In fact, we human beings have much fewer visual sensors (just our two eyes) but can drive at least as well as current autonomous vehicles. Why are we able to do this? We believe, one of the primary reasons is that, although we are limited in our visual perception, we know that it's limited. That is, we are fully aware of when and where we can see and when and where we cannot see. We know that we can't see well at night so we drive cautiously; we know that we cannot see beyond a couple of hundred meters, so we don't drive too fast; we know that we cannot see behind us, so we use side and room mirrors. Most important, we know that we cannot see behind objects, *i.e.*, that our vision can be obstructed by other objects in the scene. We are fully aware of these blind spots, so that we pay attention to those areas on the road with anticipation that something may spring out from them. That is, we know where to expect the unforeseen and we preemptively prepare for those events by attending our vision and mind to those blind spots.

Can we make computers also "see," *i.e.*, find, blind spots on the road? One approach would be to geometrically reason occlusions caused by the static and dynamic objects on the road. This requires full 3D scene reconstruction and localization of the moving camera, which is computationally expensive and prone to errors as it requires fragile ray traversal. Instead, it would be desirable to directly estimate blind spots in the 2D images without explicit 3D reconstruction or sensing. This is particularly essential for driving safety, as we would want to detect blind spots on the fly as we drive down the road. Given a frame from a live video, we would like to mark out all the occluded road regions, so that a driver or the autonomous system would know where to anticipate the unforeseen.

How can we accomplish 2D blind spot detection? Just like we likely do, we could learn to estimate blind spots directly in 2D. The challenge for this lies in obtaining the training data—the ground truth. Again, explicit 3D reasoning is unrealistic as a perfect 3D reconstruction of every frame of a video would be prohibitively expensive and error prone, especially for a dynamic road scene in which blind spots change every frame. On the other hand, labeling blind spots in video frames is also near impossible, as even to the annotators, reasoning about the blind spots from 2D images is too hard a task, particularly for a dynamic scene. How can we then, formulate blind spot estimation as a learning task?

In this paper, we introduce a novel dataset and network for estimating occluded road regions in a driving video (Fig. 1). We refer to the dataset as Road Blind Spot (RBS) Dataset and the network as BlindSpotNet (BSN). Our major contributions are two-fold: an algorithm for automatic generation of blind spot training data and a simple network that learns to detect blind spots from that training data. The first contribution is realized by implicitly reasoning 3D occlusions in a driving video from its 2D image frames by fully leveraging depth, localization, and semantic segmentation networks. This is made possible by our key idea of defining blind spots as areas on the road that are currently invisible but visible in the future. We refer to these as T-frame blind spots, $i.e.$, those 2D road regions that are currently occluded by other objects that become visible T frames later. Clearly, these form a subset of all true 2D blind spots; we cannot estimate the blind spots that never become seen in our video. They, however, cover a large portion of the blind spots ($cf.$ Fig. 6) that are critical in a road scene including those caused by parked cars on the side, oncoming cars on the other lane, street corners, pedestrians, and buildings. Most important, they allow us to derive an automatic means for computing blind spot regions for arbitrary driving videos.

Our offline automatic training data generation algorithm computes T-frame blind spots for each frame of a driving video by playing it backwards, and by applying monocular depth estimation, SLAM, and semantic segmentation to obtain the depth, camera pose, semantic regions, and road regions. By computing the visible road regions in every frame, and then subtracting the current frame's from that of T-frames ahead, we can obtain blind spots for the current frame. A suitable value for T can be determined based on the speed of the car and the frame rate of the video. Armed with this simple yet effective algorithm for computing blind spot maps, we construct the RBS Dataset. The dataset consists of 231 videos with blind spot maps computed for 21,662 frames.

For on-the-fly 2D blind spot estimation, we introduce BlindSpotNet, a deep neural network that estimates blind spot road regions for an arbitrary road scene directly for a single 2D input video frame, which fully leverages the newly introduced dataset. The network architecture is a fully convolutional encoder-decoder with Atrous Spatial Pyramid Pooling which takes in a road scene image as the input. We show that blind spot estimation can be implemented with a light-weight network by knowledge distillation from a semantic segmentation network. Through extensive experiments including the analysis of the network architectures, we show that BlindSpotNet can accurately estimate the occluded road regions in any given frame independently, yet result in consistent blind spot maps through a video.

To the best of our knowledge, our work is the first to offer an extensive dataset and a baseline method for solving this important problem of 2D blind spot estimation. These results can directly be used to heighten the safety of autonomous driving and assisted driving, for instance, by drawing attention of the limited computation resource or the human driver to those blind spots (*e.g.*, by sounding an alarm from the side with an oncoming large blind spot, if the person is looking away observed from an in-car camera). We also envision a future where BlindSpotNet helps in training better drivers both for autonomous and manual driving.

2   Related Work

Driver assistance around blind corners has been studied in the context of augmented/diminished reality [2, 20, 30]. Barnum *et al.* proposed a video-overlay system that presents a see-through rendering of hidden objects behind the wall for drivers using a surveillance camera installed at corners [2]. This system realizes realtime processing, but requires explicit modeling of blind spots beforehand. Our proposed method estimates them automatically.

Bird's-eye view (BEV) visualization [14, 33] and scene parsing [1, 11, 12, 19, 25, 29] around the vehicle can also assist drivers to avoid collision accidents. Sugiura and Watanabe [25] proposed a neural network that produces probable multi-hypothesis occupancy grid maps. Mani *et al.* [19], Yang *et al.* [29], and Liu *et al.* [14] proposed road layout estimation networks. These methods, however, require sensitive 3D reasoning to obtain coarse blind spots at runtime, while our BSN estimates blind spots directly in 2D without any 3D processing.

Amodal segmentation also handles occluded regions of each instance explicitly [8, 10, 15, 24, 27, 32, 34]. They segment the image into semantic regions while estimating occluded portions of each region. These methods, however, do not estimate objects that are completely invisible in the image. We can find a side road while its road surface is not visible at all, *e.g.*, due to cars parked in the street, by looking at gaps between buildings for example.

Understanding and predicting pedestrian behavior is also a primary objective of ADAS. Makansi *et al.* [17] trained a network that predicts pedestrians crossing in front of the vehicle. Bertoni *et al.* [3] estimated 3D locations of pedestrians around the subject by also modeling the uncertainty behind them. Our method explicitly recovers blind spots caused not only by pedestrians but also other obstacles including passing and parked cars, street medians, poles, etc.

Watson *et al.* [28] estimated free space including the area behind objects in the scene for robot navigation. They also generated a traversable area dataset to train their network. Our dataset generation leverages this footprint dataset generation algorithm to compute both visible and invisible road areas in a road scene image (*i.e.*, driving video frame). Blind spot estimation requires more than just the area behind objects as it needs to be computed and estimated across frames with a dynamically changing viewpoint.

3   Road Blind Spot Dataset

Our first goal is to establish a training dataset for learning to detect 2D blind spots. For this, we need an algorithm that can turn an arbitrary video into a

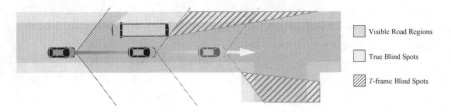

Fig. 2. We define blind spots as road regions that are currently invisible but visible in the next T-frames, which we compute by aggregating road regions across future frames and subtracting the visible road region in the current frame. These T-frame blind spots form a subset of true blind spots, cover key regions of them and, most important, can directly be computed for arbitrary driving videos (*cf.* Fig. 6 for a real example).

video annotated with 2D blind spots for each and every frame. To derive such a method, we start by defining road-scene blind spots as something that we can compute from 2D videos offline and derive a method for computing those for arbitrary driving videos. With this algorithm, we construct a large-scale dataset of road blind spot videos (RBS Dataset). This data will later be used to train a network that can estimate 2D blind spots on the fly.

3.1 T-Frame Blind Spots

In the most general form, blind spots are volumes of the 3D space that are occluded from the viewpoint by an object in the scene. Computing these "full blind volumes" would be prohibitively expensive, especially for any application that requires real-time decision making. Even though our goal in this paper is not necessarily real-time computation at this point, a full 3D reasoning of occluded volumes would be undesirable as our target scenes are dynamic. We, instead, aim to estimate 2D blind spots on the road. Dangerous traffic situations are usually caused by unanticipated movements of dynamic objects (*e.g.*, bikes, pedestrians, children, *etc.*) springing out from road areas invisible from the driver (or the camera of the car). Once we have the 2D blind spots, we can draw attention of the drivers by, for instance, extruding it perpendicularly to the road for 3D warning. By focusing on 2D blind spot estimation, we eliminate the need of explicit 3D geometric reasoning, which makes it particularly suitable for autonomous driving and ADAS applications.

As depicted in Fig. 2, given a frame of a driving video, we define our blind spots to be the road regions that are obstructed but become visible in the future. This clearly excludes blind spots that never become visible in any frame in the future, and thus the blind spots we compute and estimate are a subset of the true blind spots. That said, they have a good coverage of the true blind spots, as pedestrians have to always go through the T-frame blind spots or visible road regions to come out in front of the vehicle. These T-frame blind spots can be reliably computed from just ego-centric driving videos. Although we only investigate the estimation of these blind spots from ego-centric driving videos

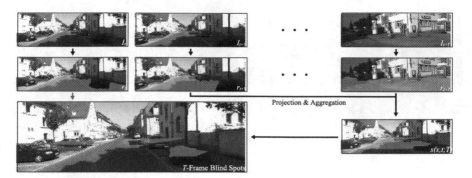

Fig. 3. Overview of algorithmic steps for generating T-frame blind spots for arbitrary driving video. Visible and invisible road maps are aggregated into the current frame from the next T frames, from which the visible road region is subtracted to obtain the T-frame blind spots for the current frame.

captured with regular perspective cameras, the wider the field of view, the better the coverage would become.

Formally, given a frame I_t, we define its blind spots as pixels $x \in \Omega_t^T$ corresponding to regions on the road Ω that are occluded but become visible in the next T frames $\{I_{t+i} | i = 1, \ldots, T\}$. Our goal is to compute the set of pixels in the blind spots Ω_t as a binary mask of the image $\omega(x, t; T) : \mathbb{R}^2 \times \mathbb{R} \mapsto \{0, 1\}$. Later BlindSpotNet will be trained to approximate this function $\omega(x, t; T)$. We refer to the blind spots of this definition as T-frame blind spots.

As depicted in Fig. 3, we compute the blind spot map ω by first computing and aggregating visible road maps at frames $I_{t:t+T}$ and then eliminating the visible road map at target frame I_t. For this, similar to Watson *et al.* [28], we compute an aggregated road map $s(x, t; T)$ by forward warping the road pixels from the next T frames $\{I_{t+i} | i = 1, \ldots, T\}$ to the target frame I_t. To perform forward warping, we assume the camera intrinsics are known, the extrinsic parameter and the depth are estimated by a visual SLAM algorithm [26] and an image-based depth estimation network [21], respectively.

Let $r(x, t)$ denote the visible road region defined as a binary mask representing the union of the road and pavement areas estimated by a semantic segmentation as introduced by Watson *et al.* [28]. We first project the visible road regions from I_{t+i} $(i = 1, \ldots, T)$ to I_t as $r'(x, t + i; t)$, and then aggregate them as

$$s(x, t; T) = r'(x, t + 1; t) \vee r'(x, t + 2; t) \vee \cdots \vee r'(x, t + T; t), \qquad (1)$$

where $\vee$ denotes the pixel-wise logical OR. As blind spots are by definition invisible regions in frame I_t, the visible road region $r(x, t)$ is subtracted from $s(x, t; T)$ to obtain the final blind spots $\omega(x, t; T)$ by

$$\omega(x, t; T) = s(x, t; T) \wedge \bar{r}(x, t), \qquad (2)$$

where $\wedge$ and $\bar{}$ denote the pixel-wise logical AND and negation, respectively.

Fig. 4. Visibility mask. Left: If the vehicle makes a turn, the blind spots straight ahead across the intersection never become visible. Right: To allow networks estimate such blind spots, we prepare a mask image to indicate the visible area for each frame.

Our method relies on three visual understanding tasks, namely semantic segmentation, monocular depth estimation, and SLAM. Although existing methods for these tasks achieve high accuracy, they can still suffer from slight errors. In the transformation from r to r', we use two such estimates, those of the camera pose and the depth, whose errors can cause residuals of blind spots after Eq. (2).

We can rectify this with simple depth comparison. We first define aggregated depth mask d_a

$$d_a(x, t; T) = \frac{1}{M} \sum_{i=1}^{T} r'(x, t+i; t) d'(x, t+i; t) \qquad M = \sum_{i=1}^{T} r'(x, t+i; t), \quad (3)$$

where $d'(x, t+i; t)$ is a depth map of frame I_{t+i} projected onto frame I_t, which is calculated in a similar way as calculation of $r'(x, t+i; t)$. M is the pixel-wise count of visible road mask over T frames. When we compare the depth map $d(x)$ of frame I_t with the aggregated depth mask d_a, the depth difference is large in the true blind spot region because true blind spots are occluded by foreground objects. On the other hand, it is small in erroneous blind spot regions because the compared depth values come from nearby pixels. Based on this observation, we remove erroneous blind spots by setting $\omega(x, t; T) = 0$ for the pixel x that satisfy $|d(x) - d_a(x)| < l_d$. Here, l_d is a threshold value determined empirically.

We may be left with small blind spots caused by, for instance, a shadow of a tire. These small blind spots are not important for driving safety. We remove these blind spots of less than 100 pixels from the final blind spot Ω_t.

For building our RBS Dataset, we opt for MiDAS [21] as the monocular depth estimator, OpenVSLAM [26] for SLAM, and Panoptic-DeepLab [6] for semantic segmentation. The scale of the depth estimated by MiDAS is linearly aligned with least squares fitting to the sparse 3D landmarks recovered by SLAM.

We use KITTI [9], BDD100k [31], and TITAN [18] datasets to build our RBS Dataset. By excluding videos for which the linear correlation coefficient in the MiDAS-to-SLAM depth alignment is less than 0.7, they provide 51, 62, and 118 videos, respectively. We obtain blind spot masks for approximately 51, 34, and 12 min of videos in total, respectively. The videos are resampled to 5 fps, and we set $T = 5$ seconds for each video. We refer to them as KITTI-RBS, BDD-RBS, and TITAN-RBS Datasets, respectively.

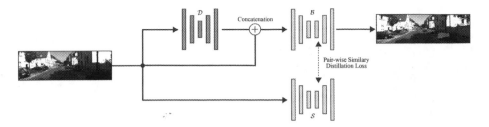

Fig. 5. Overall architecture of BlindSpotNet. We leverage the fact that blind spot estimation bears similarity to semantic segmentation by adopting a light-weight network trained with knowledge distillation from a semantic segmentation teacher network.

3.2   Visibility Mask

T-frame blind spots clearly do not cover blind spots that do not become visible through the video. For example, as illustrated in Fig. 4, consider a frame where the vehicle is making a right turn. The blind spots straight ahead across the intersection never become visible and hence are not included in the dataset, while BlindSpotNet should identify them once trained. To disambiguate such invisible regions from non-blind spots in training BSN, as shown in Fig. 4, we generate a binary mask V_t called *visibility mask* for each frame in addition to the blind spots Ω_t.

We use semantic segmentation and the distance from the camera to define the visibility mask V_t. For each pixel x, we first classify x as visible, if the semantic segmentation label is "sky." For non-sky pixels, we classify x as visible if the minimum distance from the 3D point at distance $d(x, t)$ behind x to the camera is less than a certain threshold L. In our implementation, we set $L = 16$ m.

4   BlindSpotNet

Now that we have (and can create limitless) abundant video data with per-frame blind spot annotations, we can formulate 2D blind spot detection as a learning problem. We derive a novel deep neural network for estimating blind spots in an image of a road scene. We refer to this network as BlindSpotNet and train and test it on our newly introduced RBS Dataset.

4.1   Network Architecture

As we saw in Sect. 3, blind spots are mainly determined by the scene composition of objects and their ordering in 2D. As such, at a higher level, direct image-based estimation of blind spots shares similarity in its task to semantic segmentation. The task is, however, not necessarily easier, as it is 2D dense labeling but requires implicit 3D reasoning. Nevertheless, the output is a binary map (or its probability map), which suggests that a simpler network but with a similar representation to semantic segmentation would be suitable for blind spot detection.

Figure 5 depicts the network architecture of BlindSpotNet. BlindSpotNet consists of three components: a depth estimator $\mathcal{D}$, a semantic segmentation teacher network $\mathcal{S}$, and a blind spot estimator $\mathcal{B}$. Given an input image I of size $W \times H \times 3$, the depth estimator $\mathcal{D}$ estimates the depth map D of size $W \times H$ from I. The blind spot estimator takes both the RGB image I and the estimated depth map D as inputs and generates the blind spot map $B = \{b \in [0,1]\}$.

The semantic segmentation network $\mathcal{S}$ serves as a teacher network to help train the blind spot estimator $\mathcal{B}$. The blind spot estimator $\mathcal{B}$ should be trained to reason semantic information of the scene similar to semantic segmentation, but its output is abstracted as simple as a single-channel map B. This implies that training the blind spot estimator $\mathcal{B}$ only with the T-frame blind spots can easily bypass the semantic reasoning of the scene and overfit. To mitigate this shortcut, we introduce the semantic segmentation network $\mathcal{S}$ pretrained on road scenes as a teacher and use its decoder output as a soft target of a corresponding layer output in the blind spot estimator $\mathcal{B}$.

4.2   Knowledge Distillation

Blind spot regions are highly correlated with the semantic structure of the scene. For instance, blind spots can appear behind vehicles and buildings, but never in the sky. The blind spot estimator $\mathcal{B}$ should thus be able to learn useful representations from semantic segmentation networks for parsing road scenes. For this, we distill knowledge from a pretrained semantic segmentation network to our blind spot estimator $\mathcal{B}$. Based on the work of Liu *et al.* [16], we transfer the similarity between features at an intermediate layer of each network.

Suppose we subdivide the feature map of an intermediate layer of size $W' \times H' \times C$ into a set of $w' \times h'$ patches. By denoting the spatial average of the features in the ith patch by $\boldsymbol{f}_i \in \mathbb{R}^C$, we define the similarity of two patches i and j by their cosine distance $a_{ij} = \frac{\boldsymbol{f}_i^\top \boldsymbol{f}_j}{\|\boldsymbol{f}_i\|\|\boldsymbol{f}_j\|}$. Following Liu *et al.* [16], given this pairwise similarity for patches in both the teacher semantic segmentation network $\mathcal{S}$ and the student network, *i.e.*, the blind spot estimator $\mathcal{B}$, as $a_{ij}^{\mathcal{S}}$ and $a_{ij}^{\mathcal{B}}$, respectively, we introduce a pair-wise similarity distillation loss l_{KD} as

$$l_{\mathrm{KD}} = \frac{1}{(w' \times h')^2} \sum_{i \in \mathcal{R}} \sum_{j \in \mathcal{R}} \left(a_{ij}^{\mathcal{S}} - a_{ij}^{\mathcal{B}} \right)^2, \tag{4}$$

where $\mathcal{R} = \{1, 2, \ldots, w' \times h'\}$ denotes the entire set of patches. In our implementation, we opted for DeepLabV3+ [5] as the teacher network $\mathcal{S}$.

4.3   Loss Function

In addition to the similarity distillation loss l_{KD} in Eq. (4), we employ a binary cross entropy loss l_{BCE} between the output of the blind spot estimator $\mathcal{B}$ and the T-frame blind spots given by our RBS Dataset as

$$l_{\mathrm{BCE}} = -\frac{1}{|V|} \sum_{x \in V} \left(\omega(x) \log b(x) + (1 - \omega(x)) \log(1 - b(x)) \right), \tag{5}$$

Table 1. Quantitative evaluation of RBS Datasets. The two numbers in each cell indicate the recall and the false-negative rate w.r.t.the ground truth blind spot. The results show that our T-frame blind spots approximate true blind spots well.

	CARLA		KITTI	
	Rec.↑	FN rate↓	Rec.↑	FN rate↓
T-frame BS-GT	0.372	0.013	N/A[a]	
T-frame BS (ours)	0.297	0.015	0.169	0.056

[a] T-frame BS-GT is not available since KITTI does not have ground truth semantic segmentation for most of the frames.

where x denotes the pixels in the visibility mask V, $\omega(x)$ and $b(x)$ denote the T-frame blind spots and the estimated probabilities at pixel x. $|V|$ is the total number of the pixels in V. The total loss function L is defined as a weighted sum of these two loss functions $L = l_{\mathrm{BCE}} + \lambda l_{\mathrm{KD}}$, where λ is an empirically determined weighting factor.

5   Experimental Results

We evaluate the validity of RBS Dataset and the effectiveness of BlindSpotNet (BSN), qualitatively and quantitatively with a comprehensive set of experiments.

5.1   RBS Dataset Evaluation

We first validate our T-frame blind spots with synthetic data generated by CARLA [7] and with real data from KITTI [9]. How well do they capture true blind spots? We use 360° LiDAR scans to obtain ground-truth blind spots (BS-GT). Note again that for real use such ground truth computation will be prohibitively expensive and would require LiDAR. We also use the ground-truth depth maps to compute T-frame blind spots without noise (T-frame BS-GT). In computing BS-GT, we find road regions in LiDAR points by fitting the road plane manually in 3D. For T-frame BS-GT, we used the ground truth semantic labels. We compare the T-frame blind spots generated by our data generation algorithm (T-frame BS) and T-frame BS-GT with BS-GT, and evaluate its quality in terms of the recall and the false-negative rate. Table 1 and Fig. 6 show the results. Since BS-GT is defined by sparse LiDAR points while T-frame BS-GT and T-frame BS use dense depth-maps, the precision and the false-positive (type-I error) rate do not make sense. These results show that our T-frame blind spots approximate the ground-truth blind spot well.

5.2   BlindSpotNet Evaluation

BlindSpotNet. We use MiDAS [21] and DeepLabV3+ [5] as the depth estimator $\mathcal{D}$ and the semantic segmentation subnetwork $\mathcal{S}$, respectively. We use

Fig. 6. Comparison of T-frame blind spot and "ground truth" obtained from LiDAR data. Top-left and bottom-left figures show the ground truth and T-frame blind spots, respectively. Right figure shows the 3D projection of ground truth (yellow) and T-frame blind spots (red) to LiDAR scan (blue). The green cross indicates the camera position. (Color figure online)

ResNet-18 and Atrous Spatial Pyramid Pooling [5] for the blind spot estimator $\mathcal{B}$ following DeepLabV3+. BlindSpotNet is trained by Adam [13] with $\beta_1 = 0.9, \beta_2 = 0.999, \epsilon = 1 \times 10^{-8}$, and weight decay 5×10^{-4}. The learning rate is initialized to 0.001 and a polynomial scheduler is applied. The knowledge distillation coefficient λ in Sect. 4.3 is fixed to 1.0.

We divided the 10,135 frames from KITTI-RBS into training, validation, and test sets by 8 : 1 : 1, and the 8,872 frames from BDD-RBS into training and test sets by 8 : 2. We used all the 2,655 frames from TITAN-RBS for evaluation only since each video is too short (10 to 20 s) to be used for training.

Metrics. Blind spot estimation is a binary segmentation problem. For each input frame, our BlindSpotNet outputs the probability map of blind spots. We threshold this probability map to obtain the final binary blind spot mask, and compare it with the T-frame blind spots by IoU, recall, and precision. The threshold is determined empirically for each dataset as it depends on the road scene. We plan to learn this threshold as part of the network in future work. Notice that our RBS Datasets include blind spots that become visible in T frames only. For IoU, recall, and precision, we only consider pixels in the visibility mask.

Baselines. As discussed earlier, our work is the first for image-based 2D blind spot estimation, and there are no other existing methods to the best of our knowledge. For comparison, we adapt state-of-the-art traversable region estimation [28], 2D vehicle/pedestrian/cyclist detection by semantic segmentation [5], and 3D vehicle/pedestrian/cyclist detection [4] for 2D blind spot estimation as baselines and refer to them as *Traversable*, *Detection-2D*, and *Detection-3D*, respectively.

For *Traversable* we use the hidden traversable regions, estimated by the original implementation of Watson *et al.* [28], as blind spots. *Detection-2D* is a

Table 2. Quantitative results on the test sets from KITTI-RBS, BDD-RBS, and TITAN-RBS. The lines with "w/ KITTI-RBS" and "w/ BDD-RBS" report the results by the networks trained with KITTI-RBS and BDD-RBS Datasets, respectively. These results show that depth estimation and knowledge distillation contribute independently to the final accuracy, and our BSN successfully estimates blind spots across different scenes (*i.e.*, datasets).

Model	KITTI-RBS			BDD-RBS			TITAN-RBS		
	IoU↑	Rec.↑	Prec.↑	IoU↑	Rec.↑	Prec.↑	IoU↑	Rec.↑	Prec.↑
Traversable based on [28]	0.176	0.462	0.222	0.088	0.198	0.136	0.135	0.303	0.196
Detection-2D based on [6]	0.129	0.581	0.142	0.184	**0.652**	0.204	0.142	0.435	0.182
Detection-3D based on [4]	0.182	0.368	0.265	0.059	0.067	0.316	0.048	0.057	0.216
BSN-D w/ KITTI-RBS	0.296	0.601	0.368	0.295	0.438	0.475	0.250	0.474	0.345
BSN-KD w/ KITTI-RBS	0.305	**0.646**	0.367	0.225	0.249	**0.700**	0.168	0.249	0.342
BSN (Ours) w/ KITTI-RBS	**0.330**	0.563	0.444	0.283	0.349	0.599	0.187	0.280	0.360
BSN-D w/ BDD-RBS	0.270	0.629	0.321	0.360	0.478	0.593	0.244	0.420	0.367
BSN-KD w/ BDD-RBS	0.187	0.210	**0.633**	0.350	0.443	0.624	0.253	0.529	**0.326**
BSN (Ours) w/ BDD-RBS	0.314	0.599	0.398	**0.364**	0.533	0.535	**0.257**	**0.554**	0.324

simple baseline that detects vehicle, pedestrian, and cyclist regions estimated by DeepLabV3+ [5] as blind spots. *Detection-3D* utilizes a single-image 3D detection of vehicles, pedestrians, and cyclists by Brazil *et al.* [4]. Given their detected 3D bounding boxes, *Detection-3D* returns the intersection of the projection of their far-side faces and the results by *Detection-2D* as blind spots.

We also compare with BSN without depth estimation (BSN-D), and BSN without knowledge distillation (BSN-KD) for ablation studies. In BSN-D, the depth estimator $\mathcal{D}$ is removed from BSN, and the blind spot estimator $\mathcal{B}$ is modified to take the original RGB image directly. BSN-KD disables the knowledge distillation loss by setting $\lambda = 0$ in Sect. 4.3 in BSN.

Quantitative Evaluations. Table 2 shows the results on the test sets from KITTI-RBS, BDD-RBS, and TITAN-RBS Datasets. The lines with "w/ KITTI-RBS" and "w/ BDD-RBS" indicate the results of the networks trained with KITTI-RBS and BDD-RBS, respectively. Each network, after pre-training, was fine-tuned using 20% of the training data of the target dataset to absorb scene biases. It is worth mentioning that this fine-tuning is closer to self-supervision as the T-frame blind spots can be automatically computed without any external supervision for arbitrary videos. As such, BlindSpotNet can be applied to any driving video without suffering from domain gaps, as long as a small amount of video can be acquired before running BlindSpotNet for inference. The 20% training data usage of the target scene simulates such a scenario. Note that none of the test data were used and this fine-tuning was not done for TITAN-RBS.

BSN outperforms *Traversable*, *Detection-2D*, and *Detection-3D*. These results show that blind spot estimation cannot be achieved by simply estimating footprint or "behind-the-vehicle/pedestrian/cyclist" regions. The full BSN

Fig. 7. Blind spot estimation results. BlindSpotNet successfully achieves high precision and recall for complex road scenes (KITTI, BDD, and TITAN from left to right) by estimating nuanced blind spots caused by parked and moving cars, intersections, buildings, poles, gates, *etc.*. BlindSpotNet also estimates intersection blind spots that are even not in the "ground-truth" T-frame blind spots. This demonstrates the effectiveness of the visibility mask and the advantage of BlindSpotNet of being able to train on a diverse set of scenes thanks to the fact that T-frame blind spots can be easily computed on arbitrary driving videos.

also performs better than BSN-D and BSN-KD. This suggests that both the depth estimator and knowledge distillation contribute to its performance independently. Furthermore, the performance of BSN w/ KITTI-RBS on BDD-RBS and TITAN-RBS and that of BSN w/ BDD-RBS on KITTI-RBS and TITAN-RBS demonstrate the ability of BSN to generalize across datasets.

Qualitative Evaluations. Figure 7 shows blind spot estimation results for the test sets. Compared with baseline methods, our method estimates the complex blind spots arising in these everyday road scenes with high accuracy. It is worth noting that BlindSpotNet correctly estimates the left and right blind spots in the left column example, even though the "ground-truth" T-frame blind spots do not capture them due to the visibility mask. These results clearly demonstrate the effectiveness of the visibility mask and the training on diverse road scenes whose T-frame blind spots can be automatically computed. BlindSpotNet can be trained on arbitrary road scenes as long as T-frame blind spots can be computed, *i.e.*, SLAM, semantic segmentation, and depth estimation can be applied. In this sense, it is a self-supervised method. Figure 8 shows a failure case example. By definition of T-frame blind spots, BlindSpotNet cannot estimate intersection blind spots in videos that do not have any turns. We plan to explore the use of wider perspective videos, including full panoramic views, to mitigate this issue.

Fig. 8. Failure example. Left: T-frame blind spot ("ground truth") from a frame in BDD-RBS. Right: Blind spot estimation results from BlindSpotNet trained on BDD-RBS. Due to the bias of BDD-RBS, which lacks left and right turns at intersections, BSN cannot estimate the blind spots caused by the intersection. We plan to address these issues by employing a panoramic driving video for pre-training.

Table 3. Network architecture comparison. Our model achieves comparable performance to larger models and its frame-rate is promising for realtime processing.

Architecture	IoU	Recall	Precision	# of params	GMACS	FPS
U-Net based [22]	0.289	0.484	0.417	17.3M	280.2	139.9
Small (ours)	0.330	0.563	0.444	18.1M	47.1	37.5
Medium	0.315	0.478	0.482	32.0M	93.0	19.5
Large	0.337	0.508	0.500	59.3M	160.9	11.3

Network Architecture Evaluation. We compare large/medium-sized blind spot estimators as well as a simple U-Net [22] baseline with our small-sized (lightweight) blind spot estimator. The differences between the large/medium-sized blind spot estimators and the small-sized one are backbone, channel size, and the number of decoder layers. The backbones of the large/medium-sized ones are ResNet101 and ResNet50, respectively. Table 3 lists the IoU, recall, precision, the number of the parameters, the computational complexity in GMACS, and the inference speed with a single NVIDIA TITAN V 12 GB. The results show that our model achieves comparable performance to larger models with much smaller cost and runs sufficiently fast for real-time use.

6   Conclusion

We introduced a novel computer vision task for road scene understanding, namely 2D blind spot estimation. We tackle this challenging and critical problem for safe driving by introducing the first comprehensive dataset (RBS Dataset) and an end-to-end learnable network which we refer to as BlindSpotNet. By defining 2D blind spots as road regions that are invisible from the current viewpoint but become visible in the future, we showed that we can automatically compute them for arbitrary driving videos, which in turn enables learning to detect them with a simple neural network trained with knowledge distillation from a pre-trained semantic segmentation network. We believe these results offer a promising means for ensuring safer manual and autonomous driving and open new approaches to extending self-driving and ADAS with proactive visual perception.

Acknowledgements. This work was in part supported by JSPS 20H05951, 21H04893, JST JPMJCR20G7.

References

1. Afolabi, O., Driggs-Campbell, K., Dong, R., Kochenderfer, M.J., Sastry, S.S.: People as sensors: imputing maps from human actions. In: 2018 IEEE/RSJ International Conference on Intelligent Robots and Systems (IROS), pp. 2342–2348 (2018). https://doi.org/10.1109/IROS.2018.8594511
2. Barnum, P., Sheikh, Y., Datta, A., Kanade, T.: Dynamic seethroughs: synthesizing hidden views of moving objects. In: ISMAR, pp. 111–114 (2009)
3. Bertoni, L., Kreiss, S., Alahi, A.: Monoloco: monocular 3D pedestrian localization and uncertainty estimation. In: ICCV, pp. 6861–6871 (2019)
4. Brazil, G., Liu, X.: M3D-RPN: monocular 3D region proposal network for object detection. In: ICCV (2019)
5. Chen, L.C., Zhu, Y., Papandreou, G., Schroff, F., Adam, H.: Encoder-decoder with atrous separable convolution for semantic image segmentation. In: ECCV (2018)
6. Cheng, B., et al.: Panoptic-DeepLab: a simple, strong, and fast baseline for bottom-up panoptic segmentation. In: CVPR (2020)
7. Dosovitskiy, A., Ros, G., Codevilla, F., Lopez, A., Koltun, V.: CARLA: an open urban driving simulator. In: Conference on Robot Learning, pp. 1–16. PMLR (2017)
8. Ehsani, K., Mottaghi, R., Farhadi, A.: Segan: segmenting and generating the invisible. In: CVPR (2018)
9. Geiger, A., Lenz, P., Stiller, C., Urtasun, R.: Vision meets robotics: the KITTI dataset. Int. J. Robot. Res. (2013)
10. Guo, R., Hoiem, D.: Beyond the line of sight: labeling the underlying surfaces. In: Fitzgibbon, A., Lazebnik, S., Perona, P., Sato, Y., Schmid, C. (eds.) ECCV 2012. LNCS, vol. 7576, pp. 761–774. Springer, Heidelberg (2012). https://doi.org/10.1007/978-3-642-33715-4_55
11. Hara, K., Kataoka, H., Inaba, M., Narioka, K., Hotta, R., Satoh, Y.: Predicting appearance of vehicles from blind spots based on pedestrian behaviors at crossroads. IEEE Trans. Intell. Transp. Syst. (2021)
12. Itkina, M., Mun, Y.J., Driggs-Campbell, K., Kochenderfer, M.J.: Multi-agent variational occlusion inference using people as sensors (2021). https://doi.org/10.48550/ARXIV.2109.02173. https://arxiv.org/abs/2109.02173
13. Kingma, D.P., Ba, J.: Adam: a method for stochastic optimization. In: ICLR (2015)
14. Liu, B., Zhuang, B., Schulter, S., Ji, P., Chandraker, M.: Understanding road layout from videos as a whole. In: CVPR (2020)
15. Liu, C., Kohli, P., Furukawa, Y.: Layered scene decomposition via the occlusion-CRF. In: CVPR (2016)
16. Liu, Y., Shu, C., Wang, J., Shen, C.: Structured knowledge distillation for dense prediction. IEEE TPAMI (2020)
17. Makansi, O., Çiçek, O., Buchicchio, K., Brox, T.: Multimodal future localization and emergence prediction for objects in egocentric view with a reachability prior. In: CVPR, pp. 4353–4362 (2020)

18. Malla, S., Dariush, B., Choi, C.: TITAN: future forecast using action priors. In: CVPR, pp. 11186–11196 (2020)
19. Mani, K., Daga, S., Garg, S., Narasimhan, S.S., Krishna, M., Jatavallabhula, K.M.: MonoLayout: amodal scene layout from a single image. In: WACV (2020)
20. Rameau, F., Ha, H., Joo, K., Choi, J., Park, K., Kweon, I.S.: A real-time augmented reality system to see-through cars. IEEE Trans. Visual Comput. Graphics **22**(11), 2395–2404 (2016)
21. Ranftl, R., Lasinger, K., Hafner, D., Schindler, K., Koltun, V.: Towards robust monocular depth estimation: mixing datasets for zero-shot cross-dataset transfer. IEEE TPAMI (2020)
22. Ronneberger, O., Fischer, P., Brox, T.: U-Net: convolutional networks for biomedical image segmentation. In: Navab, N., Hornegger, J., Wells, W.M., Frangi, A.F. (eds.) MICCAI 2015. LNCS, vol. 9351, pp. 234–241. Springer, Cham (2015). https://doi.org/10.1007/978-3-319-24574-4_28
23. Shivdas, S., Kelly, T.: Toyota halts all self-driving e-Palette vehicles after Olympic village accident. Reuters (2021). https://www.reuters.com/business/autos-transportation/toyota-halts-all-self-driving-e-pallete-vehicles-after-olympic-village-accident-2021-08-27/
24. Song, S., Yu, F., Zeng, A., Chang, A.X., Savva, M., Funkhouser, T.: Semantic scene completion from a single depth image. In: CVPR (2017)
25. Sugiura, T., Watanabe, T.: Probable multi-hypothesis blind spot estimation for driving risk prediction. In: 2019 IEEE Intelligent Transportation Systems Conference (ITSC), pp. 4295–4302 (2019)
26. Sumikura, S., Shibuya, M., Sakurada, K.: Openvslam: a versatile visual slam framework. In: ACM MM, pp. 2292–2295 (2019)
27. Tighe, J., Niethammer, M., Lazebnik, S.: Scene parsing with object instances and occlusion ordering. In: CVPR (2014)
28. Watson, J., Firman, M., Monszpart, A., Brostow, G.J.: Footprints and free space from a single color image. In: CVPR (2020)
29. Yang, W., et al.: Projecting your view attentively: monocular road scene layout estimation via cross-view transformation. In: CVPR, pp. 15536–15545 (2021)
30. Yasuda, H., Ohama, Y.: Toward a practical wall see-through system for drivers: how simple can it be? In: ISMAR, pp. 333–334 (2012)
31. Yu, F., et al.: BDD100K: a diverse driving dataset for heterogeneous multitask learning. In: CVPR, pp. 2636–2645 (2020)
32. Zhang, Y., Song, S., Tan, P., Xiao, J.: PanoContext: a whole-room 3D context model for panoramic scene understanding. In: Fleet, D., Pajdla, T., Schiele, B., Tuytelaars, T. (eds.) ECCV 2014. LNCS, vol. 8694, pp. 668–686. Springer, Cham (2014). https://doi.org/10.1007/978-3-319-10599-4_43
33. Zhu, X., Yin, Z., Shi, J., Li, H., Lin, D.: Generative adversarial frontal view to bird view synthesis. In: 3DV, pp. 454–463 (2018)
34. Zhu, Y., Tian, Y., Metaxas, D., Dollar, P.: Semantic amodal segmentation. In: CVPR (2017)

Gesture Recognition with Keypoint and Radar Stream Fusion for Automated Vehicles

Adrian Holzbock[1]([✉]), Nicolai Kern[2], Christian Waldschmidt[2], Klaus Dietmayer[1], and Vasileios Belagiannis[3]

[1] Institute of Measurement, Control and Microtechnology, Ulm University, Albert-Einstein-Allee 41, 89081 Ulm, Germany
{adrian.holzbock,klaus.dietmayer}@uni-ulm.de
[2] Institute of Microwave Engineering, Ulm University, Albert-Einstein-Allee 41, 89081 Ulm, Germany
{nicolai.kern,christian.waldschmidt}@uni-ulm.de
[3] Department of Simulation and Graphics, Otto von Guericke University Magdeburg,Universitätsplatz 2, 39106 Magdeburg, Germany
vasileios.belagiannis@ovgu.de

Abstract. We present a joint camera and radar approach to enable autonomous vehicles to understand and react to human gestures in everyday traffic. Initially, we process the radar data with a PointNet followed by a spatio-temporal multilayer perceptron (stMLP). Independently, the human body pose is extracted from the camera frame and processed with a separate stMLP network. We propose a fusion neural network for both modalities, including an auxiliary loss for each modality. In our experiments with a collected dataset, we show the advantages of gesture recognition with two modalities. Motivated by adverse weather conditions, we also demonstrate promising performance when one of the sensors lacks functionality.

Keywords: Camera radar fusion · Gesture recognition · Automated driving

1 Introduction

The safe interaction of traffic participants in urban environments is based on different rules like signs or the right of way. Besides static regulations, more dynamic ones like gestures are possible. For example, a police officer manages the traffic [37] by hand gestures, or a pedestrian waves a car through at a crosswalk [26]. Although the driver intuitively knows the meaning of human gestures, an autonomous vehicle cannot interpret them. To safely integrate the autonomous vehicle into urban traffic, it is essential to understand human gestures.

To detect human gestures, many related approaches only rely on camera data [3,22]. Camera-based gesture recognition is not always reliable, for instance, due to the weather sensitivity of the camera sensor [21]. One way to mitigate the drawbacks of camera-based approaches is the augmentation of the gesture

L. Karlinsky et al. (Eds.): ECCV 2022 Workshops, LNCS 13801, pp. 570–584, 2023.
https://doi.org/10.1007/978-3-031-25056-9_36

recognition system by non-optical sensor types. A sensor with less sensitivity to environmental conditions is a radar sensor [24] that has been shown to be suited for gesture recognition. Furthermore, radar sensors are not limited to detecting fine-grained hand gestures but can also predict whole body gestures [10]. Hence, radar sensors are a promising candidate to complement optical sensors for more reliable gesture recognition in automotive scenarios. This has been demonstrated for gesture recognition with camera-radar fusion in close proximity to the sensors, as presented in [17]. Furthermore, depending not only on one sensor can increase the system's reliability in case of a sensor failure. In contrast to the prior work, we propose a method for whole-body gesture recognition at larger distances and with a camera-radar fusion approach.

We present a two-stream neural network to realizing camera and radar fusion. Our approach extracts independently a representation for each modality and then fuses them with an additional network module. In the first stream, the features of the unordered radar targets are extracted with a PointNet [23] and further prepared for fusion with an spatio-temporal multilayer perceptron (stMLP) [8]. The second stream uses only a stMLP to process the keypoint data. The extracted features of each stream are fused for the classification in an additional stMLP. We train the proposed network with an auxiliary loss function for each modality to improve the feature extraction through additional feedback. For training, we use data containing radar targets of three chirp sequence (CS) radar sensors and human keypoints extracted from camera images. The eight different gestures are presented in Fig. 1 and represent common traffic gestures. The performance of our approach is tested in a cross-subject evaluation. We improve performance by 3.8% points compared to the keypoint-only setting and 8.3% points compared to the radar-only setting. Furthermore, we can show robustness against the failure of one sensor.

Overall, we propose a two-stream neural network architecture to fuse keypoint and radar data for gesture recognition. To process the temporal context in the model, we do not rely on recurrent models [28] but instead propose the stMLP fusion. In the training of our two-stream model, we introduce an auxiliary loss for each modality. To the best of our knowledge, we are the first to fuse radar and keypoint data for gesture recognition in autonomous driving.

2   Related Work

In the following, we discuss other methods regarding gesture recognition in general as well as in autonomous driving. We give an overview of approaches relying only on keypoint or radar data. Besides the single sensor gesture recognition, we present methods using a combination of different sensors.

Gesture Recognition with Camera Data. Gesture recognition is applied to react to humans outside the vehicle [22, 40] and to the passenger's desires [8, 36]. For gesture recognition of police officers, related work processes camera image snippets [3] directly. Due to the advances in human body pose estimation [4, 5],

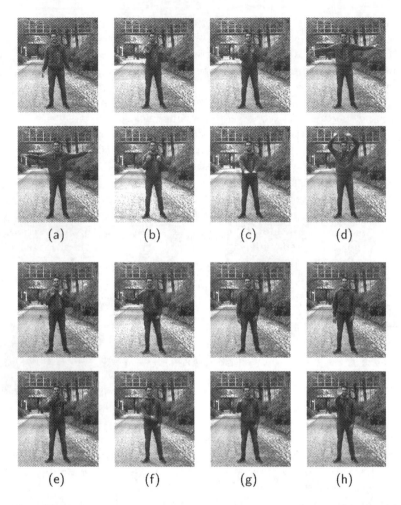

Fig. 1. Visualization of the characteristic poses of the eight gestures. (a) Fly. (b) Come closer. (c) Slow down. (d) Wave. (e) Push away. (f) Wave through. (g) Stop. (h) Thank you.

related approaches extract the body skeleton data from the images and perform gesture recognition on it [15,22,35]. For processing skeletons to predict the gesture, recurrent neural networks [35] or convolutional neural networks [22] are applied. Besides the police officer gesture recognition, the actions of other human traffic participants like cyclists [40] or pedestrians [6] are also analyzed in literature. Similar to pedestrian gesture recognition is the pedestrian intention prediction [1,25], where the pedestrian's intention to cross the street should be recognized. Changing the view to the interior of the car, there are approaches to recognize the driver's activities in order to check if the driver is focused on the traffic. For this purpose, methods like attention-based neural networks [36] or models only built on multi-layer perceptrons [8] are developed. Compared to our work, lighting and weather conditions influence the performance of camera-based approaches. Due to the fusion with the radar sensor data, our method mitigate these factors.

Gesture Recognition with Radar Data. Besides radar sensors' insensitivity to adverse weather and lighting, they also evoke fewer privacy issues than cameras. As a result, radar-based gesture recognition has received increased attention in recent years, with research efforts mainly devoted to human-machine interaction in the consumer electronics area [13]. For the control of devices with hand gestures, gesture recognition algorithms based on a wide range of neural networks have been proposed, involving, e.g., 2D-CNNs [12], 2D-CNNs with LSTMs [34], or 3D-CNNs with LSTM [42]. These approaches exploit spectral information in the form of micro-Doppler spectrograms [12] or range-Doppler spectra [34], but it is also possible to distinguish between gestures [10] and activities [29] using radar point clouds. The latter are a more compact representation of the radar observations and are obtained by finding valid targets in the radar data. Point clouds facilitate the application of geometrical transformations [29] as well as the inclusion of additional information [10]. While most research considers small-scale gesture recognition close to the radar sensor, reliable macro gesture recognition at larger distances has been also shown to be feasible with radar sensors for applications such as smart homes [18] or traffic scenarios [10,11]. While radar-only gesture recognition has shown promising results, augmenting it by camera data can further improve classification accuracy, as demonstrated in this paper. This is particularly important in safety-critical applications such as autonomous driving.

Gesture Recognition with Sensor Fusion. By combining data of multiple sensors, sensor fusion approaches can overcome the drawbacks of the individual sensor types, like the environmental condition reliance of the camera or its missing depth information. Sensor fusion has been applied e.g. for gesture-based human machine interaction in vehicles, where touchless control can increase safety [16]. Besides the image information for gesture recognition, the depth data contains essential knowledge. Therefore, other methods fuse camera data with the data of a depth sensor [19] and process the data with a 3D convolutional neural network [16,41]. Molchanov et al. [17] run a short-range radar sensor next to the RGB camera and depth camera and process the recorded data with a 3D convolutional neural network for a more robust gesture recognition. Another approach fuses the short-range radar data with infrared sensor data instead of RGB images [30]. Current fusion approaches are not only limited to short ranges and defined environments but can also be applied in more open scenarios like surveillance [9] or smart home applications [32]. Here, camera images and radar data, converted to images, are used to classify the gesture, while we use the skeletons extracted from the images and the radar targets described as unstructured point clouds. Moreover, contrary to [32], our fusion approach enhances the gesture recognition accuracy not only in cases where one modality is impaired but also in normal operation.

3   Fusion Method

We present a fusion technique for radar and keypoint data for robust gesture recognition. To this end, we develop a neural network $\hat{\mathbf{y}} = f((\mathbf{x}_R, \mathbf{x}_K), \theta)$ defined by its parameters θ. The output prediction $\hat{\mathbf{y}}$ is defined as one-hot vector $\hat{\mathbf{y}}_i \in \{0, 1\}^C$, such that $\sum_{c=1}^{C} \hat{\mathbf{y}}_i(c) = 1$ for a C-category classification problem. The radar input data is represented by $\mathbf{x}_R \in \mathbb{R}^{T \times 5 \times 300}$, where T is the number of time steps and 300 is the number of sampled radar targets in each time step, each of which is described by 5 target parameters. For the keypoint stream, 17 2D keypoints are extracted and flattened in each time step, such that the keypoint data is described by $\mathbf{x}_K \in \mathbb{R}^{T \times 34}$. We use the ground truth class label $\mathbf{y}$ for the training of $f(.)$ to calculate the loss of the model prediction $\hat{\mathbf{y}}$. In the following, we first present the architecture of our neural network and afterward show the training procedure.

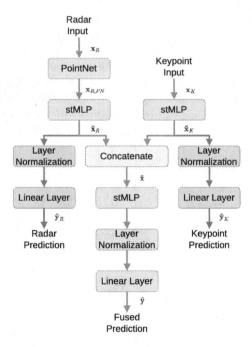

Fig. 2. Architecture of the proposed neural network for gesture recognition with radar and keypoint data. The blue layers correspond to the inference neural network while the red layers are only used to compute the auxiliary losses. (Color figure online)

3.1   Neural Network Architecture

The neural network inputs are the radar data $\mathbf{x}_R$ and the keypoint data $\mathbf{x}_K$. The network consists of two different streams, one for each modality, to extract the features of the different modalities. The information of the two streams is concatenated and fed to a joint network which does the fusion and the gesture classification. Additionally, auxiliary outputs are added to the model for the training procedure. An overview of the proposed network architecture is given in Fig. 2.

The feature extraction from the unordered radar data $\mathbf{x}_R$ is done with a PointNet [23] that extracts the features for each time step $t \in \{1, 2, \ldots, T\}$. The features of the time steps are concatenated to one feature tensor $\mathbf{x}_{R,PN} \in \mathbb{R}^{T \times 512}$. The PointNet does not process the data in the temporal dimension but only in the point dimension. For the temporal processing, we use the stMLP model [8], which replaces a standard method for temporal data processing, e.g. a long short-term memory model (LSTMs) [7]. Unlike LSTMs, the stMLP is based solely on multilayer perceptrons (MPLs) and does not have any recurrent parts. The stMLP processes the radar data $\mathbf{x}_{R,PN}$ extracted with the PointNet in the temporal and feature dimensions and outputs mixed radar features $\tilde{\mathbf{x}}_R \in \mathbb{R}^{T \times H/2}$, where H is the hidden dimension. Due to the ordered structure of the keypoint data $\mathbf{x}_K$, the mixed keypoint features $\tilde{\mathbf{x}}_K \in \mathbb{R}^{T \times H/2}$ are extracted only with an stMLP model, which processes the extracted keypoint features in both dimensions, namely the spatial and the temporal dimension. The extracted features of both modalities, $\tilde{\mathbf{x}}_R$ and $\tilde{\mathbf{x}}_K$, are concatenated to a single feature tensor $\tilde{\mathbf{x}} \in \mathbb{R}^{T \times H}$ and fed into another stMLP model. This model performs a spatial and temporal fusion of the radar and keypoint features to produce a meaningful representation for the gesture classification step. The gesture classification is performed by a Layer Normalization [2] and a single linear layer for each time step.

Additionally, we add for each modality an auxiliary output during training [31]. The auxiliary output of both modalities is built on a Layer Normalization and a linear layer for the auxiliary classification. In Fig. 2, the parts of the auxiliary outputs and the corresponding layers are drawn in red. For the gesture prediction, each auxiliary output only uses one modality and no information from the other. This means that the auxiliary radar output $\hat{\mathbf{y}}_R$ gets the extracted radar features $\tilde{\mathbf{x}}_R$ for the prediction and the auxiliary keypoint output $\hat{\mathbf{y}}_K$ the keypoint data features $\tilde{\mathbf{x}}_K$. For the model, the auxiliary outputs give additional specialized feedback for updating the network parameters in the radar and the keypoint stream. After training, the layers of the auxiliary outputs can be removed.

3.2   Model Training

For the training of the neural network introduced in Sect. 3.1 we utilize the data presented in Sect. 4.1. During training, we use the fused output of the model $\hat{\mathbf{y}}$ and the auxiliary outputs derived from the radar $\hat{\mathbf{y}}_R$ and keypoint data $\hat{\mathbf{y}}_K$. For each output, we calculate the cross entropy loss $\mathcal{L}_{CE}$, which can be formulated as

$$\mathcal{L}_{CE} = -\sum_{c=0}^{C} y_c \log(\hat{y}_c), \qquad (1)$$

where y is the ground truth label, $\hat{y}$ the network prediction, and C the number of gesture classes. We use a weighted sum of the different sub-losses to get the overall loss $\mathcal{L}$ which can be expressed by

$$\mathcal{L} = \mathcal{L}_F + \mu * (\mathcal{L}_R + \mathcal{L}_K). \tag{2}$$

In the overall loss, $\mathcal{L}_F$ is the loss of the fused output, $\mathcal{L}_R$ of the auxiliary radar output, and $\mathcal{L}_K$ of the auxiliary keypoint output. The auxiliary losses are weighted with the auxiliary loss weight μ.

4   Results

We evaluate our approach in two different settings. First, we test our approach with two modalities, assuming no issue with the sensors and getting data from both. Second, we use only one modality to evaluate the model. Only having one modality can be motivated by adverse weather conditions, e.g. the camera sensor is completely covered by snow or technical problems with one sensor.

4.1   Dataset

For the development of robust gesture recognition with the fusion of radar and keypoint data in the context of autonomous driving, we need a dataset that contains both modalities captured synchronously and at sufficient ranges. Consequently, the custom traffic gesture dataset introduced in [10] with both camera and radar data is used. Following, we specify the setup and the data processing of the dataset.

Setup. The gesture dataset comprises measurements of eight different gestures shown in Fig. 1 for 35 participants. The measurements are conducted on a small street on the campus of Ulm University as well as inside a large hall resembling a car park. For each participant, data recording is repeated multiple times under different orientations, and the measurements are labeled by means of the camera data. For the measurements, a setup consisting of three chirp sequence (CS) radar sensors and a RGB camera as illustrated by the sketch in Fig. 3 are used. The radar sensors operate in the automotive band at $77\,\mathrm{GHz}$. Each sensor has a range and velocity resolution of $4.5\,\mathrm{cm}$ and $10.7\,\mathrm{cm\,s^{-1}}$, respectively, and eight receive channels for azimuth angle estimation. The camera in the setup has a resolution of 1240×1028, and keypoints obtained from the camera serve as input to the keypoint stream of the proposed gesture recognition algorithm. All sensors are mounted on a rail and synchronized with a common trigger signal with 30 fps.

Data Processing. From the gesture recordings, per-frame radar point clouds and 2D skeletal keypoints are computed. The radar responses recorded by the sensors are processed sensor-wise to obtain range-Doppler maps [38]. The Ordered Statistics CFAR algorithm [27] is applied to extract valid targets and

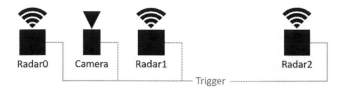

Fig. 3. Measurement system consisting of the RGB camera and three CS radar sensors. All sensors receive a common trigger signal for time synchronization.

thereby compress the information in the range-Doppler maps. For each radar target, its azimuth angle is estimated by digital beamforming [33]. The target parameters are normalized by the radar sensors' unambiguous ranges. Each target in the resulting target lists is described by its range, velocity, azimuth angle, its reflected power in dB, and the index $i_n \in 0, 1, 2$ of the radar sensor that detected it. Since the number of detected targets varies from frame to frame, target lists with a constant number N_R of targets are sampled randomly for frames whose target count exceeds N_R. Contrary, for frames where the number of targets is less than N_R, zero-padding is applied to fill the target lists. The target lists of the radar sensors are stacked, such that the final target list contains $3N_R$ targets. After repeating the processing for all T_M frames in the measurement, the radar observations are described by input data of shape $\mathbf{x}_R \in \mathbb{R}^{T_M \times 5 \times 300}$, when setting N_R to 100.

The camera data is processed by Detectron2 [39] to extract 17 keypoints in the COCO keypoint format [14]. The extracted keypoints are normalized with the image width and height to restrict the keypoint values for the neural network to the range between 0 and 1. The keypoints' 2D pixel positions over the frames are summarized in the camera observation tensor $\mathbf{o}_K \in \mathbb{R}^{T_M \times 2 \times 17}$, which is flattened to the keypoint data $\mathbf{x}_K \in \mathbb{R}^{T_M \times 34}$. After the signal processing, the radar and camera data of the measurements are downsampled to 15 fps and segmented into smaller snippets with $T = 30$ time steps each, corresponding to 2 s of observation. Finally, 15700 samples are available for training the gesture recognition model. In the cross-subject evaluation, we use the data of 7 subjects for testing and the data of the remaining subjects as training and validation data. We train the model 5 times, using different subjects for testing in each run and average over the 5 runs for the final performance.

4.2 Experimental Setup

The neural network is implemented with the PyTorch [20] framework, and we use PointNet[1] and stMLP[2] as a base for our new fusion architecture. In the PointNet, we apply the feature transformation and deactivate the input transformation. In

---

[1] https://github.com/fxia22/pointnet.pytorch.
[2] https://github.com/holzbock/st_mlp.

each stMLP structure we use 4 mixer blocks and set the stMLP hyperparameters as follows: the hidden input dimension to 256, the hidden spatial-mixing dimension to 64, and the hidden temporal-mixing dimension to 256. The loss is calculated with the function defined in Eq. 2, where we set the auxiliary weight μ to 0.5. We train our model for 70 epochs with a batch size of 32 and calculate the gradients with the SGD optimizer that uses a learning rate of 0.003, a momentum of 0.95, and a weight decay of 0.001. To get the best training result, we check the performance during the training on the validation set and use the best validation epoch for testing on the test set. The model performance is measured with the accuracy metric. To prepare the neural network for missing input data, we skip in 30% of the training samples the radar or the keypoint data, which we call skipped-modality training (SM-training). When reporting the results with a single-modality model, we remove the layers for the other modality.

4.3   Results

During testing with optimal data we assume that we get the keypoint data from the camera sensor and the targets from the radar sensors, i.e., we have no samples in the train and test set that only contain one modality's data. The results are shown in Table 1, where we compare with a model using an LSTM instead of an stMLP for temporal processing. In the *Single Modality* part of the table, we show the performance of our architecture trained and tested on only one modality. In these cases, the layers belonging to the other modality are removed. As it can be seen, the LSTM and the stMLP model are on par when training only with the radar data, and the stMLP model is better than the LSTM model trained only with the keypoint data.

In the *Fusion* part of Table 1, we show the fusion performance of our architecture. We first train the stMLP and the LSTM fusion with all the training data and then apply SM-training with a ratio of 30%. The SM-training means that we skip the radar or the keypoint data in 30% of the training samples. The stMLP fusion performs better in both cases compared to the LSTM fusion. Furthermore, skipping randomly one modality in 30% of the training data slightly benefits the test performance. Overall, the stMLP architecture performs better in the fusion of the radar and the keypoint data than the LSTM architecture. Compared to the single-modality model, the fusion improves the performance in both architectures (LSTM and stMLP) by over 4% points. This shows that the fusion of the keypoint and radar data for gesture recognition benefits the performance compared to single-modality gesture recognition.

4.4   Results with Single Modality

When we test our architecture with single modality, during testing only the radar or the keypoint data are fed into the model, but not both. This can be compared with one fully corrupted sensor due to adverse weather or a technical problem. Contrary, during training we either use all training data (no SM-training) or set the SM-training to a ratio of 30%, such that in 30% of samples one modality

Table 1. Performance of our model trained with both modalities. *Single Modality* is a model that only contains layers for one modality and is trained and tested only with this modality. The *Fusion* rows use our proposed model. *SM-training* means that we skip one modality in 30% of the training samples during the training.

Model	Temporal processing	Test data		SM-training	Accuracy in %
		Radar	Keypoint		
Single Modality	LSTM	✗	✓	✗	86.2
	stMLP	✗	✓	✗	89.9
	LSTM	✓	✗	✗	85.7
	stMLP	✓	✗	✗	85.4
Fusion	LSTM	✓	✓	✗	90.2
	stMLP	✓	✓	✗	93.7
	LSTM	✓	✓	✓	90.7
	stMLP	✓	✓	✓	93.8

is skipped. In the first part of Table 2, results are shown for the model trained without the skipped modalities, which means that the model has not learned to handle single-modality samples. Despite the missing modality in the evaluation, the model can still classify the gestures, but with a decreased accuracy. Comparing the LSTM and stMLP variants, the LSTM variant performs better with only the radar data and the stMLP with only the keypoint data.

In the second part of Table 2, we show the performance of our model trained with skipped modalities in 30% of the training samples (SM-training is 30%). When the model is trained with the skipped modalities, it learns better to handle missing modalities. This results in a better performance in the single-modality evaluation, and we can improve the accuracy by a minimum of 3% points compared to the model trained without the skipped modalities. Overall, the fusion can improve the reliability of gesture recognition in cases where one sensor fails e.g. due to adverse weather conditions or a technical sensor failure.

5   Ablation Studies

The performance of our model depends on different influence factors, which we evaluate in the ablation studies. As standard setting, we choose the hyperparameters defined in Sect. 4.2 and change the ratio of the SM-training and the auxiliary loss weight during the ablations.

5.1   Amount of Single-Modality Training Samples

The training with missing data teaches the model to handle singe modality input data, as shown in Sect. 4.4. In our standard setting for the SM-training, we randomly skip one modality in 30% of the training samples. In this ablation

Table 2. Performance of our model with single-modality data. The *Fusion* rows use our proposed model. *SM-training* means that we skip one modality in 30% of the training samples during the training.

Model	Temporal processing	Test data Radar	Test data Keypoint	SM-training	Accuracy in %
Fusion	LSTM	✗	✓	✗	69.7
	LSTM	✓	✗	✗	82.6
	stMLP	✗	✓	✗	87.7
	stMLP	✓	✗	✗	78.3
Fusion	LSTM	✗	✓	✓	80.8
	LSTM	✓	✗	✓	85.0
	stMLP	✗	✓	✓	91.2
	stMLP	✓	✗	✓	83.0

study, the SM-training ratio ranges from 0% to 60% during training and shows the influence of this parameter on the performance.

We test the model with different SM-training ratios with the test data containing both modalities and show the result in the *Fusion* row of Table 3. Additionally, we test with the data of only one modality and present the results in the *Only Keypoints* and *Only Radar* row. The experiment shows that the SM-training does not significantly influence the performance when testing with both modalities. This is explainable because in the testing set no samples have lacking modalities. Testing with only one modality, the amount of skipped-modality samples during the training influences the accuracy. Here, the performance increases until the SM-training ratio reaches 30% and then stays constant. In this case, the amount of skipped-modality samples during training influences the adaptation of the model to single-modality samples.

Table 3. Influence of the SM-training ratio on the overall accuracy.

Skip modality in %	0	10	20	30	40	50	60
Fusion	93.7	92.8	92.1	93.8	94.1	93.2	93.6
Only Keypoints	87.7	89.9	89.8	91.2	91.4	91.0	91.5
Only Radar	78.3	81.7	81.1	83.0	83.0	80.7	83.2

5.2   Loss Function

Besides the ratio of the SM-training, the auxiliary loss weight μ is an essential parameter in training. In our standard evaluation, we set μ to 0.5, while we vary the loss weight in this ablation study from 0 to 3.

The results of the different μ are shown in Table 4. In the first row, we deliver the results when testing with both modalities (*Fusion*). In the *Only Keypoints* and *Only Radar* row, we present the performance when evaluating only one modality. As we can see in Table 4, the fusion results behave equally to the single-modality results. For the Fusion as well as for the single modality, a higher μ increases the performance, and the best accuracy is reached with $\mu = 0.8$. Further increasing μ leads to a decreasing performance in gesture recognition. Comparing the results of the best $\mu = 0.8$ with the model without the auxiliary loss ($\mu = 0.0$) shows that with the auxiliary loss, the fusion performance stays equal but the single-modality accuracy increases. This indicates that the model benefits from the additional feedback of the auxiliary loss during training.

Table 4. Influence of the loss weights on the overall performance.

Loss weight	0.0	0.2	0.5	0.8	1.0	2.0	3.0
Fusion	94.2	92.9	93.8	94.1	93.7	92.4	93.4
Only Keypoints	89.2	90.8	91.2	92.1	91.0	90.1	91.7
Only Radar	84.3	81.2	83.0	84.7	82.2	81.3	82.8

6   Conclusion

We present a novel two-stream neural network architecture for the fusion of radar and keypoint data to reliably classify eight different gestures in autonomous driving scenarios. The proposed fusion method improves the classification accuracy over the values obtained with single sensors, while enhancing the recognition robustness in cases of technical sensor failure or adverse environmental conditions. In the model, we first process the data of each modality on its own and then fuse them for the final classification. We propose a stMLP fusion which applies besides the fusion of the features of both modalities also temporal processing. Furthermore, for a better overall performance of our approach, we introduce an auxiliary loss in the training that provides additional feedback to each modality stream. The evaluation of our method on the radar-camera dataset, and we show that even with missing modalities, the model can reach a promising classification performance. In the ablation studies, we demonstrate the influence of the SM-training ratio and the auxiliary loss weight.

Acknowledgment. Part of this work was supported by INTUITIVER (7547.223-3/4/), funded by State Ministry of Baden-Württemberg for Sciences, Research and Arts and the State Ministry of Transport Baden-Württemberg.

References

1. Abughalieh, K.M., Alawneh, S.G.: Predicting pedestrian intention to cross the road. IEEE Access **8**, 72558–72569 (2020)
2. Ba, J.L., Kiros, J.R., Hinton, G.E.: Layer normalization. arXiv preprint arXiv:1607.06450 (2016)
3. Baek, T., Lee, Y.G.: Traffic control hand signal recognition using convolution and recurrent neural networks. J. Comput. Des. Eng. **9**(2), 296–309 (2022)
4. Belagiannis, V., Amann, C., Navab, N., Ilic, S.: Holistic human pose estimation with regression forests. In: Perales, F.J., Santos-Victor, J. (eds.) AMDO 2014. LNCS, vol. 8563, pp. 20–30. Springer, Cham (2014). https://doi.org/10.1007/978-3-319-08849-5_3
5. Bouazizi, A., Wiederer, J., Kressel, U., Belagiannis, V.: Self-supervised 3D human pose estimation with multiple-view geometry. In: 2021 16th IEEE International Conference on Automatic Face and Gesture Recognition (FG 2021), pp. 1–8 (2021). https://doi.org/10.1109/FG52635.2021.9667074
6. Geng, K., Yin, G.: Using deep learning in infrared images to enable human gesture recognition for autonomous vehicles. IEEE Access **8**, 88227–88240 (2020)
7. Hochreiter, S., Schmidhuber, J.: Long short-term memory. Neural Comput. **9**(8), 1735–1780 (1997)
8. Holzbock, A., Tsaregorodtsev, A., Dawoud, Y., Dietmayer, K., Belagiannis, V.: A spatio-temporal multilayer perceptron for gesture recognition. In: 2022 IEEE Intelligent Vehicles Symposium (IV), pp. 1099–1106 (2022). https://doi.org/10.1109/IV51971.2022.9827054
9. de Jong, R.J., de Wit, J.J., Uysal, F.: Classification of human activity using radar and video multimodal learning. IET Radar Sonar Navig. **15**(8), 902–914 (2021)
10. Kern, N., Grebner, T., Waldschmidt, C.: PointNet+ LSTM for target list-based gesture recognition with incoherent radar networks. IEEE Trans. Aerosp. Electron. Syst. (2022). https://doi.org/10.1109/TAES.2022.3179248
11. Kern, N., Steiner, M., Lorenzin, R., Waldschmidt, C.: Robust doppler-based gesture recognition with incoherent automotive radar sensor networks. IEEE Sens. Lett. **4**(11), 1–4 (2020)
12. Kim, Y., Toomajian, B.: Hand gesture recognition using micro-doppler signatures with convolutional neural network. IEEE Access **4**, 7125–7130 (2016)
13. Lien, J., et al.: Soli: ubiquitous gesture sensing with millimeter wave radar. ACM Trans. Graph. (TOG) **35**(4), 1–19 (2016)
14. Lin, T.-Y., et al.: Microsoft COCO: common objects in context. In: Fleet, D., Pajdla, T., Schiele, B., Tuytelaars, T. (eds.) ECCV 2014. LNCS, vol. 8693, pp. 740–755. Springer, Cham (2014). https://doi.org/10.1007/978-3-319-10602-1_48
15. Mishra, A., Kim, J., Cha, J., Kim, D., Kim, S.: Authorized traffic controller hand gesture recognition for situation-aware autonomous driving. Sensors **21**(23), 7914 (2021)
16. Molchanov, P., Gupta, S., Kim, K., Kautz, J.: Hand gesture recognition with 3D convolutional neural networks. In: Proceedings of the IEEE Conference on Computer Vision and Pattern Recognition Workshops, pp. 1–7 (2015)
17. Molchanov, P., Gupta, S., Kim, K., Pulli, K.: Multi-sensor system for driver's hand-gesture recognition. In: 2015 11th IEEE International Conference and Workshops on Automatic Face and Gesture Recognition (FG), vol. 1, pp. 1–8. IEEE (2015)
18. Ninos, A., Hasch, J., Zwick, T.: Real-time macro gesture recognition using efficient empirical feature extraction with millimeter-wave technology. IEEE Sens. J. **21**(13), 15161–15170 (2021)

19. Ohn-Bar, E., Trivedi, M.M.: Hand gesture recognition in real time for automotive interfaces: a multimodal vision-based approach and evaluations. IEEE Trans. Intell. Transp. Syst. **15**(6), 2368–2377 (2014)
20. Paszke, A., et al.: Pytorch: an imperative style, high-performance deep learning library. In: Advances in Neural Information Processing Systems 32, pp. 8024–8035. Curran Associates, Inc. (2019). http://papers.neurips.cc/paper/9015-pytorch-an-imperative-style-high-performance-deep-learning-library.pdf
21. Pfeuffer, A., Dietmayer, K.: Robust semantic segmentation in adverse weather conditions by means of sensor data fusion. In: 2019 22th International Conference on Information Fusion (FUSION), pp. 1–8. IEEE (2019)
22. Pham, D.T., Pham, Q.T., Le, T.L., Vu, H.: An efficient feature fusion of graph convolutional networks and its application for real-time traffic control gestures recognition. IEEE Access **9**, 121930–121943 (2021)
23. Qi, C.R., Su, H., Mo, K., Guibas, L.J.: Pointnet: deep learning on point sets for 3D classification and segmentation. In: Proceedings of the IEEE Conference on Computer Vision and Pattern Recognition, pp. 652–660 (2017)
24. Qian, K., Zhu, S., Zhang, X., Li, L.E.: Robust multimodal vehicle detection in foggy weather using complementary lidar and radar signals. In: Proceedings of the IEEE/CVF Conference on Computer Vision and Pattern Recognition, pp. 444–453 (2021)
25. Quintero, R., Parra, I., Lorenzo, J., Fernández-Llorca, D., Sotelo, M.: Pedestrian intention recognition by means of a hidden Markov model and body language. In: 2017 IEEE 20th International Conference on Intelligent Transportation Systems (ITSC), pp. 1–7. IEEE (2017)
26. Rasouli, A., Tsotsos, J.K.: Autonomous vehicles that interact with pedestrians: a survey of theory and practice. IEEE Trans. Intell. Transp. Syst. **21**(3), 900–918 (2019)
27. Rohling, H.: Radar CFAR thresholding in clutter and multiple target situations. IEEE Trans. Aerosp. Electron. Syst. **AES-19**(4), 608–621 (1983). https://doi.org/10.1109/TAES.1983.309350
28. Schreiber, M., Belagiannis, V., Gläser, C., Dietmayer, K.: Motion estimation in occupancy grid maps in stationary settings using recurrent neural networks. In: 2020 IEEE International Conference on Robotics and Automation (ICRA), pp. 8587–8593 (2020). https://doi.org/10.1109/ICRA40945.2020.9196702
29. Singh, A.D., Sandha, S.S., Garcia, L., Srivastava, M.: Radhar: human activity recognition from point clouds generated through a millimeter-wave radar. In: Proceedings of the 3rd ACM Workshop on Millimeter-Wave Networks and Sensing Systems, pp. 51–56 (2019)
30. Skaria, S., Al-Hourani, A., Huang, D.: Radar-thermal sensor fusion methods for deep learning hand gesture recognition. In: 2021 IEEE Sensors, pp. 1–4. IEEE (2021)
31. Szegedy, C., et al.: Going deeper with convolutions. In: Proceedings of the IEEE Conference on Computer Vision and Pattern Recognition, pp. 1–9 (2015)
32. Vandersmissen, B., Knudde, N., Jalalvand, A., Couckuyt, I., Dhaene, T., De Neve, W.: Indoor human activity recognition using high-dimensional sensors and deep neural networks. Neural Comput. Appl. **32**(16), 12295–12309 (2020)
33. Vasanelli, C., et al.: Calibration and direction-of-arrival estimation of millimeter-wave radars: a practical introduction. IEEE Antennas Propag. Mag. **62**(6), 34–45 (2020). https://doi.org/10.1109/MAP.2020.2988528

34. Wang, S., Song, J., Lien, J., Poupyrev, I., Hilliges, O.: Interacting with soli: exploring fine-grained dynamic gesture recognition in the radio-frequency spectrum. In: Proceedings of the 29th Annual Symposium on User Interface Software and Technology, pp. 851–860 (2016)
35. Wang, S., Jiang, K., Chen, J., Yang, M., Fu, Z., Yang, D.: Simple but effective: upper-body geometric features for traffic command gesture recognition. IEEE Trans. Hum.-Mach. Syst. **52**(3), 423–434 (2021)
36. Wharton, Z., Behera, A., Liu, Y., Bessis, N.: Coarse temporal attention network (CTA-Net) for driver's activity recognition. In: Proceedings of the IEEE/CVF Winter Conference on Applications of Computer Vision, pp. 1279–1289 (2021)
37. Wiederer, J., Bouazizi, A., Kressel, U., Belagiannis, V.: Traffic control gesture recognition for autonomous vehicles. In: 2020 IEEE/RSJ International Conference on Intelligent Robots and Systems (IROS), pp. 10676–10683. IEEE (2020)
38. Winkler, V.: Range doppler detection for automotive FMCW radars. In: European Radar Conference, pp. 166–169. IEEE, Piscataway (2007). https://doi.org/10.1109/EURAD.2007.4404963
39. Wu, Y., Kirillov, A., Massa, F., Lo, W.Y., Girshick, R.: Detectron2 (2019). https://github.com/facebookresearch/detectron2
40. Xu, F., Xu, F., Xie, J., Pun, C.M., Lu, H., Gao, H.: Action recognition framework in traffic scene for autonomous driving system. IEEE Trans. Intell. Transp. Syst. (2021)
41. Zengeler, N., Kopinski, T., Handmann, U.: Hand gesture recognition in automotive human-machine interaction using depth cameras. Sensors **19**(1), 59 (2018)
42. Zhang, Z., Tian, Z., Zhou, M.: Latern: dynamic continuous hand gesture recognition using FMCW radar sensor. IEEE Sens. J. **18**(8), 3278–3289 (2018)

An Improved Lightweight Network Based on YOLOv5s for Object Detection in Autonomous Driving

Guofa Li[1], Yingjie Zhang[2], Delin Ouyang[2], and Xingda Qu[2(✉)]

[1] College of Mechanical and Vehicle Engineering, Chongqing University,
Chongqing 400044, China
[2] Institute of Human Factors and Ergonomics, College of Mechatronics and Control
Engineering, Shenzhen University, Shenzhen 518060, China
{zhangyingjie2020,ouyangdelin2021}@email.szu.edu.cn, quxd@szu.edu.cn

Abstract. Object detection with high accuracy and fast inference speed based on camera sensors is important for autonomous driving. This paper develops a lightweight object detection network based on YOLOv5s which is one of the most promising object detection networks in the current literature. Our proposed network not only strengthens the object positioning ability in Cartesian coordinates without using space transformation, but also reinforces the reuse of feature maps from different scales of receptive fields. To show its robustness, different driving datasets are used to evaluate the effectiveness of our proposed network. The detection performances on objects with different sizes and in different weather conditions are also examined to show the generalizability of our proposed network. The results show that our proposed network has superior object detection performances in the conducted experiments with a very high running speed (i.e., 75.2 frames per second). This implies that our proposed network can serve as an effective and efficient solution to real-time object detection in autonomous driving.

Keywords: Camera sensors · Object detection · Deep learning · Lightweight · Autonomous vehicle

1 Introduction

Object detection is a critical research topic for the development of autonomous driving technologies [1] Effectively detecting various road objects in different environments would greatly improve driving safety and travel efficiency of autonomous vehicles (AVs) [2]. To date, various deep learning based approaches have been proposed to address object detection problems for autonomous driving [3–5]. These approaches can be generally categorized into two categories, i.e., two-stage detectors and one-stage detectors [6].

The two-stage detectors are mainly based on the region convolutional neural network (R-CNN) [7]. The algorithms in these detectors attempt to extract region proposals (RPs) in the first step, and the extracted RPs are then refined

© The Author(s), under exclusive license to Springer Nature Switzerland AG 2023
L. Karlinsky et al. (Eds.): ECCV 2022 Workshops, LNCS 13801, pp. 585–601, 2023.
https://doi.org/10.1007/978-3-031-25056-9_37

and assigned for different classes based on inferred probability values. Based on R-CNN, a faster version (i.e., Fast R-CNN) was developed in [8]. Later, Ren et al. [9] proposed the idea of RP network (RPN) that shares convolutional layers with Fast R-CNN [8]. By merging RPN with Fast R-CNN into one unified network, Faster R-CNN [9] was proposed for more computationally efficient detection. Under Faster R-CNN, an input image is fed to a feature extractor, such as the ZF (Zeiler & Fergus) model [10] or VGG-16 (Visual Geometry Group-16) [11], to produce a feature map. Then, the RPN utilizes this feature map to predict RPs that potentially include objects of interest.

Another way to realize accurate object detection is the use of one-stage detectors, which typically treat object detection as a regression problem. YOLO (you only look once) [12] is about one of the most common one-stage detectors for object detection. Under YOLO, an input image is divided into a specific set of grid cells, and each cell is responsible for detecting objects whose centers are located within that cell. To that end, each cell predicts a certain number of bounding boxes, and it also predicts the confidence scores for these boxes in terms of the likelihoods with objects. Given that the box has an object, the corresponding conditional class probabilities can be predicted [13]. To filter the wrongly predicted bounding boxes out, a threshold is used on the confidence scores of the predicted bounding boxes. Incorporating the techniques of anchor box and batch normalization, YOLOv2 was developed to improve the speed for object detection while keeping the detection accuracy [14]. The YOLO model was then further improved in a version called YOLOv3 [15] that adopted the residual block and feature pyramid network (FPN) [16] to increase the detection accuracy while speeding up the detection process. More recently, YOLOv4 [17] and YOLOv5 [18] have been proposed that employ training tricks to enhance the learning accuracy.

In the development process of YOLO based networks, the performance improvement of each generation is closely related to the adjustment of backbone module of the network. The backbone of YOLOv1 is a custom structure based on GoogLeNet [19]. The backbone of YOLOv2 is Darknet-19, while YOLOv3 selects Darknet-53 as the backbone. In YOLOv4, the backbone is further adjusted as CSPDarknet53 to achieve better performance than the previous generations. The backbone of YOLOv5 still uses CSPDarknet53, but YOLOv5 integrates a variety of advanced training methods developed in the field of computer vision to improve the performance of YOLO object detection. Current YOLOv5 networks have four different structures, that are YOLOv5s, YOLOv5m, YOLOv5l and YOLOv5x. The main difference between these structures lies in the depth and width of the network controlled by a configuration file. In this study, we try to further improve the performance of the YOLO family. YOLOv5s is selected as the basic framework of our proposed network since it is a lightweight network that can facilitate real-time applications in autonomous driving [20].

In this study, a new feature extraction structure suitable for object detection network is proposed, and the backbone network of YOLOv5s is improved. By adding the coordinate information to the ordinary convolution, the object positioning accuracy in images is enhanced. At the same time, the multi-scale separation convolution model is used to better reuse the features, concentrate on the global information, and determine the location and category of the detected object. The main contribution of this study is incorporating the coordinate information and reusing features of the detection backbone network in the development of the object detection model. Without increasing the network complexity, these contributing measures can enable the network to more accurately output the frame on the Cartesian coordinates and improve the network learning performance for more accurate detection.

2 Methods

2.1 YOLOv5s

The anchor-based YOLOv5s [18] is used as the baseline of this study since it is a state-of-the-art algorithm of the YOLO family with lightweight performance and has the potential to meet real-time requirements for autonomous driving. YOLOv5s consists of a backbone network, a detection neck, and a detection head. In YOLOv5s, several image enhancement methods, such as Mosaic and CutMix, are used before putting images into the backbone network. The pre-processed images are then put into the backbone network to extract feature maps. The neck part adopts a series of network layers to mix and combine image features, and transmits images to the head module to predict the image features, generate boundary boxes and predict categories.

Backbone plays a crucial role in object detection networks. Its multi-layer structure can extract the feature information from the input images, which is then transmitted to the subsequent parts of the network for prediction. Thus, the information extracted by the backbone directly affects the prediction of the network. In this paper, we reconstruct the backbone of YOLOv5s with novel network modules including CoordConv and Hierarchical-Split block (HS block). The details of these modules and the structure of our newly proposed network are introduced in the following subsections.

2.2 CoordConv Module

Convolutional neural network (CNN) has enjoyed immense success as a key tool for effective deep learning in a broad array of applications, mainly due to its attractive structural characteristics, such as local connection, weight sharing, and down-sampling. However, [21] reveals the generic inability of CNNs to transform spatial representations between a dense Cartesian representation and a sparse pixel-based representation. To address this, the CoordConv layer (Fig. 1) is designed to help the convolution know the relative coordinate position of the processed cells in the input feature map.

The CoordConv layer can be implemented as a simple extension of standard convolution. In this extension, additional channels are instantiated and filled with coordinate information, and then they are concatenated with the input representation to form a standard convolutional layer. For convolution over two

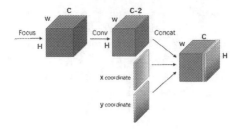

Fig. 1. Graphical representation of the CoordConv module

dimensions, two (i, j) coordinates are sufficient to completely specify an input pixel, but if desired, further channels can be added as well to bias models towards learning particular solutions. Concretely, the X coordinate channel is a H × W rank-1 matrix with its first row filled with 0's, its second row with 1's, its third row with 2's, etc. The Y coordinate channel is similar, but with columns filled in with constant values instead of rows.

Compared with the traditional CNN module, the CoordConv layer keeps the properties of few parameters and efficient computation, but it allows the network to learn to keep or to discard the translation invariance depending on the necessity of the task being learned. As the CoordConv allocates a small amount of network capacity to model non-translation invariant aspects of a problem, it can enable far more trainable models with better generalizability. If CoordConv's coordinate channel does not learn any information, CoordConv is equivalent to a traditional convolution and has complete translation invariance of traditional convolution. If the coordinate channel learns certain information, CoordConv has certain translation dependence. It can be seen that CoordConv's translation invariance and translation dependence can be dynamically adjusted according to different task requirements.

In this paper, the CoordConv module is employed in our network to solve the coordinate transform problem of input images. Since the coordinate information of these pictures is two-dimensional, only two coordinate tensors (X coordinate and Y coordinate) are added in the CoordConv module.

2.3   Hierarchical-Split Block (HS Block)

There are three important problems worthy of our consideration in object detection research: (i) how to avoid producing redundant information contained in the feature maps, (ii) how to promote the network to learn stronger feature presentations without any computational complexity, and (iii) how to achieve better performance and maintain competitive inference speed. The Hierarchical-Split block (HS block) [22] provides a flexible and efficient way to generate multi-scale feature representations (See Fig. 2).

In the HS block, an ordinary feature map in deep neural networks is split into s groups, each with w channels. Only the first group of filters can be directly connected to the next layer. The second group of the feature maps are sent to a convolution of 3 × 3 filters to extract features. The output feature maps are then split into two sub-groups in the channel dimension. One sub-group is directly

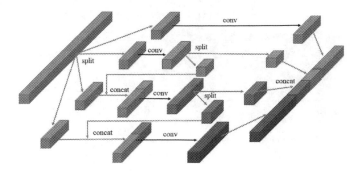

Fig. 2. A detailed view of Hierarchical-Split block with $s = 4$

connected to the next layer, while the other sub-group is concatenated with the next group of input feature maps in the channel dimension. The concatenated feature maps are operated by a set of 3×3 convolutional filters. This process is repeated until the rest of input feature maps are all processed. Finally, the feature maps from all input groups are concatenated and sent to another layer with 1×1 filters to rebuild the features. HS block adopts the method of multi-scale separation and convolution, which allows different groups of feature maps to own different scales of receptive fields. The front groups require less convolution computational times. Thus, the corresponding receptive fields of a feature cell is smaller, and the feature cells can contain detailed information of the input image. On the other hand, the back groups would capture larger objects. In this case, HS block can capture receptive fields with different sizes.

However, the computational complexity of HS block increases with the channel number. In fact, not all feature maps are necessary, since some feature maps can be generated on the existing ones. Thus, we can splice part of the feature maps obtained from the shallow modules into the deep modules(which was firstly proposed in ResNet [23]). This can help reduce the computational complexity without sacrificing the capability of the existing modules. In our research, HS blocks are employed in the Cross Stage Partial (CSP) [24] module to replace the bottleneck residual block. The developed module is called Hierarchical-Split Cross Stage Partial (HSCSP).

2.4   Proposed Network

A coordinate HS-YOLOv5s network is newly proposed in this study to enhance the learning ability for detection tasks while maintaining the lightness of the network as much as possible. Since this network mainly improves the backbone based on YOLOv5s, its main structure is consistent with the network structure of YOLO series, which can be roughly divided into three parts: backbone, neck and head. The main structure of coordinate HS-YOLOv5s is shown in Fig. 3. The backbone network is mainly used to aggregate image features on different image granularity, which is usually formed by convolution structures. The first layer of backbone is the focus module, which preprocesses the input image, aiming

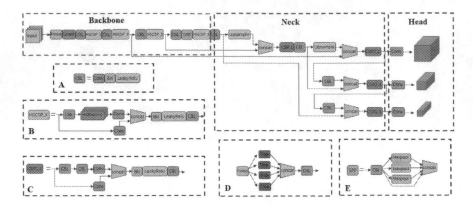

Fig. 3. The flowchart to build our stacking model

to periodically extract pixel points from high-resolution images to reconstruct low-resolution images by: (1) stacking the four adjacent points on the way, (2) focusing the W dimension and H dimension information onto the C-channel space, and (3) improving the receptive field of each pixel. Compared with the traditional down-sampling operation, the focus module can retain more complete down-sampling information [25].

In this paper, the CoordConv module is employed after the focus module in the backbone to enhance the coordinate spatial perception ability of the convolution network so that the network can better locate the target position without adding too many parameters in the learning process. In our experiments, the coordinate tensors are normalized to $[-1, 1]$. Instead of directly adding two coordinate channels to the feature map output by the focus module, we firstly reduce the number of channels by 2 through a convolution calculation with a 1×1 convolution kernel. Then, the coordinate information is added to make the subsequent network structure parameters unchanged while not increasing the amount of calculation.

Another big change we have made to the backbone is to replace the original Residual unit in the traditional cross stage partial (CSP) structure of the original YOLOv5 backbone with a HS block. The developed CSP structure is named HSCSP, which can facilitate the network to reuse the features of different scales. To better fit the structure of our network, we set the number of split groups as $s = 4$ (Fig. 2). In this case, the feature map input into the HSCSP_x module can be divided into two branches, where x denotes the reuse time of HS block in the module. The feature layer processed by the HS block conducts residual concatenation operation with an original feature map (Fig. 3(b)). In this research, the HSCPS_x is adopted repeatedly in the different stages of the backbone network to realize feature reuse in different feature scales.

In the neck and head parts, we follow the YOLOv5s structure, which uses the FPN [26] and pyramid attention network (PAN) [27] structures to detect objects with different scaling scales. FPN can transfer deep semantic features to shallow ones so as to enhance semantic expression on multiple scales. Differently, PAN transmits the positioning information from the shallow layer to the deep layer and enhances the positioning ability on multiple scales. The bidirectional pyramid combination enhances the information dissemination and has the ability to accurately retain spatial information, which helps to properly locate the pixels to form a mask. This combination can also make different feature map groups with different scale receptive fields [28]. The head part uses the previously extracted features to generalize prediction anchors and predict target class. Part of the generated bounding box is filtered out through the predefined confidence threshold, and the rest is operated by soft non maximum suppression (Soft NMS) [29] to obtain the final target box. The loss function of our network is

$$Loss = L_{cls} + L_{box} + L_{obj} \tag{1}$$

where L_{cls} is classification loss, L_{box} is localization loss and L_{obj} is confidence loss. The loss function is based on binary cross entropy (BCE) loss function.

$$L_{cls} = -\sum_{n=1}^{N} \hat{y}_i * \log(y_i) + (1 - \hat{y}_i) \log(1 - y_i) \tag{2}$$

$$y_i = sigmoid(x_i) = \frac{1}{1 + e^{-x_i}} \tag{3}$$

where N represents the total number of categories, x_i is the predicted value of the current category, y_i is the probability of the current category obtained by the activation function, and $\hat{y}_i$ is the current real value (0 or 1). The localization loss L_{box} is based on CIoU (Complete IoU) loss [30]. It is defined as,

$$L_{box} = L_{CIoU}(B, B_{gt}) = 1 - \text{CIoU}(B, B_{gt}) \tag{4}$$

$$\text{CIoU}(B, B_{gt}) = \text{IoU}(B, B_{gt}) - \frac{\rho^2(B, B_{gt})}{c^2} - \alpha v \tag{5}$$

$$v = \frac{4}{\pi}(\arctan(\frac{w^{gt}}{h^{gt}}) - \arctan(\frac{w}{h}))^2 \tag{6}$$

$$\alpha = \frac{v}{1 - \text{IoU}(B, B_{gt}) + v} \tag{7}$$

where v is the normalization of the length width ratio difference between the prediction box and the ground truth, α is a balance factor between the loss caused by the length width ratio and the loss caused by the IoU part, B and B_{gt} represent the prediction box and ground truth box, respectively, w^{gt} and h^{gt} represent the width and height of the ground truth, respectively, and w and h are the width and height of the prediction box.

The confidence loss is calculated by,

$$\begin{aligned}
L_{obj} = &-\lambda_{noobj} \sum_{i=0}^{S^2} \sum_{j=0}^{B} \mathbb{1}_{i,j}^{noobj} \left[C_i^j \log(\hat{C}_i^j) + (1 - C_i^j) \log(1 - \hat{C}_i^j) \right] \\
&-\lambda_{obj} \sum_{i=0}^{S^2} \sum_{j=0}^{B} \mathbb{1}_{i,j}^{obj} \left[C_i^j \log(\hat{C}_i^j) + (1 - C_i^j) \log(1 - \hat{C}_i^j) \right]
\end{aligned} \tag{8}$$

where S is the grid size, and B is box size. If there exists an objection, $1_{i,j}^{obj} = 1$ and $1_{i,j}^{noobj} = 0$; otherwise, $1_{i,j}^{obj} = 0$ and $1_{i,j}^{noobj} = 1$. C_i^j and $\hat{C}_i^j$ represent the ground truth and prediction of confidence for the jth prior box in the ith grid, respectively.

3   Experiment

We compared our Coordinate HS-YOLOv5s with the YOLOv3 tiny [31], which is the lightweight version of YOLOv3. Besides, the ablation study is also conducted in this paper, thus the original YOLOv5s and the modified YOLOv5s with CoordConv/HS block module were also adopted as comparison models. The source code of our proposed method is available at: https://github.com/Cyyjenkins/Coordinate_HS_YOLOv5s

3.1   Datasets

To validate our proposed network, a large scale naturalistic driving dataset, i.e., BDD100K [32], was used which is composed by 100k high resolution (1280 × 720) images. This large-scale dataset includes various traffic situations in naturalistic driving and has been widely used in environment perception studies for AVs. The BDD100K dataset consists of ten classes of objects, including bus, light, sign, person, bike, truck, motor, car, train, and rider. The aim of this study is to effectively detect cars and persons in the collected traffic environment images. As the multi-objects road detection network was typically used to accurately detect common objects in natural driving scenes, all the cars, trucks, and buses were categorized into the car class, and the persons and riders were categorized into the person class. Note that there are 1,021,857 instances of labeled cars, but only 179 instances of labeled trains in the images of BDD100K dataset. Given the very small proportion of labeled trains in the dataset which could become noise when the network learns other features, we remove the train instances from model training and testing [33]. In total, 70,000 and 10,000 images were respectively used as the training set and test set. Respectively, there are 455,010 cars, 53,335 people in the training set and 64,890 cars, 7,730 people in the test set.

To verify the robustness of the network proposed in this paper, two urban traffic datasets, namely KITTI [34] and Cityscape [35], were also utilized in our experiment. The KITTI dataset is commonly used in autonomous driving research which contains real image data collected from urban, rural and expressway scenes. There are up to 15 cars and 30 pedestrians in each image, as well as various degrees of occlusion and truncation [36]. The Cityscapes is a semantic understanding picture dataset of urban street scenes. It mainly contains street scenes from 50 different cities, and has 5,000 high-quality pixel level annotation images of driving scenes in urban environment. The Cityscapes contains a total of 19 categories, and they are further grouped into the car class and the person class as described earlier. Table 1 shows the numbers of images and instances selected from the three datasets in our experiment.

Table 1. Image number and instance number in each dataset

Dataset	Dataset type	Image num.	Car num.	Person num.
BDD100K	Training set	70,000	455,010	53,335
	Test set	1,000	64,890	7,730
KITTI	Training set	5,984	11,929	2,532
	Test set	1,497	2,935	627
Cityscapes	Training set	2,975	19,253	10,315
	Test set	500	3,299	2,236

In addition, the existing research is mainly focused on object detection in normal weather. The lack of attention to extreme weather will affect the detection performance in the corresponding scenes, which is prone to accidents in practical applications. To promote the application of autonomous driving, it is essential to ensure driving safety in extreme weather. Typical extreme weather while driving includes rain and snow. Thus, we are also interested in examining the detection ability of the proposed network in rainy and snowy weather. In our experiment, we extracted the images in rainy and snowy weather from BDD100K and conducted network training accordingly. In total, 6,831 images in these extreme weather conditions were incorporated into the training set and 901 images were into the test set. Respectively, there are 49,125 cars, 8,694 people in the training set of extreme weather and 6,545 cars, 1,180 people in the test set of extreme weather.

All the input images were resized to a resolution of 640 × 640. A percentage threshold (0.1%) [34] of occupation area in the corresponding image was used to exclude extremely tiny objects which might be generated by the tagger's error and are difficult to distinguish even by human eyes. From the perspective of network performance, these tiny objects cannot contribute to network training, but instead are causes of interference and noise.

3.2 Evaluation Criteria

This study uses the unified evaluation system of COCO API [37] to quantitatively analyze the performance of different networks and make a fair comparison between different networks in different datasets. The COCO API mainly takes the average precision (AP) and average recall (AR) as the main evaluation indicators. In this research, AP is calculated based on the IOU off value of 0.5. Since the area of our resized picture is 640 × 640, objects with an area of less than 32 × 32 (i.e., with area proportion less than 0.33%) are defined as small objects. Meanwhile, objects with an area between 32 × 32 and 96 × 96 (with area proportion less than 3%) are defined as medium objects. Objects with an area greater than 96 × 96 are defined as large objects. Generally speaking, the distance between the subject car and the target object affects the size of the target object in the collected image, and the size of the area occupied by the target object is one of the important factors affecting the difficulty of object detection. Therefore, it is reasonable for us to define the size of the target object by the area occupied by the target object in the image in this study.

3.3   Implementation Details and Hardware for Training

In our experiment, the training was carried out for 150 epochs with a learning rate of 0.01 and a batch size of 64. SGD optimizer with momentum at 0.937 and weight decay at 4e-5 was used for training the network. When using soft NMS to delete redundant bounding boxes, the IoU value to determine whether two boxes are overlap was set to 0.45. The total loss was calculated on the basis of classification and regression losses. The network was randomly initialized under the default setting of Pytorch with no pretraining on any additional dataset or any pretrained network. All the experiments were conducted on Intel-i9 9900X with two NVIDIA 3090TI.

4   Results

4.1   Qualitative Results

The qualitative results of detecting target objects of BDD100K when using different networks are shown in Fig. 4. The results show that the network with added coordinate information could detect more targets on the road than the original version of YOLOv5s and YOLOv3 tiny, but it was slightly inferior to the Coordinate HS-YOLOv5s. The Coordinate HS-YOLOv5s could better identify faraway cars and pedestrians with small sizes. At the same time, the Coordinate HS-YOLOv5s also had good detection performance on blocked objects, which is of great significance for autonomous driving, since most of the cars ahead block each other in the actual road conditions. Similar results can also be found in Fig. 5 and Fig. 6, which show qualitative detection performance of different networks in KITTI and Cityscape, from which we can see that the Coordinate HS-YOLOv5s had the best training performance in all the three datasets.

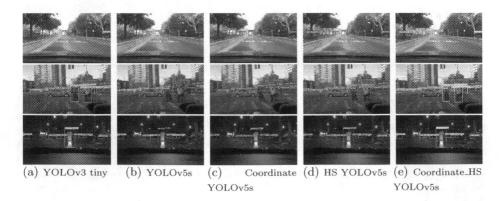

(a) YOLOv3 tiny   (b) YOLOv5s   (c)   Coordinate   (d) HS YOLOv5s   (e) Coordinate_HS
                                          YOLOv5s   YOLOv5s

Fig. 4. Qualitative detection performance between different methods in BDD100K

(a) YOLOv3 tiny   (b) YOLOv5s   (c)   Coordinate   (d) HS YOLOv5s   (e) Coordinate_HS
                                         YOLOv5s   YOLOv5s

Fig. 5. Qualitative detection performance between different methods in KITTI

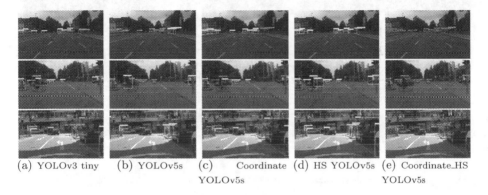

(a) YOLOv3 tiny   (b) YOLOv5s   (c)   Coordinate   (d) HS YOLOv5s   (e) Coordinate_HS
                                         YOLOv5s   YOLOv5s

Fig. 6. Qualitative detection performance between different networks in CITYSCAPES

4.2   Quantitative Results

One of the key indicators we focus on is the detection speed of different network, which is shown in Table 2. See Table 2, YOLOv3 tiny has reached the fastest detection speed, which achieved 6.2ms per image. The detection speed of our method has decreased after increasing the complexity of the network structure, but it can still meet the real-time detection requirements. Table 3 shows the results from quantitative comparisons of different networks performance. As can be seen, Coordinate HS-YOLOv5s achieved the best quantitative performance in all of the three datasets. Its APs for the overall detection, person detection and car detection were all the largest among the five examined networks. The detection accuracy of YOLOv5s adding CoordConv/HS-block was better than the original version of YOLOv5s and YOLOv3 tiny. In addition, the five examined networks were all better at detecting cars than detecting persons. Specifically, compared to the APs of person detection, the APs of car detection was 15.0% higher in BDD100K, 15.0% higher in KITTI, and 14.5% higher in Cityscapes.

Table 2. Detection speed of different networks

Networks	Speed (ms per image)
YOLOv3 tiny	**6.2**
YOLOv5s	9.7
Coordinate YOLOv5	11.2
HS-YOLOv5s	13.3
Coordinate HS-YOLOv5s	13.8

(a) YOLOv3 tiny (b) YOLOv5s (c) Coordinate (d) HS YOLOv5s (e) Coordinate_HS
YOLOv5s   YOLOv5s

Fig. 7. Qualitative detection performance between different networks in RAINY and SNOWY DATASETS

4.3 Detection Performance on Objects with Different Sizes

In this study, we also examined the performance of our network in detecting objects with different sizes. See Table 4 for the numbers of objects with different sizes in BDD100K. The detection accuracies for objects with different sizes are shown in Table 5. Regarding the overall performance, the APs of the four networks based on YOLOv5 for large object detection all exceeded 92%, but none of them could reach 65% while detecting small objects. The Coordinate HS-YOLOv5s achieved the best performance in all the three size categories. In addition, the detection performance of all the examined networks for cars was better when detecting cars versus detecting persons. Compared with the other networks, Coordinate HS-YOLOv5s significantly improved the detection ability of the network on small target objects.

4.4 Model Evaluation in Extreme Weather Situations

In the present study, we also examined the detection accuracy of our network with the extracted rainy and snowy weather dataset. The number of images in the rainy and snowy weather dataset is given in Table 6. Compared to the normal weather conditions, the detection accuracies of all the five examined networks were decreased when detecting objects in the snowy and rainy weather conditions. However, as show in Table 7 and Fig. 7, the best detection performance was

Table 3. APs (%) of different targets by different networks in different datasets

Dataset	Object	YOLOv3 tiny	YOLOv5s	Coordinate YOLOv5s	HS-YOLOv5s	Coordinate HS-YOLOv5s
BDD100K	AP-overall	68.2	83.0	83.2	83.5	**83.6**
	AP-person	55.4	75.6	75.7	**76.2**	76.1
	AP-car	81.0	90.7	90.8	90.8	**91.1**
KITTI	AP-overall	74.3	87.5	88.0	88.1	**88.3**
	AP-person	60.8	78.9	80.4	80.7	**80.8**
	AP-car	87.8	**96.1**	95.7	95.5	95.8
Cityscacpes	AP-overall	60.6	72.0	72.3	72.7	**72.8**
	AP-person	50.7	64.5	65.1	**65.8**	65.5
	AP-car	70.5	79.4	79.5	79.6	**80.0**

Table 4. Number of objects with different sizes for each object class in BDD100K

Dataset type	Object class	Small	Medium	Large
Training set	Person	32,569	19,242	1,524
	Car	179,452	205,017	70,541
	Overall	212,021	224,259	72,065
Test set	Person	4,774	2,744	212
	Car	25,761	29,094	10,035
	Overall	30,535	31,831	10,247

Table 5. APs (%) of different size of objects by networks in BDD100K

Networks	AP-overall			AP-person			AP-car		
	S	M	L	S	M	L	S	M	L
YOLOv3 tiny	42.5	65.2	78.2	38.9	58.3	73.5	47.5	74.6	88.4
YOLOv5s	58.2	82.0	92.5	55.3	76.2	86.9	61.1	87.7	**98.1**
Coordinate YOLOv5s	62.6	82.1	92.6	58.7	76.4	87.4	66.6	87.8	98.1
HS-YOLOv5s	62.8	82.4	**92.8**	**59.1**	76.6	**87.5**	66.8	**87.9**	98.1
Coordinate HS-YOLOv5s	**63.0**	**82.3**	**92.8**	59.0	**76.7**	**87.5**	**67.0**	87.9	98.1

Table 6. Number of objects in the rainy and snowy datasets

Dataset type	Object class	Small	Medium	Large
Training set	Person	5,874	2,590	230
	Car	23,017	19,015	7,093
	Overall	28,891	21,605	7,323
Test set	Person	802	351	27
	Car	3,035	2,549	961
	Overall	3,837	2,864	988

still obtained with the Coordinate HS-YOLOv5s. When using the Coordinate HS-YOLOv5s, the overall AP and APs for person detection and car detection were 15% lower, 18.6% lower, and 11.4% lower, respectively, than those when

Table 7. APs (%) of different objects by different networks in rainy and snowy datasets

Networks	AP-overall	AP-person	AP-car
YOLOv3 tiny	50.1	41.3	58.9
YOLOv5s	67.4	55.6	79.1
Coordinate YOLOv5s	68.4	57.2	**79.7**
HS-YOLOv5s	68.4	57.3	79.5
Coordinate HS-YOLOv5s	**68.6**	**57.5**	**79.7**

Table 8. APs (%) of different size of objects by networks in BDD100K

Networks	AP-overall			AP-person			AP-car		
	S	M	L	S	M	L	S	M	L
YOLOv3 tiny	35.3	50.7	66.8	30.1	44.6	56.4	37.6	53.4	82.1
YOLOv5s	42.0	68.4	80.6	34.4	61.0	66.0	49.5	75.8	95.3
Coordinate YOLOv5s	**43.4**	69.4	81.4	**36.2**	62.5	67.3	50.5	**76.4**	95.4
HS-YOLOv5s	42.8	69.5	82.0	35.0	63.0	71.6	**50.8**	76.2	95.6
Coordinate HS-YOLOv5s	43.0	**69.8**	**84.2**	35.4	**63.6**	**72.6**	50.6	76.1	**95.7**

the dataset was not limited by extreme weather conditions. As shown in Table 8, Coordinate YOLOv5s was better at detecting small objects, while Coordinate HS-YOLOv5s was better at detecting medium and large objects. Based on the results in Table 3 and 7, we conclude that rainy and snowy weather could have a negative impact on the detection accuracy for objects in different size categories.

To sum up, the Coordinate HS-YOLOv5s showed high-precision detection performance in all the selected datasets. The algorithm comparison shows that this newly proposed network can improve detection accuracy for objects with different sizes, especially for small objects. In the rainy and snowy weather, the detection accuracy of our network is also significantly improved compared with the state-of-art algorithms. Future work includes continuous efforts to improve the accuracy of the network. Some special driving scenarios, such as night and strong light scenarios, are also worth investigation.

5   Conclusions

Object detection is the virtual technology that can help improve the cognitive ability of intelligent systems. In autonomous driving, cars need object detection techniques for enhanced environmental perception. This paper proposed a novel object detection network, i.e., Coordinate HS-YOLOv5s, based on the YOLO framework. In this network, the CoordConv module is employed to enhance coordinate information for the feature map in the ordinary convolution module. Thus, the network can locate objects in Cartesian coordinates without using space transformation, which can enhance the accuracy of object positioning. HS-block CSP is also used to share map features from different scales of receptive fields and different convolution groups by using multi-scale separation and convolution. This make it possible to reuse different scales of features without increasing the computational complexity, which is beneficial for extracting the information in feature maps and detecting the location and category of target objects. We verified the effectiveness of our network by executing experiments on three datasets including BDD, KITTI, and Cityscape. The results show that our network improved the detection accuracy of YOLOv5s, especially for the small objects. In addition, our network also had higher detection precision in rainy and snowy weather. Based on the findings from the present study, we may argue that the Coordinate HS-YOLOv5s could serve as an effective solution to road object detection, and contribute to future autonomous driving applications.

Acknowledgment. This study is supported by the National Natural Science Foundation of China (52272421) and the Shenzhen Fundamental Research Fund (JCYJ201908081426- 13246).

References

1. Teichmann, M., Weber, M., Zoellner, M., Cipolla, R., Urtasun, R.: Multinet: real-time joint semantic reasoning for autonomous driving. In: 2018 IEEE Intelligent Vehicles Symposium (IV), pp. 1013–1020. IEEE (2018)
2. Dai, X.: Hybridnet: a fast vehicle detection system for autonomous driving. Signal Process. Image Commun. **70**, 79–88 (2019)
3. Wu, B., Iandola, F., Jin, P.H., Keutzer, K.: Squeezedet: unified, small, low power fully convolutional neural networks for real-time object detection for autonomous driving. In: Proceedings of the IEEE Conference on Computer Vision and Pattern Recognition Workshops, pp. 129–137 (2017)
4. Vishwakarma, A., Bhuyan, M.K.: Image fusion using adjustable non-subsampled shearlet transform. IEEE Trans. Instrum. Meas. **68**(9), 3367–3378 (2018)
5. Li, G., Ji, Z., Qu, X.: Stepwise domain adaptation (SDA) for object detection in autonomous vehicles using an adaptive centernet. IEEE Trans. Intell. Transp. Syst. (2022). https://doi.org/10.1109/TITS.2022.3164407
6. Wang, H., Yu, Y., Cai, Y., Chen, X., Chen, L., Liu, Q.: A comparative study of state-of-the-art deep learning algorithms for vehicle detection. IEEE Intell. Transp. Syst. Mag. **11**(2), 82–95 (2019)
7. Girshick, R., Donahue, J., Darrell, T., Malik, J.: Rich feature hierarchies for accurate object detection and semantic segmentation. In: Proceedings of the IEEE Conference on Computer Vision and Pattern Recognition, pp. 580–587 (2014)
8. Girshick, R.: Fast R-CNN. In: Proceedings of the IEEE International Conference on Computer Vision, pp. 1440–1448 (2015)
9. Ren, S., He, K., Girshick, R., Sun, J.: Faster R-CNN: towards real-time object detection with region proposal networks. In: Advances in Neural Information Processing Systems, vol. 28 (2015)
10. Zeiler, M.D., Fergus, R.: Visualizing and understanding convolutional networks. In: Fleet, D., Pajdla, T., Schiele, B., Tuytelaars, T. (eds.) ECCV 2014. LNCS, vol. 8689, pp. 818–833. Springer, Cham (2014). https://doi.org/10.1007/978-3-319-10590-1_53

11. Simonyan, K., Zisserman, A.: Very deep convolutional networks for large-scale image recognition. arXiv preprint arXiv:1409.1556 (2014)
12. Redmon, J., Divvala, S., Girshick, R., Farhadi, A.: You only look once: unified, real-time object detection. In: Proceedings of the IEEE Conference on Computer Vision and Pattern Recognition, pp. 779–788 (2016)
13. Hnewa, M., Radha, H.: Object detection under rainy conditions for autonomous vehicles: a review of state-of-the-art and emerging techniques. IEEE Signal Process. Mag. **38**(1), 53–67 (2020)
14. Redmon, J., Farhadi, A.: YOLO9000: better, faster, stronger. In: Proceedings of the IEEE Conference on Computer Vision and Pattern Recognition, pp. 7263–7271 (2017)
15. Redmon, J., Farhadi, A.: YOLOV3: an incremental improvement. arXiv preprint arXiv:1804.02767 (2018)
16. Lin, T.Y., Dollár, P., Girshick, R., He, K., Hariharan, B., Belongie, S.: Feature pyramid networks for object detection. In: Proceedings of the IEEE Conference on Computer Vision and Pattern Recognition, pp. 2117–2125 (2017)
17. Bochkovskiy, A., Wang, C.Y., Liao, H.Y.M.: Yolov4: optimal speed and accuracy of object detection. arXiv preprint arXiv:2004.10934 (2020)
18. Jocher, G., et al.: ultralytics/YOLOv5: V5. 0-yolov5-p6 1280 models AWS supervise. LY and youtube integrations. Zenodo **11** (2021)
19. Szegedy, C., et al.: Going deeper with convolutions. In: Proceedings of the IEEE Conference on Computer Vision and Pattern Recognition, pp. 1–9 (2015)
20. Li, G., Ji, Z., Qu, X., Zhou, R., Cao, D.: Cross-domain object detection for autonomous driving: a stepwise domain adaptive yolo approach. IEEE Trans. Intell. Veh. (2022). https://doi.org/10.1109/TIV.2022.3165353
21. Liu, R., et al.: An intriguing failing of convolutional neural networks and the coordconv solution. In: Advances in Neural Information Processing Systems, vol. 31 (2018)
22. Yuan, P., et al.: HS-ResNet: hierarchical-split block on convolutional neural network. arXiv preprint arXiv:2010.07621 (2020)
23. He, K., Zhang, X., Ren, S., Sun, J.: Deep residual learning for image recognition. In: Proceedings of the IEEE Conference on Computer Vision and Pattern Recognition, pp. 770–778 (2016)
24. Wang, C.Y., Liao, H.Y.M., Wu, Y.H., Chen, P.Y., Hsieh, J.W., Yeh, I.H.: CSPNet: a new backbone that can enhance learning capability of CNN. In: Proceedings of the IEEE/CVF Conference on Computer Vision and Pattern Recognition Workshops, pp. 390–391 (2020)
25. Yao, J., Qi, J., Zhang, J., Shao, H., Yang, J., Li, X.: A real-time detection algorithm for kiwifruit defects based on YOLOv5. Electronics **10**(14), 1711 (2021)
26. Liu, S., Qi, L., Qin, H., Shi, J., Jia, J.: Path aggregation network for instance segmentation. In: Proceedings of the IEEE Conference on Computer Vision and Pattern Recognition, pp. 8759–8768 (2018)
27. Huang, Z., Zhong, Z., Sun, L., Huo, Q.: Mask R-CNN with pyramid attention network for scene text detection. In: 2019 IEEE Winter Conference on Applications of Computer Vision (WACV), pp. 764–772. IEEE (2019)
28. Wang, H., Tong, X., Lu, F.: Deep learning based target detection algorithm for motion capture applications. In: Journal of Physics: Conference Series, vol. 1682, p. 012032. IOP Publishing (2020)
29. Bodla, N., Singh, B., Chellappa, R., Davis, L.S.: Soft-NMS-improving object detection with one line of code. In: Proceedings of the IEEE International Conference on Computer Vision, pp. 5561–5569 (2017)

30. Zheng, Z., Wang, P., Liu, W., Li, J., Ye, R., Ren, D.: Distance-IoU loss: faster and better learning for bounding box regression. In: Proceedings of the AAAI Conference on Artificial Intelligence, vol. 34, pp. 12993–13000 (2020)
31. Adarsh, P., Rathi, P., Kumar, M.: YOLO v3-tiny: object detection and recognition using one stage improved model. In: 2020 6th International Conference on Advanced Computing and Communication Systems (ICACCS), pp. 687–694. IEEE (2020)
32. Yu, F., et al.: BDD100K: a diverse driving dataset for heterogeneous multitask learning. In: Proceedings of the IEEE/CVF Conference on Computer Vision and Pattern Recognition, pp. 2636–2645 (2020)
33. Cai, Y., et al.: YOLOv4-5D: an effective and efficient object detector for autonomous driving. IEEE Trans. Instrum. Meas. **70**, 1–13 (2021)
34. Geiger, A., Lenz, P., Urtasun, R.: Are we ready for autonomous driving? The kitti vision benchmark suite. In: 2012 IEEE Conference on Computer Vision and Pattern Recognition, pp. 3354–3361. IEEE (2012)
35. Cordts, M., et al.: The cityscapes dataset for semantic urban scene understanding. In: Proceedings of the IEEE Conference on Computer Vision and Pattern Recognition, pp. 3213–3223 (2016)
36. Li, G., Xie, H., Yan, W., Chang, Y., Qu, X.: Detection of road objects with small appearance in images for autonomous driving in various traffic situations using a deep learning based approach. IEEE Access **8**, 211164–211172 (2020)
37. Lin, T.-Y., et al.: Microsoft COCO: common objects in context. In: Fleet, D., Pajdla, T., Schiele, B., Tuytelaars, T. (eds.) ECCV 2014. LNCS, vol. 8693, pp. 740–755. Springer, Cham (2014). https://doi.org/10.1007/978-3-319-10602-1_48

Plausibility Verification for 3D Object Detectors Using Energy-Based Optimization

Abhishek Vivekanandan[1]([✉]), Niels Maier[2], and J. Marius Zöllner[1,2]

[1] FZI Research Center for Information Technology, Karlsruhe, Germany
vivekana@fzi.de
[2] Karlsruhe Institute of Technology, Karlsruhe, Germany

Abstract. Environmental perception obtained via object detectors have no predictable safety layer encoded into their model schema, which creates the question of trustworthiness about the system's prediction. As can be seen from recent adversarial attacks, most of the current object detection networks are vulnerable to input tampering, which in the real world could compromise the safety of autonomous vehicles. The problem would be amplified even more when uncertainty errors could not propagate into the submodules, if these are not a part of the end-to-end system design. To address these concerns, a parallel module which verifies the predictions of the object proposals coming out of Deep Neural Networks are required. This work aims to verify 3D object proposals from MonoRUn model by proposing a plausibility framework that leverages cross sensor streams to reduce false positives. The verification metric being proposed uses prior knowledge in the form of four different energy functions, each utilizing a certain prior to output an energy value leading to a plausibility justification for the hypothesis under consideration. We also employ a novel two-step schema to improve the optimization of the composite energy function representing the energy model.

Keywords: Plausibility · Safety · 3D Object detection · SOTIF

1 Introduction

Adversarial attacks on object detection networks are making the real-world deployment of Neural Networks (NN) susceptible to safety violations, hindering the approval and conformance of vehicles to SOTIF standards. This is attributed mainly to the black box nature of NN themselves. Often the perception module, with the object detector at its core, plays a key part in situations where these errors occur. An object detector's (OD) failure to perceive an object (e.g., another car or a pedestrian crossing the street) can immediately result in an unsafe situation both for the vehicle's passengers and other traffic participants.

However, detections which are misclassified or falsely proposed by the OD also constitutes considerable risk under the Operational Design Domain (ODD) of an automated vehicle. These artifacts are called as *ghost detections* or *false*

L. Karlinsky et al. (Eds.): ECCV 2022 Workshops, LNCS 13801, pp. 602–616, 2023.
https://doi.org/10.1007/978-3-031-25056-9_38

positives often appear due to perception gaps or sensor noises. False positives not only cause sudden jerks, contributing to an uncomfortable driving experience, but could also lead to rear end collisions when braking is applied without the need for it. Motivated by the dangers and risk posed, we propose to develop a parallel checker module following the architecture design of Run-Time Assurance (RTA) [6] tests. Through experiments, we extensively show that our module checks for the plausibility of an object (in this work we chose the category car) using combinations of energy functions, thereby significantly reducing the number of False Positives relative to the base NN. The energy functions are made up of simple priors, which conforms to our definition of world knowledge [23,25]. The base network under consideration is MonoRUn [5] which uses RGB images to predict 3D Bounding Boxes defining the position and orientation of an object, forming an initial hypothesis which needs to be checked for. As shown in Fig. 1, the checker modules (marked through dashed rectangle boxes) uses raw LiDAR point clouds and Camera streams with the assumption that the sensors are synchronized accurately to provide valuable environmental information to different parts of the module.

In summary, we: 1) Design a parallel checker module for plausibility checks for the outputs of a 3D-OD network. 2) Propose a novel two-step optimization schema for composite energy function, which depends on 3D shape priors. 3) Developed computationally light rendering module to obtain 2D segmentation masks from the optimized 3D shape priors represented through the notation (y^*, z^*). 4) Finally, we propose a simple empirically evaluated threshold based False Positive filter with the help of an energy-based model.

2   Related Work

Several works aim to verify the detected objects' existence in a fusion system. Often, the Dempster-Shafer theory of evidence (DST) [22] is used to implement and combine plausibility features on various system levels. In [1], authors generate object existence probability using a high-level architecture defining different sensor fusion for their autonomous test vehicle. Each individual sensor assigns existence probabilities to the objects it detects. By following the DST rules of combination, existence probabilities of each sensor are merged by constructing basic belief assignments. In their work [9], the authors present a similar fusion system applied to the detection of cars from a roadside sensor infrastructure at a highway section consisting of various radar sensors. Different plausibility checks for individual sensors are encoded as basic belief assignments through *a priori* assumptions on geometric constraints (such as a cameras' field of view or occlusions) and parameters such as the trust into a sensors' performance, which yields a Bayesian like probability of an object's existence. The authors from [14] present a different approach for plausibility evaluation about an object's existence by employing a serial implementation checks for the consensus between two detectors from different sensor streams. A simple measurement of Latency-based threshold against a pre-defined distance threshold checks for implausibility. Works from [16,21], propose False Negatives and False positive reduction algorithm on semantic segmentation tasks. Energy-based models (EBM) have played an important part throughout the history of pre-modern machine learning. In [15], LeCun et al. gives an extensive tutorial and introduction to energy-based learning methods. They describe the concept of EBMs as capturing dependencies between variables by associating a scalar valued energy to each configuration of the variable. Energy functions are a way to encode certain priors over a set of variable states (defined through properties and entities of a system), which yields a net-zero energy value for a perfect compatibility between variables during inference and high-energy value for incompatible variables. In [18], for the discriminative task of object detection, the pose is modelled as a latent variable, and all parameters are obtained by training with a contrastive loss function from a facial pose dataset by minimizing an energy function.

Engelmann et al. [7] used a set of hand designed energy functions together with their proposed shape manifold for object segmentation task as well as pose and shape estimation. For the energy functions, they combine a Chamfer-Distance Energy, measuring the distance from the stereo points to the object's surface with some prior constraints, punishing deviations of the shape from the mean pose and the object's height over ground. From the works of [19,20,24], they apply latent shape space approaches to recover 3D shapes of a car using EBMs.

Differing from the previous presented data-driven approaches, Gustafsson et al. [11] aim to refine 3D object detection without relying on shape priors. Their work builds upon their previous work [10], in which confidence-based regression is performed on 2D detection results using an energy-based model. In their more recent work, they extend this idea by proposing to use a Deep Neural Network (DNN) to train a conditional energy-based model for probabilistic regression of 3D bounding boxes on point cloud data.

In comparison to the presented approaches here, we propose a novel approach of encoding priors into an energy-based system, from which we measure compatibility by forming a decision rule through energy value thresholds ensuring the plausibility of an object.

3   Concepts and Proposed Method

The main idea is to encode prior knowledge into the energy function such that a comparison could be made between the priors ($\boldsymbol{X} \in \mathcal{X}_i$) and the predictions from the 3D object detector ($\boldsymbol{y} \in \mathcal{Y}$). The predictions from a 3D-OD from here on would be expressed as a hypothesis, as we aim to justify whether there exists an object inside the proposed space. Individual energy functions encode a specific prior knowledge and map the observed compatibility to a single scalar value. In mathematical terms, this can be expressed in a general definition of the energy functions as, $E : \mathcal{X}_i \times \mathcal{Y} \to \mathbb{R}$ and $E(\boldsymbol{X}, \boldsymbol{y}) \geq 0 \; \forall \boldsymbol{X}, \boldsymbol{y} \in \mathcal{X}_i \times \mathcal{Y}$.

The energy functions output energy values which exhibit the property of compatibility. For a perfect compatibility, $E = E^* = 0$; while for any deviations, $E > 0$. In this paper, we use four different energy functions, where each of the function constitutes a certain prior. The energy function E_{Sil} measures silhouette

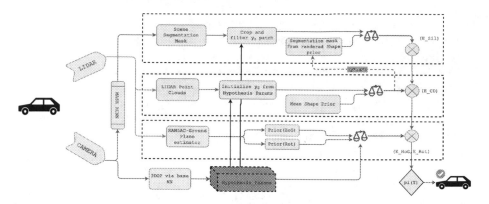

Fig. 1. General architecture describing the flow of energy values which are combined to argue about an object's presence. y_0 represents the initial hypothesis. LiDAR and Camera inputs are primarily used to provide cross sensor data streams. Top: E_{Sil}, compares segmentation mask from MaskRCNN [12] and rendered segmentation mask from the optimized outputs of CD energy function. Middle: E_{CD}, compares and optimizes a mean 3D shape prior with raw segmented point clouds within the hypothesis space. Bottom: E_{HoG}, E_{Rot}, compares ground estimates from a RANSAC regressor with the height and rotation of an initial hypothesis

alignment; E_{CD}, measures the alignment of a point cloud and the remaining two, E_{HoG} and E_{Rot} are based on ground estimates which measures the height over ground and rotational consistency of the bounding boxes. The first two energy functions depend on additional prior knowledge in the form of 3D shape priors represented as a *Truncated Signed Distance Function* (TSDF) through CAD models based on ShapeNet [4]. The two energy functions based on the ground estimate do not depend on any additional input, and are referred to as energy priors throughout this work. Each energy function can be interpreted as an expert, having expertise about one aspect that contributes to the plausibility measure. To be able to obtain feasible plausibility evaluations, the individual energy functions can be combined into an energy-based model defined as:

$$E = \alpha_0 E_{Sil} + \alpha_1 E_{CD} + \alpha_2 E_{HoG} + \alpha_3 E_{Rot} \tag{1}$$

The concatenation of such uncalibrated experts are called as product-of-experts [13].

3.1 Chamfer Distance Energy Function

Inspired from the works of [7] and [20], we construct our first energy function based on the *Chamfer Distance* (CD) to evaluate the alignment/compatibility of the shape priors and acquired point cloud data through optimization. The *Chamfer Distance* is the summation of the closest point distances between two sets of points. For two point clouds $\mathcal{A}$ and $\mathcal{B} \in \mathbb{R}^3$:

$$CD(\mathcal{A}, \mathcal{B}) = \sum_{a \in \mathcal{A}} \min_{b \in \mathcal{B}} \|a - b\|_2^2 + \sum_{b \in \mathcal{B}} \min_{a \in \mathcal{A}} \|a - b\|_2^2 \tag{2}$$

This distance measure as such satisfies the requirements of an energy function. However, [7] proposed further modification to the existing energy function to better handle the robustness and to make it compatible with second order optimizations. To reduce the effect of outliers points during optimization and also for a better signal-to-noise ratio; the *Huber loss function* $\rho : \mathbb{R} \to \mathbb{R}$ is applied to the squared TSDF values. In order to make the function independent of the point cloud size, we take the mean of all points (summed distance of the point's distance values are divided by the total number of points in the point cloud $N = | \mathcal{X}_{PC} |$). With these considerations, the Chamfer distance energy function measures the compatibility between a shape prior $\Phi = \Phi(\tau, z)$ (parameterized by the 3D pose τ and shape weights z) and the raw point clouds inside the hypothesis space. Such a measure, throughout this work, is defined as:

$$E_{CD}(X_{PC}, \Phi) = \frac{1}{N} \sum_{x_i \in \mathcal{X}_{pc}} \rho(\Phi(x_i, z)^2) \ni \rho(x) = \begin{cases} x & , x \leq \varepsilon \\ 2\sqrt{x} - \varepsilon & , \text{otherwise} \end{cases} \quad (3)$$

As our main goal is to verify that an object of class car is present inside the proposals from the 3D-OD, we fetch this information via a query (Fig. 1 black solid arrows from *Hypothesis_Params* block). A query contains pose parameters representing the hypothesis's bounding box coordinates along with the mean shape manifold (Φ_{mean}, obtained from the set of objects from ShapeNet) forming the initial hypothesis (y_0).

3.2 Silhouette Alignment Energy Function (SAEF)

The SAEF measures the consistency between the silhouette alignment of a given segmentation mask and a rendered silhouette mask of an object hypothesis from the projection function. While the matched segmentation mask of an object M is obtained from the instance segmentation network (MaskRCNN), the object hypothesis silhouette is obtained by projecting the optimized 3D TSDF shape prior Φ into the image space based on its current shape and pose estimate. As shown in the Fig. 1, dotted red line from the optimized CD energy function going into the *segmentation mask creation* block. This rendering of the shape prior is expressed through the projection function $\pi(\Phi, p) : \Phi, p \mapsto (0, 1)$, assigning each pixel p a value close to 1 inside the object and close to zero outside the object.

$$E_{Sil}(\Phi) = \frac{1}{| \Omega |} \sum_{p \in \Omega} r_{sil}(p, \Phi) \quad (4)$$

where Ω is the set of pixels (i.e., the region of interest in the image) and r_{Sil} is the residual comparing the segmentation masks per pixel:

$$r_{Sil}(p, \Phi) = -\ln(p_{fg}(p)\pi(\Phi, p) + p_{bg}(p)(1 - \pi(\Phi, p)) \quad (5)$$

Here p_{fg} and p_{bg} are the foreground and background probabilities of each pixel derived from the segmentation mask M. The residual function emits large positive values if there exist inconsistencies between the segmentation mask and object hypothesis silhouette mask. On the other hand, for consistencies' between the pixels, the ln() function becomes close to 1, resulting in a residual value close to 0.

Differentiable Rendering, inspired from the works of [19,20], about silhouette masking, we design our projection function $\pi(\mathbf{\Phi},\boldsymbol{p})$ through the following equation.

$$\pi(\mathbf{\Phi},\boldsymbol{p}) = 1 - \Phi(\boldsymbol{x}_i^o) \forall \boldsymbol{x}_i^o \in \mathcal{X}_{ray}^p \frac{1}{\exp^{\Phi(\boldsymbol{x}_i^o)\xi}+1} \tag{6}$$

where $\mathcal{X}_{ray}^p$ are 3D points sampled along a ray that is cast from the camera center through the pixel $\boldsymbol{p}$ in the pinhole camera model. The super-script ^{o} denotes the transformation from camera coordinate system into object coordinate system for evaluation in the TSDF shape grid. The function inside the product is the sigmoid function, where ξ controls the sharpness of the inflection, translating to the smoothness of the projection contours.

The idea of this projection function is, that if a point sample along the ray cast through a pixel falls into the object shape in the 3D space, the point will be assigned a negative value through the shape's TSDF. For negative values, the sigmoid function, acting as a continuous and steadily differentiable approximation of the *Heaviside step function*, will take a value close to 0. Points falling outside the shape will be assigned positive signed distance values, leading to values close to 1 in the sigmoid function.

Through this definition of the projection function, the rendering process of the shape prior silhouette masks becomes a fully differentiable function. This allows to analytically calculate desired gradients and Jacobians that can be used for optimizing the energy function. By taking into consideration only the point along each ray with the minimal signed distance value, the number of points in the further evaluation is reduced significantly. This helps us to significantly speed up the rendering process. A schematic of this process for one pixel is shown in Fig. 2a. The red-point is the sampled point along the cast ray that is closest to the shape and therefore determines the pixel value. The example rendered silhouette mask using this approach shown in Fig. 2b establishes how through this modification, the purpose of the projection function is preserved

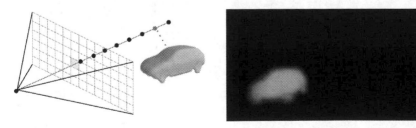

(a) Differentiable rendering for a single pixel

(b) Rendered Mask from optimized shape prior

Fig. 2. Schematic example for the differentiable rendering process obtained from the modified projection function. (Color figure online)

when compared with [20]. The point with the minimal TSDF value sampled along the ray for a pixel containing the objects' projection will still be a negative value, such that the projection function returns a value close to 1 (green in Fig. 2b). Similar, for those pixels that do not contain the object's projection, the function will still return a value close to 0.

3.3   Height over Ground Energy Function

One simple requirement for a plausible hypothesis is, for the object to be on the ground. Especially for detections of the class *Car,* this is a strong requirement. Through ground plane estimation (RANSAC regressor), the ground level of each coordinate in the x-y plane of a scene can be approximated through the function:

$$g(\boldsymbol{t}(x,y)) = \frac{1}{c}(d - ax - by) \tag{7}$$

where $\boldsymbol{t} \in \mathbb{R}^3$ is the pose of an object hypothesis in the ego-vehicle coordinate system. This allows to calculate the height over ground of an object hypothesis as

$$d_{HoG}(\boldsymbol{t}, h) = \boldsymbol{t}(z) - \frac{h}{2} - g(\boldsymbol{t}(x,y)) \tag{8}$$

where h denotes the hypothesis height, accounting for the offset in height of the 3D pose parametrization chosen as the 3D bounding box center of the object.

Using the height over ground distance, an energy function encoding this requirement can be formulated in a parabolic form as proposed in [24] and [7]:

$$E_{HoG}(\boldsymbol{y}) = (d_{HoG}(\boldsymbol{t}, h))^2 = (\boldsymbol{t}(z) - \frac{h}{2} - g(\boldsymbol{t}(x,y)))^2 \tag{9}$$

For objects being close to the ground plane, the energy function will be close to 0, while deviations are punished quadratically.

3.4   Rotation Consistent Energy Function

Similarly to the height-over-ground energy prior, another prior assumption for the orientation of an object detection hypothesis based on a ground estimate can be formulated as an energy prior. Following a similar line of reasoning for objects of class *Car* to be able to touch the ground, this requirement specifies that they should touch the ground with the wheels. This requirement of the hypothesis's orientation to align with the ground can be formulated as the hypothesis's z-axis being parallel to the ground normal vector $\boldsymbol{n}_g$. Given the orientation of the hypothesis w.r.t. the ego vehicle coordinate system through the rotation matrix R and the estimated ground plane normal, $\boldsymbol{n}_0 = [a, b, c]^T$ gives the scalar product that can be used to evaluate the alignment of the hypothesis axis and the normal vector.

The scalar product of two vectors has the property of yielding values close to 0 for orthogonal vectors, while for two normal parallel vectors it evaluates to 1. The rotation energy prior can be defined as:

$$E_{rot} = (1 - (R \cdot (0, 0, 1)^T)^T \cdot n_g)^2 \tag{10}$$

The dot product of the rotation matrix R and the vector $(0, 0, 1)^T$ is equal to taking the last row of the rotation matrix as it is describing the object's z-axis rotated to the ego-vehicle coordinate system.

4   Method

The energy-based model now describes an energy surface, on which each point in the object hypothesis space $\mathcal{Y}$ (parameterized by object 3D pose and shape prior primary components) is assigned with a value $E \geq 0$ that directly expresses the compatibility of the hypothesis through the observed raw data under the assumption of a certain shape prior. Optimal plausible hypotheses can be understood as being close to a local minimum (considering noise, etc.) on the energy surface defined by the energy-based model. The most difficult challenge to using the energy value directly for plausibility evaluation is that the energy-based model is a combination of uncalibrated individual experts. To search for the local minimum, close to an initially proposed hypothesis, optimization can be used. The goal of the optimization method is to find an optimal hypothesis y^* and a shape z^* that minimizes the energy-surface defined through the cost function:

$$_{y,z}(\alpha_0 E_{Sil} + \alpha_1 E_{CD} + \alpha_2 E_{HoG} + \alpha_3 E_{Rot}) \tag{11}$$

where α_n is a scalar value which denotes the importance factor. For our evaluation we choose two different sets of configurations for α_n, where $C_1 = [0.5, 10, 5, 5]$ and $C_2 = [10, 0.1, 1, 50, 10^4]$. This configuration hyperparameter controls the impact of individual energy functions on the overall optimization time it takes to find the local minima. An advantage of the chosen energy functions and formulations is their disposition for optimization as the individual energy functions are made differentiable. Inspired by the requirement to find the optimal value of the hypothesis parameterized by the pose and a latent variable parameterizing the shape, we propose a novel two-step optimization to find the minima which reflects true compatibility with the observed priors.

The **first optimization step,** as can be seen from line 13 of Algorithm 1 consists of solving a 3D rigid body pose combined with a search for the optimal shape problem. The 3D rigid body pose of the object hypothesis can be described through a translation and rotation of the object w.r.t. the ego vehicle coordinate system. A minimal representation of such a problem is given through 6 parameters for the 6 possible degrees-of-freedom. However, a non-minimal representation consisting of the 3 translational components and 4 quaternion parameters is chosen for the flat euclidean spaces following [2]. The pose state vector is therefore given as:

$$\boldsymbol{\tau} = \boldsymbol{\tau}_O^{EV} = [\boldsymbol{t}_O^{EV}, \boldsymbol{q}_O^{EV}] = \begin{bmatrix} t_x\ t_y\ t_z\ q_w\ q_x\ q_y\ q_z \end{bmatrix} \in \mathbb{R}^7 \tag{12}$$

Algorithm 1. Pseudocode for obtaining energy values

1: **input:** hypothesis' pose params from the prediction of a base NN
2: **output:** energy value
3: **for** samples in hypothesisList: **do**
4:      Apply Threshold filtering   ▷ filter objects based on max distance from the ego_vehicle
5:      **if** checkMinMaxPoseValid(samples): **then**
6:          fetch validBB; ▷ Validate whether the BB is plausible or not derived from ϕ_{mean}
7:          **for** validBB: **do**
8:              check removeRadiusOutlier $\leftarrow minNbrPoints$ ▷ For a hypothesis space there should exist a min number of points to satisfy further optimization criterias
9:              Identify and match box from MaskRCNN scenes ▷ False Positive when no segmentation mask is found
10:             **if** $(E_{HoG}, E_{Rot}) \leq$ minEnergyThreshold: **then**
11:                 init $\phi_{mean} \leftarrow (PoseStateVector)$ ▷ From NN
12:                 costFunction$(E_{Sil}, E_{CD}, E_{HoG}, E_{Rot})$ ▷ apply config C_1
13:                 run secondOrderOptim(costFunction, LBFGSB);
14:                 collect jointEnergyValue, (y^*, z^*);
15:                 costFunction$(E_{Sil}, E_{CD}, E_{HoG}, E_{Rot})$ ▷ apply config C_2
16:                 run secondOrderOptim(costFunction, BFGS);
17:                 return energy value w.r.t E_{CD};
18:             **else**
19:                 push up the energy value quadratically;   ▷ False Positives
20:         **else**
21:             return sqrt(calculateDeltaBetweenBB)   ▷ push up the energy value of implausible bounding box

---

where, $\boldsymbol{t}_O^{EV} = \begin{pmatrix} t_x\ t_y\ t_z \end{pmatrix}^\top$ are the components of the translation vector describing the shift of points from the object-centered coordinate system (O) to the ego-vehicle-centered coordinate system (EV) and $\boldsymbol{q}_O^{EV} = q_w + q_x \rangle + q_y| + q_z\|$ describes the quaternion parameters of the transformation. For the shape parameters z_n, the 5 shape weights corresponding to the 5 first primary components of the shape manifold are chosen. The state vector is therefore a concatenation of the pose and shape parameters:

$$\boldsymbol{\xi} = [\boldsymbol{\tau}_O^{EV}; \boldsymbol{z}] = \begin{bmatrix} t_x & t_y & t_z & q_w & q_x & q_y & q_z & z_0 & z_1 & z_2 & z_3 & z_4 \end{bmatrix} \in \mathbb{R}^{12} \tag{13}$$

Each shape weight is bounded to an interval of $[-1, 1]$. The bounded version of the 2nd order quasi-newton optimization method, L-BFGSB [17], is chosen for the current optimization step. A lightweight optimization method is chosen, as the objective of this step is to find the optimal shape parameter which fits the point cloud observation. The optimization step uses C_1 configuration as a design choice. The weight vector of the configuration expresses relatively a higher weight towards E_{CD}, as this pushes the energy function to find a better shape to fit to the observed point cloud, which otherwise suffers from translation and rotation errors. The E_{Sil} helps to attain a faster optimization by guiding the

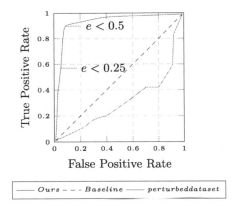

Fig. 3. Quantitative evaluation of the synthetic dataset to find an optimal energy threshold to act as a decision rule based filter.

energy function towards minima. The optimized (y^*, z^*) is the best fit to the proposal from the NN. The resulting energy from this step is in itself sufficient to argue against the plausibility of the object. But there exist cases similar to Fig. 5 where a misclassification with objects from other classes could provide a false fit, contributing to False Negatives situations. To mitigate such cases, a second optimization is needed.

The **second optimization step,** as can be seen from line 16 of Algorithm 1 uses C_2 configuration parameters with BFGS optimization method. The goal of this optimization step is to collect the energy values given the strong requirements for the object to be on the ground while the wheels touching the road surface capture through E_{HoG} and E_{Rot}. This design choice yields us an additional check to monitor whether there exists a significant gradient change in E_{CD} while the overall function pushes the rest of the priors towards minima. For TP detections, this optimization step would result in a non-significant gradient change to the energy value, as the function has already attained local minima on the energy surface. In case of an FP, arising out of misclassification, e.g., as shown in Fig. 5 where a truck is classified with high probability of being a car, the gradient change w.r.t to E_{CD} should be positive indicating an incompatibility between the priors and the observed hypothesis. In both optimization steps, the Jacobian, and the Hessian, are numerically approximated to obtain better results.

4.1   Metric

At the end of the optimization process, each proposal in the queue to be verified receives an associated uncalibrated energy value. Due to the uncalibrated nature of the energy values, a decision rule based on a threshold over the energy value needs to be chosen. To search for the cutoff threshold, we created a synthetic dataset from NuScenes Validation dataset [3]. The synthetic dataset contains a balanced mix of TP and FP created by random perturbation to the bounding boxes, including some boxes being pushed higher from the ground, simulating an adversarial attack. The baseline has a probability value of 0.5 as our perturbed dataset is balanced. The red line on the ROC chart Fig. 3 represents the amount of FP for different IOU thresholds. Each energy values contribute to the plausibility $pl(\boldsymbol{y}) \in \{1, 0\}$ and is evaluated to be either plausible or implausible based on an empirical threshold κ as defined by the following equation.

$$pl(\boldsymbol{y}) = \begin{cases} 1, & E(\boldsymbol{y}) \leq \kappa \\ 0, & E(\boldsymbol{y}) > \kappa \end{cases} \tag{14}$$

Our plausibility verification was then done for each of the samples on this perturbed dataset, and the results can be seen from the graph. The blue line from the graph is obtained for different thresholds' of energy values. Do note that the energies defined are uncalibrated. TPR has the interesting property of capturing the impact of TNs and FPs. When FPs are detected, this classification gets converted to TNs leading to a shift towards the left side of the graph. An energy threshold value of 0.5 is chosen, since this threshold converts most of the FPs present within the synthetic dataset to TN.

5 Experiments

As our work focuses towards reducing false positives and acts as a parallel module which verifies proposals from the NN we need to choose a model which suits the evaluation criteria. Our module requires no training components besides the MaskRCNN module, which needs to be retrained for the ODD. To justify the capabilities of our parallel module, we choose MonoRUn as our base network and measure its performance with the KITTI [8] test dataset. For each of the hypothesis/proposals in the scene, we have an initial plausibility check against Height-over-Ground and Rotation consistent priors. If the 3DOP produces a high-energy value (E > 0.5) we attribute the hypothesis as a *false positive*. In

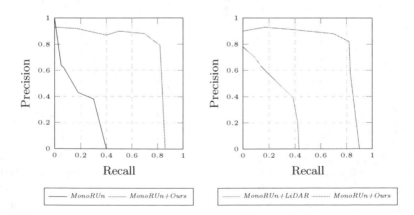

Fig. 4. Quantitative evaluation of MonoRUn network and ours on KITTI easy evaluation. The left chart shows the baseline for the network which uses no LiDAR supervision when compared with the right

addition to the height and rotation checks, we also filter the proposals based on a distance threshold to limit our verification space and this is chosen as 30 m in front and back, 15 m to the left and right of the ego vehicle's camera position. We also apply checks based on the number of points in the LiDAR space and remove proposals which have implausible Bounding box shapes. These simple checks negate the need to validate the hypothesis against further energy values, saving valuable computation time. The implication stays consistent with our world view (object belonging to class car can't float in free space) about an object's position, w.r.t our prior knowledge.

In the Fig. 4 we qualitatively evaluated our energy-based threshold filter against MonoRUn network with and without the LiDAR supervision. The AP of our filter exceeds the base NN by a wide margin. From the plots, the precision, which measure the total positivity of the observed samples, seems to achieve moderately high value which showcases that our method is effective in reducing the false positives while maintaining a significantly reduced False Negatives relative to the base NN.

In Fig. 6 for a given ground estimate, the height-over-ground and rotation energy prior can be calculated. Fig. 6 shows one of the samples which were randomly shifted and rotated. Figure 6a shows the initial hypothesis, floating above the ground plane, with the orientation being not aligned with the ground plane. This deviation from the requirements are encoded into the energy priors (e.g., a car should touch the ground with its four wheels) and is reflected in the energy values for the particular sample. Given a segmentation mask and a rendered silhouette mask of the optimal hypothesis, the E_{Sil} evaluates the compatibility of the two. For this experiment, the differentiable renderer was set to a down sample factor of 8, meaning that only every 64th (8^2) pixel the projection function was evaluated. This downsampling helps to reducing the computational intensity whilst preserving the silhouette details. Along each ray, points were sampled every 0.3 m to a maximum distance of 30 m. The parameter ξ, controlling the steepness of the sigmoid function in the projection function, was chosen to be 25. Figure 7 section (a) shows how the initial hypothesis (green mask) barely

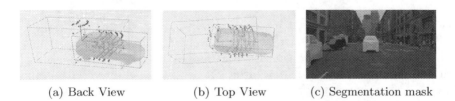

(a) Back View   (b) Top View   (c) Segmentation mask

Fig. 5. Quantitative evaluation of misclassification. In this case, the truck was misclassified as a car and point cloud observations doesn't fit the optimized shape leading to a high-energy value. (a) and (b) shows the back view and top view of the sample. Green BB, ground truth proposal. Blue BB = Optimized pose and shape vector (c) shows the segmentation mask from MaskRCNN which was trained on cityscapes to exhibit worst-case performance (Color figure online)

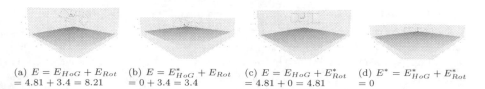

(a) $E = E_{HoG} + E_{Rot}$ (b) $E = E^*_{HoG} + E_{Rot}$ (c) $E = E_{HoG} + E^*_{Rot}$ (d) $E^* = E^*_{HoG} + E^*_{Rot}$
$= 4.81 + 3.4 = 8.21$ $= 0 + 3.4 = 3.4$ $= 4.81 + 0 = 4.81$ $= 0$

Fig. 6. Qualitative example of Height-over-Ground and Rotation energy prior-based optimization for a given ground plane and a hypothesis (blue bounding box) with random position and orientation, as shown in (a). The result of optimizing the priors individually and jointly is shown in (b), (c) and (d). (Color figure online)

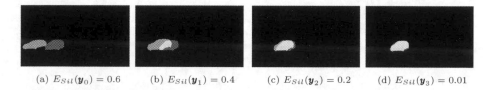

(a) $E_{Sil}(\mathbf{y}_0) = 0.6$ (b) $E_{Sil}(\mathbf{y}_1) = 0.4$ (c) $E_{Sil}(\mathbf{y}_2) = 0.2$ (d) $E_{Sil}(\mathbf{y}_3) = 0.01$

Fig. 7. Depicted are the overlap (yellow) of a segmentation mask (red) obtained from a NuScenes sample and a silhouette rendering (green) of the mean shape prior of an initial hypothesis, the optimized hypothesis and intermediate steps of the optimization process. (Color figure online)

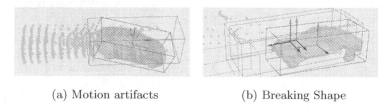

(a) Motion artifacts   (b) Breaking Shape

Fig. 8. Failure cases: True positives are being misclassified as False positives due to sensor noise. Green box represents the ground truth, blue box represents the optimized shape while red represents the initial hypothesis (Color figure online)

coincides with the segmentation mask (obtained from MaskRCNN) leading to an $E_{Sil}(y_0) = 0.6$. As the energy is the mean over all the residual values of each pixel, describing the agreement with values in the interval $[0, 1]$, the energy function is bounded $E_{Sil} \in [0, 1]$. During optimization, as shown in Fig. 7b and 7c, one can see that the energy decreases, leading to an optimal energy value in Fig. 7d. During our experiments we found some failure cases as can be seen from Fig. 8 being cause due to motion artifacts and sensor noise from the LiDAR.

Computation time required for the optimization schema as measured on Intel-Core-i7 comes around at an average of 680ms per sample without any parallelization. 70% of the compute time is spent for the optimization of the shape weights in E_{CD}. Since CD is a pointwise comparison between 2 different point clouds, this computation time could significantly be reduced by pushing the distance estimation to a GPU.

6   Conclusion

We propose a new schema which functions as a parallel checker module to verify the predictions from a 3D Object Detector. The different energy function we propose utilizes cross sensor data flows with simple priors to validate an object. We also demonstrated the viability of a decision rule base threshold filter through the synthetic dataset. From our experiments, we extensively showcase the viability of the two-step optimization schema and modified renderer towards effectively utilizing the priors in reducing the amount of *false positives* relative to the proposals. In our future work, we are planning to use motion priors along with map information to further argue about the validity of an object in space.

Acknowledgements. The research leading to these results is funded by the German Federal Ministry for Economic Affairs and Climate Action within the project "KI Wissen". The authors would like to thank the consortium for the successful cooperation.

References

1. Aeberhard, M., Paul, S., Kaempchen, N., Bertram, T.: Object existence probability fusion using dempster-shafer theory in a high-level sensor data fusion architecture, pp. 770–775. IEEE (2011). https://doi.org/10.1109/IVS.2011.5940430
2. Blanco, J.L.: A tutorial on SE(3) transformation parameterizations and on-manifold optimization. Technical report, 012010, University of Malaga (2010). http://ingmec.ual.es/jlblanco/papers/jlblanco2010geometry3D_techrep.pdf
3. Caesar, H., et al.: nuScenes: a multimodal dataset for autonomous driving. In: CVPR (2020)
4. Chang, A.X., et al.: Shapenet: an information-rich 3D model repository. CoRR abs/1512.03012 (2015). http://arxiv.org/abs/1512.03012
5. Chen, H., Huang, Y., Tian, W., Gao, Z., Xiong, L.: Monorun: monocular 3D object detection by reconstruction and uncertainty propagation. CoRR abs/2103.12605 (2021). https://arxiv.org/abs/2103.12605
6. Cofer, D., Amundson, I., Sattigeri, R., Passi, A.: Run-Time Assurance for Learning-Enabled Systems Run-Time Assurance for Learning-Enabled Systems (2020). https://doi.org/10.1007/978-3-030-55754-6
7. Engelmann, F., Stückler, J., Leibe, B.: Joint object pose estimation and shape reconstruction in urban street scenes using 3D shape priors (2016). https://doi.org/10.1007/978-3-319-45886-1_18. https://github.com/VisualComputingInstitute/ShapePriors_GCPR16
8. Geiger, A., Lenz, P., Urtasun, R.: Are we ready for autonomous driving? The kitti vision benchmark suite. In: Conference on Computer Vision and Pattern Recognition (CVPR) (2012)
9. Geissler, F., Unnervik, A., Paulitsch, M.: A plausibility-based fault detection method for high-level fusion perception systems. https://doi.org/10.1109/OJITS.2020.3027146
10. Gustafsson, F.K., Danelljan, M., Bhat, G., Schön, T.B.: Energy-based models for deep probabilistic regression. In: Vedaldi, A., Bischof, H., Brox, T., Frahm, J.-M. (eds.) ECCV 2020. LNCS, vol. 12365, pp. 325–343. Springer, Cham (2020). https://doi.org/10.1007/978-3-030-58565-5_20

11. Gustafsson, F.K., Danelljan, M., Schön, T.B.: Accurate 3D object detection using energy-based models. https://github.com/fregu856/ebms_3dod
12. He, K., Gkioxari, G., Dollár, P., Girshick, R.B.: Mask R-CNN. In: IEEE International Conference on Computer Vision, ICCV 2017, Venice, Italy, 22–29 October 2017, pp. 2980–2988. IEEE Computer Society (2017). https://doi.org/10.1109/ICCV.2017.322
13. Hinton, G.E.: Products of experts, vol. 1, pp. 1–6 (1999)
14. Khesbak, M.S.: Depth camera and laser sensors plausibility evaluation for small size obstacle detection. In: 18th International Multi-Conference on Systems, Signals & Devices, SSD 2021, Monastir, Tunisia, 22–25 March 2021, pp. 625–631. IEEE (2021). https://doi.org/10.1109/SSD52085.2021.9429373
15. LeCun, Y., Chopra, S., Hadsell, R., Ranzato, M., Huang, F.: A tutorial on energy-based learning. In: Bakir, G., Hofman, T., Schölkopf, B., Smola, A., Taskar, B. (eds.) Predicting Structured Data. MIT Press, Cambridge (2006)
16. Maag, K.: False negative reduction in video instance segmentation using uncertainty estimates. CoRR abs/2106.14474 (2021). https://arxiv.org/abs/2106.14474
17. Nocedal, J., Wright, S.: Numerical Optimization (2006). https://doi.org/10.1007/978-0-387-40065-5
18. Osadchy, M., Cun, Y.L., Miller, M.L.: Synergistic face detection and pose estimation with energy-based models (2006). https://doi.org/10.1007/11957959_10
19. Prisacariu, V.A., Segal, A.V., Reid, I.: Simultaneous monocular 2D segmentation, 3D pose recovery and 3D reconstruction. In: Lee, K.M., Matsushita, Y., Rehg, J.M., Hu, Z. (eds.) ACCV 2012. LNCS, vol. 7724, pp. 593–606. Springer, Heidelberg (2013). https://doi.org/10.1007/978-3-642-37331-2_45
20. Rao, Q., Krüger, L., Dietmayer, K.: 3D shape reconstruction in traffic scenarios using monocular camera and lidar. In: Chen, C.-S., Lu, J., Ma, K.-K. (eds.) ACCV 2016. LNCS, vol. 10117, pp. 3–18. Springer, Cham (2017). https://doi.org/10.1007/978-3-319-54427-4_1
21. Rottmann, M., Maag, K., Chan, R., Hüger, F., Schlicht, P., Gottschalk, H.: Detection of false positive and false negative samples in semantic segmentation. CoRR abs/1912.03673 (2019). http://arxiv.org/abs/1912.03673
22. Shafer, G.: A Mathematical Theory of Evidence. Princeton University Press, Princeton (1976)
23. Von Rueden, L., et al.: Informed machine learning - a taxonomy and survey of integrating knowledge into learning systems. arXiv, pp. 1–20 (2019)
24. Wang, R., Yang, N., Stuckler, J., Cremers, D.: Directshape: direct photometric alignment of shape priors for visual vehicle pose and shape estimation (2020). https://doi.org/10.1109/icra40945.2020.9197095
25. Wörmann, J., et al.: Knowledge augmented machine learning with applications in autonomous driving: a survey (2022). https://doi.org/10.48550/ARXIV.2205.04712. https://arxiv.org/abs/2205.04712

Lane Change Classification and Prediction with Action Recognition Networks

Kai Liang$^{(\boxtimes)}$, Jun Wang, and Abhir Bhalerao

Department of Computer Science, University of Warwick, Coventry, UK
{kai.liang,jun.wang.3,abhir.bhalerao}@warwick.ac.uk

Abstract. Anticipating lane change intentions of surrounding vehicles is crucial for efficient and safe driving decision making in an autonomous driving system. Previous works often adopt physical variables such as driving speed, acceleration and so forth for lane change classification. However, physical variables do not contain semantic information. Although 3D CNNs have been developing rapidly, the number of methods utilising action recognition models and appearance feature for lane change recognition is low, and they all require additional information to pre-process data. In this work, we propose an end-to-end framework including two action recognition methods for lane change recognition, using video data collected by cameras. Our method achieves the best lane change classification results using only the RGB video data of the PREVENTION dataset. Class activation maps demonstrate that action recognition models can efficiently extract lane change motions. A method to better extract motion clues is also proposed in this paper.

Keywords: Autonomous driving · Action recognition · 3D CNN · Lane change recognition

1 Introduction

Driving is one of the most common and necessary activities in daily life. It can however be often seen as a boring and inefficient experience because of traffic congestion and the tedium of long journeys. In addition, finding a parking area in a city centre can be difficult. As a newly innovative technique, autonomous driving has the potential to alleviate these problems by taking place alongside manually driven vehicles. For instance, shared autonomous vehicles are more economical than most modes of transportation, e.g., taxi and ride-hailing services, and require fewer parking spaces [14]. Furthermore, autonomous vehicles can dwindle the number of traffic accidents caused by infelicitous human behaviours or subjective factors, e.g., fatigue and drunk driving [19]. Autonomous vehicles can also operate in a coordinated mode by sharing their trajectories and behavioural intentions to improve energy efficiency and safety [11]. Consequently, autonomous vehicles are slowly becoming a new part of the infrastructure in cities [14].

© The Author(s), under exclusive license to Springer Nature Switzerland AG 2023
L. Karlinsky et al. (Eds.): ECCV 2022 Workshops, LNCS 13801, pp. 617–632, 2023.
https://doi.org/10.1007/978-3-031-25056-9_39

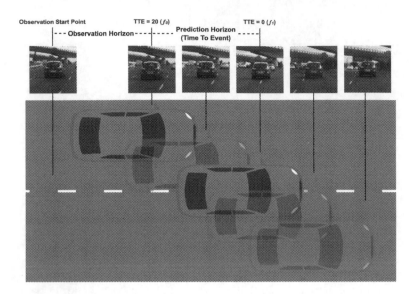

Fig. 1. A lane change event where a vehicle performs a right lane change. f_0 denotes the frame at which lane change starts and f_1 is the frame at which the rear middle part of the target vehicle is just between the lanes. The Observation Horizon is defined as 40 frames (4 s at 10 FPS before f_0). The Prediction horizon or Time To Event (TTE) (on average of length 20 frames at 10 FPS) is defined as the time from f_0 to f_1.

Developing an applicable driving system is faced however with huge difficulties, since it not only requires recognizing object instances in the road, but also requires understanding and responding to a complicated road environment, such as the behaviours of other vehicles and pedestrians, to ensure safety and efficiency. Lane change recognition, aiming to anticipate whether a target vehicle performs left lane change, right lane change or remains in its lane, plays an important role in the autonomous driving system. Autonomous vehicles and human-driven vehicles will need to coexist on the road for some time to come [8], thus lane change prediction is a tricky task since it has to cope with the complicated road environment shared with human drivers. For instance, in highway scenarios, surrounding vehicles may perform cut-in and cut-out manoeuvres unexpectedly, sometimes at high speeds. In a cut-in lane change scenario, a vehicle from one of the adjacent lanes merges into the lane right in front of the ego vehicle. A cut-out lane change scenario typically happens when the vehicle leaves its current lane to avoid slower vehicles ahead. Autonomous vehicles need to respond safely to these actions of human-driven vehicles while planning their trajectories. Hence, predicting and understanding the possible lane changes should be done at the *earliest* moment. Figure 1 illustrates an example of a right lane change where the target vehicle has its manoeuvre at f_0, the start of the Prediction Horizon, but is observed for several seconds before this point. The challenge then is to correctly predict the lane change event as early as possible.

Most of current methods [1,3,4,16,23,29] use physical variables, e.g., driving speed, acceleration, time-gap, heading angle, yaw angle, distances, etc. for lane change recognition. Nevertheless, physical variables cannot represent the type of target objects as they do not contain enough semantic information, whereas knowing the type of road agents can help autonomous vehicles make decisions accordingly. Experienced drivers predict the potential lane changes of surrounding vehicles via pure visual clues, and adjust their speed accordingly, or even change lane if necessary. This ability enables them to anticipate potentially dangerous situations and respond to the environment appropriately. Recently, action recognition models in computer vision [2,5,6,21,26–28] have demonstrated improvements on human activity recognition. 3D action recognition models efficiently extract spatial and temporal clues by only using visual information such as video data. A few works utilising action recognition methods for intelligent vehicle systems have been proposed. Nonetheless, these methods all require additional annotations to pre-process data, e.g., identifying a target vehicle in the video stream [9,11,24].

To address this problem, we propose a novel end-to-end framework involving two approaches for lane change recognition. Specifically, our method takes advantage of the recent powerful action recognition models and mimics a human drivers' behaviours by only utilizing the visual information. The first approach (RGB+3DN) only utilises the similar visual information that humans use to recognise actions, i.e. raw video data collected by cameras. This method is tested by 3D action recognition networks including I3D networks [2], SlowFast networks [6], X3D networks [5] and their variants. The second approach (RGB+BB+3DN) employs the same action recognition models, but uses information of vehicle bounding boxes to improve the detection performance. The vehicle bounding box information is embedded into each frame of the RGB video data to enhance the detection and prediction and passed to a pre-trained action recognition model. We further study the spatial and temporal attention region of activated by the recognition models by generating class activation map [31]. Furthermore, we propose a better way to extract motion clues by optimizing the pooling stride.

Our main contributions can be summarized as follows:

1. We introduce an end-to-end framework for lane change recognition from front-facing cameras involving two action-recognition approaches.
2. We perform an extensive set of experiments and achieve state-of-the-art lane classification results using raw RGB video data.
3. We generate Class Activation Maps (CAMs) to investigate the temporal and spatial attention region of the trained 3D models.
4. We propose to utilize a smaller temporal kernel size and demonstrate it can better extract motion clues.
5. We compare the performance of seven CNN action recognition networks to perform lane change classification and prediction of surrounding vehicles in highway scenarios.

Following an introduction, Sect. 2 gives an in-depth review previous works for lane change recognition and the recent development of action recognition methods, such as SlowFast networks [6]. In Sect. 3, an overview of the problem formulation is presented. Section 4 describes in detail the implementation all the proposed recognition approaches. In Sect. 5, the performance and the evaluation metrics of the different methods is assessed. We conclude with a summary of our finding and make proposals for further work.

2   Related Work

Lane Change Recognition with Physical Variables. To handle lane change recognition problems, many existing works [1,3,4,16,17,29] normally employ physical variables to represent the relative dynamics of a target vehicle with its surrounding vehicles, e.g., driving speed, acceleration, time-gap, heading angle, yaw angle, distances, etc. Lane change events are predicted by analysing physical variables with classical Machine Learning methods [8,13,17,20,22,23]. Environment information is also introduced in some works [1,22,30]. For example, Bahram et al. [1] utilise road-level features such as the curvature, speed limit and distance to the next highway junction. Lane-level features such as type of lane marking, the distance to lane end and number of lanes are also employed in their work.

Action Recognition for Lane Change Classification. Owing to the improvements in computational hardware, CNN based prediction algorithms have been adopted to understand images and videos, and even obtain superior performance than humans in some fields such as image recognition, especially for fine-grained visual categorization. To this end, some research explores the potential of the CNN for lane change detection and classification. Autonomous vehicles can make use of these algorithms to anticipate situations around them [9]. In 2014, Karpathy et al. [12] proposed the first Deep Learning action recognition framework, a single stream architecture that takes video clips as input. Simonyan et al. [24] proposed a two-stream based approach called C2D where a spatial stream takes still frames as input and a temporal stream takes multi-frame dense optical flow as input. Later, 3D convolution for action recognition was introduced by Tran et al. in 2015 [25], and since, 3D convolution methods have dominated the field of action recognition because they can efficiently extract spatial and temporal features simultaneously. Notably, the following sate-of-the-art action recognition models such as R3D [26], I3D [2], S3D [28], R(2+1)D [27], P3D [21], SlowFast [6] and X3D [5] employ 3D convolutions.

I3D [2] is a widely adopted 3D action recognition network built on a 2D backbone architecture (e.g., ResNet, Inception), expanding all the filters and pooling kernels of a 2D image classification architecture, giving them an additional temporal dimension. I3D not only reuses the architecture of 2D models, but also bootstraps the model weights from 2D pre-trained networks. I3D is able to learn spatial information at different scales and aggregate the results. It achieved the best performance (71.1% top1 accuracy) on the Kinetics-400 dataset compared against previous works [2].

Action recognition models can extract spatio-temporal information efficiently, however, require huge computational power. To address this problem,

Feichtenhofer et al. [6] proposed SlowFast, an efficient network with two pathways, i.e., a slow pathway to learn static semantic information and a fast pathway focusing on learning temporal cues. Each pathway is specifically optimised for its task to make the network efficient. SlowFast 16 × 8, R101+NonLocal outperforms the previous state-of-the-art models, yielding 79.8% and 81.8% top-1 accuracy on the Kinetics-400 dataset and the Kinetics-600 dataset respectively [6].

Feichtenhofer [5] introduced X3D networks to further improve the efficiency. X3D, a single pathway architecture also extended from a 2D image network. The most complex X3D model X3D-XL only has 11 million parameters, which is over 5 times less than SlowFast. Therefore, X3D is remarkably efficient. However, X3D-XL achieves better performance (81.9% top1 accuracy) than SlowFast on the Kinetics-600 dataset and slightly lower performance (79.1% top-1 accuracy) on the Kinetics-400 dataset with much lower computational cost [5].

Human drivers predict lane change intentions mainly use visual clues rather than physical variables. However, existing works that utilize appearance features for lane change are surprisingly few. In [18], two appearance features, the state of brake indicators and the state of turn indicators are used for lane change recognition. In [7], the authors applied two action recognition methods, the spatio-temporal multiplier network and the disjoint two-stream convolutional network, to predict and classify lane change events on the PREVENTION dataset. Their method achieved 90.3% classification accuracy. Izquierdo et al. [11] considered lane change classification as an image recognition problem and utilised 2D CNNs. Their method achieved 86.9% classification and 84.4% prediction accuracy. Laimona et al. [9] employed GoogleNet+LSTM to classify lane changes. This method yield 74.5% classification accuracy. Although the methods stated above achieved decent performance, they all require additional information of the target change vehicle such as its contour and bounding box coordinates to pre-process input data. In contrast, our proposed method, RGB+3DN, does not require any extra information.

3   Methods

In this work, we propose an end-to-end framework involving two approaches for lane change recognition classification and prediction of surrounding vehicles in highway scenarios. Seven state-of-the-art 3D action recognition models are investigated including one I3D model, two SlowFast models and four X3D models for both our first and second approaches to address the lane change recognition problem. Different from all the existing methods, the first approach does not rely on extra information. Moreover, we further investigate the temporal and spatial attention region of 3D models by using a class activation map technique [31], see Sect. 4. In this section, we firstly give the problem formulation of lane change recognition. The detail of our proposed framework are described in Sects. 3.2 and 3.3. Figure 2 illustrates the overall architecture of our method.

3.1   Problem Formulation

We consider lane change classification and prediction as an action recognition problem. Given an input video clip $V \in \mathbb{R}^{F \times C \times H \times W}$, the goal of the task is to predict the possible lane change $y \in \{0, 1, 2\}$ of this video where $\{0, 1, 2\}$ denotes lane keeping, left lane change and right lane change labels. Note that F, C, H, W are the number of frames, channels, height and width of an input video respectively and there is only one type of lane change in a video clip. The input video will then be sampled to a specific shape for different models and fed to the first convolution layer. As illustrates in Fig. 2, for I3D and SlowFast models, the features will then be sent to a max pooling layer and ResNet blocks. Whereas, for X3D models, the features will be directly sent to ResNet blocks. The features are convolved to $\mathbb{R}^{C \times T \times S^2}$, where C, T and S^2 denote channel, temporal and spatial dimension of the features. After residual convolution, the features are fed to one global average pooling layer and two convolution layers before fully connected layer. Softmax is finally applied for prediction, $\hat{y}$. The loss function is simply the categorical cross-entropy:

$$L = -\frac{1}{N} \sum_{i=1}^{N} y_i \log(\hat{y}_i), \qquad (1)$$

where N is the number of classes.

For SlowFast networks, lateral connection is performed to fuse the temporal feature of the fast pathway and spatial feature of the slow pathway by concatenation after convolution and pooling layers, as shown in Fig. 2. Based on the methodology described above, we propose the following two approaches for utilising the same seven action recognition models but with different input data:

RGB+3DN: The first method utilises *only* the visual information collected by the front-facing cameras, which is the same kind of information and approach that human drivers would use to predict manoeuvres. We test this approach with seven 3D action recognition networks involving I3D networks, SlowFast networks, X3D networks and their variants.

RGB+BB+3DN: The second method is designed over the first one. It uses the same 3D action recognition networks as the first method. Bounding box information is embedded to each frame of the RGB video data to improve classification and prediction accuracy. This method assumes that a separate vehicle prediction method has been used on the RGB input frames, prior to our lane change prediction.

In order to assess the classification and prediction ability of the 3D CNNs, and following the practice given in [7], the concept of Observation Horizon (N) and Time To Event (TTE) is used to parameterize the temporal information fed to the models. As Fig. 1 illustrates, the Observation Horizon is the time window before a target vehicle starts to perform a lane change event. Based on the investigation of Izquierdo et al. [9], the average length of an lane change event is typically 40 frames (4 s). TTE is defined as the time period from the

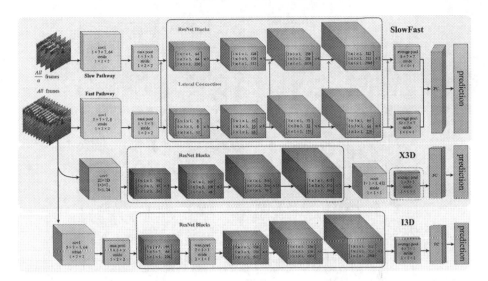

Fig. 2. Architecture of models employed. This figure takes SlowFast−R50, X3D−S and I3D−R50 as example. I3D and X3D take all the frames of a video clip as input. The number of input frames of the fast pathway is α (α = 8) times higher than the slow pathway. The fast pathway has a ratio of β (β = 1/8) channels (underlined) of the slow pathway. The red rectangle shows the temporal information extraction experiments conducted on the global average pooling layer of X3D−S. (Color figure online)

specific time point to the time point of f_1. In other words, when TTE equals 20 (half average length of an lane change event, 2 s), the specific time point is f_0, which is when the target vehicle just starts performing lane change. The following parameters of input data with respect to Observation Horizon, and Time To Event are investigated in this work:

N40-TTE00: Observation horizon: 40 frames (4 s), Time To Event: 0 frame (0 s), total data length of one sample is 60 frames (6 s). This dataset is designed to investigate the classification ability of the action recognition models, as the labeled lane change events have already occurred.

N40-TTE10: Observation horizon: 40 frames (4 s), Time To Event: 10 frames (1 s), total data length of one sample is 50 frames (5 s). This dataset is designed to investigate the prediction ability (1 s ahead) of the action recognition models.

N40-TTE20: Observation horizon: 40 frames (4 s), Time To Event: 20 frames (2 s), total data length of one sample is 40 frames (4 s). This dataset is designed to investigate the prediction ability (2 s ahead) of the action recognition models.

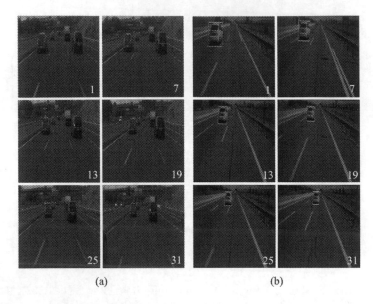

Fig. 3. Input data visualisation (a) RGB video frame data of event; (b) video combined with bounding box data. Only the 1th, 7th, 13th, 19th, 25th and 31st frames are shown. A vehicle in the right lane in (a) and the left lane in (b) perform left and right lane change manoeuvres respectively. The frame data is resized to have aspect ratio of 1 ready for input to a classification CNN.

3.2 RGB+3DN: 3D Networks and RGB Video Data

This section describes our first method, RGB+3DN. Different from all the existing methods, it only requires the original RGB video as input to perform prediction.

We extract the samples of each class using the annotation provided by the PREVENTION dataset. The samples are initially centre-cropped from 1920×600 to 1600×600 pixels, then resized to 400×400 pixels in spatial resolution. In order to reduce the computational cost, the data is further downsampled from 10 FPS to 32/6 FPS in frame rate. Data augmentations are applied to expand the datasets. Figure 3 (a) illustrates a sample of RGB video data.

3.3 RGB+BB+3DN: 3D Networks and Video Combined with Bounding Box Data

In this second approach, we employ the same seven CNNs as the first approach, however, the input video data used for this method is different and incorporates vehicle bounding box information. The method of processing video combined bounding box data is inspired by [9], where they employed 2D image classification models and still images combined with contours of vehicle motion histories.

To generate video combined bounding box data, we firstly use the same method of generating RGB video data to extract raw video clips of each class.

Then the vehicle bounding boxes of vehicles in each frame are rendered into colour channels of each frame: the red channel is used to store the scene appearance as a gray scale image; the green channel to draw a bounding box of the target vehicle; and the blue channel to draw the bounding boxes of surrounding vehicles. These frames of each sample are then converted into video clips. Figure 3 (b) illustrates a sample of video combined bounding box data. Note that the assumption here is that an estimated bounding boxes of vehicles in the scene and the location of the target vehicle (the vehicle undertaking the lane-change) is available with the RGB input frames, perhaps by running a separate vehicle detection and tracking algorithm.

4   Experiments

We evaluate and compare seven action recognition models on the PREVENTION dataset [10] for lane change classification and prediction. This section describes the experiments in detail.

4.1   Dataset

The PREVENTION dataset consists of video footage, object bounding boxes, lane change events, trajectories, lane marking annotations, LiDAR and radar data. The dataset has 5 records. In each record, there is an annotation file named *detections_filtered.txt*, which contains the bounding box and contour information of all objects detected in the visual scene. The annotation information is stored in the form $[frame, ID, class, xi, yi, xf, yf, conf, n]$ where *class* is the variable used to denote the types of the objects tracked, and n is a sequence of x and y coordinates denote the contour of the tracked vehicle. The *lane_changes.txt* file contains the information with regard to the detected vehicle lane change events. The information is denoted in the format $[ID, ID - m, LC - type, f0, f1, f2, blinker]$ where $ID - m$ is the unique ID of each vehicle performing the lane change, $LC - type$ denotes the type of each detected lane change event: 3 stands for left lane change and 4 stand for right lane change, $f0$ refers to the frame number at which lane change starts, $f1$ denotes the frame number at which the rear middle part of the vehicle is just between the lanes, $f2$ is the number of the end frame.

The data used for the first approach, RGB+3DN, and the second approach, RGB+BB+3DN, are RGB video data and video combined bounding box data respectively. Because of the limited annotation available in the PREVENTION data, training and validation data used for the second approach is more limited than used for the first approach. Data augmentation methods such as random cropping, random rotation, colour jittering, horizontal flipping are performed to the all data. After data augmentation, the total number of each class of RGB video data and video combined bounding box data is 2420 and 432 respectively. For RGB video data, 1940 samples of each class are used for training and the rest 480 samples are used for validation. For video combined bounding box data, sample numbers of each class in the training set and the test set are 332 and 100.

4.2   Evaluation Metrics

We perform 4-fold cross validation to all methods. As a three-class classification problem, the accuracy was considered as the main metric to assess the performance of the networks employed in each scheme. The same as the evaluation metrics used in [24], the model accuracy was calculated by

$$Acc = \frac{TP}{Total\ No.\ of\ Samples} \tag{2}$$

i.e. the number of true positive samples (TP) for the three classes divided by the total number of samples. The final top-1 accuracy is calculated by averaging the results of each fold.

Table 1. Top-1 classification and prediction accuracy of RGB+3DN method on the PREVENTION dataset.

Model	GFLOPs	TTE-00	TTE-01	TTE-20
X3D-XS	**0.91**	82.78	75.69	64.10
X3D-S	2.96	**84.79**	75.00	63.82
X3D-M	6.72	82.36	69.72	63.19
X3D-L	26.64	77.57	69.30	59.51
I3D	37.53	83.05	73.75	63.54
SF, R50	65.71	81.04	**76.67**	**65.00**
SF, R101	127.20	71.80	74.79	62.08

Table 2. Top-1 classification and prediction accuracy of RGB+BBS+3DN method on the PREVENTION dataset.

Model	TTE-00	TTE-01	TTE-20
X3D-XS	98.33	**99.17**	**98.86**
X3D-S	**99.33**	98.33	97.82
X3D-M	98.17	98.67	97.19
X3D-L	98.43	98.34	97.51
I3D	98.67	98.84	96.54
SF, R50	98.50	98.50	98.67
SF, R101	97.33	97.67	96.33

4.3   Lane Change Classification and Prediction with RGB+3DN

Lane Change Classification. Table 1 presents the top-1 accuracy of lane change classification and prediction on RGB video data. Among all the models, SlowFast-R101 (with 127.2 GFLOPs) is the most complex model. It outperforms all the other models on the Kinetics-400 dataset [6]. However, SlowFast-R101 networks only yield 71.80% top-1 accuracy on RGB video data, which is the lowest. Whereas, the relatively lightweight SlowFast-R50 networks, with nearly half GFLOPs, achieve 81.04% top-1 accuracy. Surprisingly, although all the four X3D models are much more lightweight than the SlowFast models, they outperform all the SlowFast models. The second most lightweight model X3D-S achieves the best performance (84.79% top-1 accuracy). The I3D model is the second best model. It yields 83.05% top-1 accuracy, which is only 1.74% lower than the best model.

An interesting finding can be observed from the results described above, i.e., in general, for lane change lane change classification, the lighter the model, the better the accuracy is. This could be explained because the capacity of the models employed is large, but the dataset does not have sufficient variation.

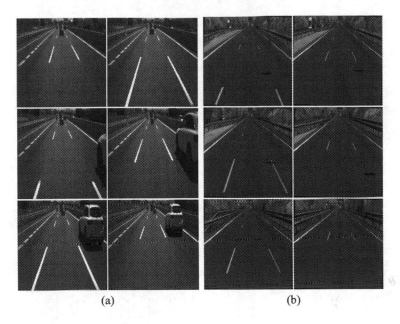

(a)   (b)

Fig. 4. TP and FP examples: (a) TP example, the vehicle in the middle performing right lane change is correctly classified. (b) FP example, the vehicle in the right lane performing left lane change is classified as lane keeping incorrectly due to small target.

Lane Change Prediction. Models can anticipate 1 s and 2 s ahead on TTE-10 and TTE-20 data respectively. The prediction accuracy of all the models decrease more significantly than their classification accuracy. This can be explained as TTE-10 and TTE-20 data provide less information than TTE-00 data. As can be seen in Table 1, the best model on both TTE-10 and TE-20 data is the SlowFast networks, which yield top-1 accuracy of 76.67% and 65% for 1 and 2 s anticipation respectively. On the prediction experiments, a similar finding as on the classification experiments is observed, i.e. higher model complexity leads to lower prediction accuracy. For instance, SlowFast-R50 is more lightweight than SlowFast-R101. However, SlowFast-R50 outperforms SlowFast-R101 by nearly 2%. This finding also applies to the all X3D models.

By analysing classification and prediction confusion metrics of the RGB+3DN method, we observe that, for lane change classification, the accuracy of right lane change class is higher than left lane change. This can be explained as many of the Left Lane Change events take place in areas far from the ego vehicle and the target vehicles are relatively small, as can be seen in Fig. 4 (b). Whereas, right lane changes normally take place in front of the ego vehicle, which is close to the cameras, as Fig. 4 (a) illustrates: clearer footage results in better image representations and better recognition accuracy. Lane keeping is always the best predicted class for TTE-00, TTE-10 and TTE-20 data, as its scene is relatively simple and easy to predict. For lane change prediction, the accuracy of lane keeping class is significantly higher than the other two classes. This may

Table 3. Comparison of different methods on PREVENTION dataset.

Method	Extra information	TTE-00	TTE-01	TTE-20
GoogleNet + LSTM [9]	ROI + Contour	74.54	–	–
2D Based [11]	Contour	86.90	84.50	–
Two stream based [7]	ROI	90.30	85.69	91.94
VIT [15]	Non	81.23	–	–
Ours (RGB + 3DN)	**Non**	84.79	76.67	65.00
Our (RGB + BB+ 3DN)	BBS	**99.33**	**99.17**	**98.86**

be due to there being fewer frames given to these events, some lane and right lane changes also resemble lane keeping events. Therefore, left lane and right lane changes can be sometimes miss-predicted as lane keeping.

4.4   Lane Change Classification and Prediction with RGB+BB+3DN

Table 2 illustrates the classification and prediction results on video combined with bounding box data. As can be observed, regardless of the classification or prediction results, the performance of each method does not vary much. The best accuracy is only 3.00% higher than the lowest one, which is very different from the experiments of the RGB+3DN method on the RGB video data. The X3D-L and SlowFast-R101 are always the two lowest performing models. Whereas, the best performing models are always from the X3D family.

The temporal anticipation accuracy is not affected much in any of the models, although there is still a 1% to 1.5% drop in performance between TTE-00 and TTE-20 in all cases. The prediction accuracy however is considerably enhanced by knowing the location of vehicles by their bounding boxes giving much higher performance than with RGB information alone.

4.5   Comparison to Previous Methods

As Table 3 illustrates, Simonyan et al.'s two stream based method obtains better accuracy than our RGB+3DN method, however, their method requires bounding box coordinates of the target vehicle for Region of Interest (ROI) cropping. Furthermore, their validation data of each class is highly unbalanced. Although our RGB+3DN only uses the original video data collected by cameras and does not require additional information, we still outperform some methods, e.g., GoogleNet + LSTM and VIT.

The method for processing the data of our RGB+BB+3DN method is inspired by Izquierdo et al.'s work [9]. In [9], their best performing model, GoogleNet + LSTM yields 74.4% for lane change classification. Because 3D models can better extract spatio-temporal features than 2D CNNs, our RGB+BB+3DN method achieves top-1 classification accuracy of 99.33%, which is significantly higher. With extra bounding box information, our RGB+BB+3DN method out-performs all the methods compared.

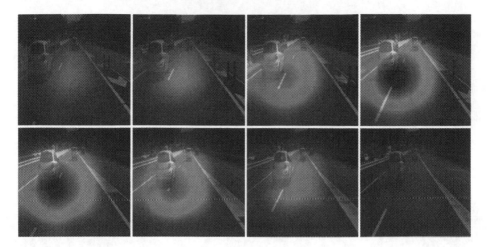

Fig. 5. Class activation maps. Only the 25th to 32nd frames of the input video are shown. The model mainly focuses on the frames where lane change happens, as well as the edge of the target vehicle and lane marking which it is about to cross.

4.6   Class Activation Maps

To investigate whether our models recognize the motion clues and learn spatio-temporal information, we generated class activation maps on X3D-S and RGB video data. The features of the last convolution layer were used to calculate the CAMs. The calculated scores were normalised across all patches and frames before visualisation. Experimental results show that action recognition models can efficiently extract both spatial and temporal information. As Fig. 5 illustrates, in the spatial domain, the model focuses on the edge of the target vehicle and lane marking which the target vehicle is about to cross. Whereas, in the temporal domain, the X3D-S only focuses on those frames that the lane change event happens, specifically, frames 27 to 30.

4.7   Optimizing Temporal Information Extraction

The 3D CNNs we employ in this work are initially designed for recognising general human behaviours and trained on human behaviours datasets such as Kinetics-400 and Kinetics-600. These datasets are formed by video clips with relatively high frame rates (25 fps) [2]. Therefore, in order to efficiently extract motion clues, the temporal kernel size of the global layer (as shown in the red rectangle of Fig. 2) of X3D is originally designed as 16. Whereas, the frame rate of our data is much less, 32/6. Moreover, while we inspected class activation maps, an interesting finding was revealed, i.e., the model only pays attention to the 3rd to 5th frames where the target vehicle approaches the lane marking and overlay it, as shown in Fig. 5. Therefore, we postulated that a larger temporal kernel size could potentially introduce noisy information, and decreasing the temporal dimension size of the kernels might improve the model performance. This seems

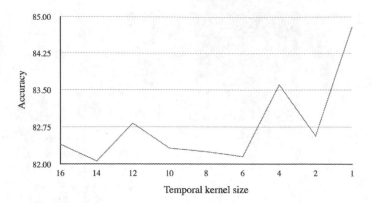

Fig. 6. Experiments on temporal kernel size. Smaller kernel size results in better accuracy.

to be the case. As Fig. 6 shows, as we decrease the temporal kernel size from 16 to 1, the model classification accuracy increases by 2.39% from 82.40% to 84.79%.

5   Conclusions

Two approaches involving seven 3D action recognition networks are adopted to classify and anticipate lane change events on the PREVENTION dataset. For our RGB+3DN method, the lane change recognition problem is formulated as an action recognition task only utilising visual information collected by cameras. The best performing model, X3D-S achieves state-of-the-art top-1 accuracy (84.79%) for lane change classification using the original RGB video data of the PREVENTION dataset. Our RGB+BB+3DN method achieves significant accuracy improvement (TTE-20 98.86%) by taking the advantage of 3D CNNs and additional bounding box information. Furthermore, we generated class activation maps to investigate the spatial and temporal attention region of 3D CNNs. These CAMs demonstrated that action recognition models are able to extract lane change motions efficiently. We proposed a way to better extract relevant motion clues by decreasing the dimension of the temporal kernel size.

As further work, more data needs to be added for training and evaluation to expand the diversity of the dataset and prevent overfitting. Introducing vehicle and lane detectors to the 3D networks to further exploit appearance clues of the data and improve model performance is one potential strategy. Although still challenging, in this way, the model might achieve the performance of RGB+BB+3DN method of this work without separate bounding box/target vehicle information being required.

References

1. Bahram, M., Hubmann, C., Lawitzky, A., Aeberhard, M., Wollherr, D.: A combined model-and learning-based framework for interaction-aware maneuver prediction. IEEE Trans. Intell. Transp. Syst. **17**(6), 1538–1550 (2016)
2. Carreira, J., Zisserman, A.: Quo vadis, action recognition? A new model and the kinetics dataset. In: Proceedings of the IEEE Conference on Computer Vision and Pattern Recognition, pp. 6299–6308 (2017)
3. Deo, N., Trivedi, M.M.: Convolutional social pooling for vehicle trajectory prediction. In: Proceedings of the IEEE Conference on Computer Vision and Pattern Recognition Workshops, pp. 1468–1476 (2018)
4. Deo, N., Trivedi, M.M.: Multi-modal trajectory prediction of surrounding vehicles with maneuver based LSTMS. In: 2018 IEEE Intelligent Vehicles Symposium (IV), pp. 1179–1184. IEEE (2018)
5. Feichtenhofer, C.: X3D: expanding architectures for efficient video recognition. In: Proceedings of the IEEE/CVF Conference on Computer Vision and Pattern Recognition, pp. 203–213 (2020)
6. Feichtenhofer, C., Fan, H., Malik, J., He, K.: Slowfast networks for video recognition. In: Proceedings of the IEEE/CVF International Conference on Computer Vision, pp. 6202–6211 (2019)
7. Fernández-Llorca, D., Biparva, M., Izquierdo-Gonzalo, R., Tsotsos, J.K.: Two-stream networks for lane-change prediction of surrounding vehicles. In: 2020 IEEE 23rd International Conference on Intelligent Transportation Systems (ITSC), pp. 1–6. IEEE (2020)
8. Izquierdo, R., Parra, I., Muñoz-Bulnes, J., Fernández-Llorca, D., Sotelo, M.: Vehicle trajectory and lane change prediction using ANN and SVM classifiers. In: 2017 IEEE 20th International Conference on Intelligent Transportation Systems (ITSC), pp. 1–6. IEEE (2017)
9. Izquierdo, R., Quintanar, A., Parra, I., Fernández-Llorca, D., Sotelo, M.: Experimental validation of lane-change intention prediction methodologies based on CNN and LSTM. In: 2019 IEEE Intelligent Transportation Systems Conference (ITSC), pp. 3657–3662. IEEE (2019)
10. Izquierdo, R., Quintanar, A., Parra, I., Fernández-Llorca, D., Sotelo, M.: The prevention dataset: a novel benchmark for prediction of vehicles intentions. In: 2019 IEEE Intelligent Transportation Systems Conference (ITSC), pp. 3114–3121. IEEE (2019)
11. Izquierdo, R., et al.: Vehicle lane change prediction on highways using efficient environment representation and deep learning. IEEE Access **9**, 119454–119465 (2021)
12. Karpathy, A., Toderici, G., Shetty, S., Leung, T., Sukthankar, R., Fei-Fei, L.: Large-scale video classification with convolutional neural networks. In: Proceedings of the IEEE Conference on Computer Vision and Pattern Recognition, pp. 1725–1732 (2014)
13. Kasper, D., et al.: Object-oriented Bayesian networks for detection of lane change maneuvers. IEEE Intell. Transp. Syst. Mag. **4**(3), 19–31 (2012)
14. Kato, S., Takeuchi, E., Ishiguro, Y., Ninomiya, Y., Takeda, K., Hamada, T.: An open approach to autonomous vehicles. IEEE Micro **35**(6), 60–68 (2015). https://doi.org/10.1109/MM.2015.133
15. Konakalla, N., Noor, A., Singh, J.: CNN, CNN encoder-RNN decoder, and pretrained vision transformers for surrounding vehicle lane change classification at future time steps (2022). https://cs231n.stanford.edu/reports/2022/pdfs/105.pdf

16. Lee, D., Kwon, Y.P., McMains, S., Hedrick, J.K.: Convolution neural network-based lane change intention prediction of surrounding vehicles for ACC. In: 2017 IEEE 20th International Conference on Intelligent Transportation Systems (ITSC), pp. 1–6. IEEE (2017)

17. Li, J., Lu, C., Xu, Y., Zhang, Z., Gong, J., Di, H.: Manifold learning for lane-changing behavior recognition in urban traffic. In: 2019 IEEE Intelligent Transportation Systems Conference (ITSC), pp. 3663–3668. IEEE (2019)

18. Li, J., Dai, B., Li, X., Xu, X., Liu, D.: A dynamic Bayesian network for vehicle maneuver prediction in highway driving scenarios: framework and verification. Electronics **8**(1), 40 (2019)

19. Litman, T.: Autonomous vehicle implementation predictions. Victoria Transport Policy Institute Victoria, BC, Canada (2017)

20. Liu, P., Kurt, A., Özgüner, Ü.: Trajectory prediction of a lane changing vehicle based on driver behavior estimation and classification. In: 17th International IEEE Conference on Intelligent Transportation Systems (ITSC), pp. 942–947. IEEE (2014)

21. Qiu, Z., Yao, T., Mei, T.: Learning spatio-temporal representation with pseudo-3D residual networks. In: Proceedings of the IEEE International Conference on Computer Vision, pp. 5533–5541 (2017)

22. Schlechtriemen, J., Wedel, A., Hillenbrand, J., Breuel, G., Kuhnert, K.D.: A lane change detection approach using feature ranking with maximized predictive power. In: 2014 IEEE Intelligent Vehicles Symposium Proceedings, pp. 108–114. IEEE (2014)

23. Schlechtriemen, J., Wirthmueller, F., Wedel, A., Breuel, G., Kuhnert, K.D.: When will it change the lane? A probabilistic regression approach for rarely occurring events. In: 2015 IEEE Intelligent Vehicles Symposium (IV), pp. 1373–1379. IEEE (2015)

24. Simonyan, K., Zisserman, A.: Two-stream convolutional networks for action recognition in videos. In: Advances in Neural Information Processing Systems, vol. 27 (2014)

25. Tran, D., Bourdev, L., Fergus, R., Torresani, L., Paluri, M.: Learning spatiotemporal features with 3D convolutional networks. In: Proceedings of the IEEE International Conference on Computer Vision, pp. 4489–4497 (2015)

26. Tran, D., Ray, J., Shou, Z., Chang, S.F., Paluri, M.: Convnet architecture search for spatiotemporal feature learning. arXiv preprint arXiv:1708.05038 (2017)

27. Tran, D., Wang, H., Torresani, L., Ray, J., LeCun, Y., Paluri, M.: A closer look at spatiotemporal convolutions for action recognition. In: Proceedings of the IEEE Conference on Computer Vision and Pattern Recognition, pp. 6450–6459 (2018)

28. Xie, S., Sun, C., Huang, J., Tu, Z., Murphy, K.: Rethinking spatiotemporal feature learning for video understanding. arXiv preprint arXiv:1712.04851 (2017)

29. Yao, W., et al.: On-road vehicle trajectory collection and scene-based lane change analysis: Part II. IEEE Trans. Intell. Transp. Syst. **18**(1), 206–220 (2016)

30. Yoon, S., Kum, D.: The multilayer perceptron approach to lateral motion prediction of surrounding vehicles for autonomous vehicles. In: 2016 IEEE Intelligent Vehicles Symposium (IV), pp. 1307–1312. IEEE (2016)

31. Zhou, B., Khosla, A., Lapedriza, A., Oliva, A., Torralba, A.: Learning deep features for discriminative localization. In: Proceedings of the IEEE Conference on Computer Vision and Pattern Recognition, pp. 2921–2929 (2016)

Joint Prediction of Amodal and Visible Semantic Segmentation for Automated Driving

Jasmin Breitenstein[(✉)], Jonas Löhdefink, and Tim Fingscheidt

Institute for Communications Technology, Technische Universität Braunschweig,
Schleinitzstrasse 22, 38106 Braunschweig, Germany
{j.breitenstein,j.loehdefink,t.fingscheidt}@tu-bs.de

Abstract. Amodal perception is the ability to hallucinate full shapes of (partially) occluded objects. While natural to humans, learning-based perception methods often only focus on the visible parts of scenes. This constraint is critical for safe automated driving since detection capabilities of perception methods are limited when faced with (partial) occlusions. Moreover, corner cases can emerge from occlusions while the perception method is oblivious. In this work, we investigate the possibilities of joint prediction of amodal and visible semantic segmentation masks. More precisely, we investigate whether both perception tasks benefit from a joint training approach. We report our findings on both the Cityscapes and the Amodal Cityscapes dataset. The proposed joint training outperforms the separately trained networks in terms of mean intersection over union in amodal areas of the masks by 6.84% absolute, while even slightly improving the visible segmentation performance.

1 Introduction

Reliable environment perception is an important part of automated driving to ensure safety of all traffic participants. Visual perception methods rely for this on camera image or video data to extract the desired environment information. For dependable environment perception furthermore the detection of corner cases is necessary. Corner cases are unpredictable and potentially dangerous situations in automated driving and can range from pixel errors and unknown objects to unusual trajectories only visible in coherent video data [2,5,9]. Corner cases can also be inherent to the methods themselves [9].

In this work, we focus on a special type of corner cases resulting from unavailable amodal perception. Amodal perception describes the ability to imagine the full shape of objects that are completely or partially occluded. Standard visual perception methods have strong deficits when it comes to recognizing occluded objects [21], which is potentially dangerous, e.g., when occluded pedestrians close to the road are not properly detected. Hence, there is an interest to develop methods being able to perform amodal perception. For different visual perception tasks, the task in amodal perception varies slightly however the goal is always to perceive the occluded regions correctly: In object detection the aim

L. Karlinsky et al. (Eds.): ECCV 2022 Workshops, LNCS 13801, pp. 633–645, 2023.
https://doi.org/10.1007/978-3-031-25056-9_40

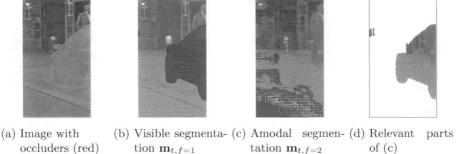

(a) Image with occluders (red)   (b) Visible segmenta- tion $\mathbf{m}_{t,f=1}$ (c) Amodal segmen- tation $\mathbf{m}_{t,f=2}$ (d) Relevant parts of (c)

Fig. 1. Example from the Amodal Cityscapes test dataset (a) with corresponding visible (b) and amodal (c) semantic segmentation of our proposed joint training approach hallucinating the full shape of the partially occluded pedestrian.

is to correctly identify occluded objects, in instance segmentation the aim is to predict the entire (visible and occluded) shape of each instance. We consider the task of amodal semantic segmentation where we aim to predict the class label of the occluded class on pixel level. Figure 1 visualizes our task on an amodal semantic segmentation dataset [4]: The inserted occluders shaded in red are synthetically placed in an image. We obtain both a visible (b) and an amodal (c) semantic segmentation showing what can be directly seen in the image, and what is occluded in the scene, in this case, e.g., one half of a pedestrian.

While there is a multitude of works on semantic segmentation, e.g., focusing on improving the performance [3,23], efficiency [14,19], or robustness [1,12], amodal semantic segmentation has not received the same amount of attention in the research community. Existing prior works have separated the tasks of visible and amodal semantic segmentation from each other [24], or used a simplified version of semantic segmentation by predicting groups and not separating semantic classes [4,15].

However, thinking about human perception, amodal perception is linked to visible perception as the visible areas give cues about the occluded object and its shape [18,22]. Thus an investigation of methods that can jointly predict amodal and visible semantic segmentation labels appears beneficial. For the task of amodal instance segmentation, we have already seen methods following that paradigm and reaching the current state of the art [10,11].

In this work, we investigate the task of joint amodal and visible semantic segmentation. For that, we use the Amodal Cityscapes [4] dataset to train and evaluate our methods. We provide guidelines as to where to fuse amodal and visible semantic segmentation. Additionally, our proposed approach outperforms the baseline methods on Amodal Cityscapes by a large margin, while also relaxing the assumptions on amodal semantic segmentation.

The rest of the paper is structured as follows: We first summarize the related work for amodal segmentation, while paying special attention to methods relying on joint training approaches for visible and amodal segmentations. Section 3 then describes our method for amodal semantic segmentation, and, finally, Sect. 4 reports and discusses the experimental results of our approach.

2 Related Work

2.1 Amodal Segmentation

So far, the task of amodal semantic segmentation has received only few attention in the research community. It was first defined by Zhu et al. [24], who provide two baseline methods for class-agnostic object segmentation next to a non-automotive amodal object segmentation dataset. We build upon their `AmodalMask` method in our approach where we use a deep neural network for semantic segmentation to directly predict the amodal semantic segmentation masks. An adaptation builds upon `AmodalMask` as well by directly predicting the entire object mask [7]. Sun et al. [20] predict object boundaries for partially occluded objects by fitting a Bayesian model on the neural network features. Separate amodal segmentation methods using a standard method for visible segmentation trained on amodal masks are also used to report baseline results on the SAIL-VOS dataset [10]. There are indications, however, that separate training of visible and amodal segmentation is suboptimal, e.g., due to the more limited amount of amodal training data which interferes with learning meaningful representations. Accordingly, we propose a joint training method in this work.

2.2 Joint Amodal and Visible Semantic Segmentation

Joint amodal and semantic segmentation has also rarely been studied so far. We build upon ideas from amodal *instance* segmentation, where a few joint training methods have been proposed before. Hu et al. [10] published a method which extends the standard `Mask R-CNN` [8] by an additional branch for the amodal mask. Qi et al. [16] use a similar approach applying multi-level coding to improve the amodal instance perception. Another method models both occluder and occludee mask, and hence also their relationship, in parallel [11]. A similar approach joins features from an occluder and occludee network branch in an amodal object branch to predict the desired masks [17]. Mohan et al. [13] train a panoptic segmentation that combines methods for joint amodal and visible instance segmentation and background segmentation to predict panoptic labels.

While joint amodal and visible semantic segmentation has been approached by previous works [4,15], those approaches are limited to predicting occluded groups of semantic classes and cannot identify when objects of the same group occlude each other, e.g., when a person occludes another person. This is because both visible and amodal semantic segmentation are learnt by adapting the final softmax layer to predict the visible semantic group and the present class by group, resulting in additional amodal labelling, and restricting the visible class at each pixel to be in a different group than the respective amodal class. In our

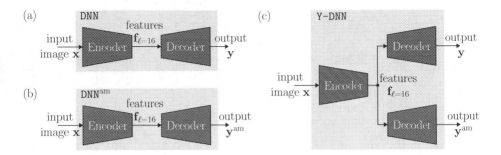

Fig. 2. Method overview of (a) the segmentation DNN, (b) the amodal DNN (DNN^{am}), (c) the joint amodal and visible semantic segmentation Y-DNN.

work we follow the paradigm that both tasks can benefit from each other, but we eliminate the restrictions by and the need for semantic grouping. Instead, we directly predict the visible and amodal semantic class per pixel. Hence, we also drop the assumption that classes of the same group cannot occlude each other.

3 Proposed Method for Joint Amodal and Visible Semantic Segmentation

Figure 2 shows the overall functioning of the proposed method. Given an input image $\mathbf{x}_t \in [0,1]^{H \times W \times C}$ for $C = 3$ color channels, $t \in \mathcal{T} = \{1, \ldots, T\}$, the standard semantic segmentation deep neural network (DNN) (Fig. 2(a)), predicts the (visible) output $\mathbf{y}_t = (\mathbf{y}_{t,s}) \in [0,1]^{H \times W \times S}$, where S is the number of semantic segmentation classes, i.e., in our setting $S = 19$, and $\mathbf{y}_{t,s} \in [0,1]^{H \times W}$. The set of semantic classes is given by $\mathcal{S} = \{1, \ldots, S\}$. For standard semantic segmentation, we obtain a semantic segmentation mask

$$\mathbf{m}_t = \operatorname*{argmax}_{s \in \mathcal{S}} \mathbf{y}_{t,s}, \tag{1}$$

where the argmax decision is done for each pixel i in $\mathbf{y}_{t,s} = (y_{t,s,i})$ with $i \in \mathcal{I} = \{1, \ldots, H \times W\}$. In our experiments, we use the ERFNet (=DNN) consisting of layers $\ell \in \{1, \ldots, L\}$ as basis for our experiments with $L = 23$, and $\ell = 16$ being the last layer of the encoder.

In contrast, the amodal DNN (DNN^{am}, Fig. 2(b)), predicts the amodal output of the softmax layer $\mathbf{y}_t^{\mathrm{am}} \in [0,1]^{H \times W \times S}$. The DNN and DNN^{am} provide different outputs due to their training targets: The DNN is trained using the (visible) ground truth $\overline{\mathbf{y}}_t \in \{0,1\}^{H \times W \times S}$, while the DNN^{am} is trained using ground truth amodal labels $\overline{\mathbf{y}}_t^{\mathrm{am}} \in \{0,1\}^{H \times W \times S}$.

Finally, in Fig. 2(c), the Y-DNN for joint amodal and visible semantic segmentation is shown. While in Figs. 2(a) and (b) we have two separate networks and two separate trainings, the Y-DNN is able to predict both $\mathbf{y}_t \in [0,1]^{H \times W \times S}$ and $\mathbf{y}_t^{\mathrm{am}} \in [0,1]^{H \times W \times S}$ at the same time. This is done by sharing the encoder and then using two separate decoders to predict both visible and amodal labels.

In all three cases above, we obtain the semantic segmentation mask $\mathbf{m}_t = (\mathbf{m}_{t,f}) = (m_{t,i,f}) \in \mathcal{S}^{H \times W \times F}$ with

$$\mathbf{m}_{t,f=1} = \underset{s \in \mathcal{S}}{\operatorname{argmax}}\, \mathbf{y} \qquad (2)$$

$$\mathbf{m}_{t,f=2} = \underset{s \in \mathcal{S}}{\operatorname{argmax}}\, \mathbf{y}^{\mathrm{am}}, \qquad (3)$$

where we consider only a visible layer ($f = 1$) and an occluded layer ($f = 2$), i.e., $F = 2$. Of course, for the DNN (Fig. 2(a)), we only obtain a mask $\mathbf{m}_{t,f=1}$ (2), and for the DNN^{am} (Fig. 2(b)), we obtain only a mask $\mathbf{m}_{t,f=2}$ (3).

To summarize, our method and investigations are novel in three ways: First, by allowing joint amodal and visible semantic segmentation training to share the same encoder, the *invisible* mIoU will be shown to improve vs. the state of the art. Second, to the best of our knowledge, we are the first in proposing a class-specific joint network structure and training for visible and amodal semantic segmentation, hence allowing all classes to occlude all classes. Third, we investigate both an efficient and a high-performing network for amodal semantic segmentation with the aim to show the specific effect the joint training has on the overall performance.

4   Experimental Validation and Discussion

4.1   Datasets and Metrics

For our experimental evaluation, we use two datasets. The first is the well-known Cityscapes dataset [6] ($\mathcal{D}_{\mathrm{CS}}$). It is often used for visual perception tasks on automotive data. While we could use the Cityscapes dataset to train a (visible) semantic segmentation, it does not contain amodal labels. Hence, we cannot train amodal semantic segmentations with it. For this task, we use an adapted version of the Cityscapes dataset, namely the Amodal Cityscapes dataset [4] ($\mathcal{D}_{\mathrm{amCS}}$). This dataset is generated by inserting instances from other Cityscapes source images from the same dataset split into the target image. This way, we obtain amodal ground truth knowledge about the now occluded areas. As this amodal ground truth is based on the insertion of instances, i.e., objects of the classes **person, rider, car, truck, motorcycle, bicycle, bus, train** (mobile objects), we only evaluate the proposed amodal segmentation methods wherever those known insertions are present.

We use the Amodal Cityscapes training data split ($\mathcal{D}_{\mathrm{amCS}}^{\mathrm{train}}$) for training, and the Amodal Cityscapes validation data split ($\mathcal{D}_{\mathrm{amCS}}^{\mathrm{val}}$) to monitor the training and select the best-performing model. Evaluation results are then reported on the available Amodal Cityscapes test dataset ($\mathcal{D}_{\mathrm{amCS}}^{\mathrm{test}}$), and also on the Cityscapes validation dataset ($\mathcal{D}_{\mathrm{CS}}^{\mathrm{val}}$). Evaluation on $\mathcal{D}_{\mathrm{CS}}^{\mathrm{val}}$ shall allow to check to what extent the original task of semantic segmentation suffers by including amodality. We report mIoU and mIoU^{inv} and follow the definitions by Breitenstein et al. [4] for the latter metric:

Table 1. Visible and invisible mIoU (%) performance of the ERFNet and its amodal variations on the original and the Amodal Cityscapes dataset. Fields marked with $*$ cannot be calculated due to missing amodal ground truths. Best results are shown in bold.

Method	$\mathcal{D}_{amCS}^{test}$		$\mathcal{D}_{CS}^{val}$	
	mIoU	mIoU^{inv}	mIoU	mIoU^{inv}
ERFNet	62.99	5.00	67.21	$*$
ERFNet^{am}	20.16	36.48	21.00	$*$
Y-ERFNet	**63.32**	**43.32**	**68.35**	$*$

$$\text{mIoU}^{inv} = \frac{1}{S} \sum_{s \in \mathcal{S}} \frac{\text{TP}_s^{inv}}{\text{TP}_s^{inv} + \text{FP}_s^{inv} + \text{FN}_s^{inv}}, \tag{4}$$

where true positives for class s (TP_s^{inv}) are given if $\overline{m}_{t,i,2} = m_{t,i,2} = s$ holds for a pixel position i and an image with index t. Note that the ground truth mask $\overline{m}_{t,2}$ corresponds to the one-hot amodal target mask $\overline{y}_t^{am}$. False positives FP_s^{inv} are present if $\overline{m}_{t,i,2} \neq s = m_{t,i,2}$, and false negatives FN_s^{inv} if $\overline{m}_{t,i,2} = s \neq m_{t,i,2}$. Naturally, on the amodal Cityscapes dataset in *training*, the pixel position i has to be restricted to areas where an amodal ground truth is available to measure mIoU^{inv}, i.e., $i \in \{j \in \mathcal{I} | \overline{m}_{t,j,f=2} = s, s \in \mathcal{S}\}$ [4]. This restriction to amodal ground truth pixels is also done for the quantitative *evaluation* in Tables 1, and 3, since we can only quantify performance where the ground truth is known. In a real application however, amodal ground truth is unknown. Accordingly, in a practical application we should evaluate $m_{t,f=2}$ only where mobile objects were detected in the visible mask $m_{t,f=1}$. This setting is shown in our qualitative evaluation results, e.g., in Figs. 1, 3, 4, 5, where other pixels in the amodal mask are simply set to white background color.

4.2   Training Details

We train our methods following the original training protocol of the ERFNet [19]. This also holds for the amodal variants. For the ERFNet, we train our networks for 120 epochs using the Adam optimizer with an exponential learning rate schedule, and learning rate 0.01. We use a batch size of 6 for all trainings. The single-task trainings use the cross-entropy loss. For the Y-ERFNet method, we calculate cross-entropy losses for the two decoder heads, respectively, being weighted equally during training.

4.3   Quantitative Results

The results for the ERFNet variants can be found in Table 1. An mIoU of 62.99% is reached on $\mathcal{D}_{amCS}^{test}$, and 67.21% on $\mathcal{D}_{CS}^{val}$ using the plain ERFNet. The low mIoU^{inv} of 5.00% compares visible predictions with amodal ground truth, naturally not exceeding chance level since the predicted visible $\mathbf{m}_{t,1}$ is no good replacement for the desired amodal $\mathbf{m}_{t,2}$. However, it also indicates that for a relevant amount of pixels the visible and the amodal class coincide, supporting our decision and novelty to omit the *groups*. Considering the *cross-task performance* between amodal and visible semantic segmentation, the decent value of the *visible* mIoU of ERFNet^{am} (20.16%) is higher than the mIoU^{inv} of ERFNet (5.00%), even though in both cases we regard tasks that are not explicitly learned by the respective network. In both cases, mIoU and mIoU^{inv} contributions different from zero stem from false positives and amodal predictions coinciding with the visible ground truth. However, the better mIoU results of the ERFNet^{am} are likely due to ignoring all areas without available amodal ground truth during training, meaning that the network has only seen mobile classes in training. This leads to noisy predictions wherever classes unknown to the network appear, as can be seen, e.g., in Fig. 1(c).

More importantly, for its designated task of amodal semantic segmentation, the mIoU^{inv} of ERFNet^{am} (36.48%) is significantly higher compared to ERFNet (5.00%). Moreover, we observe that our proposed Y-ERFNet even outperforms both: While on both datasets the visible mIoU slightly improves by 0.33% and 1.14% absolute, respectively, mIoU^{inv} increases significantly by 6.84% absolute (43.32% vs. 36.48%), due to the proposed joint amodal and visible training without the concept of groups.

Visible mIoU Details: To analyze the effect of the joint training on the specific visible semantic and occluded classes, we additionally report both mIoU and mIoU^{inv} per class on the Amodal Cityscapes test dataset ($\mathcal{D}_{\text{amCS}}^{\text{test}}$). Table 2 shows the (visible) mIoU per class. This includes the segmentation performance on the visible occluders. We observe that the ERFNet performs slightly better than the Y-ERFNet for some static classes, e.g., the mIoU of the class sidewalk is 80.09% vs. 79.62%. However, for most of the mobile classes, the Y-ERFNet outperforms the ERFNet, e.g., for class person (67.26% vs. 65.04%), rider (28.62% vs. 26.65%) or truck (49.71% vs. 46.22%). Hence, the joint amodal and visible semantic segmentation training helps in learning more informative visible features of mobile objects, and thus, our approach reflects better the vulnerable road users.

Invisible mIoU Details: Table 3 shows the per-occludee-class mIoU^{inv} for ERFNet, ERFNet^{am}, and Y-ERFNet. Occludee classes are the semantic classes appearing in the amodal ground truth at a pixel position $i \in \mathcal{I}$, i.e., $s = \overline{m}_{t,i,f=2}$. We see that in this case, the Y-ERFNet outperforms the other methods for almost all classes. The only exception is for the bicycle class, where the ERFNet^{am} performs just slightly better. For so-called vulnerable road users (person, rider), the Y-ERFNet improves the mIoU^{inv} in comparison to the ERFNet^{am} by 6.14% absolute (person) and by 7.72% absolute (rider). This indicates that *the joint training process not only improves perception of visible mobile classes and, especially, vulnerable road users, but also leads to large improvements concerning their amodal perception.*

Table 2. Visible mIoU (%) $s = \overline{m}_{t,i,f=1}$ per class on the Amodal Cityscapes test dataset ($\mathcal{D}_{\mathrm{amCS}}^{\mathrm{test}}$) for the ERFNet-based semantic segmentation method. Best results in bold, total results from Table 1.

	road	sidewalk	traffic light	wall	fence	pole	building
ERFNet	**97.27**	**80.09**	**59.18**	**42.03**	**42.07**	55.33	**89.29**
ERFNet^{am}	42.30	34.66	1.42	12.78	8.67	30.93	51.51
Y-ERFNet	97.15	79.62	**59.18**	39.32	40.61	**55.34**	89.10
	terrain	vegetation	traffic sign	sky	person	rider	car
ERFNet	60.64	90.29	68.51	92.92	65.04	26.65	87.04
ERFNet^{am}	15.86	58.30	15.33	11.27	20.53	8.11	37.10
Y-ERFNet	**60.81**	**90.33**	**69.11**	**93.16**	**67.26**	**28.62**	**87.52**
	truck	motorcycle	bicycle	bus	train	total	
ERFNet	46.22	**28.86**	68.08	55.12	42.18	62.99	
ERFNet^{am}	8.31	3.19	20.38	0.69	1.73	20.16	
Y-ERFNet	**49.71**	26.54	**68.82**	**55.59**	**45.28**	**63.32**	

Table 3. mIoU^{inv} (%) per occludee class $s = \overline{m}_{t,i,f=2}$ on the Amodal Cityscapes test dataset ($\mathcal{D}_{\mathrm{amCS}}^{\mathrm{test}}$) for the ERFNet-based semantic segmentation method. Best results in bold, total results from Table 1.

Occludee:	road	sidewalk	traffic light	wall	fence	pole	building
ERFNet	1.67	2.42	2.31	3.57	3.33	4.70	5.69
ERFNet^{am}	84.74	54.32	0.01	33.11	28.37	29.18	62.14
Y-ERFNet	**86.87**	**58.92**	**0.05**	**35.96**	**35.69**	**34.21**	**66.65**
Occludee:	terrain	vegetation	traffic sign	sky	person	rider	car
ERFNet	3.70	5.37	3.82	6.53	6.00	7.26	17.09
ERFNet^{am}	35.40	62.35	11.86	43.47	38.29	14.94	72.34
Y-ERFNet	**40.39**	**65.98**	**21.63**	**49.17**	**44.43**	**22.66**	**73.44**
Occludee:	truck	motorcycle	bicycle	bus	train	total	
ERFNet	0.94	2.08	4.29	4.09	10.13	5.00	
ERFNet^{am}	27.40	5.01	**35.35**	30.22	23.86	36.48	
Y-ERFNet	**36.88**	**7.47**	34.78	**51.89**	**50.70**	**43.32**	

4.4 Qualitative Results

We show a visualization of our results on images from the Amodal Cityscapes test dataset ($\mathcal{D}_{\mathrm{amCS}}^{\mathrm{test}}$) [4].

Figure 3 shows qualitative results of the ERFNet, ERFNet^{am}, and Y-ERFNet for the given input image $\mathbf{x}_t$ (Fig. 3(a)). We observe that the visible semantic segmentation is well predicted for both the ERFNet (Fig. 3(b)) and the Y-ERFNet (Fig. 3(f)), as also indicated by the results in Table 1. The predictions of the

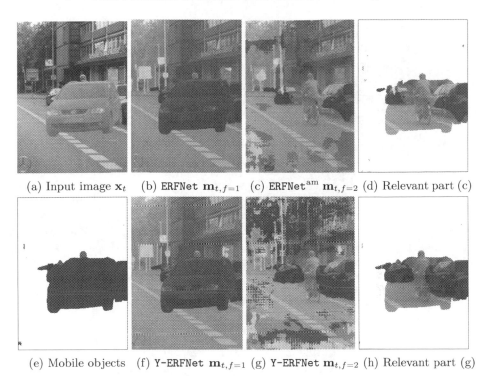

(a) Input image $\mathbf{x}_t$ (b) ERFNet $\mathbf{m}_{t,f=1}$ (c) ERFNet^{am} $\mathbf{m}_{t,f=2}$ (d) Relevant part (c)

(e) Mobile objects   (f) Y-ERFNet $\mathbf{m}_{t,f=1}$ (g) Y-ERFNet $\mathbf{m}_{t,f=2}$ (h) Relevant part (g)

Fig. 3. Qualitative results of the ERFNet (b) and ERFNet^{am} (c) visualized on the Amodal Cityscapes test dataset ($\mathcal{D}_{\text{amCS}}^{\text{test}}$) for an input image (a). Subfigure (d) visualizes the predictions of ERFNet^{am} where mobile objects (c) are predicted. The bottom row further shows the results of the Y-ERFNet with the predicted mobile objects (e), the visible semantic segmentation (f), the amodal semantic segmentation (g), and the amodal semantic segmentation where the visible semantic segmentation predicted mobile objects (*relevant parts*) (h).

amodal semantic segmentation $\mathbf{m}_{t,f=2}$ are shown in Fig. 3(c) for the ERFNet^{am}, and in Fig. 3(g) for the Y-ERFNet. Note that at first glance these predictions look quite chaotic, but the amodal decoders have only been trained to "look behind" mobile objects, not "behind" the road, buildings, or vegetation. This is why in practice these masks shall only be evaluated behind pixels, where any of the mobile objects have been detected. Figure 3(e) shows the visible mobile objects as predicted by the Y-ERFNet. Figure 3(d) shows the relevant parts of the amodal semantic segmentation in Fig. 3(c), inserted only where the visible semantic segmentation Fig. 3(b) predicted mobile objects. Figure 3(h) shows the same visualization for the Y-ERFNet, inserting the amodal semantic segmentation of Fig. 3(g) wherever the semantic segmentation in Fig. 3(f) predicted mobile objects. Comparing Figs. 3(d) and (h), we see that the occluded person (in this case a rider) is anticipated by both methods, however, the Y-ERFNet's prediction (Fig. 3(h)) has a larger segmentation overlap as is also indicated by the quantitative results in Table 1.

(a) Ground truth visible segmentation $\overline{\mathbf{m}}_{t,f=1}$ (b) Ground truth amodal segmentation $\overline{\mathbf{m}}_{t,f=2}$ (c) Amodal segmentation 3(g) inserted in the visible ground truth (a).

Fig. 4. Qualitative results of the Y-ERFNet visualized using the **ground truth masks** of the Amodal Cityscapes test dataset ($\mathcal{D}_{\mathrm{amCS}}^{\mathrm{test}}$) for the same input image as in Fig. 3(a). Subfigure (a) shows the visible ground truth corresponding to the input image. Subfigure (b) shows the amodal ground truth behind the occluder. Subfigure (c) shows the amodal prediction of the Y-ERFNet inserted in the visible ground truth (a) wherever an amodal ground truth (b) is available.

In Fig. 4 we show the ground truth for (a) visible and (b) amodal semantic segmentation for the respective input image $\mathbf{x}_t$ from Fig. 3(a). We insert the amodal semantic segmentation $\mathbf{m}_{t,f=2}$, shown in Fig. 3(g), into the ground truth semantic segmentation $\overline{\mathbf{m}}_{t,f=1}$, Fig. 4(a), wherever $\overline{\mathbf{m}}_{t,f=2}$ (Fig. 4(b)) is existing. Figure 4(c) shows the result. We see a semantic segmentation mask that sufficiently recovers the amodal ground truth including the footpoint of the person/bicycle, although the bicycle itself is erroneously represented by **person** pixels. However, note that this misclassification is indeed helpful for further perception functions, as distance estimation, e.g., heavily relies on correct footpoints. Thus, the amodal prediction yields sufficient results not only when inserted in predicted mobile objects, but even compared with the ground truth.

Figure 5 shows the same setting as Fig. 3 for a different input image $\mathbf{x}_t$, inserting the amodal semantic segmentation $\mathbf{m}_{t,f=2}$ into the predicted visible mobile objects (Fig. 5(b)) from the semantic segmentation $\mathbf{m}_{t,f=1}$ (Fig. 5(a)), hence visualizing the relevant parts. In Fig. 5(c), we insert the predicted amodal semantic segmentation as replacement for all predicted visible mobile objects, as this is the setting our methods have been trained on using the Amodal Cityscapes dataset. It can be observed that indeed the mobile objects are replaced by the prediction of the amodal segmentation method. For the car in the middle of the road, the amodal segmentation yields background, i.e., road, building, cars, and especially the fully anticipated shape of the partially visible person on the right behind the car. While this is all predicted reasonably, we observe that three pedestrians were fully occluded by the car, and therefore could not be predicted by the amodal decoder. However, in this case, also a human would struggle to predict fully occluded pedestrians behind the occluding car *without* temporal

(a) Predicted visible seg- (b) Predicted mobile   (c) Relevant parts of the
    mentation $\mathbf{m}_{t,f=1}$ objects in $\mathbf{m}_{t,f=1}$ amodal segmentation
                                                                $\mathbf{m}_{t,f=2}$ in (a)

Fig. 5. Qualitative results of the `Y-ERFNet` with the visible semantic segmentation (a), the predicted mobile objects (b), and the amodal semantic segmentation where the visible semantic segmentation predicted mobile objects (*relevant parts*) (c).

context. We leave the question of integrating temporal context open for future work.

Overall, we observe that a separate prediction of the specific visible and occluded semantic classes outperforms the previous groupwise setting [4,15]. Moreover, from our experiments we also derive that the joint training of visible and amodal semantic segmentation outperforms the separate training approach. Specifically, we show better results in both (visible) mIoU and mIoU^{inv}, establishing a new baseline method for amodal segmentation in general as well as setting a new benchmark for amodal segmentation on the Amodal Cityscapes dataset $\mathcal{D}^{\mathrm{toot}}_{\mathrm{amCS}}$ [4].

5 Conclusion

In this paper, we proposed a novel joint training approach for visible and amodal semantic segmentation as opposed to single-task training. In particular, the proposed method is predicting all semantic classes separately instead of only predicting groupwise labels. The single-task training predicting each semantic class separately using one network for each task already allows predicting both segmentation masks and yields an invisible mIoU of 36.48%. The joint training approach leads to a significant improvement, achieving an invisible mIoU of 43.32% on the Amodal Cityscapes test dataset. For the visible mIoU, the separate training achieves an mIoU of 62.99%, while the joint approach leads to a slightly improved visible mIoU of 63.32%. We observe that for the `ERFNet`, both visible and amodal semantic segmentation can benefit from the joint training approach.

Acknowledgements. We mourn the loss of our co-author, colleague and friend Jonas Löhdefink. Without his valuable input this work would not have been possible. The research leading to these results is funded by the German Federal Ministry for Economic Affairs and Energy within the project "KI Data Tooling - Methods and tools for the generation and refinement of training, validation and safeguarding data for AI functions in autonomous vehicles." The authors would like to thank the consortium for the successful cooperation.

References

1. Bär, A., Klingner, M., Varghese, S., Hüger, F., Schlicht, P., Fingscheidt, T.: Robust semantic segmentation by redundant networks with a layer-specific loss contribution and majority vote. In: Proceedings of CVPR - Workshops, Seattle, WA, USA, pp. 1348–1358 (2020)
2. Bogdoll, D., Nitsche, M., Zöllner, J.M.: Anomaly detection in autonomous driving: a survey. In: Proceedings of CVPR - Workshops, New Orleans, LA, USA, pp. 4488–4499 (2022)
3. Bolte, J.A., et al.: Unsupervised domain adaptation to improve image segmentation quality both in the source and target domain. In: Proceedings of CVPR - Workshops, Long Beach, CA, USA, pp. 1404–1413 (2019)
4. Breitenstein, J., Fingscheidt, T.: Amodal cityscapes: a new dataset, its generation, and an amodal semantic segmentation challenge baseline. In: Proceedings of IV, Aachen, Germany, pp. 1–8 (2022)
5. Breitenstein, J., Termöhlen, J.A., Lipinski, D., Fingscheidt, T.: Systematization of Corner Cases for Visual Perception in Automated Driving. In: Proc. of IV. pp. 986–993. Las Vegas, NV, USA (Oct 2020)
6. Cordts, M., et al.: The cityscapes dataset for semantic urban scene understanding. In: Proceedings of CVPR, Las Vegas, NV, USA, pp. 3213–3223 (2016)
7. Follmann, P., König, R., Härtinger, P., Klostermann, M.: Learning to see the invisible: end-to-end trainable amodal instance segmentation. In: Proceedings of WACV, Waikoloa Village, HI, USA, pp. 1328–1336 (2019)
8. He, K., Gkioxari, G., Dollár, P., Girshick, R.: Mask R-CNN. In: Proceedings of ICCV, Venice, Italy, pp. 2980–2988 (2017)
9. Heidecker, F., et al.: An application-driven conceptualization of corner cases for perception in highly automated driving. In: Proceedings of IV, Nagoya, Japan, pp. 644–651 (2021)
10. Hu, Y.T., Chen, H.S., Hui, K., Huang, J.B., Schwing, A.G.: SAIL-VOS: semantic amodal instance level video object segmentation - a synthetic dataset and baselines. In: Proceedings of CVPR, Long Beach, CA, USA, pp. 3105–3115 (2019)
11. Ke, L., Tai, Y.W., Tang, C.K.: Deep occlusion-aware instance segmentation with overlapping BiLayers. In: Proceedings of CVPR, Nashville, TN, USA, pp. 4019–4028 (2021)
12. Klingner, M., Bär, A., Fingscheidt, T.: Improved noise and attack robustness for semantic segmentation by using multi-task training with self-supervised depth estimation. In: Proceedings of CVPR - Workshops, Seattle, WA, USA, pp. 1299–1309 (2020)
13. Mohan, R., Valada, A.: Amodal panoptic segmentation. In: Proceedings of CVPR, New Orleans, LA, USA, pp. 21023–21032 (2022)
14. Poudel, R.P., Liwicki, S., Cipolla, R.: Fast-SCNN: Fast Semantic Segmentation Network. arXiv preprint arXiv:1902.04502, pp. 1–9 (2019)
15. Purkait, P., Zach, C., Reid, I.D.: Seeing behind things: extending semantic segmentation to occluded regions. In: Proceedings of IROS, Macau, SAR, China, pp. 1998–2005 (2019)

16. Qi, L., Jiang, L., Liu, S., Shen, X., Jia, J.: Amodal instance segmentation with KINS dataset. In: Proceedings of CVPR, Long Beach, CA, USA, pp. 3014–3023 (2019)
17. Reddy, N.D., Tamburo, R., Narasimhan, S.: WALT: Watch and learn 2D amodal representation using time-lapse imagery. In: Proceedings of CVPR, New Orleans, LA, USA, pp. 9356–9366 (2022)
18. Rensink, R.A., Enns, J.T.: Early completion of occluded objects. Vision Res. **38**(15), 2489–2505 (1998)
19. Romera, E., Álvarez, J.M., Bergasa, L.M., Arroyo, R.: ERFNet: efficient residual factorized ConvNet for real-time semantic segmentation. IEEE Trans. Intell. Transp. Syst. (T-ITS) **19**(1), 263–272 (2018)
20. Sun, Y., Kortylewski, A., Yuille, A.: Amodal segmentation through out-of-task and out-of-distribution generalization with a Bayesian model. In: Proceedings of CVPR, New Orleans, LA, USA, pp. 1215–1224 (2022)
21. Wang, A., Sun, Y., Kortylewski, A., Yuille, A.: Robust object detection under occlusion with context-aware CompositionalNets. In: Proceedings of CVPR, Seattle, WA, USA, pp. 12645–12654 (2020)
22. Weigelt, S., Singer, W., Muckli, L.: Separate cortical stages in amodal completion revealed by functional magnetic resonance adaptation. BMC Neurosci. **8**(1), 1–11 (2007)
23. Yuan, Y., Chen, X., Wang, J.: Object-contextual representations for semantic segmentation. In: Vedaldi, A., Bischof, H., Brox, T., Frahm, J.-M. (eds.) ECCV 2020. LNCS, vol. 12351, pp. 173–190. Springer, Cham (2020). https://doi.org/10.1007/978-3-030-58539-6_11
24. Zhu, Y., Tian, Y., Metaxas, D., Dollár, P.: Semantic amodal segmentation. In: Proceedings of CVPR, Honolulu, HI, USA, pp. 1464–1472 (2017)

Human-Vehicle Cooperative Visual Perception for Autonomous Driving Under Complex Traffic Environments

Yiyue Zhao[ID], Cailin Lei[ID], Yu Shen[✉][ID], Yuchuan Du[ID], and Qijun Chen[ID]

Tongji University, Shanghai 201804, China
{1952430,2010762,yshen,ycdu,qjchen}@tongji.edu.cn

Abstract. Human-vehicle cooperative driving has become one of the critical stages to achieve a higher level of driving automation. For an autonomous driving system, the complex traffic environments bring great challenges to its visual perception tasks. Based on the gaze points of human drivers and the images detected from a semi-automated vehicle, this work proposes a framework to fuse their visual characteristics based on the Laplacian Pyramid algorithm. By adopting Extended Kalman Filter, we improve the detection accuracy of objects with interaction risk. This work also reveals that the cooperative visual perception framework can predict the trajectory of objects with interaction risk better than simple object detection algorithms. The findings can be applied in improving visual perception ability and making proper decisions and control for autonomous vehicles.

Keywords: Human-vehicle cooperative visual perception · Image fusion · Object detection · Complex traffic environments

1 Introduction

With the rapid development of autonomous vehicles (AVs), human-vehicle cooperative driving has become a critical stage of fully autonomous driving. On the one hand, it could improve driving safety and efficiency in urgent and risky situations, such as distraction or incorrect operations of drivers [16,21,29]. On the other hand, AVs could assist in completing accurate vehicle control and reduce the workload of human drivers [2,5,7].

Existing studies pay more attention to automatic engineering and drivers' behaviors [16,21]. In the field of automatic engineering, there are several traditional solutions to assist manual driving from the vehicle perspective. For example, the authors of [9] designed an autonomous-mode steering wheel, which could emit a signal to drivers to switch to manual mode. In addition, the authors of [17] proposed a switched cooperative driving model from a cyber-physical perspective to minimize the impact of the transitions between automated and manual driving on traffic operations.

L. Karlinsky et al. (Eds.): ECCV 2022 Workshops, LNCS 13801, pp. 646–662, 2023.
https://doi.org/10.1007/978-3-031-25056-9_41

In addition, drivers' behaviors play an important role in traffic safety. An increasing number of researchers utilize visual, auditory, tactile, and other models to stimulate drivers in the automated mode [11,25,39]. The authors of [28] studied tactile and multi-sensory spatial warning signals on distracted drivers perform better than a simple sensor.

However, existing research seldom studies human-vehicle cooperative driving in the visual perception field. Besides, endless changes occur in the complex traffic scenarios, such as complex infrastructures, mixed traffic flow, etc. They bring great challenges to the cooperative system, especially the reliability of visual perception [13]. Without accurate detection, the system could hardly make proper decisions and control strategies [31]. Thus, accurate and efficient visual perception is essential to the safety of human-vehicle cooperative driving.

1.1   Visual Perception of Autonomous Driving

Existing autonomous driving visual perception studies mainly include semantic segmentation and object detection of road scenes. The former perceives the scenes from a global perspective, and it does not care about the specific kinematic states of each object. But the latter cares more about changes in objects' position and kinematic states within a continuous time series. So object detection provides more details for subsequent decision-making and control [22].

Since the PASCAL Visual Object Classes competition (PASCAL VOC) in 2012, deep learning has become the most powerful technique to extract objects' features from a massive amount of raw data [12,15]. One-stage detectors like You Only Look Once (YOLO) and Single Shot MultiBox Detector (SSD) have the advantages of fast detection and generalized feature extraction [26,33]. Two-stage detectors such as Regional Convolutional Neural Network (RCNN) and Feature Pyramid Network (FPN) have higher accuracy through region proposal [10,18]. However, these algorithms cannot adapt to complex traffic environments directly. The adaptability to complex road infrastructures (like large flyovers, overhead, etc.) and mixed traffic flow needs to be improved.

1.2   Cooperative Visual Perception Methods

Experiments About Cooperative Perception. The above mentioned visual perception algorithms almost predict objects with the same weight so that they cannot detect the objects with interaction risk. Thus, there is a large barrier between AVs' visual perception systems and human drivers' visual features.

Combined with visual sensors and perception algorithms, some scientists designed a human-vehicle cooperative navigation system and simulated visual perception in SCANeR^{TM} Studio [3,36]. Through the precise perception by LIDAR and dynamic command of drivers, AVs can adjust the edge and center of object detection. The experiment revealed that cooperative perception assisted proper decision-making and control in risky scenarios. However, the above process still needed dynamic feedback from human drivers. It decreased the efficiency and did not reduce the cognitive load of the human agent.

Image Fusion Method for Cooperative Visual Perception. Traditional image fusion methods include crop and paste operations by Pillow, Matplotlib, Scikit-image Packages, etc. [30]. However, these methods might lose the features of original images and add signal noise to images during the transformation.

In 1987, Gaussian Pyramid was utilized to fuse images. The original image was the bottom layer of Gaussian Pyramid. It was executed with Gaussian fuzzy filtering and down-sampling three to six times to get the images with the minimal size, namely the top layer of the pyramid [38]. Then the $(n+1)^{th}$ layer of Gaussian Pyramid was upsampled and reconstructed as the nth layer of Laplacian Pyramid [4]. By applying Laplacian Pyramid, the pixel information and features could be extracted and the noise could be diminished.

To bridge the above mentioned research gaps, this paper proposed a visual fusion model for the human-vehicle cooperative perception system. The precision and effect of object detection and image fusion were both verified in the real-world road environment. Based on the transfer learning method and the image fusion algorithm, human-vehicle cooperative visual perception performed better than simple visual perception algorithms. It could predict the trajectory of objects with interaction risk more accurately than the simple object detection algorithms under complex traffic environments, including the left-turning collision, lane change, and urgent avoidance scenarios [6,14,35].

The rest of this paper is arranged as follows. Section 2 introduces the critical methodology of human-vehicle cooperative visual perception. The detailed process of the experiment is illustrated in Sect. 3; Sect. 4 concludes this paper with contributions and future studies.

2   Methodology

To improve the perception ability and driving safety for AVs under complex traffic environments, this study proposed a visual perception fusion approach for human-vehicle cooperative driving. It included two steps. First, an object detection algorithm was used to detect the positions and kinematic state of objects under complex traffic environments. Due to the uncertainty of traffic flow, current object detection algorithms cannot adapt to urgent conflict scenarios. Therefore, this study utilized the transfer learning method to enhance the detection precision in complex traffic environments. Second, this study proposed an image fusion algorithm to fuse drivers' gaze points with in-vehicle camera screens which were predicted by the object detection algorithm. By analyzing the distribution of drivers' gaze points, the object with interaction risk could be determined. The framework of this study is shown in Fig. 1.

2.1   Data Acquisition Under Complex Traffic Environments

To obtain the visual characteristics of manual driving in real-world scenarios and vehicles' real-time kinematic states, this study collected multi-source data, including three aspects:

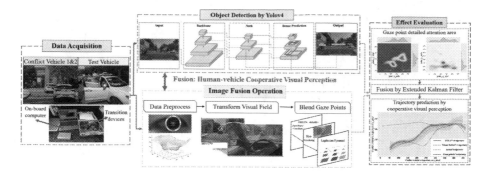

Fig. 1. Framework of human-vehicle cooperative perception.

- Drivers' eye-tracking data
- Scenario tracking data
- Real-time kinematic data

Drivers' eye-tracking data recorded the location information of drivers' pupils. It could be extracted to obtain the gaze points' characteristics, namely the object of the driver's gaze at any given moment. The location coordinate (x, y) was ranging from 1920×1080 pixel. And the scenario tracking data was derived from the screen of in-vehicle cameras, which recorded the scenarios in front of the ego vehicle. And the real-time kinematic data returned the high-accuracy information of motion, including relative distance, velocity, acceleration, etc.

2.2 Data Preprocessing

The scenario tracking data and real-time kinematic data would be transformed subsequently. This part introduced the processing of eye-tracking data. By the Canny operator (1), the boundaries of pupils were determined rapidly [34].

$$\hat{I} = \sum_{\omega} a_{guid} I_{guid} + b_{guid}$$

$$a_{guid} = \frac{\Phi_{\omega}^2}{\Phi_{\omega}^2 + \varepsilon} \tag{1}$$

$$b_{guid} = (1 - a_{guid}) M_{\omega}$$

where

$I_{guid}=$ bootstrap image of the bootstrap filter
$\hat{I}=$ images I_{guid} after filtering
$a_{guid}, b_{guid}=$ linearity coefficient
$\omega=$ filter window
$M_{\omega}=$ the means of I_{guid} in filter window ω
$\Phi_{\omega}=$ the variance of I_{guid} in filter window ω
$\varepsilon=$ regular term to adjust values of a_{guid}

After detecting the boundary of pupils, this study found the location of the pupil center. According to equation $(x - a)^2 + (y - b)^2 = r^2$, Hough Transform found the circle (a, b, r) that covered most of the boundary point (x, y) in the binary image. Then the pupil center coordinate was determined as (a, b) [1]. The Hough Transform algorithm is shown below.

Algorithm 1: Pupil Center Coordinate Determining by Hough Transform

---

Require: $\boldsymbol{L}$, $\boldsymbol{W}$, $\boldsymbol{R}$, $\boldsymbol{x}$, $\boldsymbol{y}$, $\boldsymbol{r}$

1: Determine the sizes of input images: $\boldsymbol{L}$=image.length, $\boldsymbol{W}$=image.width, $\boldsymbol{R}=\frac{1}{2} \cdot min(\boldsymbol{L}, \boldsymbol{W})$.
2: Establish a three-dimension array $(\boldsymbol{L}, \boldsymbol{W}, \boldsymbol{R})$ and calculate the number of boundary points $(\boldsymbol{x}, \boldsymbol{y})$ of each parameter circle.
3: Iterate all boundary points of images, compute circle equations of each boundary point, and record the results into the array: $(\boldsymbol{x}, \boldsymbol{y}) \rightarrow (\boldsymbol{L}, \boldsymbol{W}, \boldsymbol{R})$.
4: Find the circle containing the most boundary points (i.e., the Hough circle): $(x - a)^2 + (y - b)^2 = r^2$.

5: Return the pupil center coordinate $(\boldsymbol{L}, \boldsymbol{W}, \boldsymbol{R})$.

By Algorithm 1, the location of the pupil's center was obtained. Taking the geometric center point of two pupils, the trajectory of gaze points was simplified.

2.3   Data Fusion Algorithm

To achieve the human-vehicle cooperative visual perception, this paper established a data fusion algorithm between object detection screen and human drivers' gaze points.

Object Detection of Complex Traffic Environments. The existing object detection algorithms have difficulty identifying the objects in complex traffic environments. When the traffic environment becomes complex (like more traffic elements or interaction risk), the accuracy of object detection declines sharply [27]. So this study improved the algorithm by transfer learning [24].

Transfer learning migrates hyper-parameters from the source domain to the target domain, which effectively utilizes the learning results of the pre-trained model. Thus, the training process requires a smaller volume of datasets and improves efficiency rapidly. The method was divided into two steps:

Step 1: migrate data in the Microsoft Common Objects in Context dataset (COCO) to the complex traffic environment dataset (CFE) [19]. Because of the simple samples of COCO, the pre-trained model cannot adapt to complex traffic environments well. Thus, this study established the mapping relationship of object classes between COCO and CFE.

Step 2: optimize the hyper-parameters of the pre-trained model. Due to the large size of objects in the COCO, the pre-trained model cannot detect tiny objects under complex traffic environments. Therefore, this study optimized the parameters to improve the efficiency and accuracy [20,23], as shown in Table 1.

Table 1. Optimized hyper-parameters and effect

Optimized Hyper-parameter	Effect
Input size: 608 × 608	Improve boundary accuracy
Regularization parameter: $\alpha = 10^{-4}$	Prevent overfitting
Learning rate: Cosine anneal	Find the optimal solution fast
Image augmentation: Mosaic	Detect small-size objects
Label smoothing: 0.005	Reduce misclassification
Loss function: Complete Intersection over Union	Stabilize the target anchor

Image Fusion Algorithms. Through data preprocessing, this study obtained the center coordinate (x_i, y_i) of gaze points of human drivers. And the gaze point could be represented by a circle that the center coordinate was (x_i, y_i) and the radius was 35 pixels. With the Crop function in Pillow (a package of Python), gaze points of human drivers could be extracted frame by frame. And the crop box would be transformed into coordinate telescoping as shown in Eq. (2).

$$x_j = \frac{(x_i - 480)x_1}{x_0 - x_1} + \frac{35x_1}{2x_0}(\text{min takes -,max takes +})$$

$$y_j = \frac{(y_i - 10)y_1}{y_0 - y_1} \pm \frac{35y_1}{2y_0}(\text{min takes -,max takes +})$$

(2)

where

x_0=horizontal axis pixel value of eye-tracking screen
x_1=horizontal axis pixel value of in-vehicle screen
y_0=vertical axis pixel value of eye-tracking screen
y_1=vertical axis pixel value of in-vehicle screen
x_i=horizontal coordinate of the gaze point
y_i=vertical coordinate of the gaze point

Then this study captured the area of eye-tracking images ranging from $[x_{jmin} : x_{jmax}, y_{jmin} : y_{jmax}]$ and applied Laplacian of Gaussian (LoG) to smooth the signals n_r, n_c (the length and width of images) from the eye-tracking images. The detailed Gaussian Filter [38] is shown as Eq. (3).

$$LoG(x, y) = \frac{\partial^2 G_\sigma(x, y)}{\partial x^2} + \frac{\partial^2 G_\sigma(x, y)}{\partial y^2} = -\frac{1}{\pi\sigma^4}e^{-\frac{x^2+y^2}{2\sigma^2}}(1 - \frac{x^2 + y^2}{2\sigma^2}) \quad (3)$$

Based on Algorithm 2 below, this study constructed Gaussian and Laplacian Pyramid [4]. Drivers' gaze points of eye-tracking image B were fused with in-vehicle camera screen A after object detection.

Algorithm 2: Image fusion Algorithm

Require: n_c, n_r, A, B

1: Down-sample B until the Dots Per Inch (DPI) reaches the minimum to construct Gaussian Pyramid of B: $n_r^{'} = \frac{n_r}{2^n}, n_c^{'} = \frac{n_c}{2^n}$.

2: Up-sample and smooth the minimal images of the top layer in the Gaussian Pyramid until the image's size is equal to the initial image: $LB = B$.laplacian_pyramid

3: Repeat the above operation to obtain the Laplace pyramid of A: LA

4: Operate binary mask for B to get gaze zone): $M = B$.mask

5: Repeat Step 1 to get the Gaussian Pyramid of M: GM

6: Use GM nodes as weights and synthesize Laplace Pyramid LS from LA and LB: $LS(i,j) = GM(i,j) \cdot LA(i,j) + (1 - GM(i,j)) \cdot LB(i,j)$

7: Reconstruct pyramid LS to return fused images S.

2.4 Data Postprocessing

The precise object detection of traffic scenarios and fused images could be further applied to analyze manual eye-tracking characteristics in the process of driving. The object covered by the drivers' gaze points would be given a larger weight. It made the visual perception system better detect objects with interaction risk in complex traffic environments, such as left-turning collisions, lane changes, etc. Through Extended Kalman Filter algorithm [32], the trajectory of drivers' gaze points and object detection algorithm could be fused. And the fusion trajectory was closer to the real trajectory of interaction objects.

3   Experiment

3.1   Setup

The experiment recruited nineteen volunteers with different driving experiences to drive the ego vehicle. It was produced by DongFeng and achieved level 2 automation. This paper collected data in Tongji University on 8th and 9th May 2021, covering sunny and cloudy conditions. Participants were required to wear the eye tracking device Dikablis Professional Glasses 3.0, as shown in Fig. 2. It could track pupils' trajectory and return the pupils' position per 0.117s.

In addition, various sensors were installed on the ego vehicle. In-vehicle camera from Robosense recorded the scenario in front of the ego vehicle [37]. Real-time kinematic (RTK) from Robosense could return the vehicles' kinematic state

 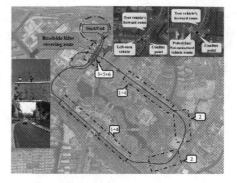

Fig. 2. Eye tracking device.   **Fig. 3.** Experimental scenarios.

in real-time including relative distance, velocity and acceleration at the longitudinal direction, and the distance of deviation from the lane centerline.

To obtain manual eye-tracking features under various complex traffic environments, the experiment set up six driving scenarios. They contained three conflict scenarios: left-turning collision, lane change, and urgent avoidance for pedestrians. Table 2 shows the description of experimental scenarios.

Table 2. Experimental scenario design

Number	Experimental Scenarios
1	Straight driving
2	Turn left or right without obstacles
3	Conflict between motorized vehicle and motorized vehicle
4	Two-vehicle or three-vehicle following
5	Conflict between pedestrian and vehicle
6	Conflict between non-motorized vehicle and motorized vehicle

The experimental scenario was located at Tongji University, Shanghai. Based on the scenarios in Table 2, the experiment constructed a closed route on the campus. Figure 3 illustrates the driving route and conflict scenarios.

The nineteen participants included nine males and ten females(mean ages (M) = 21 years, Standard deviation (SD) = 1.1 years). The driving experience of participants was ranging from 0.5 to 5 years (M = 1.875 years, SD = 0.75 years).

Participants were divided into three groups, and there were no obvious demographic differences between participants among these three groups. The first group needed to tackle the vehicle-vehicle conflict in Experiment 3. The second group would meet conflicts between pedestrians and vehicles in Experiment 5.

And the third group needed to deal with the conflict between non-motorized vehicles and motorized vehicles in Experiment 6.

3.2   Object Detection of Vehicle's Visual Perception

Dataset. Due to the advantages of transfer learning, the required scale of the dataset was small. This paper extracted 505 images of complex traffic scenarios like left-turning traffic collision, lane change, and urgent avoidance for pedestrians. Through drawing rectangles, classes of objects were labeled manually.

Training Detail. The dataset was divided into train-set and test-set by 9:1 randomly. And this paper executed 10-fold division and cross-validation. The improved backbone of the model was CSPDarkNet53 [3] and the activate function was replaced by Mish, as shown in Eq. (4).

$$Mish = x \cdot tanh(ln(1 + e^x)) \tag{4}$$

The feature enhancement extraction network added Spatial Pyramid Pooling (SPP) and PANet, which could greatly increase the receptive field and isolate the most significant contextual features [8].

Evaluation Metrics. The metrics for evaluation included: F1-score, Precision Rate, Recall Rate, and Average Precision (AP). Mean Average Precision (mAP) has become a common metric to evaluate the results of multi-classification tasks.

Loss Study. To assess the validity of the pre-trained model, this paper applied the pre-trained weight of the COCO dataset [19] to train. And the loss result in the process of iterating is shown in Fig. 4.

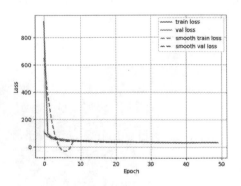

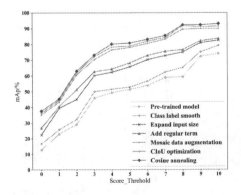

Fig. 4. Loss of the pre-trained model.

Fig. 5. Detection accuracy improvement.

Figure 4 shows the pre-trained model converged fast and the mAP was 69.3%. It ranked at the top level among the open-source algorithms.

Object Detection Algorithm Improvement by Transfer Learning. This part would improve the framework of the pre-trained model by using transfer learning, making it better adapt to complex traffic scenarios.

Then, this study utilized the pre-trained model to predict our dataset directly. The mAP of twelve classes remained at 41.7%, and the problems of missing or wrong predictions were exposed. To improve the accuracy and efficiency of the pre-trained model, this study set control experiments including six experimental groups and one control group. Through adjusting hyper-parameters of the pre-trained model (according to the order in Table 1), the predicted mAP of each group was recorded. Figure 5 shows the comparison result.

It revealed the effect of transfer learning. Training YOLOv4 on the collected dataset using transfer learning achieved 75.52%. Mosaic data augmentation and input size expansion improved the predicted precision most effectively. Then the improved model could distinguish different objects in complex traffic environments. It could lay a solid foundation for visual fusion in scenarios with interaction risk.

3.3 Gaze Point Fusion

Instead of regarding the conflict object as a particle, visual perception of AVs needs to capture the riskiest object in complex traffic environments. So this study fused drivers' gaze points with in-vehicle camera screens after object detection. The process was divided into three steps.

Gaze Point Preprocessing. During the experiment, the eye-tracking device recorded pupils' changes dynamically. And this paper applied Canny Operator to gain the edge information of pupils [34]. Compared with Laplacian of Gaussian Operator (LoG), Canny Operator applies Non-Maximum Suppression to detect the optimal edges of pupils. Through Canny Operator, the signal noise was suppressed and it remained the complete characteristic of pupils.

After detecting the pupils' edge, this study determined the center of the pupil. Through Hough Transform and edge points detected by Canny Operator, the center point and area of pupils could be obtained completely.

Coordinate Conversion of Visual Field. The visual field of the in-vehicle camera screen was different from gaze points, and the eye-tracking device was shaking in the driving process. So this study unified the coordinate between the in-vehicle camera screen and drivers' gaze points.

Through coordinate transforming Eq. (2), this study transformed coordinate of gaze points (size = 1920×1080) to the coordinate of the in-vehicle screen (size = 480×270). Then human driver's visual field was identical to the vehicle.

Gaze Point Extraction and Fusion. Based on the coordinate information of pupils, this study collected the center of gaze points and set 70×70 pixels squares to crop gaze points from the eye-tracking screen. By Laplacian Pyramid, the signal noise was diminished and the characteristics were reserved. Finally, this study applied Eq. (2) to fuse the gaze points with the in-vehicle camera screen. Figure 6 illustrates an example of visual fusion.

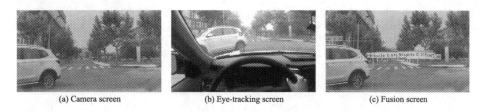

| (a) Camera screen | (b) Eye-tracking screen | (c) Fusion screen |

Fig. 6. The example of visual fusion.

3.4   Ground-Truth Processing

To verify the advantages of human-vehicle cooperative visual perception for autonomous driving, this paper utilized Real-time Kinetic (RTK) data to analyze the accuracy of the visual fusion method. Because the results of human-vehicle visual perception are Two-Dimensional images or videos, this paper transformed relative distance in the real-world road into a pixel-level trajectory to validate in the same coordinate. The transforming formula is shown as Eq. (5).

$$
\begin{aligned}
p_{x_j} &= \frac{s_{x_2} - s_{x_1}}{p_{x_2} - p_{x_1}} \cdot \frac{(x_i - 480)x_1}{x_0 - x_1} \cdot s_{x_j} \\
p_{y_j} &= \frac{s_{y_2} - s_{y_1}}{p_{y_2} - p_{y_1}} \cdot \frac{(y_i - 10)y_1}{y_0 - y_1} \cdot s_{y_j}
\end{aligned}
\tag{5}
$$

where

p_{x_j} =horizontal value of trajectory at the pixel level
p_{y_j} =vertical value of trajectory at the pixel level
s_{x_j} =horizontal value at the geographic coordinate
s_{y_j} =vertical value at the geographic coordinate

3.5   Validation of Human-Vehicle Cooperative Visual Perception

Furthermore, this study validated the effect of human-vehicle cooperative visual perception and explored differences in visual characteristics between AVs' visual perception systems and human drivers.

Experimental Scenarios. This study selected the left-turn conflict scenario to validate the effect of cooperative visual perception. The ego vehicle moved forward while another vehicle turned left suddenly at the intersection.

And RTK returned the states of the ego vehicle and the closest conflict object dynamically. Additionally, RTK also provided the kinematic states of the closest conflict object and the relative distance.

Evaluation Metrics. Time to Collision (TTC) measures the urgent degree of the conflict scenario. The First order of TTC can be calculated by Eq. (6):

$$TTC = \frac{\Delta S}{\Delta v} \tag{6}$$

where:

$\Delta S=$ Distance between test vehicles and conflict object
$\Delta v=$ Relative velocity derivation in the longitudinal direction

In the left-turning conflict, $\Delta v = v_{1y} + v_{2y}$. v_{1y}, v_{2y} indicated the velocity at the longitudinal direction of the ego vehicle and conflict vehicle. Then this study obtained the spatial distribution of gaze points under different TTC.

Trajectory and Attentive Zone of Gaze Points. This study collected 36 left-turning conflict scenarios from 19 participants. And the distribution of drivers' gaze points and the center of anchors were obtained. Figure 7 shows the comparison between gaze points of human drivers and AVs' visual perception.

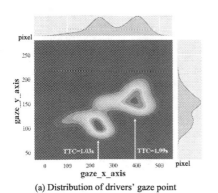

(a) Distribution of drivers' gaze point

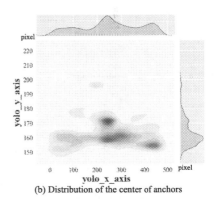

(b) Distribution of the center of anchors

Fig. 7. Distribution of drivers' gaze points and object detection algorithm.

Figure 7 suggested that when TTC was beyond 2.0s, drivers' gaze points were nearly out of the closest conflict vehicles. When TTC declined below 2.0s, drivers tended to care about the farther zone of the conflict vehicle (like the rear tire of vehicles). While TTC $\leq$ 1.03s, drivers paid more attention to the closer zone away from them, as shown in Fig. 7(a). In contrast, the distribution of YOLO almost remained at the same value at the y axis. It could not represent the real motion of conflict objects.

Unlike existing object detection algorithms determining conflict objects' positions by the center point of predicted anchors, human-vehicle cooperative visual perception cares about the detailed contour position of conflict objects. It indicates the riskiest zone in urgent conflict scenarios. To improve the accuracy of visual perception for AVs, it is essential to fuse gaze points and object detection.

Trajectory Prediction. Based on the result above, this paper further assessed the validation of cooperative visual perception by comparing the actual ground-truth trajectory and the trajectory after visual fusion.

– Actual trajectory
– Gaze points' trajectory
– YOLO's trajectory

Due the nonlinear and continuously changing features of vehicles' motion, this paper applied Extended Kalman Filter (EKF) to fuse the trajectories of gaze points and YOLOv4 [32]. The fused trajectory was determined as the predicted trajectory of conflict objects. The EKF algorithm is shown as Eq. (7).

$$
\begin{aligned}
\hat{z}_{k+1} &= h(\hat{x}_{k+1|k}) \\
\hat{x}_{k+1|k+1} &= \hat{x}_{k+1} + K_{k+1}(z_{k+1} - \hat{z}_{k+1|k}) \\
\hat{x}_{k+1} &= f(\hat{x}_{k|k}) \\
K_{k+1} &= P_{k+1|k} H_{k+1}^T (H_{k+1} P_{k+1|k} H_{k+1}^T + R_{k+1})^{-1} \\
P_{k+1|k} &= F_k P_{k|k} F_k^T + Q_k \\
P_{k+1|k+1} &= (I - K_{k+1} H_{k+1}) P_{k+1|k}
\end{aligned}
\tag{7}
$$

where

$x_{k|k}$: state estimate value at the moment k
z_{k+1}: measured vector value at the moment $k + 1$
$f(x_k)$: non-linear function of the state vector
$h(x_k)$: non-linear function of the measured vector
$P_{k+1|k}$: state estimate covariance at the moment $k + 1$
K_{k+1}: state matrix at the moment $k + 1$
F_k: partial derivative of $f(x_k)$ to x at the moment k
H_k: partial derivative of $h(x_k)$ to x at the moment k

Based on the non-linear function $f(x_k)$ of state vector x_k (including gaze point trajectory and YOLO's trajectory) and non-linear function $h(x_k)$ of the measured vector z_k (fusion trajectory), this study could approximate the next state estimate z_{k+1} of the system through Kalman Filter algorithm. Figure 8 shows the prediction result of the conflict object.

It revealed that the visual fusion's trajectory was closer to the actual trajectory in green than YOLO's trajectory and gaze points' trajectory alone. The Root Mean Squared Error (RMSE) of the predicted trajectory was 0.88 pixel square.

So human-vehicle cooperative visual perception could better predict the trajectory and kinematic states of conflict objects. The accurate prediction could support the decision-making and control of AVs.

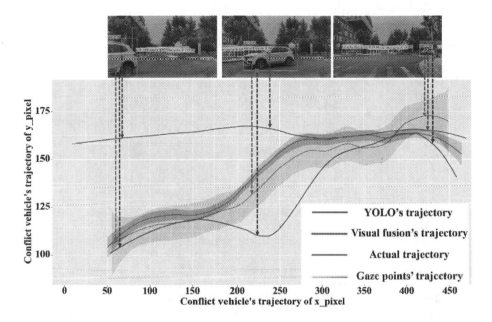

Fig. 8. Actual trajectory and predicted trajectory after visual fusion.

4   Conclusion

This study proposed a visual fusion perception framework for human-vehicle cooperative driving under complex traffic environments. Based on the transfer learning method, this study improved the object detection algorithm YOLOv4 and made it better adapt to detect the critical elements in complex traffic environments. The mAP of critical traffic elements was improved to 75.52%. Then, this paper proposed a visual fusion perception method between drivers' gaze points and in-vehicle camera screens after object detection. It was evaluated in the left-turning conflict experiment. Results revealed that cooperative visual perception could predict the conflict object's trajectory more accurately than the simple object detection algorithm. The findings could detect the objects with interaction risk and provide accurate data for decision-making and control for AVs.

For future research, this study can add a gyroscope to the eye-tracking devices to record the rotation angle of the driver's head. It can obtain more precise positions and changes of drivers' gaze points. By calculating the deviation between drivers' visual field and vehicle direction forward, the model can adjust fusion coordinate automatically. It improves the automation and efficiency of human-vehicle cooperative visual perception.

Acknowledgments. This work is supported by the National Key Research and Development Program of China under Grant No. 2020AAA0108101.

References

1. Ballard, D.H.: Generalizing the hough transform to detect arbitrary shapes. Pattern Recogn. **13**(2), 111–122 (1981)
2. Bennett, R., Vijaygopal, R., Kottasz, R.: Attitudes towards autonomous vehicles among people with physical disabilities. Transport. Res. Part A: Policy Pract. **127**, 1–17 (2019)
3. Bochkovskiy, A., Wang, C., Liao, H.M.: Yolov4: optimal speed and accuracy of object detection. CoRR abs/2004.10934 (2020). https://arxiv.org/abs/2004.10934
4. Burt, P., Adelson, E.: The Laplacian pyramid as a compact image code. IEEE Trans. Commun. **31**(4), 532–540 (1983)
5. Correia, G., Arem, B.V.: Solving the user optimum privately owned automated vehicles assignment problem (uo-poavap): a model to explore the impacts of self-driving vehicles on urban mobility. Transport. Res. Part B: Methodol. **87**, 64–88 (2016)
6. Dabbour, E., Easa, S.: Proposed collision warning system for right-turning vehicles at two-way stop-controlled rural intersections. Transport. Res. Part C: Emerg. Technol. **42**(42), 121–131 (2014)
7. Endsley, M.R.: Toward a theory of situation awareness in dynamic systems. Human Fact. J. Human Factors Ergon. Soc. **37**(1), 32–64 (1995)
8. Gai, W., Liu, Y., Zhang, J., Jing, G.: An improved tiny yolov3 for real-time object detection. Syst. Sci. Control Eng. Open Access J. **9**(1), 314–321 (2021)
9. Gazit, R.Y.: Steering wheels for vehicle control in manual and autonomous driving (2015)
10. Girshick, R., Donahue, J., Darrell, T., Malik, J.: Rich feature hierarchies for accurate object detection and semantic segmentation. In: 2014 IEEE Conference on Computer Vision and Pattern Recognition, pp. 580–587 (2014)
11. Gold, C., Bengler, K.: Taking over control from highly automated vehicles. In: Human Factors and Ergonomics Society Annual Meeting Proceedings, vol. 8, no. 64 (2014)
12. Hinton, G.E., Salakhutdinov, R.R.: Reducing the dimensionality of data with neural networks. Science **313**(5786), 504–507 (2006)
13. Huang, T., Fu, R., Chen, Y.: Deep driver behavior detection model based on human brain consolidated learning for shared autonomy systems. Measurement **179**, 109463 (2021)
14. Kalra, N., Paddock, S.M.: Driving to safety: how many miles of driving would it take to demonstrate autonomous vehicle reliability? Transport. Res. Part A: Policy Pract. **94**, 182–193 (2016)
15. LeCun, Y., Bengio, Y.: Deep learning. Nature **521**(7553), 436–444 (2015)

16. Li, J., Yao, L., Xu, X., Cheng, B., Ren, J.: Deep reinforcement learning for pedestrian collision avoidance and human-machine cooperative driving. Inf. Sci. **532**, 110–124 (2020)
17. Li, Y., Sun, D., Zhao, M., Dong, C., Cheng, S., Xie, F.: Switched cooperative driving model towards human vehicle copiloting situation: A cyberphysical perspective. J. Adv. Transp. **2018**, 1–11 (2018)
18. Lin, T.Y., Dollar, P., Girshick, R., He, K., Hariharan, B., Belongie, S.: Feature pyramid networks for object detection. In: 2017 IEEE Conference on Computer Vision and Pattern Recognition (CVPR) (2017)
19. Lin, T.-Y., et al.: Microsoft COCO: common objects in context. In: Fleet, D., Pajdla, T., Schiele, B., Tuytelaars, T. (eds.) ECCV 2014. LNCS, vol. 8693, pp. 740–755. Springer, Cham (2014). https://doi.org/10.1007/978-3-319-10602-1_48
20. Loshchilov, I., Hutter, F.: SGDR: stochastic gradient descent with warm restarts. In: International Conference on Learning Representations (2017)
21. Ma, R., Kaber, D.B.: Situation awareness and workload in driving while using adaptive cruise control and a cell phone. Int. J. Ind. Ergon. **35**(10), 939–953 (2005)
22. Murthy, C., Hashmi, M.F., Bokde, N., Geem, Z.W.: Investigations of object detection in images/videos using various deep learning techniques and embedded platforms-a comprehensive review. Appl. Sci. **10**(9), 3280 (2020)
23. NgoGia, T., Li, Y., Jin, D., Guo, J., Li, J., Tang, Q.: Real-time sea cucumber detection based on YOLOv4-tiny and transfer learning using data augmentation, pp. 119–128 (2021). https://doi.org/10.1007/978-3-030-78811-7_12
24. Pan, S.J., Qiang, Y.: A survey on transfer learning. IEEE Trans. Knowl. Data Eng. **22**(10), 1345–1359 (2010)
25. Politis, I., Brewster, S.: Language-based multimodal displays for the handover of control in autonomous cars. In: Proceedings of the 7th International Conference on Automotive User Interfaces and Interactive Vehicular Applications (2015)
26. Redmon, J., Divvala, S., Girshick, R., Farhadi, A.: You only look once: unified, real-time object detection. In: 2016 IEEE Conference on Computer Vision and Pattern Recognition (CVPR), pp. 779–788 (2016). https://doi.org/10.1109/CVPR.2016.91
27. Son, T., Mita, S., Takeuchi, A.: Road detection using segmentation by weighted aggregation based on visual information and a posteriori probability of road regions. In: 2008 IEEE International Conference on Systems, Man and Cybernetics, pp. 3018–3025 (2008). https://doi.org/10.1109/ICSMC.2008.4811758
28. Spence, C., Ho, C.: Tactile and multisensory spatial warning signals for drivers. IEEE Trans. Haptics **1**(2), 121–129 (2008)
29. Stender-Vorwachs, J., Steege, H.: Legal aspects of autonomous driving. Internationales verkehrswesen **70**(MAYSPEC.), 18–20 (2018)
30. Walt, S., Schnberger, J.L., Nunez-Iglesias, J., Boulogne, F., Yu, T.: Scikit-image: image processing in python. arXiv e-prints (2014)
31. Wang, J., Huang, H., Zhi, P., Shen, Z., Zhou, Q.: Review of development and key technologies in automatic driving. Application of Electronic Technique (2019)
32. Wang, Y., Papageorgiou, M.: Real-time freeway traffic state estimation based on extended kalman filter: a general approach. Transport. Res. Part B: Methodol. **39**(2), 141–167 (2007)
33. Liu, W., et al.: SSD: single shot multibox detector. In: Leibe, B., Matas, J., Sebe, N., Welling, M. (eds.) ECCV 2016. LNCS, vol. 9905, pp. 21–37. Springer, Cham (2016). https://doi.org/10.1007/978-3-319-46448-0_2
34. Xu, J., Wang, H., Huang, H.: Research of adaptive threshold edge detection algorithm based on statistics canny operator. In: MIPPR 2015: Automatic Target Recognition and Navigation, vol. 9812, p. 98121D (2015)

35. Yang, W., Zhang, X., Lei, Q., Cheng, X.: Research on longitudinal active collision avoidance of autonomous emergency braking pedestrian system (AEB-P). Sensors **19**(21), 4671 (2019)
36. Yue, K., Victorino, A.C.: Human-vehicle cooperative driving using image-based dynamic window approach: System design and simulation. In: 2016 IEEE 19th International Conference on Intelligent Transportation Systems (ITSC) (2016)
37. Zhang, J., Xiao, W., Coifman, B., Mills, J.P.: Vehicle tracking and speed estimation from roadside lidar. IEEE J. Selected Top. Appl. Earth Observ. Remote Sens. **13**, 5597–5608 (2020)
38. Zhang, T., Mu, D., Ren, S.: Information hiding algorithm based on gaussian pyramid and geronimo hardin massopust multi-wavelet transformation. Int. J. Digit. Content Technol. Appl. **5**(3), 210–218 (2011)
39. Ziemke, T., Schaefer, K.E., Endsley, M.: Situation awareness in human-machine interactive systems. Cogn. Syst. Res. **46**, 1–2 (2017)

MCIP: Multi-Stream Network for Pedestrian Crossing Intention Prediction

Je-Seok Ham$^{(\boxtimes)}$ ⓘ, Kangmin Bae ⓘ, and Jinyoung Moon ⓘ

Artificial Intelligence Research Laboratory, Electronics and Telecommunications Research Institute (ETRI), Daejeon, Republic of Korea
{jsham,kmbae,jymoon}@etri.re.kr

Abstract. Predicting the crossing intention of pedestrian is an essential task for autonomous driving systems. Whether or not a pedestrian will cross a crosswalk is a significantly inevitable skills for safety driving. Although many datasets and models are proposed to precisely predict the intention of pedestrian, they lack the ability to integrate different types of information. Therefore, we propose a Multi-Stream Network for Pedestrian Crossing Intention Prediction (MCIP) based on our novel optimal merging method. The proposed method consists of integration modules that takes two visual and three non-visual elements as an input. We achieved state-of-the-art performance on accuracy of pedestrian crossing intention, F1-score, and AUC with both public standard pedestrian datasets, PIE and JAAD. Furthermore, we compared the performance of our MCIP with other networks quantitatively by visualizing the intention of the pedestrian. Lastly, we performed ablation studies to observe the effectiveness of our multi-stream methods.

Keywords: Pedestrian crossing intention · Semantic segmentation · Autonomous vehicle · Pedestrian detection

1 Introduction

Recently, autonomous driving technology has been highly studied in variable fields. Many researchers produced datasets based on various environments related to pedestrian crossing, such as PIE (Pedestrian Intention Estimation) [12,13,21,24–28], JAAD (Joint Attention in Autonomous Driving) [25,26], STIP (Stanford-TRI Intent Prediction) [15], Euro-PVI (Pedestrian Vehicle Interactions in Dense Urban Centers) [1], TITAN (Trajectory Inference using Targeted Action priors Network) [20], and etc. Each dataset has its own prediction model and most of the datasets are filmed by an ego-vehicle. Each prediction model applied its own neural networks method, for example, the prediction model based on PIE dataset is composed of long short-term memory (LSTM) unit. Also, in case of model based on STIP dataset, they adopted scene graph convolutional neural networks, and for Euro-PVI dataset, they suggested Joint-*beta*-cVAE model for anticipating future pedestrian trajectories.

L. Karlinsky et al. (Eds.): ECCV 2022 Workshops, LNCS 13801, pp. 663–679, 2023.
https://doi.org/10.1007/978-3-031-25056-9_42

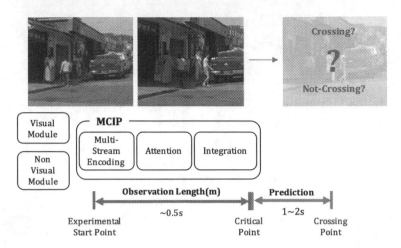

Fig. 1. Proposed pedestrian crossing intention prediction algorithm: They accept two modules (visual module and non visual module) and are composed of MSE (Multi-Stream Encoding) unit, attention unit, bonding unit. There are three points and we set the interval between experimental start point and critical point to observation length and the interval between critical point and crossing point to prediction time.

In terms of efficiency and safety of driverless driving environments, it is an essential factor that the autonomous vehicle correctly recognize the objects around the vehicle such as pedestrian, traffic sign, or neighbor vehicle. Because the safety of pedestrian is susceptible issue, so the self-driving system has no choice but to keep an eye on the stream of pedestrian behavior. Prediction is a significantly challenging topic because there may be unusual cases where the intention of pedestrian is suddenly changed unlike past frame information. However, from the standpoint of the autonomous driving system, the technology to predict in advance whether a pedestrian will cross the crosswalk or not is very important. In other words, as the behavior of pedestrian would significantly affect the driverless vehicle safety, the prediction of future pedestrian's action is considered to be an important issue. Especially, the intention prediction of pedestrian, that is, crossing or not-crossing is a highly dominant factor for self-driving system.

Therefore, we propose a novel algorithm for predicting future crossing intention of pedestrian referred to as Multi-Stream Network for Pedestrian Crossing Intention Prediction (MCIP). The MCIP consists of two modules: 1) Non-Visual Module (NVM) and 2) Visual Module (VM). In non-visual module (NVM), there are three inputs with bounding box and pose keypoints of pedestrians and speed of ego-vehicle. The bounding box includes the coordinate information (top left and bottom right) of pedestrian on image which is a local tracking trajectory data of pedestrian who has the intention of crossing. We used the 18 joints with the 36 dimension vectors of 2D coordinates generated by OpenPose for pose keypoints input. The speed of ego-vehicle is included in the PIE dataset, but not the JAAD dataset. For visual module (VM), two inputs are passed through MCIP, those are local context and global context. We cropped the image to twice the size of the bounding box for local context input. For global context input, we applied semantic segmentation to the image.

To summarize, our three main contributions are as below:

- We propose a novel multi-stream network for pedestrian crossing intention prediction (MCIP) based on 5 inputs.
- Our proposed method employs diverse inputs based on CNN unit for visual module and RNN unit for encoder, which is composed of Multi-Stream Encoding (MSE) unit, attention unit, integration unit.
- We verify the effectiveness of MCIP, achieving the state-of-the-art performance on the three metrics, crossing intention accuracy (CIA), F1-Score and AUC of pedestrian using two public pedestrian behavior benchmark datasets, PIE and JAAD. We also demonstrate the prediction aspects of pedestrian crossing intention through quantitative and qualitative results.

2   Related Work

Pedestrian Detection. Many researchers have studied detection of pedestrian behaviors. Abdul Hannan Khan et al. [11] proposed F2DNet, which is their new version of two-stage detection algorithms. They used focal detection network based on HRNet backbone to detect pedestrian on the crosswalk or driveway. Irtiza Hasan et al. [10] studied comprehensive experiments for crossing datasets. They tested their method on three state-of-the-art autonomous driving datasets, Caltech [8], CityPersons [37], and EuroCity Persons (ECP) [3] and they compared various datasets using cascade R-CNN [4]. Zebin Lin et al. [14] proposed Exemplar-guided contrastive learning network (EGCL) which included faster R-CNN to detect pedestrian. Fang Qingyun et al. [23] suggested a Cross-Modality Fusion Transformer (CFT), which consists of RGB input module and thermal input module and each output of two modules are fused based on transformer model.

Pedestrian Intention Prediction. There are many datasets that can be used for predicting pedestrian intention of crossing the streets and each dataset has its own prediction networks. Most of the datasets are collected using an ego-vehicle. PIE dataset is the standard crossing pedestrian dataset for predicting pedestrian crossing intention. Rasouli et al. [24] used LSTM and convolutional LSTM unit for two inputs(bounding box and visual context), but it achieved low accuracy of pedestrian intention estimation. Rasouli et al. [27] studied another network named as SF-GRU (Stacked with multilevel fusion GRU), which operates with five sources including pedestrian appearance, surrounding context, poses, bounding boxes, and ego-vehicle speed. Additionally, some researchers accept transformer models [9,16,17,22,31,34,36] for faster training than LSTM based models [2,12,18,28,33], but they still tend to perform poorly in terms of accuracy. Another approach for predicting pedestrian crossing intention is to use GAN

(Generative Adversarial Networks) [7,19]. Lukas Neumann et al. [21] employed self-supervised learning for vehicle motion prediction and view normalization. For STIP (Stanford-TRI Intent Prediction) dataset, Bingbin et al. [15] studied the graph convolutional approach. They integrated pedestrian-centric graph and location-centric graph. Sparse graphic convolutional methods also used for predicting pedestrian trajectory [29]. Apratim Bhattacharyya et al. [1] proposed Joint-*beta*-cVAE model using Euro-PVI dataset, which is for not only pedestrian, but also bicyclist. They used joint trajectory model and predicted the future trajectory of pedestrian. Srikanth Malla et al. [20] introduced TITAN (Trajectory Inference using Targeted Action priors Network) which consists of variable actions of person and vehicle. Amir Rasouli et al. [24] introduced JAAD (Joint Attention in Autonomous Driving) dataset, and it does not contain pedestrian crossing intention contrary to PIE dataset.

3   MCIP: Multi-Stream Network for Pedestrian Crossing Intention Prediction

3.1   Model Architecture

We introduce new algorithms for forecasting pedestrian crossing intention. The overall architecture of Multi-Stream Network for Pedestrian Crossing Intention Prediction (MCIP) is shown in Fig. 2. It consists of two modules which are Non Visual Module (NVM) with three inputs (pose, bounding box, ego-vehicle speed) and Visual Module (VM) with two inputs (local context, global context). Also, our network includes CNN unit, RNN unit, Multi-Stream Encoding (MSE) unit, attention unit, and integration unit.

In this work, we concentrate on the growth of performance of pedestrian crossing intention accuracy (CIA). From the observation of past frames of pedestrian, we predict the pedestrian crossing intention in future frames. We set the three points, experimental start point, critical point, and crossing point (Fig. 1). When we start to observe the target pedestrian, we call that point as experimental start point. The critical point refers to the point at which the observation ends and prediction starts. The interval from experimental start point to critical point is defined as the observation length (m) which means collecting the pedestrian information during past frames. We set the observation length as 15 frames ($\sim$0.5s) on this experiment. After critical point, the prediction starts. The crossing point means a future point in time when the pedestrian actually initiates a real action (crossing or not crossing). We defined the prediction time as the interval between crossing point and critical point. In this experiment, we obtained the crossing intention accuracy by changing the prediction time from 1s to 4s.

3.2   Input Acquisition

Non Visual Module (NVM). NVM receives features and not visual images as input and sequentially passes through the RNN unit. m is denoted as the

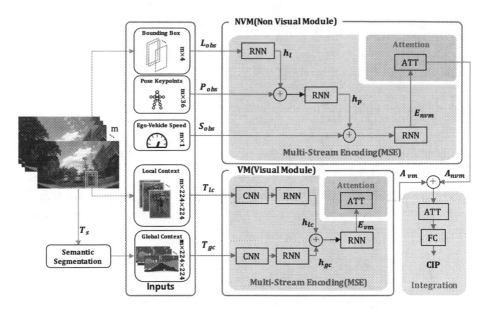

Fig. 2. The proposed architecture of Multi-Stream Network for Pedestrian Crossing Intention Prediction (MCIP). The MCIP consists of two modules: 1) non visual module (NVM), and 2) visual module (VM). For NVM, there are three inputs, bounding box (L_{obs}), pose keypoints (P_{obs}), and ego-vehicle speed (S_{obs}). For VM, there are two inputs, local context (T_{lc}), global context (T_{gc}). The NVM and VM modules are composed of multi-stream encoding (MSE), attention unit, and integration unit. The outputs of each MSE unit in NVM and VM, E_{nvm} and E_{vm}, go into input of attention unit. The integration unit merges the two output feature A_{nvm} and A_{vm} from attention unit to predict the final intention of the pedestrian (CIP, Crossing Intention Prediction).

observation length (frames), and we set $m = 15$ at which we achieved highest intention accuracy. t is the critical point $60 \sim 240$ frames before the crossing event. The NVM accepts 3 different inputs: 1) bounding boxes L_{obs}, 2) pose keypoints P_{obs} and, 3)ego-vehicle speed S_{obs}. The bounding boxes provides the exact location of the pedestrians to predict the intention. The bounding boxes L_{obs} is given as:

$$L_{obs} = \{l_i^{t-m}, l_i^{t-m+1}, ..., l_i^t\}, \tag{1}$$

where $l_i = [x_1, y_1, x_2, y_2] \in \mathbb{R}^4$ is a 2D bounding box described by top-left ($[x_1, y_1]$) and bottom-right ($[x_2, y_2]$) coordinates of pedestrian. The dimension of L_{obs} is $m \times 4$, which means 4 coordinates. We also applied pose keypoints P_{obs} to understand the state of the pedestrian. The pose keypoints P_{obs} are given as:

$$P_{obs} = \{p_i^{t-m}, p_i^{t-m+1}, ..., p_i^t\}, \tag{2}$$

where p_i is a 36 dimension vector which consists of $2D$ coordinates of 18 joints detected by OpenPose [5,6,30,32]. So, the dimension of P_{obs} is $m \times 36$. The ego-vehicle speed S_{obs} is given as:

$$S_{obs} = \{s_i^{t-m}, s_i^{t-m+1}, ..., s_i^t\}, \tag{3}$$

where s_i is the speed of ego-vehicle, which is included in PIE dataset, but not JAAD dataset, and the dimension of S_{obs} is $m \times 1$.

We established RNN unit as Gated Recurrent Unit (GRU). While Rasouli et al. [24] used LSTM and Convolutional LSTM for predicting pedestrian intention crossing, we use GRU which is faster and efficient than LSTM. Each of the three NVM inputs passes through each RNN unit. First, bounding box L_{obs} passes through 1st RNN unit, and the output is h_l. This is concatenated with pose keypoints input P_{obs} and this output goes into 2nd RNN unit. h_p, which is the output of 2nd RNN, is concatenated with ego-vehicle speed S_{obs} as last NVM input. The output of concatenation passes through 3rd RNN unit and the final result of MSE (Multi-Stream Encoding) unit of NVM is E_{nvm}, which goes into attention unit.

Visual Module (VM). VM receives visual image features and passes through CNN unit. The VM accepts two different visual inputs: 1) local context T_{lc}, 2) global context T_{gc}. The local visual context T_{lc} is given as:

$$T_{lc} = \{t_{lc}^{t-m}, t_{lc}^{t-m+1}, ..., t_{lc}^t\}, \tag{4}$$

where local visual context is cropped the image to twice the size of the bounding box of pedestrian. We resized the image as 224×224.

The global context T_{gc} is given as:

$$T_{gc} = \{t_{gc}^{t-m}, t_{gc}^{t-m+1}, ..., t_{gc}^t\}, \tag{5}$$

where global context is developed from semantic segmentation. For semantic maps, we use DeepLabV3 model with pretraining on Cityscapes Dataset and resized the image to 224×224.

According to the overall flow of MSE on VM, each of the two visual module inputs passes through CNN unit, and the corresponding features pass through the RNN unit. We use VGG16 model pretrained on ImageNet to encode image inputs on visual module. For a detailed procedure of MSE_{VM}, after each visual input is resized to a dimension of $[224 \times 224]$, T_{lc} and T_{gc}, both of them are passed through the CNN unit. The output of CNN features go into RNN unit, and the last hidden states of each RNN are h_{lc} and h_{gc}. After concatenating and passing them through the last RNN, the output is the E_{nvm}, which is the input to the attention unit of visual module.

3.3 Implementation Unit

Attention Unit. α is the attention vector and m is the observation length. The weight of attention unit is defined as below:

$$\alpha = \frac{\exp(s(h_m, \tilde{h}_p))}{\sum_{p^t=1}^{T} \exp(s(h_m, \tilde{h}_{p^t}))} \tag{6}$$

where h_m is the end of hidden state of the encoder and h_{p^t} is the previous hidden state of the encoder. $score(h_m, h_{p^t})=h_m^T W_s h_p$ and W_s is a trainable weight matrix.

$$\beta_{attention} = \tanh(W_x[h_x : h_m]) \tag{7}$$

where W_x is a weight matrix, and h_m is the last hidden state. h_x is the concatenated sum of total attention vectors and weighted hidden states are specified as $h_x = \sum_{p^t} \alpha h_{s^t}$.

Integration Unit. Each output of non-visual module (A_{nvm}) and visual module (A_{vm}) is concatenated in integration unit. The output of concatenation is passed through last attention module and FC layer. The final output is the crossing intention prediction.

4   Experiments

4.1   Dataset

Table 1. Comparison of two public pedestrian datasets, PIE and JAAD.

	PIE [24]	JAAD [25]
# of video clips	48	346
Length of each video clips	10 min	5–10 s
FPS(Frames Per Second)	30	30
Total # of frames	909 K	82 K
Total # of annotated frames	293 K	82 K
# of pedestrians with behavior annotations	1.8 K	686
Total # of pedestrians	1,842	2,786
# of pedestrian bounding boxes	739 K	379 K
Intend to cross and cross	519	–
Intend to cross and don't cross	894	–
Do not intend to cross	429	–
# of crossing pedestrian	–	495
# of not-crossing pedestrian	–	191

We verify the strength of our proposed method for pedestrian crossing intention prediction on two public pedestrian behavior benchmark datasets, Pedestrian Intent Estimation (PIE) [12,13,21,24–28] and Joint Attention in Autonomous Driving (JAAD) [25,26]. The properties of both datasets are compared on Table 1.

Pedestrian Intention Estimation (PIE): PIE dataset [24,27] is filmed on sunny clear day in Toronto, Canada. It was captured during 6 h with 30fps HD format(1920 × 1080) and each video is about 10 min, with 6 sets in total. A total of 1,842 samples containing pedestrians are used, out of which 48% (880 samples) are used for training, 39% (719 samples) for testing and the remaining 13% (243 samples) for validation. 1842 pedestrians are grouped into 4 classes. 1) They intend to cross, and they actually cross the crosswalk (512 samples). 2) They intend to cross, but they do not cross the crosswalk (898 samples). 3) They do not intend to cross, but they actually cross the crosswalk (2 samples). 4) They do not intend to cross, and they actually do not cross the crosswalk (430 samples). When object is occluded 25%~75%, it is defined as partial occlusion. When object is occluded 75% or more, it is defined as full occlusion. OBD (On-Board Diagnostics) sensor is installed inside the ego-vehicle and it measured vehicle speed, heading direction and GPS coordinates.

Joint Attention for Autonomous Driving (JAAD): JAAD dataset [25,26] is filmed in various locations in North America and Eastern Europe. It was captured in different weather conditions and total of 2,786 pedestrians appear in the JAAD dataset. This dataset has 346 short video clips which are recorded at 30 FPS, so the total number of frames is 82,032. Attributes consists of gender, age, direction of motion and etc. $JAAD_{beh}$ includes video clips of pedestrians who actually cross (495 samples) or intend to cross (191 samples). $JAAD_{all}$ contains additional video clips of pedestrians (2100 samples) who do not cross the street. There are no vehicle information such as speed and GPS, as opposed to PIE dataset.

4.2 Experimental Settings

Implementation Details. The proposed novel algorithm was implemented using Tensorflow. We trained our network using 8 GPU server with Titan RTX 24GB and the CPU core is Intel Xeon ® gold 6248. Also, MCIP is trained during 400 epochs using a 258 batch size with L2 regularization of 0.001 and a dropout of 0.5 is added after attention module. For the decoder, we use gated recurrent units (GRU) with 256 hidden units for encoding non-visual module with sigmoid (σ) activation. MSE unit is trained with RMSprop optimizer with a learning rate of 5×10^{-5} on both PIE dataset and JAAD dataset.

Evaluation Metrics. We compared our method with others using the following three evaluation metrics. First, crossing intention accuracy (CIA) refers to how accurately the model prediced. The output of the MCIP is binary classification of crossing intention. Second, AUC(Area Under Curve) is the calculated value of the base area of the ROC(Receiver Operation Characteristic) curve. A higher AUC means that the model can discriminate the classes well. Finally, F1 score consists of a combination of precision and recall.

Ablations. We performed two ablation studies and one prediction time study. The first ablation study was conducted to compare the effect of different 5 inputs: local context, global context, pose keypoints, bounding box, and vehicle speed for MCIP. The second ablation was performed to contrast three different version of MCIP. We compared the performance of not only our proposed MCIP model, but also the MCIP-ATT model in which the order of the attention unit and integration unit is changed, and the MCIP-POSE model in which only the pose input is separated and passed through the attention unit. Finally, prediction time study is an experiment comparing the performance of two datasets, both PIE and JAAD, while changing the prediction time form 1 s to 4 s.

5   Results

5.1   Qualitative Results

We demonstrate the qualitative results of our proposed network in Fig. 3. The pedestrian with crossing intention is denoted as a blue bounding box and pedestrian with no crossing as a red bounding box. We showed the past frame{{t-3} frame to {t} frame} from t, which is the critical point. While the frame is changing, the prediction result of the pedestrian's crossing intention continues to change. Some pedestrians are predicted to continue crossing, or some pedestrians are predicted to be not-crossing to crossing or vice versa depending on the direction or situations of the pedestrian. So, we predicted the {t+1} frame of the pedestrian, which corresponds to the Fig. 3 on the far right.

5.2   Quantitative Results

Results Achieved Using Our Proposed Model. The quantitative comparison results of our proposed model are shown in Table 2. We tested on two public pedestrian datasets, PIE and JAAD. We compared the performance of each algorithm using three metrics, crossing intention accuracy (CIA), AUC, F1-score. This proposed method achieved best performance on all metrics. We also show separate results on the JAAD dataset, which was divided into $JAAD_{beh}$ and $JAAD_{all}$ which included not-crossing pedestrian. First, PIE_{int} [24] model is included combination of LSTM and Convolutional LSTM. This model accepts only two inputs, bounding box coordinates and visual context image, not ego-vehicle speed which is used for predicting trajectory prediction, in order to extract pedestrian crossing intention. So, when PIE_{int} model used total two inputs, the final accuracy of pedestrian intention estimation is 79%. Second, SF-GRU(Stacked with multilevel fusion GRU) [27] model used GRU networks including five sources {ego-vehicle speed, bounding box, pose, visual features, and their surrounding images}. They verified the effect of observation length and they achieved 84% accuracy of crossing intention when observation length was 0.5s. Third, ARN(Attentive Relation Network) [35] model applied not only bounding box and visual images but also six different traffic inputs such as features of traffic neighbor, traffic light, traffic

t-3   t-2   t-1   t   t+1

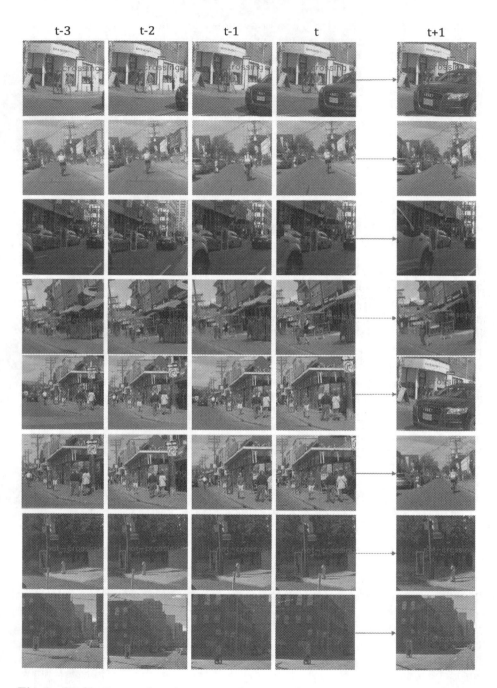

Fig. 3. Qualitative results of our network. The pedestrian with crossing intention is denoted as blue bounding box and not-crossing intention pedestrian as red bounding box. We demonstrated the past frame $t - 3 \sim t$ and prediction frame $t + 1$. (Color figure online)

sign, crosswalk, station, and ego motion. When they tested with full model including ARN networks, they achieved 84% accuracy using "event-to-crossing" setting. Fourth, PCPA(Pedestrian Crossing Prediction with Attention) [13] model uses four inputs: bounding box, pose, speed, and local context. This model uses GRU with 256 hidden units, and its performance in terms of accuracy is 87%. Finally, our model (MCIP) is verified using two modules separately; Firstly, using only NVM and secondly, using both NVM and VM. When we tested with only NVM, the accuracy of pedestrian crossing intention is 85%. However, when we added VM, the accuracy increased to 89% which is the best performance compared to other models. AUC and F1 score were also the highest in comparison with other approaches on the PIE dataset. In case of $JAAD_{beh}$ and $JAAD_{all}$ datasets, our model was also the best performing one. The performance was better when the VM was included than when it was not included, showing a similar trend as on the PIE dataset.

Table 2. Model Performance comparison between PIE and JAAD datasets. Our proposed model achieved the state-of-the-art performance as the crossing intention accuracy (CIA) on all datasets. PIE_{int} model used only two inputs(bounding box and local context). SF-GRU (Stacked with multilevel fusion GRU) model used five sources including two CNN features(pedestrian appearance and surrounding context). ARN (Attentive Relation Network) model added six different traffic inputs. PCPA (Pedestrian Crossing Prediction with Attention) took four inputs(bounding box, pose, vehicle speed, local context). We verified our model in two ways depending on whether or not VM are included. The crossing intention accuracy of the model including the VM was higher for all datasets than the accuracy when the VM was not included.

	PIE			$JAAD_{beh}$			$JAAD_{all}$		
Approach	ACC↑	AUC↑	F1↑	ACC↑	AUC↑	F1↑	ACC↑	AUC↑	F1↑
PIE_{int} [24]	0.79	0.73	0.87	–	–	–	–	–	–
SF-GRU [27]	0.84	0.83	0.72	–	–	–	–	–	–
ARN [35]	0.84	0.88	0.90	–	–	–	0.87	**0.92**	**0.70**
PCPA [13]	0.87	0.86	0.77	0.58	0.50	0.71	0.85	0.86	0.68
MCIP NVM	0.85	0.86	0.76	0.61	0.53	0.72	0.84	0.82	0.62
NVM+VM	**0.89**	**0.87**	**0.81**	**0.64**	**0.55**	**0.78**	**0.88**	0.84	0.66

Ablation Studies I. The results of ablation studies are shown in Table 3. We performed ablation studies while omitting one input from the five input features from non-visual module (NVM) and visual module (VM). The five inputs are: {Bounding Box, Pose, Vehicle Speed, Local Context, Global Context}. We experimented ablation studies while subtracting each one input among five inputs. The input that had significant effect on accuracy is vehicle speed. When we did not include the vehicle speed input, the performance was decreased by 12% compared to result when every inputs were considered. Among the five inputs, the input that had the least effect on the overall results was the local context. When we excluded the local context input, the accuracy of crossing intention decreased only by 2%. Especially, AUC scores highest when all inputs are included as well as when global context is excluded.

In summary, the order of inputs that have the most influence on the crossing intention accuracy is as follows: ego-vehicle speed, pose keypoints, bounding box, global context, and local context. The vehicle speed information is the most important factor for achieving higher performance because the JAAD dataset does not involve vehicle annotation, and the average performance of on the JAAD dataset is lower than on the PIE dataset.

Table 3. Ablation studies with 5 inputs for MCIP. When the vehicle speed is only excluded, the accuracy of pedestrian crossing intention is greatly decreased to a value of 79%. On the other hand, when we subtract the input of local context, the crossing intention accuracy is the least decreased by only 1%. The order of inputs that affect the accuracy in decreasing magnitude is vehicle speed, pose keypoints, bounding box, global context, and local context. When we use all five inputs, the other metrics scores (AUC and F1 score) are also the highest. But, it does not affect the AUC value when we excluded the global context input.

Bounding box	Pose keypoints	Vehicle speed	Local context	Global context	ACC↑	AUC ↑	F1↑
✓	✓	✓	✓	✓	**0.89**	**0.87**	**0.81**
-	✓	✓	✓	✓	0.87	0.86	0.78
					(−0.02)	(−0.01)	(−0.03)
✓	−	✓	✓	✓	0.86	0.81	0.74
					(−0.03)	(−0.06)	(−0.07)
✓	✓	−	✓	✓	0.79	0.76	0.65
					(−0.10)	(−0.11)	(−0.16)
✓	✓	✓	−	✓	0.88	0.84	0.77
					(−0.01)	(−0.03)	(−0.04)
✓	✓	✓	✓	−	0.87	**0.87**	0.79
					(−0.02)	(−0.00)	(−0.02)

Ablation Studies II. The second group of ablation studies consists of the results obtained from three versions of MCIP models. We compared the results using three metrics (ACC, AUC, F1) on each MCIP model as shown in Table 4. Compared to our proposed MCIP model, the model of MCIP-ATT reverses the order of attention unit and integration unit, but MSE unit is same. That is, in the MCIP model, two outputs that pass through RNN unit in NVM and VM are combined after passing through the attention unit, whereas in the MCIP-ATT model, two outputs that passed RNN unit in NVM and VM are first combined and then the merged output passes through a single RNN unit. The output of RNN unit passes through the attention unit last. The performances of MCIP-ATT are overall lower than those of the proposed MCIP model on PIE dataset. The accuracy of pedestrian crossing intention is 83%, which is 6% lower than the proposed MCIP model. AUC was 83%(4% lower) and F1-score was 73%(8% lower).

In case of MCIP-POSE model, the most distinctive feature is that it separates the pose keypoints input and passes it through the RNN unit and the attention unit alone. So, two inputs of bounding box and ego-vehicle speed are concatenated to each other and VM is the same as MCIP. A total of three RNN outputs pass through each of the three attention units (in contrast to 2 attention units in the proposed MCIP model). Next, three attention unit outputs are finally concatenated. The performance of MCIP-POSE was higher than that of MCIP-ATT, but lower than that of proposed MCIP model on PIE dataset. Pedestrian crossing intention accuracy was 86%, which is 3% lower than the proposed MCIP. AUC was 85% (2% lower) and F1-score was 77% (4% lower).

Table 4. Ablation studies for three MCIP models. First method, MCIP, is our proposed model. The second model, MCIP-ATT, changes the order of attention unit and integration unit. The order of MCIP-ATT: MSE unit-Integration unit-Attention unit ↔ The order of proposed MCIP: MSE unit-Attention unit-Integration unit. The last model, MCIP-POSE, is different in the sense that only the pose keypoints input is separated and passed through the RNN unit and the attention unit. So, total three outputs of each three attention unit are concatenated to each other. The performance was the highest for all three metrics in the proposed MCIP model, followed by MCIP-POSE and MCIP-ATT.

Method	Model architecture	ACC↑	AUC↑	F1↑
MCIP	MSE → ATT → Integration unit	**0.89**	**0.87**	**0.81**
MCIP-ATT	MSE → Integration unit → ATT	0.83(−0.06)	0.83(−0.04)	0.73(−0.08)
MCIP-POSE	P_{obs} → ATT → Integration unit	0.86(−0.03)	0.85(−0.02)	0.77(−0.04)
	MSE($\notin P_{obs}$) → ATT → Integration unit			

Prediction Time. The crossing intention prediction results of variable prediction time are shown in Table 5. When we predicted the pedestrian intention of crossing, we observed 15 frames before critical point which starts observing pedestrian. We experimented with 4 prediction frames (30 frames to 120 frames). The highest accuracy was obtained when the pedestrian's crossing intention was predicted after 1 s which is the best performance for all datasets as well as AUC and F1 score. When our model predicts the intention after 2 s, the accuracy was decreased by only 1.13%, which follows the same trend as after 1 s on PIE dataset. However, when the model predicted the pedestrian intention after 3 s, the accuracy was decreased by 10% compared to after 1 s. When the number of prediction frames is 120, the intention prediction rate was dramatically decreased to 64%. The other two metrics were also decreased when increasing the number of prediction frames.

In JAAD dataset, the flow of intention accuracy is similar as PIE dataset. Overall, three metric results of $JAAD_{beh}$ was lower than those of $JAAD_{all}$. If we set the prediction frame to 30, $JAAD_{beh}$ accuracy is 64%, and the accuracy of $JAAD_{all}$ dataset is 88%. It can be seen that as the number of prediction frames increases, that is, the more distant future was predicted, the crossing intention accuracy was decreased gradually on both JAAD datasets. For the remaining two metrics (AUC and F1 score), the values decreased as the prediction time increases.

Table 5. The prediction results of our model for longer period of time. For all datasets, as the prediction time increases, the prediction accuracy continues to decrease. For PIE dataset, up to $2s$ our model successfully predicts the intention with relatively small amount of performance degradation. However, in case of $3s$ and $4s$, the accuracy decreased gradually. In the JAAD dataset, the decrease in accuracy with prediction time was similar.

Prediction frame	t	PIE			$JAAD_{beh}$			$JAAD_{all}$		
		ACC↑	AUC↑	F1↑	ACC↑	AUC↑	F1↑	ACC↑	AUC↑	F1↑
30	1	**0.89**	**0.87**	**0.81**	**0.64**	**0.55**	**0.78**	**0.88**	**0.84**	**0.66**
60	2	0.88	0.84	0.80	0.59	0.45	0.74	0.64	0.59	0.35
90	3	0.80	0.79	0.68	0.45	0.43	0.55	0.56	0.51	0.29
120	4	0.64	0.65	0.52	0.38	0.50	0.36	0.44	0.52	0.27

6 Conclusions

This paper proposed a novel algorithm, named Multi-Stream Network for Pedestrian Crossing Intention Prediction(MCIP) based on two public pedestrian datasets, PIE and JAAD. We used the driving scene semantic segmentation and results of pedestrian detection, so we predicted the pedestrian's future crossing intention. We showed our proposed model achieved the state-of-the-art performance compared to other networks through quantitative and qualitative experiments. The key idea of our proposed model is an effective multi-stream network for integrating of several inputs. Our multi-stream network consists of both visual modules and non-visual modules. The network is composed of three units, MSE(Multi-Stream Encoding) unit, attention unit and integration unit. We also studied the effect of five different inputs{bounding box, pose keypoints, ego-vehicle speed, local context image, global context segmentation map}, model types{MCIP, MCIP-ATT, MCIP-POSE}, and prediction times. We predicted the pedestrian crossing intention up to 4 s. We anticipate that our algorithm can be used to prevent traffic road accidents by predicting the trajectory of nearby pedestrians.

Acknowledgement. This work was supported by the Institute of Information and Communications Technology Planning and Evaluation (IITP) Grant through the Ministry of Science and ICT (MSIT), Government of Korea (Development of Previsional Intelligence Based on Long-Term Visual Memory Network) under Grant 2020-0-00004.

References

1. Bhattacharyya, A., Reino, D.O., Fritz, M., Schiele, B.: Euro-PVI: pedestrian vehicle interactions in dense urban centers. In: Proceedings of the IEEE/CVF Conference on Computer Vision and Pattern Recognition (CVPR) (2021)
2. Bouhsain, S.A., Saadatnejad, S., Alahi, A.: Pedestrian intention prediction: a multitask perspective. ArXiv preprint arXiv:2010.10270 (2020)
3. Braun, M., Krebs, S., Flohr, F., Gavrila, D.M.: Eurocity persons: a novel benchmark for person detection in traffic scenes. In: IEEE Transactions on Pattern Analysis and Machine Intelligence (TPAMI) (2019)
4. Cai, Z., Vasconcelos, N.: Cascade R-CNN: high quality object detection and instance segmentation. In: IEEE Transactions on Pattern Analysis and Machine Intelligence (TPAMI) (2019)
5. Cao, Z., Hidalgo Martinez, G., Simon, T., Wei, S., Sheikh, Y.A.: OpenPose: realtime multi-person 2D pose estimation using part affinity fields. In: IEEE Transactions on Pattern Analysis and Machine Intelligence (TPAMI) (2019)
6. Cao, Z., Simon, T., Wei, S.E., Sheikh, Y.: Realtime multi-person 2D pose estimation using part affinity fields. In: Proceedings of the IEEE/CVF Conference on Computer Vision and Pattern Recognition (CVPR) (2017)
7. Dendorfer, P., Elflein, S., Leal-Taixé, L.: MG-GAN: a multi-generator model preventing out-of-distribution samples in pedestrian trajectory prediction. In: Proceedings of the IEEE/CVF International Conference on Computer Vision (ICCV) (2021)
8. Dollar, P., Wojek, C., Schiele, B., Perona, P.: Pedestrian detection: an evaluation of the state of the art. In: IEEE Transactions on Pattern Analysis and Machine Intelligence (TPAMI) (2011)
9. Giuliari, F., Hasan, I., Cristani, M., Galasso, F.: Transformer networks for trajectory forecasting. In: 2020 25th International Conference on Pattern Recognition (ICPR) (2021)
10. Hasan, I., Liao, S., Li, J., Akram, S.U., Shao, L.: Generalizable pedestrian detection: the elephant in the room. In: Proceedings of the IEEE/CVF Conference on Computer Vision and Pattern Recognition (CVPR) (2021)
11. Khan, A.H., Munir, M., van Elst, L., Dengel, A.: F2DNet: fast focal detection network for pedestrian detection. ArXiv preprint arXiv:2203.02331 (2022)
12. Kim, K., Lee, Y.K., Ahn, H., Hahn, S., Oh, S.: Pedestrian intention prediction for autonomous driving using a multiple stakeholder perspective model. In: 2020 IEEE/RSJ International Conference on Intelligent Robots and Systems (IROS) (2020)
13. Kotseruba, I., Rasouli, A., Tsotsos, J.K.: Benchmark for evaluating pedestrian action prediction. In: Proceedings of the IEEE/CVF Winter Conference on Applications of Computer Vision (WACV) (2021)
14. Lin, Z., Pei, W., Chen, F., Zhang, D., Lu, G.: Pedestrian detection by exemplar-guided contrastive learning. ArXiv preprint arXiv:2111.08974 (2021)
15. Liu, B., et al.: Spatiotemporal relationship reasoning for pedestrian intent prediction. IEEE Robot. Autom. Lett. (RA-L) **PP**(99), 1 (2020)
16. Lorenzo, J., et al.: CAPformer: pedestrian crossing action prediction using transformer. Sensors **21**(17), 5694 (2021)
17. Lorenzo, J., Parra, I., Sotelo, M.: IntFormer: predicting pedestrian intention with the aid of the transformer architecture. ArXiv preprint arXiv:2105.08647 (2021)

18. Lorenzo, J., Parra, I., Wirth, F., Stiller, C., Llorca, D.F., Sotelo, M.A.: RNN-based pedestrian crossing prediction using activity and pose-related features. In: IEEE Intelligent Vehicles Symposium (IV) (2020)
19. Lv, Z., Huang, X., Cao, W.: An improved GAN with transformers for pedestrian trajectory prediction models. Int. J. Intell. Syst. **36**(12), 6989–7962 (2021)
20. Malla, S., Dariush, B., Choi, C.: Titan: future forecast using action priors. In: Proceedings of the IEEE/CVF Conference on Computer Vision and Pattern Recognition (CVPR) (2020)
21. Neumann, L., Vedaldi, A.: Pedestrian and ego-vehicle trajectory prediction from monocular camera. In: Proceedings of the IEEE/CVF Conference on Computer Vision and Pattern Recognition (CVPR) (2021)
22. Postnikov, A., Gamayunov, A., Ferrer, G.: Transformer based trajectory prediction. ArXiv preprint arXiv:2112.04350 (2021)
23. Qingyun, F., Dapeng, H., Zhaokui, W.: Cross-modality fusion transformer for multispectral object detection. ArXiv preprint arXiv:2111.00273 (2021)
24. Rasouli, A., Kotseruba, I., Kunic, T., Tsotsos, J.K.: Pie: a large-scale dataset and models for pedestrian intention estimation and trajectory prediction. In: Proceedings of the IEEE/CVF International Conference on Computer Vision (ICCV) (2019)
25. Rasouli, A., Kotseruba, I., Tsotsos, J.K.: Agreeing to cross: how drivers and pedestrians communicate. In: IEEE Intelligent Vehicles Symposium (IV) (2017)
26. Rasouli, A., Kotseruba, I., Tsotsos, J.K.: Are they going to cross? a benchmark dataset and baseline for pedestrian crosswalk behavior. In: Proceedings of the IEEE/CVF International Conference on Computer Vision Workshops (ICCVW) (2017)
27. Rasouli, A., Kotseruba, I., Tsotsos, J.K.: Pedestrian action anticipation using contextual feature fusion in stacked RNNs. In: Proceedings of The British Machine Vision Conference (BMVC) (2019)
28. Razali, H., Mordan, T., Alahi, A.: Pedestrian intention prediction: a convolutional bottom-up multi-task approach. Transport. Res. Part C: Emerg. Technol. **130**, 103259 (2021)
29. Shi, L., et al.: SGCN: sparse graph convolution network for pedestrian trajectory prediction. In: Proceedings of the IEEE/CVF Conference on Computer Vision and Pattern Recognition (CVPR) (2021)
30. Simon, T., Joo, H., Matthews, I., Sheikh, Y.: Hand keypoint detection in single images using multiview bootstrapping. In: Proceedings of the IEEE/CVF Conference on Computer Vision and Pattern Recognition (CVPR) (2017)
31. Sui, Z., Zhou, Y., Zhao, X., Chen, A., Ni, Y.: Joint intention and trajectory prediction based on transformer. In: 2021 IEEE/RSJ International Conference on Intelligent Robots and Systems (IROS) (2021)
32. Wei, S.E., Ramakrishna, V., Kanade, T., Sheikh, Y.: Convolutional pose machines. In: Proceedings of the IEEE/CVF Conference on Computer Vision and Pattern Recognition (CVPR) (2016)
33. Yang, D., Zhang, H., Yurtsever, E., Redmill, K., Ozguner, U.: Predicting pedestrian crossing intention with feature fusion and spatio-temporal attention. IEEE Transactions on Intelligent Vehicles (T-IV) (2022)
34. Yao, H.Y., Wan, W.G., Li, X.: End-to-end pedestrian trajectory forecasting with transformer network. ISPRS Int. J. Geo-Inf. **11**(1), 44 (2022)
35. Yao, Y., Atkins, E., Johnson-Roberson, M., Vasudevan, R., Du, X.: Coupling intent and action for pedestrian crossing behavior prediction. In: Proceedings of 30th International Joint Conference on Artificial Intelligence (IJCAI) (2021)

36. Yin, Z., Liu, R., Xiong, Z., Yuan, Z.: Multimodal transformer networks for pedestrian trajectory prediction. In: Proceedings of 30th International Joint Conference on Artificial Intelligence (IJCAI) (2021)
37. Zhang, S., Benenson, R., Schiele, B.: Citypersons: a diverse dataset for pedestrian detection. In: Proceedings of the IEEE/CVF Conference on Computer Vision and Pattern Recognition (CVPR) (2017)

SimpleTrack: Understanding and Rethinking 3D Multi-object Tracking

Ziqi Pang[1(✉)], Zhichao Li[2], and Naiyan Wang[2]

[1] UIUC, Urbana-Champaign, USA
ziqip2@illinois.edu
[2] TuSimple, California, USA

Abstract. 3D multi-object tracking (MOT) has witnessed numerous novel benchmarks and approaches in recent years, especially those under the "tracking-by-detection" paradigm. Despite their progress and usefulness, an in-depth analysis of their strengths and weaknesses is not yet available. In this paper, we summarize current 3D MOT methods into a unified framework by decomposing them into four constituent parts: pre-processing of detection, association, motion model, and life cycle management. We then ascribe the failure cases of existing algorithms to each component and investigate them in detail. Based on the analyses, we propose corresponding improvements which lead to a strong yet simple baseline: SimpleTrack. Comprehensive experimental results on Waymo Open Dataset and nuScenes demonstrate that our final method could achieve new state-of-the-art results with minor modifications. Furthermore, we take additional steps and rethink whether current benchmarks authentically reflect the ability of algorithms for real-world challenges. We delve into the details of existing benchmarks and find some intriguing facts. Finally, we analyze the distribution and causes of remaining failures in SimpleTrack and propose future directions for 3D MOT. Our code is at https://github.com/tusen-ai/SimpleTrack.

Keywords: 3D multi-object tracking · Autonomous driving

1 Introduction

Multi-object tracking (MOT) is a composite task in computer vision, combining both the aspects of localization and identification. Given its complex nature, MOT systems generally involve numerous interconnected parts, such as the selection of detections, the data association, the modeling of object motions, etc. Each of these modules has its special treatment and can significantly affect the system performance as a whole. Therefore, we would like to ask *which components in 3D MOT play the most important roles, and how can we improve them?*

Z. Pang—This work is complete during the first author's internship at TuSimple.

Supplementary Information The online version contains supplementary material available at https://doi.org/10.1007/978-3-031-25056-9_43.

Bearing such objectives, we revisit the current 3D MOT algorithms [3,10,13,14, 30,32,40,46,47]. These methods mostly adopt the "tracking by detection" paradigm, where they directly take the bounding boxes from 3D detectors and build up tracklets across frames. We first break them down into four individual modules and examine each of them: pre-processing of input detections, motion model, association, and life cycle management. Based on this modular framework, we locate and ascribe the failure cases of 3D MOT to the corresponding components and discover several overlooked issues in the previous designs.

First, we find that inaccurate input detections may contaminate the association. However, purely pruning them by a score threshold will sacrifice the recall. Second, we find that the similarity metric defined between two 3D bounding boxes need to be carefully designed. Neither distance-based nor simple IoU works well. Third, the object motion in 3D space is more predictable than that in the 2D image space. Therefore, the consensus between motion model predictions and even poor observations (low score detections) could well indicate the existence of objects. Illuminated by these observations, we propose several simple yet non-trivial solutions. The evaluation on Waymo Open Dataset [37] and nuScenes [8] suggests that our final method "SimpleTrack" is competitive among the 3D MOT algorithms (in Table 4 and Table 5).

Besides analyzing 3D MOT algorithms, we also reflect on current benchmarks. We emphasize the need for high-frequency detections and the proper handling of output tracklets in evaluation. To better understand the upper bound of our method, we further break down the remaining errors based on ID switch and MOTA metrics. We believe these observations could inspire the better design of algorithms and benchmarks.

In brief, our contributions are as follow:

- We analyze each component in 3D MOT methods and their failure cases, based on a decomposition of "tracking-by-detection" 3D MOT framework.
- We propose corresponding treatments for each module and combine them into a simple baseline. The results are competitive on the Waymo Open Dataset and nuScenes.
- We also analyze existing 3D MOT benchmarks and explain the potential influences of their designs. We hope that our analyses could shed light for future research.

2   Related Work

Most 3D MOT methods [3,10,13,14,30,32,40,46,47] adopt the "tracking-by-detection" framework because of the strong power of detectors. We first summarize the representative 3D MOT work and then highlight the connections and distinctions between 3D and 2D MOT.

2.1   3D MOT

Many 3D MOT methods are composed of rule-based components. AB3DMOT [40] is the common baseline of using IoU for association and a Kalman filter as the motion model. Its notable followers mainly improve on the association part: Chiu *et al.* [10] and CenterPoint [46] replace IoU with Mahalanobis and L2 distance, which performs better on nuScenes [8]. Some others notice the importance of life cycle management, where

CBMOT [3] proposes a score-based method to replace the "count-based" mechanism, and Pöschmann *et al.* [30] treats 3D MOT as optimization problems on factor graphs. Despite the effectiveness of these improvements, a systematic study on 3D MOT methods is in great need, especially where these designs suffer and how to make further improvements. To this end, our paper seeks to meet the expectations.

Different from the methods mentioned above, many others attempt to solve 3D MOT with fewer manual designs. [2,9,18,41] leverage rich features from RGB images for association and life cycle control, [32] manages to fuse the object proposals from different modalities, and Chiu *et al.* [9] specially uses neural networks to handle the feature fusion, association metrics, and tracklet initialization. Recently, OGR3MOT [47] follows Guillem *et al.* [7] and solves 3D MOT with graph neural networks (GNN) in an end-to-end manner, focusing on the data association and the classification of active tracklets, especially. SDVTracker [13] systematically investigates 3D MOT, and proposes to extract descriptors of agents for association and update the tracks in an interaction-aware manner. Compared to SDVTracker, which is innovative analysis on 3D MOT, SimpleTrack focuses on the priors of 3D MOT and rule-based 3D MOT systems, and our analyses are based on public datasets Waymo Open Dataset and nuScenes.

2.2   2D MOT

2D MOT shares the common goal of data association with 3D MOT. Some notable attempts include probabilistic approaches [1,19,33,35], dynamic programming [12], bipartite matching [6], min-cost flow [4,49], convex optimization [29,38,39,48], and conditional random fields [45]. With the rapid progress of deep learning, many methods [7,15–17,22,43] learn the matching mechanisms and others [20,23,24,26,28] learn the association metrics.

Similar to 3D MOT, many 2D trackers [5,25,36,51] also benefit from the enhanced detection quality and adopt the "tracking-by-detection" paradigm. However, the objects on RGB images have varied sizes because of scale variation; thus, they are harder for association and motion models. But 2D MOT can easily take advantage of rich RGB information and use appearance models [21,22,36,42], which is not available in LiDAR based 3D MOT. In summary, the design of MOT methods should fit the traits of each modality.

3   3D MOT Pipeline

In this section, we decompose 3D MOT methods into the following four parts. An illustration is in Fig. 1.

Pre-processing of Input Detections. It pre-processes the bounding boxes from detectors and selects the ones to be used for tracking. Some exemplar operations include selecting the bounding boxes with scores higher than a certain threshold. (In "Pre-processing" Fig. 1, some redundant bounding boxes are removed.)

Motion Model. It predicts and updates the states of objects. Most 3D MOT methods [3,10,40] directly use the Kalman filter, and CenterPoint [46] uses the velocities predicted by detectors from multi-frame data. (In "Prediction" and "Motion Model Update" Fig. 1.)

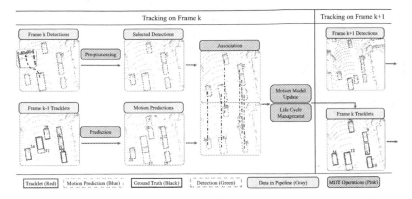

Fig. 1. 3D MOT pipeline. For simplicity, we only visualize the steps between frame k and frame k+1. Best view in color.

Association. It associates the detections with tracklets. The association module involves two steps: similarity computation and matching. The similarity measures the distance between a pair of detection and tracklet, while the matching step solves the correspondences based on the pre-computed similarities. AB3DMOT [40] proposes the baseline of using IoU with Hungarian algorithm, while Chiu *et al.* [10] uses Mahalanobis distance and greedy algorithm, and CenterPoint [46] adopts the L2 distance. (In "Association" Fig. 1.)

Life Cycle Management. It controls the "birth", "death" and "output" policies. "Birth" determines whether a detection bounding box will be initialized as a new tracklet; "Death" removes a tracklet when it is believed to have moved out of the attention area; "Output" decides whether a tracklet will output its state. Most of the MOT algorithm adopts a simple count-based rule [10,40,46], and CBMOT [3] improves birth and death by amending the logic of tracklet confidences. (In "Life Cycle Management" Fig. 1.)

4 Analyzing and Improving 3D MOT

In this section, we analyze and improve each module in the 3D MOT pipeline. For better clarification, we ablate the effects of every modification by removing it from the final variant of SimpleTrack. By default, the ablations are all on the validation split with the CenterPoint [46] detection. We also provide additive ablation analyses and the comparison with other methods in Sect. 4.5.

4.1 Pre-processing

To fulfill the recall requirements for detection AP, current detectors usually output a large number of bounding boxes with scores roughly indicating their quality. However, if these boxes are treated equally in the association step of 3D MOT, the bounding boxes with low quality or severe overlapping, which is not fully addressed by a loose NMS in 3D detectors, may deviate the trackers to select the inaccurate detection for extending or

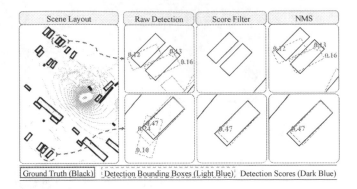

Fig. 2. Comparison between score filtering and NMS. To remove the redundant bounding boxes on row 2, score filtering needs at least a 0.24 threshold, but this will eliminate the detections on row 1. However, NMS can well satisfy both by removing the overlapping on row 2 and maintaining the recall on row 1.

Table 1. Left: ablation for NMS on nuScenes. Right: ablation for NMS on WOD.

NMS	AMOTA↑	AMOTP↓	MOTA↑	IDS ↓
✗	0.673	0.574	0.581	557
✓	**0.687**	**0.573**	**0.592**	**519**

NMS	Vehicle			Pedestrian		
	MOTA↑	MOTP↓	IDS(%)↓	MOTA↑	MOTP↓	IDS(%)↓
✗	0.5609	**0.1681**	0.09	0.4962	**0.3090**	5.00
✓	**0.5612**	**0.1681**	**0.08**	**0.5776**	0.3125	**0.42**

forming tracklets (as in the "raw detection" of Fig. 2). Such a gap between the detection and MOT task needs careful treatment.

3D MOT methods commonly use confidence scores to filter out the low-quality detections and improve the precision of input bounding boxes. However, such an approach may be detrimental to the recall as they directly abandon the objects with poor observations (top row in Fig. 2). It is also especially harmful to metrics like AMOTA, which needs the tracker to use low score bounding boxes to fulfill the recall requirements.

To improve the precision without significantly decreasing the recall, our solution is simple and direct: we apply *stricter* non-maximum suppression (NMS) to the input detections. Please note that 3D detectors [46] already applies NMS prior to 3D MOT, however, we emphasize that the IoU threshold for NMS should be higher in 3D MOT, compared to detection. As shown in the right of Fig. 2, the NMS operation alone can effectively eliminate the overlapped low-quality bounding boxes while keeping the diverse low-quality observations, even on regions like sparse points or occlusion. *Therefore, by adding NMS to the pre-processing module, we could roughly keep the recall, but greatly improves the precision and benefits MOT.*

On WOD, our stricter NMS operation removes 51% and 52% bounding boxes for vehicles and pedestrians and nearly doubles the precision: 10.8% to 21.1% for vehicles, 5.1% to 9.9% for pedestrians. At the same time, the recall drops relatively little from 78% to 74% for vehicles and 83% to 79% for pedestrians. According to Table 1, this largely benefits the performance, especially on the pedestrian (right part of Table 1), where the object detection task is harder.

Table 2. Left: comparison of motion models on Waymo Open Dataset. "KF" denotes Kalman filters; "CV" denotes constant velocity model; "KF-PD" denotes the variant using Kalman filter only for motion prediction. Right: comparison of motion models on nuScenes. Details in Sect. 4.2.

Method	Vehicle			Pedestrian		
	MOTA↑	MOTP↓	IDS(%)↓	MOTA↑	MOTP↓	IDS(%)↓
KF	**0.5612**	**0.1681**	**0.08**	**0.5776**	**0.3125**	**0.42**
CV	0.5515	0.1691	0.14	0.5661	0.3159	0.58
KF PD	0.5516	0.1691	0.14	0.5654	0.3158	0.63

Method	AMOTA↑	AMOTP↓	MOTA↑	IDS↓
KF	0.687	0.573	**0.592**	519
CV	**0.690**	**0.564**	**0.592**	516

4.2 Motion Model

Motion models depict the motion status of tracklets. They are mainly used to predict the candidate states of objects in the next frame, which are the proposals for the following association step. Furthermore, the motion models like the Kalman filter can also potentially refine the states of objects. In general, there are two commonly adopted motion models for 3D MOT: Kalman filter (KF), *e.g.* AB3DMOT [40], and constant velocity model (CV) with predicted speeds from detectors, *e.g.* CenterPoint [46]. The advantage of KF is that it could utilize the information from multiple frames and provide smoother results when facing low-quality detection. Meanwhile, CV deals better with abrupt and unpredictable motions with its explicit speed predictions, but its effectiveness on motion smoothing is limited. In Table 2, we compare the two of them on WOD and nuScenes, which provides clear evidence for our claims.

In general, these two motion models demonstrate similar performance. On nuScenes, CV marginally outperforms KF, while it is the opposite on WOD. The advantages of KF on WOD mainly come from the refinement for the bounding boxes. To verify this, we implement the "KF-PD" variant, which uses KF only for providing motion predictions prior to association, and the outputs are all original detections. Eventually, the marginal gap between "CV" and "KF-PD" in Table 2 left supports our claim. On nuScenes, the CV motion model is slightly better due to the lower frame rates on nuScenes (2 Hz). To prove our conjecture, we apply KF and CV both under a higher 10 Hz setting on nuScenes[1], and KF marginally outperforms CV by 0.696 versus 0.693 in AMOTA.

To summarize, *the Kalman Filter fits better for high-frequency cases because of more predictable motions, and the constant velocity model is more robust for low-frequency scenarios with explicit speed prediction.* Since inferring velocities is not yet common for detectors, we adopt the Kalman filter for without loss of generality.

4.3 Association

Association Metrics: 3D GIoU IoU based [40] and distance based [10, 46] association metrics are the two prevalent choices in 3D MOT. As in Fig. 3, they have typical but different failure modes. IoU computes the overlapping ratios between bounding boxes, so it cannot connect the detections and motion predictions if the IoU between them are all zeros, which are common at the beginnings of tracklets or on objects with abrupt

---

[1] Please check Sect. 5.1 for how we 10 Hz settings on nuScenes.

motions (the left of Fig. 3). The representatives for distance-based metrics are Mahalanobis [10] and L2 [46] distances. With larger distance thresholds, they can handle the failure cases of IoU based metrics, but they may not be sensitive enough for nearby detection with low quality. We explain such scenarios on the right of Fig. 3. On frame k, the blue motion prediction has smaller L2 distances to the green false positive detection, thus it is wrongly associated. Illuminating by such example, we conclude that the distance-based metrics lack discrimination on orientations, which is just the advantage of IOU based metrics.

To get the best of two worlds, we propose to generalize "Generalized IoU" (GIoU) [34] to 3D for association. Briefly speaking, for any pair of 3D bounding boxes B_1, B_2, their 3D GIoU is as Eq. 1, where I, U are the intersection and union of B_1 and B_2. Furthermore, the convex hull of B_1, B_2 is denoted by C (short for *convex hull*). V represents the volume of a polygon. We set GIoU > −0.5 as the threshold for every category of objects on both WOD and nuScenes for this pair of associations to enter the subsequent matching step.

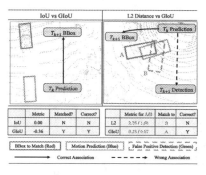

Fig. 3. Illustration of association metrics. IoU (left)/L2 Distance (right) versus GIoU. Details are in Sect. 4.3. (Color figure online)

$$V_U = V_{B_1} + V_{B_2} - V_I,$$
$$\mathbf{GIoU}(B_1, B_2) = V_I/V_U - (V_C - V_U)/V_C. \tag{1}$$

As in Fig. 3, the GIoU metric can handle both patterns of failures. The quantitative results in Fig. 4 also show the ability of GIoU for improving the association on both WOD and nuScenes.

Matching Strategies. Generally speaking, there are two approaches for the matching between detections and tracklets: 1) Formulating the problem as a bipartite matching

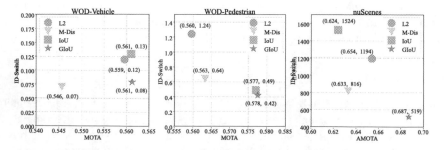

Fig. 4. Comparison of association metrics on WOD (left & middle) and nuScenes (right). "M-Dis" is the short for Mahalanobis distance. The best method is closest to the bottom-right corner, having the lowest ID-Switches and highest MOTA/AMOTA. IoU and GIoU use Hungarian algorithm for matching, while L2/M-Dis use greedy algorithm (explained in Sect. 4.3).

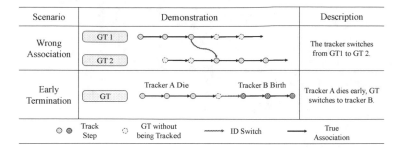

Fig. 6. Illustration for two major types of ID-Switches.

problem, and then solving it using Hungarian algorithm [40]. 2) Iteratively associating the nearest pairs by greedy algorithm [10,46].

We find that these two methods heavily couples with the association metrics: IoU based metrics are fine with both, while distance-based metrics prefer greedy algorithms. We hypothesize that the reason is that the range of distance-based metrics are large, thus methods optimizing global optimal solution, like the Hungarian algorithm, may be adversely affected by out-

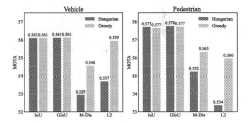

Fig. 5. Comparison of matching strategies on WOD.

liers. In Fig. 5, we experiment with all the combinations between matching strategies and association metrics on WOD. As demonstrated, IoU and GIoU function well for both strategies, while Mahalanobis and L2 distance demand greedy algorithm, which is also consistent with the conclusions from previous work [10].

4.4 Life Cycle Management

We analyze all the ID-Switches on WOD[2], and categorize them into two groups as in Fig. 6: wrong association and early termination. Different from the major focus of many work, which is association, we find that the early termination is actually the dominating cause of ID-Switches: 95% for vehicle and 91% for pedestrian. Among the early terminations, many of them are caused by point cloud sparsity and spatial occlusion. To alleviate this issue, we utilize the free yet effective information: consensus between motion models and detections with low scores. *These bounding boxes are usually of low localization quality, however they are strong indication of the existence of objects if they agree with the motion predictions.* Then we use these to extend the lives of tracklets.

Bearing such motivation, we propose "Two-stage Association." Specifically, we apply two rounds of association with different score thresholds: a low one T_l and a

---

[2] We use py-motmetrics [11] for the analysis.

Detection with Scores		0.8	0.6	0.4	0.2	0.8
Frame Number		1	2	3	4	5
One-stage	Action	Initialize	Match	Predict	Death	Initialize
	Object ID	A	A	A	A	B
	ID-Switch	0	0	0	0	1
Two-stage	Action	Initialize	Match	Extend	Extend	Match
	Object ID	A	A	A	A	A
	ID-Switch	0	0	0	0	0

Fig. 7. Comparison for "One-stage" and "Two-stage" association with a hypothetical example. "Extend" means "extending the life cycles," and "Predict" means "using motion predictions due to no association." Suppose $T_h = 0.5$ and $T_l = 0.1$ are the score thresholds, the "one-stage" method early terminates the tracklet because of consecutively lacking associations. Details in Sect. 4.4.

high one T_h (e.g. 0.1 and 0.5 for pedestrian on WOD). In stage one, we use the identical procedure as most current algorithms [10,40,46]: only the bounding boxes with scores higher than T_h are used for association. In stage two, we focus on the tracklets unmatched to detections in stage one and relax the conditions on their matches: detections having scores larger than T_l will be sufficient for a match. If the tracklet is successfully associated with one bounding box in stage two, it will still keep being alive. However, as the low score detections are generally in poor quality, we don't output them to avoid false positives, and they are also not used for updating motion models. Instead, we use motion predictions as the latest tracklet states, replacing the low quality detections.

We intuitively explain the differences between our "Two-stage Association" and traditional "One-stage Association" in Fig. 7. Suppose $T = 0.5$ is the original score threshold for filtering detection bounding boxes, the trackers will then neglect the boxes with scores 0.4 and 0.2 on frames 3 and 4, which will die because of lacking matches in continuous frames and this eventually causes the final ID-Switch. In comparison, our two-stage association can maintain the active state of the tracklet.

Table 3. Ablation for "Two-stage Association" on WOD. "One" and "Two" denotes the previous one-stage association and our two-stage association methods. Details in Sect. 4.4.

Method	Vehicle			Pedestrian		
	MOTA↑	MOTP↓	IDS(%)↓	MOTA↑	MOTP↓	IDS(%)↓
One	0.5567	0.1682	0.46	0.5718	**0.3125**	0.96
Two	**0.5612**	**0.1681**	**0.08**	**0.5776**	**0.3125**	**0.42**

In Table 3, our approach greatly decreases the ID-Switches without hurting the MOTA. This proves that SimpleTrack is effective in extending the life cycles by using detections more flexibly. Parallel to our work, a similar approach is also proven to be useful for 2D MOT [50].

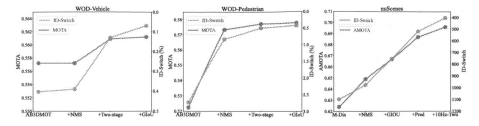

Fig. 8. Improvements from SimpleTrack on WOD (left & middle) and nuScenes (right). We use the common baselines of AB3DMOT [40] on WOD and Chiu *et al.* [10] on nuScenes. For nuScenes, the improvements of "10 Hz-Two" (10 Hz detection and two-stage association) is in Sect. 5.1, and "Pred" (outputting motion model predictions) is in Sect. 5.2. The names for modifications are on the x-axis. Better MOTA and ID-Switch values are higher on the y-axis for clearer visualization.

Table 4. Comparison on WOD test split (L2). CenterPoint [46] detections are used. We list the methods using public detection. For AB3DMOT [40] and Chiu *et al.* [10], we report their best leaderboard entries.

Method	Vehicle			Pedestrian		
	MOTA↑	MOTP↓	IDS(%)↓	MOTA↑	MOTP↓	IDS(%)↓
AB3DMOT [40]	0.5773	**0.1614**	0.26	0.5380	0.3163	0.73
Chiu *et al.* [10]	0.4932	0.1689	0.62	0.4438	0.3227	1.83
CenterPoint [46]	0.5938	0.1637	0.32	0.5664	0.3116	1.07
SimpleTrack	**0.6030**	0.1623	**0.08**	**0.6013**	**0.3114**	**0.40**

4.5   Integration of SimpleTrack

In this section, we integrate the techniques mentioned above into the unified Simple-Track and demonstrate how they improve the performance step by step.

In Fig. 8, we illustrate how the performance of 3D MOT trackers improve from the baselines. On WOD, although the properties of vehicles and pedestrian are much different, each technique is applicable to both. On nuScenes, every proposed improvement is also effective for both the AMOTA and ID-Switch.

We also report the test set performance and compare with other 3D MOT methods. SimpleTrack could achieve new state-of-the-art results with nominal cost, running at 120 FPS after detectors. (in Table 4, Table 5).[3]

5   Rethinking NuScenes

Besides the techniques mentioned above, we delve into the design of benchmarks. The benchmarks greatly facilitate the development of research and guide the designs of algorithms. Contrasting WOD and nuScenes, we find that despite more than 70% of vehicles

---

[3] Validation split comparisons are in the supplementary.

Table 5. Comparison on nuScenes test split. CenterPoint [46] detections are used. We list the methods using public detection. For CBMOT [3] and OGR3MOT [47], we report their numbers with CenterPoint [46] detection. Our numbers using 2 Hz and 10 Hz frame rate detections are reported (details of 10 Hz setting are in Sect. 5).

Methods	AMOTA↑	AMOTP↓	MOTA↑	IDS ↓
AB3DMOT [40]	0.151	1.501	0.154	9027
Chiu *et al.* [10]	0.550	0.798	0.459	776
CenterPoint [46]	0.638	0.555	0.537	760
CBMOT [3]	0.649	0.592	0.545	557
OGR3MOT citech43graphmot	0.656	0.620	0.554	**288**
SimpleTrack (2 Hz)	0.658	0.568	0.557	609
SimpleTrack (10 Hz)	**0.668**	**0.550**	**0.566**	575

staying static all the time, which may lead to biases to the evaluation, WOD is closer to real-world scenarios. As for nuScenes, two aspects are critical: 1) The frame rate of nuScenes 2 Hz, while WOD 10 Hz. Such low frequency adds unnecessary difficulties to 3D MOT (Sect. 5.1). 2) The evaluation of nuScenes requires high recalls with low score thresholds. And it also pre-processes the tracklets with interpolation, which encourages trackers to output the confidence scores reflecting the entire tracklet quality, but not the frame quality (Sect. 5.2).

We hope these two findings could inspire the community to rethink the benchmarks and evaluation protocols of 3D tracking. *And such modifications may help: a. support 10 Hz setting; b. discard the "interpolation" in evaluation API.*

Table 6. Frequency comparison of benchmarks.

Benchmark	Data	Annotation	Model
Waymo Open Dataset	10 Hz	10 Hz	10 Hz
nuScenes	20 Hz	2 Hz	2 Hz

5.1   Detection Frequencies

Tracking generally benefits from higher frame rates, because motion is more predictable in short intervals.

We compare the frequencies of point clouds, annotations, and common MOT frame rates on the two benchmarks in Table 6. On nuScenes, it 20 Hz point clouds but 2 Hz annotations. This leads to most common detectors and 3D MOT algorithms work 2 Hz, even they actually utilize all 20 Hz LiDAR data and operate faster 2 Hz. Therefore, we investigate the effect of high-frequency data as follows. Although the information is more abundant with high frequency (HF) frames, it is non-trivial to incorporate them because nuScenes only evaluates on the low-frequency frames, which we refer to as "evaluation frames." In Table 7, simply using all 10 Hz frames does not improve the performance. This is because the low-quality detection on the HF frames may deviate the trackers and hurt the performance on the sampled evaluation frames. To overcome this issue, we explore by first applying the "One-stage Association" on HF frames, where only the bounding boxes with scores larger than $T_h = 0.5$ are considered and used for motion model updating. We then adopt the "Two-stage Association" (described in Sect. 4.4) by using the boxes with scores larger than $T_l = 0.1$ to extend the tracklets. As in Table 7, our approach significantly improves both the AMOTA and ID-Switches. We also try to even increase the frame rate 20 Hz, but this barely leads to further improvements due to the deviation issue. So SimpleTrack uses 10 Hz setting in our final submission to the test set.[4]

5.2   Tracklet Interpolation

We start with the evaluation protocol on nuScenes, where they interpolate the input tracklets to fill in the missing frames and change all the scores with their tracklet-average scores as illustrated in Fig. 9. Such interpolation on nuScenes encourages the trackers to treat tracklet quality holistically and output calibrated quality-aware scores. However the quality of boxes may vary a lot across frames even for the same tracklet, thus we suggest depicting the quality of a tracklet by only one score is imperfect. Moreover, future information is also introduced in this interpolation step and it changes the tracklet results. This could also bring the concern on whether the evaluation setting is still a fully online one which is crucial for autonomous driving.

To prove this argument, we output the motion model predictions for frames and tracklets without associated detection bounding boxes, and empirically assign them lower scores than any other detection. In our case, their scores are $0.01 \times S_P$, where S_P is the confidence score of the tracklet in the previous frame. As shown in Fig. 9, Our approach can explicitly penalize the low-quality tracklets, which generally contain more missing boxes replaced by motion model predictions. In Table 8, this simple experiment improves the overall recall and

Table 7. MOT with higher frame rates on nuScenes. "10 Hz" is the vanilla baseline of using all the detections on high frequency (HF) frames. "-One" denotes "One-stage," and "-Two" denotes "Two-stage." Details in Sect. 5.1.

Setting	AMOTA↑	AMOTP↓	MOTA↑	IDS↓
2 Hz	0.687	0.573	0.592	519
10 Hz	0.687	0.548	0.599	512
10 Hz - One	**0.696**	0.564	**0.603**	450
10 Hz - Two	**0.696**	0.547	0.602	**403**
20 Hz - Two	0.690	**0.547**	0.598	416

AMOTA. Moreover, some attempts [3,47] have changed tracklet scores without explicitly declaring the "interpolation" issue which should be aware to ensure the effectiveness of the benchmark.

---

[4] Because of the submission time limits to nuScenes test set, we are only able to report the "10 Hz-One" variant in Table 5. It will be updated to "10 Hz-Two" once we had the chance.

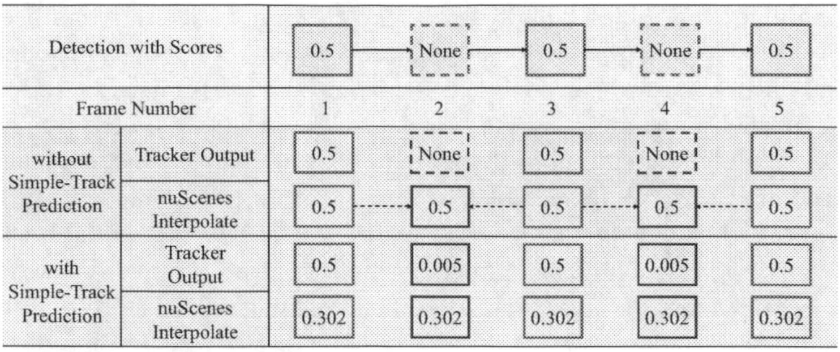

Detection with Scores		0.5	None	0.5	None	0.5
Frame Number		1	2	3	4	5
without Simple-Track Prediction	Tracker Output	0.5	None	0.5	None	0.5
	nuScenes Interpolate	0.5	0.5	0.5	0.5	0.5
with Simple-Track Prediction	Tracker Output	0.5	0.005	0.5	0.005	0.5
	nuScenes Interpolate	0.302	0.302	0.302	0.302	0.302

Fig. 9. How the motion predictions and nuScenes interpolation changes tracklet scores. Dashed arrows are the directions for interpolation. On Frame 2 and 4 the boxes with score 0.05 are our motion predictions. The "0.5" and "0.302" are the tracklet-average scores with or without motion predictions. Details in Sect. 5.2.

Table 8. Improvement from "outputting motion model predictions" on nuScenes (2 Hz detections for ablation).

Predictions	AMOTA↑	AMOTP↓	MOTA↑	IDS ↓	RECALL↑
×	0.667	0.612	0.572	754	0.696
✓	**0.687**	**0.573**	**0.592**	**519**	**0.725**

6   Error Analyses

In this section, we conduct analyses on the remaining failure cases of SimpleTrack and propose potential future directions for improving "tracking by detection" paradigm. Without loss of generality, we use WOD as an example.

6.1   Upper Bound Experiment Settings

To quantitatively evaluate the causes of failure cases, we contrast SimpleTrack with two different oracle variants. The results are summarized in Table 9.

GT Output erases the errors caused by "output" policy. We compute the IoU between the bounding boxes from SimpleTrack with the GT boxes at the "output" stage, then use the IoU to decide if a box should be output instead of the detection score. [5]

GT All is the upper bound of tracking performance with CenterPoint boxes. We greedily match the detections from CenterPoint to GT boxes, keep the true positive and assign them ground-truth ID.

---

[5] The ID-Switch increases because we output more bounding boxes and IDs. The 0.003 false positives in pedestrians are caused by boxes matched with the same GT box in crowded scenes.

6.2   Analyses for "Tracking by Detection"

ID-Switches. We break down the causes of ID-Switches as in Fig. 6. Although early termination has been greatly decreased by the scale of 86% for vehicle and 70% for pedestrian with "Two-stage Association," it still takes up 88% and 72% failure cases in the remaining ID-Switches in SimpleTrack for vehicle and pedestrian, respectively. We inspect these cases and discover that most of them result from long-term occlusion or the returning of objects from being temporarily out of sight. Therefore, in addition to improving the association, potential future work can develop appearance models like in 2D MOT [21,22,36,42] or silently maintain their states to re-identify these objects after they are back.

Table 9. Oracle experiments on WOD.

Method	Vehicle				Pedestrian			
	MOTA↑	IDS(%)↓	FP↓	FN↓	MOTA↑	IDS(%)↓	FP↓	FN↓
SimpleTrack	0.561	0.078	0.104	0.334	0.578	0.425	0.109	0.309
GT Output	0.741	0.104	**0.000**	0.258	0.778	0.504	0.003	0.214
GT All	**0.785**	**0.000**	**0.000**	**0.215**	**0.829**	**0.000**	**0.000**	**0.171**

FP and FN. The "GT All" in Table 9 shows the upper bound for MOT with Center-Point [46] detection, and we analyze the class of vehicle for example. Even with "GT All" the false negatives are still 0.215, which are the detection FN and can hardly be fixed under the "tracking by detection" framework. Comparing "GT All" and Simple-Track, we find that the tracking algorithm itself introduces 0.119 false negatives. We further break them down as follows. Specifically, the difference between "GT Output" and "GT ALL" indicates that the 0.043 false negatives are caused by the uninitialized track-lets resulting from NMS and score threshold in pre-processing. The others come from life-cycle management. The "Initialization" requires two frames of accumulation before outputting a tracklet, which is same as AB3DMOT [40]. This yields a marginal 0.005 false negatives. Our "Output" logic uses detection score to decide output or not, taking up the false negatives number 0.076. Based on these analyses, we can conclude that the gap is mainly caused by the inconsistency between the scores and detection quality. By using historical information, 3D MOT can potentially provide better scores compared to single frame detectors, and this has already drawn some recent attention [3,47].

7   Conclusions and Future Work

In this paper, we decouple the "tracking by detection" 3D MOT algorithms into several components and analyze their typical failures. With such insights, we propose corresponding enhancements of using *NMS*, *GIoU*, and *Two-stage Association*, which lead to our SimpleTrack. In addition, we also rethink the frame rates and interpolation pre-processing in nuScenes. We eventually point out several possible future directions for "tracking by detection" 3D MOT.

However, beyond the "tracking by detection" paradigm, there are also branches of great potential. For better bounding box qualities, 3D MOT can refine them using long term information [27,31,44], which are proven to outperform the detections based only on local frames. The future work can also transfer the current manual rule-based methods into learning-based counterparts, *e.g.* using learning based intra-frame mechanisms to replace the NMS, using inter-frame reasoning to replace the 3D GIoU and life cycle management, etc.

References

1. Bar-Shalom, Y., Fortmann, T.E., Cable, P.G.: Tracking and data association (1990)
2. Baser, E., Balasubramanian, V., Bhattacharyya, P., Czarnecki, K.: FANTrack: 3D multi-object tracking with feature association network. In: IV (2019)
3. Benbarka, N., Schröder, J., Zell, A.: Score refinement for confidence-based 3D multi-object tracking. In: IROS (2021)
4. Berclaz, J., Fleuret, F., Turetken, E., Fua, P.: Multiple object tracking using k-shortest paths optimization. IEEE Trans. Pattern Anal. Mach. Intell. **33**(9), 1806–1819 (2011)
5. Bergmann, P., Meinhardt, T., Leal-Taixe, L.: Tracking without bells and whistles. In: ICCV (2019)
6. Bewley, A., Ge, Z., Ott, L., Ramos, F., Upcroft, B.: Simple online and realtime tracking. In: ICIP (2016)
7. Braso, G., Leal-Taixe, L.: Learning a neural solver for multiple object tracking. In: CVPR (2020)
8. Caesar, H., et al.: nuScenes: a multimodal dataset for autonomous driving. In: CVPR (2020)
9. Chiu, H., Li, J., Ambrus, R., Bohg, J.: Probabilistic 3D multi-modal, multi-object tracking for autonomous driving. In: ICRA (2021)
10. Chiu, H.k., Prioletti, A., Li, J., Bohg, J.: Probabilistic 3D multi-object tracking for autonomous driving. arXiv:2001.05673 (2020)
11. py-motmetrics Contributors: py-motmetrics. https://github.com/cheind/py-motmetrics
12. Fleuret, F., Berclaz, J., Lengagne, R., Fua, P.: Multicamera people tracking with a probabilistic occupancy map. IEEE Trans. Pattern Anal. Mach. Intell. **30**(2), 267–282 (2007)
13. Gautam, S., Meyer, G.P., Vallespi-Gonzalez, C., Becker, B.C.: Sdvtracker: real-time multi-sensor association and tracking for self-driving vehicles. arXiv preprint arXiv:2003.04447 (2020)
14. Genovese, A.F.: The interacting multiple model algorithm for accurate state estimation of maneuvering targets. J. Hopkins APL Tech. Dig. **22**(4), 614–623 (2001)
15. He, J., Huang, Z., Wang, N., Zhang, Z.: Learnable graph matching: incorporating graph partitioning with deep feature learning for multiple object tracking. In: CVPR (2021)
16. Hornakova, A., Henschel, R., Rosenhahn, B., Swoboda, P.: Lifted disjoint paths with application in multiple object tracking. In: ICML (2020)
17. Jiang, X., Li, P., Li, Y., Zhen, X.: Graph neural ased end-to-end data association framework for online multiple-object tracking. arXiv preprint arXiv:1907.05315 (2019)
18. Kim, A., Osep, A., Leal-Taixé, L.: EagerMOT: 3D multi-object tracking via sensor fusion. arxiv:2104.14682 (2021)
19. Kim, C., Li, F., Ciptadi, A., Rehg, J.M.: Multiple hypothesis tracking revisited. In: ICCV (2015)
20. Lan, L., Tao, D., Gong, C., Guan, N., Luo, Z.: Online multi-object tracking by quadratic pseudo-boolean optimization. In: IJCAI (2016)

21. Leal-Taixé, L., Canton-Ferrer, C., Schindler, K.: Learning by tracking: siamese CNN for robust target association. In: CVPR Workshops (2016)
22. Li, J., Gao, X., Jiang, T.: Graph networks for multiple object tracking. In: WACV (2020)
23. Liang, T., Lan, L., Luo, Z.: Enhancing the association in multi-object tracking via neighbor graph. arXiv preprint arXiv:2007.00265 (2020)
24. Liu, Q., Chu, Q., Liu, B., Yu, N.: GSM: graph similarity model for multi-object trackin. In: IJCAI (2020)
25. Lu, Z., Rathod, V., Votel, R., Huang, J.: RetinaTrack: online single stage joint detection and tracking. In: CVPR (2020)
26. Pang, J., et al.: Quasi-dense similarity learning for multiple object tracking. In: CVPR (2021)
27. Pang, Z., Li, Z., Wang, N.: Model-free vehicle tracking and state estimation in point cloud sequences. In: IROS (2021)
28. Peng, J., et al.: TPM: multiple object tracking with tracklet-plane matching. Pattern Recogn. **107**, 107480 (2020)
29. Pirsiavash, H., Ramanan, D., Fowlkes, C.C.: Globally-optimal greedy algorithms for tracking a variable number of objects. In: CVPR (2011)
30. Pöschmann, J., Pfeifer, T., Protzel, P.: Factor graph based 3D multi-object tracking in point clouds. In: IROS (2020)
31. Qi, C.R., et al.: Offboard 3D object detection from point cloud sequences. In: CVPR (2021)
32. Rangesh, A., Trivedi, M.M.: No blind spots: full-surround multi-object tracking for autonomous vehicles using cameras and lidars. In: IV (2019)
33. Reid, D.: An algorithm for tracking multiple targets. IEEE Trans. Autom. Control **24**(6), 843–854 (1979)
34. Rezatofighi, H., Tsoi, N., Gwak, J., Sadeghian, A., Reid, I., Savarese, S.: Generalized intersection over union. In: CVPR (2019)
35. Rezatofighi, S.H., Milan, A., Zhang, Z., Shi, Q., Dick, A., Reid, I.: Joint probabilistic data association revisited. In: ICCV (2015)
36. Sadeghian, A., Alahi, A., Savarese, S.: Tracking the untrackable: learning to track multiple cues with long-term dependencies. In: ICCV (2017)
37. Sun, P., et al.: Scalability in perception for autonomous driving: Waymo Open Dataset. arxiv:1912.04838 (2019)
38. Tang, S., Andres, B., Andriluka, M., Schiele, B.: Subgraph decomposition for multi-target tracking. In: CVPR (2015)
39. Tang, S., Andres, B., Andriluka, M., Schiele, B.: Multi-person tracking by multicut and deep matching. In: Hua, G., Jégou, H. (eds.) ECCV 2016. LNCS, vol. 9914, pp. 100–111. Springer, Cham (2016). https://doi.org/10.1007/978-3-319-48881-3_8
40. Weng, X., Wang, J., Held, D., Kitani, K.: 3D multi-object tracking: a baseline and new evaluation metrics. In: IROS (2020)
41. Weng, X., Wang, Y., Man, Y., Kitani, K.: GNN3DMOT: graph neural network for 3D multi-object tracking with 2D-3D multi-feature learning. In: CVPR (2020)
42. Wojke, N., Bewley, A., Paulus, D.: Simple online and realtime tracking with a deep association metric. In: ICIP (2017)
43. Xu, Y., et al.: How to train your deep multi-object tracker. In: CVPR (2020)
44. Yang, B., Bai, M., Liang, M., Zeng, W., Urtasun, R.: Auto4D: learning to label 4D objects from sequential point clouds. arxiv:2101.06586 (2021)
45. Yang, B., Huang, C., Nevatia, R.: Learning affinities and dependencies for multi-target tracking using a CRF model. In: CVPR (2011)
46. Yin, T., Zhou, X., Krähenbühl, P.: Center-based 3D object detection and tracking. In: CVPR (2021)
47. Zaech, J., Dai, D., Liniger, A., Danelljan, M., Gool, L.V.: Learnable online graph representations for 3D multi-object tracking. arXiv:2104.11747 (2021)

48. Roshan Zamir, A., Dehghan, A., Shah, M.: GMCP-tracker: global multi-object tracking using generalized minimum clique graphs. In: Fitzgibbon, A., Lazebnik, S., Perona, P., Sato, Y., Schmid, C. (eds.) ECCV 2012. LNCS, vol. 7573, pp. 343–356. Springer, Heidelberg (2012). https://doi.org/10.1007/978-3-642-33709-3_25
49. Zhang, L., Li, Y., Nevatia, R.: Global data association for multi-object tracking using network flows. In: CVPR (2008)
50. Zhang, Y., et al.: ByteTrack: multi-object tracking by associating every detection box. arXiv preprint arXiv:2110.06864 (2021)
51. Zhang, Y., Wang, C., Wang, X., Zeng, W., Liu, W.: FairMOT: on the fairness of detection and re-identification in multiple object tracking. Int. J. Comput. Vis. **129**, 1–19 (2021)

Ego-Motion Compensation
of Range-Beam-Doppler Radar Data
for Object Detection

Michael Meyer[1,2]([✉])[ID], Marc Unzueta[1][ID], Georg Kuschk[1][ID],
and Sven Tomforde[2][ID]

[1] Cruise, Munich, Germany
{michael.meyer,marc.unzueta,georg.kuschk}@getcruise.com
[2] Christian-Albrecht University, Kiel, Germany
st@informatik.uni-kiel.de

Abstract. With deep learning based perception tasks on radar input data gaining more attention for autonomous driving, the use of new data interfaces, specifically range-beam-doppler tensors, are explored to maximize the performance of corresponding algorithms. Surprisingly, in past publications, the Doppler information of this data has only played a minor role, even though velocity is considered a powerful feature. We investigate the hypothesis that the sensor ego-velocity, induced by the ego vehicle motion, increases the data generalization complexity of the range-beam-doppler data and propose a phase shift of the electromagnetic wave to normalize the data by compensating for the ego vehicle motion. We show its efficacy versus non-compensated data with an improvement of 8.7% average precision (AP) for object detection tasks.

1 Introduction

One of the main tasks that autonomous vehicles need to perform is the perception of their surroundings. Cameras, lidars, and radars are common sensors used to perform a variety of sub-tasks, including free space detection [1], detection of dynamic traffic participants [2,3], and scene classification [4].

While most deep learning based research in the area of autonomous driving perception focuses on the use of lidar and camera, recently there have been also some more publications on radar perception [5–7].

The direct measurement of an object's (relative radial) velocity is one of the main strengths of radar sensors. Although this is expected to be a strong feature, in past publications the influence of the Doppler dimension was often not investigated [8–10], or it only led to minor improvements in the object detection task [11]. One reason for this might be the dependence of the measured velocity value on the ego-velocity of the radar sensor: The same scene will result in a different radar signal when the sensor (and ego vehicle) has a different velocity.

L. Karlinsky et al. (Eds.): ECCV 2022 Workshops, LNCS 13801, pp. 697–708, 2023.
https://doi.org/10.1007/978-3-031-25056-9_44

(a) Camera image with 3D annotations.

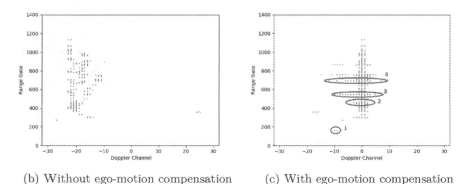

(b) Without ego-motion compensation   (c) With ego-motion compensation

Fig. 1. Ego motion compensation effect on the radar Doppler channel. Corresponding objects are marked both in camera and Range-Doppler data.

Ideally, features extracted by machine learning techniques such as neural networks are invariant to abstract input variables like translation, rotation, illumination, and scale. Deep neural networks have shown that they are capable of doing so with a sufficiently large dataset. However, in most cases, it is favorable to align variations in the data, rather than training the network to learn it implicitly. Feature pyramid networks (FPNs) [12] for example, introduce a mechanism to detect objects at different scales, and spatial transformers [13] increase invariance against rotations and non-rigid deformations.

In this work, we present a procedure to compensate for the sensor ego-motion through phase shifts in the complex signal. We then investigate the hypothesis that ego-motion compensation leads to an improvement in the performance of an object detection task on these tensors.

Pallfy *et al.* [14] use both the range-beam-doppler tensor and the radar point cloud but only perform the compensation of the ego-motion on the point cloud. For radar point clouds, the ego-motion compensation is quite common [14–16], but for the tensor data, so far it has been neglected.

Our main contributions are as follows:

- We show a mathematical formulation on how to perform the ego-motion compensation for the range-beam-doppler tensors through phase shifts in the complex signal (see Sect. 3 and Fig. 1).
- We demonstrate that this ego-motion compensation leads to an improvement of +8.7% points in the AP performance of an object detection network that is using radar data (see Sect. 4).

2   Radar Processing Theory

2.1   Radar Foundations

In this work, a frequency modulated continuous wave (FMCW) radar is used. FMCW radars emit a signal with a linear modulated frequency and record the signal that is reflected back from objects in the scene (see Fig. 2). The received signal is mixed with the transmitted signal to obtain the frequency difference Δf between the two. This beat frequency is a function of the object range. The speed of light c_0 and the time the signal traveled Δt can be used to calculate the range R:

$$R = \frac{c_0 |\Delta t|}{2} = \frac{c_0 |\Delta f|}{2 \left| \frac{df}{dt} \right|} \tag{1}$$

This range is usually calculated from the frequency with a Fast-Fourier-Transform (FFT) which is referred to as Range-FFT.

In case an object is moving, the frequency difference is (1) a superposition of the frequency shift related to range and (2) a Doppler component due to the radial velocity between object and radar sensor.

From a single chirp, the Doppler information cannot be extracted as it is overlapping with the range component, but through sending and receiving multiple consecutive chirps and processing them together, it is possible to resolve both range and velocity.

An object moving with a velocity v measured by two consecutive chirps separated by time T_c will result in a phase difference $\Delta\phi$ corresponding to a displacement $r = v \cdot T_c$ [17]:

$$\Delta\phi = \frac{4\pi r}{\lambda} = \frac{4\pi v T_c}{\lambda} = \frac{4\pi f v T_c}{c_0} \tag{2}$$

where $\lambda = \frac{c_0}{f}$ is the wavelength of the radar signal.

Equation 2 can be used to derive the velocity v:

$$v = \frac{c_0 \Delta\phi}{4\pi T_c f} \tag{3}$$

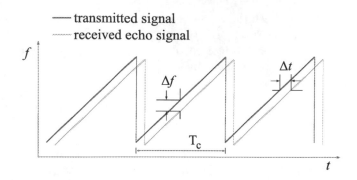

Fig. 2. Transmitted and received frequency over time of a radar sensor.

Here the velocity measurement is based on a phase difference and consequently this velocity will be ambiguous. The measurement is unambiguous only if $|\Delta\phi| < \pi$. Therefore, the maximum unambiguous velocity depends on the chirp transmission time:

$$v_{max} = \frac{\lambda}{4T_c} \tag{4}$$

With multiple chirps, the radar sensor is also able to distinguish between multiple different velocities at the same range. Usually, a set of N chirps is processed through a second FFT, which is often called Doppler-FFT.

The result is N radial velocity / Doppler measurements for each range.

Multiple receiver channels are used to infer object angles (beams) through digital beamforming. Two antennas receive the reflected signal from the same object, with a phase difference caused by the different distances the signal travels between the object and each antenna. Therefore, the phase shift depends on the relative position of the antennas and the position of the object.

For two receive antennas being distance d apart (see Fig. 3), the phase difference of an object at angle θ is:

$$\Delta\phi = \frac{2\pi f d \sin(\theta)}{c_0} \tag{5}$$

Hence, the angle can be determined as:

$$\theta = \sin^{-1}\left(\frac{c_0 \Delta\phi}{2\pi f d}\right) \tag{6}$$

In practice, radar sensors nowadays can use much more complex antenna layouts than one transmit and two receive antennas [18], nevertheless, the underlying principle of inferring the angle through phase differences between the antennas remains the same, and most radar sensors that are used today work based on the concepts described above. At one point in their data processing chain, these radar sensors use a data tensor with dimensions ranges, beams, and Doppler

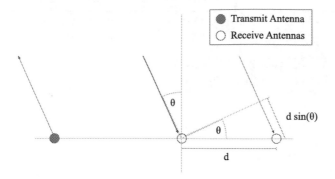

Fig. 3. Angle estimation with two receive antennas.

channels to detect data points of interest. In traditional radar processing, a constant false alarm rate (CFAR) algorithm is often used to extract a point cloud from this tensor, whereas in this paper we work directly with this tensor.

For a more thorough introduction to radar fundamentals and its classical signal processing, we refer to [19, 20].

3 Methods and Experiments

3.1 Ego-Motion Compensation

For autonomous driving, radar sensors are mounted on a (moving) vehicle, where the ego-velocity then influences the resulting Doppler/velocity measurement. In the following paragraphs, we derive how the radar signal, specifically the Doppler measurement, can be made invariant to the ego-velocity through shifting the phase of the original signal.

The measured phase difference between two consecutive chirps is actually a superposition of (a) the radial velocity of objects in the scene and (b) the component of the ego-velocity in the direction of the objects. Thus, Eq. 2 can be extended to:

$$\Delta\phi = \Delta\phi_{Object} + \Delta\phi_{Ego\text{-}Vehicle} \tag{7}$$

$$= \frac{4\pi T_c f v_{Object}}{c_0} + \frac{4\pi T_c f v_{Ego\text{-}Vehicle}}{c_0} \tag{8}$$

Therefore, the signal from each consecutive chirp has to be shifted by $\Delta\phi_{Ego\text{-}Vehicle}$, where $v_{Ego\text{-}Vehicle}$ stands for a speed in a particular direction and thus depends on the beam.

For one particular beam b the phase difference between the first and the n-th chirp caused by the ego-motion $\Delta\phi_{b_{Ego\text{-}Vehicle}}$ is proportional to the time difference $t = nT_c$:

Algorithm 1. Overview of the ego-motion compensation procedure with phase shifts. Bold variables represent vectors, the circumflex/hat indicates that it is normalized.

1: **procedure** EGO MOTION COMPENSATION
2:      determine ego-velocity vector of radar sensor $\mathbf{v}$
3:
4:      **for** each beam b **do**
5:
6:          calculate ego-speed v_b in beam direction $\hat{\mathbf{u}}_b$:
7:          $v_b \leftarrow \mathbf{v} \cdot \hat{\mathbf{u}}_b$
8:
9:          **for** each chirp n **do**
10:
11:             calculate phase shift $\Delta\phi_{bn}$:
12:             $\Delta\phi_{bn} \leftarrow \frac{4\pi f n T_c v_b}{c_0}$
13:
14:             shift the radar signal Ψ_{bn} by phase $\Delta\phi_{bn}$:
15:             $\Psi_{bn} \leftarrow \Psi_{bn} e^{i\Delta\phi_{bn}}$

$$\Delta\phi_{bn\,Ego\text{-}Vehicle} = \frac{4\pi f n T_c v_b}{c_0} \tag{9}$$

During the processing of the radar signal from measured frequencies to range-beam-doppler tensor, Eq. 9 is used to calculate which phase shifts were induced into the signal through the radial motion of the radar sensor as described above. The original signal is then adjusted so that these phase shifts are eliminated, and the resulting range-beam-doppler tensor is invariant to the ego-velocity. The procedure is displayed in Algorithm 1.

In order to explain the influence of this method, a scene with moving and static objects is selected, as seen in Fig. 1a. The range gate and doppler channel indices are plotted for high magnitude values of the range-beam-doppler tensor. Without ego motion compensation, the doppler channel indices are dependent on the ego-velocity (Fig. 1b). When the ego-motion compensation method is applied, all returns from static objects are visible in a vertical line at doppler channel 0, while all moving objects result in points outside the vertical line (Fig. 1c). In this plot, the points that are not encircled in one of the object clusters and have a doppler channel $\neq 0$ belong to moving objects outside the field of view of the camera. This shows how the proposed method makes the range-beam-doppler data invariant to changes in ego-velocity.

3.2 Radar Object Detection

To determine the influence of the ego-motion compensation on perception tasks, we train an object detection network (similar to *RT-Net* [21]) in birds-eye-view (BEV) on range-beam-doppler tensors. The same network is trained twice, once with ego-motion compensated data and once on the same data, which has not been ego-motion compensated during the data preprocessing. The networks'

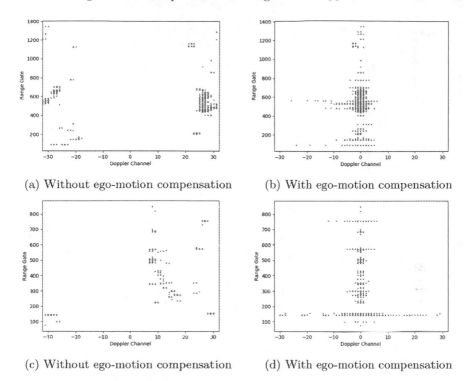

(a) Without ego-motion compensation   (b) With ego-motion compensation

(c) Without ego-motion compensation   (d) With ego-motion compensation

Fig. 4. Scatterplots (a) without and (b) with ego-motion compensation for two randomly selected frames. The range gate indices are plotted over doppler channel indices for all magnitudes in the tensor above a certain threshold. The purpose of these plots is to show the distribution of high-magnitude values in the range-beam-doppler tensor. As expected, with ego-motion compensation, all static objects have the same doppler channel.

6D object detection predictions (x, y, length, width, yaw, score) are evaluated against manually labeled ground truth of the class *car*.

The architecture of the RT-Net model is shown in Fig. 5. First, the data is converted from a polar representation with dimensions $1360 \times 64 \times 128$ to Cartesian space using bilinear interpolation, into a grid of 680×533 pixels, which is fed into a ResNet50 backbone which extracts the Radar features which is then fed into a FPN to extract features at different scales. The 6D object detection output tensor is regressed using a single head which computes deltas with respect to a set of anchors for the location and dimensions, altogether with a L1 loss function.

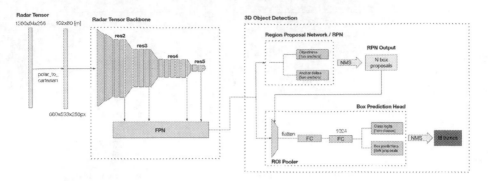

Fig. 5. RT-Net model architecture.

In case of the Radar+Camera architecture, the camera features are projected into BEV using an Orthographic Feature Transform as in [22] and the features camera/radar features are concatenated. The model is trained using Adam optimizer for 80 epochs on a 8×4 distributed T4 GPU cluster.

3.3   Dataset

To better understand the impact of the ego-motion invariance, we provide details on the dataset used. The data was recorded in a highly dense urban environment of the inner city of San Francisco with a recording rate of 0.25 Hz. It contains 9697 frames, which for the object detection task are split as 70% for training, 20% for validation, and 10% for testing.

A 77 GHz radar sensor with a maximum range of 102m is used to record radar data. The recorded radar tensor is a $1360 \times 64 \times 128$ complex-valued tensor, which contains the range, azimuth, and Doppler information.

As the paper results are focused on cars, only details on that class are provided. There is a total of 224,011 cars in all frames, with a mean of about 23 cars per scene. The ego-motion mean speed of the car is 6.05 m/s and a standard deviation of 3.67 m/s. Having an evenly distributed ego vehicle speed is important to validate our experiments' validity under all conditions. The distribution of the ego vehicle forward velocity is shown in Fig. 6.

The location of the manually annotated 3D objects is limited to [0, 102] m towards the front and [−40, 40] m to the side.

4   Results

To test if the ego-motion compensation is working, the indices of high magnitudes in the range-beam-doppler tensor are determined and the range index is plotted over the Doppler channel. Since in big cities' downtown areas there are numerous static objects, such as buildings and traffic lights, it can be expected that most of these high magnitude objects are static. The plots (see Fig. 4) are created for

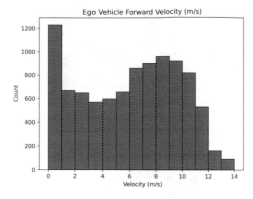

Fig. 6. Ego vehicle forward velocity dataset distribution

the same frame twice - once with ego-motion compensation and once without. One can clearly see that with ego-motion compensation, the large majority of signals have a zero Doppler velocity, whereas previously they were distributed in the various Doppler channels seemingly random depending on the ego-velocity.

The performance of an object detection network trained on the data that had been ego-motion compensated is presented in Table 1. The ego-motion compensation leads to an 8.7% AP improvement for object detection on radar data. For object detection on radar and camera data, the improvement was 3.5% AP.

Table 1. Average precision (Intersection over Union threshold of 0.5) for object class *car* based on radar (+camera) data, with and without ego-motion compensation.

	Average Precision	
	Radar + Camera	Radar
Without radar ego-motion compensation	63.5%	34.9%
With radar ego-motion compensation	67.0%	43.6%
Difference	+3.5%	+8.7%

5 Discussion

Compensating for the ego-motion of the sensor leads to an immense improvement in the object detection performance on radar data. For a network based on radar and camera data, the radar data alone is less important, so it is no surprise that in that case, the improvement is smaller. Nonetheless, the ego-motion compensation of the radar data also results in a significant improvement for object detection based on radar and camera.

The presented method can be used for all deep learning tasks on radar data, which use range-azimuth-Doppler tensor data. Certainly, in the near future there will be more and more work on solving autonomous driving tasks with radar data and so far this data level has shown promising results.

Even though some publications (e.g. [11]) reported better average precision values, these are very dependent on the dataset and strongly depend on the considered detection range and evaluation parameters, and thus are not comparable to this work.

Representing the radar data with a raw, unprocessed radar tensor is a new research domain, which has its own complications. The raw tensors need a large network bandwidth switch in order to collect and store all the information at a reasonable frame rate. At the moment, our dataset is recorded at only 0.25 Hz. As a potential solution, some preprocessing in the form of radar compression could be applied to reduce the bandwidth load while maintaining the model performance, as mentioned in [23]. On the other hand, deployability could be a road blocker as well as long as the bandwidth issues are not addressed. With that said, we propose the initial improvements required to standardize such a new radar representation.

Our work shows the potential of data normalization of radar tensors w.r.t. the Doppler dimension and the effect on deep neural network based object detection. However, the task of object detection is just an example for the importance of this data normalization. Other tasks, like velocity estimation of moving objects, are likely to benefit even more from this ego-motion compensation.

6   Conclusion

In this paper, we present a method to compensate the ego-velocity on the radar data, achieving velocity invariance, through shifting the phase of the original signal, as the measured phase difference between two consecutive chirps is a superposition of the object's radial velocity and the ego-velocity component in the direction of the object.

While ego-motion compensation is a common approach for radar point clouds, we demonstrate that it is also important when training neural networks directly on the range-beam-doppler data.

We provide the mathematical formulation used and the algorithm description to show the approach's efficacy with an 8.7% AP performance improvement on a dataset with a balanced ego-velocity distribution versus the non-invariant data experiment.

In future work, we will investigate the impact of the proposed ego-motion compensation for velocity estimation.

References

1. Scheck, T., Mallandur, A., Wiede, C., Hirtz, G.: Where to drive: free space detection with one fisheye camera. In: Twelfth International Conference on Machine Vision (ICMV 2019) (2020)

2. Lang, A.H., Vora, S., Caesar, H., Zhou, L., Yang, J., Beijbom, O.: PointPillars: fast encoders for object detection from point clouds. In: Proceedings of the IEEE/CVF Conference on Computer Vision and Pattern Recognition, pp. 12697–12705 (2019)
3. Meyer, G.P., Laddha, A., Kee, E., Vallespi-Gonzalez, C., Wellington, C.K.: Laser-Net: an efficient probabilistic 3D object detector for autonomous driving. In: CVF Conference on Computer Vision and Pattern Recognition (CVPR), pp. 12669–12678 (2019). IEEE (2019)
4. Meyer, M., Kuschk, G., Tomforde, S.: Complex-valued convolutional neural networks for automotive scene classification based on range-beam-doppler tensors. In: 2020 IEEE 23rd International Conference on Intelligent Transportation Systems (ITSC) (2020)
5. Lim, T.Y., et al.: Radar and camera early fusion for vehicle detection in advanced driver assistance systems. In: Conference on Neural Information Processing Systems Workshops (2019)
6. Orr, I., Cohen, M., Zalevsky, Z.: High-resolution radar road segmentation using weakly supervised learning. In: Nature Machine Intelligence, pp. 1–8 (2021)
7. Scheiner, N., et al.: Seeing around street corners: non-line-of-sight detection and tracking in-the-wild using doppler radar. In: Proceedings of the IEEE/CVF Conference on Computer Vision and Pattern Recognition, pp. 2068–2077 (2020)
8. Dong, X., Wang, P., Zhang, P., Liu, L.: Probabilistic oriented object detection in automotive radar. In: Conference on Computer Vision and Pattern Recognition (CVPR) Workshops, pp. 458–467 (2020)
9. Kothari, R., Kariminezhad, A., Mayr, C., Zhang, H.: Object detection and heading forecasting by fusing raw radar data using cross attention. CoRR (2022). https://doi.org/10.48550/arXiv.2205.08406
10. Zhang, A., Nowruzi, F.E., Laganiere, R.: Raddet: range-azimuth-doppler based radar object detection for dynamic road users. In: 2021 18th Conference on Robots and Vision (CRV), pp. 95–102 (2021)
11. Major, B., et al.: Vehicle detection with automotive radar using deep learning on range-azimuth-doppler tensors. In: International Conference on Computer Vision Workshops (2019)
12. Lin, T.Y., Dollár, P., Girshick, R., He, K., Hariharan, B., Belongie, S.: Feature pyramid networks for object detection. In: IEEE Conference on Computer Vision and Pattern Recognition (CVPR) (2017)
13. Jaderberg, M., Simonyan, K., Zisserman, A., kavukcuoglu, k.: Spatial transformer networks. In: Cortes, C., Lawrence, N., Lee, D., Sugiyama, M., Garnett, R. (eds.) Advances in Neural Information Processing Systems, vol. 28. Curran Associates, Inc. (2015)
14. Palffy, A., Dong, J., Kooij, J.F., Gavrila, D.M.: CNN based road user detection using the 3D radar cube. IEEE Robot. Autom. Lett. $5(2)$, 1263–1270 (2020)
15. Niederlöhner, D., et al.: Self-supervised velocity estimation for automotive radar object detection networks (2022)
16. Ulrich, M., et al.: Improved orientation estimation and detection with hybrid object detection networks for automotive radar (2022)
17. Iovescu, C., Rao, S.: The fundamentals of millimeter wave sensors. In: Texas Instruments, pp. 1–8 (2017)
18. Albagory, Y.: An efficient conformal stacked antenna array design and 3D-beamforming for UAV and space vehicle communications. Sensors $21(4)$, 1362 (2021)
19. Skolnik, M.I.: Radar handbook. McGraw-Hill Education (2008)

20. Stoica, P., Li, J., Xie, Y.: On probing signal design for MIMO radar. IEEE Trans. Sig. Process. **55**(8), 4151–4161 (2007)
21. Meyer, M., Kuschk, G., Tomforde, S.: Graph convolutional networks for 3D object detection on radar data. In: IEEE/CVF International Conference on Computer Vision (ICCV) Workshops (2021)
22. Roddick, T., Kendall, A., Cipolla, R.: Orthographic feature transform for monocular 3D object detection. CoRR (2018). http://arxiv.org/abs/1811.08188
23. Meyer, M., Nekkah, S., Kuschk, G., Tomforde, S.: Automotive object detection on highly compressed range-beam-doppler radar data. In: EuRAD (2022)

RPR-Net: A Point Cloud-Based Rotation-Aware Large Scale Place Recognition Network

Zhaoxin Fan[1]([⊠]), Zhenbo Song[3], Wenping Zhang[1], Hongyan Liu[2], Jun He[1], and Xiaoyong Du[1]

[1] Key Laboratory of Data Engineering and Knowledge Engineering of MOE, School of Information, Renmin University of China, Beijing 100872, China
{fanzhaoxin,wpzhang,hejun,duyong}@ruc.edu.cn
[2] Department of Management Science and Engineering, Tsinghua University, Beijing 100084, China
hyliu@tsinghua.edu.cn
[3] School of Computer Science and Engineering, Nanjing University of Science and Technology, Nanjing 210094, China
songzb@njust.edu.cn

Abstract. Point cloud-based large scale place recognition is an important but challenging task for many applications such as Simultaneous Localization and Mapping (SLAM). Taking the task as a point cloud retrieval problem, previous methods have made delightful achievements. However, how to deal with catastrophic collapse caused by rotation problems is still under-explored. In this paper, to tackle the issue, we propose a novel Point Cloud-based Rotation-aware Large Scale Place Recognition Network (RPR-Net). In particular, to solve the problem, we propose to learn rotation-invariant features in three steps. First, we design three kinds of novel Rotation-Invariant Features (RIFs), which are low-level features that can hold the rotation-invariant property. Second, using these RIFs, we design an attentive module to learn rotation-invariant kernels. Third, we apply these kernels to previous point cloud features to generate new features, which is the well-known SO(3) mapping process. By doing so, high-level scene-specific rotation-invariant features can be learned. We call the above process an Attentive Rotation-Invariant Convolution (ARIConv). To achieve the place recognition goal, we build RPR-Net, which takes ARIConv as a basic unit to construct a dense network architecture. Then, powerful global descriptors used for retrieval-based place recognition can be sufficiently extracted from RPR-Net. Experimental results on prevalent datasets show that our method achieves comparable results to existing state-of-the-art place recognition models and significantly outperforms other rotation-invariant baseline models when solving rotation problems.

J. He—111.

Supplementary Information The online version contains supplementary material available at https://doi.org/10.1007/978-3-031-25056-9_45.

Keywords: Point cloud · Place recognition · Rotation-invariant · Dense network architecture

1   Introduction

Autonomous Driving (AD) and Simultaneous Localization and Mapping (SLAM) are becoming increasingly more important in some practical applications. Moreover, large scale place recognition often acts as the spine of them. It plays an indispensable role in a SLAM system or an autonomous driving system. Specifically, on the one hand, place recognition could provide a self-driving car with accurate localization information in a high definition map (HD Map). On the other hand, the recognition result is always used for loop-closure detection [2,17] in a SLAM system.

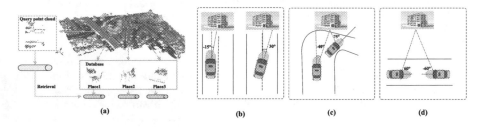

Fig. 1. Task definition and motivation. (a) an illustration of point cloud-based large scale place recognition. (b) change lanes causes rotation problems. (c) turning a corner causes rotation problems. (d) driving from different direction causes rotation problems.

Image and point cloud are the two kinds of most frequently used data formats for large scale place recognition. However, the former is proven to be sensitive to illumination changes, weather changes, etc., making methods [1,31] based on it difficult to achieve robust performance. In contrast, point cloud scanned by LiDAR is more reliable because it is calculated from reflected lasers, which is less sensitive to the above environmental changes. Therefore, for performance purpose, we study point cloud-based large scale place recognition in this paper.

Figure 1(a) illustrates a pipeline of point cloud-based large scale place recognition methods: 1) A HD Map of an area is firstly pre-built as a strong prior. 2) Then, the HD Map is uniformly divided into a database of submaps with accurate localization information. 3) When a car or a robot travels the area, the scanned point cloud at hand should be compared with each of these submaps to find the most similar location-available submap to determine the location of the car. The first two steps are easy to do offline, while the third step, to find the most similar sample from a large amount of topologically variable point clouds in an on-the-fly retrieval manner, is harder to achieve. Apparently, the place recognition problem is essentially the foundation of such a challenging retrieval problem.

For the point cloud retrieval problem, challenges lay in how to encode point clouds into discriminative and robust global descriptors for similarity calculation. In recent years, many deep learning-based point cloud encoding models [7,8,11, 15,23,25,28,33] have been proposed and have achieved acceptable recognition performances. However, we observe that nearly all existing methods are facing a serious common defect: they are suffered from catastrophic collapse caused by rotation problems, i.e., when the scanned point clouds are rotated compared to the pre-defined submaps, their performance would significantly drop (evidenced in Fig. 4 and Table 1). Undoubtedly, this would make the algorithm unreliable and cause ambiguities for decision making. What's more, safety risks would rise if these models are deployed to self-driving cars and robots. What makes things worse is that rotation problems are very common in practical scenarios. As shown in Fig. 1(b), (c) and (d), when a vehicle or robot changes lanes, turns a corner or drives from different directions, the scanned point cloud is equivalent to being rotated relative to the submaps in the HD map. Hence, to increase reliability and safety, it is very necessary and important to take the *rotation problems* into consideration when designing point cloud-based place recognition models.

In this paper, we propose to solve the problem by learning rotation-invariant global descriptors. In particular, we propose a novel point cloud-based rotation-aware large scale place recognition model named RPR-Net. To power RPR-Net with the rotation-invariant ability, we first design a kind of Attentive Rotation-Invariant Convolution named ARIConv, which learns rotation-invariant features in three steps:

First, taking a point cloud as input, three kinds of Rotation-Invariant Features (RIFs) named Spherical Signals (SS), Individual-level Local Rotation-Invariant Features (ILRIF) and Group-level Local Rotation-Invariant features (GLRIF) are extracted. The three kinds of features hold the rotation-invariant property of the point cloud from three different perspectives: invariance of individual points in spherical domain, invariance of individual-level relative position in a local region, and invariance of point distribution in a local region. All the three kinds of features are low-level features which are representative. Second, taking RIFs as input, we propose an attentive module to learn rotation-invariant convolutional kernels. In the attentive module, momentous RIFs that are more relevant to a specific scene are adaptively highlighted, such that the most significant rotation-invariant related hidden modes can be enhanced. To do so, we implement a RIF-wise attention in the high-level latent space. Through this way, the model could learn how to re-weight different RIFs' mode to generate more representative convolutional kernels to achieve the scene-specific SO(3) mapping process. Third, we apply the learned convolutional kernels to features of previous layers to obtain high-level features of the current layer. Due to that the above process is a SO(3) mapping process that can effectively inherit the rotation-invariant property of both RIFs and the input features, both the learned kernels and the corresponding convolved point cloud features would perpetually be strictly rotation-invariant.

To complete the large scale place recognition task, we further propose a dense network architecture to build RPR-Net using ARIConvs, which benefited from learning rotation-invariant features as well as strengthening semantic character-istic of scenes. After trained, RPR-Net would predicted robust and discriminative global rotation-insensitive descriptors from raw point clouds. Then, recognition-by-retrieving can be achieved.

We have conducted extensive experiments following the evaluation setting of top rotation-invariant related papers [29, 30]. Experimental results show that our proposed model can achieve comparable results to existing state-of-the-art large scale place recognition models. Meanwhile, our model also significantly outper-forms other strong rotation-invariant baselines. What's more, the advantage of our novel proposed model is even more remarkable when dealing with rotation problems of point clouds. Specifically, when the point clouds are rotated at dif-ferent rotation levels, recognition accuracy of our model is almost constant, yet all state-of-the-art place recognition competitors fail to work.

Our contributions can be summarized as:

- We propose a novel model named RPR-Net, which, to the best of our knowl-edge, is the first strictly rotation-invariance dense network designed for point cloud-based large scale place recognition.
- We propose an Attentive Rotation-Invariant Convolution operation, which learns rotation-invariant by mapping low-level rotation-invariant features into attended convolutional kernels.
- We achieve the state-of-the-art performance when dealing with rotation prob-lems and achieve comparable results with most of existing methods on several original non-rotated benchmarks.

2   Related Works

2.1   Rotation-Invariant Convolution

In our work, the attentive rotation-invariant convolution is a key component. In early years, deep learning models like PoinetNet [19], PointNet++ [20] and DGCNN [26] try to use a T-Net [19] module to make models less sensitive to rota-tion. It learns a rotation matrix to transform a raw point cloud into canonical coordinates. Though operating on raw point clouds, T-Net is not strictly rotation-invariant. In fact, it is challenging to achieve strictly rotation-invariant using a normal convolution operation [3, 5, 22]. Methods like Spherical cnns [4, 6] propose to project 3D meshes onto their enclosing spheres for learning global rotation-invariant features. They are not suitable for raw point cloud-based methods and in-efficient. To solve the issue, Sun et al. [24] propose SRINet to learn point projec-tion features that are rotation-invariant. Zhang et al. [34] design RIConv, which utilizes low-level rotation-invariant geometric features like distances and angles to learn convolutional weights for learning rotation-invariant features. Kim et al. [10] propose to utilize graph neural networks to learn rotation-invariant represen-tations. Li et al. [12] introduce a region relation convolution to encode both local and non-local informations. Their main goal is alleviating the inevitable global information loss caused by rotation-invariant representations. Recently, You et al.

[29,30] propose PRIN, where an adaptive sampling strategy and a 3D spherical voxel convolution operation are introduced to learn point-wise rotation-invariant features. These works, though are strictly rotation-invariant, are tailored for point cloud classification or segmentation. They either cannot learn sufficient scene-level features, or are too memory expensive, making them not applicable to the large scale place recognition task. In our work, the ARIConv is designed to construct the dense RPR-Net, which can learn both powerful geometric features and powerful semantic features as well as keeping the rotation-invariant property. To our knowledge, we are the first to design strictly rotation-invariant convolution for large scale place recognition.

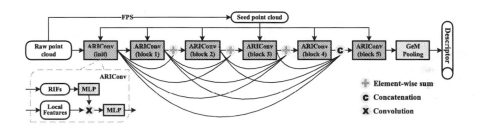

Fig. 2. Illustration of network architecture.

2.2   Large Scale Place Recognition

According to available data formats, large scale place recognition methods can be categorized into image-based methods and point cloud-based methods. Both methods share the same scheme: first encode scenes into global descriptors and then use K-Nearest-Neighbors (KNNs) algorithm to retrieve similar location-available scenes. Commonly, designing better global descriptors attracts more research attention. As described before, image-based methods [1,31] are sensitive to changes of illumination, weather, season, etc. Thus, they are not reliable. Compared with them, point cloud-based methods are more reliable. Point-NetVLAD [25] is the first point cloud-based deep learning method that achieves successful place recognition results. It uses PointNet [19] and NetVLAD [1] to learn global descriptors. PointNetVLAD only takes advantage of global features and neglects the importance of extracting local/non-local features, therefore, a line of works propose different ways to improve it. Methods like [7,15,23] employ Graph Convolutional Networks [26] to capture better local features, while methods like [7,23,28,33] use self-attention mechanism [27] to capture better non-local features to improve performance. There are also some methods [8,11] propose to voxelize unstructured point clouds into regular voxels for learning global descriptors, which also achieve acceptable place recognition results. Evidenced by our experimental results, these existing methods are greatly suffered from rotation problems. In this paper, we propose RPR-Net and ARIConv, which demonstrate notable advantages over existing point cloud-based large scale place recognition methods.

3   Methodology

3.1   Overview

Problem Statement: Our goal is to achieve place recognition when a car or robot travels around an area as shown in Fig. 1(a). Given a HD Map M of an area, we first divide it into a database of submaps represented as point clouds, denoted by $D = \{s_1, s_2, \cdots, s_m\}$, where m is the number of submaps. Each submap is attached with its unique localization information. When the car/robot travels around the area, it would scan a query point cloud q. We need to find where the car/robot is in the HD Map utilizing D and q. To achieve the goal, we formulate it as a retrieval problem, defined as: $s^* = \phi(\varphi(D|\Theta), \varphi(q|\Theta))$, where φ is the deep learning model encoding all point clouds into global descriptors, Θ is model parameters, ϕ is a KNN searching algorithm, and s^* is the most similar submap to q. After s^* is retrieved, we regard q and s^* share the same localization information, i.e., the place recognition task at current location is achieved. In this paper, we focus on designing a novel deep learning model φ to learn robust and discriminative rotation-invariant point cloud descriptors. We formulate φ as our RPR-Net.

Network Architecture: We first introduce the network architecture of RPR-Net. To learn representative descriptors, we expect RPR-Net be equipped with the following powers: 1) It should be rotation-invariant. 2) Descriptors learned from it can represent the scene geometry well. 3) High-level scene semantics can be effectively extracted. With all three aspects being satisfied, the final learned features would be powerful enough for scene description and simlarity calculation. To achieve the three goals, we propose ARI-Conv (detailed later) and a dense network architecture as shown in Fig. 2. Specifically, 6 ARIConvs and one GeM Pooling [21] layer are used to constitute the skeleton of the network. The former is used to learn high-level point-wise rotation-invariant features, while the later is used to aggregate the these features into a global descriptor. Note that ARI-Conv mainly explores how to learn rotation-invariant utlizing geometric characteristics. Therefore, the skeleton is principally equipped the two first two kinds powers, while learning high-quality scene semantics is still hard to achieve if ARIConv layers are simply stacked. To this end, to learn the third power, we adopt the *densely connected* idea [9,14,32] to design a dense network architecture, whose benefits has been demonstrated in many works. To implement, for ARIConv block 1 to 4, we choose to take the element-wise sum of all previous layers' features as their input. Then, we concatenate outputs of the 4 layers and regard it as the input of the last ARIConv layer (ARIConv block 5), which learns the final point-wise rotation-invariant features. We use element-wise sum in block 1 to 4 rather than use concatenation like in [7] mainly due to that element-wise sum is much more efficient. Through such a process, powerful and discriminative rotation-invariant global descriptors can be obtained and later used for efficient retrieval-based place recognition. Next, we introduce the ARIConv and its attributes in detail.

3.2   Low Level Rotation Invariant Features

ARIConv's duty is to learn point-wise rotation-invariant features. As mentioned before, its process consists of three steps. The first step is to extract low-level Rotation-Invariant Features (RIFs) from a point cloud, which will be used to learn rotation-invariant kernels and features later. In our work, each point's RIFs are extracted in a local neighbourhood level. Specifically, given a input point cloud $p \in R^{N \times 3}$, we first use Farthest-Point-Sampling (FPS) to sample a seed point cloud $p_s \in R^{N_s \times 3}$. Then, for each point of p_s, we search the K nearest neighbor of it from p, which can be defined as a matrix of dimension $N_s \times K \times 3$. And the RIFs are extracted from each $K \times 3$ local group. For each points, we extract three kinds of RIFs: Spherical Signals, Individual-level Local Rotation-Invariant Features and Group-level Local Rotation-Invariant Features. We choose these RIFs due to three reasons.: 1) They are extracted in a local region, which can better represent scene details and defend outliers. 2) Most of them are low-level geometry scalers, so that they can preserve the scene geometry better. 3) They are strictly rotation-invariant and are easy to be calculated. These advantages would benefit the final retrieval task. Next, we introduce them in detail.

Spherical Signals. Suppose x_i is any point in point cloud p. The spherical signals are obtained by transforming x_i into its spherical coordinate: $(s(\alpha, \beta), h)$, where $s(\alpha, \beta) \in S^2$ denotes its location in the unit sphere, and $h \in H = [0, 1]$ denotes the radial distance to the coordinate origin, as shown in Fig. 3(a). It has been proven in many works [4,6,30] that the spherical signals constructed from the unit spherical space is rotation-invariant. Therefore, we also adopt spherical signals as features for learning rotation-invariant kernels in our work. For convenient, we use the method proposed in SPRIN [30] to build the spherical signal f operating on the raw sparse point cloud. Besides, considering that the radial distance h is also rotation-invariant since it is a scalar and the coordinate origin will never be changed, we also adopt it as part of the rotation-invariant spherical signals. In total, for an input point cloud, we can get a dimension of $N_s \times K \times 2$ spherical signals.

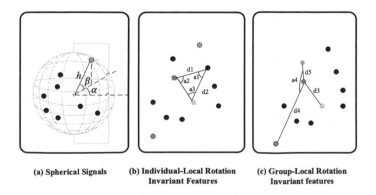

(a) Spherical Signals   (b) Individual-Local Rotation   (c) Group-Local Rotation
                               Invariant Features   Invariant features

Fig. 3. Illustration of three kinds of RIFs.

Individual-Level Local Rotation-Invariant Features. Besides spherical signals, another representation that is constant rotation-invariant is the relative distance between two points as well as the angles between two vectors. Therefore, inspired by RIConv [34], we extract five features for each point as shown in Fig. 3(b), where the red point is defined as the center point (or seed point) x_i of a $K \times 3$ local group, the yellow point is the mean center of all points in the local group, and black points are the K-nearest neighbors of x_i. It is obviously that for each individual black point, no matter how the point cloud is rotated, features $[d_1, d_2, a_1, a_2, a_3]$ will never be changed, i.e., they are rotation-invariant, where d_i means distance and a_i means angles. In total, for the whole point cloud, we can get a dimension of $N_s \times K \times 5$ individual-level local rotation-invariant features. Note that, for each individual black point (neighbor point) in Fig. 3(b), the five features are unique and different from that of other points.

Group-Level Local Rotation-Invariant Features. Individual-level local rotation-invariant features have described the invariance of each individual points in a local group, while the global invariance of the local group itself has not been characterized. For a group, we find that what insensitive to rotation is the *mean/min/max* property between all points. Therefore, as shown in Fig. 3(c), we extract $[d_3, d_4, a_5, a_4]$ as our group-level local rotation-invariant features to describe the characteristic of the local group itself. In Fig. 3(c), the red point is the center point (or seed point) x_i of a $K \times 3$ local group, the yellow point is the mean center of all points in the local group and the blue and purple point are the closest and farthest neighbors of the seed point x_i among the K neighbors, respectively. For N_s local groups, we can get $N_s \times 4$ group-level local rotation-invariant features. In order to keep the dimension of it to be consistent with the other two kinds of RIFs, we repeat it K times and finally form a $N_s \times K \times 4$ feature matrix. From another point of view, it means all points in a local group share the same GLRIFs. Note both the ILRIFs and GLRIFs are strong in expressing scene geometry besides learning rotation-invariant features, benefited from the inherent attributes of relative distances and relative angles between different points.

3.3   Attentive Rotation-Invariant Convolution Operation

Through the above process, we can extract a $N_s \times K \times 11$ RIFs matrix. Taking the matrix as input, the second and the third steps are to generate rotation-invariant kernels and apply these kernels to input features, i.e., to conduct the convolution operation.

Kernels Generation: We take three aspects into mind when learning convolutional kernels. First, they should be rotation-invariant. Second, they should be equipped with strong scene representation ability. Third, they should be computational-efficient. To meet these requirements, we choose to use a shared MLP [19] to map the RIFs into convolutional kernels. Since the MLP processes each point individually, it is obviously rotation-invariant and it has been proven in many works [19] that MLP is both effective and efficient. Utilizing the shared MLP, we can get a $N_s \times K \times C_{in}C_{out}$ kernel matrix κ. The kernel matrix can be then used for convolving input features of the seed point cloud p_s. However, simply using kernels learned by a shared MLP neglects the fact that each RIF may play a different role in generating the kernel. This motivates us that we should treat different RIFs with different importance. To this end, we decide to introduce a RIF-wise attention mechanism in the latent space to solve this problem. Specifically, in κ, we propose to multiply each element of $C_{in}C_{out}$ channels with a learned attention score to highlight more important RIFs in an implicit mode. The attention mechanism can be easily implemented by:

$$\kappa = \hat{\kappa} \cdot sigmoid(fc(avg\_pool(\hat{\kappa}))) \tag{1}$$

where $\hat{\kappa}$ is the kernels before being attended. Finally, we get a $N_s \times K \times C_{in}C_{out}$ attentive kernel matrix κ. We reshape it into $N_s \times K \times C_{in} \times C_{out}$ and use it to obtain more high-level point-wise rotation-invariant point cloud features later. We call the above process an attentive module. Note again, the purpose of the attentive module is not to attend individual RIF, but to promote the model to learn how to re-weight different RIFs' latent mode, so that more representative convolutional kernels can be generated.

Convolution Operation: After the convolutional kernels are learned, high-level output features are obtained by applying the kernels to the input features using a convolution operation. Suppose the input features of the seed point cloud is $f_s \in R^{N_s \times C_{in}}$. We use the K-nearest neighbor indices at the previous step to construct a feature matrix f_{sg}. The dimension of f_{sg} is $N_s \times K \times C_{in}$. Taking f_{sg} and κ as input, the convolution operation for each point can be implemented as:

$$f_n = \sum_{k=1}^{K} \sum_{c_{in}=1}^{C_{in}} \kappa(n, k, c_{in}) f_{sg}(n, k, c_{in}) \tag{2}$$

where $\kappa(n, k, c_{in}) \in C_{out}$ is vector, $f_{sg}(n, k, c_{in})$ is scalar and f_n is the convolved features. Totally, we will get a $N_s \times C_{out}$ final convolved features matrix. Then, we use a shared MLP to get the final output rotation-invariant features $f_{so} \in R^{N_s \times C_{out}}$. Since the kernels are rotation-invariant, the above convolutional process is actually a SO(3) mapping process.

Till now, we have introduced the RPR-Net's dense network achitecture and the proposed ARIConv. As stated before, we use a GeM Pooling [21] layer to aggregate learned point-wise features into a final global descriptor. Since GeM Pooling only operating on each point individually and its average pooling operation is channel wise, there is no doubt that it is rotation-invariant. Therefore, our whole network architecture is rotation-invariant. One may also notice that each ARIConv layer takes a seed point cloud as its input. For efficiency, we let all ARIConv layers share the same seed point cloud.

4   Experiments

4.1   Dataset

To verify the effectiveness of our method, we use the benchmark datasets proposed by [25], which are now the most authoritative and representative datasets

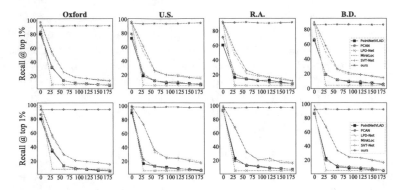

Fig. 4. Comparison of our method with state-of-the-art place recognition methods at different rotation levels. The top row is the result at the baseline training setting; the bottom row is the result at the refined training setting. The values along the horizontal axis represents different maximal rotation degrees.

for point cloud-based large scale place recognition. The benchmark consists of four datasets: one outdoor dataset called Oxford generated from Oxford Robot-Car [16] and three in-house datasets: university sector (U.S.), residential area (R.A.) and business district (B.D.). Each point cloud in these datasets contains 4096 points after removing the ground plane. Each point cloud is shifted and rescaled to be zero mean and inside the range of [–1, 1]. Therefore, the impact of translation is filtered. Each dataset contains 21,711/400/320/200 submaps for training and 3,030/80/75/200 scans for testing for Oxford., U.S., R.A. and B.D., respectively. During training, point clouds are regarded as correct matches if they are at maximum 10m apart and wrong matches if they are at least 50m

Table 1. Quantitative comparison of our methods with sate-of-the-art place recognition models at the baseline training setting. Top rows are results on the original datasets. Bottom rows are results when both submaps and query point clouds are randomly rotated along three axis.

	Avg recall at top 1%(w/o rotation)				Avg recall at top 1 (w/o rotation)			
	Oxford	U.S.	R.A.	B.D.	Oxford	U.S.	R.A.	B.D.
PointNetVLAD	80.3	72.6	60.3	65.3	81.0	77.8	69.8	65.3
PCAN	83.8	79.1	71.2	66.8	83.8	79.1	71.2	66.8
LPD-Net	94.9	96.0	90.5	89.1	86.3	87.0	83.1	82.3
Minkloc3D	**97.9**	95.0	91.2	88.5	93.0	86.7	80.4	81.5
SVT-Net	97.8	**96.5**	**92.7**	**90.7**	**93.7**	**90.1**	**84.3**	**85.5**
Ours	92.2	94.5	91.3	86.4	81.0	83.2	83.3	80.4
	Avg recall at top 1%(w/ rotation)				Avg recall at top 1 (w/ rotation)			
	Oxford	U.S.	R.A.	B.D.	Oxford	U.S.	R.A.	B.D.
PointNetVLAD	5.0	5.8	5.8	4.0	1.6	2.1	2.6	2.5
PCAN	5.2	5.7	4.5	4.4	1.7	1.7	1.6	2.3
LPD-Net	7.2	4.8	5.2	4.2	2.2	1.4	1.1	2.2
Minkloc3D	11.9	14.1	11.7	13.3	4.7	6.8	6.5	8.5
SVT-Net	12.1	14.0	9.6	12.9	4.8	6.4	5.1	8.6
Ours	**92.2**	**93.8**	**91.4**	**86.3**	**81.1**	**83.3**	**82.0**	**80.0**

apart. During testing, the retrieved point cloud is regarded as a correct match if the distance is within 25m between the retrieved point cloud and the query scan.

4.2   Implementation Details

We use Triplet loss to train our model due to its effectiveness, following [8,11]. The output feature dimension of all ARIConv layers is 64 except the final layer *ARIConv block 5*, which is 256. Therefore, the dimension l of the final global descriptor is also 256. The number of neighbors K in each local group is set as 32. Following previous works [8,11], we adopt two training settings. The baseline training setting only uses the training set of the Oxford dataset to train models and the models are trained for 40 epochs. The refined training setting addtionally adds the training set of U.S. and B.D except using the Oxford dataset and the models are trained for 80 epochs. We use average recall at top 1% and average recall at top 1 as evaluation metrics following all previous works. We implement our model using PyTorch [18] platform and optimize it using the Radam optimizer [13]. Random jitter, random translation, random points removal and random erasing augmentation are adapted for data augmentation during training. To verify our model's superiority in solving problems caused by rotation, we also follow existing top rotation-invariant related papers [29,30] to randomly rotate point clouds for evaluation, which is a fair setting since LiDAR point cloud is collected around. More details can be found in the **supplementary material**.

4.3   Results

In this subsection, to verify the effectiveness and robostness of our model, we first quantitatively compare our method with state-of-the-art large scale place recognition methods including PointNetVLAD [25], PCAN [33], LPD-Net [15], MinkLoc3D [11] and SVT-Net [7]. We also compare our method with strong rotation-invariant baseline models RIConv [34] and SPRIN [30]. Qualitative verification and complexity analysis are also included.

Quantitative Comparison with Place Recognition Models: In Fig. 4, we compare our method with existing top place recognition methods at both the baseline training setting and the refined training setting. The point clouds are randomly rotated along the gravity axis by different rotation levels. It is the most common case that point clouds are rotated along the gravity axis compared to submaps in practical scenarios. The compared methods are trained by using random rotation as data augmentation. Note that methods like LPD-Net have tried to used T-Net to solve the rotation problems. It can be seen that all compared methods fail even the point clouds are only rotated by a small degree, while performance of our method is almost constant, which significantly demonstrates that our model is more robust towards rotation compared to other methods. The poor performance of other methods verifies that data augmentation and T-Net are not sufficient enough to make the trained models be strong enough to deal with rotation problems. In contrast, our RPR-Net, for the first time, learns strictly rotation-invariant global descriptors, which are much more robust and reliable.

Table 2. Quantitative comparison of our method with sate-of-the-art rotation-invariant baseline models at the baseline training setting. Top rows are results on the original datasets. Bottom rows are results when both submaps and query point clouds are randomly rotated along three axis.

	Avg recall at top 1% (w/o rotation)				Avg recall at top 1 (w/o rotation)			
	Oxford	U.S.	R.A.	B.D.	Oxford	U.S.	R.A.	B.D.
RIConv	88.2	90.3	87.3	82.6	73.3	75.5	74.8	75.2
SPRIN	89.5	92.1	88.2	83.7	79.2	78.5	78.7	80.0
Ours	**92.2**	**94.5**	**91.3**	**86.4**	**81.0**	**83.2**	**83.3**	**80.4**
	Avg recall at top 1%(w/ rotation)				Avg recall at top 1 (w/ rotation)			
	Oxford	U.S.	R.A.	B.D.	Oxford	U.S.	R.A.	B.D.
RIConv	88.3	90.2	86.8	83.2	73.4	76.1	74.9	76.2
SPRIN	89.2	91.8	89.2	83.6	79.0	78.8	79.0	76.2
Ours	**92.2**	**93.8**	**91.4**	**86.3**	**81.1**	**83.3**	**82.0**	**80.0**

To see if our method can deal with more challenging cases, we randomly rotate the point clouds along the three axis following the evaluation setting of [29,30]. Results on the baseline training setting are shown in Table 1. The models' performances under non-rotated situation are also included. We can find that when point clouds are not rotated (the original datasets), our method achieves competitive results with existing methods, which means except to keep the rotation-invariant property, our model also retains the ability of representing as much scene geometries and semantics in the global descriptor as possible. Then, when all point clouds are randomly rotated, our method achieves the state-of-the-art performance. Compared to the non-rotated situation, there is almost no performance drop. However, all other methods almost don't work at all. The above analysis again demonstrates that our model is superior to other competitors. Note again that rotation problems would frequently happen in practical scenarios as shown in Fig. 1, therefore, according the above experimental results, our model has inherent advantages in ensuring recognition reliability.

Quantitative Comparison with Rotation-Invariant Baseline Models: Since existing place recognition models all fail to handle rotation problems. To further demonstrate the superiority of RPR-Net, we adopt RIConv [34] and SPRIN [30], two state-of-the-art rotation-invariant backbones for classification and segmentation, as φ to learn rotation-invariant global descriptors. Both baseline models are carefully tuned to achieve their best performance. Results on the baseline training setting are shown in Table 2. Though all models achieve rotation-invariant, our model significantly outperforms RIConv and SPRIN. We believe the superiority of our method comes from three aspects: 1) We propose three kinds of RIFs to inherit the rotation-invariant property as well as extracting local geometry. 2) We propose an attentive module to learn powerful rotation-invariant convolutional kernels. 3) We propose a dense network architecture to better capture scene semantics. For results on the refined training setting, please see the **Supplementary Materials**.

Qualitative Verification: To verify if our model indeed learns point-wise rotation-invariant features, we visualize learned features of different layers using T-SNE in Fig. 5. It can be seen that no matter how we rotate the same point cloud, the output point-wise features keep unchanged. It indicates that RPR-Net indeed learns strictly rotation-invariant features. We attribute this to the strong power of the proposed RIFs and the learned rotation-invariant convolutional kernels.

Complexity Analysis: The total number of trainable parameters of our model is 1.1M, which is comparable to the current most light-weight models Minkloc3D (1.1M) and SVT-Net (0.9M). However, our model is much more robust towards rotation than them as discussed before. For running time, our model takes 0.283 s to process a point cloud scan on a V100 GPU, which is relatively slow. Therefore, there is still much room for running time reduction. We leave solving the limitation as our future work.

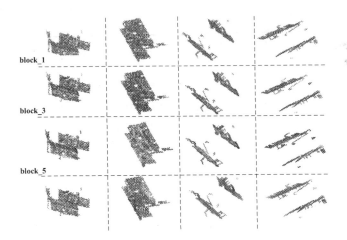

Fig. 5. T-SNE visualization of learned features.

Table 3. Results about impact of network architecture.

	Avg recall at top 1% (w/o rot)				Avg recall at top 1% (w rot)			
	Oxford	U.S.	R.A.	B.D.	Oxford	U.S.	R.A.	B.D.
w/o dense connection	90.1	91.4	90.4	84.0	90.3	91.6	89.8	83.4
w/ NetVLAD	90.4	90.8	87.6	81.3	90.6	90.1	86.2	81.4
Ours	**92.2**	**94.5**	**91.3**	**86.4**	**92.2**	**93.8**	**91.4**	**86.3**

4.4   Ablation Study

In this subsection, we experimentally analyse the necessity of each module and the correctness of the network designs. All experiments are conducted at the baseline training setting. The evaluation metric is average recall at top 1%. Both results on randomly rotated along three axes and results on non-rotated situation are shown. The impact of the network architecture and the impact of ARIConv designs are investigated in this part. For more experiments, please see the **Supplementary Materials**.

Impact of Network Architecture: The dense network architecture is one of the key componet of improving performance. It plays an important role in learning high quality high-level semantic features. As shown in Table 3, when it is removed, the performance decreases a lot. Then, expect using GeM Pooling for aggregating global descriptors, many works [15,25,33] utilize NetVLAD [1] to achieve this function. Therefore, we also replace GeM Pooling with NetVLAD to compare their performances. Obviously, from Table 3, we find that though both methods can achieve rotation-invariant, GeM Pooling is a better choice because it demonstrates better overall performance.

Table 4. Results about impact of ARIConv.

	Avg recall at top 1% (w/o rot)				Avg recall at top 1% (w/ rot)			
	Oxford	U.S.	R.A.	B.D.	Oxford	U.S.	R.A.	B.D.
w/o GLRIF	**92.7**	92	90.4	85.5	**92.7**	91.7	91.1	85.6
w/o ILRIF	46.9	59.1	59.8	48.7	46.6	59.5	58.5	49.2
w/o SS	91.4	94.1	89.7	85.1	91.4	93.6	90.0	85.3
w/o attentive	92.3	92.7	90.6	85.4	92.2	93.1	89.9	85.6
Ours	92.2	**94.5**	**91.3**	**86.4**	92.2	**93.8**	**91.4**	**86.3**

Impact of ARIConv: ARIConv is the most basic and pivotal component of our work. It core units are the three kinds of RIFs and the attentive operation. In the Table 4, we investigate the impacts of them. First, we conclude that all three kinds of RIFs can hold the rotation-invariant property of the learned features, because no matter using what kinds of RIFs, the place recognition performance doesn't decrease if we rotate the input point clouds. Second, we also find that the ILRIF is the strongest component since the performance greatly drops when it is removed. Meanwhile, the other two kinds of modules also contribute, though the contribution is not equal. Third, we note that the attentive module in ARIConv is also critical. When it is removed, the model's performances on the three indoor datasets decrease a lot, which means the generalization ability of the model is reduced. We also tried to directly mapping RIFs into high-dimensional features. However, in this case, the model fails to converge, therefore, we believe learning convolutional kernels is the right way of inheriting the rotation-invariant property of RIFs. In a word, all designs in ARIConv is necessary and effective.

5 Conclusion

In this paper, we propose a novel rotation-aware large scale place recognition model named RPR-Net, which focus on learning rotation-invariant global point cloud descriptors to solve the catastrophic collapse caused by rotation. We propose a novel ARIConv, which is equipped with three kinds of RIFs and an attentive module to conduct rotation-invariant convolution operation. Taking ARIConv as a basic unit, a dense network architecture is constructed. Experimental comparison with state-of-the-art place recognition models and rotation-invariant models has demonstrated the superiority of our method. Our research may be potentially applied to SLAM, robot navigation, autonomous driving, etc., to increase the robustness and the reliability of place recognition.

Acknowledgements. This work was supported by the National Social Science Major Program under Grant No. 20&ZD161 and National Natural Science Foundation of China (NSFC) under Grant No. 62172421.

References

1. Arandjelovic, R., Gronat, P., Torii, A., Pajdla, T., Sivic, J.: NetvLAD: CNN architecture for weakly supervised place recognition. In: Proceedings of the IEEE Conference on Computer Vision and Pattern Recognition, pp. 5297–5307 (2016)
2. Campos, C., Elvira, R., Rodríguez, J.J.G., Montiel, J.M., Tardós, J.D.: Orb-slam3: an accurate open-source library for visual, visual-inertial, and multimap slam. IEEE Trans. Rob. (2021)
3. Chen, C., Li, G., Xu, R., Chen, T., Wang, M., Lin, L.: Clusternet: deep hierarchical cluster network with rigorously rotation-invariant representation for point cloud analysis. In: CVPR, pp. 4994–5002 (2019)
4. Cohen, T.S., Geiger, M., Köhler, J., Welling, M.: Spherical CNNs. arXiv preprint arXiv:1801.10130 (2018)
5. Dym, N., Maron, H.: On the universality of rotation equivariant point cloud networks. arXiv preprint arXiv:2010.02449 (2020)
6. Esteves, C., Allen-Blanchette, C., Makadia, A., Daniilidis, K.: Learning SO(3) equivariant representations with spherical CNNs. International Journal of Computer Vision **128**(3), 588–600 (2019). https://doi.org/10.1007/s11263-019-01220-1

7. Fan, Z., Liu, H., He, J., Sun, Q., Du, X.: SRNET: A 3D scene recognition network using static graph and dense semantic fusion. In: CGF, vol. 39, pp. 301–311. Wiley Online Library (2020)

8. Fan, Z., Song, Z., Liu, H., He, J., Du, X.: SVT-Net: a super light-weight network for large scale place recognition using sparse voxel transformers. arXiv preprint arXiv:2105.00149 (2021)

9. Huang, G., Liu, Z., Van Der Maaten, L., Weinberger, K.Q.: Densely connected convolutional networks. In: CVPR, pp. 4700–4708 (2017)

10. Kim, S., Park, J., Han, B.: Rotation-invariant local-to-global representation learning for 3D point cloud. arXiv preprint arXiv:2010.03318 (2020)

11. Komorowski, J.: MinkLoc3D: point cloud based large-scale place recognition. In: WACV, pp. 1790–1799 (2021)

12. Li, X., Li, R., Chen, G., Fu, C.W., Cohen-Or, D., Heng, P.A.: A rotation-invariant framework for deep point cloud analysis. arXiv preprint arXiv:2003.07238 (2020)

13. Liu, L., et al.: On the variance of the adaptive learning rate and beyond. arXiv preprint arXiv:1908.03265 (2019)

14. Liu, Y., Fan, B., Meng, G., Lu, J., Xiang, S., Pan, C.: DensePoint: learning densely contextual representation for efficient point cloud processing. In: CVPR, pp. 5239–5248 (2019)

15. Liu, Z., et al.: LPD-Net: 3D point cloud learning for large-scale place recognition and environment analysis. In: CVPR, pp. 2831–2840 (2019)

16. Maddern, W., Pascoe, G., Linegar, C., Newman, P.: 1 year, 1000 km: the oxford robotcar dataset. The Int. J. Rob. Res. **36**(1), 3–15 (2017)

17. Memon, A.R., Wang, H., Hussain, A.: Loop closure detection using supervised and unsupervised deep neural networks for monocular slam systems. Robot. Auton. Syst. **126**, 103470 (2020)

18. Paszke, A., et al.: Automatic differentiation in pytorch . In; 31st Conference on Neural Information Processing Systems (NIPS 2017), Long Beach, CA, USA (2017)

19. Qi, C.R., Su, H., Mo, K., Guibas, L.J.: PointNet: deep learning on point sets for 3d classification and segmentation. In: CVPR, pp. 652–660 (2017)

20. Qi, C.R., Yi, L., Su, H., Guibas, L.J.: Pointnet++: deep hierarchical feature learning on point sets in a metric space. In: 30th Proceedings of Conference on Advances in Neural Information Processing Systems (2017)

21. Radenović, F., Tolias, G., Chum, O.: Fine-tuning CNN image retrieval with no human annotation. Trans. Pattern Anal. Mach. Intell. **41**(7), 1655–1668 (2018)

22. Rao, Y., Lu, J., Zhou, J.: Spherical fractal convolutional neural networks for point cloud recognition. In: CVPR, pp. 452–460 (2019)

23. Sun, Q., Liu, H., He, J., Fan, Z., Du, X.: DAGC: employing dual attention and graph convolution for point cloud based place recognition. In: ICMR, pp. 224–232 (2020)

24. Sun, X., Lian, Z., Xiao, J.: SriNet: learning strictly rotation-invariant representations for point cloud classification and segmentation. In: ACM MM, pp. 980–988 (2019)

25. Uy, M.A., Lee, G.H.: Pointnetvlad: deep point cloud based retrieval for large-scale place recognition. In: CVPR, pp. 4470–4479 (2018)

26. Wang, Y., Sun, Y., Liu, Z., Sarma, S.E., Bronstein, M.M., Solomon, J.M.: Dynamic graph CNN for learning on point clouds. Trans. Graphics **38**(5), 1–12 (2019)

27. Woo, S., Park, J., Lee, J.-Y., Kweon, I.S.: CBAM: convolutional block attention module. In: Ferrari, V., Hebert, M., Sminchisescu, C., Weiss, Y. (eds.) ECCV 2018. LNCS, vol. 11211, pp. 3–19. Springer, Cham (2018). https://doi.org/10.1007/978-3-030-01234-2_1

28. Xia, Y., et al.: SOE-Net: a self-attention and orientation encoding network for point cloud based place recognition. In: CVPR, pp. 11348–11357 (2021)
29. You, Y., et al.: Pointwise rotation-invariant network with adaptive sampling and 3d spherical voxel convolution. In: AAAI, vol. 34, pp. 12717–12724 (2020)
30. You, Y., et al.: PRIN/SPRIN: on extracting point-wise rotation invariant features. arXiv preprint arXiv:2102.12093 (2021)
31. Yu, J., Zhu, C., Zhang, J., Huang, Q., Tao, D.: Spatial pyramid-enhanced NetVLAD with weighted triplet loss for place recognition. TNNLS **31**(2), 661–674 (2019)
32. Yu, R., Wei, X., Tombari, F., Sun, J.: Deep positional and relational feature learning for rotation-invariant point cloud analysis. In: Vedaldi, A., Bischof, H., Brox, T., Frahm, J.-M. (eds.) ECCV 2020. LNCS, vol. 12355, pp. 217–233. Springer, Cham (2020). https://doi.org/10.1007/978-3-030-58607-2_13
33. Zhang, W., Xiao, C.: PCAN: 3D attention map learning using contextual information for point cloud based retrieval. In: CVPR, pp. 12436–12445 (2019)
34. Zhang, Z., Hua, B.S., Rosen, D.W., Yeung, S.K.: Rotation invariant convolutions for 3d point clouds deep learning. In: 3DV, pp. 204–213. IEEE (2019)

Learning 3D Semantics From Pose-Noisy 2D Images with Hierarchical Full Attention Network

Yuhang He[1] , Lin Chen[2](✉) , Junkun Xie[3] , and Long Chen[4]

[1] Department of Computer Science, University of Oxford, Oxford, UK
yuhang.he@cs.ox.ac.uk
[2] Institute of Photogrammetry and GeoInformation (IPI),
Leibniz University Hanover, Hanover, Germany
chen@ipi.uni-hannover.de
[3] School of Computer Science and Engineering, Sun Yat-sen University,
Guangzhou, China
xiejk3@mail2.sysu.edu.cn
[4] Institute of Automation, Chinese Academy of Sciences, Beijing, China
long.chen@ia.ac.cn

Abstract. We propose a novel framework to learn 3D point cloud semantics from 2D multi-view image observations containing pose errors. Normally, LiDAR point cloud and RGB images are captured in standard automated-driving datasets. This motivates us to conduct a "task transfer" paradigm so that 3D semantic segmentation benefits from aggregating 2D semantic cues. However, pose noises are contained in 2D image observations and erroneous prediction from 2D semantic segmentation renders the "task transfer" difficult. To consider those two factors, we perceive each 3D point using multi-view images and for every single image, a patch observation is employed. Moreover, the semantic labels of a block of neighboring 3D points are predicted simultaneously, enabling us to exploit the point structure prior. A hierarchical full attention network (HiFANet) is designed to sequentially aggregate patch, bag-of-frames and inter-point semantic cues. The hierarchical attention mechanism is tailored for different levels of semantic cues. Each preceding attention block largely reduces the feature size before feeding to the next attention block, making our framework slim. Experiment results on Semantic-KITTI show that the proposed framework outperforms existing 3D point cloud based methods significantly, requiring less training data and exhibiting tolerance to pose noise. The code is available at https://github.com/yuhanghe01/HiFANet.

Keywords: Attention network · Semantic segmentation · Pose noise

Supplementary Information The online version contains supplementary material available at https://doi.org/10.1007/978-3-031-25056-9_46.

1   Introduction

Directly learning from 3D point cloud is difficult. Challenges derive from four main aspects: First, 3D point cloud is massive and a typical Velodyne HDL-64E scan leads to millions of points. Processing such large data is prohibitively expensive for many algorithms and computation sources. Second, 3D point cloud is unstructured and unordered as well. It records neither the physical 3D world texture nor object topology information, which have often been used as important priors by image based environment perception methods [14,26,33]. Third, data imbalance issue. Due to the 3D physical world layout that particular categories conquer most of the space, captured 3D point cloud is often dominated by classes such as road, building and sidewalk. Other categories (*i.e.* traffic sign, poles, pedestrian) with minor point cloud presence but vital for self-driving driving scenario understanding and high-quality map generation are often overwhelmed by dominating classes. Lastly, capturing 3D point cloud is a dynamic process, resulting in inconsistent and nonuniform data sampling. Distant objects are much more sparsely sampled than close objects.

Most self-driving data collection platforms collect 3D point cloud and RGB images simultaneously, with the LiDAR scanner and camera being pre-calibrated and synchronized. This motivates us to transfer 3D point cloud segmentation to its 2D image based counterpart (we call "task transfer") so that the segmentation of point cloud can largely benefit from various matured 2D image semantic segmentation networks. However, such seemly-fascinating "task transfer" comes with a price: In real-scenario, LiDAR-Camera pose is often noisy so accurate 3D-2D correspondences are non-guaranteed. In addition, view-angle change easily results in distorted image observation. Moreover, 2D semantic segmentation method may also give erroneous predictions, as 2D semantic segmentation network is not designed or trained to be viewing direction invariant.

To tackle the aforementioned challenges, we first propose to perceive each 3D point from multi-view images so that bag-of-frame observations for each single 3D point are obtained. Multi-view image observation reduces the impact of the unfavoured viewing directions as it introduces extra semantic cues. Moreover, instead of looking into single-pixel of an image, we focus on a small patch-area around the pixel. The patch observation strategy mitigates 3D-2D correspondence error led by pose noise and further enables neural network to learn pose noise tolerant representation in a data-driven way. Moreover, we process a local group of spatially or temporally close 3D points at the same time, so that we can exploit 3D points structure prior (*i.e.* two points' spatial location). Actually, the local 3D point group and the corresponding 2D observation can be treated as seq2seq learning problem [35], where one sequence is 2D image and the other is 3D point cloud. To accommodate these different data representation properties, we propose a hierarchical fully attention network (HiFANet) to sequentially and hierarchically aggregate the patch observation, bag-of-frame observation and inter-point structural prior to infer the 3D semantics. Such hierarchical attention blocks design enables the neural network to learn to efficiently aggregate semantics at different levels. Moreover, the preceding attention block naturally reduces the feature representation size before feeding it to the next attention block, so the whole framework is slim by design.

In sum, our contribution is three fold: first, we propose to transfer 3D point semantic segmentation problem to its counterpart in 2D images. Second, to counteract the pose noise impact, we propose to associate each single 3D point with multi-view patch observation so that the neural network can learn to tolerate pose inaccuracy. Third, we formulate it as a seq2seq problem so that we can best exploit the structural prior arising from both 3D point cloud and 2D images to improve the performance.

2   Related Work

3D semantic segmentation can be divided into three main categories: point-based, voxel-based and 2D projection based methods [15,41].

Point based methods compute the features from points and can be categorized into three sub-classes [15]: Multi Layer Perceptron (MLP), point convolution and graph convolution based methods. MLP based method apply MLP directly on points to learn features, such as PointNet [31], HRNN [43], PointNet++ [32], PointWeb [46]. In comparison, point covolution based methods apply convolution on individual point. Representative works in this group are PointwiseCNN [19], PCNN [38], PointConv [40], RandLA-Net [17] and PolarNet [45]. In the third class, the points are connected with graph structure, graph convolution is further applied to capture more meaningful local information. Example works include DeepGCNs [24], AGCN [42], HDGCN [25] and 3DContextNet [44].

In voxel based methods, voxels divide 3D space into volumetric grids, which are used as input for 3D convolutional neural networks. The voxel used is either uniform [8,20,28] or non-uniform [12,34]. Methods in this group are restricted by the fact that the computation burden grows fast with the scale of scene. Consequently, the usage of those methods in large scale becomes impractical.

In projection based methods, point cloud is projected into synthetic but multi-view image planes and then 2D CNNs are used by each view, finally semantic results from mutliple views are aggregated [6,13,16,22], However, this idea is restricted by misinterpretation stem from sparse sampling of 3D points. Our work shares the similar idea to convert 3D point cloud to 2D plane, but we exploit 2D RGB images to assist 3D semantic segmentation and we rely on 2D semantic segmentation to predict 3D semantics.

In 2D semantic segmentation, FCN [27] is one of the first works using deep neural network for semantic segmentation by replacing the fully connected layer with fully convolution layers. The following works, e.g., SegNet [1] and [30], use more sophisticated way to encode the input image and decode the latent representation so that images are better segmented. Obtaining features at multiple scale is manipulated either at convolution kernel level or through pyramid structure. The former leads to the method of using dilated convolution and representative works are DeepLabV2 [3] and DeepLabV3 [5]. The latter is implemented in PSPN [47] and [11]. Also, attention mechanisms are used to weight features softy for semantic segmentation task in [4]. In this paper, we make use of the network proposed in [48] as our base feature extractor, since it uses synthetic predicting to scale up training data and the trained label is also robust, benefiting from the usage of the boundary relaxation strategy proposed in that paper.

This paper utilizes features from multi-view patches sampled from camera images, which are not accurately aligned with 3D point cloud, to benefit the semantic segmentation of 3D point cloud. In this context, the central issue is how to aggregate multi view image features in a sophisticated way so that 3D points can be better separated in the feature space spanned by those aggregated features.

3   Problem Formulation

We have a sequence of N 3D point cloud frames $\mathbf{P} = \{P_1, P_2, \cdots, P_N\}$, and framewise associated 3D point semantic label $\mathbf{C}$ and RGB image $\mathbf{I}$. Such data is collected by platform where LiDAR scanner and camera are carefully synchronized and pre-calibrated with noisy pose information $P_o = [R|t]$ (rotation matrix R and translation t). Moreover, the relative pose between any two neighboring point cloud frames can be obtained via IMU system. With the noisy pose, we can theoretically project any 3D point to any image plane. Off-the-shelf image semantic segmentation method [48] is adopted to get semantic result $\mathbf{S}$ for each image, each pixel of which consists of categorical semantic label and semantic-aware representation r. Our goal is to train a model $\mathcal{F}$ parameterized by θ to predict point cloud semantics from images $\mathbf{C} = \mathcal{F}(\mathbf{I}, \mathbf{S}|P_o, \theta)$.

4   Hierarchical Full Attention Network

The fundamental idea of designing our framework is two-fold: "task transfer" which learns 3D point cloud semantics from 2D images; further address accompanying challenges brought by the "task transfer" through a "learning" perspective by fully exploiting the potential of deep neural networks in a hierarchical way. Specifically, given the pose information between any 3D point cloud frame and any 2D image, we can obtain N patch observations $\{\mathcal{P}_1, \cdots, \mathcal{P}_N\}$ for each 3D point by projecting it to its neighboring image frames (we call bag-of-frames), where a patch observation $\mathcal{P}_i$ indicates a $k \times k$ squared patch centered at the pixel $[u_x, u_y]$ of the 3D point's i-th observation image frame. $[u_x, u_y]$ corresponds to the 3D point projection location with noisy pose information. In the meantime, a pre-trained 2D image semantic segmentation model is available, so we can get both the categorical semantic label s_j and the semantic-guided feature representation r_j for the j-th pixel in the patch. The feature representation r_j can be easily obtained by taking the penultimate layer activation of the model trained on 2D images. So the patch observation can be expressed as,

$$\mathcal{P} = \{(s_1, r_1)_i, \cdots, (s_{k^2}, r_{k^2})_i\}_{i=1}^{N} \tag{1}$$

Introducing patch observation instead of single-pixel observation in 2D image is to address pose noise challenge, which we will give detailed discussion in next

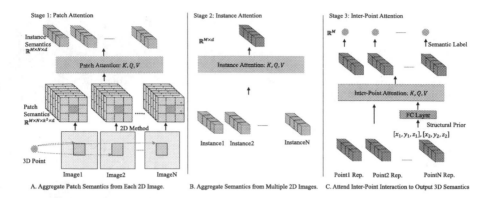

Fig. 1. HiFANet pipeline: Given the pose, we project M 3D points to their nearest top-N RGB images to get $k \times k$ patch observations. Off-the-shelf 2D image segmentation model is trained to get each patch's semantic feature representation as well as categorical semantic labels. HiFANet is a three-stage hierarchical fully attentive network. It first learns to aggregate patch representation into an instance representation (left image), then aggregates multiple image instances into one point-wise representation (middle image), and finally an inter-point attention module to attend structural and feature interaction among 3D points to output per-point elegant semantic feature representation, which is then used to predict the ultimate semantic label.

section. Instead of learning 3D semantic for each 3D point separately, we model M neighboring 3D points simultaneously, which benefits us to use 3D points structure prior to escalate the performance. For example, an intuitive spatial prior is that two spatially-close 3D points are much more likely to share the same semantic label than those lie far apart. In sum, our model takes M 3D points' Cartesian coordinates as well as each 3D point's N patch observations as input and outputs each 3D point's semantic label. It is worth noting that M 3D points forms a point sequence and $M \times N$ image observation forms another image sequence, the whole framework can be treated as a seq2seq task, either spatially or temporally. The framework input simply consists of image-learned semantic information (categorical label or feature presentation), no extra constraint is involved and we do not directly process 3D point cloud.

With "task transfer", the main task of our framework is to efficiently aggregate semantic clues arising from bag of 2D image frames. To this end, we propose a hierarchical full attention three-stage aggregation mechanism, in which we first learn to aggregate patch observation into an instance observation (i.e., single pixel observation in an image), and then learn to aggregate multiple instances in the bag-of-frames for each 3D point into 3D point wise observation, and finally attend all the structure prior and interaction between 3D points to output the target semantic label for each single 3D point. Our framework is fully attentive and invariant to images observation order permutation. The hierarchical attention mechanism design has two advantages: it first enables the neural network to fully learn specified attention tailed for different semantic representation, second it aggressively reduces the feature size so that we keep the whole framework slim.

4.1   Patch Attention for Patch Aggregation

Patch attention tends to aggregate the patch observation into a single-pixel observation. Within each $k \times k$ patch, we call the centered point the principle point and the remaining points are neighboring points. The basic idea behind the patch attention is to attend all points in the patch with a trainable weight before weighted-adding them together to generate one feature. Since the principle point records the most-confident 3D point semantic related feature representation, we add a short-cut connection between the principle point feature and the attended to feature representation. To minimize the computation cost, we adopt criss-cross like attention module [21] to attend all neighboring points to the principle point. Specifically, given the feature representation $r_i \in \mathbb{R}^{k \times k \times d}$, the principle points lies in $[\frac{k}{2}, \frac{k}{2}]$ and has feature representation f_p of length d, the output feature f_{pa} after patch attention can be expressed as,

$$f_{pa} = \sum_{j=1}^{k \times k} w_j \cdot V_j + f_p \tag{2}$$

where w_j is the learned weight for the j-th point in the patch. To learn the attention weight w, we draw inspiration from self-attention module [37] to learn a patch Key $K = \mathbb{R}^{k \times k \times d_1}$ and patch Query $Q = \mathbb{R}^{k \times k \times d_1}$ and a patch Value $V = \mathbb{R}^{k \times k \times d}$. The three parts can be efficiently learned via 1×1 2D convolution on the patch observation. To reduce the computation cost (usually $d_1 \ll d$), we set $d_1 = 64$ and $d = 256$. With K and Q we can further compute the scaled dot-product attention where the attended weight w can be obtained by,

$$w = \text{softmax}(\frac{Q_p K}{\sqrt{d_1}}) \tag{3}$$

Q_p is the principle point query. With Eq.(3), we can get the weight of each point to the principle point. The patch attention is a self-attention module, it requires no extra supervision and can efficiently attend the final single-pixel observation in with paralleling computation.

4.2   Instance Attention for Image Aggregation

Instance attention module takes $\mathbb{R}^{M \times N \times d}$ semantic feature as input, and aims to aggregate bag-of-frames features to get 3D point wise feature. We call the aforementioned patch-attention aggregated pixel-wise semantic representation in each image frame as an instance, because it represents an independent observation towards a 3D point. The multiple instances arising from bag-of-frames form an *Instance Set* [23,39], which means these instances are orderless, the final accurate semantic label may derive from an individual instance or multiple instances combination. To satisfy the instance set property, the instance attention module has to be order-permutation invariant. Commonly seen set-operators include max-pooling and average-pooling. In HiFANet, we first apply a self-attention layer like the patch attention block does to attend each instance by all the remaining instances. Finally, we apply average pooling to merge multiple instances into one instance representation.

4.3 Inter-point Attention for 3D Points Aggregation

Inter-point attention take $\mathbb{R}^{M \times d}$ semantic feature learned by instance attention module as input. Unlike the previous two attention modules that just focus on per-point semantic feature learning, inter-point attention module fully considers the interaction between 3D points, including the spatial structure interaction and semantic feature interaction. We adopt a Transformer [37] multi-head self-attention like network to construct the inter-point attention module. Specifically, the input feature is fed to learn per-point Key $K = \mathbb{R}^{M \times d_2}$ and per-point Query $Q = \mathbb{R}^{M \times d_2}$ as well as per-point Value $V = \mathbb{R}^{M \times d}$. To involve structural prior, we encode the relative Cartesian position difference between any two 3D points $p_i - p_j$. The Cartesian position difference is further fed to two consecutive fully connection layers to get the structural prior encoding K_{pe}, which is the same size of K. The original Key K is then updated by adding K_{pe},

$$K = K + K_{pe} \tag{4}$$

The updated K in Eq. (4) naturally contains the structural prior. With the Q and updated K, we can compute the attention weight for each single 3D point w.r.t the remaining 3D points, as is shown in Eq. (3). The attention weight is further applied to combine value V to get the final per-point semantic representation, which is further concatenated with a classification layer for semantic classification.

In sum, HiFANet sequentially and hierarchically aggregates patch semantics, instance semantics and inter-point semantics to learn semantic representation for each 3D point. It is fully attentive and learns compartmentalized and certain attention blocks w.r.t. different aggregation granularity separately. The preceding attention layer largely reduce the feature size before feeding it to the next layer, so the whole neural network is slim. Detailed HiFANet pipeline is shown in Fig. 1.

5 Discussion on HiFANet Design Motivation

The feasibility of such "task transfer" lies in the availability of the pose information between LiDAR scanner and the camera, which enables us to project 3D point cloud onto the image plane to get each 3D point's correspondence in the image plane. We hereafter call such correspondence as a 2D observation. The "task transfer" poses three main challenges that may jeopardize the performance.

1. Pose noise. Sensor calibration often suffers from internal and external noise. Noisy pose information leads to inaccurate 2D observations. This stays as the most prominent challenge.

2. View-angle. Projecting a cluster of point cloud belonging to a specific category (*i.e.* car) to an image plane often leads to distorted 2D observation. In severe cases, it leads to wrong observation due to the occlusion caused by view-angle difference.
3. Void projection. While LiDAR scanner scans in 360°, pinhole camera simply captures the forward-facing view. This mismatch of perception field inevitably leads to void projection in which point cloud cannot find observation in any image.

Addressing the above three challenges leads to our proposed framework. To mitigate the pose noise impact, we propose to use patch observation to replace pixel observation. Pixel-wise observation is fragile and sensitive to pose noise, a small change leads to totally different observation. Patch-wise, on the contrary, becomes much more resilient to pose noise because it covers possible observations potentially led by noisy pose. Moreover, introducing patch-wise observation avoids us directly optimizing $[R|t]$ in an iterative way. To address the view-angle and void projection issue, we propose to involve multiple observation arising from different view-angles. With the multi-view observations, we naturally obtain multiple clues for each 3D point.

5.1   Pose Noise and Patch Observation

The pose between LiDAR scanner coordinate system and camera coordinate system can be formulated as a rotation matrix R and translation T. A 3D point $[x, y, z]$ projects onto an image plane, the corresponding observation location $[u, v]$ in the 2D image plane is computed by,

$$[u, v, 1] = K[R|T] \cdot [x, y, z, 1]^T \tag{5}$$

Please note that the projected pixel location is normalized by its 3rd dimension. The pose noise of sensor calibration (between laser scanner and camera) renders the location of true projected point uncertain. However, in our approach, a patch is extracted and then the attention is learned to focus on the pixel closest to the true projected points.

In order to investigate the influence of the pose noise on the location of projected points, a toy simulation experiment is provided and illustrated in Fig. 2. As can be observed in Fig. 2, given the translation noise for the calibration between the LiDAR scanner and camera as 10cm and the rotation angle noise as 1°, the projection error on the image plane (1024 × 512 pixels) is around 40 pixels for near camera object points. Since the patch extracted on each camera view is within a $k \times k$ patch in the downsized feature maps (normally at 1/16 or 1/32 resolution), the information encoded in the image is then well preserved for the attention module to discover, although the pose noise exist.

5.2   View-Angle and Void Projection

View-angle easily leads to titled, occluded and even erroneous observation. A 3D point that is observed in one viewpoint (an RGB image) can be obstructed

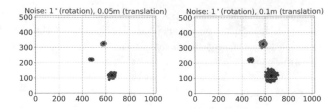

Fig. 2. The influences of pose noise to the projected point coordinates on image plane (in pixel) for 3D world points that are various distant from the camera plane (green: 5 m; yellow: 10 m; blue:20 m). The simulated noises of rotation angles are 1° (for each rotation angle) for both cases and the translation noise are 0.05 m (left), 0.1 m (right) for each of the three axes in world coordinate system. (Color figure online)

in another neighboring viewpoint. Traditional 3D reconstruction framework like structure-from-motion (SfM [9]) suffer from the same dilemma. The void projection jeopardizes the "task transfer" proposal because it causes large number of 3D points being 2D image unobserved.

To mitigate the two challenges, we propose to observe a single 3D point from multiple view-angles. On the one hand, it reduces the risk of one 3D point being observed at an unfavored view angle. On the other hand, it maximally ensures each 3D point cloud to be observed by at least one 2D image. Moreover, this strategy brings us the advantage of aggregating semantic clues arising from multiple images to better estimate semantics. Multiple view-angles observation can be efficiently aggregated in parallel in HiFANet.

6   Experiments and Results

We conduct experiment on the Semantic-KITTI dataset [2]. Since we need the inter-frame odometry information to project each 3D point to multiple RGB frames but the official provided test dataset (sequence 11–20) does not provide such information, we do not follow the official split but instead create the train/test/val split by ourselves and further train the methods involved in comparison with the split dataset from scratch. The same problem applies to other relevant datasets such as Waymo and CityScapes [7], and we only run experiment on Semantic-KITTI dataset in this paper.

Data Preparation. We run experiment on sequence 00–10 because the inter-frame odometry information is available for the 11 sequences, with which we can register all point cloud frames from a sequence to a uniform 3D coordinate so that each 3D point can be freely projected to any image plane. There are 13 semantic categories in total: *road, side-walk, building, fence, pole, traffic sign, vegetation, terrain, person, bicyclist, car, motorcycle* and *bicycle*. Some categories like *road, building, vegetation* and *terrain* dominate most of the points, whereas the others' portion is very small. An extra *unlabelled* background category is added. Sequence 06 is selected as test set as it contains all semantic categories and

Table 1. Quantitative result on semantic-KITTI [2] dataset. B, K and M mean billion, thousand and million, respectively. A.Acc: average accuracy

Category	Method	Train size	Param num	mIoU (↑)	A.Acc (↑)
Point based	PointNet [31]	2.8 B	3.53 M	0.036	0.105
	PointNet++ [32]	2.8 B	0.97 M	0.055	0.156
	RangeNet++(CRF) [29]	2.8 B	50.38 M	0.500	0.878
	RangeNet++(KNN) [29]	2.8 B	50.38 M	0.512	0.899
	KPConv [36]	2.8 B	18.34 M	0.466	0.868
	RandLANet [18]	2.8 B	1.24 M	0.578	0.913
Aggregation	BoF Num = 1	23 K	137 M	0.422	0.845
	BoF Num = 3	23 K	137 M	0.437	0.852
	BoF Num = 5	23 K	137 M	0.436	0.852
	Patch Size = 1	23 K	137 M	0.436	0.850
	Patch Size = 3	23 K	137 M	0.436	0.851
	Patch Size = 5	23 K	137 M	0.436	0.852
Multi-View	AvgPool_FC	0.5 M	0.04 M	0.451	0.872
	HiFANet_noPA	0.5 M	2.5 M	0.537	0.891
	HiFANet_noSP	0.5 M	2.7 M	0.561	0.920
	HiFANet	0.5 M	2.7 M	**0.620**	**0.933**

account for 20% data of the whole dataset. Sequence 08 is selected for validation and the remaining 9 sequences serve as training set. To get each 3D point's N neighboring image observations, we project it to its closest N image planes. N is set as 5 because it then covers 64% of the whole point cloud dataset with patch size $k = 5$ and 3D points number size $M = 10$. Those 3D points that fail to find N image observations are discarded during test but left for training point cloud based models. The image based semantic representation and semantic label are obtained from VideoProp [48] model pre-trained on KITTI dataset [10]. The semantic representation is a 256-d feature. Therefore, the size of patch semantics representation feeds to HiFANet is $5 \times 5 \times 256$. For the evaluation metric, we adopt the standard mIoU and average accuracy [2].

Methods to Compare. The first method category we tend to compare is pure 3D point cloud based semantic segmentation method. It helps us to gain an understanding of how far our proposed "task transfer" strategy goes, comparing with directly learning from 3D points. The second method category we compare with is the semantic result giving by deterministically aggregating the category semantic labels predicted by 2D image aggregation method, it gives us an understanding of how good image based semantic prediction methods can perform, by varying the observation number like image number and patch size. The third category is multi-view learning method which means designing neural network to learn from image semantic representations, as our proposed HiFANet does.

Ablation Study. we want to figure out the impact of the involvement of patch feature representation, structural prior on the performance. We thus test two

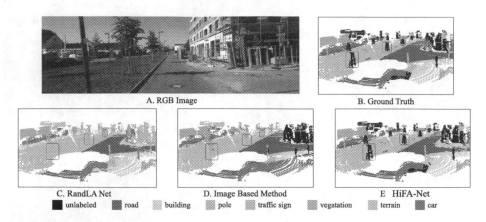

Fig. 3. Close-up visualization of various methods on unlabelled tree stake. While point based method erroneously classifies them as pole and image based method as terrain, HiFANet accurately recognizes it by fully combing 2D image based semantics and 3D structural priors.

HiFANet variants: reduce the patch size to 1 so no patch attention module is applied (**HiFANet_noPA**), no structural prior involvement in inter-point attention module (**HiFANet_noSP**). Moreover, to test the effectiveness of our proposed full attention network, we train another simple semantic aggregation network, in which we simply average-pool all the input feature (patch and instance feature) to get per-point feature, and further concatenate two full connection layer (of size 256, 128) to directly predict the semantic label (**AvgPool_FC**). Please note that AvgPool_FC is a simple neural network and it is order-permutation invariant.

Five recent 3D point cloud based methods: PointNet [31], PointNet++ [32], RangeNet [29] (two variants, with KNN and CRF), KPConv [36] and Rand-LANet [18] are selected for comparison study. For image aggregation methods, we simply deterministically choose the semantic label with maximum occurrence times. Within multi-view learning methods, all HiFANet variants are trained with the same hyper-parameter setting as HiFANet.

Quantitative Result is shown in Table 1. We can observe that point cloud based methods training requires much larger number of training dataset than both image aggregation methods and our proposed multi-view learning methods. This shows the advantage of learning semantics from 2D images. The compactly-organized and topology-preserving RGB images enables neural network to learn meaningful semantic representations with much fewer training samples. Within image aggregation methods, involving extra bag-of-frame observations increases the performance, but the performance gain is not prominent due to the view-angle and occlusion challenges. Moreover, expanding the patch size also improves the performance, which shows capability of introducing patch-wise observation in mitigating the dilemma caused by observation uncertainty. In sum, aggregating image-predicted semantics can achieve comparable performance than point based

A. RGB Image   B. Ground Truth

C. RandLA Net   D. Image Based Method   E. HiFANet

■ unlabeled   ■ road   ■ building   ■ pole   ■ traffic sign   ■ vegatation   ■ terrain   ■ car

Fig. 4. Global visualization of various methods comparison. While point based method fails to classify traffic sign and image based method generates spatially distributed prediction, HiFANet successfully avoids these dilemmas and gives the right semantics.

methods. It further shows the potential of designing neural network to learn from image learned semantic representations, instead of simply voting them.

Within multi-view learning methods, we can observe that all methods outperform image aggregation methods, showing the advantage of neural network learning over deterministic semantic aggregation. Simply adding several fully connection layers (AvgPool_FC) generates inferior performance than the other three HiFANet variants. This result shows that more advanced semantic aggregation strategy is needed to better aggregate semantic cues arising from multiple image observations. At the same time, either removing the patch attention module or the structural prior module inevitably reduces the performance. Patch observation introduces extra semantic cues in a pose noise sensitive way and structural prior regularizes the whole network training. Finally, HiFANet generates the best performance over all methods, far outweighing other methods by a large margin.

Qualitative Result is shown in Fig. 3 and Fig. 4. In the close-up comparison of tree stakes in Fig. 3, as it is a category falls out of our consideration, it should be regarded as *unlabelled* category. However, 3D point based method RandLANet [18] (sub-figure B.) mixes it with *pole* due to their point cloud representation similarity. Image aggregation method (sub-figure D.) directly predicts it as terrain because of its color similarity and connection with the tree leaves. HiFANet (sub-figure E.), however, fully exploits 3D point structural prior information to predict the correct semantics. For example, the tilted angle of tree stakes over the ground makes it unlikely to be a pole (which is usually vertical to ground), nor terrain (no angle information).

The global comparison of various methods is shown in Fig. 4. We can observe that point based method (C. RandLA-Net) failed to predict the large traffic sign (red box in the RGB image) because such samples are rarely seen in training dataset. At the same time, due to the pose noise, image based method distributes

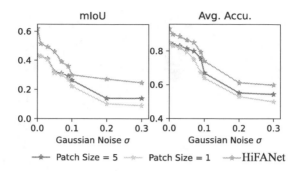

Fig. 5. Pose noise test: performance variation trend under various Gaussian pose noise level.

car 3D points to large area (see the largely distributed red points in sub-figure D., near the light blue). Our proposed HiFANet can maximally avoid these dilemmas. It obtains semantic representation from RGB images, so it does not require massive training dataset and large presence of all classes. The hierarchical attention design and the involvement of 3D structural prior equip HiFANet with capability to dynamically alleviate the erroneous prediction led by pose noise. In sum, our proposed HiFANet achieves promising performance with relatively small training dataset. It also exhibits pose noise tolerance capability, which is a common challenge in real scenario.

6.1   More Experimental Result

We report the detailed mIoU and mAP score for each individual class in Table 2. We can see from the table that our proposed HiFANet achieves the best performance on most categories. Image based method (with BoF = 5) obtains inferior performance on some categories such as car, rider and traffic sign, due to the pose noise. Our proposed HiFANet maximally resists the negative impact of pose noise and thus is capable of obtaining promising performance.

6.2   Discussion on Pose Noise

We further want to test our proposed HiFANet performance under various pose noise level. To this end, we add Gaussian pose noise to the point-to-image projection matrix in Eq. 5. The Gaussian noise level is controlled by the Gaussian deviation σ (the mean value is set 0). We compare HiFANet with two image aggregation variants: with patch size 1 and 5. Since the introduction of patch observation is to handle pose noise, it helps us to understand patch observation (patch size = 5) resistance to pose noise against the original observation (patch size = 1), and against HiFANet.

The Gaussian pose noise σ is linearly spaced from 0 to 0.3. The result is shown in Fig. 5, from which we can observe that adding more pose noise reduces the performance of all methods. The variant with patch size 1 suffers most while

Table 2. Detailed IoU score for each category on semantic-KITTI [2] dataset

Method	Road	Sidewalk	Building	Fence	Pole	Traffic-sign	Vegetation
PointNet [31]	0.031	0.069	0.113	0.043	0.036	0.022	0.041
PointNet++ [32]	0.066	0.023	0.079	0.042	0.112	0.014	0.036
RangeNet++(CRF) [29]	0.878	0.745	0.742	0.232	0.252	0.313	0.612
RangeNet++(KNN) [29]	0.895	0.769	0.819	0.258	0.333	0.291	0.648
KPConv [36]	0.738	0.574	0.653	0.244	0.469	**0.400**	0.533
RandLANet [18]	0.883	0.760	0.883	0.323	0.537	0.319	0.731
Image Based BoF = 5	0.888	0.710	0.378	0.154	0.189	0.362	0.598
HiFANet	**0.910**	**0.790**	**0.903**	**0.349**	**0.540**	0.374	**0.755**

Method	Terrain	Person	Rider	Car	Motor-cycle	Bicycle	
PointNet [31]	0.054	0.000	0.000	0.052	0.002	0.003	
PointNet++ [32]	0.183	0.000	0.002	0.133	0.010	0.000	
RangeNet++(CRF) [29]	0.875	0.088	0.356	0.853	0.375	0.176	
RangeNet++(KNN) [29]	0.896	0.114	0.414	0.856	0.178	0.183	
KPConv [36]	0.767	0.249	**0.696**	0.739	0.360	0.000	
RandLANet [18]	0.910	0.216	0.572	0.909	0.470	0.003	
Image Based BoF = 5	0.889	0.210	0.055	0.563	0.533	0.146	
HiFANet	**0.912**	**0.247**	0.577	**0.933**	**0.547**	0.169	

HiFANet maximally mitigates the pose noise impact. It thus shows the advantage of involving patch observation in tackling pose noise and our carefully designed HiFANet is capable of learning pose noise tolerant feature representation.

For more implementation details and the network architecture of our method, please refer to the supplementary material.

7 Conclusion

We propose a three-stage hierarchical fully attentive network, HiFANet, to label the point cloud semantically. The patch observation strategy and bag-of-frames multi-view observation enable HiFANet to handle point-image projection pose noise. Compared to point cloud based methods, HiFANet requires significantly less amount of data and outperforms point based methods by a large margin. The downside our method is that HiFANet's good performance still depends relatively on the LiDAR-camera pose accuracy. If the pose accuracy drops significantly, HiFANet's performance reduces accordingly. Designing more pose-noise tolerant method thus forms a potential future research direction. Another point is that HiFANet only builds on 2D image observations, a joint learning from both the image and point cloud may further improve the performance.

References

1. Badrinarayanan, V., Kendall, A., Cipolla, R.: SegNet: a deep convolutional encoder-decoder architecture for image segmentation. IEEE Trans. Pattern Anal. Mach. Intell. **39**(12), 2481–2495 (2017)

2. Behley, J., et al.: SemanticKITTI: a dataset for semantic scene understanding of LiDAR sequences. In: Proceedings of the IEEE/CVF International Conference on Computer Vision (ICCV) (2019)

3. Chen, L.C., Papandreou, G., Kokkinos, I., Murphy, K., Yuille, A.L.: DeepLab: semantic image segmentation with deep convolutional nets, atrous convolution, and fully connected CRFs. IEEE Trans. Pattern Anal. Mach. Intell. **40**(4), 834–848 (2017)

4. Chen, L.C., Yang, Y., Wang, J., Xu, W., Yuille, A.L.: Attention to scale: scale-aware semantic image segmentation. In: Proceedings of the IEEE Conference on Computer Vision and Pattern Recognition, pp. 3640–3649 (2016)

5. Chen, L.-C., Zhu, Y., Papandreou, G., Schroff, F., Adam, H.: Encoder-decoder with Atrous separable convolution for semantic image segmentation. In: Ferrari, V., Hebert, M., Sminchisescu, C., Weiss, Y. (eds.) ECCV 2018. LNCS, vol. 11211, pp. 833–851. Springer, Cham (2018). https://doi.org/10.1007/978-3-030-01234-2_49

6. Chen, L., He, Y., Chen, J., Li, Q., Zou, Q.: Transforming a 3-D lidar point cloud into a 2-D dense depth map through a parameter self-adaptive framework. IEEE Trans. Intell. Transp. Syst. **16**, 165– 176 (2017)

7. Cordts, M., et al.: The cityscapes dataset for semantic urban scene understanding. In: Proceedings of the IEEE Conference on Computer Vision and Pattern Recognition (CVPR) (2016)

8. Dai, A., Ritchie, D., Bokeloh, M., Reed, S., Sturm, J., Nießner, M.: Scancomplete: large-scale scene completion and semantic segmentation for 3D scans. In: Proceedings of the IEEE Conference on Computer Vision and Pattern Recognition (CVPR), pp. 4578–4587 (2018)

9. Furukawa, Y., Ponce, J.: Accurate, dense, and Robust multi-view stereopsis. In: Proceedings of the IEEE Computer Vision and Pattern Recognition (CVPR) (2007)

10. Geiger, A., Lenz, P., Urtasun, R.: Are we ready for autonomous driving? The KITTI vision benchmark suite. In: Proceedings of the IEEE Conference on Computer Vision and Pattern Recognition (CVPR), pp. 3354–3361 (2012)

11. Ghiasi, G., Fowlkes, C.C.: Laplacian pyramid reconstruction and refinement for semantic segmentation. In: Leibe, B., Matas, J., Sebe, N., Welling, M. (eds.) ECCV 2016. LNCS, vol. 9907, pp. 519–534. Springer, Cham (2016). https://doi.org/10.1007/978-3-319-46487-9_32

12. Graham, B., Engelcke, M., Van Der Maaten, L.: 3D semantic segmentation with submanifold sparse convolutional networks. In: Proceedings of the IEEE Conference on Computer Vision and Pattern Recognition (CVPR), pp. 9224–9232 (2018)

13. Guerry, J., Boulch, A., Le Saux, B., Moras, J., Plyer, A., Filliat, D.: SnapNet-r: consistent 3D multi-view semantic labeling for robotics. In: Proceedings of the IEEE International Conference on Computer Vision Workshops (ICCVW), pp. 669–678 (2017)

14. He, K., Zhang, X., Ren, S., Sun, J.: Deep residual Learning for Image Recognition. In: Proceedings of the IEEE Computer Vision and Pattern Recognition (CVPR) (2016)

15. He, Y., et al.: Deep learning based 3D segmentation: a survey. arXiv preprint arXiv:2103.05423 (2021)

16. He, Y., Chen, L., Li, M.: Sparse depth map upsampling with RGB image and anisotropic diffusion tensor. In: 2015 IEEE Intelligent Vehicles Symposium (IV) (2015)

17. Hu, Q., et al.: Randla-net: efficient semantic segmentation of large-scale point clouds. In: Proceedings of the IEEE/CVF Conference on Computer Vision and Pattern Recognition (CVPR), pp. 11108–11117 (2020)

18. Hu, Q., et al.: Randla-Net: efficient semantic segmentation of large-scale point clouds. In: Proceedings of the IEEE Conference on Computer Vision and Pattern Recognition (CVPR) (2020)

19. Hua, B.S., Tran, M.K., Yeung, S.K.: Pointwise convolutional neural networks. In: Proceedings of the IEEE Conference on Computer Vision and Pattern Recognition (CVPR), pp. 984–993 (2018)

20. Huang, J., You, S.: Point cloud labeling using 3D convolutional neural network. In: 2016 23rd International Conference on Pattern Recognition (ICPR), pp. 2670–2675. IEEE (2016)

21. Huang, Z., et al.: CCNet: Criss-Cross attention for semantic segmentation. IEEE Trans. Pattern Analy. Mach. Intell. (TPAMI) (2020)

22. Lawin, F.J., Danelljan, M., Tosteberg, P., Bhat, G., Khan, F.S., Felsberg, M.: Deep projective 3D semantic segmentation. In: Felsberg, M., Heyden, A., Krüger, N. (eds.) CAIP 2017. LNCS, vol. 10424, pp. 95–107. Springer, Cham (2017). https://doi.org/10.1007/978-3-319-64689-3_8

23. Lee, J., Lee, Y., Kim, J., Kosiorek, A., Choi, S., Teh, Y.W.: Set transformer: a framework for attention-based permutation-invariant neural networks. In: Proceedings of the 36th International Conference on Machine Learning (ICML), pp. 3744–3753 (2019)

24. Li, G., Muller, M., Thabet, A., Ghanem, B.: Deepgcns: Can GCNs go as deep as CNNS? In: Proceedings of the IEEE/CVF International Conference on Computer Vision (CVPR), pp. 9267–9276 (2019)

25. Liang, Z., Yang, M., Deng, L., Wang, C., Wang, B.: Hierarchical depthwise graph convolutional neural network for 3d semantic segmentation of point clouds. In: 2019 International Conference on Robotics and Automation (ICRA), pp. 8152–8158. IEEE (2019)

26. Liu, W., et al.: SSD: single shot multibox detector. In: Leibe, B., Matas, J., Sebe, N., Welling, M. (eds.) ECCV 2016. LNCS, vol. 9905, pp. 21–37. Springer, Cham (2016). https://doi.org/10.1007/978-3-319-46448-0_2

27. Long, J., Shelhamer, E., Darrell, T.: Fully convolutional networks for semantic segmentation. In: Proceedings of the IEEE Conference on Computer Vision and Pattern Recognition, pp. 3431–3440 (2015)

28. Meng, H.Y., Gao, L., Lai, Y.K., Manocha, D.: Vv-Net: Voxel VAE net with group convolutions for point cloud segmentation. In: Proceedings of the IEEE/CVF International Conference on Computer Vision (ICCV), pp. 8500–8508 (2019)

29. Milioto, A., Vizzo, I., Behley, J., Stachniss, C.: RangeNet++: fast and accurate LiDAR semantic segmentation. In: IEEE/RSJ International Conference on Intelligent Robots and Systems (IROS) (2019)

30. Noh, H., Hong, S., Han, B.: Learning deconvolution network for semantic segmentation. In: Proceedings of the IEEE International Conference on Computer Vision, pp. 1520–1528 (2015)

31. Qi, C.R., Su, H., Mo, K., Guibas, L.J.: PointNet: deep learning on point sets for 3D classification and segmentation. In: IEEE Conference on Computer Vision and Pattern Recognition (CVPR) (June 2017)

32. Qi, C.R., Yi, L., Su, H., Guibas, L.J.: Pointnet++: deep hierarchical feature learning on point sets in a metric space. In: Conference on Neural Information Processing Systems (NeurIPS) (2017)

33. Ren, S., He, K., Girshich, R., Sun, J.: Faster r-CNN: towards real-time object detection with region proposal networks. In: Advances in Neural Information Processing Systems (NeurIPS) (2015)

34. Riegler, G., Osman Ulusoy, A., Geiger, A.: Octnet: learning deep 3D representations at high resolutions. In: Proceedings of the IEEE Conference on Computer Vision and Pattern Recognition (CVPR), pp. 3577–3586 (2017)

35. Sutskever, I., Vinyals, O., V. Le, Q.: Sequence to sequence learning with neural networks. In: Conference on Neural Information Processing Systems (NeurIPS) (2014)

36. Thomas, H., Qi, C.R., Deschaud, J.E., Marcotegui, B., Goulette, F., Guibas, L.J.: Kpconv: flexible and deformable convolution for point clouds. In: Proceedings of the IEEE International Conference on Computer Vision (CVPR) (2019)

37. Vaswani, A., et al.: Attention is all you need. In: Conference on Advances in Neural Information Processing Systems (NeurIPS) (2017)

38. Wang, S., Suo, S., Ma, W.C., Pokrovsky, A., Urtasun, R.: Deep parametric continuous convolutional neural networks. In: Proceedings of the IEEE Conference on Computer Vision and Pattern Recognition (CVPR), pp. 2589–2597 (2018)

39. Wang, Y., Li, J., Metze, F.: A comparison of five multiple instance learning pooling functions for sound event detection with weak labeling. In: IEEE International Conference on Acoustics, Speech and Signal Processing (ICASSP) (2019)

40. Wu, W., Qi, Z., Fuxin, L.: Pointconv: deep convolutional networks on 3D point clouds. In: Proceedings of the IEEE/CVF Conference on Computer Vision and Pattern Recognition (CVPR), pp. 9621–9630 (2019)

41. Xie, Y., Tian, J., Zhu, X.X.: Linking points with labels in 3D: a review of point cloud semantic segmentation. IEEE Geosc. Remote Sens Mag. 8(4), 38–59 (2020)

42. Xie, Z., Chen, J., Peng, B.: Point clouds learning with attention-based graph convolution networks. Neurocomputing 402, 245–255 (2020)

43. Ye, X., Li, J., Huang, H., Du, L., Zhang, X.: 3D recurrent neural networks with context fusion for point cloud semantic segmentation. In: Ferrari, V., Hebert, M., Sminchisescu, C., Weiss, Y. (eds.) ECCV 2018. LNCS, vol. 11211, pp. 415–430. Springer, Cham (2018). https://doi.org/10.1007/978-3-030-01234-2_25

44. Zeng, W., Gevers, T.: 3DContextNet: K-d tree guided hierarchical learning of point clouds using local and global contextual cues. In: Leal-Taixé, L., Roth, S. (eds.) ECCV 2018. LNCS, vol. 11131, pp. 314–330. Springer, Cham (2019). https://doi.org/10.1007/978-3-030-11015-4_24

45. Zhang, Y., et al.: PolarNet: an improved grid representation for online lidar point clouds semantic segmentation. In: Proceedings of the IEEE/CVF Conference on Computer Vision and Pattern Recognition (CVPR), pp. 9601–9610 (2020)

46. Zhao, H., Jiang, L., Fu, C.W., Jia, J.: Pointweb: enhancing local neighborhood features for point cloud processing. In: Proceedings of the IEEE/CVF Conference on Computer Vision and Pattern Recognition (CVPR), pp. 5565–5573 (2019)

47. Zhao, H., Shi, J., Qi, X., Wang, X., Jia, J.: Pyramid scene parsing network. In: Proceedings of the IEEE Conference on Computer Vision and Pattern Recognition, pp. 2881–2890 (2017)

48. Zhu, Y., et al.: Improving semantic segmentation via video propagation and label relaxation. In: IEEE Conference on Computer Vision and Pattern Recognition (CVPR) (2019)

SIM2E: Benchmarking the Group Equivariant Capability of Correspondence Matching Algorithms

Shuai Su$^{(\boxtimes)}$, Zhongkai Zhao , Yixin Fei , Shuda Li , Qijun Chen ,
and Rui Fan

Tongji University, Shanghai 201804, China
{sushuai,kanez,amyfei,qjchen,rfan}@tongji.edu.cn

Abstract. Correspondence matching is a fundamental problem in computer vision and robotics applications. Solving correspondence matching problems using neural networks has been on the rise recently. Rotation-equivariance and scale-equivariance are both critical in correspondence matching applications. Classical correspondence matching approaches are designed to withstand scaling and rotation transformations. However, the features extracted using convolutional neural networks (CNNs) are only translation-equivariant to a certain extent. Recently, researchers have strived to improve the rotation-equivariance of CNNs based on group theories. Sim(2) is the group of similarity transformations in the 2D plane. This paper presents a specialized dataset dedicated to evaluating sim(2)-equivariant correspondence matching algorithms. We compare the performance of 16 state-of-the-art (SoTA) correspondence matching approaches. The experimental results demonstrate the importance of group equivariant algorithms for correspondence matching on various sim(2) transformation conditions. Since the subpixel accuracy achieved by CNN-based correspondence matching approaches is unsatisfactory, this specific area requires more attention in future works. Our dataset is publicly available at: mias.group/SIM2E.

Keywords: Correspondence matching · Computer vision · Robotics · Rotation-equivariance · Scaling-equivariance · Convolutional neural networks

1 Introduction

Correspondence matching is a key component in autonomous driving perception tasks, such as object tracking [1], simultaneous localization and mapping (SLAM) [2], multi-camera online calibration [3], 3D geometry reconstruction [4], panorama stitching [5], and camera pose estimation [6], as shown in Fig. 1. Sim(2) transformation consists of rotation, scaling, and translation. Sim(2)-equivariant correspondence matching is significantly important for autonomous driving, as vehicles often veer abruptly. Classical algorithms leverage a detector, a descriptor, and a matcher to determine correspondences. The detector and descriptor

© The Author(s), under exclusive license to Springer Nature Switzerland AG 2023
L. Karlinsky et al. (Eds.): ECCV 2022 Workshops, LNCS 13801, pp. 743–759, 2023.
https://doi.org/10.1007/978-3-031-25056-9_47

Fig. 1. Autonomous driving perception tasks involving correspondence matching.

provide the locations and the descriptions of interest points (point-like features in an image), and the matcher produces the final correspondences. Moravec *et al.* [7] presented the concept of interest points. Harris [8] judges whether the pixel is a corner based on the local image gradient changes. The scale-invariant feature transform (SIFT) [9] is a rotation-invariant and scale-invariant algorithm that consists of a detector and a descriptor. The distance among descriptors is computed using cosine distance. The correspondences of a given image pair are determined using the nearest neighbor matching algorithms. As a hand-crafted algorithm, SIFT [9] achieves rotation-invariance by computing the main directions of local features (in an image patch of 16×16 pixels.). ASIFT [10] aimed to improve the performance of SIFT [9] on affine transformation. Oriented FAST [11] and Rotated Binary Robust Independent Elementary Features (BRIEF) [12] (ORB) [13] greatly minimize the trade-off between accuracy and speed and have been widely used in visual SLAM [13–15].

Deep learning has been applied successfully in numerous computer vision tasks in recent years. LIFT [16] is an architecture of learning-based rotation-invariant feature detection and description approach. It consists of three modules: detector, orientation estimator, and descriptor. Similar to SIFT [9], a scale-space pyramid is used to obtain multi-scale correspondence detection results. SuperPoint [17] is a self-supervised framework for correspondence detection and description. It is a two-stage method: (1) in the first stage, a feature extractor is trained on a synthetic dataset generated by rendering patterns of corners; (2) in the second stage, the descriptor is trained on images from the COCO-dataset [18] that are augmented by random homography matrices including transformations such as rotation, scaling, and translation. Unlike SuperPoint, D2Net [19] is a one-stage approach that jointly detects and describes correspondences. D2Net is trained using the correspondences obtained from large-scale structure from motion (SfM) reconstructions. R2D2 [20] proposes a framework to find more repeatable correspondences, and it can simultaneously estimate the reliability and repeatability of correspondences. DISK [21] uses reinforcement learning to realize end-to-end correspondence matching. It performs better at small angle changes but worse than SuperPoint [17] at large angle changes.

SuperGlue [22] achieves superior performance by modeling correspondence matching as a graph matching problem. The inputs of the graph neural network in SuperGlue [22] include descriptors, positions, and scores of keypoints. The Sinkhorn algorithm [23] is utilized to solve the optimal-transport problem. However, the graph edges in SuperGlue exponentially grow as the number of correspondences increases. SGMNet [24] uses a seeded graph to reduce computation and memory costs significantly. LoFTR [25] is a detector-free and end-to-end architecture. It uses convolutional neural networks (CNN) as feature extractors and a coarse-to-fine strategy to obtain more accurate pixel-level results. Similar to SuperGlue, it also fuses descriptors with the position information. Unlike LoFTR, another end-to-end correspondence matching network, referred to as MatchFormer [26], uses a hierarchical extract-and-match transformer. It is demonstrated that the correspondence matching operation can also be conducted in the encoder. RoRD [27] uses orthographic view generation to improve correspondence matching by increasing the visual overlap using orthographic projection. It also shows that rotation invariance can be improved by augmenting the training dataset with random rotation, scaling, and perspective transformations.

Group-equivariant convolutional neural networks (G-CNN) are equivariant under a specific transformation (e.g., rotation, translation, etc.) which can also be represented by a special group. Researchers have designed G-CNNs using different mathematical approximations. Cohen et al. [28] proposed the first G-CNN. Li et al. [29] use the cyclic replacement to achieve P4-group equivariance. Cohen et al. [30] use the Fast Fourier Transform (FFT) to approximate the integral of a group. E2-CNN [31] is a general G-CNN framework that analyzes and models the orientation and symmetry of images. GIFT [32] is a rotation-equivariant and scaling-equivariant descriptor based on G-CNN. It uses E2-CNN [31] rather than conventional CNNs to describe local visual features. On the other hand, SEKD [33] is a group-equivariant correspondence detector based on G-CNN, which greatly improves the performance of rotation-equivariant correspondence matching. ReF [34] is a rotation-equivariant correspondence detection and description framework. It uses a G-CNN to extract group-equivariant feature maps and a group-pooling operation to get rotation-invariant descriptors. SE2-LoFTR [35] replaces the feature extractor of LoFTR with E2-CNN, achieving significantly better results on the rotated-HPatches dataset [27]. Furthermore, it also mentioned in [35] that the position information is not rotation-equivariant while the descriptor is rotation-equivariant. The methods mentioned above only consider the equivariance of local features. Unfortunately, the equivariance of position information is rarely discussed. Cieslewski *et al.* [36] presented an algorithm to match correspondences without descriptors, namely, only position information is used. This algorithm is evaluated on the KITTI [37] dataset (containing relatively ideal scenarios), demonstrating robust performance even without descriptors. Similar to [36], ZZ-Net [38] is an algorithm for matching two 2D point clouds. It demonstrates that correspondence matching without descriptors can work in rotation-only conditions. Therefore, the current research on the equivariance of position information needs to be further expanded.

2   SIM2E Dataset

2.1   Data Collection and Augmentation

To ensure the pixel-level accuracy of correspondence matching ground truth, we scrape frames from online time-lapse videos. The cameras used to capture such time-lapse videos are fixed. Our SIM2E dataset contains many challenging scenarios, such as moving clouds in the sky and changing illumination conditions. We choose the first frame of each video as the reference image and use the rest of the frames as target (query) images. We also publish our data augmentation code so that interested readers can conduct sim(2) transformations on our dataset according to their own needs.

2.2   SIM2E-SO2S, SIM2E-Sim2S, and SIM2E-PersS

The rotation and scaling operations produce many black backgrounds. To increase the difficulty of correspondence matching, we generate synthetic backgrounds to fill these black areas. Our dataset is split into three subsets: SIM2E-SO2S, SIM2E-Sim2S, and SIM2E-PersS.

- **SIM2E-SO2S (Rotation and Synthetic Background)**: The target images are rotated by random angles between 0 ° and 360 °. Scaling is not applied.
- **SIM2E-Sim2S (Rotation, Scaling, Translation and Synthetic Background)**: The target images are rotated by random angles between 0 ° and 360 °. Random scaling ranging between 0.4 and 1, and random translation ranging between 0 and 0.2 are also applied.
- **SIM2E-PersS: (Perspective Transformation and Synthetic Background)**: Random perspective transformations are applied to the target images, where the perspective parameters (the two elements on the 3rd row, the 1st and 2nd columns of the homography matrix, respectively) are random values between –0.0008 and 0.0008. The shear angle is randomly set to $[-10°, 10°]$. The target images are rotated by random angles between 0 ° and 360 °. Random scaling ranging between 0.4 and 1 is applied. Random translation ranging between 0 and 0.2 is applied.

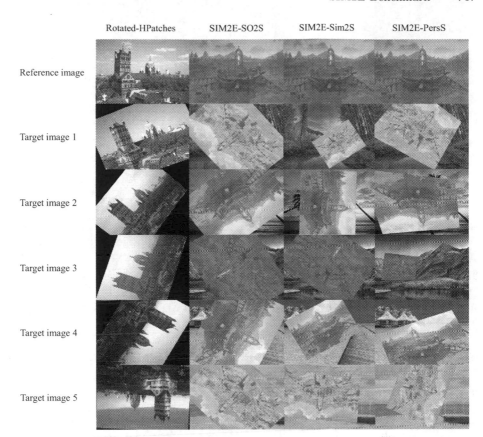

Fig. 2. Rotated-HPatches dataset and the three sub-sets of our created SIM2E dataset.

2.3 Comparison with Other Public Correspondence Matching Datasets

As shown in Table. 1, the existing datasets for correspondence matching algorithm evaluation can be grouped into two types: 3D scenes [39–44] and planar scenes [27,45].

Aachen Day-Night [39] is a public dataset designed to evaluate the performance of outdoor visual localization algorithms in changing illumination conditions (day-time and night-time). The dataset contains a scenario where images were taken with a hand-held camera at different times of the day. It is widely used to evaluate the performance of correspondence matching algorithms, especially when the illumination change is significant.

ScanNet [41] is a large-scale real-world dataset containing 2.5M RGB-D images (1513 scans acquired in 707 different places, such as offices, apartments, and bathrooms). All the scans are annotated with estimated calibration parameters, camera poses, reconstructed 3D surfaces, textured meshes, dense object-level semantic segmentations, and aligned computer-aided design (CAD) models.

Table 1. Comparison between our SIM2E dataset and other public datasets.

Dataset	Type	Illumination change	Rotation	Scaling	Dataset size
AachenDayNight [39]	3D	Significant	Small	Medium	Large
ScanNet [41]	3D	Slight	Small	Small	Large
MegaDepth [42]	3D	Slight	Small	Large	Large
Inloc [43]	3D	Medium	Small	Medium	Large
TartanAir [44]	3D	Very significant	Small	Medium	Large
Hpatches [45]	Plane	Significant	Small	Small	Small
Rotated-HPatches [27]	Plane	Significant	Large	Medium	Small
SIM2E (ours)	Plane	Very significant	Large	Large	Small

MegaDepth [42] is a large-scale dataset for the evaluation of depth estimation and/or correspondence matching algorithms. It uses SfM and multi-view stereo (MVS) techniques to acquire 3D point clouds, which can then be used to train and evaluate single-view depth estimation and/or correspondence matching networks. However, the 3D point clouds generated using SfM/MVS in the MegaDepth dataset are not sufficiently accurate and dense.

Compared to the Aachen Day-Night [39] and MegaDepth [42] datasets which were created in outdoor scenarios, the Inloc [43] dataset focuses on indoor localization problems. The Inloc dataset consists of a database of RGB-D images, geometrically registered to the floor maps and augmented with a separate set of RGB target images (annotated with manually verified ground-truth 6DoF camera poses in the global coordinate system of the 3D map).

Unlike the aforementioned datasets that are relatively ideal in terms of either motion or illumination conditions, TartanAir [44], a synthetic dataset used to evaluate visual SLAM algorithms, is collected using a photo-realistic simulator (with the presence of moving objects, changing illumination and weather conditions). Such a more challenging dataset fills the gap between synthetic and real-world datasets.

Hpatches [45] are created using other public datasets. It can be split into two subsets: illumination and viewpoint, which are two crucial aspects of correspondence matching. It can also be split into three subsets: EASY, HARD, and TOUGH, according to the sizes of the overlapping areas between reference and target images. Randomly rotating the target images in the Hpatches [45] dataset produces a new dataset, referred to as Rotated-HPatches [27].

Similar to the Hpatches and Rotated-HPatches datasets, our SIM2E dataset provides accurate subpixel correspondence matching ground truth. On the other hand, the illumination, rotation, and scaling changes are significant in our dataset. Therefore, compared to other existing public datasets, our SIM2E dataset can be used to evaluate the sim(2)-equivariant capability of correspondence matching algorithms more comprehensively. However, the size of the current version of our SIM2E dataset is small. We will therefore increase its size in our future work.

The rotation distributions of Rotated-HPatches and our SIM2E subsets are shown in Fig. 3. It can be observed that in the Rotated-HPatches dataset, slight

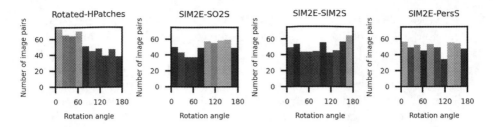

Fig. 3. Rotation distributions of the rotated-HPatches and our SIM2E datasets.

rotations ($\leq 60°$) account for a large proportion. In contrast, the three subsets of our SIM2E dataset are uniformly distributed. Most existing learning-based correspondence matching approaches have poor rotation-equivariant capabilities, and their performances are satisfactory only when there are slight rotations. Therefore, our SIM2E dataset can provide more acceptable results when evaluating the rotation-equivariant capability of a given correspondence matching algorithm.

3 Experiments

3.1 Experimental Setup

The group-equivariant capabilities of six classical and ten learning-based correspondence matching approaches are evaluated on our SIM2E dataset.

For classical correspondence matching approaches, we use the OpenCV [46] implementations of AKAZE [47], BRISK [48], KAZE [49], ORB [13], FREAK [50], and SIFT [9] in our experiments. All these classical approaches use the nearest neighbor matching algorithm for correspondence matching. The ratio test technique (threshold is set to 0.7) is also used to improve the overall performance.

For learning-based correspondence matching approaches, we use the official weights of each model. These models were trained on different datasets, as detailed below:

- **SuperPoint** [17] is trained on the MS-COCO [18] dataset, a large-scale dataset for object detection and segmentation.
- **R2D2** [20] is trained on the Aachen Day-Night [39] dataset and a retrieval dataset [51].
- **ALIKE** [52] is trained on the MegaDepth [42] dataset.
- **GIFT** [32] is trained on the MS-COCO [18] dataset and finetuned on the GL3D [53] dataset (consisting of indoor and outdoor scenes).
- **RoRD** [27] is trained on the PhotoTourism [54] dataset, where the 3D structures of scenes are obtained using SfM.
- **SuperGlue** [22] is trained with the indoor models in the ScanNet [41] dataset and the outdoor models in the MegaDepth [42] dataset.

- **SGMNet** [24] is trained on the GL3D [53] dataset. Our experiments utilize the SIFT version of SGMNet, where the detector and descriptor are rotation-invariant.
- **LoFTR** [25] is trained with the same experimental setup as SuperGlue.
- **MatchFormer** [26] is trained with the same experimental setup as Super-Glue and LoFTR. Limited by our GPU memory, the lightweight version of MatchFormer is used in our experiments.
- **SE2-LoFTR** [35] is trained on the MegaDepth dataset.

Furthermore, the mean matching accuracy (MMA) is employed to quantify the performance of the aforementioned correspondence matching algorithms, which are run on a PC with an Intel Core i7-10870H CPU and an NVIDIA RTX3080-laptop GPU (having a 16 GB DDR4 memory).

3.2   Comparison of the SoTA Approaches on the Rotated-HPatches Dataset

The Rotated-HPatches [27] dataset is generated using the Hpatches [45] dataset to evaluate rotation-equivariant capability of correspondence matching methods. Each sub-folder of the Rotated-HPatches dataset contains one reference image and five target images. The target images are obtained by rotating the reference image at a random angle. The correspondence matching ground truth is acquired using the homography matrices between each pair of reference and target images. As illustrated in Fig. 4(a), the SoTA correspondence matching algorithms demonstrate significantly different performances on the Rotated-HPatches dataset.

The classical algorithms, such as AKAZE, BRISK, KAZE, and SIFT, achieve the best overall performances on the Rotated-HPatches dataset, as they consider both the scaling and rotation invariance of visual features. Benefiting from the higher dimensional feature descriptors, these four algorithms outperform ORB and FREAK.

On the other hand, SuperPoint, R2D2, and ALIKE are developed without considering rotation invariance. Therefore, their performances are relatively poor on the Rotated-HPatches dataset. GIFT [32] uses SuperPoint as the feature detector. Its feature descriptor is developed based on G-CNN to acquire the rotation-equivariant capability. As expected, GIFT significantly outperforms SuperPoint.

As can be seen from Table 2 and Fig. 4(a), when the tolerance δ exceeds 5, the learning-based methods demonstrate better performances than classical methods. For instance, SGMNet outperforms all classical methods when $\delta > 5$ and SE2-LoFTR shows similar performance to the classical methods when $\delta > 8$. Referring to [24], SGMNet is a lightweight version of SuperGlue and demonstrates slightly worse performance than SuperGlue. However, SuperGlue with SuperPoint performs much worse than SGMNet. This is probably because SGM-Net uses SIFT as its detector and descriptor, which has the rotation-equivariant capability.

Table 2. The performance of SoTA correspondence matching approaches on the Rotated-HPatches dataset. N denotes the average number of valid matches.

Method	$\delta \leq 1$	$\delta \leq 3$	$\delta \leq 5$	$\delta \leq 10$	N
AKAZE [47]	**0.426**	0.744	0.812	0.841	203
BRISK [48]	0.419	0.745	0.816	0.841	282
KAZE [49]	0.423	**0.753**	**0.825**	0.858	505
ORB [13]	0.253	0.576	0.646	0.672	20
SIFT [9]	0.484	0.762	0.809	0.830	727
FREAK [50]	0.278	0.519	0.567	0.594	27
SuperPoint [17]	0.123	0.249	0.274	0.295	120
R2D2 [20]	0.057	0.129	0.144	0.153	158
ALIKE [52]	0.111	0.172	0.184	0.201	59
GIFT [32]	0.203	0.406	0.439	0.448	186
RoRD [27]	0.030	0.191	0.378	0.620	1077
SuperGlue [22]	0.185	0.425	0.491	0.552	479
SGMNet [24]	0.369	0.688	0.814	**0.893**	1278
LoFTR [25]	0.037	0.167	0.259	0.350	506
MatchFormer [26]	0.033	0.162	0.256	0.358	600
SE2-LoFTR [35]	0.197	0.556	0.720	0.842	1305

It can be observed that SoTA classical methods always perform better than learning-based methods when the tolerance is small. With the decrease in tolerance, the MMA scores achieved by learning-based methods drop considerably. This is probably because the learning-based methods are generally trained via self-supervised learning, where the tolerance to determine positive samples is typically set to 3. Furthermore, the model is difficult to converge in the training phase when the tolerance is too small, e.g., less than 1, while the model's accuracy degrades dramatically when the tolerance is too large. Moreover, compared to classical methods, learning-based methods can always obtain more correspondences. Therefore, improving the subpixel accuracy of learning-based approaches without reducing the number of valid matches is a research area that requires more attention.

3.3 Comparison of the SoTA Approaches on Our SIM2E Dataset

In this paper, we quantify the group-equivalent capabilities of the aforementioned algorithms on the three subsets of our SIM2E Dataset.

Experimental Results on the SIM2E-SO2S Subset. Our SIM2E-SO2S subset is created in a similar fashion to the Rotated-HPatches dataset. Since we select time-lapse videos with more challenging illumination conditions, apply more uniformly distributed random rotations, and add synthetic backgrounds to the target image, the SIM2E-SO2S subset is expected to reflect the correspondence matching algorithms' group-equivariant capabilities more comprehensively. As can be observed from Fig. 4(b), all the SoTA methods perform much worse on our SIM2E-SO2S subset because it is more challenging than

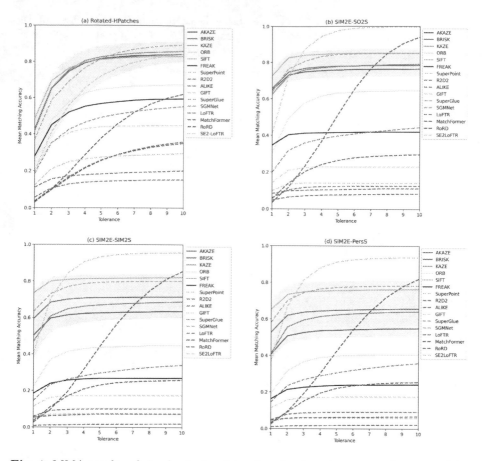

Fig. 4. MMA results of six classical and ten learning-based correspondence matching approaches on the Rotated-HPatches dataset and our SIM2E-SO2S, SIM2E-SIM2S, and SIM2E-PersS subsets. MMA results for the classical methods are shown as solid lines, while MMA results for the learning-based methods are shown as dashed lines.

the Rotated-HPatches dataset. Furthermore, similar to the experimental results in the Rotated-HPatches experiments (see Fig. 4(a)), SE2-LoFTR, SGMNet, AKAZE, BRISK, KAZE, and SIFT also achieve the best group-equivariant capabilities on our SIM2E-SO2S subset (see Fig. 4(b)). This validates the effectiveness of our SIM2E-SO2S subset in terms of evaluating a correspondence matching algorithm's group-equivariant capability. Moreover, the MMA scores achieved by these six methods differ more significantly in the SIM2E-SO2S experiments. Therefore, our SIM2E-SO2S subset more comprehensively quantifies the group-equivariant capability of a given correspondence matching algorithm.

Table 3. The performance of SoTA correspondence matching approaches on the SIM2E-SO2S subset. N denotes the average number of valid matches.

Method	$\delta \leq 1$	$\delta \leq 3$	$\delta \leq 5$	$\delta \leq 10$	N
AKAZE [47]	0.647	0.749	0.759	0.765	72
BRISK [48]	0.656	0.768	0.779	0.785	113
KAZE [49]	0.625	0.760	0.778	0.792	129
ORB [13]	0.302	0.578	0.633	0.646	24
SIFT [9]	**0.727**	0.841	0.849	0.853	264
FREAK [50]	0.346	0.412	0.416	0.421	9
SuperPoint [17]	0.098	0.140	0.141	0.142	40
R2D2 [20]	0.059	0.117	0.123	0.126	53
ALIKE [52]	0.083	0.101	0.106	0.112	22
GIFT [32]	0.142	0.224	0.229	0.230	51
RoRD [27]	0.034	0.232	0.516	0.941	630
SuperGlue [22]	0.197	0.359	0.398	0.445	192
SGMNet [24]	0.639	0.808	0.844	0.851	1226
LoFTR [25]	0.047	0.071	0.077	0.083	329
MatchFormer [26]	0.055	0.203	0.264	0.297	451
SE2-LoFTR [35]	0.410	**0.876**	**0.976**	**0.998**	1640

Table 4. The performance of SoTA correspondence matching approaches on the SIM2E-SIM2S subset. N denotes the average number of valid matches.

Method	$\delta \leq 1$	$\delta \leq 3$	$\delta \leq 5$	$\delta \leq 10$	N
AKAZE [47]	0.503	0.615	0.627	0.636	22
BRISK [48]	0.595	0.701	0.711	0.715	51
KAZE [49]	0.470	0.649	0.670	0.687	54
ORB [13]	0.222	0.414	0.437	0.449	10
SIFT [9]	**0.736**	0.809	0.814	0.821	165
FREAK [50]	0.185	0.258	0.267	0.270	3
SuperPoint [17]	0.054	0.072	0.073	0.073	20
R2D2 [20]	0.059	0.096	0.100	0.102	23
ALIKE [52]	0.053	0.066	0.071	0.075	6
GIFT [32]	0.113	0.171	0.173	0.174	32
RoRD [27]	0.027	0.195	0.444	0.854	287
SuperGlue [22]	0.145	0.262	0.296	0.340	118
SGMNet [24]	0.633	0.770	0.793	0.801	802
LoFTR [25]	0.008	0.013	0.016	0.019	155
MatchFormer [26]	0.035	0.171	0.231	0.258	232
SE2-LoFTR [35]	0.407	**0.844**	**0.933**	**0.956**	847

Experimental Results on the SIM2E-SIM2S Subset. Compared to the SIM2E-SO2S subset, the SIM2E-SIM2S subset contains scaling and translation transformations that shrink the size of the overlapping area between image pairs.

Table 5. The performance of SoTA correspondence matching approaches on the SIM2E-PersS subset. N denotes the average number of valid matches.

Method	$\delta \leq 1$	$\delta \leq 3$	$\delta \leq 5$	$\delta \leq 10$	N
AKAZE [47]	0.407	0.526	0.539	0.549	14
BRISK [48]	0.526	0.637	0.649	0.656	34
KAZE [49]	0.410	0.595	0.624	0.642	42
ORB [13]	0.190	0.366	0.398	0.406	7
SIFT [9]	**0.651**	0.749	0.756	0.764	130
FREAK [50]	0.162	0.226	0.237	0.242	3
SuperPoint [17]	0.039	0.057	0.058	0.058	13
R2D2 [20]	0.047	0.086	0.090	0.092	16
ALIKE [52]	0.046	0.060	0.062	0.067	5
GIFT [32]	0.113	0.169	0.174	0.174	24
RoRD [27]	0.025	0.189	0.425	0.821	257
SuperGlue [22]	0.145	0.268	0.306	0.358	120
SGMNet [24]	0.581	0.745	0.773	0.783	758
LoFTR [25]	0.010	0.016	0.018	0.022	120
MatchFormer [26]	0.031	0.150	0.221	0.255	161
SE2-LoFTR [35]	0.370	**0.814**	**0.912**	**0.937**	772

As illustrated in Fig. 4(c) and Table 4, the performances of SE2-LoFTR, SGM-Net, SIFT, and BRISK remain stable in the SIM2E-SIM2S experiments, while other models' performances degrade dramatically. Therefore, our SIM2E-SIM2S subset can be used to quantify not only the rotation-equivariant capability but also the scaling-equivariant capability of correspondence matching algorithms.

Experimental Results on the SIM2E-PersS Subset. Compared to the SIM2E-SO2S and SIM2E-SIM2S subsets, the SIM2E-PersS subset contains random perspective transformations. Therefore, correspondence matching on the SIM2E-PersS subset is more challenging. Similarly, in the SIM2E-PersS experiments (see Fig. 4(d) and Table 5), the performances of SE2-LoFTR, SGMNet, and SIFT remain stable, while the performances of BRISK, AKAZE, and KAZE degrade more significantly. Therefore, our SIM2E-PersS subset can be utilized to quantify sim(2)-equivariant capability of correspondence matching algorithms in a more in-depth manner.

3.4 Discussion

The experimental results presented above show that SGMNet, SE2-LoFTR, SIFT, BRISK, KAZE, and AKAZE demonstrate similar group-equivariant capabilities. As can be seen in Table 2, the performances of these algorithms are very similar, while SE2-LoFTR achieves the worst performance on the Rotated-HPatches dataset. As can be observed in Fig. 4(b)–(d), our created SIM2E

Fig. 5. Correspondence matching results on our SIM2E dataset.

dataset can reflect the group-equivariant capabilities of the existing correspondence matching algorithms more comprehensively. The compared SoTA methods achieve the best performances on the SIM2E-SO2S subset and the worst overall performances on the SIM2E-PersS subset. Therefore, we believe that our created SIM2E dataset can help users obtain more objective and in-depth evaluation results of their developed correspondence matching algorithms' group-equivariant capabilities.

4   Conclusion

This paper presented a benchmark dataset for the evaluation of sim(2)-equivariant capability of correspondence matching approaches. We first discussed the classical and learning-based methods and the mainstream of developing group-equivariant network architectures. We qualitatively and quantitatively evaluated sixteen SoTA correspondence matching algorithms on the Rotated-HPatches dataset and three subsets of our created SIM2E dataset. These results suggest that our SIM2E dataset is much more challenging than public correspondence matching datasets, and it can comprehensively reflect the group-equivariant capability of SoTA correspondence matching approaches. In summary, group-equivariant detection, group-equivariant description, and group-equivariant position information are vital for group-equivariant correspondence matching. SuperGlue, LoFTR, and SGMNet use neural networks to fuse global position information and local feature descriptors, and achieve superior performances over others. However, obtaining group equivariance of position information is still challenging, as discussed in [35]. The scaling-equivariant and rotation-equivariant capabilities of learning-based approaches are close to classical approaches. However, the sub-pixel accuracy achieved by the former is still unsatisfactory.

Acknowledgements. This work was supported by the National Key R&D Program of China, under grant No. 2020AAA0108100, awarded to Prof. Qijun Chen. This work was also supported by the Fundamental Research Funds for the Central Universities, under projects No. 22120220184, No. 22120220214, and No. 2022-5-YB-08, awarded to Prof. Rui Fan.

References

1. Zhou, H., et al.: Object tracking using SIFT features and mean shift. Comput. Vis. Image Underst. **113**(3), 345–352 (2009)
2. Yang, Yu., et al.: Accurate and robust visual localization system in large-scale appearance-changing environments. IEEE/ASME Trans. Mechatron. (2022). https://doi.org/10.1109/TMECH.2022.3177237
3. Ling, Y., Shen, S.: High-precision online markerless stereo extrinsic calibration. In: 2016 IEEE/RSJ International Conference on Intelligent Robots and Systems (IROS), pp. 1771–1778. IEEE (2016)
4. Fan, R., et al.: Road surface 3D reconstruction based on dense subpixel disparity map estimation. IEEE Trans. Image Process. **27**(6), 3025–3035 (2018)
5. Brown, M., G Lowe, D.: Automatic panoramic image stitching using invariant features. Int. J. Comput. Vis. **74**(1), 59–73 (2007)
6. Fan, R., Liu, M.: Road damage detection based on unsupervised disparity map segmentation. IEEE Trans. Intell. Transp. Syst. **21**(11), 4906–4911 (2019)
7. Moravec, H.P.: Techniques towards automatic visual obstacle avoidance. (1977)
8. Harris, C., Stephens, M., et al..: A combined corner and edge detector. In: Alvey Vision Conference, vol. 15, pp. 10–5244. Citeseer (1988)

9. G Lowe, D.: Object recognition from local scale-invariant features. In: Proceedings of the Seventh IEEE International Conference on Computer Vision, vol. 2, pp. 1150–1157. Ieee (1999)
10. Morel, J.-M., Guoshen, Yu.: ASIFT: a new framework for fully affine invariant image comparison. SIAM J. Imag. Sci. **2**(2), 438–469 (2009)
11. Rosten, E., Drummond, T.: Machine learning for high-speed corner detection. In: Leonardis, A., Bischof, H., Pinz, A. (eds.) ECCV 2006. LNCS, vol. 3951, pp. 430–443. Springer, Heidelberg (2006). https://doi.org/10.1007/11744023_34
12. Calonder, M., Lepetit, V., Strecha, C., Fua, P.: BRIEF: binary robust independent elementary features. In: Daniilidis, K., Maragos, P., Paragios, N. (eds.) ECCV 2010. LNCS, vol. 6314, pp. 778–792. Springer, Heidelberg (2010). https://doi.org/10.1007/978-3-642-15561-1_56
13. Rublee, E., Rabaud, V., Konolige, K., Bradski. G.: ORB: an efficient alternative to SIFT or SURF. In 2011 International Conference on Computer Vision, pp. 2564–2571. Ieee (2011)
14. Mur-Artal,, R., Martinez Montiel, J.M., Tardos, J.D.: ORB-SLAM: a versatile and accurate monocular SLAM system. IEEE Trans. Rob. **31**(5), 1147–1163 (2015)
15. Mur-Artal, R., Tardós, J.D.: ORB-SLAM2: an open-source SLAM system for monocular, stereo, and RGB-D cameras. IEEE Trans. Rob. **33**(5), 1255–1262 (2017)
16. Yi, K.M., Trulls, E., Lepetit, V., Fua, P.: LIFT: learned invariant feature transform. In: Leibe, B., Matas, J., Sebe, N., Welling, M. (eds.) ECCV 2016. LNCS, vol. 9910, pp. 467–483. Springer, Cham (2016). https://doi.org/10.1007/978-3-319-46466-4_28
17. DeTone, D., Malisiewicz, T., Rabinovich, A.: SuperPoint: self-supervised interest point detection and description. In: Proceedings of the IEEE Conference on Computer Vision and Pattern Recognition Workshops, pp. 224–236 (2018)
18. Lin, T., et al.: Microsoft COCO: common objects in context. In: Fleet, D., Pajdla, T., Schiele, B., Tuytelaars, T. (eds.) ECCV 2014. LNCS, vol. 8693, pp. 740–755. Springer, Cham (2014). https://doi.org/10.1007/978-3-319-10602-1_48
19. Dusmanu, M., et al.: D2-Net: a trainable CNN for joint description and detection of local features. In: Proceedings of the IEEE/CVF Conference on Computer Vision and Pattern Recognition, pp. 8092–8101 (2019)
20. Revaud, J., et al.: R2D2: repeatable and reliable detector and descriptor. arXiv preprint arXiv:1906.06195 (2019)
21. Tyszkiewicz, M., Fua, P., Trulls, E.: DISK: learning local features with policy gradient. Adv. Neural. Inf. Process. Syst. **33**, 14254–14265 (2020)
22. Edouard Sarlin, P., DeTone, D., Malisiewicz, T., Rabinovich, A.: SuperGlue: Learning feature matching with graph neural networks. In Proceedings of the IEEE/CVF Conference on Computer Vision and Pattern Recognition, pp. 4938–4947 (2020)
23. Cuturi. M.: Sinkhorn distances: lightspeed computation of optimal transport. In: 26th Advances in Neural Information Processing Systems (2013)
24. Chen, H., et al.: Learning to match features with seeded graph matching network. In: Proceedings of the IEEE/CVF International Conference on Computer Vision, pp. 6301–6310 (2021)
25. Sun, J., Shen, Z., Wang, Y., Bao, H., Zhou, X.: LoFTR: detector-free local feature matching with transformers. In: Proceedings of the IEEE/CVF Conference on Computer Vision and Pattern Recognition, pp. 8922–8931 (2021)
26. Wang, Q., Zhang, J., Yang, K., Peng, K., Stiefelhagen, R.: MatchFormer: interleaving attention in transformers for feature matching. arXiv preprint arXiv:2203.09645 (2022)

27. Parihar, U.S., et al.: RoRD: rotation-robust descriptors and orthographic views for local feature matching. In: 2021 IEEE/RSJ International Conference on Intelligent Robots and Systems (IROS), pp. 1593–1600. IEEE (2021)
28. Cohen, T., Welling. M.: Group equivariant convolutional networks. In: International Conference on Machine Learning, pp. 2990–2999. PMLR (2016)
29. Li, J., Yang, Z., Liu, H., Cai, D.: Deep rotation equivariant network. Neurocomputing **290**, 26–33 (2018)
30. Cohen, T.S., Geiger, M., Köhler, J., Welling, M.: Spherical CNNs. arXiv preprint arXiv:1801.10130 (2018)
31. Weiler, M., Cesa, G.: General E(2)-equivariant steerable CNNs. In: 32rd Proceedings of Conference on Advances in Neural Information Processing Systems (2019)
32. Liu, Y., Shen, Z., Lin, Z., Peng, S., Bao, H., Zhou, X.: GIFT: learning transformation-invariant dense visual descriptors via group CNNs. In: 32rd Proceedings of Conference on Advances in Neural Information Processing Systems (2019)
33. Lee, J., Kim, B., Cho, M.: Self-supervised equivariant learning for oriented keypoint detection. arXiv preprint arXiv:2204.08613 (2022)
34. Peri, A., Mehta, K., Mishra, A., Milford, M., Garg, S., Madhava Krishna, K.: ReF-rotation equivariant features for local feature matching. arXiv preprint arXiv:2203.05206 (2022)
35. Bökman, G., Kahl, F.: A case for using rotation invariant features in state of the art feature matchers. arXiv preprint arXiv:2204.10144 (2022)
36. Cieslewski, T., Bloesch, M., Scaramuzza, D.: Matching features without descriptors: implicitly matched interest points. arXiv preprint arXiv:1811.10681 (2018)
37. Geiger, A., Lenz, P., Urtasun, R.: Are we ready for autonomous driving? The KITTI vision benchmark suite. In: 2012 IEEE Conference on Computer Vision and Pattern Recognition, pp. 3354–3361. IEEE (2012)
38. Bökman, G., Kahl, F., Flinth, A.: ZZ-Net: a universal rotation equivariant architecture for 2D point clouds. arXiv preprint arXiv:2111.15341 (2021)
39. Sattler, T., Weyand, T., Leibe, B., Kobbelt, L.: Image retrieval for image-based localization revisited. In: BMVC, vol. 1, p. 4 (2012)
40. Sattler, T., et al.: Benchmarking 6DOF outdoor visual localization in changing conditions. In: Proceedings of the IEEE Conference on Computer Vision and Pattern Recognition, pp. 8601–8610 (2018)
41. Dai, A., Chang, A.X., Savva, M., Halber, M., Funkhouser, T., Nießner, M.: ScanNet: richly-annotated 3D reconstructions of indoor scenes. In: Proceedings of the IEEE Conference on Computer Vision and Pattern Recognition, pp. 5828–5839 (2017)
42. Li, Z., Snavely, N.: MegaDepth: learning single-view depth prediction from internet photos. In: Proceedings of the IEEE Conference on Computer Vision and Pattern Recognition, pp. 2041–2050 (2018)
43. Taira, H., et al.: InLoc: indoor visual localization with dense matching and view synthesis. In: Proceedings of the IEEE Conference on Computer Vision and Pattern Recognition, pp. 7199–7209 (2018)
44. Wang, W., et al.: TartanAir: a dataset to push the limits of visual SLAM. In: 2020 IEEE/RSJ International Conference on Intelligent Robots and Systems (IROS), pp. 4909–4916. IEEE (2020)
45. Balntas, V., Lenc, K., Vedaldi, A., Mikolajczyk, K: HPatches: a benchmark and evaluation of handcrafted and learned local descriptors. In: Proceedings of the IEEE Conference on Computer Vision and Pattern Recognition, pp. 5173–5182 (2017)

46. Bradski, G.: The OpenCV library. Dr. Dobb's J. Softw. Tools Prof. Program. **25**(11), 120–123 (2000)
47. Alcantarilla, P.F., Solutions, T.: Fast explicit diffusion for accelerated features in nonlinear scale spaces. IEEE Trans. Patt. Anal. Mach. Intell, **34**(7), 1281–1298 (2011)
48. Leutenegger, S., Chli, M., Siegwart, R.Y.: BRISK: Binary robust invariant scalable keypoints. In: 2011 International Conference on Computer Vision, pp. 2548–2555. IEEE (2011)
49. Alcantarilla, P.F., Bartoli, A., Davison, A.J.: KAZE features. In: Fitzgibbon, A., Lazebnik, S., Perona, P., Sato, Y., Schmid, C. (eds.) ECCV 2012. LNCS, vol. 7577, pp. 214–227. Springer, Heidelberg (2012). https://doi.org/10.1007/978-3-642-33783-3_16
50. Alexandre Alahi, Raphael Ortiz, and Pierre Vandergheynst. FREAK: Fast retina keypoint. In 2012 IEEE conference on computer vision and pattern recognition, pages 510–517. Ieee, 2012
51. Radenović, F., Iscen, A., Tolias, G., Avrithis, Y., Chum, O.: Revisiting oxford and Paris: large-scale image retrieval benchmarking. In: Proceedings of the IEEE Conference on Computer Vision and Pattern Recognition, pp. 5706–5715 (2018)
52. Zhao, X., Wu, X., Miao, J., Chen, W., Chen, P.C.Y., Li, Z.: ALIKE: accurate and lightweight keypoint detection and descriptor extraction. IEEE Trans. Multim. (2022)
53. Shen, T., Luo, Z., Zhou, L., Zhang, R., Zhu, S., Fang, T., Quan, L.: Matchable image retrieval by learning from surface reconstruction. In: Jawahar, C.V., Li, H., Mori, G., Schindler, K. (eds.) ACCV 2018. LNCS, vol. 11361, pp. 415–431. Springer, Cham (2019). https://doi.org/10.1007/978-3-030-20887-5_26
54. Snavely, N., Seitz, S.M., Szeliski, R.: Photo tourism: Exploring photo collections in 3D. In: ACM SIGGRAPH 2006 Papers, pp. 835–846 (2006)

Author Index

Printed in the United States
by Baker & Taylor Publisher Services